Bioactive Polymeric Systems

An Overview

Bioactive Polymeric Systems

An Overview

Edited by

Charles G. Gebelein
Youngstown State University
Youngstown, Ohio

and

Charles E. Carraher, Jr.
Wright State University
Dayton, Ohio

Plenum Press • New York and London

Library of Congress Cataloging in Publication Data

Main entry under title:

Bioactive polymeric systems.

Includes bibliographies and index.

1. Polymers and polymerization—Physiological effect. 2. Polymers in medicine. 3. Controlled release preparations. I. Gebelein, Charles G. II. Carraher, Charles E.

QP801.P64B53 1985 661'.8 85-6399

ISBN 978-1-4757-0407-5 ISBN 978-1-4757-0405-1 (eBook)

DOI 10.1007/978-1-4757-0405-1

Softcover reprint of the hardcover 1st edition 1985
A Division of Plenum Publishing Corporation
233 Spring Street, New York, N.Y. 10013

Contributors

James M. Anderson Departments of Pathology and Macromolecular Science, Case Western Reserve University, Cleveland, OH 44106.

Chye H. Ang School of Chemistry, The University of New South Wales, Kensington, New South Wales, Australia 2033.

Bernadette M. Cardarelli Unique Technologies Inc., 71 South Cleveland Avenue, Mogadore, OH 44260.

Nate F. Cardarelli Unique Technologies Inc., 71 South Cleveland Avenue, Mogadore, OH 44260.

Charles E. Carraher, Jr. Department of Chemistry, Wright State University, Dayton, OH 45435.

Y. D. Clonis The Biotechnology Centre, University of Cambridge, Downing Street, Cambridge CB2 3EF, England.

Klaus Dorn Johannes Gutenberg-Universität, Fachbereich Chemie, Institut für Organische Chemie, Joh.-Joachim-Becher-Weg 18-22, D-6500 Mainz, West Germany.

David L. Gardner Batelle-Columbus Laboratories, 505 King Avenue, Columbus, OH 43201.

John L. Garnett School of Chemistry, The University of New South Wales, Kensington, New South Wales, Australia 2033.

Charles G. Gebelein Department of Chemistry, Youngstown State University, Youngstown, OH 44555.

David J. Giron Department of Immunology, Wright State University, Dayton, OH 45435.

David I. Gustafson Shell Development Company, P.O. Box 4248, Modesto, CA 95352.

Gerhard Hoerpel Johannes Gutenberg-Universität, Fachbereich Chemie, Institut für Organische Chemie, Joh.-Joachim-Becher-Weg 18-22, D-6500 Mainz, West Germany.

Yukio Imanishi Department of Polymer Chemistry, Kyoto University, Yoshida Honmachi, Sakyo-ku, Kyoto, 606, Japan.

Melvin H. Keyes Owens-Illinois, Inc., One SeaGate, Toledo, OH 43666.

Sung Wan Kim Department of Pharmaceutics, University of Utah, Salt Lake City, UT 84112.

Ronald G. Levot School of Chemistry, The University of New South Wales, Kensington, New South Wales, Australia 2033.

Hilton B. Levy National Institute of Allergy and Infectious Diseases, Bethesda, MD 20205.

Mervyn A. Long School of Chemistry, The University of New South Wales, Kensington, New South Wales, Australia 2033.

C. R. Lowe The Biotechnology Centre, University of Cambridge, Downing Street, Cambridge CB2 3EF, England.

Barbara A. Metz Battelle-Columbus Laboratories, 505 King Avenue, Columbus, OH 43201.

Jocelyn M. Miller College of Forest Resources, University of Washington, Seattle, WA 98040.

Raphael M. Ottenbrite Department of Chemistry and the Massey Cancer Center, Virginia Commonwealth University, Richmond, VA 23284.

Robert V. Petersen Department of Pharmaceutics, College of Pharmacy, University of Utah, Salt Lake City, UT 84112.

Thomas Quinn Johns Hopkins Hospital, Baltimore, MD.

Helmut Ringsdorf Johannes Gutenberg-Universität, Fachbereich Chemie Institut für Organische Chemie, Joh.-Joachim-Becher-Weg 18-22, D-6500 Mainz, West Germany.

James Samanen Peptide Chemistry Department, Smithkline Beckman, Inc., 1050 Page Mill Road, Palo Alto, CA 94304.

Seshaiyer Saraswathi Owens-Illinois, Inc., One SeaGate, Toledo, OH 43666.

Conrad Schuerch Department of Chemistry, State University of New York, College of Environmental Science and Forestry, Syracuse, NY 13210.

William J. Scott Department of Microbiology, Wright State University, Dayton, OH 45435.

Kiichi Takemoto Faculty of Engineering, Osaka University, Yamadaoka, Suita, Osaka 565, Japan.

John M. Whitely Division of Biochemistry, Department of Basic Clinical Research, Scripps Clinic and Research Foundation, La Jolla, CA 92037.

Anthony Winston Department of Chemistry, West Virginia University, Morgantown, WV 26506.

Amar Yahianoui College of Forest Resources, University of Washington, Seattle, WA 98040.

Preface

The vast array of libraries in the world bear mute witness to the truth of the 3000-year-old observation of King Solomon who stated ". . . of making many books there is no end, and much study is a weariness of the flesh." Yet books are an essential written record of our lives and the progress of science and humanity. Here is another book to add to this huge collection, but, hopefully, not just another collection of pages, but rather a book with a specific purpose to aid in alleviating the "weariness of the flesh" that could arise from much studying of other journals and books in order to obtain the basic information contained herein.

This book is about polymeric materials and biological activity, as the title notes. Polymeric materials, in the broad view taken here, would include not only synthetic polymers (e.g., polyethylene, polyvinyl chloride, polyesters, polyamides, etc.), but also the natural macromolecules (e.g., proteins, nucleic acids, polysaccharides) which compose natural tissues in humans, animals and plants. In the broad sense used here, biological activity is any type of such action whether it be in medication, pest control, plant-growth regulation, and so on. In short, this book attempts to consider, briefly, the use of any type of polymeric material system with essentially any kind of biological activity.

Many books have been written about various segments of the vast spectrum of biological activity and polymers. This book is, however, unique in its conception in that what exists in the 22 chapters are brief, introductory reviews of a wide variety of bioactive polymeric systems which are written by experts in their fields for a scientist who is *not* an expert in this specific field. The purpose of these chapters is to provide a scientist with some interest in a particular bioactive polymeric system with the basic, and relevant, information available in fields other than his or her own specialty. This will enable a specialist to examine how someone in a different field might attack a problem relevant to their own and rapidly assess how these other techniques, materials, or approaches might apply to their own research or development problems. The resulting cross-fertilization should help advance all of the various types of bioactive polymeric materials. These 22 chapters contain an aggregate of about 3000 references to other reviews and more detailed papers in the area of bioactive polymeric systems which will enable the reader to obtain further background in any desired area.

The organization of this book is centered mainly around the category of the polymeric system used to achieve the biological activity result, rather than on the end usage. The book is organized in the following manner:

Overview	Chapter 1
Controlled-Release Systems	Chapters 2–8
Special Experimental Techniques	Chapters 9 and 10
Natural Polymer Systems	Chapters 11–15
Pseudonatural Polymer Systems	Chapters 16 and 17
Synthetic Polymer Systems	Chapters 18–22

By far, the largest emphasis in this book is on medically related biological activity and various aspects of this topic can be found in Chapters 1, 2, 4, 6, 7, 12–15, and 18–22. Chapters 3, 5, 8–11, and 16–17 are mainly concerned with nonmedical applications. In actual fact, information relating to either medical or nonmedical applications can be found in nearly all the chapters and the various approaches and/or techniques delineated are often usable in either of these main areas. This is to be expected since the desired end result can often be achieved in several ways. While this theme is developed more fully in Chapter 1, the following three examples illustrate the interrelationships between the bioactive polymeric systems considered in this book.

Enzymes are obviously a biologically active system which is macromolecular. Several diseases (e.g., phenylketonuria, tyrosinosis, etc.) arise from the lack of a specific enzyme and can be treated by the administration of the appropriate enzyme. Sometimes an enzyme can be used medically in other types of disease treatment. For example, the enzyme L-asparaginase has been shown to suppress the growth of certain tumors. The catalytic activity of enzymes is not, however, limited to medical applications. Enzymes are used in various industrial processes, such as fermentation, and in laundry detergents. Some of these above applications might be done more effectively by a bound enzyme. However, since an enzyme normally contains a specific site on the macromolecule which gives rise to the bioactivity, might not this same end result be achievable with a completely synthetic polymer with the same type of site—a pseudoenzyme or an enzyme-mimetic polymer? Could the same result, potentially, be achieved by an apparently unrelated polymer structure or system?

While cancer is actually a multiplicity of diseases, we can generalize and note that the treatment of this disease often includes chemical means—the chemotherapy approach. Obviously, these agents can be enclosed in a polymeric matrix to control the release of the drug agent to the cancer and possibly make the treatment more effective. A given anticancer agent could also be readily bound to a polymer and either be released to the body or be active in the bound form. The polymeric material in either case could be a synthetic one or a natural material. Carrying this approach one step further, one could attach the anticancer agent to a bioactive polymer (natural or synthetic), such as an enzyme or an antibody, and possibly enhance the drug activity. The possibilities seem endless. Only time will reveal what approach might prove best for cancer treatment. It may well be that various kinds of cancer will respond better to different approaches.

Finally, consider the problem of the control of algae growth in a pond. The end goal is merely to prevent algae buildup in any way possible, at a reasonable expense (including labor) and considering the effect of the chemical agent on the total ecosystem. In that light, it would hardly matter whether the system was the controlled release of an algicide, an antialgae synthetic polymer, or an enzyme that destroyed the algae. The problem is how to find a method of control that would be more efficient than the periodic hand addition of an algicide.

The chapters in this book will, hopefully, bring these interrelationships into clearer focus and aid in the solution of numerous apparently unrelated problems. We especially encourage the reader to consider the use of the other more unfamiliar techniques in their own problems, where appropriate, and also to consider the use of their own approaches to other areas as might be suggested by this book.

The editors wish to thank each author for their excellent chapters. We also thank our families for their special form of assistance with this book. It would have been nearly impossible to complete this volume without their encouragement and warm smiles.

Charles G. Gebelein
Department of Chemistry,
Youngstown State University,
Youngstown, OH 44555
and
Department of Pharmacology,
Northeastern Ohio Universities,
College of Medicine

and

Charles E. Carraher, Jr.
Department of Chemistry,
Wright State University,
Dayton, OH 45435

Contents

3. Controlled-Release Pesticides: A Historical Summary and State of the Art

Nate F. Cardarelli and Bernadette M. Cardarelli

4. Controlled Release of Antifertility Agents

David L. Gardner and Barbara A. Metz

5. Controlled Release and Plant-Growth Regulators

Jocelyn M. Miller and Amar Yahiaoui

6. Hydrogels for Controlled Drug Release

Sung Wan Kim

7. Biodegradable Drug Delivery Systems Based on Polypeptides

Robert V. Petersen

8. Controlled-Release Animal Repellents in Forestry

David I. Gustafson

9. Affinity Chromatography

C. R. Lowe and Y. D. Clonis

10. Application of Radiation Grafting in Reagent Insolubilization

Chye H. Ang, John L. Garnett, Ronald G. Levot, and Mervyn A. Long

11. Immobilized Enzymes

Melvin H. Keyes and Seshaiyer Saraswathi

12. Biomedical Polypeptides—A Wellspring of Pharmaceuticals

James Samanen

13. Drug Delivery with Protein and Peptide Carriers

John M. Whiteley

14. Biomedical Applications of Polysaccharides

Conrad Schuerch

15. Interferon Induction by Polymers

Hilton B. Levy and Thomas Quinn

16. Functionality and Applicability of Synthetic Nucleic Acid Analogues

Kiichi Takemoto

17. Enzyme-Mimetic Polymers

Yukio Imanishi

18. Bioactive Carboxylic Acid Polyanions

Raphael M. Ottenbrite

19. Polymeric Antitumor Agents on a Molecular and Cellular Level

Klaus Dorn, Gerhard Hoerpel, and Helmut Ringsdorf

20. Biological Activities of *cis*-Dichlorodiamineplatinum II and Its Derivatives

Charles E. Carraher, Jr., William J. Scott, and David J. Giron

21. Iron-Complexing Bioactive Polymers

Anthony Winston

22. Biological Activities and Medical Applications of Metal-Containing Macromolecules

Charles E. Carraher, Jr.

1

Bioactive Polymeric Systems, An Overview

Charles G. Gebelein

Abstract. The breadth of the bioactive polymeric systems is illustrated in this chapter. The area includes controlled-release systems, such as erodible systems, diffusion-controlled systems, mechanical systems, and microcapsules. Bioactive polymeric systems would also include biologically active polymers, such as natural polymers, synthetic polypeptides, pseudoenzymes, pseudonucleic acids, and polymeric drugs. In addition, the area can include immobilized bioactive materials, such as immobilized enzymes, antibodies, and other bioactive agents. Some potential examples of these systems and approaches are given.

1. Introduction

The major purpose of this chapter is to define and to illustrate the term "bioactive polymeric systems" and to serve as an introduction for the balance of this book. In this chapter, most of the various types of systems that involve both polymers and bioactivity will be discussed briefly, and many of these will receive further elaboration in subsequent chapters. No attempt is made here to be encyclopedic in scope or exhaustive in documentation. The area of bioactive polymeric systems is too broad and the number of available applications is too large for this to be possible in the limited size of this chapter, or even this book. This chapter is primarily a smorgasbord designed to whet the appetite of the reader and to challenge the reader in attempting new approaches to solving problems in bioactive polymeric systems.

2. Bioactive Polymeric Systems

2.1. What Is Bioactivity?

Bioactivity can be defined in several ways. For the purpose of this chapter, and the entire book, we will consider bioactivity to be the interaction

Charles G. Gebelein • Department of Chemistry, Youngstown State University, Youngstown, Ohio 44555.

of some agent, usually chemical, on a biological system. Examples of such bioactivity would include: (1) the action of a drug on a disease center; (2) the action of a herbicide on weeds; (3) the action of an insecticide on insects; and (4) the prevention of conception by an antifertility chemical agent.

2.2. What Are Bioactive Polymeric Systems?

In the sense that we employ this term in this book, a bioactive polymeric system is any bioactive system that utilizes any kind of polymeric material in producing, enhancing, or controlling bioactivity. This is, admittedly, a broad definition, but the basic plan of this book is to present this general topic in its manifold aspects. Elaboration of this definition will become apparent in the examples and sections that follow.

3. Classes of Bioactive Polymeric Systems

This broad area could be subdivided many different ways. In this chapter, we shall subdivide into three basic areas: (1) controlled-release systems; (2) biologically active polymers; and (3) immobilized bioactive materials. Naturally, there will be some overlap between these areas, but this will occur in any system of division utilized in this field. Each of these major categories will be further subdivided into more specific areas for the convenience of discussion and for the benefit of the readers.

4. Polymeric Controlled-Release Systems

A controlled-release system is simply any polymeric system that regulates or controls the release of some type of bioactive agent. Generally, these systems are developed to restrict the concentration of this agent to some fairly narrow range in order to elicit the desired activity while reducing the other potentially dangerous side effects to acceptable levels. Examples of such systems exist in nature. Many polypeptides consist of large molecules which release, on demand by the body's chemistry, a controlled amount of a subsection of this polypeptide which has some specific biological activity. Many hundreds of such systems have been developed synthetically in the past several decades. We shall consider examples of (1) erodible systems, (2) diffusion-controlled systems, (3) mechanical systems, and (4) microcapsules. In all these cases, the bioactive agent is entrapped in some manner in the polymeric system. In actual practice these modes of approach are frequently combined. Systems in which this agent is attached chemically to the polymer will be considered in the next two sections.

4.1. Erodible Systems

The basic approach in an erodible system is to contain the bioactive agent within some polymeric system that will be dissolved away or destroyed by the biological media in which it is to be placed. One of the oldest medical applications of this methodology is the enteric-coated drugs in which the drug is protected from the stomach acids and enzymes and is then released in the intestines. Placing copper salts in biodegradable, natural polymers, to control barnacle formation on ships, would be another old application of this approach. Some examples of a more recent vintage would include the "timed capsule" cold and/or headache remedies.

Several types of natural and synthetic polymers are of potential utility for bioerodible systems. These include polysaccharides, polypeptides, polyesters, poly(ortho esters), polyacetals, polyketals, and some complexation systems. These topics have been discussed in several recent papers[1-5] and will be treated further in Chapters 3, 5, 7, 8, and 14 of this book. The most commonly used synthetic polymers are the poly(lactic acids) and the poly(glycolic acids). The bioagent is placed within the polymer either as a depot or a reservoir, or dispersed fairly uniformly throughout the polymer (monolithic system). In either case, the polymeric material is eroded away by solvent, chemical, and/or biological action and the bioactive agent is released. The precise kinetics of release would depend on the exact polymeric system used, the mode of bioagent dispersal, and the configuration of the entire system. These factors have been discussed in the literature.[3]

The advantage of the erodible systems in medical applications would be that the polymeric material would disappear and not have to be removed from the patient. Aside from the enteric coatings and the "timed capsules" noted above, these erodible systems are being explored for potential medical application in drug-addiction control, contraception, cancer treatment, and a wide variety of drug release systems. The use of erodible polymeric systems in nonmedical applications continues, but the special advantage of a "noncontaminating," biodegradable polymer is usually less important compared to the longer, more precise control offered by some other systems.

4.2. Diffusion-Controlled Systems

Diffusion-controlled systems are of two basic types: (1) the reservoir or depot type; and (2) the monolithic type. These systems normally employ a nondegradable polymer, such as poly(dimethylsiloxane), natural or synthetic rubber, a vinyl polymer, or an acrylate (methacrylate) polymer, although devices have been developed based on erodible polymers. In both types of diffusion-controlled systems, the rate of release of the bioactive agent is regulated or controlled by the diffusion of this agent through the polymeric matrix. This is, in turn, dependent on the composition of this polymeric material and on the chemical nature of the bioactive agent. The major difference arises, however, from the nature of the two systems themselves. In the reservoir type, the bioactive agent is enclosed as a depot or reservoir within the polymeric material, while in the monolithic type the bioactive agent is dispersed throughout the polymeric matrix in a relatively uniform fashion.

The kinetics of the release from a reservoir system usually approximates zero order, which means that the release rate is independent of time. This zero order kinetics prevails until the reservoir is depleted, and then the rate of release decreases rapidly. In the case of the monolithic system, the kinetics approximate first order, which means that the release rate shows a steady decrease with time in a manner similar to radioactive decay. Both types usually give an initial high release of the bioactive agent in what can be called a "burst effect." In real systems, deviations from these ideals are common. It is also possible to combine these types by placing a second polymeric layer around a monolithic device as sort of a membrane which usually changes the kinetic pattern to approximately zero order for much of the time interval involved.[1,6–8]

Either system has potential utility in medical and nonmedical applications; these will be discussed more fully in Chapters 3–6, and 8. Some of the nonmedical applications include animal flea-repellent collars, insecticide-releasing strips, and various agricultural applications. Some medical uses include the release of pilocarpine from a ethylene–vinyl acetate copolymer (Ocusert®, Alza Corp.) for the control of glaucoma, and the release of progesterone from the same polymeric system to control fertility (Progestasert®, Alza Corp.).[9,10] Other polymeric systems have been studied for contraceptive devices.[11] Diffusion systems have also been utilized to release narcotic antagonists and anticancer agents.[12]

4.3. Mechanical Devices

By the definition used here, a mechanical device would be one that utilizes any form of design control that regulates the release of the bioactive agent in a manner other than direct diffusion control or erosion. This would, therefore, include devices "powered" by osmotic pressure and the transdermal membrane systems, in addition to the more obvious pumps and extracorporeal devices.

Possibly the most widely studied external pump device would be the insulin infusion pump or the "artificial pancreas" through which better control can be obtained over the insulin concentrations, and therefore the glucose concentrations, in the body. A number of such systems have been devised using some form of a polymeric pump or reservoir to contain and circulate the insulin solution into the bloodstream of the patient. Some devices can vary the rate of administration of the insulin and implantable infusion pumps have been developed using poly(dimethylsiloxane). Considerable effort is in progress to couple these pumps with a miniature sensor and this could closely approximate the behavior of the healthy pancreas.[13,14]

When a device that is constructed using a semipermeable membrane, enclosing a solution of a bioactive agent, is fitted with a small orifice, the diluted solution of this agent can be driven from the device by osmotic pressure as water diffuses through the membrane into the more concentrated interior of the device. These devices (e.g., the Oros® System of Alza Corp.) can be used for the controlled release of various drugs, including insulin.[9,10,15]

Transdermal devices utilize a semipermeable membrane to regulate the release of a drug directly onto the skin; this drug then permeates the skin into the specified region of the body for the desired physiological activity. The devices are held in place by an adhesive layer, which should not be the rate-limiting unit in a properly constructed system, or are taped in place, although this would be less desirable and more bulky. The rate of drug release is controlled, of course, by the semipermeable membrane and should not be so great that the drug accumulates on the skin surface (i.e., the skin should not be the rate-controlling part of the system). Devices have been developed for the administration of nitroglycerine (for treating angina pectoris) and scopolamine (for treating motion sickness).[9,10,16]

The parenteral administration of drugs using plastic tubing, with or without an external pump, could also be considered a mechanical, controlled-release system, but these are outside the scope of this chapter.

4.4. Microcapsules

Essentially, a microcapsule is a small particle-sized membrane system. The combination of high membrane permeability and the very large surface area that can be achieved with these systems confer some unusual properties on the devices and have made them of interest for various artificial organs, as well as drug release systems.

Microcapsules have been used to enclose essentially every type of bioactive agent. Sometimes, the polymeric material has also been bioerodible, which combines the microcapsule and erodible systems. These systems have been used to enclose various enzymes, such as catalase or L-asparaginase (which can be used to treat some forms of cancer), and have been shown to maintain the enzymatic activity for prolonged periods of time at room temperature. They have also been used to encapsulate material such as the beta-cells of the Islets of Langerhans and function as a source of insulin production (artificial pancreas).[17–21]

5. Biologically Active Polymers

5.1. Natural Polymers

Obviously the natural proteins, enzymes, polysaccharides, and nucleic acids are biologically active. Entire books have been written for each of these kinds of bioactive polymers.[22] Many of these macromolecules have been synthesized and a wide range of related derivatives have been prepared, often with even greater activity. We will only cite the endophins and enkephalins here and note that some synthetic analogues have shown over 28,000 times the analgesic activity of the natural materials without a corresponding increase in the addictive properties.[23]

Natural polymers have also been widely used as the polymeric matrix for controlled-release systems (usually biodegradable) and as a carrier macromolecule for other bioactive agents.[24] Part of this wide usage is due to the fact that these macromolecules are already biologically active and this fact can sometimes be utilized to make a modified system more specific. These topics will be covered in more detail in Chapters 7, 12, and 13.

5.2. Synthetic Polypeptides

In addition to the example of the endophins cited in the preceding section, hundreds of other examples of modified, natural polypeptides have appeared in the literature.[25,26] Some of these studies are described further in Chapter 12, and no further elaboration will be made here.

5.3. Pseudoenzymes

The biological activity of enzymes need not be documented since the highly specific, catalytic activity of these polypeptides is well known. This specific activity is usually attributed to a "lock-in-key" mechanism of action. Many researchers have attempted to duplicate this specific activity by duplicating the section of the enzyme macromolecule that was believed to be responsible for these interactions and these synthetic polymers can be called pseudoenzymes, enzyme-analogues, synzymes, or enzyme-mimetic polymers.[27–29]

Although enzymelike activity has been demonstrated on many occasions with these pseudoenzymes, they usually show much less catalytic activity than a true enzyme. Nevertheless, the activity is real even when the polymers are vinyl derivatives rather than polypeptides. The possible medical applications of such synthetic pseudoenzymes is great. There are many diseases which are caused by an enzyme deficiency, such as phenylketonuria, tyrosinosis, or histidinemia. When these diseases are treated by the administration of the missing enzyme, the body rapidly destroys this foreign protein. A pseudoenzyme might not be recognized in this manner by the body's defense system and might be able to treat these enzyme-deficient diseases effectively. While this is an exciting possibility, this appears unlikely in the near future.

These pseudoenzymes might, of course, be used to manufacture some chemical agent that is currently produced using an enzyme. The advantage that might accrue here is that the pseudoenzymes might be more stable than the natural enzyme, even when compared with the bound forms of the enzymes. This type of application appears more likely than the medically related ones. Some of the knowledge acquired may, however, prove valuable in developing other types of bioactive synthetic polymers. The topic of enzyme-mimetic polymers will be covered in more detail in Chapter 17.

5.4. Pseudonucleic Acids

The bioactivity of the nucleic acids DNA and RNA in genetic replication and protein synthesis is well established and does not need elaboration here. As with the enzymes considered previously, the specific activity of the nucleic acids is believed due to certain portions of the macromolecule and many researchers have attempted to synthesize pseudonucleic acids, usually based on vinyl polymer backbones, although some studies have used polypeptides.[30–33] Some of these pseudonucleic acids have exhibited biological activity (usually antiviral or antitumor activity). This subject is considered in more detail in Chapter 16.

5.5. Polymeric Drugs

In its broadest sense, a polymeric drug is any macromolecule that contains a drug unit attached to or within the backbone chain or exhibits biological activity without such a unit. Thousands of such polymers have been prepared and several reviews of this area exist.[34–41] Chapters 18–22 will consider several different types of polymeric drugs in greater detail; here we will only overview this area.

For many years it was debated whether or not synthetic polymeric materials could elicit biological activity. While active research programs in this general area predate the 1950s, demonstrations of actual bioactivity from these materials have been more recent. Nevertheless, the fact that so many of the biological activities in living systems are controlled by or utilize macromolecules made the prospects seem promising. One could, for example, note the hundreds of known, natural bioactive polymers. In addition, derivatives were made for many of these which still showed activity, sometimes even greater than the "natural" polymer. The fact that even simple dimers, trimers, and tetramers of a single amino acid showed less antibacterial activity than the corresponding homopolymer[42] also suggested the feasibility of polymeric drugs even though they were not developed immediately.

Polymers that contain a known therapeutic unit have now been prepared for many different types of drugs. Some of these do show biological activity, but it is not always clear how this occurs. In many cases, the drug unit cleaves off of the polymer and becomes available in the biological media. In this respect, these polymeric drugs are a type of drug release system, but the kinetics would be expected to be more complex since a hydrolytic or enzymatic cleavage reaction must occur in addition to any diffusion phenomena. Some polymeric drugs do not appear to cleave readily but still exhibit biological activity. In these cases, the polymer is assumed to enter directly into the cells *via* endocytosis or indirectly *via* piggyback endocytosis using a natural polypeptide, or other natural polymer, as the carrier. (Considerations of this latter effect have led many researchers to use various natural polymers as carriers for drugs, either monomeric or polymeric.) There are, however, a few examples of polymeric drugs, which have the therapeutic units attached to objects that are clearly too large to enter the cell (e.g., agarose beads), and still exhibit biological activity by a mechanism that is unclear at this time.[43]

There are now many polymers that exhibit biological activity without the presence of any known druglike unit. The best example of this type of polymeric drug is probably the divinyl ethermaleic anhydride cyclocopolymer (DIVEME or "Pyran" copolymer).[44] These polymers could elicit their bioactivity in a variety of ways. The DIVEME polymer, for example, appears to interact with the immune system to promote a larger number of macrophages which then cause the bioactivity that is observed.[45] Other polymeric materials, usually polyanions, elicit their activity by causing an increase in the interferon concentration.[46] (See also Chapter 15.) The pseudoenzymes (Section 5.3) and the pseudonucleic acids (Section 5.4) could also be considered polymeric drugs in which the molecule as a whole is bioactive to a much greater extent than the individual subunits.

Polymeric drugs, partly because of their high molecular weight, are capable of modifications that could completely destroy the bioactivity in low-molecular-weight systems. Thus, portions of the macromolecule would be dedicated to changing the solubility or "targeting" the drug without gross effects on the bioactivity of the segments that contain the drug unit. This concept of multifunctional polymeric drugs has been discussed in the literature[34,38,39,47] and the potential use of such polymers in cancer therapy has been reviewed.[48] (See also Chapters 18, 19, and 22.)

6. Immobilized Bioactive Materials

6.1. Immobilized Enzymes

Enzymes, while they are powerful catalysts with many industrial and medical applications, have their limitations. These include cost, difficulty of recovery from a chemical reaction, temperature sensitivity, pH limitations, and the possibility of inactivation by a variety of agents. Therefore, the concept of an immobilized enzyme system that might alleviate many of these difficulties was developed. This subject has been reviewed[49–53] and is discussed in more detail in Chapter 11. Here we will only consider the basic features of these systems.

Enzymes can be immobilized or made insoluble in water, the usual reaction media, by five basic methods which have many variations in each case. These include: (1) adsorption [ADE]; (2) covalent bonding [CBE]; (3) cross-linking [CLE]; (4) matrix entrapment [MEE]; and (5) membrane encapsulation [EIM]. Each system has its advantages and disadvantages. The ADE systems can be prepared easily on a wide variety of carrier surfaces and often show enhanced enzymatic activity; but they are reversible, have a high carrier-to-enzyme ratio, and the enzyme is still exposed to any deactivating influence. The CBE system is not reversible and is usually more stable, but the attachment reaction sometimes deactivates the enzyme. In addition, the loading of the enzyme on the surface is still not very high. The CBE system is fairly simple, however, and not very expensive to make in many cases.

The CLE system only involves the enzyme and some low-molecular-weight cross-linking agent, but some of these can render the entire system inactive. In addition, the reactants might not reach the catalytic portion of the enzyme in highly cross-linked systems. The MEE systems can be made readily and there are a variety of polymeric materials that can be used to entrap the enzyme. Unfortunately, these systems, while stable, tend to have very low enzyme-to-carrier ratios. In addition, much of the catalytic activity seems to occur near the surface of the particles, which means much of the enzyme content is not used. The EIM systems, in which the enzyme is closed in a thin polymeric membrane, appear to overcome some of these problems. These microcapsules of the enzyme, usually in the form of a solution, are often very resistant to attack by proteolytic enzyme, antibodies, and leukocytes, and have high thermal stability. The combination of high surface area and ultrathin membranes (0.02 μ) results in good reactivity, and potential medical applications have been cited for these materials.[(50)]

In addition to the potential therapeutic applications, immobilized enzymes find use in analysis (enzyme electrodes) and a variety of industrial processes, such as the preparation of high-fructose syrups, milk treatment, resolution of DL-amino acid pairs, and the preparation of some antibiotics. These uses will probably increase in the future.

6.2. Other Immobilized Bioagents

The same basic techniques that have been used to immobilize enzymes could also be used to immobilize other biologically active agents, such as antibodies, antigens, chloroplasts, or mitochondria. Although the research to date in this area is less than for the enzymes, the results appear promising.[(49,50,54)] It would appear possible to combine some of these agents to achieve special effects. For example, an antibody could be immobilized on a polymer that also contained an immobilized enzyme and the combination would promote a targeted result. Since these binding reactions could be done on a wide variety of polymers, a soluble polymer could be chosen and this would permit the immobilized antibody:enzyme to move about and accomplish the desired result.

7. Examples of Bioactive Polymeric Systems

What we are going to examine in this section are some examples or illustrations of how bioactive polymeric systems can be used. The preceding sections also contained many examples of these varied applications. The main point here is to illustrate the versatility of the various types of bioactive polymeric systems; no attempt will be made at completeness—to do this would only belabor the obvious.

7.1. Many Uses for Any System

The multiple use of some systems is obvious since these applications already exist. Diffusion-controlled release systems are currently being used in a very wide variety of medical, industrial, agricultural, and home applications. Alternate uses of some other systems might not be as obvious, however. Such a system would be the polymeric drugs. This is in part a function of the name, and in part a function of an incomplete, unclear understanding of the nature of these bioagents. Polymeric drugs are merely a single grouping of a larger class which could be termed "synthetic bioactive polymers."

Some attempts to develop polymeric food colorings, antioxidants, and related food additives have been reported,[55] but apparently have been too difficult to put into the marketplace because of cost factors and government regulations. Only time will tell whether or not this situation will ever change.

A synthetic bioactive polymer would not have its activity restricted to medical applications and could run the gamut of potential biological activities. One could conceive of a polymeric insecticide, herbicide, plant-growth regulator, food additive, or any other bioagent. The limiting factors to designing such systems would include cost and a lack of knowledge of the basic, detailed biochemistry involved in the end use. Yet the possibility exists to design a herbicide, for example, that would only attack certain types of weeds—a level of specificity beyond the realm of today's herbicide industry. But to do this might require more detailed knowledge about this plant, and others as well, than is available.

In a similar manner, one could conceive of a polymeric insecticide that only attacked mosquitoes. Armed with this weapon, malaria could become a thing of the past. What's more, this could be achieved without polluting our environment as is often done with the usual insecticide, such as DDT. Obviously, we would need to know many details about mosquito biochemistry as well as similar information for other insects, animals, and humankind. The possibilities for precise control exist, but the knowledge, and perhaps the desire, do not.

Finally, we might note that a bioactive polymer might make an excellent choice for a tissue, or organ replacement after surgery. If the correct units were present, they might reduce rejection problems.

These novel uses of synthetic bioactive polymers are future possibilities. For now, we still need to know more details about human biochemistry in order to treat our own diseases and problems. Perhaps soon we may see a "cure" for cancer using polymeric drugs. The sad fact remains that the present anticancer polymer drugs have a long way to go to achieve the ultimate finesse of which these systems are capable. This level of perfection is unlikely ever to be achieved by simple, low-molecular-weight therapeutic agents.

7.2. Many Solutions for Any Problem

Often a researcher can only see a single approach to a problem—the one on which he or she is working. Obviously this is seldom the case. In the field of bioactive polymeric systems we again see the truth that multiple solutions often exist for a single problem. True, some solutions may be better, or less expensive, but sometimes we do not always recognize the almost infinite possibilities that exist in this field. We will consider possible multiple approaches to cancer treatment and to mosquito control.

Cancer can be treated by a variety of methods, including chemotherapy. The antineoplastic agent could be placed into a controlled-release system, as is being tried by researchers worldwide. Likewise, the use of antitumor polymers, such as pseudonucleic acids, DIVEMA, and many others, is also being studied by many. It is known that some forms of cancer can be treated by certain enzymes (e.g., L-asparaginase) and this approach is being utilized in microcapsules as well as direct administration. Could modifications of L-asparaginase work better? Often an enzyme achieves its effect using only a small portion of the total molecule. If this is the case for L-asparaginase, it might be possible to isolate this subunit and utilize it as part of a polymeric drug.

Chemotherapy is only one of many methods used to treat cancer. Some polymers are affected by radiation in different ways than others. Could these differences be used to develop polymers that might make radiation therapy more effective? Sometimes surgical removal of part of the body is necessary. If the replacement polymeric materials were designed to include some permanently attached antitumor agents, it might be possible to prevent any recurrence of the cancer.

Compared to cancer treatment, mosquito control seems almost trivial, unless, of course, you live in an area where malaria reaches epidemic proportions. Controlled-release systems are already used here and need not be detailed. The concept of an antimosquito polymer is discussed in Section 7.1 and also need not be repeated. Mosquitoes, like every other living creature, have a set of specific enzymes and nucleic acids that control their biochemistry. Could these bioagents be utilized to control mosquitoes, and therefore malaria? Could an artificial "DNA" be designed that would cause mosquitoes to become sterile or otherwise unable to compete successfully with other, more desirable, life forms? Could an enzyme, or derivative thereof, be developed that could alter the biochemistry of these insects and destroy them? Naturally, such agents could be incorporated into another bioactive polymeric system for better ease of use and more effective action.

8. Summary

This chapter has outlined some of the various types of bioactive polymeric systems that presently exist, and we have also speculated on some possible new areas that might be developed in the future. The field of bioactive polymeric systems already exists in the marketplace and is growing.

Much of this development has been in what could be termed "one-dimensional" pathways wherein only a single approach was utilized. Much of the future development in this fertile field will be "multidimensional" and will utilize several of the approaches described in this chapter and this book at the same time. This will be both a challenge and a source of frustration since the many approaches are often not familiar to those in other areas where a completely different terminology is sometimes used. Nevertheless, this is the wave of the future, and only those whose "ship of knowledge" is large enough to span several crests will successfully navigate these waters.

9. References

1. S. D. Bruck, ed., *Controlled Drug Delivery*, Vols. I & II, CRC Press, Boca Raton, Fla. (1983).
2. C. G. Pitt, T. A. Marks, and A. Schindler, in: *Controlled Release of Bioactive Materials* (R. Baker, ed.), pp. 19–43, Academic Press, New York (1980).
3. J. Heller and R. Baker, in: *Controlled Release of Bioactive Materials* (R. Baker, ed.), pp. 1–17, Academic Press, New York (1980).
4. N. S. Mason, C. S. Miles, and R. E. Sparks, in: *Biomedical and Dental Applications of Polymers* (C. G. Gebelein and F. F. Koblitz, eds.), pp. 279–291, Plenum Press, New York (1981).
5. S. Yolles, R. M. Roat, M. F. Sartori, and C. L. Washburne, in: *Biological Activities of Polymers* (C. E. Carraher, Jr. and C. G. Gebelein, eds.), pp. 233–241, American Chemical Society, Washington, D.C. (1982).
6. A. F. Kydonieus, *Controlled Release Technologies, Methods, Theory and Applications*, Vols. I & II, CRC Press, Boca Raton, Fla. (1980).
7. W. R. Good and K. F. Mueller, in: *Controlled Release of Bioactive Materials* (R. Baker, ed.), pp. 155–175, Academic Press, New York (1980).
8. W. D. Rhine, V. Sukhatme, D. S. T. Hsieh, and R. S. Langer, in: *Controlled Release of Bioactive Materials* (R. Baker, ed.), pp. 177–187, Academic Press, New York (1980).
9. A. Zaffaroni, in: *Biomedical and Dental Applications of Polymers* (C. G. Gebelein and F. F. Koblitz, eds.), pp. 293–313, Plenum Press, New York (1981).
10. A. Zaffaroni, *CHEMTECH* **6**, 756–761 (1976).
11. G. W. Duncan, D. R. Kalkwarf, and J. T. Veal, in: *Polymers in Medicine and Surgery* (R. L. Kronenthal, Z. Oser, and E. Martin, eds.), pp. 205–212, Plenum Press, New York (1975).
12. S. Yolles, in: *Polymers in Medicine and Surgery* (R. L. Kronenthal, Z. Oser, and E. Martin, eds.), pp. 245–261, Plenum Press, New York (1975).
13. J. V. Santiago, A. H. Clemens, W. L. Clarke, and D. M. Kipnis, *Diabetes* **28**(1), 71–84 (1979).
14. H. J. Sanders, *Chem. Eng. News March 2, 1981*, 30–45.
15. F. Theeuwes and B. Eckenhoff, in: *Controlled Release of Bioactive Materials* (R. Baker, ed.), pp. 61–82, Academic Press, New York (1980).
16. S. K. Chandrasekaran and J. E. Shaw, in: *Controlled Release of Bioactive Materials* (R. Baker, ed.), pp. 99–106, Academic Press, New York (1980).

17. J. A. Bakan, in: *Polymers in Medicine and Surgery* (R. L. Kronenthal, Z. Oser, and E. Martin, eds.), pp. 213–235, Plenum Press, New York (1975).
18. F. Lim and A. M. Sun, *Science* **210**, 980 (1980).
19. T. Kondo, in: *Surface and Colloid Science* (E. Matijevic, ed.), Vol. 10, pp. 1–43, Plenum Press, New York (1978).
20. T. M. S. Chang, ed., *Artificial Kidney, Artificial Liver, and Artificial Cells*, Plenum Press, New York (1978).
21. T. M. S. Chang, in: *Biomedical Polymers. Polymeric Materials and Pharmaceuticals for Biomedical Use* (E. P. Goldberg and A. Nakajima, eds.), pp. 171–187, Academic Press, New York (1980).
22. C. H. Li, ed., *Hormonal Proteins and Peptides*, Vols. I & II, Academic Press, New York (1973).
23. S. H. Snyder, *Chem. Eng. News November 28, 1977*, 26–35.
24. A. G. Walton, in: *Biomedical Polymers. Polymeric Materials and Pharmaceuticals for Biomedical Use* (E. P. Goldberg and A. Nakajima, eds.), pp. 53–83, Academic Press, New York (1980).
25. C. H. Li, ed., *Hormonal Proteins and Peptides*, Vol. 8, Academic Press, New York (1980).
26. J. Ramachandran, in: *Biological Activities of Polymers*, (C. E. Carraher, Jr. and C. G. Gebelein, eds.), pp. 119–132, American Chemical Society, Washington, D.C. (1982).
27. J. A. Pavlisko and C. G. Overberger, in: *Biomedical and Dental Applications of Polymers* (C. G. Gebelein and F. F. Koblitz, eds.), pp. 257–278, Plenum Press, New York (1981).
28. Y. Imanishi, *J. Polym. Sci., Macromol. Rev.* **14**, 1–205 (1979).
29. A. S. Lindsey, in: *Reviews in Macromolecular Chemistry*, Vol. 4, (G. B. Butler and K. F. O'Driscoll, eds.), pp. 1–47, Marcel Dekker, New York (1970).
30. K. Takemoto, *J. Polym. Sci., Polym. Symp. 55*, 105–125 (1976).
31. K. Takemoto, in: *Polymeric Drugs* (L. G. Donaruma and O. Vogl, eds.), pp. 103–129, Academic Press, New York (1978).
32. J. Pitha, in: *Biomedical and Dental Applications of Polymers* (C. G. Gebelein and F. F. Koblitz, eds.), pp. 203–213, Plenum Press, New York (1981).
33. J. Pitha, M. Akashi, and M. Draminski, in: *Biomedical Polymers. Polymeric Materials and Pharmaceuticals for Biomedical Use* (E. P. Goldberg and A. Nakajima, eds.), pp. 271–297, Academic Press, New York (1980).
34. C. G. Gebelein, *Polym. News* **4**, 163–171 (1978).
35. L. G. Donaruma and O. Vogl, eds., *Polymeric Drugs*, Academic Press, New York (1978).
36. L. G. Donaruma, R. M. Ottenbrite, and O. Vogl, eds., *Anionic Polymeric Drugs*, Wiley–Interscience, New York (1980)..
37. C. M. Samour, *CHEMTECH* **8**, 494–501 (1978).
38. H. Ringsdorf, *J. Polym. Sci., Polym. Symp. 51*, 135–153 (1975).
39. H. G. Batz, in: *Advances in Polymer Science*, Vol. 23 (H. J. Cantrow, ed.), pp. 25–53, Springer-Verlag, New York (1977).
40. D. S. Breslow, *Pure Appl. Chem.* **46**, 103–113 (1976).
41. V. A. Kropachev, *Pure Appl. Chem.* **46**, 355–361 (1976).
42. E. Katchalski, L. Bichowski-Slomnitzki, and B. F. Volcani, *J. Biochem.* **55**, 671 (1953).
43. D. C. LaPorte, K. S. Rosenthal, and D. R. Sturm, *Biochemistry* **16**, 1642–1648 (1977).
44. G. B. Butler, in: *Anionic Polymeric Drugs* (L. G. Donaruma, R. M. Ottenbrite, and O. Vogl, eds.), pp. 49–141, Wiley–Interscience, New York (1980).
45. L. G. Baird and A. M. Kaplan, in: *Anionic Polymeric Drugs* (L. G. Donaruma, R. M. Ottenbrite and O. Vogl, eds.), pp. 185–210, Wiley–Interscience, New York (1980).
46. M. C. Breinig, A. E. Munson, and P. S. Morahan, in: *Anionic Polymeric Drugs* (L. G. Donaruma, R. M. Ottenbrite, and O. Vogl, eds.), pp. 211–226, Wiley–Interscience, New York (1980).

47. C. G. Gebelein, R. M. Morgan, R. Glowacky, and W. Baig, in: *Biomedical and Dental Applications of Polymers* (C. G. Gebelein and F. F. Koblitz, eds.), pp. 191–201, Plenum Press, New York (1981).
48. C. G. Gebelein, in: *Biological Activities of Polymers* (C. E. Carraher, Jr. and C. G. Gebelein, eds.), pp. 193–203, American Chemical Society, Washington, D.C. (1982).
49. H. H. Weetal, ed., *Immobilized Enzymes, Antigens, Antibodies and Peptides*, Dekker, New York (1975).
50. T. M. S. Chang, ed., *Biomedical Applications of Immobilized Enzymes and Proteins*, Vols. 1 & 2, Plenum Press, New York (1977).
51. M. D. Trevan, *Immobilized Enzymes*, Wiley, New York (1980).
52. M. H. Keyes, *Kirk-Othmer, Encyclopedia of Chemical Technology*, 3rd ed., Vol. 9, pp. 148–172, Wiley, New York (1980).
53. S. P. O'Neill, *Rev. Pure Appl. Chem.* **22**, 133–143 (1972).
54. T. Fujimura, F. Yoshii, I. Kaetsu, Y. Inoue, and K. Shibata, *Z. Naturforsch.* **35c**, 477–482 (1980).
55. N. Weinshenker, in: *Polymeric Drugs* (L. G. Donaruma and O. Vogl, eds.), pp. 17–37, Academic Press, New York (1978).

2

Biocompatibility of Bioactive Polymeric Systems

James M. Anderson

It is obvious that the general area of bioactive polymeric systems is an extremely broad area of science which includes systems intended for human use, animal use, agriculture applications, and diagnostic assays. While biocompatibility considerations are important to all of these areas, the determination of biocompatibility is of utmost importance to those areas where bioactive polymeric systems are considered for human or animal use. In particular, the evaluation of biocompatibility of systems intended for implantation or injection in humans or animals is a necessary step in the research and development of such systems. It is the intent of this paper to briefly outline and describe biocompatibility procedures and methods for the evaluation of systems intended for implantation or injection into humans or animals.

The term "biocompatibility" is commonly used to provide an indication of the host and material responses which occur in a specific application of a given material or device. Black has recently addressed biocompatibility from the perspective of the living-system physiology and has defined biocompatibility as the biological performance of a given material in a specific application that is judged suitable to that situation.[1] Biological performance is defined as the interaction between materials or devices and living systems, and the two aspects of the biological performance are the host response and the material response. The host response is the local and systemic responses, other than the intended therapeutic response, of living systems to the material or device. The material response is the response of the material or device to living systems. A clear, specific, and absolute definition of biocompatibility does not exist at this time. This is not surprising considering the numerous and interdisciplinary factors which must be used to describe the biocompatibility of a given material in a given application for a given duration. In general, the biocompatibility of a given material or device is dependent on the material/host or tissue interactions which occur

---

James M. Anderson • Departments of Pathology and Macromolecular Science, Case Western Reserve University, Cleveland, Ohio 44106.

Table 1. Factors Influencing the Biocompatibility of Bioactive Polymeric Systems

Type and form of polymeric system
Surface and bulk composition of polymeric system
Chemical, physical, and mechanical properties of polymeric system
Influence of service conditions on the properties of the polymeric system
Animal model
Implant site: tissue and/or organ
Anticipated duration of the bioactive polymeric system
Influence of service conditions on the tissues of the implant site

over the anticipated duration of the use of the material or device under appropriate service conditions. Table 1 contains interrelated factors which can influence the biocompatibility of bioactive polymeric systems or devices.

1. Types of Bioactive Polymeric Systems

The general types of bioactive polymer systems are presented in Table 2. Polymers are commonly used as carriers for bioactive agents and the bioactive agents are released from the polymeric system by a variety of physical and chemical mechanisms.

In general, physical carriers of bioactive agents involve matrix or monolithic devices and the release of the drug is through diffusion mechanisms. Polymers used in these systems are considered to be non-biodegradable and include silicon rubber, poly-2-hydroxyethyl methacrylate and its copolymers, and so on. Matrix systems may also include biodegradable polymers such as polylactic acid, polyglycolic acid, polycaprolactone, and the polyorthoesters. It is generally believed that these biodegradable matrix systems release their bioactive agents during the biodegradation of the matrix polymer. Diffusion mechanisms may be operative in these systems and this would be a function of the unique bioactive-agent/polymer system under consideration. Chemical carriers of bioactive agents may include soluble and insoluble polymers. Biodegradable matrix systems may be considered to be an example of a chemical carrier, as the bioactive agent is released through

Table 2. General Types of Bioactive Polymeric Systems

1. Polymers as physical carriers of bioactive agents: matrix and monolithic devices
2. Polymers as chemical carriers of bioactive agents: soluble and insoluble polymers
3. Entrapped and immobilized bioactive species

the hydrolysis of labile bonds in the main chain of the polymer. The general mode of application for these systems is implantation or injection into muscle or adipose tissue. Soluble polymers which carry bioactive agents present a special set of circumstances in regard to biocompatibility. If chemical carriers of bioactive agents utilizing soluble polymers are injected directly into the blood or following injection into tissue the polymer bioactive-agent system diffuses into blood, the bioactive-agent polymer system must be treated as a drug with regard to its biocompatibility. The three main phases of drug action must be considered. These include: the pharmaceutical phase, which is the disintegration of the dosage form or dissolution of the bioactive-agent polymer system; the pharmacokinetic phase, which involves adsorption, distribution, metabolism, and excretion; and the pharamacodynamic phase, which involves biocompatibility of these systems in their interaction with the various cells and organs in the reticuloendothelial system.

Entrapped and immobilized bioactive species differ from physical and chemical carriers of bioactive agents in that they function without the release of the bioactive agent. Examples include entrapped glucose oxidase, within a highly permeable polymer matrix, for use in the detection and quantitation of glucose for insulin release systems. Immobilized systems involve the interfacial immobilization of enzymes, antigens, and antibodies, Immobilized systems differ from entrapped systems in that the bioactive agent is at the interface with blood or tissue.

2. Biocompatibility—General Considerations

Any test utilized to determine the biocompatibility of a given bioactive-agent polymer system must be well-defined and capable of reproduction. It is also generally accepted that standard biocompatibility tests are also accurate, significant, specific, and economical. These criteria, coupled with the multi- and interdisciplinary factors (Table 1) which control biocompatibility, illustrate the task which must be faced in determining the biocompatibility of a given bioactive-agent polymer system in a given application for a given implant time. In determining the tissue biocompatibility or the blood biocompatibility of a bioactive-agent polymer system, the limitations of the test methodology must be considered. This is an extremely important consideration as there is a tendency to extrapolate results from a static test to the dynamic situation, and the compatibility under nonfunctional conditions to that of functional conditions.

A comprehensive evaluation of the biocompatibility of a bioactive-agent polymer system would include the biocompatibility evaluation of the bioactive agent, the polymer system, and the bioactive-agent/polymer system. If the bioactive-agent/polymer system is to be considered for injection, a determination of the biocompatibility of the vehicle must be done. The vast majority of bioactive-agent/polymer systems under consideration involve the use of well-known drugs whose physicochemical and biological properties are relatively well known. Lee and Robinson have addressed these considerations in a review on physicochemical and biological factors influencing the design and performance of bioactive-agent/polymer systems.[2] An in-depth analysis of these considerations is beyond the scope of this paper, however, abundant information is available in the literature regarding various biocompatibility aspects of bioactive agents.

In general, biocompatibility testing protocols involve *in vitro* or *in vivo* interaction of the candidate material or device with cells, tissues, and organs. Both *in vitro* and *in vivo* tests are usually designed to mimic the in-use interactions and behavior of bioactive polymeric systems. *In vitro* tests commonly involve cell-culture techniques, and the primary purpose of these tests is to identify the toxic behavior of a material or device which may lead to toxicity and cell death. *In vivo* tests obviously use implantation procedures and provide a more realistic appraisal of the behavior of the material or device. *In vivo* tests also permit the examination of both the local-tissue response and the systemic responses, if present. Table 3 identifies the three major approaches to biocompatibility testing of bioactive polymeric systems. Primary acute toxicity screening involves both *in vitro* and *in vivo* methods. Short- and long-term implantation of bioactive polymeric systems provide information on the interaction of the system with cells, tissues, and organs. As seen in Table 1, numerous factors can influence the biocompatibility of bioactive polymeric systems. Functional assays for biocompatibility may also be carried out on candidate bioactive polymeric systems. These tests usually measure in a quantitative fashion some interaction parameter of the material with the cells or tissues. Functional assays may also involve the study of the release behavior and pharmacokinetics of the candidate bioactive polymer system. Numerous reviews are available on various aspects of the biocompatibility of biomaterials and implantable devices.[3,4] These studies are important as they provide prospectives on the identification and testing of various factors involved in the tissue/material interaction.

Table 3. Approaches to Biocompatibility Testing of Bioactive Polymeric Systems

1. Primary acute toxicity screening
2. Tissue-interaction studies: short- and long-term implantation
3. Functional assays for biocompatibility

3. Primary Acute Toxicity Screening

Autian and co-workers have developed a primary acute toxicity screening program for the toxicological evaluation of biomaterials.[5,6] As Autian points out, the two most important criteria for an implantable material or device are the biofunctionability and biocompatibility during the anticipated life span of the material or device. If one or the other fails, the patient may be at risk. Biofunctionability refers to the ability of the device to perform the purpose for which it was designed and this is of the utmost importance in considering bioactive-agent/polymer release systems. Biocompatibility refers to the ability of the material or device to remain biologically inert during the period of implantation. Biocompatibility from Autian's perspective is assessed in terms of the toxicity or lack of toxicity of the material or device.

Although Autian and his group have not specifically addressed the question of biocompatibility of bioactive-agent/polymer release systems, it is apparent that the rationale for the development of the acute toxicity screening program considers functional aspects of bioactive-agent/polymer sustained-release systems. The acute toxicity screening program has been primarily designed to assess toxicological manifestations of an implanted biomaterial or device which results primarily from the release of chemical constituents from the material. These chemical constituents, if released from the material upon implantation, may lead to a local-tissue response; a systemic toxicological response; an allergic response; and carcinogenic, teratogenic, and/or mutagenic responses. One or more of these toxic consequences may result following the implantation of a device or material.

The release of chemical constituents from materials following implantation is exactly what bioactive-agent/polymer sustained-release systems are designed to provide. It is anticipated that the chemical constituent which has been released will lead to a desirable local-tissue response or systemic response. Thus, Autian's primary acute toxicity screening program can be considered to be an excellent preliminary test for the biocompatibility of bioactive-agent/polymer sustained-release systems. Important in this consideration is the fact that many release systems show a burst effect immediately following implantation. In these cases, the bioactive agent may be present in the local-tissue environment in toxic concentrations. An examination of the acute release behavior coupled with the primary acute toxicity screening tests will lead to an appropriate evaluation of the material.

Table 4 describes briefly the tests carried out on material and material extracts in the primary acute toxicity screening program. The tests on extracts of the material may be redundant when bioactive-agent/polymer

Table 4. Primary Acute Toxicity Screening: Material and Material Extracts

1. Tissue-culture/agar overlay responses for the material and extracts: detects the response of monolayer-cell culture to readily diffusible components from materials, 1 day.
2. Intramuscular rabbit implant test—gross observation and histopathology evaluation: detects the response of tissue to leachable components from materials, 7 days.
3. Rabbit blood hemolysis test: detects the ability of the material to lyse red blood cells, 1 h.
4. Intracutaneous test in rabbits on extracts: detects the irritant response elicited by extracts, 3 days.
5. Systemic toxicity in mice on extracts: measures the systemic toxicity of extracts, 7 days.
6. Inhibition of cell-growth test on aqueous extracts: measures the ability of distilled-water extracts to inhibit cell growth in culture, 3 days.

sustained-release systems are being tested. The toxicity characteristics of drugs or bioactive agents commonly used in sustained-release systems are generally known, but the primary acute toxicity screening program provides for an appreciation of the cell and tissue responses at the local level. Results from the primary acute toxicity screening program provide for a cumulative toxicity response (CTI). Fourteen biological responses are used to determine the cumulative toxicity response.

The primary acute toxicity screening program may be utilized in a number of different ways to examine the behavior of bioactive-agent/polymer sustained-release systems. First, burst effects with changing rates of release and variable concentrations of the bioactive agent in the local-tissue environment may be examined. Second, this program provides an easy screening test for the comparison of the parent polymer system and the bioactive-agent/polymer system. Finally, biodegradable polymers may be examined and the degradation or hydrolysis products from these materials may be tested. Thus, the primary acute toxicity screening program can provide a substantial amount of information regarding the biocompatibility of bioactive-agent/polymer release systems.

4. Tissue-Interaction Studies

The short- and long-term implantation of candidate materials and systems in animals can provide information relevant to the biocompatibility of these materials or systems. These tests commonly provide an indication of the tissue interaction with the implanted material or system. Numerous reviews are available on this general topic.[7–12] For the most part, these tests are subjective in nature and it is difficult to ascertain subtle differences in biocompatibility.

Gourlay and co-workers have described an *in vivo* method for screening polymeric materials for biocompatibility.[13] Their test is based on grading

Table 5. Biocompatibility Testing: *In Vivo* Implantation Studies on Polymer Powders

1. Indicators of toxicity
    - Degree of muscle cell damage
    - Thickness of tissue response
    - Overall cell density
    - Number of polymorphonuclear leukocytes and erythrocytes
    - Number of eosinophils, lymphocytes, and foreign-body giant cells
    - Number of fibrocytes and mononuclear phagocytes
2. Mean standard toxicity scores (MSTS)
    - Nontoxic: Polymers with MSTS's equal to or significantly greater than that for poly(glycolic acid), but significantly less than that of poly(isobutyl 2-cyanoacrylate).
    - Moderately toxic: Polymers with MSTS's not significantly different than that of poly(isobutyl 2-cyanoacrylate).
    - Toxic: Polymers with MSTS's significantly greater than that of poly(isobutyl-2-cyanoacrylate) and equal to or greater than the value of poly(methyl-2-cyanoacrylate).

acute and subacute tissue reactions at 7 and 28 days, respectively, of polymeric powders following implantation in rats. Tissue reactions to implanted polymer particles were assessed by a modification of the method to grade the tissue response to gut sutures. In the Gourlay test, differential weighted scores were used to grade different indicators of the tissue response to the implanted polymers. Table 5 lists the various indicators of toxicity used by Gourlay *et al.*, and also the immune standard toxicity scores relative to standard polymers. The various indicators of toxicity were graded and the grade for each of the six indicators was multiplied by an empiric weighting factor listed in Table 6. The sum of the scores of the six tissue-response indicators was the tissue reaction score for a given polymeric material. In each case, the higher the tissue reaction score, the more toxic the response in the tissue examined. Mean standard tissue reaction scores were derived from the six different tissue response scores at days 7 and 28.

Table 6. Tissue-Response Weighting Factors

Weighting factor	Tissue response
7	Muscle cell damage
6	Number of polymorphonuclear leukocytes and erythrocytes
5	Thickness of tissue response
3	Overall cell density
2	Number of lymphocytes, eosinophils, and foreign-body giant cells
1	Number of fibrocytes and mononuclear phagocytes

Langer *et al.* have utilized the rabbit cornea as an *in vivo* site for determining the biocompatibility of polymers to be used for the sustained release of macromolecules.[14] The rabbit cornea has a number of advantages over other implant sites for determining biocompatibility. *In vivo* observations can be made easily and frequently. The cornea is convenient to view and implanted polymers can be observed without sacrificing the animal. The cornea is a sensitive indicator of inflammatory stimuli because it is transparent and avascular. Variations in the biocompatibility as reflected by different inflammatory characteristics, such as edema, cellular infiltration, and neovascularization, can be observed with the use of a stereomicroscope.[15] Langer's studies showed that poly(2-hydroxyethyl methacrylate) and alcohol-washed ethylene–vinyl acetate copolymer were noninflammatory in the corneas of rabbits. Other polymers such as polyacrylamide and poly(vinyl pyrrolidone) produced significant information. Implantation of polymers and bioactive-agent polymer systems in rabbit corneas can provide useful information, although not quantitative, regarding biocompatibility and toxicity.

The relatively recent development of methods to prepare polymeric microcapsules and microspheres, especially for biodegradable polymers, has provided the opportunity to consider injectable bioactive-agent/polymer systems. In considering the biocompatibility of injectable systems, the injection vehicle must also be evaluated for biocompatibility. Fortunately, injection injury has been well studied. Gray has provided an excellent review article on the pathological evaluation of injection injury.[16] Topics covered in this review article include factors affecting tolerance injection sites, injection injury in muscle, evaluation of muscle injury in animals, injury injection in subcutaneous tissue, intravenous injection injury, and injection sequelae and artifacts. Although Gray does not address the biocompatibility of injected polymeric systems, his review provides a basis for understanding injection injury. Unfortunately, there is a paucity of studies in the literature which address the biocompatibility of injected bioactive-agent/polymer systems. Gurny and co-workers have provided some information on the histopathologic evaluation of biodegradable polyester lattices injected into the thigh muscles of rats.[17] In addition, they have provided information on the injection injury provided by saline, Tween 85, sodium laurel sulfate, and Pluronic F68 vehicles.

Soluble polymers as chemical carriers for bioactive agents present unique considerations in regard to their respective biocompatibilities. These unique considerations involve the systemic effects of the soluble bioactive-agent/polymer system. While soluble polymers carrying bioactive agents may be injected directly into the bloodstream and may contain antibodies for targeting purposes, organs and tissues other than those to

Table 7. The Mononuclear-Phagocyte System

Bone marrow	Committed Monoblasts Promonoblasts
Blood	Monocytes
Tissues	Inflammatory macrophages Liver (Kupffer cells) Lung (alveolar macrophages) Connective tissue (histocytes) Bone marrow (macrophages) Spleen and lymph nodes (fixed and free macrophages) Serous cavities (pleural and peritoneal macrophages) Nervous system (?microglial cells)

which the bioactive-agent/polymer system is directed may be involved in interaction with the bioactive-agent/polymer system. While the bioactive-agent/polymer system may be directed to a specific tissue or organ, the fact that the bioactive-agent/polymer system may interact with other tissues or organ systems requires that these tissues or organs also be investigated for their respective biocompatibilities. In this regard, the chemical and physical properties of the bioactive-agent/polymer system may play a more important role in determining biocompatibility than that usually described for implantable systems.

Variable organ effects must be considered when the bioactive polymer system is designed to be injected directly into the bloodstream. These systems usually utilize soluble polymer as chemical carriers for the bioactive agent. Numerous reviews have been published on these systems.[18–21] Systemic aspects of biocompatibility must be considered with these systems. Defense mechanisms common to the organs and tissues of the body must be considered in addressing the biocompatibility of these bioactive polymer systems. In this regard, the mononuclear-phagocyte system (reticulo-endothelial system) is of primary importance. Table 7 provides a list of the numerous cells and organs or tissues which are involved in this system. The mononuclear-phagocyte system continuously monitors the vascular system for foreign substances. The cells of this system are highly specialized for the function of endocytosis and intracellular digestion and remove foreign substances from the bloodstream.

Salthouse and co-workers have utilized enzyme histochemical techniques to appreciate cell function at the implant site.[22–24] This effort to examine the cellular enzyme activity at the polymer–tissue interface has led to a new and different approach toward the evaluation of biocompatibility. The study of cellular enzyme activity at the polymer–tissue interface has led to a greater appreciation of the effects of different materials on the cellular response and the function of the cells associated with the implanted materials. Utilizing the enzyme histochemical technique, Salthouse *et al.* have shown that the presence of different implant materials can exert a modifying effect on both the cellular and enzyme response to the material with a greater variation in the pattern and level of enzyme activity being found at the polymer–tissue interface. Using microspectrophotometry and standard implant and tissue specimens, these investigators have shown the quantitative variations, measured as optical density, in acid phosphatase and aminopeptidase with implant duration for various types of polymers.

To place biocompatibility on an objective basis, we have investigated the use of the cage implant system to determine the dynamic nature of cell function occurring at the implant site. This cage system containing the biomedical polymer or the drug–polymer system permits the formation of an inflammatory exudate which surrounds the material.[25,26] The cage implant system provides a simple means by which the inflammatory exudate which bathes the biomedical polymer within the cage can be monitored in a serial fashion without sacrificing the animal. Following the withdrawal of a small fraction of the inflammatory exudate which surrounds the biomedical polymer or drug–polymer delivery system, total white-cell concentrations, polymorphonuclear-leukocyte concentrations, mononuclear-leukocyte concentrations, extracellular alkaline phosphatase activity, and acid phosphatase activity can be determined in a quantitative fashion. Thus, assays which are sensitive, reliable, reproducible, and quantifiable can be used to determine the temporal variations in cell function in the inflammatory exudate which interacts with the implanted biomedical polymer or drug–polymer delivery system. In addition, the cage system also permits the determination of the release characteristics of polymeric drug delivery systems with time by appropriate quantitation of the drug present in the inflammatory exudate at various time intervals. Thus, the cage system

Table 8. Cage Implant System and Biocompatibility Assays

Material changes
Surface properties
Bulk properties
Biodegradation
Inflammatory exudate
Quantitative and differential white-cell counts
Extracellular enzyme analysis
Total protein and protein electrophoresis
Quantitative drug analysis

provides correlative, comparative, and quantitative information on the release characteristics and cellular response and activation, that is, biocompatibility, of polymeric delivery systems. Table 8 summarizes various assays that may be carried out to determine biocompatibility of parent polymer and drug–polymer sustained-release systems. (Details on the cage implant system have been provided elsewhere.)

The cage system permits an analysis of the parent polymer and the drug–polymer sustained-release system allows comparisons to be made and the local response to the released drug to be determined. We have carried out such a comparison using silicone rubber and gentamicin. In the following study, silicone rubber/gentamicin implants containing 40% gentamicin by weight were used. Control specimens were empty cages that were implanted.[26]

Implantation of the stainless-steel cages provokes an acute inflammatory response that is characterized by large numbers of polymorphonuclear (PMN) leukocytes or granulocytes present in the exudate within the cage. The accumulation of white cells and the predominance of PMN leukocytes or granulocytes, principally neutrophils, in the exudates over the first four days following implantation are clearly demonstrated in Table 9 for the control, silicone rubber, and gentamicin/silicone rubber implants.

Resolution of the acute inflammatory response was followed by a mild chronic response that occurred between the first and second weeks of implantation. The variation in the total white-cell concentration in the exudate as a function of implantation time for each system is presented in Table 9. In the acute phase of the response, days 4 and 7, no difference was noted between silicone rubber and the respective control values. On the other hand, the total white-cell concentrations for the gentamicin/silicone rubber system at days 4 and 7 were statistically different than the corresponding

Table 9. The Variation in Total White-Cell Concentration in the Exudate as a Function of Implantation Time

	Total white cells per μl		
Time (days)	Control	Silicone	Gentamicin/Silicone
4	4160 ± 520^a	3640 ± 510	6960 ± 510^b
7	820 ± 180	920 ± 140	2550 ± 260^b
14	260 ± 50	700 ± 180^b	750 ± 10^b
21	120 ± 50	300 ± 80	350 ± 290

[a] $\pm$ items represent mean value $\pm$ standard deviation.

[b] Statistically different at the 95% confidence level ($p < 0.05$) when compared to the mean control value. Student's t test for unpaired samples.

Table 10. Gentamicin Release from Silicone Rubber into the Exudate with Time

Time (days)	Gentamicin release (μg/ml)
4	40.7 ± 20.8
7	8.7 ± 1.5
14	4.5 ± 1.9
21	3.5 ± 0.9

control values, indicating an increased inflammatory response. Since the drug-free silicone rubber system was not different, this would suggest that the gentamicin released into the exudate over the first week of implantation elicited this variation in the inflammatory response. Table 10 shows the concentration of gentamicin present in the exudate at times comparable to those for the cell and enzyme studies. As expected, a large burst effect had occurred over the first four days.

The variation in the extracellular alkaline phosphatase activity in the exudate is a function of implantation time for the control, silicone rubber, and gentamicin/silicone rubber systems as shown in Table 11. The silicone rubber system was comparable to the controls at each time period indicating that cells in the exudate in the silicone system were not activated to release their granular constituents. Statistical differences with the control were noted for the gentamicin/silicone rubber system at day 4 following implantation. This suggests that specific interaction between the cells in the exudate and the gentamicin/silicone rubber system resulted in exocytosis of alkaline phosphatase from the cells in the exudate. As alkaline phosphatase is found primarily in polymorphonuclear-leukocytes in the acute inflammatory response, it is not remarkable that there were no significant differences for the

Table 11. The Variation in Extracellular Alkaline Phosphatase Activity in the Exudate as a Function of Implantation Time

	Alkaline phosphatase activity (U/dl)		
Time (days)	Control	Silicone	Gentamicin/Silicone
4	20.9 ± 3.2[a]	21.4 ± 7.4	46.0 ± 9.0[b]
7	14.7 ± 1.9	17.5 ± 3.0	21.2 ± 4.0
14	11.0 ± 1.2	9.8 ± 0.7	12.9 ± 3.5
21	8.4 ± 2.1	9.8	8.4 ± 0.7

[a] ± items represent mean value ± standard deviation.

[b] Statistically different at the 95% confidence level ($p < 0.05$) when compared to the control value. Student's t test for unpaired samples.

silicone rubber and gentamicin/silicone rubber systems when compared to the control values at days 14 and 21.

The silicone rubber system, when compared to the controls, did not show an increased inflammatory response, as measured by the white-cell concentration in the exudate, or evidence of white-cell activation in the acute stage, as measured by the concentration of alkaline phosphatase in the exudate. This suggests that the silicone rubber/polymer system is biocompatible. On the other hand, the gentamicin/silicone rubber system exhibited both an increased inflammatory response and evidence of cell activation in the acute stage when compared to the controls. As the same silicone rubber was used in both the silicone rubber and gentamicin/silicone rubber systems, it is suggested that the differences in biocompatibility are the result of the gentamicin which is being released from the silicone rubber.

5. Summary

Bioactive polymeric systems presented varying degrees of complexity in the evaluation of their respective biocompatibilities. Not only must the chemical and physical properties of the polymer be appreciated, but the biocompatibility of the bioactive agent under consideration must be determined in each unique situation. The unique combination of a bioactive agent and the polymer must be evaluated for its inherent biocompatibility. Important in these biocompatibility considerations is the function of the device and the specific mechanism by which the bioactive-agent functions. This paper has presented, in a general manner, those variables which may control biocompatibility and methods or protocols which may be used to determine the biocompatibility of bioactive polymeric systems.

References

1. J. Black, *Biological Performance of Materials: Fundamentals of Biocompatibility*, Marcel Dekker, New York (1981).
2. V. H.-L. Lee and J. R. Robinson in: *Sustained and Controlled Release Drug Delivery Systems* (J. R. Robinson, ed.), Chapter 3, pp. 71–121, Marcel Dekker, New York (1978).
3. D. F. Williams, ed., *Fundamental Aspects of Biocompatibility*, Vols. I to II, CRC Press, Boca Raton, Fla. (1981).
4. D. F. Williams, ed., *Biocompatibility in Clinical Practice*, Vols. I & II, CRC Press, Boca Raton, Fla. (1982).
5. J. Autian, *Artif. Organs* **1**, 53 (1977).
6. J. E. Turner, W. H. Lawrence, and J. Autian, *J. Biomed. Mater. Res.* **7**, 39 (1973).
7. D. L. Coleman, R. N. King, and J. D. Andrade, *J. Biomed. Mater. Res.* **8**, 199 (1974).
8. D. L. Coleman, R. N. King, and J. D. Andrade, *J. Biomed. Mater. Res. Symp.* **5**(1), 65 (1974).

9. R. H. Rigdon, *CRC Crit. Rev. Food Sci. Nutrit.* **7**, 435 (1975).
10. L. Marion, E. Haugen, and I. A. Mjor, *J. Biomed. Mater. Res.* **14**, 343 (1980).
11. R. I. Leininger, *CRC Crit. Rev. Bioeng.* **1**, 333 (1972).
12. C. A. Homsy, *J. Biomed. Mater. Res.* **4**, 341 (1970).
13. S. J. Gourlay, R. M. Rice, A. F. Hegyeli, C. W. R. Wade, J. G. Dillon, H. Jaffe, and R. K. Kulkarni, *J. Biomed. Mater. Res.* **12**, 219 (1978).
14. R. Langer, H. Brem, and D. Tapper, *J. Biomed. Mater. Res.* **15**, 267 (1981).
15. S. B. Aaronson and R. C. Horton, *Arch. Ophthalmol.* **85**, 306 (1971).
16. J. E. Gray in: *Sustained and Controlled Release Drug Delivery Systems* (J. R. Robinson, ed.), Chapter 5, pp. 351–410, Marcel Dekker, New York (1978).
17. R. Gurny, N. A. Peppas, D. D. Harrington, and G. S. Banker, *Drug Dev. Ind. Pharm.* **7**, 1 (1981).
18. L. G. Donaruma, R. M. Ottenbrite, and O. Vogl, eds., *Anionic Polymeric Drugs*, Wiley, New York (1981).
19. G. Gregoriadis, *Drug Carriers in Biology and Medicine*, Academic Press, London (1979).
20. H. Ringsdorf, *J. Polym. Sci., Polym. Symp.* **51**, 135 (1975).
21. E. Chiellini and P. Guisti, eds., *Polymers in Medicine: Biomedical and Pharmacological Applications*, Plenum Press, New York (1983) (in press).
22. T. N. Salthouse and B. F. Matlaga in: *Fundamental Aspects of Biocompatibility* (D. F. Williams, ed.), Volume II, pp. 233–257, CRC Press, Boca Raton, Fla. (1981).
23. T. N. Salthouse and B. F. Matlaga, *Ann. Histochim.* **21**, 105 (1976).
24. T. N. Salthouse, B. F. Matlaga, and R. K. O'Leary, *Toxicol. Appl. Pharmacol.* **25**, 201 (1973).
25. R. Marchant, A. Hiltner, C. Hamlin, A. Rabinovitch, R. Slobodkin, and J. M. Anderson, *J. Biomed. Mater. Res.* **17**, 301 (1983).
26. J. M. Anderson, R. Marchant, and M. McClurken, in: *Long-Acting Contraceptive Delivery Systems* (G. I. Zatuchni *et al.*, eds.), p. 248, Harper & Row, New York (1984).

3

Controlled-Release Pesticides

A Historical Summary and State of the Art

Nate F. Cardarelli and Bernadette M. Cardarelli

Abstract. Controlled-release pesticide technology began in prehistoric times with the advent of antifouling paintlike composition for ship hulls. Such leaching-type systems were gradually improved upon over the centuries. In 1964 it was discovered that very long-term release of an antifouling agent from an elastomeric matrix could be achieved through a diffusion–dissolution phenomenon. In the same decade, novel systems based upon the incorporátion of a volatile insecticide in a noncompatible carrier within a plastic matrix were developed that could provide three or more months' emission of the agent.

The diffusion–dissolution concept was extended to molluscicide emission in 1966 and herbicide release in 1969. Carrier-type systems were applied to solid-pesticide release in the 1960s and early 1970s. In 1976 the discovery of new methods of achieving requisite porosity in plastic-based compositions permitted the development of long-lasting molluscicides, aquatic insecticides, aquatic and terrestrial herbicides, and plant nutrients.

In parallel development work, the principle of coacervation was used to form toxic microcapsules which found a ready market in the 1970s for control of crop insects. Variations of microcapsular and other systems were developed in the mid- and late 1970s for use in insect control through pheromone release.

This report briefly describes controlled-release pesticides in their historical context from earliest time to the present state of the art up to late 1982.

1. Antifouling

1.1. Toxic Paints

The attachment and accumulation of marine organisms on ship hulls and other surfaces exposed to temperate and tropical waters results in the destruction of streamlines, increased weight, and general deterioration of structures. The problem is ancient and the first solution was discovered by mariners in prehistoric times. At some point in history prior to 1400 B.C., seafaring men of the Mediterranean area, and presumably elsewhere,

Nate F. Cardarelli and Bernadette M. Cardarelli • Unique Technologies Inc., 71 South Cleveland Avenue, Mogadore, OH 44260.

discovered that arsenic compounds and sulfur in wax or asphaltum would discourage barnacles and other foulants for a period of time. Also lead sheathing was used with some degree of effectiveness in preventing both fouling and the penetration of marine worms.[1] In each instance the slow release of toxic ions provided an interface poisonous to larval foulants. Both methods were used throughout the ages with gradual improvement noted in antifouling coatings. Prior to 1625 A.D., copper sheathing was also introduced.

The advent of metal hulls negated the use of copper sheathing which had been popular through medieval times. The copper–iron galvanic cell created led to rapid corrosion. Sandwich-type constructions were attempted whereby the iron and copper were separated by a wooden barrier, but this too failed and the marine engineers' attention was focused on the development of long-lasting paint coatings.[2]

In 1854 McInnes patented a metallic soap containing copper sulfate which was applied hot over the already painted hull.[3] McInnes also recommends the use of arsenic and oxides of mercury. The "modern age" of antifouling paint technology began when Tarr and Wonson patented a composition containing copper oxide as the agent suspended in a tar base using naphtha or benzene as the solvent.[4] Copper oxide gradually transplanted other antifoulants such as salts and oxides of mercury and arsenic. By 1943, the U.S. Navy had developed and standardized several formulations based upon copper oxide in a resinous film former. However, the use of copper oxide has several sizable problems. A galvanic cell would develop unless a barrier existed between the antifouling paint and a metallic hull thus requiring an anticorrosive intermediate paint. Another major problem lies in the poor longevity arising from the nature of the paint film.

Antifouling paints in the 1940 to 1960 era functioned through two mechanisms. The basic need was to maintain a toxic interface between the paint film and the surrounding water. Proper emission of copper ions (or mercury, arsenic, lead, etc.) will do this; the trick is to establish and maintain the emission over a reasonable period of time. The interfacial concentration must be greater than a certain threshold value where the toxicity is inadequate to prevent attachment. One type of antifouling paint, now rarely seen, functions through an exfoliating mechanism wherein the film-forming material is slightly water soluble. As solvation occurs, toxic ions or molecules are gradually released. The second mechanism relies upon leaching wherein the loss of the toxicant through gradual solution in water creates porosity within the matrix allowing water ingress and contact with further toxic molecules. In order to develop an adequate porosity, no less than 85% of the paint-film dry weight must be copper oxide with superior paints ranging as high as 92%. Consequently, the extremely high essential toxicant loadings lead to films having low tensile and tear strengths, generally mediocre adhesion, and poor resistance to the marine environment. Details regarding the nature of fouling, toxic-paint formulation, antifouling mechanisms, and research and development efforts prior to 1951 are aptly described in Reference 5.

The patent literature describes many paint developments over the three decades from 1930 to 1960. In general, the appearance of new polymers – vinyls, chlorinated natural rubber, chloroprenes, and so on, provided the paint chemist with an opportunity to broaden the art. The clear thrust was in two directions: increased longevity under marine-use conditions and the use of nonionic and/or organic antifoulants. As early as 1946, Young described certain low-molecular-weight aldehyde/phenolics (e.g., the condensation product of *p*-cresol with cinnamic aldehyde), as antifouling agents.[(6)]

1.2. Organotin Antifoulants

The organotins first appeared in the patent literature in 1960 wherein various triphenyl and tributyltins were shown to have antifouling properties.[(7)] Miller examined a number of pesticidal materials against several major categories of fouling and discovered that bis(tri-*n*-butyltin)oxide (TBTO) showed broad-spectrum activity against fouling organisms indigenous to southern Floridian waters.[(8)]

Antifouling paints containing TBTO appeared on the commercial market in 1962 using a vinyl or epoxy polymer as the film-forming element. They were clearly inferior in terms of longevity to the copper oxide paints.[(9)] They did possess some advantages – the major advantage lying in their nonionic nature thus avoiding the need for an anticorrosive undercoat. A number of organotins have been found useful as antifoulants: bis (tributyltin) adipate,[(10)] tributyltin pentachlorophenate,[(11)] tributyltin sulfide,[(12)] and tributyltin fluoride,[(13)] as well as combinations of organotins and other pesticidal agents.[(14)]

1.3. Antifouling Rubber

In 1964 one of the authors (N.F.C.) was presented with a sonar-dome problem. The housing of several types of domes is composed of reinforced plied rubber. Antifouling paints are ineffective due to their rapid loss through cavitation phenomena associated with high-intensity sound-wave transmission. One solution was to incorporate an antifouling agent within the rubber negating the use of a toxic paint. Copper and/or mercury salts can be formulated in the elastomers involved, in this case various chloroprenes, but do not release into the interface and thus provide no fouling protection.

It was discovered that TBTO and other organotins would slowly emit from an elastomeric surface upon exposure to water. The mechanism involved is based upon a diffusion phenomenon rather than upon leaching. Organotins are soluble to varying extents within not only chloroprene, but also natural rubber, acrylonitrile rubber, butyl rubber, and, in general, those materials thermodynamically and kinetically classified as elastomers. Rates of solubility vary from 2% to over 12%.[9] Some degree of solubility is also noted in the so-called plastomers which are polymers and copolymers possessing both an elastomeric and a plastic nature, such as the ethylene–propylene–diene polymers. True plastics are not solvents for the antifouling organotins. Upon formulating an organotin in elastomers, wherein the agent content is held below the solubility limit, a condition of solution equilibrium is established. The elastic and strength properties of the polymeric matrix are not affected to any appreciable degree. Organotins are highly soluble in elastomers and of very limited solubility in water – from under 1 ppm to not much greater than 3 ppm.* Upon immersion in water, agent molecules on or very near the elastomer surface move into the interface through mathematically describable dissolution processes. This, in turn, creates a localized solution disequilibrium and internal agent molecules, the solute, migrate towards the depleting surface driven by solution pressure. The cycle repeats and long-term agent emission is observed.

The process is termed diffusion–dissolution and is aptly described in the technical literature.[9,15,16] The first commercial TBTO/neoprene coating was used on several U.S. Coast Guard maintained buoys.[17] Effective control of barnacles and other foulants was noted for over nine years. Various small-ship hulls and other objects were similarly covered and, in general, efficacy was maintained for over 10 years.[18] The product, marketed by the B.F. Goodrich company of Akron, Ohio as NOFOUL™, showed a decrease in tin content from 2.4% initially to 0.61% at the 10-year mark with a drop in tensile strength from 1600 to 664 psi.

This invention led directly to the development of insecticidal, herbicidal, and molluscicidal controlled-release rubber formulations which will be described in the succeeding sections of this chapter.

In order to improve agent-emission longevity (because in 1964/65 we did not know that the product would exceed 10 years), several other developments were made. The thrust of this effect was to find a method of increasing organotin content without seriously affecting physical properties – as is the case when the solution limit is exceeded. It was found

* The literature, including the handbooks, claim water solubilities to 30 ppm. They are *incorrect*. Difficulties involved in measuring solubility of extremely hydrophobic materials are well known. Water "solutions" invariably contain molecule aggregates in colloidal suspension and are not true solutions as scientifically defined.

possible to use a membrane–reservoir system wherein the reservoir was literally a sponge mechanically holding up to 30% organotin and laminated to a rubber sheet which would slowly transmit the agent from the reservoir to a watery interface.[19] A second approach was to first add the organotin, which must be liquid, TBTO, or tri (*n*-butyltin) sulfide, to a hollow microcapsule, such as the phenolic "microballoons," and then mechanically bind the filled microcapsules within a polymeric matrix.[20,21]

NOFOUL™ elastomers are now recognized for their long-term biological efficacy[22] and have been used, basically by the U.S. Navy, in specialty applications. Whereas antifouling rubber coatings are far superior in terms of biological lifetime to antifouling paints, whose efficacy in tropical waters seldom exceed eight months,[23] the expense usually does not justify the results. Antifouling rubber is applied using several adhesive coats and a fair degree of skill on the part of the applicator.

1.4. Extension of the Concept

The need to preserve wooden structures, such as pilings, utility poles, and fence posts, exposed not only to damage by marine organisms but also subject to attack by fungal organisms, boring worms, insects, and even birds have long been recognized. Creosote and creosote-type materials have been utilized by themselves, and with pesticidal additives for some time.[24] Organotin impregnation of pilings has been found useful.[25] Lower-alkyl organotins combine readily with cellulosic materials providing a cellulosic–organotin inimical to worm and fungus attack. The fungicidal and bactericidal nature of the R_3SnX compounds, wherein R = alkyl groups having two to five carbon atoms, is well recognized.[26–28] Unfortunately, the use of organotin impregnates on marine wood will control worm and borer attack but is of little value in the prevention of sessile fouling.[29]

Artifacts submerged in tropical and temperate waters are not only subject to fouling but are also attacked by various marine worms, such as the teredo or "shipworm"; the *limnoria* species; the "gribble" or "marine termite," whose sphere of activity, being air breathers, is at the water/air interface; and pholad clams which will burrow into concrete pilings and sea walls. Penetration of the lead sheathing of marine cables has been reported[30] and active boring attack on objects at depths as low as 6000 ft.[31] Tributyltins and triphenyl tins are quite toxic to teredo larvae, far more so than creosote and various conventional pesticides.[32]

Edelstein *et al.* added oxybisphenoxarsine to neoprene and showed that sonar domes could be protected against fouling.[33] However, the use of arsenic-bearing compounds has not progressed, presumably due to potential environmental concerns.

The Australians have done considerable work with organotin-bearing antifouling elastomers with results similar to those seen with the B.F. Goodrich product.[34] A $2\frac{1}{2}$-year effective life was found using 10% tributyltin fluoride in neoprene.[35]

1.5. Further Developments in Antifouling Paint Technology

Antifouling sheet rubber use has been essentially confined to special-use areas, such as sonar domes, propellers, rudders, and so on, where conventional antifouling paints fail rapidly. Economics and other practicalities dictate the continued use of paint-type coatings for most marine objects. Antifouling paint development in the last decade has taken two basic approaches: improved toxic agents, with most studies keyed to organotin and organolead compounds, and improved paint vehicles.

Wolf and Van Londen were the first to update the antifouling paint compounds "bible," Reference 5, by further description of the copper-based systems.[36] In 1970 it was shown that poor antifouling capability arises from certain basic incompatibilities between copper compounds and typical paint vehicles, whereas such problems can be overcome using organotin, organomercurials, and combinations of the two.[37] Various new tin compounds were discovered having not only unique antifouling capabilities, but also improved compatibility with polymers used as the film-forming element in marine paints.[38–42]

Considerable effort has been expended in the development of antifouling paints based upon triphenyllead acetate and other organoleads.[43]

Improved polymeric vehicles for organotins and other antifoulants have tremendously increased antifouling paint efficacy in terms of coatings integrity and foul-free life. The literature is replete with hundreds of reports. Certain discoveries stand out as milestones of the technology. In 1963 Friedsham noted a bivalent-metal treatment of modified alkyl resins improved physical properties.[44] One objectionable quality of organotin paints lies in the need for relatively high agent loads, 45–55% based on dry weight. Fahlstrom showed that chlorinated methanobenzene-based compounds added to the paint formulation allowed the reduction of agent loading to 7% or less.[45] Proper resin/rosin proportions also improved antifouling paint performance.[46] Improved methods of maleinization and catalyzation of alkyd–melamine resins led to enhanced antifouling paint quality.[47]

1.6. Mixed-Agent Paints

In 1969 Beers noted that the addition of zinc powder or zinc oxide to paints containing an organotin or an organoarsenic enhanced performance.[48] Similarly, a synergistic action between tributyltin fluoride and triazine A (which is rubber soluble) was observed.[49] Also organoleads in combination with copper oxide improved antifouling longevity,[50] as well as combinations of organoleads and organotins.[51]

1.7. Notes on Antifouling Paint Technology

Whereas hundreds of major articles have been published concerning antifouling paint research and development over the last decade, there are a few, hitherto unmentioned, of considerable value. Antifouling mechanisms are ably discussed by Phillip,[50] Carr and Kronstein,[52] Bishop and Silva,[53] and the now-dated review by Saroyan.[54] Monaghan *et al.* have described the organotin release mechanism from paints.[55] Beiter has provided a comprehensive review of the factors affecting toxicant emission from paint emphasizing the organotins.[56]

1.8. Organometallic Polymers

It has been known since 1965 that organotins can be polymerized with acrylic and methacrylic acid into film-forming polymers (author's, N.F.C., unpublished work). In 1972 Kanakkanatt pointed out that TBTO would readily chemically combine with and cross-link neoprene rubber.[57] Akagane and Matsuura reacted triethyltin hydroxide and other organotins with a copolymer of styrene–maleic anhydride to create an antifouling polymer having appended an organotin moiety.[58] This discovery extended the Montermoso technology describing organotin acrylic polymers.[59] Release was brought about, in the Akagane and Matsuura development, through hydrolysis of the pendent moiety according to the equation. $L = K_0(e^{-kt} - e^{-Dt})$, where L is the emission rate, K_0 is an experimental constant, D is the diffusion coefficient, and k is the hydrolysis constant.

In 1972 organotin polymers were copolymerized with fluoromonomers to create antisliming optical materials for underwater use.[60] A number of reports have been published by the U.S. Navy describing the methods of preparing OMPs (organometallic polymers) wherein several organotins have been appended to polystyrene, polymethacrylate, polyester, polyvinyl, and cellulose polymer backbones.[61–63] Test results are provided which indicate several year's antifouling potential.[62,63] The nature of the invention is thoroughly discussed in the pertinent patent.[64] The OMPs can be incorporated in a paint vehicle or in elastomeric compounds. Upon exposure to the marine environment, the toxic moiety is slowly emitted, through hydrolytic cleavage, rendering the interface repellent to fouling organisms.

The excessively high leach rates observed with conventional organotin paints is avoided thus substantially reducing environmental impact.

The OMPs are film formers, but the relatively poor strength characteristics of those OMPs examined by the authors will probably preclude their direct use as the film-forming element of an antifouling paint. It has been observed, and confirmed, by the senior author that OMPs, which are extremely hydrophobic, when formulated with a hydrophilic polymer provide lacquers which are not only antifouling but also reduce turbulent drag by 3–4% under dynamic laboratory conditions. Three field trials have been confirmatory with maximum velocity increases of 1.5–2.5 knots. This work is still in the experimental stage, but will be published in due time.

Dunn and Oldfield covulcanized tributyltin acrylate with a number of elastomers, natural rubber, styrene–butadiene rubber, polybutadiene, polychloroprene, and so on, and also plastomers and true plastics, such as polyethylene, using a peroxide cure to produce elastomeric backbones with an appended organotin group.(65) Upon immersion, long-term antifouling was observed.

In his work with organolead antifouling paints, Kronstein discovered that the antifoulant released contained not only the lead component but also the polymeric component of the paint.(66) For instance, when triphenyllead acetate is incorporated in a VYHH (87% vinyl chloride–13% vinyl acetate) base, the released antifoulant is the triphenyllead acetate–VYHH reaction product.(67,68)

2. Molluscicidal Elastomers

The development of a successful long-term antifouling rubber based upon agent emission through a diffusion–dissolution mechanism encouraged the search for other areas, wherein an aquatic pest might be controlled through application of this technology. One promising area was freshwater snail control.

2.1. Snail-Borne Parasitic Disease

A number of species of tropical snails serve as the intermediate host for various parasites affecting man. Schistosomiasis, a dread disease afflicting in excess of 300 million human victims, is relatively well known. However, there are five other parasitic diseases, each afflicting millions of humans, wherein various snail species are implicated. Considering just schistosomiasis, where the impact in terms of human mortality and suffering, and economic costs, are the greatest, the clear need for long-term control methods is evident.

Whereas a number of drugs are available for human treatment, the major ones are all carcinogenic and two are mutagenic; no immunity is conferred nor is the disease cured. Destruction of the parasite, a trematode worm, in the human body can be accomplished; but the damage done to the liver and other organs, basically through trematode ova blocking minor blood vessels and cutting off the nutrient and oxygen supply to the cells, is not repairable. Consequently, the disease can be arrested but a "cure" probably never occurs. Since immunization does not exist, the "cured" human can be, and usually is, reexposed to the disease upon contact with infested water. Thus therapy alone is inadequate.

The schistosome life cycle can be described as follows. Adult worms within the human body reproduce sexually, the female laying thousands of fertile eggs several times each hour. In the normal sequence, ova are excreted and often enter a freshwater habitat containing susceptable snails. The eggs hatch upon contact with freshwater and the microscopic miracidia are attracted, following a chemical gradient, to the host snail. They penetrate and form a sporocyst wherein asexual reproduction occurs. In time the cercarian larval form leave the sporocyst, entering freshwater, and seek a human target. Once found, penetration occurs in a few seconds and the cercaria enter the body, mature, find a partner of the opposite sex, and the cycle repeats itself. An infected snail sheds hundreds to thousands of cercaria daily.

To halt the disease, one must intervene at some point in the life cycle. Though ova, cercaria, and miracidia are far less tolerant of toxicants, the conventional point of attack has been through the destruction of the snail using molluscicides.

The basic problem is that snails are quite difficult to find and kill. A 99% reduction in population is of no long-term avail and the "missed" snails are extremely prolific so that within a few months population resurgence recreates the same snail density. In any case, snails from nontreated areas soon appear in the treated sites. Consequently, public-health teams must continually re-treat the same area. Conventional molluscicide formulations create a toxic-water condition for at best a few weeks, and usually only a few days. Since economic resources for labor and mollusciciding chemicals are always inadequate, the nations where schistosomiasis is endemic – and there are over 60 of them in Africa, South America, the Caribbean, and tropical Asia – lack the resources necessary for wide-area prophylaxis and therapeutic treatment. Installing modern sanitation, fresh water, and the like to halt human pollution of their own water supply is of such staggering cost as to be beyond the world's combined excess capital.

Obviously, a controlled-release molluscicide would be of great merit. By keeping the infected-water courses continuously toxic to snails by one treatment lasting a number of years, the same economic resources would allow broader geographical scope to the control efforts and at least the potential for eradicating the target snail populations over large territorial extents.

Schistosomiasis, and the other human snail-borne diseases are of a debilitating nature. As the worm burden in a given victim increases, liver function decreases, and the physical ability to work decreases. Lowered productivity thus reduces the afflicted nations' available capital for public health. Thus a disease–poverty cycle, one breeding the other, is established. The degree of human infections run as high, or higher than, 60% in a number of the nations, which means that that portion of the population is producing foodstuffs and other goods below the normal rate.

Although temperate climates are not suitable for the specific nonrefractory snail species, the United States and elsewhere have a problem area in which humans inadvertently intervene in a snail–fish or snail–bird cycle. In such cases, cercaria attempt penetration of the human target creating an allergic reaction arising from the chemical penetrants secreted by the parasite. One such schistosome dermatitis, termed "swimmers itch," not only is an irritant to swimmers, but causes a number of deaths per year in the Great Lakes States and Ontario.

There are other parasitic infections of economic importance where various snail species serve as the intermediate host, such as fascioliasis, which infects sheep and cattle.

2.2. Controlled-Release Organotin Molluscicides

The successful development of a controlled-release antifouling material based upon organotin emission from a sheet elastomer through a diffusion–dissolution mechanism led to the extension of the art into the molluscicide area. In 1965 it was well known that several organotins were effective molluscicides based upon laboratory studies.[69–72] A long-term residual activity had been noted.[73] Field trials with tributyltin oxide applied in a conventional manner showed promise.[74,75]

Antifouling rubber compounds, generally loaded with TBTO, were found effective under laboratory conditions in repeat-challenge bioassay against various snail species.[76–78] A number of small-scale field trials using various TBTO/elastomers were performed in several tropical sites. Table 1 is a compendium of results.

Results were generally good, the long-term activity being of great interest. Analysis of molluscicidal pellets for residual tin after return from 14 and 22 months' service at field sites in Rhodesia indicated over three and possibly five years' release life.[79]

The initial materials evaluated were essentially various antifouling

Table 1. Field Evaluations of Controlled-Release Organotin Compounds[79]

Location	Site description	Compound	Agent
Tanzania	Gravel pit, 42,000 liters	633B	3% TBTO
	Swampy pool, 28,000 liters	633B	3% TBTO
St. Lucia	Marsh, 28 m^2	BioMet™ SRM	6% TBTO
	Ravine, 1 × 20 m	BioMet™ SRM	6% TBTO
Iran	Ponds (six total)	BioMet™ SRM	6% TBTO
Phillipines	Swamp, 373 m^2	CBL-9B	20% TBTF
	Irrigation canal (150 m)	BioMet™ SRM	6% TBTO
Brazil	Ditch	443A	5.8% TBTO
	Irrigated truck farm	443A	5.8% TBTO
	Pool (2 total)	443A	5.8% TBTO
	Pit (2 total)	BioMet™ SRM	6% TBTO
	Ponds (4 total)	BioMet™ SRM	6% TBTO
	Canals (2 total)	BioMet™ SRM	6% TBTO
	Well (1 total)	BioMet™ SRM	6% TBTO
Rhodesia	Reservoir, 950-m margin	NOFOUL™	6% TBTO
	Dam, 480 m^2	CBL-9B	20% TBTF
	Ponds: 2092 m^2, 850 m^2, 768 m^2	CBL-9B1	20% TBTF
	Dam, 1,320 m^2	BioMet™ SRM	6% TBTO
	Dam, 810 m^2	BioMet™ SRM	6% TBTO

rubber compounds based upon TBTO, tri-*n*-butyltin fluoride (TBTF) and bis (tri-*n*-butyltin) sulfide (TBTS). The first commercial material, BioMet™ SRM (M & T Chemicals Company, Rahway, N.J.) became available in 1974. This material is based upon natural rubber rather than the more costly chloroprenes. The controlling art is given by Cardarelli.[80] In a separate development, formulation CBL-9B was the result of a study for the World Health Organization based on a natural rubber/TBTF compound as follows:

Natural rubber	67.70%
Carbon black	6.77%
Zinc oxide	1.35%
PBNA (phenyl-β-naphthylamine)	0.68%
Sulfur	1.69%
Benzothiazyl disulfide	1.35%
TBTF	20.31%
Stearic acid	0.14%

While CBL-9B is probably superior, in terms of lower application dosage, over BioMet™ SRM, the higher cost per pound of the former makes both materials about equal in effectiveness.[81]

There are two major problems mitigating the use of elastomer-based organotin molluscicides – processing costs and lack of knowledge regarding long-term toxicological effects and environmental burden. The latter is gradually being overcome; however, reduction in the multistep processing requirements is not likely with the present state of the art. However, scrap materials, such as rubber reclaim, or inexpensive polymers, such as asphalt, used in the binding element may significantly reduce manufacturing costs.[82–84]

2.3. Technology Based upon Organotin/Elastomer Development

Several organotins have played a key role in the development of controlled-release pesticides used initially on elastomeric-based formulations and later with thermoplastic-based dispensers. To a large extent the antifouling and molluscicide studies showed that the controlled-release concept was workable in these two areas. A natural technological sequence led to the development of copper sulfate/elastomer molluscicides, herbicides,

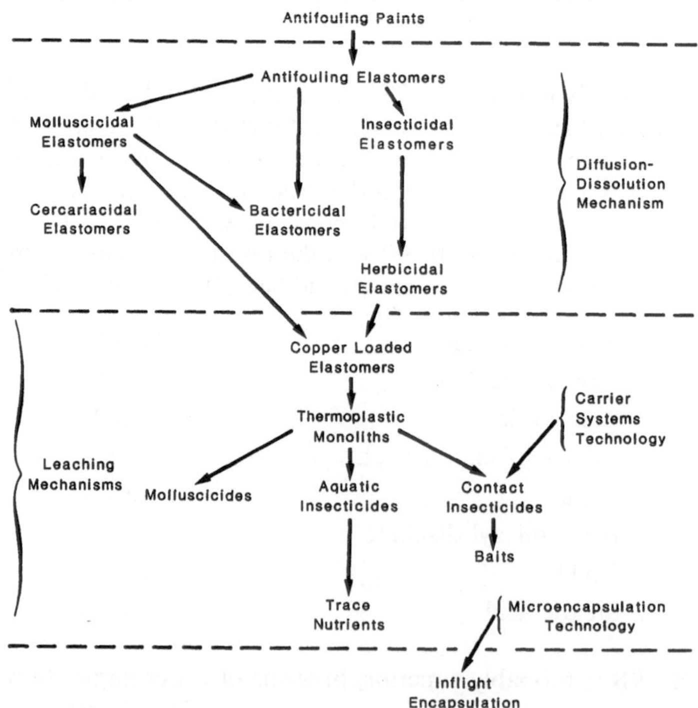

Figure 1. Technological sequence.

and algaecides and extension of the diffusion–dissolution mechanisms to elastomer-based bactericides, herbicides, and aquatic insecticides. A technological sequence chart is shown in Figure 1. The input of parallel technologies is also shown. Each area is described in the following sections.

2.4. Organotin Properties: Toxicology and Chemodynamics

Due to the leading and unique role of organotins in the development of controlled-release pesticide systems, and their predictably continuing place in the technology, a section on their properties is in order.

The alkyl organotins, R_2SnX_2 and R_3SnX, show selective toxicity to animal life depending upon the length of the R group. The methyl and ethyl alkyls use as pesticides is unlikely. The dibutyl and tributyl tins show *relatively* low mammalian toxicity and a high degree of toxicity to freshwater aquatic snails, amphibious snails, and select insecta. Susceptibility in aquatic insect larva, such as the mosquito, is possibly a function of the degree of acidity of the gastrointestinal tract. Under moderate- to high-acidity *in vitro* environments, absorption is high, whereas with a slightly acidic to alkaline gut, absorption is probably much lower.

A wealth of information exists on the acute toxicology of the tributyltins to mammals, piscine life, shellfish, insects, plankton, bacteria, and vascular plants. It is conventionally summarized in Reference 79. At the dosages necessary for snail (and mosquito) control no acute effects are observed with rodents, fish, and plants. In all the previously listed field evaluations of controlled-release organotin molluscicides, there has been *no* fish kill observed and relatively minor suppression of phytoplankters – with a subsequent population resurgence. The intoxication threshold for fish lies in the 0.1–0.5 ppm concentration range, whereas the lethal dose for snails is in the 0.003 or lower ppm range, depending on the length of exposure.[76,79]

There are essentially four competing molluscicidal products: organotin (TBTO or TBTF), copper sulfate, *n*-tritylmorpholine (Frescon™), and niclosamide.

In a comparison of toxicological properties, organotins appear to pose less of an *acute* toxic hazard to nontarget biota. Frescon, for instance, is a highly toxic nerve poison to fish.[85] The snail LC_{90} for niclosamide is 3–8 ppm/h, whereas the fish LC_{90} is 0.05–0.3 ppm/h.[96]

The salient feature of organotin molluscicides and mosquito larvicides is the long-term effect – desired for its economy in both chemical use and labor savings. Although no field tests have, to date, been sufficiently long term as to permit the experimental determination of release life, it must be measured in years. Various field observers have reported 21 months,[83] 389 days,[78] 12 months,[87] and 14 months.[88] However, this advantage could be offset by the accumulative effects, if any, of organotins on nontarget species. It must be kept in mind that organotins at the ultra-low-use concentrations create a *chronic* intoxication syndrome in snails-requiring two to three weeks for mortality, as opposed to the acute intoxication observed using molluscicides in the conventional manner. Organotins are accumulative poisons and it is eminently reasonable to inquire as to the possibilities of long-term effects on nontargets exposed continuously to the very low organotin content in the treated-water course. The nature of the snail kill mechanism is critical to answering the query.

TBTO and TBTF (which probably hydrolyses to TBTO within a few hours of water contact) are extremely hydrophobic. Whether any true water solubility exists is even questionable. Agent molecules are released from the elastomeric dispenser through a dissolution process based upon system kinetics. Once released the dwell time in water is extremely low, probably a few hours at most, with rapid absorption by cellulosic materials, suspended soil particles, and the water-course bottom soil. Detection of organotin in water is difficult indeed, although reasonably accurate techniques are known to concentrations as low as 10 ppb.[(89)] If other organic matter is present, generally the analyst cannot find organotin in the water. Elaborate isolation studies have shown that the snail contacts the toxicant from ingestion of the soil and organic matter and not through simple exposure to the water.[(79,90,91)] The snail, in browsing the bottom soil, ingests grit containing absorbed TBTO. Once in the gut, desorption occurs and the agent moves through the stomach and intestinal wall. The toxic moiety, which is almost certainly not TBTO but an unknown tin-bearing biotoxin, accumulates at cidal sites. The vital process affected may be blockage of transamination, the probable binding sites being the thiol groups of cysteine and the =NH of histidine.[(92)] The gross effort is destruction of membrane cells leading to permeation of the haemolymph and massive internal hemorrhage.[(79)]

The selective mode of snail contact leads to chronic terminal intoxication. Nontarget species, such as fish, which do not ingest soil particles, at least to the same extent, are unaffected.

In organotin studies with mice, 92–94% of the ingested agent is excreted unchanged in the feces, with about 2% noted in the urine with short-term storage of the residue in a few organ systems.[(93)] Similar results have been noted with bovines.[(79)]

TBTO and TBTF and their suspected decomposition products are nonteratogenic, at least over four generations of mice continuously orally exposed.[(94,95)] Few carcinogenicity studies have been performed with organotins. Triphenyltin hydroxide has been shown to be noncarcinogenic,[(96)] and also dibutyltin diacetate.[(97)] In a six-month carcinogenic study, TBTF applied dermally to Swiss white mice resulted in no neoplastic lesions.[(93)]

Triphenyltin acetate is neither carcinogenic nor mutagenic.[98,99] Recent work by the authors have shown that both TBTO and TBTF are antitumorgenic in the mouse, and that some natural mouse anticarcinogens contain tin in their molecular structure.[100] The anticarcinogenic nature of natural organotins in mice, with presumptive evidence for their presence in humans, has been observed and will be published in 1984 by the authors.

TBTO and TBTF degrade rapidly in water through a series of increasingly less toxic intermediates with stannic oxide as the terminal product.[101] It has been presumed that TBTF converts to TBTO through an intermediate hydroxide within a few days, then to tributyltin hydroxide, followed by dealkylation to dibutyltin dihydroxide, dibutyltin oxide, monobutyltin trihydroxide, butylstannoic acid, and finally, stannic oxide, a ubiquitous nontoxic natural constituent of the earth's crust.[101] There is considerable supporting evidence[102,100,101] that bacteria enhance the decomposition rate.[104] However, several steps in this process have been challenged.[105] Photolytic cleavage by ultraviolet radiation also leads to rapid degradation of various organotins.[106,107]

2.5. Controlled-Release Cercariacides

The general approach to intervention in the schistosome transmission cycle has been to destroy the snail using conventional molluscicides and administer various vermifuges to infected humans that kill or render the adult schistosome sterile. However, interdiction of the miracidium or the cercaria would also be a valid approach. It has been known that organotins will destroy the adult trematode at relatively low concentration.[108] Ritchie *et al.* evaluated TBTO against both larval forms and found that 10 ppb suppresses infectivity of cercariae with only 5-min exposure and miracidia within 40-min.[109] The effect is reliable enough to allow the determination of TBTO and TBTF concentration in water through noting effects on cercaria after 30-min exposure.[110] Santos *et al.* noted that cercariae of amphibious snails were destroyed within an hour of exposure to BioMet™ SRM and CBL-9B.[111] It has been suggested that controlled-release organotins be used in schistolarva control since the larval stages succumb rapidly to organotin concentrations $\frac{1}{10}$ to $\frac{1}{100}$ of that necessary for snail control.[79,112] However, the difficulties involved in conducting field tests with schistolarva as the target have likely discouraged further investigation. Techniques have improved[113] and one can predictably see future effort along this line.

2.6. Other Controlled-Release Molluscicides

The basic problem retarding the acceptance of organotin molluscicides in the 1960s and early 1970s lies in the lack of toxicological data. Once recognized, one apparent method of circumventing the problem would be to develop controlled-release formulations of known conventionally used molluscicides which presumably have the blessings of public-health agencies.

Niclosamide/elastomer compounds were formulated in 1968. Laboratory evaluation indicated that a chloroprene-based material was effective against snails for 36 weeks.(114)

N-tritylmorpholine, sold as Frescon™ by the Shell Chemical Company Ltd., was similarly compounded in elastomers. However, the necessary evaluated temperature needed for cross-linking led to thermal degradation and loss of the molluscicidal property.

Tetrachloroisophthalonitrile, a known antifoulant,(115) was formulated in natural rubber, styrene–butadiene copolymers, chloroprene, and other elastomers as a controlled-release molluscicide. It was shown to be effective in repeat-challenge bioassay against host snails, snail eggs, and cercaria.(116)

Organolead materials, especially triphenyllead acetate, known antifoulants were found to be molluscicidal with LC_{50}'s in the 0.05–0.01 ppm range under laboratory conditions, showing considerable efficacy in repeat-challenge bioassays.(118)

2.7. Controlled-Release Copper Sulfate Elastomers

Copper sulfate has long been used in snail control. It is far less expensive than organotins, niclosamide, and *n*-tritylmorpholine, although effective snail control requires much higher concentration.(119) However, a controlled-release version might be suitable. Since copper sulfate lacked solubility in all elastomers evaluated, it was not possible to create a controlled-release formulation based upon a diffusion–dissolution mechanism. Thus the researchers involved keyed their work towards the development of a leaching-type system. Large agent loadings were necessary to create proper porosity and the only effective binding elastomer discovered were various EPDM (ethylene–propylene–diene) polymers that would hold a loading as high as 86% with sufficient pellet integrity.(120,121) An effective biocidal life of four to six months was observed with the commercialized version.(122) One unique formulation feature was the need for a secondary leachant to adjust interfacial pH. Copper sulfate emitted from a dispensing pellet or granule in alkaline water tends to form an insoluble copper carbonate surface film that retards further leaching action. The addition of ammonium sulfate which coleaches is ameliorative. Formulation E-51 (Incracide™ E-51) is shown as follows.(122)

Epcar 5465	100.0 parts (by wt)

Sulfur	1.0 parts
Mercaptobenzothiazole	0.2 parts
Zinc oxide	3.0 parts
Tetramethylthiuram disulfide	1.0 parts
HAF carbon black	0.0 parts
Ammonium sulfate	2.0 parts
Copper sulfate monohydrate	157.2 parts

Copper sulfate monohydrate or anhydrous copper sulfate must be used. The common pentahydrate decomposes at curing temperature (290–300°F) releasing superheated liquid H_2O which combines with sulfur, the curing agent, to give sulfuric and sulfurous acids which attack and degrade the binding elastomer.

The continuous exposure of snails to copper ion, at concentrations far below those conventionally used, 20–60 ppm, leads to a terminal chronic intoxication with mortality ensuing in several weeks.[123,124]

2.8. Critique

Comparison of controlled-release elastomer-based molluscicides is difficult in that relatively few field trials have been undertaken with copper-releasing materials and niclosamide, and none at all, to our knowledge, with controlled-release organoleads and tetrachloroiso-phthallonitrile. Various field reports have shown that TBTO elastomers compare well with conventionally applied niclosamide, and *n*-tritylmorpholine.[82] Triphenyllead acetate has been reported as less effective than tributyltin oxide as a molluscicide.[125] Controlled-release niclosamide was found ineffective in marshes.[126]

Controlled-release copper sulfate does not provide the usually requisite long-term release and several investigators have concluded that it is less effective than organotin/elastomers.[78,88,127]

Organotin controlled-release compounds are believed to be the most cost effective, especially when used for focal control of host snails.[87]

2.9. Bait Molluscicides

Baits are a form of controlled release wherein the agent is firmly bound within a polymeric matrix, with little or no seepage into the environment, and an attractant material is slowly emitted. Snail baits containing organotins, niclosamide, and later copper sulfate, were made in the laboratory and were found to be repellent to most snails. However, *n*-tritylmorpholine in a polyacrylic thermoplastic, using finely powdered clay as a diluent, was found effective under laboratory conditions in 1969.[128] A large number of kairomones (food-type attractants) were evaluated and considerable species specificity found.[129] The initial work with aquatic freshwater and marine snails were extended to amphibious snails.[76,130] Tropical field evaluations showed efficacy.[131]

Cercaria are attracted to plasticized polyacrylates if specific amino acids are slowly emitted.[131] Penetration and mortality were noted. Possibly miracidia could likewise be attracted using miraxone or perhaps a water-soluble magnesium salt as the attractant.[132]

The superior snail attractants were wheat germ, beef-heart infusion agar, and powdered lettuce for aquatic freshwater snails, and fish flour for marine snails.[79] Recent work by Thomas indicates that maltose serves as both an attractant and an arrestant for one important schistosome-bearing snail.[133]

A simple formulation for aquatic snails is as follows:

Carboset 526*	Binder	64.0%
Carboset 515*	Plasticizer	4.0%
Wheat germ	Attractant	26.4%
Calcium chloride	Curative	0.1%
TCPN†	Fungicide	0.5%
N-tritylmorpholine	Agent	5.0%

3. Controlled-Release Insecticidal Elastomers

In 1965 it was noted that TBTO- and TBTF-loaded antifouling rubber destroyed mosquito larva under laboratory conditions. However, repeat-challenge bioassay showed little long-term efficacy.[134] Large dosage rates would be required for adequate kill.[135] By 1967 the senior author had incorporated chlordane, dieldrin, endrin, heptachlor, lindane, dibrom, abate (temephos), dursban (chlorpyrifos), and baytex in natural rubber, polybutadiene, EPDM, and other elastomers.

Each exhibited a degree of solubility in the polymeric matrix and thus a diffusion–dissolution phenomenon was seen upon water immersion. Abate/EPDM exhibited superior larvicidal activity and longevity against mosquitoes.[134,136] In 26 repeat-challenge bioassays of 72 hours' duration, 100% larva mortality was noted.[137]

* Trademark, B.F. Goodrich Chemical Company, Cleveland, Ohio.

† Tetrachloroisophthallonitrile, Diamond Shamrock Company, Cleveland, Ohio.

4. Bactericidal and Fungicidal Elastomers

It was well established by 1963 that organotins were fungicidal and bactericidal, and indeed several are used as mildewcides and bactericides. Little study has been done in the area of controlled-release organotin formulations for protection against microbial organisms. In the one known published study, 17 organotins containing elastomers were evaluated against bacteria and fungi as a means of disinfecting potable water.[138] One TBTO/chloroprene compound was effective against bacteria in potable water and also in human urine.

Tri-*n*-butyltin fluoride and tributyltin laurate release from elastomeric sheet compounds will destory various pathogenic and nonpathogenic fungi through vapor intoxication at 0.001 ppm in air.[139]

5. Herbicidal Elastomeric Formulations

The development of controlled-release herbicides using the diffusion–dissolution emission concept began in 1969 under the aegis of the U.S. Army Corps of Engineers. Early studies showed that the butoxy-ethanol ester of 2,4-dichlorophenoxy-acetic acid, "2,4-D BEE," was soluble in a number of elastomers at levels up to 30%. Various 2,4-D amines, also evaluated, had much lower solubilities. Formulations developed were keyed towards phytozone treatment that is, releasing the herbicide only in the water strata where the particular aquatic weed had its biological niche. Consequently, floating weeds would be targeted using a floating dispenser, rooted vascular plants by a sinking pellet, and nonrooted submerged targets by use of anchored strands. A number of formulations based upon 2,4-D BEE in natural rubber were evaluated against *Eichornia crassipes* (water hyacinth), *Elodea canadensis*, *Myriophyllum spicatum* (Eurasian water milfoil), and other aquatic biota.[140] The superior compound, 14ACE-B, showed a 21–27 month effective release period.[140,141]

The rather simple 14ACE-B recipe is as follows in both the sinking and floating versions:

Ingredient	14ACE-B Sinker	14ACE-B Floater
Natural rubber	100.0	100.0
2,4-D BEE	30.0	30.0
Zinc oxide	60.0	1.0
Sulfur	0.5	0.5

CBTS (*N*-cyclohexyl-2-benzothiazyl sulfenamide)	2.0	2.0
Stearic acid	0.5	0.5
TMTDS (tetramethylthiuram disulfide)	1.0	1.0
HAF (carbon black)	20.0	20.0

Later, iron oxide was substituted for 59 parts of zinc oxide in the sinking version.

Early studies indicated that the amount and type of carbon black used moderated the release rate; increase the amount of black and the diffusion path length increases thus reducing the emission rate. Also going from fine-structured blacks to coarse-structured carbon blacks increased the loss rate. It was also found in laboratory studies that 2,4-D BEE "layered out" in still water with considerable horizontal movement but relatively little vertical translocation.[(142)]

In 1970 it was noted that "Habers" rule $CT = C'T'$, was not valid for aquatic plants continuously exposed to very low herbicide dosages. This discovery, later termed the "chronicity effect," led to the evaluation of a number of common herbicides against various plant species. The concept is of great importance; normal herbicide treatment at 2–40 ppm led to acute intoxication within a few hours' exposure, whereas continuous dosing, that is, release from a dispenser, resulted in a chronic intoxication at dosages as low as 0.001 ppm with mortality ensuing in two to five weeks. The end result was the same in each case; however, the amount of available toxicant in the water course is dramatically reduced. Thus the environmental impact is lessened, and the chemical costs reduced.

The chronicity phenomenon has been fully reported in the literature.[(76,143–145)] Table 2 is a compendium of observations as to the presence or absence of the effect.

The chronic intoxication phenomenon is easily noted by comparing the LT_{100} (lethal time to 100% population mortality) at various dosages. For instance, using Diquat against Eurasian water milfoil it is observed that at a 1 ppm/day release rate the LT_{100} is 14 days; at 0.1 ppm/day it is also 14 days; at 0.01 ppm/day, 19 days; and at 0.001 ppm/day, 38 days.[(76)]

Statistical evaluation of thousands of bioassays shows, beyond doubt, the existence of the chronicity effect, with regard to aquatic plants.[(146)] Matrix analysis indicates that this particular effect is determinable under laboratory conditions and is independent of the agent used and the plant species.

It is believed that the chronicity effect is based upon the target-plant's inability to distinguish the presence of xenobiotics at ultralow concentrations

Table 2. Plant–Herbicide Evaluations Showing the Chronicity Phenomenon[a]

	Plant Species					
Herbicide	Eurasian water milfoil[b]	Vallisneria[c]	Cabomba[d]	Elodea[e]	Water Hyacinth[f]	Southern Naiad[g]
2,4-D BEE[h]	+	+	+	+	+	+
2,4-D Butyl ester[i]	+	+	N	+	0	N
Silvex acid[j]	+	+	0	+	0	N
Fenac[k]	+	+	+	0	N	+
Dichlobenic[l]	+	+	N	+	N	+
2,4-D Dimethylamine	N	N	N	N	0	N
2,4-D Oleylamine	+	N	N	N	+	N
Silvex, sodium salt[m]	+	+	0	+	0	N
Diquat[n]	+	+	+	+	+	+
Acrolein	0	0	0	0	0	0
Hydrothol[o]	N	N	N	N	0	N
Endothall[p]	0	0	+	0	N	+
Fenuron[q]	N	0	+	+	N	N

[a] + = Phenomenon present; 0 = Phenomenon absent; N = not evaluated.
[b] *Myriophylium spicaum.*
[c] *Vallisneria americana.*
[d] *Cabomba caroliniana.*
[e] *Elodea canadensis.*
[f] *Eichornia crassipes.*
[g] *Najas quadrilupensis.*
[h] Butoxyethanol ester of 2,4-dichlorophenoxyacetic acid.
[i] Butyl ester of 2,4-dichlorophenoxyacetic acid.
[j] 2-(2,4,5-Trichlorophenoxy) propionic acid.
[k] Sodium 2,3,6-trichlorophenyl acetate.
[l] 2,6-Dichlorobenzonitrile.
[m] Sodium [2(2,4,5-trichlorophenoxy)] propionate.
[n] 6,7-Dihydrodipyrido (1,2-a:2,1-c) pyrazidiium dibromide.
[o] Di(*N,N*-dimethyl alkylamine) salt of endothall.
[p] 7-Oxabicyclo (2,2,1) heptane-2,3-dicarboxylic acid.
[q] 3-Phenyl-1,1-dimethyl urea.

and thus the physiological defensive mechanisms are not triggered. Laboratory studies and stochastic considerations lend credence to this view.[147,148] It is noteworthy that this phenomenon, basically the results of continuous target exposure to ultralow pesticide concentrations, appears to hold true with molluscicides and larvicides, with regard to the mosquito.

Since copper sulfate is not only molluscicidal but also herbicidal and algaecidal,[149] the controlled-release version, Incracide E-51, has been evaluated against both algae and vascular aquatic plants.[150]

Laboratory studies indicated considerable efficacy for a number of controlled-release herbicide compounds based upon the diffusion-dissolution emission mechanism.[141,151–153] Although a number of artificial outdoor pool tests were performed with good results,[142] only one field trial has been attempted to date. Results have shown several years' water hyacinth control with one application.[154]

The basic problem is an economic one concerning usage of controlled-release aquatic herbicides based upon elastomeric monoliths. Such systems require complex processes in manufacturing and are thus costly.

6. Controlled-Release Thermoplastic Systems: An Overview

The first known plastic-based release systems were developed in the early 1960s using a carrier-type mechanism; that is, a plasticized polyvinyl chloride base incorporated an agent, nonsoluble in that base, but soluble in the plasticizer. As the plasticizer molecules migrated to the surface—a phenomenon known as "blooming" to the compounders' art—they carried with them agent molecules which volatilized. If the agent was a volatile insecticide then toxic vapors were released and a method of flying-insect control resulted. This material was, within a few months of commercialization, properly sized, a buckle attached, and the first pet flea collar developed. Such carrier systems, modified with a moderating laminated membrane, were also used to deliver a nonvolatile contact insecticide to a polymer–air interface and the "roach tape" was created. Other slight modifications in agent choice led to such reservoir and membrane systems being marketed with an antimicrobial agent for hospital uses.

Such known pest-control agents were not soluble in thermoplastic or thermosetting polymers, the diffusion–dissolution system was inoperable, and thus a carrier mechanism was essential. This concept dominated the scene through most of the 1960s.

The next advance was in the creation of simple leaching-type systems; however, both longevity and efficacy were low. The one commercialized product required relatively huge application dosages for reasonable mosquito larva control. Following this, the direct attachment of the toxic moiety, usually an aquatic herbicide, to a monomer, was attempted, then polymerizing so that the toxic element was appended to a polymer backbone. Upon exposure to water, hydrolytic cleavage occurred. However, the lack of suitable porosity rendered water penetration of the dispenser difficult and long-term efficacy dubious in many applications. Use of an organotin moiety appended to various polymers did provide a useful antifouling agent.

In 1975 it was found that the use of a secondary coleachant whose function was to create porosity made possible the development of long-term controlled-release leaching-type systems based upon a thermoplastic binder.

7. Early Concepts

In tracing the development of controlled-release pesticide technology, it should be noted that there are several distinct pathways involved. These in turn developed from the formulators' art of the 1940s and 1950s, which we can term "conventional" systems, and thus preclude it from our general discussion. However, there are a number of developments that bridge the gap between the conventional methodologies of the 1950s and the controlled-release development of the middle and late 1960s. Those especially germane are discussed in this section.

It was common practice in the era from 1930 to 1960 to absorb a pesticide on some finely powdered particulate, such as clays, talc, and chalk wherein the absorbant served as an extender and the necessity for desorption, a time-dependent phenomenon, to occur gave a degree of slow release of the agent. In 1963 Geary patented the use of an impregnated system, using mica, clay, and so on, as the absorbant for various organic pesticides, then coating the particle formed with the urea–formaldehyde polymer.[(155)] Polymerization was *in situ*. Several years later, Ugi showed that certain isonitriles are fungicidal and acaricidal, and can be polymerized to a low-molecular-weight pesticide.[(156)]

Sanders and Knoke demonstrated that liquid plastics, used as adjuvants, would prolong the persistency of endosulfan, endrin, linedane, and malathion insecticides.[(157)] Similarly, latex-coated pesticidal compactions have been used to enhance release longevity.[(158)] Insoluble polymeric binders are also used and release time extended, in one case from 3 days to 28 days.[(159)] In one interesting development an agricultural agent is absorbed on silica or diatomaceous earth and bound to an inert nucleus using a water-soluble polymeric adhesive such as carboxymethyl cellulose.[(160)] This process is the reverse of the microencapsulation technique to be described in a later section. In another instance the use of a short oil resin modified alkyd resin coating extended the release life of pirimicarb granules from 5 to 120 hours under field conditions.[(161)]

The Balassa patent provides a thorough description of the use of various waxes and resins for the coating of herbicidal granules.[(162)] It is taught that one adds the herbicide to an inert porous carrier, with a bactericidal/fungicidal preservative, and then mixes it with waxes, wood resins, gum resins, terpene resins, cellulosics, creosote, and so on, to increase in-use release life.

8. Carrier Systems: Insecticides

The earliest known development work leading to the creation of a long-term controlled-release pesticidal formulation were keyed to the use of a carrier system. Basically these are three-phase systems in which the pesticide is soluble in an additive, usually a plasticizer, and the additive, in turn, is soluble in the plastic.

8.1. Insecticidal Strip

It was observed that dichlorvos (dimethyl-2,2-dichlorovinyl phosphate, DDVP) can be incorporated in polyvinyl chloride and slowly released as a vapor.[163,164] The well-known commercial product, Shell Chemical Company's No-Pest™ strip, resulted. This is, or was, 20% dichlorvos in a polyvinyl chloride matrix using dioctylphthalate or other plasticizers as the carrier material. Efficacy has been amply demonstrated, for instance, in long-term control of catch-basin mosquitoes.[165]

In the period 1970/71, a series of patents were issued to Aries which described the use of such three-phase systems[166–170] and extended usage to parasiticides. Aries showed that the following composition provided controlled release[167]:

DDVP	20 g
Dioctylphthalate	18 g
Methoxy-4-styrene	0.4 g
Vinyl styrene	1.6 g
Porous alumina	360 g

DDVP-impregnated cellulose fibers were also noted as controlled-release materials.[166] These systems used a coevaporant to regulate agent loss; that is, the rate of DDVP loss depends upon the evaporation rate of the incorporated DDVP solvent.[168] Aries reported that the coevaporant/DDVP system was utilizable with polyethylene, polypropylene, polystyrene, polyvinyl esters, polyesters, and others, using dimethyl-, diethyl-, dipropyl-, dibutyl-, and dioctylphthalate.[169] Nicotanate stabilizers or other stabilizers were essential.[170]

8.2. Flea Collars

The perceived need for a method of protecting domestic animals from ectoparasites is old. Incorporation of a pesticide-impregnated pad between dog-collar plies was suggested in 1940.[171] Peo suggested using a pesticide containing fabric tube attachment to the collar.[172] In 1952 Coash discovered that DDT could be added to film-forming compositions and fly repellency achieved.[173] Several pertinent inventions were added to the art in 1957. Leeds showed that a porous plastic material could be filled with a fluid, such as an ink, which would release slowly into the ambient environment.[174] Polyvinyl chloride, polyethylene, and other plastics were suggested. The art was extended by Bracey, who incorporated not only insecticides but also bactericides in aminoplastics such as butylated urea formaldehyde and melamine formaldehyde.[175] Doyle described a hollow pet collar composed of a flexible-tube construction with an insecticide or deodorant enclosed.[176]

In 1965 Menn *et al.* incorporated DDVP in polyvinyl chloride and showed that resulting granular formulations fed to sheep and other animals would destroy internal parasites.[177] The art was extended by Feinberg to cover a large number of insecticides incorporated in various vinyl resins.[178] Although this invention was keyed to paints and floor polishes, it is noted that the insecticides evaluated are solid, high-boiling chemicals as well as DDVP and the other volatile agents. In 1968 Hackney showed that solid insecticides, such as various carbamates, could be incorporated in nonvinyl, nonplasticized thermoplastics, such as polyethylene, rendering the contemplated product insect resistant. Such formulations are not controlled release, but do show compatibility of insecticides and various polymers.[179] Alfes *et al.* have not only thoroughly described the necessity for a carrier in vinyl compounds but went on to show that a polyester–vinyl thermosetting copolymer with DDVP *and* a select filler, such as kaolin or quartz, will create the essential porosity for vapor release.[180]

The initial flea collars were no more than the No-Pest™ strip cut to size and a buckle attached. In 1969 the senior author prepared a number of experimental formulations using both liquid and solid insecticides incorpoated in polyvinyl chloride plastisol with various plasticizers. In 1974 Duffey and Coleman patented a plastic cylinder containing *O*-(2-isopropoxyphenyl)-*N*-methylcarbomate, a solid insecticide in a fatty acid – polyethylene glycol – for topical application to the head area of the pet.[181] Whereas this is not a collar, the art first taught in the subject patent has application to the general domestic-animal insect-protection area.

In 1974 Grubb and Baxter patented a pet collar containing a solid carbamate insecticide in a plasticized vinylic resin.[182] Again dioctylphthalate is used as the carrier. In the same year, Aries *et al.* described a novel collar for cats, dogs, horses, and sheep.[183] This collar has laminated to it, or otherwise attached, a container having a flexible plastic sheath. The sheath has an imperforate surface directed towards the animal and a perforated surface directed away from the wearer. A DDVP/polyvinyl chloride/

plasticizer compound in granular form is loaded in the hollow portions of the sheath.

Potter extended the art to include DDVP dispersed in a blend of styrene block polymer with plasticized vinyl chloride.[184] DDVP in urethane rubber has also been described as being useful in pet collars.[185] This system is not of the carrier type, but probably, since DDVP is soluble in various polyurethanes, relies upon a diffusion–volatilization release mechanism. Pasarela also extended the art to acaricidal resin compositions of plasticized polyvinyl chloride.[186] Although the release of a deodorant, a fragrance from a pet collar, has yet to reach a practical stage, it has been suggested in at least one known patent.[187] In a recent patent, Dick advocates the use of diazinone and/or diazoxone in plasticized vinyl or vinylidene polymers.[188]

Greenberg and Cloyd have developed a pet collar releasing both a liquid and a solid agent (namely dimethyl-1,2-dibromo-2,2-dichloroethyl phosphate) and a nonvolatile carbamate simultaneously.[189] Two polyvinyl chloride homopolymers are used with a plasticizer carrier, such as di-2-ethylhexylphthalate, and a porosity-control additive. Efficacy appears to be superior over other insecticidal flea collars. The release of both a vapor and a solid insecticide is novel. Present published technology is capped by this invention, which represents the curent state of the art.

8.3. Insecticidal Roach Tape and Related Inventions

In 1948 Engel reported that a plasticized polyvinyl chloride containing medicaments, disinfectants, and insecticides could be put up in a paste form which would solidify into an elastic film upon cooling.[190] Glycerine, or another substance, was used as an additive to carry the agent to the surface where release occurred. Watson noted that DDT (dichlorodiphenyltrichloroethane) and benzene hexachloride insecticides could be incorporated in paints using a suitable plasticizer, that is, a "carrier," to bring the agents to the film surface.[191] This technological sequence was followed by a Shell International Company patent noting that DDVP could be incorporated in natural or synthetic waxes and vegetable or animal fats. A diluent (actually a carrier), such as dibutylphthalate, cetyl alcohol, and so on, was added to provide long-term release of insecticidal vapors.[192] This concept was extended to include DDVP in various plastics – both thermoplastic and thermosetting materials.[193] This included the polyolefins, polymethylmethacrylate, and similar substances, with emphasis on polyvinyl chloride. The referenced patent describes the basic carrier-type release mechanism underlying the insecticidal strips, early pet collars, and the subject matter of this section. The art was extended to other insecticides such as lindane, dieldrin, chlordane, DDT, and malathion in plasticized paint films, the plasticizing element serving as a solvent for the agent.[(194)] Boullenger demonstrated that not only polyvinyl chloride could be used, but also ethylene–vinyl acetate copolymers and polyethylene – with an appropriate additive to act as the carrier material.[(195)]

A considerable advance in the art was made by Hyman *et al.* wherein an insecticidal-containing reservoir was laminated to a polymeric membrane, the latter serving as a release-rate regulant.[(196)] Again, plasticized polyvinyl chloride was used. Migration of the agent from the reservoir to and through the membrane was continuous. Antibacterial and antifungal agents were used, generally organotins, such as tributyltin neodecanoate, but also antibiotics, organoarsenics, organosulfur, and organomercuric compounds. Several compositions have been marketed as surface disinfectants (antibacterial) for hospital use.

The concept was extended to the use of diazinon, dursban (chlorpyrifos), and other insecticides in laminated structures, manufactured as a tape, for household and restaurant cockroach control.[(197,198)] The latest advance in this technology consists of a dual-purpose additive to the thermoplastic – a material that will serve not only as a carrier of the agent, but also as an attractant.[(199)] Table 3 depicts several formulations.

Soy oil and lecithin serve as an attractant as well as a carrier for the insecticide. A 10-second exposure of the German cockroach, *blatella germanica*, to the active surface results in a terminal acute intoxication. Such materials are attractive, as well as lethal, to various ant species and the larval crawling stage of agricultural insects.

Table 3. Bait Contact Insecticides

	Recipe				
Ingredient	A	B	C	D	E
Ethylene–Propylene copolymer[a]	69.0%	–	84.0%	74.0%	–
Ethylene–Vinyl Acetate copolymer[b]	–	68.8%	–	–	84.0%
Zinc stearate	1.0%	1.0%	1.0%	1.0%	1.0%
Diazinon[c]	20.0%	–	10.0%	20.0%	10.0%
Baygon[d]	–	15.2%	–	–	–
Soy oil	10.0%	10.0%	–	–	5.0%
Lecithin	–	–	5.0%	5.0%	–

[a]Vistalon 702, Exxon Chemical Company.
[b]Microthene MU 763, USI Chemicals.
[c]50% Wettable powder, *O,O*-diethyl-*O*-(2-isopropyl-6-methyl-5-pyrimidinyl) phosphorothioate, Ciba Geigy Corporation.
[d]70% powder, 2-(1-methylethoxy) phenol methylcarbamate, Mobay Chemical Corporation.

9. Insecticidal and Molluscicidal Monoliths

The monolithic incorporation of a pesticide in a polymeric material is a concept as old as the antifouling paints. As previously described, there is a technological sequence from pesticidal paint-type coatings to the diffusion–dissolution type release mechanisms seen with the organotin elastomers, and later organophosphate elastomers. Studies leading to the direct incorporation of a toxic agent in a nonplasticized plastic material, wherein the release mechanism could not be diffusion–dissolution, due to lack of solubility, and no carrier material is present, began in the 1960s. It is believed that the major preliminary work in this direction was instituted at the U.S. Army Environmental Hygiene Agency at Edgewood Arsenal, Maryland. Early work with insecticide-impregnated plaster briquettes[(200,201)] was followed by the incorporation of temephos, chlorpyrifos, and other mosquito larvicides in plastic pellets using polyamides and polyvinyl chloride.[(202)] Briquettes did not provide long-term control.[(203)] Chlorpyrifos–polyethylene and chlorpyrifos–polyvinyl chloride proved superior.[(204,205)] The latter relied upon di(2-ethylhexyl) phthalate, a plasticizer, as the carrier. However, the chlorpyrifos–polyethylene compound was not plasticized, the agent being released through a simple leaching mechanism. Over 21 weeks of biologically effective agent release with this formulation was noted in artificial pools.[(206)]

In a second series of tests, two formulations of chlorpyrifos–polyethylene and one of chlorpyrifos–polyvinyl chloride provided mosquito larva control for 18 months in outdoor artificial pools.[(207)] Whereas the long-term insect-control feature was never in doubt, the necessary quantities of material to achieve this were great. One-time application at 5 ppm active required 131–153 kg of material per hectare – an impractically large amount.

The above tests, importantly, clearly showed that continuous larvae exposure to the agent at much lower levels than short-term conventional exposure provided the same lethality – that is, 0.23 ppb continuous as compared to the conventional over 0.9 ppb.[(208)]

The chlorpyrifos–polyethylene material, using a chlorinated polyethylene base to achieve a density greater than that of water, was marketed by the Dow Chemical Company as Dursban™ CR-10. The product was not very successful, probably due to the inordinately high treatment levels required. Examination of the commercial granule by the author demonstrated poor release efficacy; although 18–24 months' control was observed in laboratory repeat-challenge bioassay, at the point where emission rate was below the kill threshold, over 50% of the agent remained entrapped in the pellet. The inability of a thermoplastic monolithic dispenser to deliver all or most of its pesticide content to the ambient environment constituted a basic difficulty.

9.1. The Porosigen Concept and Controlled Release

9.1.1. Molluscicides and Aquatic Larvicides

Deschiens and Flock had tried incorporating an organotin in polyethylene by absorption in 1970; however, field release from the granule produced was but a few days.[209] The apparent problem was lack of sufficient porosity to allow water ingress into the dispensing unit. Extended-term agent release from an open cell foam was possible by the mechanical washing out of the agent once immersed, or through the application of compression as with long-term lubricating plastics.[210,211] Such systems were not useful to the senior author who was seeking a long-lasting molluscicide dispenser based upon the slow emission of copper ion and tributyltin fluoride from a thermoplastic. Nonporous polymers would retain and release organotins, but depended upon a carrier.[211,212] Carrier systems, by their nature and as previously discussed, have poor efficiency due to the inability to provide complete agent release.

A method of developing thermoplastic porosity was needed wherein the development was slow so that water ingress into an aquatic dispenser would gradually contact and remove the pesticide. Agents of low water solubility did not leach at practical rates due to the extremely poor rate of porosity.[213] Copper sulfate, a highly water-soluble molluscicide, was leached from polyethylene at too high a rate thus providing only four or five months' release in the laboratory. Also copper ion has a number of serious disadvantages as a molluscicide – though widely used as it is inexpensive. The preferred agent, tri-*n*-butyltin fluoride (TBTF), has a true water solubility of *less* than 1 ppm, although in molecular aggregates arising from adding to water by serial dilution, or from emulsified systems, it may be somewhat over 3 ppm. However, TBTF was far preferred due to its nonpersistent nature, very low kill threshold, and lethal mechanism.

In 1975 the senior author incorporated water-soluble salts and oxides in polyethylene and other thermoplastics as additives. As the soluble component leached out of a dispensing pellet, a porosity system was gradually developed allowing water ingress and contact with the TBTF agent. This in turn created a gradual emission of the agent and a new mechanism of release.[214] Through the use of calcium carbonate, ammonium sulfate, and other soluble entities, it was possible to achieve long-term release of TBTF. Low-density polyethylene, ethylene–propylene copolymers, and ethylene–vinyl acetate were found to be appropriate binding elements. By

Table 4. Molluscicidal and Larvicidal TBTF Thermoplastic Formulations

	Parts by weight			
Ingredient	1	2	3	4
Ethylene–Vinyl Acetate copolymer[a]	56.2	28.6	22.0	43.0
Zinc stearate	2.4	2.4	1.0	2.0
Calcium carbonate	17.0	17.0	25.0	17.0
TBTF	24.4	24.4	30.0	30.0
Polyethylene[b]	—	27.6	22.0	—
Silicon dioxide	—	—	—	8.0

Note: Formulation No. 3 is presently marketed as Ecopro™ 1330 by the Environmental Chemicals Company of Barrington, Illinois.
[a]Microthene MU 763, U.S. Chemicals, NY.
[b]Microthene MN 718, U.S. Chemicals, NY.

fusing two polymers of diverse rheological character, for example, melt index, into a polymer alloy, free volume was increased thus allowing greater efficiency. Several molluscicidal and larvicidal recipes are shown in Table 4.

Long-term repeat-challenge bioassay of Ecopro™ 1330 against host snails of schistosomiasis and mosquito larva shows effective kill at ultralow water concentration for over 1100 days under laboratory conditions.[215,216] The concept was extended to include a number of soluble or slightly soluble additives that upon leaching into the ambient water would create a porous condition within the polymer matrix.[217] The term "porosigen" was coined, and acceptable to the scientific if not the literary community, meaning a porosity-enhancing agent—"porosity generator."

The obvious success of the porosigen-type leaching systems encouraged investigation of the incorporation of more conventional materials in this type of dispenser. Also the wide-scale use of an organotin in potable waters was *not* enthusiastically greeted by the Environmental Protection Agency. A new formulation of an *acceptable*, *approved*, and *conventional* mosquito larvicide was needed.

DDVP, chlorpyrifos, and temephos all lend themselves to incorporation in plastic matrices. The first two show a slight degree of volatile loss during processing-at temperature, and the safety of DDVP for some uses has been called into question.[218] Temephos, under the trade name Abate®, was known to be an effective larval insecticide somewhat superior to chlorpyrifos.[219] Incorporation in polyvinyl chloride had been done by the U.S. Army, and reasonably good mosquito control in woodland pools was demonstrated.[220]

Temephos was consequently incorporated in various polymers with a porosigen additive and field and laboratory tested.[221] Granules were observed to silt over in mud-bottom water courses and consequently various floating versions were made. However, this form was not suitable for use in

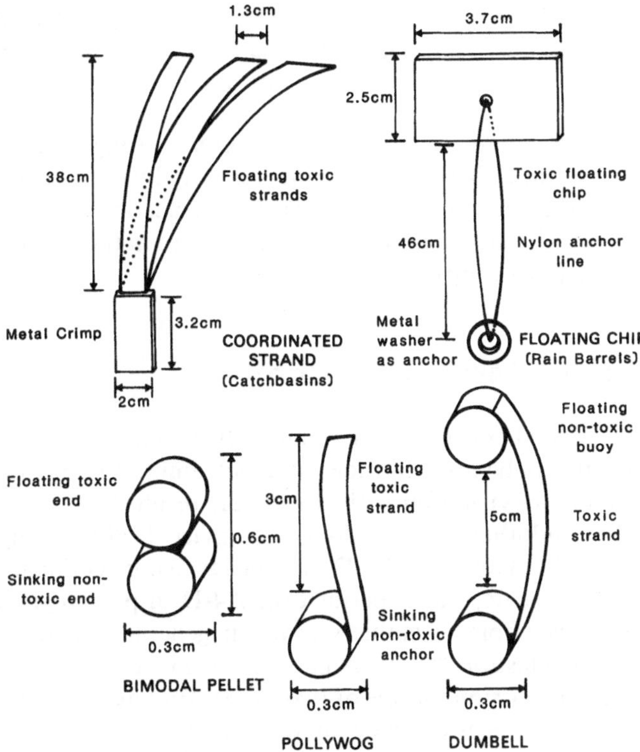

Figure 2. Thermoplastic dispensers for aquatic systems.

flowing-water systems. Consequently, various anchored versions were made.[222] Types of dispensers are depicted in Figure 2.

The following anchored chip formulation is in use in potable-water systems, such as rain barrels and stoneware jars, and has shown longer-term control than is achievable with conventional larvicides and several briquette release systems.[223] The Ecopro 1707 recipe for chips and anchored strands is as follows:

EPM*	34.0
LDPE†	34.0
Zinc stearate	0.2
Calcium carbonate	16.5
Ammonium sulfate	1.0
Silicon dioxide	6.1
Temephos (90%)	8.2

* Ethylene–propylene copolymer, Vistalon 702, manufacturer Exxon Company.
† Low-density polyethylene, MN 718, manufacturer U.S. Chemicals Company.

The plastic alloy creates an increased free volume for better mobility of the temephos. The $CaCO_3$, $(NH_4)_2SO_4$, and SiO_2 mix of porosigens induces the proper porosity to provide in excess of one-year continuous temephos emission.

The bimodel pellets (Figure 2) float on mud, raising as descending silt raises the mud surface. The upper portion is essentially the 1707 recipe, while the anchored end is an iron-oxide-loaded thermoplastic providing a density of around 1.2 g/cm^3. The anchored strands are in use in the United States for catch-basin mosquito control.[(224)]

10. Thermoplastic Aquatic Herbicide Systems

In 1970 several forms of 2,4-D and silvex acid were formulated in polyvinyl acetate, polyvinyl chloride, and polyamide (nylon) plastics and evaluated against water milfoil with reasonable results.[(225)] The same group also noted that endothall and dichlobenil in plasticized polyvinyl chloride was also effective.[(226)] However, 2,4-D on attaclay, a conventional method of dispensing the herbicide, was superior over 2,4-D in polyvinyl chloride.[(227)] Also, 2,4-D release from elastomers gave, in general, superior long-term results over 2,4-D plastics.[(141,153,228)] A basic problem with thermoplastics was a rapid, high-level emission for one or two weeks, trailing off to an ineffectual loss rate.[(229)] Further work lagged in the aquatic control area until the development of the porosigen-containing thermoplastic herbicide dispensers.

A number of aquatic herbicidal materials utilizing a thermoplastic binder with a porosigen additive and one of several agents were developed in

Table 5. Aquatic Herbicide Recipes

	Parts by (wt %)							
Ingredient	1A	1F	2F	2G	3A	3E	3H	4D
LDPE[a]	29.0	24.5	31.0	31.0	74.0	31.0	27.5	29.0
EPM[b]	29.0	24.5	31.0	31.0	—	31.0	27.5	29.0
Zinc stearate	1.0	1.0	1.0	1.0	1.0	1.0	1.0	0.1
Calcium carbonate	—	5.0	10.0	5.0	—	10.0	5.0	5.0
Ammonium sulfate	10.0	6.0	—	5.0	—	—	8.0	—
Silicon dioxide	—	4.0	—	—	—	—	4.0	5.0
Ethylene glycol	—	4.0	2.0	2.0	—	2.0	4.0	—
Diuron	31.0	31.0	—	—	—	—	—	—
2,4-D acid	—	—	25.0	25.0	—	—	—	—
Diquat	—	—	—	—	25.0	25.0	25.0	—
Atrazine	—	—	—	—	—	—	—	31.0

[a] Microthene MN 718.
[b] Vistalon 702.

1980. These were laboratory tested against a number of aquatic weeds. Agents evaluated were Diuron [3-(3,4-dichlorophenyl)-1,1-dimethylurea]; Simazine [2-chloro-4,6-bis(ethylamino)-*S*-triazine]; Diquat [6,7-dihydrodipyrido (1,2a'2'1'-c) pyrazidiinium dibromide]; Fenac (2,3,6-trichlorophenylacetic acid); Bromacil (5-bromo-3-*sec*-butyl-6-methyluracil); Atrazine (2-chloro-4-ethylamino-6-isopropylamino-*S*-triazine); Dichlobenil (2,6-dichlorobenzonitrile); 2,4-Dacid; 2,4-D esters; and 2,4-D amines. Those bind-

Table 6. Bioassay of Controlled-Release Aquatic Herbicides

Compound	Agent	Concentration (ppm-p)	Plant	LT_{90} (days)
1A	Diuron	1	Elodea	32
	Diuron	1	Cabomba	35
	Diuron	10	Cabomba	26
	Diuron	10	Vallisneria	53
	Diuron	10	Myriophyllum	20
1F	Diuron	1	Elodea	38
	Diuron	10	Elodea	28
	Diuron	1	Vallisneria	50
	Diuron	10	Vallisneria	50
	Diuron	1	Cabomba	37
	Diuron	10	Cabomba	17
	Diuron	10	Myriophyllum	19
2F	2,4-D acid	1	Cabomba	27
2G	2,4-D acid	10	Elodea	19
	2,4-D acid	10	Cabomba	19
3A	Diquat	1	Elodea	10
	Diquat	1	Vallisneria	18
	Diquat	10	Vallisneria	14
	Diquat	1	Cabomba	28
3F	Diquat	1	Elodea	12
	Diquat	1	Vallisneria	17
	Diquat	10	Vallisneria	13
	Diquat	1	Cabomba	19
	Diquat	10	Cabomba	18
	Diquat	1	Myriophyllum	19
3H	Diquat	1	Elodea	8
	Diquat	1	Vallisneria	23
	Diquat	10	Vallisneria	17
	Diquat	1	Cabomba	25
	Diquat	10	Cabomba	18
	Diquat	1	Myriophyllum	13
	Diquat	10	Myriophyllum	9
4D	Atrazine	10	Vallisneria	35
	Atrazine	1	Cabomba	24

ing polymers found useful were low-density polyethylene, high-density polyethylene, ethylene–vinyl acetate copolymer, polypropylene, polystyrene, polyesters, polyamides, polyvinyl acetate. polyurethane, and ethylene–propylene copolymer.

Typical recipes are shown in Table 5. These materials were evaluated against *Elodea canadensis*, *Vallisneria americana*, *Cabomba caroliniana*, and *Myriophyllum spicatum* water weeds at 1 ppm-p* and 10 ppm-p, as seen in Table 6.

Controlled-release materials are not evaluated using conventional parameters such as LC_{50}, LC_{90}, LD_{50}, and so on, since such values are meaningless when a given dispenser releases its agent continually, and intoxication is chronic rather than acute. The value used is the LT_{90} or LT_{100} – lethal time until 90 or 100% of the target population succumbed.

Since Diquat is highly water soluble it has a dual function, acting both as a herbicide and a porosigen.

11. Controlled Release through Pendent Substitution

In 1965 McFadden *et al.* patented a method of chemically combining several insecticides with ethenoid polymers.[230] Ethyl acrylate, styrene–ethyl acrylate, and others were used as monomers. Parathion, methyl parathion, and so on, were the agents. The end product was a film-forming latex or solvent soluble system. Geary demonstrated that DDT, chlordane, and other chlorinated hydrocarbons could be polymerized with amidoaldehyde monomers.[231] A little later Baltazzi patented several polymerizable monomer/herbicide systems such as[232]:

Glycerol + phthalic anhydride + 2,4-D acid	herbicidal solid resin
Glycerol + phthalic anhydride + TCA	herbicidal solid resin
Glycerol + human acid + 2,4-D acid	herbicidal solid resin

He also noted that diethanolamine can be substituted for glycerol.

Direct polymerization of a given pesticide to a higher-molecular-weight version, thus producing less water solubility, has had little study. Beasley and Collins did report that 2,4-D could be self-polymerized using ferric nitrate as a catalyst[233]; however, further study of this system, if attempted, remains unknown to the authors.

The work of Allan and his colleagues is especially pertinent to the entire

* 1ppm-p indicates 1 ppm of the active agent in the dispensing pellet and *not* in the water. Pellets release for upwards of two years.

controlled-release area. It was shown that 11 herbicides could be chemically bonded to sawdust, tree bark, and lignin, each having hydroxyl groups, by esterification.[234] The generalized equation is

$$\left.\begin{matrix}\text{sawdust}\\\text{bark}\\\text{lignin}\end{matrix}\right\} - \text{OH} + \text{Herbicide} + \text{COCl} \rightarrow \left.\begin{matrix}\text{sawdust}\\\text{bark}\\\text{lignin}\end{matrix}\right\} - \text{OCO} - \text{herbicide} + \text{HCl}$$

Herbicide release occurs through degradation of the CO linkage, probably by hydrolytic cleavage and bacterial activity. The resulting products were shown to be less toxic to germinating nontarget-tree seeds.[235,236]

While target-weed control results are impressive, it is also reported that only 37% of the available herbicide, 2,4-dichlorophenoxy-butyric acid, was actually releaseable.[237] Allan *et al.* also demonstrated that a chemosterilant can be appended to forest waste products.[238]

McCoy and Swanson chemically altered oil-insoluble pesticides by combination with a hydroxylated and alkylated benzene nucleus.[239] This would enhance usage in oil-based formulations.

In 1971 Harris and co-workers incorporated Fenac acid, a herbicide, in polyethylene and partially coated the prepared pellets with wax so that only a portion were open to water contact.[240] It was noted that release rate was dependent upon the degree of porosity and tortuosity of the matrix.[241] In order to increase porosity, Harris *et al.* added sucrose to polyethylene, leached out the additive with water leaving a pore structure, and then filled that structure with Fenac.[242] The next step was to directly attach the herbicide to the polymeric backbone.

Monomers containing hydrolyzable herbicide moieties were prepared and polymerized.[243] Over 80% herbicide content as pendent side chains was achieved. Polymerization methods for various vinyl/2,4-D and vinyl/silvex polymers have been described.[244] Fenac/polyethylene and Fenac/ethylcellulose have been prepared,[245] and also polymerized vinyl and acryloyloxyethyl esters of 2,4-D and 2-(2,4,5-trichlorophenoxy) propionic acid.[246] It was found that the normally slow release rate of 2,4-D from such polymers could be increased by also appending hydrophilic groups as well as the polymer backbone.[247]

Others have used the same or similar approach to herbicide release from polymers. Sinclair evaluated lactic and glycolic acid polymers.[248] Mehltretter *et al.* reacted 2,4-D esters with starch to get insoluble compounds and, in the presence of water, glucoside bond cleavage occurred releasing the herbicidal moiety.[249]

12. Controlled-Release Juvenile Hormones

It is well known that juvenile hormones and their analogues are potent agents for use against insect larva.[252–254] Morphogenetic agents for nine insect species have been identified.[255] Methoprene, the major in-use juvenile hormone, shows greater activity to many insects as compared to conventional insecticides,[256] and is essentially nontoxic to mammals.[257] However, resistance can be acquired,[258] including insecticide cross-resistance with juvenile-hormone mimics.[259] Pyrethrin juvenile-hormone combinations are synergistic.[260]

Methoprene, showing a number of uses in control of mosquitoes and other insects, is photolytically unstable thus greatly limiting its value. However, incorporation in a plastic that shields ultraviolet radiation provides a relatively long-lasting release technique. Incorporated in a polyurethane foam dispenser at 3% active, good catch-basin mosquito control has been demonstrated.[261] Slow release formulations show greater activity than methoprene conventionally applied.[262]

Altosid™ SR-10, the commercial controlled-release version of methoxy – isopropyl(*E*,*E*)-11-methoxy-3,7,11-trimethyl-2,4-dodecadienoate – has been marketed and successfully used in mosquito control. Very low concentrations are adequate.[263] Methoprene in a controlled-release version has also been found effective in horn fly control.[264]

13. Monoliths in Agriculture

Controlled-release concepts developed basically in the antifouling and public-health areas have been extended to the control of pest forms in various aspects of agriculture. The literature is replete with descriptions of conventional pesticides and their usages. Most reports are not relevant to controlled release, where dosages, application methodologies, and so on, differ greatly. The description of the physicochemistry of pesticide release from conventional granular formulations by Furmidge *et al.*[265] and the report of Scott and Phillips on herbicide diffusion coefficients[266] have been found useful by the authors.

The monumental studies in the mid-1960s by Allen and his colleagues are probably the first reduction to practice of the concepts regarding controlled-release pesticides. Various pesticides were incorporated in polyamides, polyurea, polystyrene, cellulose, ethylene–vinyl acetate copolymers, and others.[267] A large reduction, perhaps as high as 1000 : 8, conventional *vs*. controlled release, in chemical usage was predicted.[268] Allen *et al.* found that the crystallinity of polymers was a disadvantage, and used additives to reduce it.[269] Polyamide preparations using carbofuran as the agent showed release up to 340 days with organophosphate materials not doing quite so well.[270]

By 1970 a number of investigators were active in the agricultural controlled-release pesticide area and literature on new techniques began to proliferate. For convenience we have divided the agricultural usages into several broad areas: monoliths, laminated structures, microencapsulation, pendent moieties, hollow fibers, and so on. The division is artificial in that an advance in the art in one area cross-fertilizes with the others. As of November 1982, there were approximately 1500 literature references referring to controlled release in agriculture and the akin areas of forestry, animal husbandry, lawn care, and so on. The discovery and synthesis of pheromones, and insect juvenile hormones, led to the need for special methods of prolonging persistency and thus provided emphasis for the development of controlled-release devices.

This section deals with monoliths, that is, dispensing media where the agent is uniformly dispersed in a polymeric matrix.

Stokes *et al.* prepared monoliths of aldicarb and dimethoate in urea–formaldehyde, polyamide, and other polymers.[271] For adequate release it was necessary to incorporate a plasticizing agent which served as a carrier. Results were generally poor, excepting aldicarb in cellulose acetate. Model systems coupled with laboratory evaluation suggested that a practical controlled-release system was feasible.[272] Tall oil–acid polymers were suggested as binders for various insecticides and herbicides absorbed on diatomaceous earth – a method of extending the life of conventional granular formulations.[273] A patent issued to Campbell in 1973 describes the use of Douglas fir wood chips as the absorber carrier for DDT, carbamates, and so on; a wax coating is used to reduce emission rate.[274]

The era from 1974 to 1976 saw the initiation of a number of new controlled-release areas. The 1976 Controlled Release Symposium saw the introduction of CR (controlled-release) devices in cattle[275]; extension of monoliths and pendent work in the forestry areas, especially with herbicides[276,277]; use of silicate-based monoliths[278]; the first report on controlled-release nematicides[279]; and a general extension of the use of insecticides in CR form.[280]

The use of controlled-release monolithic devices in pest control for livestock is a growing area of investigation. Miller *et al.* report that a plaster of Paris bolus containing methoprene inserted in the bovine rumen would provide good horn fly larvae control in the manure.[281] Implants of methoprene in poly(lactic acid) have been successfully used against cattle grubs.[282] The damage by insects, ticks, and mites are estimated to cost the U.S. livestock industry three billion dollars per year in decreased meat and milk production and two-thirds of that loss is borne out by beef cattle.[283] Rumen boluses of diflubenzuron are reported to provide 14 weeks of manure larvae control of the horn and stable fly.[283] Cattle ear tags and leg bands of 15% Stirofos or 20% dichlorvos in polyvinyl chloride will provide some duration of adult horn fly and tick control.[281,283]

Recently, antihelmintic agents in polypropylene matrices used as ear implants in Hereford steers have been shown to provide a degree of protection against parasitic worms and fly larva developing in manure.[284]

Continued work with starch and starch–xanthide-based controlled-release matrices has indicated considerable merit to this approach. DBCP (1,2-dibromo-3-chloropropane) nematicide in a starch–xanthide matrix provided 95–100% reduction in root-knot infections in the laboratory.[285] The nematicide work was extended to NNDD, *N*,*N*-dimethyl dodecanamine in polyvinyl acetate and ethylene (vinyl acetate) copolymers, the latter giving good laboratory results.[286] Ethoprop nematicide also has been formulated in a lignan matrix.[287]

Various pesticides have been incorporated in starch matrices and new techniques of trapping the agent in the cellular structure of the matrix developed.[288,289] EPTC has been dispersed in gelatinized starch followed by a boric acid treatment to enhance controlled-release properties.[290]

Methoprene has been incorporated in the cell structure of a polyamide matrix to provide fly control in manure through feeding of the granule to poultry.[291]

The development of herbicide-bearing monoliths continued during the past few years. Dellicolli demonstrated that 2,4-D/lignin matrices were practical.[287] Dichlobenil and chlorthiamid herbicides have been incorporated in urea–formaldehyde resins and good weed control in rice paddies reported.[292] The work in aquatic herbicide monoliths has continued and been extended to terrestrial weed control.[293,294] The interesting question of long-term (100+ years) control of vegetation over radioactive-waste burial sites has been raised. In the first known work along this line a 10%-trifluralin-loaded polypropylene pellet was developed and evaluated.[295]

In other recent advances, Williams has shown that chlorpyrifos or Reldan, loaded 1% in polyethylene film packaging material, provides 12–18 months' protection of stored fruit and grains against the rice weevil, lesser grain borer, and other insect pests.[296] It has also been reported that the incorporation of methyl parathion in ethylcellulose pseudolatex and polymethyl methacrylate provides superior insect control over microencapsulated forms of that agent.[297]

New polymers have been developed possessing properties useful to controlled-release technology. These include hydrophilic materials,[298] cotton cellulose,[299] and the bacteria-degradable poly(3-hydroxybutylate), a thermoplastic ester.[300] The use of controlled-release methods of releasing plant-growth stimulants is a relatively new area and one of considerable future potential.[301]

14. Agricultural Uses of Porosigen-Containing Monoliths

Leaching-type dispensers using a porosigen additive have been prepared with select herbicides, insecticides, and trace nutrients.

It has long been recognized that considerable fertilizer loss, up to 80% of that applied, occurs with conventional modes of application. Various attempts have been made to reduce the loss arising from groundwater mechanical washing or percolation through some sort of coating of the granule.[76,302] Sulfur-coated ureas constitute one method.[303] In work initiating in 1977 the authors compounded various salts of zinc, iron, copper, boron, manganese, cobalt, selenium, and molybdenum in alloys of polyethylene with ethylene (vinyl acetate copolymer) and/or ethylene-propylene copolymers with porosigens, such as calcium carbonate, to create long-term controlled-release systems dispensing ions of the respective trace nutrients.[304] Evaluation under laboratory conditions indicated a three- to five-year release life with leach rates adjustable through the mix of various porosigens of differing water solubilities.[305]

Monolithic porosigenic systems using chlorpyrifos and turbufos as the agent have shown soil persistency in excess of 28 weeks in the laboratory.[306] Materials are presently undergoing large-scale tests against soil insect pests of sugar cane.

A number of thermoplastic monoliths containing a herbicide and a porosigen additive were initially prepared and tested by the senior author against aquatic weeds, and then terrestrial weeds. Evaluated herbicides were Diuron, 2,4-D acid, Diquat, Fenac, Simazine, Bromocil, Dichlobenil, the dimethylamine of 2,4-D, and the isooctyl ester of 2,4-D. All show varying degrees of control of aquatic weeds. More importantly, several compounds each of 2,4-D acid, Diuron, and Dichlobenil were field tested against the common dandelion and ragweed. One-seasonal control was achieved with Diuron and 2,4-D acid controlled-release formulation, and essentially two-year control with the controlled-release Dichlobenil. Materials evaluated against terrestrial weeds are shown in Table 7.

Herbicides were also formulated not only in various thermoplastics, such as polyethylene, polypropylene, polystyrene, polyurethane, polyamide, but also in polyesters, epoxies, and other thermosetting resins. In general the binding matrix is of lesser importance than the choice of porosigens in

Table 7. Controlled-Release Terrestrial Herbicides Recipes

Ingredient	2508F	2508J	2508H	2501F	2501H	2501J	2502F	2502H	2502J
EPM[a]	33.5	30	29.5	28	24.5	24	31	27.5	27
LDPE[b]	33.5	30	29.5	28	24.5	24	31	27.5	27
Zinc stearate	1	1	1	1	1	1	1	1	1
Calcium carbonate	10	5	10	10	5	10	10	5	10
Silicon dioxide	—	4	—	—	4	—	—	4	—
Ammonium sulfate	—	6	10	—	6	10	—	6	10
Ethylene glycol	2	4	—	2	4	—	2	4	—
Diuron (80%)[c]	—	—	—	31	31	31	—	—	—
2,4-D acid	—	—	—	—	—	—	25	25	25
Dichlobenil (50%)	20	20	20	—	—	—	—	—	—
Dosages, a.i.	4	4	4	5	5	5	2	2	2
Pounds/Acre	12	12	12	20	20	20	10	10	10

[a] Vistalon 702.
[b] Low-density polyethylene MN 718.
[c] 3-(3,4-dichlorophenyl)1,1-dimethylurea.

regulating release rate. (This information will be published in detail later. A patent is pending.)

15. Microencapsulation

Methods of forming polymeric microcapsules containing an enveloping substance and a central core of an agent have been known since 1939. Over 280 U.S. patents have been issued describing various portions of the art. Methodology has been described in a number of recent reports and texts.(307–312) A large variety of polymers and macromolecules have been used as enveloping membranes: gelatin gum arabic, starch, sugar, ethylcellulose, carboxymethylcellulose, paraffin, polyvinyl acetate, polyethylene, polypropylene, polystyrene, polyacrylamide, polyethers, polyesters, polyamides, polyureas, polybutadiene, polyisoprene, polysiloxane, polyurethane, epoxy resins, inorganic silicates, and others.(308)

One of the early microencapsulation successes was with the insecticide methyl parathion. Encapsulation did not greatly extend field life, but dramatically reduced the handling hazard.(313) The commercialized material, Pencap M® (Pennwalt Corp.), showed lower mammalian toxicity, longer persistence, and lower-dosage requirements for control of various insects attacking cotton, beans, apples, tomatoes, corn, and other crops.(314) Persistence increased from one to two days, conventional, to five to seven days using a polyamide microcapsule.(315) The concept was extended to the use of diazinon in similar cross-linked polyamide microcapsules for household insect control. It was shown that it required four to eight times as much diazinon emulsifiable concentrate to achieve the same level of control of roaches, ants, flies, and silverfish.[316]

Controlled-release microcapsules of various insecticides have been prepared and evaluated against a variety of agricultural insect pests. This includes formulation of diazinon, chlorpyrifos, and acephate tested against various soil insects, such as cabbage maggot, corn borer, cabbage looper, cabbage worm, and Japanese beetle.[317] A 1% temephos in a polyethylene microcapsule has shown over 14 weeks of mosquito larvae control.[318] Foliar insect control with microcapsules has been demonstrated.[319] Tice *et al.* have shown efficacy of chlorpyrifos microcapsules.[320] Sumithion has been encapsulated in polyurethane and shown to be efficacious against mosquitoes and cockroaches.[321,322]

Relatively little work has been done with microencapsulated herbicides. The herbicidal performance of encapsulated chlorpropham is improved over that seen with conventional formulations.[323,324] Silicate capsules containing 2-(2,4,5-trichlorophenoxy) propionic acid showed promise.[325]

Microcapsules containing dimilin, a chitin synthesis inhibitor, have been shown promising against mosquito larva.[326] Tributyltin-chloride-containing microcapsules are noted as extending the antifouling life of marine paints.[327] Microencapsulation of nutrients and veterinary drugs are also encouraging new controlled-release areas. For instance, encapsulation of fish feed eliminates eutrophication.[328] Microencapsulated fasciolicides, such as diamfenetide and albendazole, may reduce parasitic infections which cause a 500 million dollar annual loss to animal producers.[329] Miocrocapsules of Famfur and Stirofos in polylactic acid injected directly into test animals provided several weeks of tick control.[330] Diflubenzuron microcapsules in chicken feed have provided fly larvae control in manure.[331]

Microencapsulation of feed additives for ruminants enhance their stability in the rumen for later release in the abomasum.[332]

Pheromone and attractant microcapsules have been evaluated in several situations. Mitchell reports on the use of encapsulated Gossyplure as a weapon against the pink bollworm of cotton[333] and Keiser *et al.* have shown the merit of using methyl eugenol against the oriental fruit fly.[334]

The heterogeneous nature of microcapsules is well known.[335] The spray drift problem arising from the wide range of particle sizes from spraying equipment presents a problem of some magnitude to agriculture. In an attempt to ameliorate both problems, a method was developed whereby pest-control agents were solvated or emulsified in a polymeric solution. Upon spraying, coacervation occurs and microcapsules of a relatively narrow-size range and wall thickness are developed as the solvent evaporates.[336] The wall matrix is a carboxylated polyacrylate and the solvent system consists of ammonia stabilized water and/or ethanol. Cross-linking is affected with a divalent metal ion, calcium being preferred, and occurs as the ammonia component of the vehicle volatilizes. A relatively wide number of insecticides and pheromones lend themselves to this system—termed "in-flight encapsulation." The first commercialized product, Pyrellin SCS (Webbe Wright Company, Fort Myers, Florida, manufacturer), used as a barn spray, was effective for 19 times as long as the nonencapsulated material.[337]

16. Pendent Systems in Agriculture

Increased activity in the study of pendent systems is noticed in the 1977–1982 period, especially with herbicides. Allan *et al.* reported that herbicides such as 2,4-D can be linked to α-cellulose through a hydrolyzable bond.[338] Metribuzin moieties have been bonded to polyvinyl alcohol, chitin, and cellulose.[339]

A six-year field study has shown that there is growth enhancement of Douglas fir trees protected from weed competition by 2,4-dichloro-phenoxy-butyric acid chemically bonded to Douglas fir bark.[340] Metribuzin has been studied by several groups as a pendent to a number of polymers.[341] Chitin is one of the inexpensive preferred backbone materials.[342] Also cellulose, amylose, and amylopectin have been used for both metribuzin and 2,4-D acid pendents.[343,344] Metribuzin can also be attached to poly(vinyl alcohol) through a diisocyanate linkage.[345] Tetrachlorophenol isocyanate will react with wood to prevent brown rot fungus attack.[346] In the continuing pendent herbicide study by Harris and Whitlock, the 2-acryloyloxyethyl ester of 2,4-D was copolymerized with 2-hydroxyethyl methacrylate to provide a hydrolyzable 2,4-D herbicidal moiety.[347]

Work with pendent insecticides is in its infancy. Kraft-lignin has been chemically bonded to carbofuran for the protection of paddy rice from insects.[348] The degree of control achieved is the same as that seen with conventional means of treatment, but an advantage of lower application rate is observed. Interesting work has been done by Meyers *et al.* where trichlorfon is bonded to a polymeric backbone using an unstable bond, such as an ester, amide, imine, or anhydride; but a diacid spacer (succinic, maleic, perfluorosuccinic, etc.) is used to enhance solubilization of the insecticide.[349] The spacer serves as a rate-controlling moiety.

17. Pheromones in Agriculture

Pheromone usage in trapping devices and space confusants has seen

Table 8. Pheromones in Agriculture

Target pest	Pheromone	References
Cabbage looper	Looplure	353, 361
Gypsy moth	Disparlure	353, 362, 363
Oriental fruit moth	Z-8-dodecenyl acetate	354
Mediterranean fruit fly	Trimedlure	355, 357, 364
Elm bark beetle	Multilure	356
Ips typographus	*cis*-Verbanol, ipsdienol, and methylbutenol mixture	358
Western pine shoot borer	Z- and E-9-dodecenyl acetate	359
Boll weevil	Grandlure	360, 364
Tobacco budworm	Virelure	360, 364
Pink bollworm	Gossyplure	360, 364
Fall army worm	Z-9-dodecen-1-ol acetate	360

increasing use in agriculture. The benefits derived from their use include: practicality; safety to nontargets (generally such materials have little or no toxicity); nonrepellent, nonpersistent, continuous, and uniform release; and no residue buildup.[(350)]

The laminated three-ply structure termed the "Hercon" dispenser has been the most popular of a number of methods of applying pheromones. Table 8 depicts the successful field and laboratory usage of pheromone release systems. The Hercon dispenser can be comminuted to granules, powders, and other configurations to aid in dispersion.[(351)] Such systems can be used as confusants for insect-mating disruption,[(352)] or as attractants for mechanical or insecticide-loaded traps.

Reider has dramatically shown the value in pheromone use as an attractant for mass trapping of insects to prevent forest damage in a three-year program in Norway as depicted in Table 9.[(358)]

Another method of slowly releasing a pheromone is through the use of an agent-loaded hollow fiber. These materials, developed and marketed by the Conrel Division of Albany International, release the attractant vapors at

Table 9. Mass Trapping of *Ips typographus* (Bark Beetle)

Year	Number of traps deployed	Number of trees destroyed by target (in millions)
1978	0	3.3
1979	605,000	3.0
1980	650,000	2.7
1981	530,000	0.9

the air/pheromone interface. Capillary action leads to a continual pheromone supply at this site. The fibers lend themselves to single component *and* multicomponent systems. For instance, the three elements of the elm bark beetle pheromone—multistriatin, 4-methyl-3-heptanol, and α-cubebene—can be simultaneously released, as is necessary, from a single dispensing unit.[365,366] The Conrel™ fibers have provided good bioassay results with pheromones against the pink bollworm,[367–369] elm bark beetle,[368] and the grape berry moth.[368] Hollow fibers have also been used to release inhibitors against the Southern pine beetle.[370]

Comparisons between the Hercon dispenser, Conrel fibers, and microencapsulated products show greater target control with the first two over the last.[362]

18. Insecticidal Carrier Systems

The Hercon® laminant functions through a carrier-type mechanism dependent upon the presence of a plasticizer in the polyvinyl chloride materials used. Baygon was incorporated in the Hercon dispenser in 1975 providing excellent cockroach control.[371] Later on, attractant–toxicant lure was made for fly control.[372,373] Hercon dispensers containing chlorpyrifos and malathion provided seven months' protection to stored rice.[374] By 1981 Hercon dispensers containing various insecticides had been evaluated against a number of insects; thimet *vs.* potato leaf hopper; diazinon and chlorpyrifos *vs.* fire ant; and malathion, diazinon, and chlorpyrifos *vs.* the rice borer.[375] In a comparison study of the Hercon dispenser in conjuction with microencapsulated and starch-loaded toxicant, better results in general were noted with the Hercon dispenser.[376] Diazinon, chlorpyrifos, and acephate in the various types of dispensers were evaluated against 13 agricultural pests (cabbage worm, diamond back moth, cabbage flea beetle, cabbage root maggot, wireworm, alfalfa blotch leafminer, potato leaf hopper, corn rootworm, corn borer, fall army worm, corn eatworm, seed corn maggot, and the anion maggot). While control of any single pest varied between the CR dispenser used, consistent superiority over conventional application was noted.

19. Other Release Systems

A number of membrane–reservoir systems have been developed for pharmaceutical dispensers. Few of these concepts have been carried into the pesticide area. Lee reported on the use of a polyvinyl alcohol–membrane system for the controlled release of chlorinated isocyanurate for bacteria control in swimming pools.[377]

In 1973 Nichols patented a method of preparing cellulose triacetate useful as a carrier for liquids due to its microcellular structure.[378] The material is trademarked as Poroplastic, a gel, or Sustrelle, a dry powder or slurry of microbeads, and marketed by Molecular Research Corporation of Massachusetts. It can be formed as a film or a fiber. The gelled product is created by solvent replacement with a pesticide or other agent loading.[379] Payloads of pesticides, pheromones, juvenile-hormone mimics, and so on, as high as 96% are possible.[380] The material is essentially a molecular "sponge." It is believed that the relatively difficult method of loading through solvent displacement has slowed market acceptance.

One of the latest, and most intriguing, controlled-release system is based upon magnetic control.[381] Unlike all other systems, the release rate is changeable on demand. Essentially, an agent is dispersed in an ethylene–vinyl acetate copolymer using magnetic steel beads as an additive. Introducing a magnetic field increases the agent release by as much as 100%. Bovine serum albumin has been released by this method and also insulin in diabetic rats.[382]

References

1. S. W. Katzev and M. L. Katzev, Last harbor for the oldest ship, *Natl. Geog.* **146**, 618 (1974).
2. J. Engelhart, C. Beiter, and A. Freiman, Recent Developments in Slow Release Organotin Antifoulants, Preceedings of the Controlled Release Pesticide Symposium, University of Akron, Ohio (September, 1974), p. 18.1.
3. J. McInnes, Composition for Coating the Bottoms of Iron Ships, British Patent 1356 (June 21, 1854).
4. J. G. Tarr and A. H. Wonson, Paint for Ships Bottoms, U.S. Patent 40,515 (November 3, 1863).
5. *Marine Fouling and Its Prevention*, U.S. Naval Institute, Annapolis, Md. (1952).
6. G. H. Young, Antifouling Composition, U.S. Patent 2,398,069 (April 9, 1946).
7. M. Varvill, Protection of Structures in Sea Water Against Fouling by Marine Growths, British Patent 851,902 (1960).
8. S. Miller, Antifouling Potentials of Pesticide Materials, Final Report, U.S. Navy Contract NOAS 59-6182-C, Marine Laboratory, Institute of Marine Science, University of Miami (March 1961).
9. N. F. Cardarelli, *Controlled Release Pesticides Formulations*, CRC Press, Boca Raton, Fla. (1976).
10. J. Robbins and D. M. Updegraf, Bis (Tributyltin) Adipate and Antifouling Materials Comprising Same, U.S. Patent 3,236,793 (1966).
11. Dai Nippon Paint Manufacturing Co., Japanese Patent 6,903,103 (1969).
12. BioMet Bis (Tri-*n*-Butyltin) Oxide Ingredient for Ship Bottom Paints, Technical Data Sheet No. 6223, M & T Chemicals Inc., Rahway, New Jersey (December 1965).
13. BioMet Tri-*n*-Butyltin Fluoride Ingredient for Shipbottom Paints, Technical Data Sheet No. 259, M & T Chemicals Inc., Rahway, New Jersey (1970).

14. R. W. Beers, Antifouling Coating Composition, U.S. Patent 3,801,534 (1974).
15. N. F. Cardarelli and H. F. Neff, Compositions de preservations contre le Incrustations Sous-Marines, France, patent 1,506,704, (November 13, 1967).
16. E. H. Bollinger, Controlled Release Antifouling Rubber Coating, Proceedings of the Controlled Release Pesticide Symposium, University of Akron, Ohio (September, 1974), p. 19.1.
17. D. R. Wuerzer, R. L. Senderling, and N. F. Cardarelli, *Rub. World* **157**(2), 77 (1967).
18. G. A. Janes and R. L. Senderling, The Diffusion–Dissolution Controlled Release Mechanism – Performance Evaluation of an Antifouling Coating After Ten Years' Service, Proceedings of the Controlled Release of Bioactive Materials Symposium, New Orleans, Louisiana (August 1979), p. II.3.
19. N. F. Cardarelli and S. J. Caprette, Antifouling Covering, U.S. Patent 3,426,473 (February 11, 1969).
20. N. F. Cardarelli and H. F. Neff, Biocidal Elastomeric Compositions, U.S. Patent 3,639,583 (February 1, 1972).
21. N. F. Cardarelli, Biocidal Elastomeric Composition and Method for Dispersing Biocides Therewith, U.S. Patent 3,851,053 (November 26, 1974).
22. "NOFOUL": Technical Background, B.F. Goodrich Company Report (Published), Aerospace and Defense Division, Akron, Ohio (January 1969).
23. Boat paints, *Consumers Reports*, pp. 223–227 (April 1969).
24. F. G. W. Smith, R. B. Bottoms, E. Abrams, and S. M. A. Miller, *Forest Prod. J.* **340** (September 1956).
25. M. Varvill, Protection of Structures in Sea Water against Fouling by Marine Growths, British Patent 851,902 (October 19, 1960).
26. H. P. Vind and H. Hochman, An Evaluation of Organotin Compounds as Preservatives of Marine Timbers, Report, American Wood Preserver's Association (1962).
27. *Tin and Its Uses, Pub. Tin*, No. 36, Research Institute (1956).
28. R. J. Zedler and C. B. Beiter, *Soap Chem. Spec.* (March 1962).
29. D. Oldfield and G. F. Samson, Potential Antifouling Coatings for Timber, Technical Note 319, Australian Defence Scientific Service, Defence Standards Laboratory, Maribyrnong, Victoria, Australia (November 1973).
30. L. R. Snoke and A. P. Richards, *Science* **124**, 443 (1956).
31. J. S. Muraoka, Biodeteriorations of Materials: Part II (AD631078), U.S. Navy Civil Engineering Laboratory, Port Hueneme, California (February 1966).
32. H. Hochman and H. P. Vind, Screening of Chemical Toxicity to Marine Borers – Final Report, TR 426, U.S. Navy Civil Engineering Laboratory, Port Hueneme, California (February 1966).
33. H. P. Edelstein, S. A. Eller, and R. G. Grunther, *Naval Eng. J.* **115** (February 1970).
34. J. M. D. Woodford, Underwater Marine Coatings: Part II: Marine Biocidal Rubbers Containing Organotin Toxics, Report 496, Defence Standards Laboratory, Maribyrnong, Victoria, Australia (March 1972).
35. A. De Forest, R. W. Pettis, and A. T. Phillip, Underwater Marine Coatings: A Detailed Examination of Elastomeric Antifouling Materials after Marine Immersion, Report 589, Defence Standards Laboratory, Maribyrnong, Victoria, Australia (May 1974).
36. P. De Wolf and A. M. Van Londen, *Nature* **209**, 272 (1966).
37. Mitsui Shipbuilding and Engineering Company and Hokko Chem. Ind. Company, Antifouling Paints, British Patent 1,211,768 (November 11, 1970).
38. I. Hechenbleikner and P. F. Thompson, Carboxymercaptal Hydrocarbon Tin Salts with Antifouling Marine Paint or Coating Biocidal Activity, U.S. Patent 3,463,644 (August 26, 1969).

39. Antifouling systems based on organotins: Australian Navy trials, *Tin and Its Uses*, No. 96 (1973).
40. Underwater Protective Coating Compositions Based on Triorganotin Compounds, British Patent 1,348,752, Brit. Pat. Abs., Vol. V, No. 12, p. 12 (April 30, 1974).
41. M. Onozuka, Y. Hayashi, and Y. Adachi, Antifouling Composition for Use in Water Comprising a Polymer and an Organotin Compound, U.S. patent 4,010,141 (1977).
42. M. H. Gitlitz and A. Frieman, Antifouling Composition, U.S. Patent 4,012,347 (March 15, 1977).
43. D. S. Carr, Organolead Antifouling Paints, Proceedings of the Controlled Release Pesticide Symposium, University of Akron, Ohio (September 1974), p. 20.1.
44. L. W. Friedsham, Bivalent Metal Hydroxide Treatment of Drying Oil Modified Alkyd Resins, U.S. Patent 3,110,690 (November 12, 1973).
45. G. B. Fahlstrom, Preservative Composition for Marine Use, U.S. patent 3,227,563 (January 4, 1966).
46. P. A. Herbert, D. F. Bowerman, and K. S. Ford, Chlorinated rubber marine coatings: Performance tests including antifouling and blends of CR with tar, *Am. Chem. Soc.* **34**(1), 594 (1974).
47. N. A. Ghanem, Research Work Pertaining to Protective Coatings and Corrosion Protection, Report, National Research Centre, Cairo, Egypt Workshop (November–December 1977).
48. R. W. Beers, Antifouling Coating Composition, U.S. Patent 3,801,534 (April 2, 1974).
49. L. V. Wake, Improved Antifouling by Marine Coatings Containing Mixed Toxics, Technical Note 363, Department of Defence, Materials Research Laboratory, Maribyrnong, Victoria, Australia (January 1975).
50. A. T. Phillip, *Progr. Org. Coat.* **2**, 159 (1973/74).
51. M. Kronstein, Simultaneous Use of Different Organometallic Toxicants in Exposed Antifouling Paints, Proceedings of the Controlled Release of Bioactive Materials Symposium, New Orleans, Louisiana (August 1979), p. II.7.
52. D. S. Carr and M. Kronstein, *Mod. Paint Coat.* **65**(12), 23 (1975).
53. J. H. Bishop and S. R. Silva, Antifouling paint film structure, with particular reference to cross sections, *Am. Chem. Soc.*, **30**(1), 364 (1970).
54. J. R. Saroyan, *Ocean Engineering*, Vol. I, Pergamon Press, Great Britain (1969).
55. C. P. Monaghan, V. H. Kulkarni, and M. Y. Good, Further Evaluation of a Diffusion Model for the Characterization of the Leaching Properties of Several Conventional Antifouling Coatings, Proceedings of the Controlled Release of Bioactive Materials Symposium, National Bureau of Standards, Gaithersburg, Maryland (August 1978), p. 7.22.
56. C. B. Beiter, Factors Influencing the Release of Organotin Antifoulants from Ship Bottom Paints, Proceedings of the Controlled Release Pesticide Symposium, University of Akron, Ohio (September 1976), p. 2.22.
57. S. V. Kanakkanatt, Schistosomiasis Control by Slow Release Molluscicides, Report to the World Health Organization, Geneva, Switzerland (September 15, 1972).
58. K. Akagane and M. Matsuura, *Shikizai Kyokaishi* **45**(2), 69 (1972).
59. J. C. Montermoso, Organotin Acrylic Polymers, U.S. Patent 3,016,369 (August 1972).
60. J. A. Montemarano and E. J. Dyckman, Antislime Coatings, Part III: Antislime Organometallic Polymers of Optical Quality, Report 3597, Naval Ship Research and Development Center, Bethesda, Maryland (September 1972).
61. J. A. Montemarano and E. J. Dyckman, Antifouling Organometallic Structural Plastics, Report 4159, Naval Ship Research and Development Center, Bethesda, Maryland (August 1973).
62. E. J. Dyckman, J. A. Montemarano, and E. C. Fischer, *Naval Eng. J.* **85**, 33 (1973).

63. J. A. Montemarano and E. J. Dyckman, Biologically Active Polymeric Materials Exhibiting Controlled Release Mechanisms for Fouling Prevention, Proceedings of the Controlled Release Pesticide Symposium, University of Akron, Ohio (September 1974), p. 21.1.
64. E. J. Dyckman, J. A. Montemarano, E. C. Fischer, and R. Ressler, Non-Polluting, Antifouling Composition and Method, U.S. Patent 3,979,354 (September 7, 1976).
65. P. Dunn and D. Oldfield, A New Concept in Antifouling Elastomers, Report, Third Australian Technical Convention Institute of the Rubber Industry, Terrigal, New South Wales (September 1974).
66. M. Kronstein, Study of Released Matter From Organometal Toxicant–Antifouling Paints Exposed Underwater, Proceedings of the Controlled Release of Bioactive Materials Symposium, Corvallis, Oregon (August 1977), p. 7.37.
67. D. S. Carr and M. Kronstein, *Mod. Coat.* **65**, 23 (1975).
68. M. Kronstein, Organolead Antifouling Paint Studies Using Atomic Absorption Analysis, Proceedings of the Controlled Release Pesticide Symposium, University of Akron, Ohio (September 1976), p. 2.4.
69. R. Deschiens and H. Floch, *Compt. Rend.* **255**, 1236 (1962).
70. H. Floch and R. Deschiens, Etude comparee de l'action molluscicide du 5,2′-dichloro-4-nitro-salicylanalide (Bayer f73) et de sels (acetate et chlorure) de triphyletain, *Bull. Soc. Pathol. Exot.* **55**, 816 (1962).
71. L. S. Ritchie, L. A. Berrios-Duran, L. P. Frick, and I. Fox, Molluscicidal time-concentration relation of organotin comounds, *Bull. WHO* **31**, 147 (1964).
72. A. H. Freiberg, Tributyltin Molluscicides, U.S. Patent 3,439,098 (April 15, 1969).
73. R. Deschiens and H. Flock, Donnes complementaires sur l'action molluscicide remanente de sels de triphenyl-etain, *Bull. Soc. Pathol. Exot.* **56**, 22 (1963).
74. R. Deschiens, H. Brottes, and L. Mvogo, Application sur le terrain au cameroun, dans le prophylaxie des bilharzioses de l'action molluscicide de l'oxide de tributyletain, *Bull. Soc. Pathol. Exot.* **59**, 968–973 (1966).
75. L. A. Berrios-Duran, L. S. Ritchie, and H. B. Wessel, Field screening tests on molluscicides against *Bimphalaria glabrata, Bull. WHO* **39**, 316 (1968).
76. L. A. Berrios-Duran and L. S. Ritchie, Molluscicidal activity of bis (tri-*n*-butyltin) oxide formulated in rubber, *Bull. WHO* **39**, 310 (1967).
77. A. Fenwick, Laboratory evaluation of biocidal rubber pellets 633B as a residual molluscicide against *Biomphalaria pfeifferi, Bull. WHO* **41**(2), 326 (1969).
78. K. Y. Chu, Effects of environmental factors on the molluscicidal activities of slow-release hexabutyldistannoxane and copper sulfate, *Bull. WHO* **54**, 417 (1976).
79. N. F. Cardarelli, *Controlled Release Molluscicides*, Monograph, Environmental Management Laboratory, University of Akron, Ohio (May 1977).
80. N. F. Cardarelli, Method for Dispersing Toxicants to Kill Disease Spreading Water-Spawned Larva, Trematodes, Molluscs, and Similar Organisms, and the Products Used in Such Methods, U.S. Patent 3,717,181 (December 17, 1968).
81. A. T. Santos, M. J. Santos, B. L. Blas, and E. A. Banez, Studies on the Effect of BioMet and CBL-9B on *O. quadrasi* and *Schistosoma japonicum* Cercariae, Proceedings of the Controlled Release Pesticides Symposium, Natural Bureau of Standards, Gaithersburg, Maryland (August 1978), p. 105.
82. B. Gilbert, L. A. Paes Leme, W. Benson, and M. Figueirido, Slow Release in Tropical Disease Control, Proceedings of the Controlled Release Pesticide Symposium, University of Akron, Ohio (September 1974), p. 4.1.
83. B. Gilbert, Slow Release Molluscicides in Schistosomiasis Control, Proceedings of the Controlled Release Pesticide Symposium, University of Akron, Ohio (September 1976), p. 5.9.

84. B. Gilbert, L. A. P. Leme, A. M. Ferreira, M. S. Bulhoes, and C. Castleton, Field tests of hexabutyldistannoxane (TBTO) in slow-release formulations against *Biomphalaria* spp., *Bull WHO* **49**, 633 (1973).
85. J. Willomitzer and Z. Lucky, *Helminthologia* **11**, 285 (1970).
86. A. Abdallah, Schistosomiasis Control. Report of a World Health Organization Expert Committee, WHO Technical Report Serial No. 515 (1973).
87. C. J. Shiff and A. C. Evans, *Cent. Afr. J. Med.* **23** (11, Suppl.), (1977).
88. E. S. Upatham, Focal Control of *Schistosoma haematobium*—Transmitting Snails, *Bulinus* (*Ph.*) *abyssinicus* Using Controlled Release Tri-*n*-Butyltin Fluoride and Copper Sulfate, Proceedings of the Controlled Release of Bioactive Materials Symposium, II.57–II.59, New Orleans, Louisiana (August 1979), p. II.57.
89. L. R. Sherman and T. Carlson, *J. Anal. Tox.* **4**, 31 (1980).
90. W. H. Evans, BioMet™ SRM Contact Studies, Proceedings of the Controlled Pesticide Symposium, University of Akron, Ohio (September 1976), p. 5.48.
91. C. J. Shiff, C. Yiannakis, and R. C. Evans, Further Trials with TBTO and Other Slow Release Molluscicides in Rhodesia, Proceedings of the International Controlled Release Pesticide Symposium, Wright State University, Dayton, Ohio (September 1975), p. 177.
92. P. J. Smith, A. J. Crowe, V. G. K. Das, and J. Duncan, *Pestic. Sci.* **10**, 419 (1979).
93. W. H. Evans, N. F. Cardarelli, and D. J. Smith, *J. Toxicol. Environ. Health.* **5**, 871 (1979).
94. N. F. Cardarelli, Laboratory and Field Evaluations of Controlled Release Molluscicides and Schistolarvicides, Annual Report, Edna McConnell Clark Foundation, New York, New York 276-0091 (July 1, 1977).
95. N. F. Cardarelli and K. E. Walker, Chronic Effects of Ultralow Toxicant Concentrations, Final Report, U.S. National Institute Health, 1-R22-AI11861, Bethesda, Maryland (July 1, 1976).
96. Bioassay of Triphenyltin Hydroxide for Possible Carcinogenicity, U.S. National Cancer Institute Technical Report, Series No. 139, (1978).
97. Bioassay of Dibutyltin Diacetate for Possible Carcinogenicity, U.S. National Cancer Institute Technical Report, Series No. 183 (1979).
98. S. R. M. Innes, *J. Natl. Cancer Inst.* **42**, 1101 (1969).
99. S. S. Epstein, *Appl. Pharmacol.* **23**, 288 (1972).
100. N. F. Cardarelli and B. M. Quitter, Organotins in Carcinogenesis, Abstract, Controlled Release of Bioactive Materials Symposium, Fort Lauderdale, Florida (July 1982), p. 125.
101. N. F. Cardarelli and W. H. Evans, in: *Controlled Release of Bioactive Materials* (R. Baker, ed.), pp. 357–385, Academic Press, New York (1980).
102. A. W. Sheldon, *Am. Chem. Soc. Ser.* **34**(1), 600 (1974).
103. V. F. Hodge, S. L. Seidel, and E. D. Goldberg, *Anal. Chem.* **51**, 1256 (1979).
104. W. B. Bollen and C. M. Tu, *Tin and Its Uses*, No. 94, 13 (1972).
105. C. P. Monaghan, J. F. Hoffman, E. J. O'Brien, L. M. Frenzel, and M. Y. Good, An Evaluation of the Leaching Mechanisms for Organotin Containing Antifouling Coatings, Proceedings of the Controlled Release Pesticide Symposium, Oregon State University, Corvallis, Oregon (August 1977) (unpublished).
106. H. Akagi and Y. Sakagamic, On degradation of organotin compounds by ultraviolet rays, *Bull. Inst. Public Health Jpn.* **20**, (1971).
107. R. D. Barnes, A. T. Bull, and R. C. Poller, *Pestic. Sci.* **4**, 305 (1973).
108. D. Y. Mitchum and T. D. Moore, *Progr. Fish Cult.* **31**(3), 143 (1969).
109. L. S. Ritchie, V. A. Lopez, and J. M. Cora, Prolonged Application of an Organotin Against *Biomphalaria glabrata* and *Schistosoma mansoni*, Report, Puerto Rico Nuclear Center, San Juan (1974).
110. V. A. Lopez and L. S. Ritchie, The Release of TBTO and Other Toxicants from

Elastomers as Measured by a Bioassay, Proceedings of the Controlled Release Pesticide Symposium, University of Akron, Ohio (September 1974), p. 38.1.

111. A. T. Santos, B. L. Blas, E. L. V. Banez, and C. Reyes, Laboratory Trials on the Effect of BioMet and CBL-9B on *Oncomelania hupensis quadrasi* and *Schistosoma japanicum* Cercaria, Proceedings of the Controlled Release Pesticide Symposium (August 1977), p. 105.
112. N. F. Cardarelli, Cercariacidal Potency of Slow Release Molluscicides, Proceedings of the Controlled Release Pesticide Symposium, University of Akron, Ohio (September 1974), p. 39.1.
113. A. Theron, A differential filtration technique for the measurement of schistosome cercarial densities in standing waters, *Bull. WHO* **57**, 971 (1979).
114. J. W. Miles, R. T. Taylor, G. C. Robbins, and T. B. Blue, Evaluation of Conventional and Slow-Release Formulations of Molluscicides Against *Biompharalaria glabrata*, Proceedings of the International Conference on Schistosomiasis, Cairo, Egypt, 469 (1978).
115. S. Miller, Evaluation of DAC 2787 as an Antifouling Toxicant for Shipbottom Coatings, Progress Report, Miami Marine Research Inc., March–April 1965 (unpublished).
116. N. F. Cardarelli, Methods of Controlling the Gastropod Vectors of Parasitic Flukes, U.S. Patent 3,767,809 (October 3, 1973).
117. P. R. Hira and G. Webbe, *J. Helminth* **46**(1), 11 (1972).
118. N. F. Cardarelli, Development and Laboratory Evaluation of Controlled Release Organolead Molluscicides, Proceedings of the Controlled Release Pesticide Symposium, University of Akron, Ohio (September 1976), p. 5.1.
119. R. C. Wall, *Sterkiana* **63/64**, 1 (1976).
120. K. E. Walker and N. F. Cardarelli, Development of Slow Release Copper Sulfate as a Molluscicide, International Copper Research Association, Report Project 203, New York, New York (July 1, 1973).
121. N. F. Cardarelli, in: *Schistosomiasis Control* (T. C. Cheng, ed.), pp. 177–240, Academic Press, New York (1974).
122. N. F. Cardarelli and K. E. Walker, Slow Release Copper Toxicant Composition, U.S. Patent 4,012,221 (March 15, 1977).
123. T. C. Cheng and J. T. Sullivan, Biological Studies on Copper Containing Molluscicides, Report No. 193, International Copper Research Association, New York, New York (June 30, 1974).
124. N. F. Cardarelli and K. E. Walker, Development of Slow Release Copper Sulfate as a Molluscicide, Report Project 203, International Copper Research Association, (July 1, 1974).
125. H. A. Floch and R. Deschiens, Etude des proprietes molluscicides du triphenylacetate de plomb, *Bull. Soc. Pathol. Exot.* **68**, 315 (1975).
126. K. Walker, Field and Laboratory Tests of Some New Molluscicides and Some Novel Formulations of Existing Ones, Fifth Annual Report, Research and Control Department, Castries, St. Lucia (February 1971–1972), p. 104.
127. E. S. Upatham, Laboratory Trials of Various Molluscicide on *Bulinus* (*Ph.*) *abyssinicus*, the Intermediate Host of *Schistosoma haematobium* in Somalia, Proceedings of the Controlled Release of Bioactive Materials Symposium, New Orleans, Louisiana (August 1979), p. II.52.
128. M. Koltnow, K. Walker, and N. Cardarelli, Destruction of Aquatic Snail by Means of a Bait Concept, WHO Report Series PD/MOL/69,6, Geneva, Switzerland (1969).
129. K. E. Walker, S. Z. Mansdorf, S. A. Quinn, and N. F. Cardarelli, Pollution control by means of target biocides, *Proc. Inst. Environ. Sci.* 324 (April 1970).
130. S. M. Bille, K. E. Walker, and N. F. Cardarelli, A Bait Method of Destroying *Oncomelania* and *Biomphaloria glabrata*, Snail Hosts of Schistosomiasis; and Related

Studies, Final Report, U.S. Army Medical Research and Development Command, DADA 17-69-C-9166, (July 1971).

131. S. M. Bille, K. E. Walker, and N. F. Cardarelli, Field Tests of Molluscicidal Baits, Final Report, U.S. Army Medical Research and Development Command, DADA 17-72-G-9351 (October 1, 1971–December 19, 1971).
132. H. H. Stibbs, E. Charnan, S. Ward, and M. L. Karnovsky, *Nature* **260**, 702 (1976).
133. J. D. Thomas and D. Assefa, *Comp. Biochem. Physiol.* **63C**, 99 (1979).
134. H. A. Schultz and A. B. Webb, *Mosq. News* **29**(1), 38 (1969).
135. A. H. Boike and C. B. Rathburn, *Mosq. News* **33**(4), 501 (1973).
136. Laboratory Bioassay of Various Biocidal Rubbers as Mosquito Larvicides, Prelimary Report, U.S. Army Environmental Hygiene Agency, Edgewood Arsenal, Maryland, Project 31-3-67/68 (January 26, 1968).
137. N. F. Cardarelli and J. C. Hess, Floating Larvicide, U.S. Patent 3,590,119 (June 29, 1971).
138. N. F. Cardarelli and C. J. Major, Biocidal Rubber for Water Reclamation Systems, AMRL-TR-69-17, Aerospace Medical Research Lab, Air Force Systems Command, Wright-Patterson AFB, Dayton, Ohio (February 1969).
139. N. F. Cardarelli, *Tin and Its Uses* No. 93, 16 (1972).
140. S. Bille, S. Z. Mansdorf, and N. F. Cardarelli, Development of Slow Release Herbicide Materials for Controlling Aquatic Plants, Annual Report Department of the Army, Office Chief Engineer, Washington, D.C., DACW 73-70-C-0030 (November 1969–September 1970) (unpublished).
141. L. L. Nelson, Evaluation of Slow Release Herbicide Formulations for Aquatic Weed Control, Report, U.S. Army Environmental Hygiene Agency, Edgewood Arsenal, Maryland, Entomological Special Study 31-004-69/72 (January 21, 1972).
142. S. Bille, S. Z. Mansdorf, and N. F. Cardarelli, Department of Slow Release Herbicidal Materials for Controlling Aquatic Plants, Final Report, Department of the Army, Office Chief Engineer, Washington, D.C., DACW 73-70-C-0030 (November 1969–July 1971) (unpublished).
143. E. A. Branum and S. M. Bille, Pest control through chronic intoxication, *Proc. Inst. Environ. Sci. 1972*, p 241.
144. N. F. Cardarelli, *Weeds, Trees and Turf* **11**, 36 (1972).
145. G. A. Janes and N. F. Cardarelli, Aquatic Herbicides Chronicity Study, Final Report, Department of the Army, Office Chief Engineer, Washington, D.C., DACW 73-73-C-0042 (February 15, 1974) (unpublished).
146. S. A. Quinn, N. F. Cardarelli, and E. O. Gangstad, *J. Aquat. Plant Manag.*, **15**, 74 (1977).
147. N. F. Cardarelli and S. V. Kanakkanatt, A Hypothesis on Chronic Intoxication by Herbicides in Ultralow Concentrations, Report, American Chemical Society's 30th NW Regional Meeting, University of Hawaii, Honolulu (June 1975) (unpublished).
148. N. F. Cardarelli, A Hypothesis Concerning the Mechanism of the Chronicity Phenomenon in Aquatic Plants, Proceedings of the Controlled Release Pesticide Symposium, Wright State University, Dayton, Ohio (September 1975), p. 349.
149. A. C. Maldonado, *Weed Sci.* **22**, 443 (1974).
150. N. F. Cardarelli, Evaluation of Incracide E-51 Against Select Aquatic Weeds and Algae, Final Report, International Copper Research Association, New York, New York (April 5, 1975).
151. S. Z. Mansdorf, Slow Release: Development of Aquatic Herbicide Carrier Systems, Proceedings of the Controlled Release Pesticide Symposium, University of Akron, Ohio (September 1974), p. 12.1.
152. N. F. Cardarelli, W. E. Thompson, and G. A. Janes, *Rub. Dev. J.* **27**, 54 (1974).
153. K. K. Steward, Weed Investigations—Aquatic and Noncrop Areas, Annual Report U.S. Department of Agriculture, ARS, Fort Lauderdale, Florida (July 1970–June 1971).

154. W. E. Thompson, Field Tests of Slow Release Herbidices, Proceedings of the Controlled Release Pesticide Symposium, University of Akron, Ohio (September 1974), p. 15.1.
155. R. J. Geary, Particulate Pesticidal Composition Coated with An Amido–Aldehyde Resin Polymerized *In situ*, U.S. Patent 3,074,845 (January 22, 1963).
156. I. Ugi, Methods for Controlling Pests and Fungi with Polyisonitriles, U.S. Patent 3,278,371 (October 11, 1966).
157. J. L. Sanders and J. K. Knoke, *J. Econ. Entomol.* **60**, 588 (1967).
158. J. T. Miles and K. Schreiber, *Can. J. Plant Sci.* **51**, 347 (1971).
159. R. A. Stokes, J. R. Copedge, and R. L. Ridgeway, *J. Agric. Food Chem.* **18**, 195 (1970).
160. Granule Destine a Absorber et Dispenser un Agent Chimique Pour L'agriculture et Son Procede de Preparation, France Patent 2,082,026 (December 26, 1969).
161. D. Seaman and R. P. Warrington, *Pestic. Sci.* **3**, 799 (1972).
162. L. L. Balassa, Herbicidal Compositions and Methods of Preparing Them, U.S. Patent 3,725,031 (April 3, 1973).
163. F. B. Folckemer, R. E. Hanson, and A. Miller, Resin Compositions Comprising Organophosphorus Pesticides, U.S. Patent 3,318,769 (May 9, 1967).
164. J. W. Miles, G. W. Pierce, and J. E. Woehst, *J. Agric. Food Chem.* **10**, 244 (1962).
165. G. D. Brooks, C. M. Elmore, H. F. Schoof, and G. T. Carmichael, Field Evaluation of Three Types of DDVP Dispensers for the Control of *Culex pipiens quinquefasciatus* Say in Catch Basins, Proceedings of the 50th Annual Meeting of the New Jersey Mosquito Extermination Association and the 19th Meeting of the American Mosquito Control Association, American Mosquito Control Association, Albany, New York (1963), p. 353.
166. R. Aries, Compositions Pesticides Stabilisees, France Patent 2,105,097 (September 25, 1970).
167. R. Aries, Compositions Pesticides de Toxicite Redvite, France Patent 2,108,927 (October 20, 1970).
168. R. Aries, Compositions Pesticides Stabilisees, France Patent 2,112,136 (November 6, 1970).
169. R. Aries, Compositions Antiparasitaires Stabilisees, France Patent 2,117,768 (December 16, 1970).
170. R. Aries, Compositions Antiparasitaires Remanentes, France Patent 2,117,769 (December 16, 1970).
171. F. K. Banks, Insecticidal Collar for Animals, U.S. Patent 2,205,711 (June 25, 1940).
172. H. J. Peo, Flea Repelling Animal Collar, U.S. Patent 2,734,483 (February 14, 1956).
173. C. V. Coash, Pest Control Coating Compositions, U.S. Patent 2,621,163 (December 9, 1952).
174. H. R. Leeds, Process for Making Micro-Reticulated Material, U.S. Patent 2,777,824 (January 15, 1957).
175. P. Bracey, Pesticidal Coating Compositions, U.S. Patent 2,788,320, (April 9, 1957).
176. T. J. Doyle, Animal Collar Having a Pocket for Insecticide, U.S. Patent 2,791,202 (May 7, 1957).
177. J. J. Menn, F. B. Flockemer, and A. Miller, Antihelmintic Compositions and Methods of Employing the Same, U.S. Patent 3,166,472 (January 19, 1965).
178. I. Feinberg, Process for Preparing a Stable Polymer Latex Containing an Insecticide, U.S. Patent 3,400,093 (September 3, 1968).
179. R. J. Hackney, Insect Resistant Thermoplastic Compositions, U.S. Patent 3,408,323 (October 29, 1968).
180. F. Alfes, W. Behrenz, K. Raichle, and K. Weirauch, Shaped Articles with Long-Term Vapour Emission, U.S. Patent 3,608,062 (September 21, 1971).
181. T. K. Duffey and W. R. Coleman, Composition and Method for the Control of Fleas on Domesticated Animals, U.S. Patent 3,826,232 (July 30, 1974).

182. L. M. Grubb and J. K. Baxter, Tick and Flea Collar of Solid Vinylic Resin–Carbamate Insecticide, U.S. Patent 3,852,416 (December 3, 1974).
183. R. Aries, A. Dechaumet, and G. Dupre, Device for Securing a Vaporizable Active Substance to an Animal Collar, U.S. Patent 3,814,061 (June 4, 1974).
184. R. C. Potter, Insecticidal Compositions Employing Certain Block Copolymers, U.S. Patent 4,065,555 (December 27, 1977).
185. M. von Bittera, H. U. Sieverking, W. Standel, and H. Vosge, Tierhalsbander mit Insektizider Wirksamkeit, Federal Republic of Germany Patent 2,715,596 (October 19, 1978).
186. N. R. Pasarela, Acaricidal Resin Composition Containing Spiro (Cyclopropane-1,1′-Indene)-2-Carboxylic Acid, 3,3-Dimethyl-Cyano-*m*-Phenoxybenzyl Ester, U.S. Patent 4,145,409 (March 20, 1979).
187. L. K. Colliard, Pet Collar, U.S. Patent 4,091,766 (May 30, 1978).
188. P. R. Dick, Devices for Protecting Animals from Ectoparasites, U.S. Patent 4,150,109 (April 17, 1979).
189. J. Greenberg and G. D. Cloyd, Pet Collar, U.S. Patent 4,158,051 (June 12, 1979).
190. W. Engel, Improvements Relating to the Therapeutic Application of Medicaments and the Utilization of Disinfectants, Insecticides and Insectifuges, Great Britain Patent 599,237 (March 8, 1948).
191. G. H. R. Watson, Improvements in or Relating to Paints, Lacquers and Like Coating Compositions, Great Britain Patent 796,368 (June 11, 1958).
192. Shell Internationale Research Maatschappij N.V., Improvements Relating to Pesticidal Compositions, Great Britain Patent 903,159 (August 15, 1962).
193. Shell Internationale Research Maatschappij N.V., An Insecticidal Composition, Great Britain Patent 955,350 (April 15, 1964).
194. Montecatini Societa Generale Per L'industria Mineraria E Chimica, Insecticidal Paints, Great Britain Patent 999,067 (July 21, 1965).
195. Y. Boullenger, Compositions Chimiques Contenant un Azote Trivalent et aidant L'evaporation des Insecticides, France Patent 2,124,776 (September 22, 1972).
196. S. Hyman, B. S. Bernstein, and R. Kapoor, Activated Polymer Materials and Processes for Making Same, U.S. Patent 3,705,938 (December 12, 1972).
197. A. F. Kydonieus and I. K. Smith, Efficacy of Hercon™ Polymeric Controlled Release Dispensers (PRCD) in Cockroach Control, Report, Herculite Protective Fabrics Corporation, New York, New York (1974).
198. A. R. Quisumbing, D. J. Lawatsch, and A. F. Kydonieus, Field Studies on the Control of the German Cockroach, *Blatella germanica* using Hercon™ "Roach Tape" and Standard Sprays, Proceedings of the International Controlled Release Pesticide Symposium, Wright State University, Dayton, Ohio (September 1975), p. 247.
199. N. F. Cardarelli, Biologically Active Insecticide Containing Polymeric Formulation, U.S. Patent 4,237,113 (December 2, 1980).
200. W. W. Barnes, A. B. Webb, and L. B. Savage, *Mosq. News* **27**, 488 (1967).
201. W. W. Barnes and A. B. Webb, *Mosq. News* **28**, 458 (1968).
202. J. T. Whitlaw and E. S. Evans, *J. Econ. Entomol.* **61**, 889 (1968).
203. J. H. Nelson, E. S. Evans, N. E. Pennington, and W. H. Young, The U.S. Army Environmental Hygiene Agency Activities in the Area of Slow Release Insecticides, Proceedings of the Controlled Release Pesticide Symposium, University of Akron, Ohio (September 1974), p. 30.1.
204. R. N. Wilkinson, W. W. Barnes, A. R. Gillogly, and C. D. Minnemeyer, *J. Econ. Entomol.* **64**, 1 (1971).
205. T. A. Parker, Effectiveness of 9.9 Percent Dursban in Polyethylene Applied as a

Pre-Season Larvicide, Report, U.S. Army Environmental Hygiene Agency Special Study No. 31-005-71 (AD 725573) (1970).
206. T. Miller, L. L. Young, L. W. Roberts, D. R. Roberts, and R. N. Wilkinson, *Mosq. News* **33**, 148 (1973).
207. T. A. Miller, L. L. Nelson, and W. W. Young, *Mosq. News* **33**, 172 (1973).
208. L. L. Nelson, T. A. Miller, and W. W. Young, *Mosq. News* **33**, 396 (1973).
209. R. Deschiens and H. A. Floch, Utilisation Molluscicide du Polyethylene Impregne par L'oxyde de Tributyletain, *Bull. Soc. Pathol. Exot.* **63**, 73 (1970).
210. *Prod. Eng.* (August 27, 1974).
211. R. I. Christy and C. R. Meeks, Polyurethane Retainers for Ball Bearings, Proceedings of the International Ball Bearing Symposium, Charles Stark Draper Laboratory, Cambridge, Massachusetts (June 1973).
212. S. Hyman, Bacteristatic Material, U.S. Patent 3,279,986 (October 19, 1966).
213. N. F. Cardarelli, Controlled Release of Compounds Utilizing a Plastic Matrix, U.S. Patent 4,400,374 (August 23, 1983).
214. N. F. Cardarelli, Method and Composition for the Long Term Controlled Release of a Non-Persistent Organotin Pesticide from An Inert Monolithic Thermoplastic Material, U.S. Patent 4,166,111 (August 28, 1979).
215. N. F. Cardarelli, *Mosq. News* **38**, 328 (1978).
216. N. F. Cardarelli, Formulation and Evaluation of Controlled Release Pesticide Materials Based Upon the Monolithic Incorporation of the Agent in Thermoplastic Alloys with Porosigen Additives, Proceedings of the Controlled Release of Bioactive Materials Symposium (Abstract), Fort Lauderdale, Florida (July 1981), p. 117.
217. N. F. Cardarelli, Method and Composition for the Long Term Controlled Release of a Non-Persistent Organotin Pesticide from an Inert Monolithic Plastic Dispenser, U.S. Patent 4,237,114 (December 2, 1980).
218. J. W. Gillett, *Res. Rev.* **44**, 115 (1972).
219. B. M. Glancy, *Mosq. News* **28**, 205 (1968).
220. L. W. Roberts and T. A. Miller, Polymer Formulation of Dursban and Abate as Mosquito Larvicides, Report, Entomological Special Study 31-006-71/72, U.S. Army Environmental Health Agency, Edgewood Arsenal, Maryland (April–October 1970).
221. T. Quick, N. F. Cardarelli, and L. Boswell, Laboratory and Field Evaluation of Controlled Release Temephes Mosquito Larvicides, Proceedings of the Controlled Release of Bioactive Materials Symposium, Fort Lauderdale, Florida (July 1980) (unpublished).
222. N. F. Cardarelli, Floating Pesticides Dispenser, U.S. Patent 4,228,614 (October 21, 1980).
223. S. T. Aminah, H. L. Mathis, and I. G. Seregeg, Persistence of Two Insect Development Inhibitors in Domestic Water Containers in Jakarta, Indonesia, Report, World Health Organization WHO/VBC/81.832 (1981).
224. T. J. Quick, N. F. Cardarelli, R. J. Ellin, and L. R. Sherman, in: *Controlled Release of Pesticides and Pharmaceuticals*, (D. H. Lewis, ed.), pp. 275–285. Plenum Press, New York (1981).
225. W. W. Barnes and J. T. Whitlaw, Preliminary Evaluation of Controlled Release Herbicide Formulations for Aquatic Weeds, Report, U.S. Army Environmental Hygiene Agency, Edgewood Arsenal, Maryland, Project 31-006-69/70 (1970).
226. W. W. Young and L. L. Nelson, Evaluation of Controlled Release Herbicide Formulations for Aquatic Weeds, Report, U.S. Army Environmental Hygiene Agency, Edgewood Arsenal, Maryland, Project 31-004-69/71 (July 1969–July 1970).
227. K. K. Steward and L. L. Nelson, Evaluations of Controlled Release PVC Formulations of 2,4-D on Watermilfoil, Report, Interagency Commission Aquatic Plant Control Program, Office Chief Engineer, Department of the Army (October 1970) (unpublished).

228. E. O. Gangstad, R. H. Scott, and R. G. Cason, Controlled Release Herbicides, Technical Report No. 1, Aquatic Plant Control Program, U.S. Army Engineer Waterways Experiment Stations, Vicksburg, Mississippi (October 1972).
229. K. K. Steward and L. L. Nelson, *Hyacinth Control J.* **10**, 35 (1972).
230. R. T. McFadden, R. R. Languer, and L. L. Wade, Biocidally-Active Mixed Phosphorothioate Ester-Containing and Mixed Phosphoroamidate Ester-Containing Polymeric Materials, U.S. Patent 3,212,967 (October 19, 1965).
231. R. J. Geary, Pesticidal Resins and Method of Preparing Same, U.S. Patent 3,223,513 (December 14, 1965).
232. E. Baltazzi, Alkyd Resins with Herbicidal Properties, U.S. Patent 3,334,941 (September 26, 1967).
233. M. L. Beasley and R. L. Collins, *Science* **169**, 767 (1970).
234. G. C. Allan, C. S. Chopra, and R. M. Wilkins, *TAPPI* **54**(8), 1293 (1971).
235. G. C. Allan, C. S. Chopra, A. Neogi, and R. M. Wilkins, *Int. Pest Control* **13**, 10 (1971).
236. G. C. Allan, C. S. Chopra, and R. M. Russel, *Int. Pest Control* **14**(2), 15 (1972).
237. G. C. Allan, C. S. Chopra, M. W. Maggi, A. N. Neogi, and R. N. Wilkins, *Int. Pest Control* **15**(1), 4 (1973).
238. G. C. Allan, A. N. Neogi, C. S. Chopra, and R. M. Wilkins, *Nature* (*London*) **234**(5328), 349 (1971).
239. F. C. McCoy and L. W. Swanson, Oil-Substituting Nitrogen Containing Pesticidal Compounds, U.S. Patent 3,681,348 (August 1, 1972).
240. F. W. Harris, S. O. Norris, and L. K. Post, Investigation of Factors Influencing Release of Herbicides from Polymer Matrices: The Polyethylene Matrix, Quarterly Report, Contract DACW 73-71-C-0053, U.S. Army Corps of Engineers (October–December 1971).
241. F. W. Harris, *Hyacinth Control J.* **11**, 61 (1973).
242. F. W. Harris, S. O. Norris, and L. K. Post, *Weed Sci.* **21**(4), 318 (1973).
243. F. W. Harris, W. A. Feld, and L. K. Post, Investigation of Factors Influencing the Release of Herbicides from Polymers Containing Herbicides as Pendent Size Chains, Annual Report, Contract DACW 73-74-C-0001, U.S. Army, Office Chief Engineer, Washington, D.C. (July 1973–June 1974).
244. F. W. Harris and L. K. Post, Synthesis of Polymers Containing Aquatic Herbicides as Pendent Substituents, Proceedings of the Controlled Release Pesticide Symposium, University of Akron, Ohio (September 1974), p. 17.1.
245. W. A. Feld, L. K. Post, and F. W. Harris, Controlled Release from Polymers Containing Pesticides as Pendent Substituents, Proceedings of the Controlled Release Pesticide Symposium, Wright State University, Dayton, Ohio (September 1975), p. 113.
246. F. W. Harris and L. K. Post, *J. Polym. Sci.* **13**, 225 (1975).
247. F. W. Harris, W. A. Feld, and B. K. Bowen, Polymers Containing Pendent Herbicides Substituents, Proceedings of the International Controlled Release Pesticide Symposium, Wright State University, Dayton, Ohio (September 1975), p. 334.
248. R. G. Sinclair, *Environ. Sci. Technol.* **7**, 955 (1973).
249. C. L. Mehltretter, W. B. Roth, F. B. Weakley, T. A. McGuire, and C. R. Russell, *Weed Sci.* **22**, 415 (1974).
250. B. S. Shasha, D. Trimmnell, and F. S. Otey, *J. Polym. Sci.* **19**, 1891 (1981).
251. B. S. Shasha, W. M. Doane, and C. R. Russell, *J. Polym. Sci.* **14**, 417 (1976).
252. P. A. Cruickshank, Some Juvenile Hormone Analogues: A Critical Appraisal, *Metteil. Schweizerischen Entomol. Gesells.* **44**(1/2), 7, (1971).
253. Y. L. Meltzer, *Hormonal and Attractant Pesticide Technology*, Noyes Data Corporation Publication, Park Ridge, New Jersey (1971).
254. L. C. Post and W. R. Vincent, *Naturwissenschaften* **60**, 431 (1973).

255. J. J. Menn and F. M. Pallos, Development of Morphogenetic Agents in Insect Control, Report, Stauffer Chemical Company, Mt. View, California, (1974).
226. D. C. Cerf and G. P. Georghiou, *Nature* (*London*) **239**, 401 (1972).
257. L. J. Hoffman, J. H. Ross, and J. J. Menn, *J. Agric. Food Chem.* **21**, 156 (1973).
258. T. M. Brown and A. W. A. Brown, *J. Econ. Entomol.* **67**, 799 (1974).
259. P. B. Vinson and F. W. Plapp, *J. Agric. Food Chem.* **22**, 356 (1974).
260. M. S. Chodnekar, A. Pfiffner, N. Rigassi, U. Schweiter, and M. Suchy, Insecticidal Pyrethrins in Combination with Juvenile Hormones, U.S. Patent 3,839,562 (October 1, 1974).
261. R. L. Dunn and F. E. Strong, *Mosq. News* **33**, 110 (1973).
262. Y. Noguchi and T. Ohtaki, *Jpn. J. Sanit. Zool.* **25**, 185 (1974).
263. C. B. Rathburn and A. H. Boike, Laboratory and Small Plot Field Tests of Altosid and Dimilin for the Control of Mosquito Larvae, Report, West Florida Ardhropod Research Laboratory, Florida Department of Health, Panama City, Florida (1975).
264. J. A. Miller, M. L. Beadles, J. S. Palmers, and M. O. Pickens, *J. Econ. Entomol.* **70**, 589 (1977).
265. C. G. L. Furmidge, A. C. Hill, and J. M. Osgerby, *J. Sci. Food Agric.* **17**, 518 (1966).
266. H. D. Scott and R. E. Phillips, Diffusion of selected herbicides in soil, *Soil Sci. Soc. Am. Proc.* **36**, 714 (1972).
267. G. G. Allan, Controlled Release Pesticide, Canada Patent 846,785 (July 14, 1970).
268. Biocides under wraps, *Chem. Week* **111**, 40 (1972).
269. G. C. Allan, C. S. Chopra, J. F. Friedhoff, R. I. Gara, M. W. Maggi, A. N. Neogi, S. C. Roberts, and R. M. Wilkins, *CHEMTECH* **4**, 1719 (1973).
270. G. C. Allan, R. I. Gara, and R. M. Wilkins, *Int. Pest Control* (*London*) **16**, (1974).
271. R. A. Stokes, J. R. Coppedge, D. Y. Bull, and R. L. Ridgeway, *J. Agric. Food Chem.* **21**, 103 (1973).
272. J. R. Coppedge, R. A. Stokes, R. L. Ridgeway, and R. E. Kinzer, Slow Release Formulations of Aldicarb: Modeling of Soil Persistence, Publication No. 20, Cotton Insects Research Laboratory, U.S. Department of Agriculture, ARS, College Station, Texas (December 1974).
273. R. L. Collins, Microgranules and Microencapsulation of Pesticides, Proceedings of the Controlled Release Pesticide Symposium, Wright State University, Dayton, Ohio (September 1975), p. 105.
274. C. C. Campbell, Pesticidal Compositions, U.S. Patent 3,740,419 (June 19, 1973).
275. R. H. Laby, Controlled Release and the Animal Producer: An Emerging Technology, Proceedings of the Controlled Release Pesticide Symposium, University of Akron, Ohio (September 1976), p. 8.36.
276. G. C. Allan, Controlled Release in Forestry, Proceedings of the Controlled Release Pesticide Symposium, University of Akron, Ohio (September 1976), p. 7.15.
277. R. M. Wilkins, Design and Evaluation of Polymer Combinations for the Controlled Release of Herbicides, Proceedings of the Controlled Release Pesticide Symposium, University of Akron, Ohio (September 1976), p. 7.1.
278. V. B. Schandle, R. D. Sjogren, and C. Thies, Silicate Capsule Formulations for Mosquito Control, Proceedings of the Controlled Release Pesticide Symposium, University of Akron, Ohio (September 1976), p. 6.13.
279. J. Feldmesser, B. S. Shasha, and W. M. Doane, Nematocides in Starch for Controlled Release, Proceedings of the Controlled Release Pesticide Symposium, University of Akron, Ohio (September 1976), p. 6.18.
280. M. S. Mulla, Slow Release Insecticide Formulations for the Control of Insect Vectors, Proceedings of the Controlled Release Pesticide Symposium, University of Akron, Ohio (September 1976), p. 6.36.

281. J. A. Miller, M. L. Beadles, and W. J. Gladney, Sustained Release of Pesticides: The Potential in Livestock Pest Control, Proceedings of the Controlled Release Pesticide Symposium, Corvallis, Oregon (August 1977), p. 253.

282. H. Jaffe, P. Giang, and J. Miller, Implants of Methoprene in Poly(Lactic Acid) against Cattle Grubs, Proceedings of the Controlled Release of Bioactive Materials Symposium, National Bureau of Standards, Gaithersburg, Maryland (August 1978), p. 5.5.

283. J. A. Miller, E. E. Kunze, and D. D. Oehler, in: *Controlled Release of Pesticides and Pharmaceuticals* (D. H. Lewis, ed.), pp. 311–318, Plenum Press, New York (1981).

284. J. A. Miller, R. O. Drummond, and D. D. Oehler, A Sustained-Release Implant for Delivery of Ivermectin for Control of Livestock Pests, Proceedings of the Controlled Release of Bioactive Materials Symposium, Fort Lauderdale, Florida (July 1981), p. 276.

285. J. Feldmesser and B. S. Shasha, Evaluation of Controlled Release DBCP-Starch Xanthide Granules in Laboratory and Greenhouse Soil Tests, Proceedings of the Controlled Release Pesticide Symposium, Corvallis, Oregon (August 1977), p. 205.

286. J. Feldmesser, H. Jaffe, P. A. Giang, and G. Wolfhard, Experimental Controlled-Release Formulations of a Nematicidal Amine, Proceedings of the Controlled Release of Bioactive Materials Symposium, Fort Lauderdale, Florida (July 1981), p. 147.

287. H. T. Dellicolli, in: *Controlled Release Pesticides* (H. B. Scher, ed.), pp. 84–93, American Chemical Society Symposium Series 53, Washington, D.C. (1977).

288. W. M. Doane, B. S. Shasha, and C. R. Russell, in: *Controlled Release Pesticides* (H. B. Scher, ed.), pp. 74–83, American Chemical Society Symposium Series 53, Washington, D.C. (1977).

289. B. S. Shasha, Encapsulation of Bioactive Materials Within a Starch Matrix, Proceedings of the Controlled Release of Bioactive Materials Symposium, Fort Lauderdale, Florida (July 1981), p. 149.

290. B. S. Shasha, Encapsulation Within a Starch–Borate Matrix, Proceedings of the Controlled Release of Bioactive Materials Symposium, Fort Lauderdale, Florida (July 1982), p. 49.

291. J. W. Young, T. M. Graves, R. Curtis, and M. M. Furniss, in: *Controlled Release Pesticides* (H. B. Scher, ed), pp. 184–199, American Chemical Society Symposium Series 53, Washington, D.C. (1977).

292. E. H. Schacht, G. E. Desmarets, and E. J. Goethals, in: *Controlled Release of Pesticides and Pharmaceuticals* (D. H. Lewis, ed.), pp. 159–170, Plenum Press, New York (1981).

293. G. James, Recent Evaluations of Controlled Release Herbicides, Proceedings of the Controlled Release of Bioactive Materials Symposium, National Bureau of Standards, Gaithersburg, Maryland (August 1978), p. 3.1.

294. K. Das, A Controlled Release Herbicide Formulation for Terrestrial Weeds, Proceedings of the Controlled Release of Bioactive Materials Symposium, National Bureau of Standards, Gaithersburg, Maryland (August 1978), p. 3.29.

295. F. G. Burton, D. A. Cataldo, J. F. Cline, and W. Skiens, The Use of Controlled Release Herbicides in Waste Burial Sites, Proceedings of the Controlled Release of Bioactive Materials Symposium, Fort Lauderdale, Florida (July 1981), p. 261.

296. M. L. Williams, Bio-Active Plastic Films, Proceedings of the Controlled Release of Bioactive Materials Symposium (Extended Abstracts), Fort Lauderdale, Florida (July 1982), p. 173.

297. B. D. Couriel, G. E. Peck, and R. M. Hollingworth, Utilization of Molecular Scale Entrapment Technology in the Chemical Stabilization and Improvement of the Biological Activity of Methyl Parathion, Proceedings of the Controlled Release of Bioactive Materials Symposium (Extended Abstract), Fort Lauderdale, Florida (July 1982), p. 151.

298. D. Lewis, D. Cowser, and M. Hamilton, Hydrophilic Polymers as Excipients for Biologically Active Agents, Proceedings of the Controlled Release of Bioactive Materials

Symposium, National Bureau of Standards, Gaithersburg, Maryland (August 1978), p. 2.6.
299. A. B. Pepperman and K. E. Savage, Preparation of Controlled-Release Formulations of Metribuzin Based on Cotton Cellulose, Proceedings of the Controlled Release of Bioactive Materials Symposium, New Orleans, Louisiana (August 1979), p. III.7.
300. R. A. Holmes, Biodegradability of Poly(3-Hydroxybutyrate) Films in Soil, Proceedings of the Controlled Release of Bioactive Materials Symposium (Extended Abstracts), Fort Lauderdale, Florida (July 1982), p. 170.
301. G. G. Allan, W. Balaba, C. Balaban, J. Dutkiewicz, A. W. Lee, W. Mikels, and R. A. Struszczyk, Plant Growth Stimulants: A New Area of Application for Controlled Release Technology, Proceedings of the Controlled Release of Bioactive Materials Symposium, New Orleans, Louisiana (August 1979), p. III.20.
302. J. P. Kealy, Low Solubility Fertilizer Compound, Proceedings of the Controlled Release of Bioactive Materials Symposium, National Bureau of Standards, Gaithersburg, Maryland (August 1978), p. 9.26.
303. J. G. A. Fiskell and R. B. Diamond, Soil Evaluation of Sulfur-Coated Ureas, Proceedings of the Controlled Release of Bioactive Materials Symposium, Corvallis, Oregon (August 1978), p. 299.
304. N. F. Cardarelli, Controlled Release Trace Nutrients, Proceedings of the Controlled Release of Bioactive Materials Symposium, National Bureau of Standards, Gaithersburg, Maryland (August 1978), p. 9.1.
305. N. F. Cardarelli, Controlled Release of Trace Nutrients, U.S. Patent 4,299,613 (November 10, 1981).
306. B. E. Hitchcock, CR Formulations Persistence, Progress Report, Bureau of Sugar Experiment Stations, Mackay, Australia (December 10, 1980).
307. H. S. Hall and T. M. Hinkes, The Wuerster Process for Controlling Pesticides, Proceedings of the Controlled Release Pesticides Symposium, University of Akron, Ohio (September 1976), p. 4.1.
308. H. B. Scher, in: *Microencapsulated Pesticides* (H. B. Scher, ed.), pp. 126–144, American Chemical Society Symposium Series 53, Washington, D.C. (1977).
309. J. E. Vandegaar, Microencapsulation as a Method for Controlled Release, Proceedings of the Release of Bioactive Materials Symposium, National Bureau of Standards, Gaithersburg, Maryland (August 1978), p. 2.1.
310. J. A. Bakan, in: *Controlled Release Technologies*, Vol. II (A. F. Kydonieus, ed.), CRC Press, Fort Lauderdale, Florida (1980).
311. R. C. Koestler, in: *Controlled Release Technology*, Vol. II (A. F. Kydonieus, ed.), pp. 117–132, CRC Press, Fort Lauderdale, Florida (1980).
312. G. R. Somerville and J. T. Goodwin, in: *Controlled Release Technologies*, Vol. II (A. F. Kydonieus, ed.), pp. 155–164, CRC Press, Fort Lauderdale, Florida (1980).
313. B. C. Pass and H. W. Dorough, *J. Econ. Entomol.* **66**, 1117 (1973).
314. C. B. DeSavigny and E. E. Ivy, Microencapsulated Pesticides, *Amer. Chem. Soc., Div. Org. Coat. Plast. Chem.* **33**(2), Meeting, Chicago, Illinois (August 1973), p. 554.
315. R. C. Koestler and E. E. Ivy, Controlled Release of Biologically Active Agents from Nylon-Type Microcapsules, Proceedings of the Controlled Release Pesticide Symposium, University of Akron, Ohio (September 1974).
316. R. C. Koestler, Knox Out 2FM Insecticide, Microencapsulated Diazinon, Proceedings of the Controlled Release of Bioactive Materials Symposium, New Orleans, Louisiana (August 1979).
317. N. L. Gauthier, Field Testing Experiments with Slow Release Insecticide Granular Formulations to Assess Insect Control, Proceedings of the Controlled Release Pesticide Symposium, Corvallis, Oregon (August 1977), p. 226.
318. K. G. Das and V. B. Tungikar, Controlled Release Larvicide, Proceedings of the

Controlled Release Bioactive Materials Symposium, New Orleans, Louisiana (August 1979), p. IV.37.

319. N. L. Gauthier, Vegetable Foliar Insect Control With Encapsulated Granular Insecticides, Proceedings of the Controlled Release of Bioactive Materials Symposium, National Bureau of Standards, Gaithersburg, Maryland (August 1978), p. 5.29.
320. T. R. Rice, D. H. Lewis, and D. R. Cowser, Controlled Release of Pesticide from Chlorpyrifos Microcapsules, Proceedings of the Controlled Release of Bioactive Materials Symposium, Fort Lauderdale, Florida (July 1981), p. 135.
321. H. Fuyama, G. Shinjo, and K. Tsuji, Encapsulated Sumithion: Inner Active Ingredient Concentration and Biological Efficacy, Proceedings of the Controlled Release of Bioactive Materials Symposium, Fort Lauderdale, Florida (July 1981), p. 141.
322. H. Fuyama, G. Shinjo, and K. Tsuji, Encapsulated Sumithion, Proceedings of the Controlled Release of Bioactive Materials Symposium (Abstracts), Fort Lauderdale, Florida (July 1982), p. 145.
323. W. A. Gentner and L. L. Danielson, The Influence of Microencapsulation on the Herbicidal Performance of Chlorpropham, Proceedings of the Controlled Release Pesticide Symposium, University of Akron, Ohio (September 1976), p. 7.26.
324. J. H. Dawson, Prolonged Control of Cuscuta with Microencapsulated Chlorpropham, Proceedings of the Controlled Release Pesticide Symposium, University of Akron, Ohio (September 1976), p. 6.30.
325. R. Helfner, Evaluation of Silicate Capsules Containing 2-(2,4,5-Trichlorophenoxy) Propionic Acid, Proceedings of the Controlled Release Bioactive Materials Symposium, New Orleans, Louisiana (August 1979), p. III.16.
326. R. D. Sjogren, Preliminary Evaluation of Capsules Containing a Chitin Synthesis Inhibitor, Proceedings of the International Controlled Release Pesticide Symposium, Wright State University, Dayton, Ohio (September 1975), p. 217.
327. R. P. Porter and J. B. Miale, Extended Control of a Microencapsulated Liquid Organometallic Biocide in Vinyl Rosin Paint, Proceedings of the Controlled Release of Bioactive Materials Symposium (Abstract), Fort Lauderdale, Florida (July 1982), p. 111.
328. R. G. Arnold, Microcapsules and Controlled Release New Animal Drugs, Proceedings of the Controlled Release of Bioactive Materials Symposium (Abstract), Fort Lauderdale, Florida (July 1982), p. 201.
329. R. S. Rew and R. H. Fetterer, Development of Controlled-Release Fasciolicides for Ruminants, Proceedings of the Controlled Release of Bioactive Materials Symposium (Abstract), Fort Lauderdale, Florida (July 1982), p. 209.
330. H. Jaffe, D. K. Hayes, P. A. Giang, R. O. Drummond, and T. M. Whetsone, Injectable Formulations for the Controlled Release of Pesticides Against Ticks on Cattle, Proceedings of the Controlled Release Pesticide Symposium, Corvallis, Oregon (August 1977), p. 272.
331. R. W. Miller and C. Corley, Poultry Feed Additives for Fly Control: Improvements in Efficiency with Controlled Release Formulations, Proceedings of the Controlled Release Pesticide Symposium, Corvallis, Oregon (August 1977), p. 264.
332. S. H. Wu, C. C. Dannelly, and R. J. Komarek, in: *Controlled Release of Pesticides and Pharmaceuticals* (D. H. Lewis, ed.), pp. 319–331, Plenum Press, New York (1981).
333. E. R. Mitchell, Recent Advances in the Use of Sex Pheromones for Control of Insect Pests, Proceedings of the Controlled Release Pesticide Symposium, Corvallis, Oregon (August 1977), p. 41.
334. I. Keiser, R. M. Koyabashi, and E. J. Harris, Enhanced Duration of Effectiveness Against the Oriental Fruit Fly and Guava Foliage Treated with Encapsulated Insecticides and Encapsulated Methyl Eugenol, Proceedings of the Controlled Release Pesticide Symposium, University of Akron, Ohio (September 1976), p. 3.1.

335. T. Dappert and C. Thies, The Heterogeneous Nature of Microcapsules, Proceedings of the Controlled Release of Bioactive Materials Symposium, National Bureau of Standards, Gaithersburg, Maryland (August 1978), p. 2.18.
336. C. H. Himel and N. F. Cardarelli, In-Flight Encapsulation of Particles, U.S. Patent 4,286,020 (August 25, 1981).
337. R. L. Lipsey, Research Report on Pyrellin SCS at the Al-Marah Registered Arabian Horse Ranch, Micanopy, Florida, Institute of Food and Agricultural Sciences, College of Agriculture, University of Florida (July 15, 1980) (unpublished).
338. G. G. Allan, J. W. Beer, and N. J. Cousin, in: *Controlled Release Pesticides* (H. B. Scher, ed.), pp. 94–101, American Chemical Society Symposium Series 53, Washington, D.C. (1977).
339. C. Y. McCormick and M. Fooladi, in: *Controlled Release Pesticides*, (H. B. Scher, ed.), pp. 112–125, American Chemical Society Symposium Series 53, Washington, D.C. (1977).
340. G. G. Allan, J. W. Beer, and M. J. Cousins, Current Status of Controlled Release Herbicides in Pacific Northwest Reforestation, Proceedings of the Controlled Release Pesticide Symposium, Corvallis, Oregon (August 1977), p. 19.
341. K. Savage, C. McCormick, and B. Hutchinson, Biological Evaluation of Polymeric Controlled Activity Herbicide Systems Containing Pendent Metribuzin, Proceedings of the Controlled Release of Bioactive Materials Symposium, National Bureau of Standards, Gaithersburg, Maryland (August 1978), p. 3.18.
342. C. McCormick, D. Lichatowich, and M. Fooladi, Synthesis and Characterization of Controlled Activity Pendent Herbicide Systems Utilizing Chitin and Other Biodegradable Polymers, Proceedings of the Controlled Release of Bioactive Materials Symposium, National Bureau of Standards, Gaithersburg, Maryland (August 1978), p. 3.6.
343. C. L. McCormick, K. W. Anderson, J. A. Pelezo, and D. K. Lichatowich, in: *Controlled Release of Pesticides and Pharmaceuticals* (D. H. Lewis, ed.), pp. 147–158, Plenum Press, New York (1981).
344. C. L. McCormick and K. W. Anderson, Synthesis and Characterization of Natural Saccharides with Attached Metribuzin, Proceedings of the Controlled Release of Bioactive Materials Symposium (Abstract), Fort Lauderdale, Florida (July 1982), p. 32.
345. C. L. McCormick and K. W. Anderson, Polymers Containing Pendently Attached Metribuzin as Potential Controlled-Release Herbicides, Proceedings of the Controlled Release of Bioactive Materials Symposium (Abstract), Fort Lauderdale, Florida (July 1981), p. 255.
346. R. M. Rowell, Bonding of Toxic Chemicals to Wood, Proceedings of the Controlled Release of Bioactive Materials Symposium, Fort Lauderdale, Florida (July 1981), p. 97.
347. F. W. Harris and M. W. Whitlock, Synthesis and Hydrolysis of Uniform Composition Copolymers Containing Pendent 2,4-Dichloro-Phenoxyacetic Acid Substituent, Proceedings of the Controlled Release of Bioactive Materials Symposium (Abstract), Fort Lauderdale, Florida (July 1981), p. 237.
348. R. M. Wilkins, The Design and Use of Lignin-Based Controlled Release Formulations for Pest Control in Tropical Rice, Proceedings of the Controlled Release of Bioactive Materials Symposium (Abstract), Fort Lauderdale, Florida (July 1981), p. 72.
349. W. E. Meyers, D. H. Lewis, R. K. Vandermeer, and C. S. Lofgren, Polymers Containing Pendent Insecticides, Proceedings of the Controlled Release of Bioactive Materials Symposium, Fort Lauderdale, Florida (July 1981), p. 171.
350. J. P. Kreig, Combatting Crawling Insects with Hercon™ Controlled-Release Pesticide Applications, Report, Health Chemical Corporation, New York, New York (1974) (unpublished).
351. A. F. Kydonieus, S. Baldwin, and S. Hyman, Hercon™ Granules and Powders for

Agricultural Applications, Proceedings of the Controlled Release Pesticide Symposium, University of Akron, Ohio (September 1976), p. 4.23.

352. J. R. Plimmer, B. A. Bierl, E. D. DeVilbiss, and B. L. Smith, Evaluation of Controlled Release Formulations of Insect Pheromones for Mating Disruption, Proceedings of the Controlled Release Pesticide Symposium, University of Akron, Ohio (September 1976), pp. 3.29.
353. M. Beroza, E. C. Paszek, E. R. Mitchell, B. A. Bierl, J. R. McLaughlin, and D. L. Chambers, *J. Environ. Entomol.* **3**, 926 (1974).
354. C. R. Gentry, B. A. Bierl, and J. L. Blythe, Air Permeation Trials with the Oriental Fruit Moth Pheromone, Proceedings of the Controlled Release Pesticide Symposium, University of Akron, Ohio (September 1976), p. 3.22.
355. I. Keiser, M. Jacobson, and J. A. Silva, in: *Controlled Release of Pesticides and Pharmaceuticals* (D. H. Lewis, ed.), pp. 245–257, Plenum Press, New York (1981).
356. R. Cuthbert, Release of the Three Pheromone Components of Elm Bark Beetle Pheromone, Multilure from Controlled Release Dispensers, Proceedings of the Controlled Release of Bioactive Materials Symposium, National Bureau of Standards, Gaithersburg, Maryland (August 1978), p. 5.49.
357. I. Keiser, R. Miyabara, J. Silva, and E. Harris, Enhanced Duration of Effectiveness of Trimedlure as an Attractant to Male Mediterranean Fruit Flies under Field Conditions, Proceedings of the Controlled Release of Bioactive Materials Symposium, Natural Bureau of Standards, Gaithersburg, Maryland (August 1978), p. 5.40.
358. R. Lie, Result of Mass Trapping of *Ips Typographus* by Use of Pheromone Baited Traps, Proceedings of the Controlled Release of Bioactive Materials Symposium (Abstract), Fort Lauderdale, Florida (July 1982), p. 82.
359. G. E. Daterman, C. Startwell, and L. L. Sower, Status of Mating Disruption as a Control Approach for Selected Forest Insects, Proceedings of the Controlled Release of Bioactive Materials Symposium, New Orleans, Louisiana (August 1979), p. IV.8.
360. A. F. Kydonieus, B. Bierl-Leonhardt, J. R. Plimmer, M. W. Berry, and A. R. Quisumbing, Cotton Insects: Dispenser Development and Disruption of Mating Trials, Proceedings of the Controlled Release of Bioactive Materials Symposium, New Orleans, Louisiana (August 1979), p. IV.13.
361. K. L. Smith, Y. Ninomiya, and R. W. Baker, Development of BioLure™ Controlled Release Pheromone, Proceedings of the Controlled Release of Bioactive Materials Symposium (Abstract), Fort Lauderdale, Florida (July 1981), p. 63.
362. J. R. Plimmer, B. A. Bierl, R. E. Webb, and C. P. Schwable, in: *Controlled Release Pheromone in the Gypsy Moth Program*, (H. B. Scher, ed.), pp. 168–183, American Chemical Society Symposium Series 53, Washington, D.C. (1977).
363. B. A. Bierl, E. D. DeVilbiss, and J. R. Plimmer, in: *Use of Pheromones in Insect Control Programs Slow Release Formulations* (D. R. Paul and F. W. Harris, eds.), pp. 265–272, American Chemical Society Symposium Series 33, Washington, D.C. (1976).
364. A. F. Kydonieus and I. K. Smith, in: *Controlled Release of Pheromones Through Multi-layered Polymeric Dispensers*, (D. R. Paul and F. W. Harris, eds.), pp. 283–294, American Chemical Society Symposium Series 33, Washington, D.C. (1976).
365. E. Ashare, T. W. Brooks, and D. W. Swenson, Controlled Release from Hollow Fibers, Proceedings of the Controlled Release Pesticide Symposium, Wright State University, Dayton, Ohio (September 1975), p. 42.
366. T. W. Brooks and D. W. Swenson, Design Considerations in the Formulation of Controlled Release Products, Proceedings of the Controlled Release Pesticide Symposium, University of Akron, Ohio (September 1976), p. 4.5
367. T. Brooks, C. Doane, D. Swensen, J. Haworth, D. Kelly, D. Osborne, and R. Kitterman, Gossyplure, A Controlled Release Sex Pheromone Formulation, Proceedings of the

Controlled Release of Bioactive Materials Symposium, National Bureau of Standards, Gaithersburg, Maryland (August 1978), p. 2.67.

368. E. Ashare, T. W. Brooks, and D. W. Swenson, in: *Controlled Release Polymeric Formulations*, (D. R. Paul and F. W. Harris, eds.), pp. 273–282, American Chemical Society Symposium Series 33, Washington, D.C. (1976).

369. T. W. Brooks, C. C. Doane, D. G. Osborn, and J. K. Haworth, Experience in Using a Hollow Fiber Controlled Release Formulation in Pheromone Mediated Suppression of *Pectinophora gossypiella* under Humid Tropical Conditions, Proceedings of the Controlled Release of Bioactive Symposium, New Orleans, Louisiana (August 1979), p. IV.1.

370. T. L. Payne, J. E. Coster, and P. C. Johnson, Hollow Fibers in Controlled-Release of Southern Pine Beetle Inhibitors, Proceedings of Controlled Release Pesticide Symposium, University of Akron, Ohio (September 1976), p. 3.9.

371. A. F. Kydonieus, I. K. Smith, and S. Hyman, A Polymeric Delivery System for the Controlled Release of Pesticides, Proceedings of the Controlled Release Pesticide Symposium, Wright State University, Dayton, Ohio (September 1975), p. 60.

372. A. R. Quisumbing, A. F. Kydonieus, D. R. Calsett, and J. B. Haus, Hercon Lure'N Kill™ Flytape: A Non-Fumigant Insecticidal Strip Containing Attractants, Proceedings of the Controlled Release Pesticide Symposium, University of Akron, Ohio (September 1976) p. 3.40.

373. A. R. Quisumbing and A. F. Kydonieus, Use of Laminated Controlled Release Contact-Action Insecticidal Strips for Cluster Fly Control, Proceedings of the Controlled Release of Bioactive Materials Symposium, National Bureau of Standards, Gaithersburg, Maryland (August 1978), p. 5.62.

374. B. L. Parker and N. L. Gauthier, Controlled Release Formulations of Chlorpyrifos and Malathion for Suppression of Rice Storage Pests, Proceedings of the Controlled Release Pesticide Symposium (Abstract), Fort Lauderdale, Florida (July 1981), p. 74.

375. H. L. Collins, N. Foster, S. Cade, N. Gauthier, and A. F. Kydonieus, Conventional Agrichemical Controlled Release Formulations for Agriculture, Forestry and Food Protection, Proceedings of the Controlled Release Pesticide Symposium, Fort Lauderdale, Florida (July 1981), p. 66.

376. N. Y. Gauthier, in: *Controlled Release of Pesticides and Pharmaceuticals* (D. H. Lewis, ed.), pp. 259–274, Plenum Press, New York, (1981).

377. P. I. Lee, Polymeric Alcohol Membrane System for the Controlled Release of Chlorinating Agents, Proceedings of the Controlled Release of Bioactive Materials Symposium, Fort Lauderdale, Florida (July 1982), p. 101.

378. L. D. Nichols, Process of Preparing Gelled Cellulose Triacetate Products and the Products Produced Thereby, U.S. Patent 3,846,404 (May 23, 1973).

379. L. D. Nichols, Poroplastic and Sustrelle: Controlled Release Vehicles Having Broad Compatibility with Dissolved and Precipitated Pesticides, Proceedings of the International Controlled Release Pesticides Symposium, Wright State University, Dayton, Ohio (September 1975), p. 95.

380. A. S. Obermayer and L. D. Nichols, in: *Controlled Release Polymeric Formulations* (D. R. Paul and F. W. Harris, eds.), pp. 303–307, American Chemical Society Symposium Series 33, Washington, D.C. (1976).

381. S. T. Hsieh and R. Langer, Magnetically Controlled Drug-Delivery System, Proceedings of the Controlled Release of Bioactive Materials Symposium, Fort Lauderdale, Florida (July 1981), p. 24.

382. D. S. T. Hsieh and R. Langer, Magnetic Modulation of Insulin Release in Diabetic Rats, Proceedings of the Controlled Release of Bioactive Materials Symposium, Fort Lauderdale, Florida (July 1982), p. 90.

4

Controlled Release of Antifertility Agents

David L. Gardner and Barbara A. Metz

Abstract. The sustained–controlled release concept is actively being investigated in many different fields, including agriculture, aquaculture, food, cosmetics, veterinary, and pharmaceutics. One particular area in the pharmaceutical field which has received considerable attention is the fertility area. In fact, some of the earliest sustained–controlled release delivery systems were associated with this area and involved the sustained–controlled release of antifertility agents. One can preselect the site at which the antifertility agent is to be delivered by developing novel delivery systems and/or by choosing the route of administration. Depending, again, upon the site selected or mode of administration chosen, one also preselects the distribution of that agent, that is, the antifertility agent will be delivered and act at a localized site (tissue or organ) or will be distributed systemically and act at a remote site.

This chapter details the various devices, for example, microspheres, IUDs, and vaginal rings, and routes of administration, for example, oral, intranasal, vaginal, being pursued for the sustained-controlled delivery of antifertility agents. In addition, the antifertility agents (steroid or nonsteroid) being selected for delivery and some of the more recent and novel approaches to delivery of these agents are discussed.

1. Introduction

The use of contraceptives and devices to regulate and control fertility has been practiced for centuries. Many of the contraceptive approaches practiced in the past would be considered crude by today's standard, for example, honey and natron (native sodium carbonate), gummy substrates, and prescriptions of a mixture containing either crocodile or elephant dung.[1] However, the need to regulate or control fertility effectively may be more critical today, especially in developing countries where population growth may be dependent upon the availability of feeding the masses.

As a result of this need, there has been a continuing effort to develop better and more novel contraceptive systems. Newer contraceptive developments have been exemplified by the "pill," intrauterine devices, and

David L. Gardner and Barbara A. Metz • Battelle-Columbus Laboratories, 505 King Avenue, Columbus, Ohio 43201.

vasectomies. The desire to develop further contraceptive systems is based upon the fact that some of these present-day contraceptives, for example, oral contraceptives and intrauterine devices, have certain inherent disadvantages. In the case of oral contraceptives these include: the requirement of daily ingestion; the subsequent daily variation in blood steroid concentrations following ingestion; and the reported increased incidence of circulatory diseases.[2] However, longer-acting, that is, sustained-release, oral contraceptives are being evaluated and the development of these products may minimize the disadvantages cited above.[3,4] Intrauterine devices, on the other hand, have been associated with an increased frequency of pelvic infection and unwanted pregnancies that are more likely to be ectopic.[5]

Sustained-release drug delivery systems are designed to overcome the above-mentioned problems. By carefully designing the drug delivery system, the amount of drug released is maintained at a relatively uniform dose for a desired time period. Some factors that must be considered in designing the delivery system are: (1) drug half-life; (2) polymers employed; (3) blood–drug concentration required per day; (4) injection site; (5) total capacity of the drug delivery system; and (6) amount or volume of the delivery system which can be implanted, injected, and so on.

The modes of administering these novel sustained–controlled release devices and systems are concentrated around oral ingestion, vaginal, cervical, or intrauterine insertion, and parenteral injection. The mode of administration sometimes dictates that the released contraceptive agent will be distributed systemically; while in other administration modes, the agent will be delivered to a localized site, that is, pretargeted delivery.

This chapter is devoted to examining the different sustained–controlled release delivery systems of antifertility agents which are actively being investigated. Advantages and disadvantages of each system are highlighted where appropriate. Since most of the sustained–controlled release delivery systems have been investigated solely in the female, with only one or two being investigated in the male, our discussion centers around the female. The last part of the chapter concentrates on some of the more novel delivery routes being examined for delivery of contraceptive agents.

2. Injectables

Injectable depot formulations considered in this section include steroids administered (1) as micronized aqueous or oil suspensions, (2) in biodegradable microspheres, and (3) as polymer complexes. Some of the advantages of an injectable formulation include ease of administration, extended duration of steroid release, an effective mode of contraception, and reduction of the compliance factor. Disadvantages include the development of irregular bleeding patterns, delayed fertility following discontinuation, possible weight gain, and irreversibility of treatment.

The injectable progestogen formulations function similarly to the combined oral contraceptives in that they can: (1) inhibit ovulation in women; (2) cause a thickening of the cervical mucus, thus reducing sperm motility or penetration through the cervix; (3) alter transport of ovum through the oviduct; and (4) reduce the likelihood of nidation, that is, prevent endometrial implantation of the fertilized egg.

2.1. Suspensions

Injectable formulations have been investigated for fertility control since the late 1950s.[6,7] The initial injectable formulations concentrated on two approaches: a once-a-month injection designed to maintain the normal bleeding pattern and the once-every-three-months' injection, which dictated that the "normal" menstrual cycle would be unpredictable.[7] The once-a-month injectable necessitated the use of an estrogen and progestogen component, whereas the once-every-three-months' injection involved only a progestogen.

Injectable formulations were developed because of the discovery that certain steroids were not degraded as rapidly as others following oral ingestion, that is, their biologic half-life was increased. Although the initial injectable formulations were not construed as controlled release in the true sense, the nature of their formulation provided for an extended duration of action. The initial two steroids so formulated were medroxyprogesterone acetate (MPA; trade name Depo-Provera) and norethindrone enanthate (NET-EN; trade name Norigest).

In initial clinical studies, MPA (150 mg) was administered as an aqueous microcrystalline suspension, while NET-EN (200 mg) was formulated in an oil. Following intramuscular administration, serum MPA concentrations ranged from 10–25 ng/ml at 5 to 20 days posttreatment, followed by a decrease to 5–10 ng/ml at 30 days. Blood levels of MPA at the end of three months had declined to 0.6 ng/ml.[8] In the case of NET-EN, serum concentrations of 4–23 ng/ml were obtained during the first 20 days following intramuscular administration. However, the serum concentrations were undetectable 46 to 110 days postinjection.[9,10]

The release pattern observed with both MPA and NET-EN intramuscular injections indicated that the blood profile was nonlinear and that the initial blood concentrations far exceeded the required minimal effective concentration, reported to be approximately 1 ng/ml for both MPA and NET-EN.[11] Thus, although the depot injection site provided for an extended duration of action, the concept of controlled release was not fully achieved in these studies, that is, a fixed amount of the steroid was not released each day.

More recently, the World Health Organization has been conducting a two-year multinational comparative trial consisting of 200 mg of NET-EN administered at 60-day intervals, of 200 mg of NET-EN given at 60-day intervals for six months followed by 200 mg of NET-EN at 84-day intervals, or 150 mg of DMPA (depot-medroxyprogesterone acetate) administered at 90-day intervals.[12] After 18 months of observation, the overall discontinuation rate of the drugs has been similar among the three groups, that is, 61.8–63.5 per 100 women. However, terminations of the drugs due to amenorrhea were significantly higher among DMPA users (12.1 and 17.4 per 100 women at 12 and 18 months) than among NET-EN users (6.8–8.2 per 100 women at 12 months and 10.4–10.9 per 100 women at 18 months). In addition, there were significant differences in the pregnancy rate among the three groups. Among NET-EN users (84 drug treatment intervals), the pregnancy rate was 1.6 per 100 women, while among the DMPA users it was 0.2 per 100 women.

The use of DMPA as an effective contraceptive agent also has been investigated in males and has been found to reduce sperm levels.[13] However, these studies also indicated that DMPA diminished endogenous testosterone production. Thus, later studies have included an androgen component to help maintain circulatory androgen levels.[14–16] The results of these studies indicated that a marked reduction in sperm count or complete azoospermia occurred. However, the failure to suppress spermatogenesis completely in all men, and the prolonged suppression and recovery of spermatogenesis following cessation of treatment, represent drawbacks to the depot injection approach.[17]

Other injectable formulations which have been investigated preclinically or clinically are listed in Table 1. The future of injectable oil or suspension-type formulations in the United States remains in doubt since only Depo-Provera is currently being produced for international distribution. Depo-Provera has not received approval in the United States because of its association with heart cancer detected in beagle dogs and because of questions raised about *in situ* human cervical carcinoma formation.[6,18,19]

2.2. Biodegradable Microspheres

One of the most exciting approaches to fertility control centers around microencapsulated formulations which can be parenterally injected. These formulations consist of small particles ($< 250\,\mu m$) composed of

Table 1. Injectable Formulations Evaluated Preclinically or Clinically[6,20]

Estrogen only	Estrogen/Progestogen combination		Progestogen only
	Estrogen component	Progestogen component	
Estradiol undecylate			Hydroxyprogesterone caproate
Quinestrol	Estradiol enanthate	Algestone acetophenide (formerly known as dihydroxyprogesterone acetophenide)	Levonorgestrel nonanoate
			Levonorgestrel undecylate
	Estradiol undecylate	Norethindrone enanthate (also known as norethisterone enanthate)	
	Estradiol benzoate	Norgestrel	
	Estradiol cypionate	Medroxyprogesterone acetate	
	Chlormadinone acetate	Hydroxyprogesterone caproate	

biodegradable materials. The biodegradable materials include *d,l*-polylactic acid, poly(lactide-co-glycolide), and albumin.[21,22] The lactide and glycolide materials have been chosen because of their known biocompatibility as suture materials, while albumin has been shown to be nonimmunogenic if cross-linked, yet will biodegrade following implantation.

Particles can be either microspheres (a nonreservoir device in which the drug is distributed homogeneously throughout the matrix, i.e., a monolithic system), or microcapsules (a reservoir device in which the drug is a true solution, a suspension, or emulsion surrounded by a thin semipermeable membrane). To date, only microspheres containing the contraceptive agent have been evaluated in fertility control.

The most fully developed formulation is that by Beck and associates in which micronized norethisterone (NET) is incorporated in *d,l*-polylactide microspheres.[21] This system has been evaluated preclinically in baboons[21,23,24] and clinically,[25] and provides for continuous release of NET over a six-month period following intramuscular injection. The microspheres used in these studies contained 25% NET by weight and ranged in size from 60 to 240 μm diameter. Following intramuscular injection (doses ranging from 29 to 370 mg of microspheres and containing 7.25–94.5 mg of NET, respectively), there was an initial small burst of NET released in the serum followed by a gradual decline over a six-month period. The amount of NET released per day approximated 0.48 mg as an average value.

These clinical studies indicated that the NET blood levels were proportional to the dose administered. In general, lower doses showed no effect on ovarian function, but the intermediate to high doses exhibited partial or complete suppression of ovarian function, that is, they inhibited ovulation.[24] Based on these initial studies, Beck and associates have determined that 1.33–3.45 mg NET/kg is necessary to suppress ovulation for six months.[25]

2.3. Polymer–Steroid Complexes

Steroids converted to ester derivatives (prodrugs) have exhibited sustained release following intramuscular injection.[26] This is based upon the fact that the prodrug itself is not active, but will release the steroid via hydrolysis by plasma esterases. Thus, the rate of hydrolysis controls the rate of steroid release. This concept recently has been extended to the covalent attachment of steroids to biodegradable polymers.[27] The steroids, norethisterone and norethindrone, have been covalently attached to a polyglutamic acid backbone or to poly(α-amino acids), respectively. Preliminary studies in rats indicate norethisterone may be released for periods greater than a year, while norethindrone, coupled to poly-N^5-(3-hydroxypropyl)-L-glutamine microparticles, and administered subcutaneously in rats, showed a near zero-order release rate for greater than 200 days.[28]

Another related prodrug approach involves forming a basic derivative of the steroid by attaching a tertiary amino group to the steroid.[29] Ethynyl estradiol and norethindrone have been derivatized by attaching a dimethylamino group via an ester or oxime linkage. These basic derivatives are then converted to either a zinc or aluminum tannate complex. Thus, the proposed release mechanism for this formulation is comprised of two processes: first, there is release of the base-substituted steroid from the complex and second, the labile linkage is cleaved, thus releasing the parent steroid.

Initial experiments in rats indicated that the zinc tannate complex of ethynyl estradiol was effective for 90 days, whereas ethynyl estradiol alone was only active for 60 days. However, continued testing in rats now suggests that a 20 mg/kg dose of the above complex may be effective for one year, that is, the appearance of the uteri was suggestive of prolonged estrogenic stimulation. These complexes were suspended in an aluminum monostearate-oil gel for injection, since this vehicle further reduces the rate of drug release.[29]

3. Subdermal Implants

3.1. Nonbiodegradable Implants

Subdermal implants have been investigated for the controlled delivery of antifertility agents since the late 1960s. Their use as a controlled-release system was based upon two significant developments. The first occurred in 1964 when Folkman and Long demonstrated that physiologically active materials would diffuse through silicone rubber.[30] The second development was the observation that small quantities of progestogen administered daily provided an antifertility effect in women without altering cyclic ovarian activity.[31] This was followed by the report of Dziuk and Cook that steroids would diffuse at a relatively constant rate through silicone rubber capsules, both under *in vitro* conditions and when implanted in ewes.[32] Segal and Croxatto then put these developments together by combining the slow diffusion of steroids through silicone rubber and the recognized antifertility effects of small quantities of progestogens.[33] Their studies confirmed earlier reports that steroids would diffuse through silicone rubber and that the rate of diffusion was affected by the thickness and total surface area of the silicone rubber capsule.

The initial PDS (polydimethylsiloxane) implants in patients were conducted by Chang and Kincl using megestrol acetate (MA) in 1968.[34] This was followed by two clinical pilot studies in humans using MA in silicone rubber capsules—one by Croxatto and collaborators in Chile,[35] and the other by Tatum and associates in Brazil.[36] The results of these initial pilot studies in patients indicated the degree of contraception was dependent upon the number of implants and that long-term contraception was possible without inhibiting ovulation.

In the studies by Tatum and associates,[36] the release of megestrol acetate from four to six silicone capsules was estimated to be 72–108 μg per day. However, this was based upon *in vitro* release data. In studies by Croxatto and associates,[35] three capsules released 78 μg per day and four capsules released 104 μg per day, again based upon *in vitro* data. They noted that the theoretical functional life of each implant would approximate 1000 days, based upon these *in vitro* release data. Thus, these patient studies demonstrated that if an appropriate number of capsules were implanted, pregnancy would be prevented.

In addition, Silastic subdermal implants offered the following advantages over oral contraceptives. These included: (1) continuous constant release of the steroid for periods of up to one year or longer; (2) easy implant insertion via a trocar; (3) surgical removal of the implant, if required, due to side effects; (4) elimination of patient compliance; and (5) rapid return of fertility following removal.

Additional clinical studies continued to confirm the efficacy of subdermal implants using megestrol acetate.[37–40] However, the use of megestrol acetate in PDS capsules for contraceptive purposes was seriously undermined when it was discovered that breast nodules had developed in female dogs when several progestogens, including megestrol acetate, were administered.[41] Thus, there was a clear incentive to find alternative progestogens, since the basic concept of controlled delivery had been demonstrated.

Table 2 lists some of the alternative steroids which have been, and continue to be, investigated for contraceptive purposes via Silastic subdermal implants. Two steroids, ST-1435 and levonorgestrel, have proven to be potent inhibitors of ovulation with ST-1435 reported to be at least seven times more potent than levonorgestrel.[53] For an excellent summary of *in vitro* and *in vivo* steroid release rates from Silastic rods and capsules, see the article by Nash *et al.*[55]

3.2. Biodegradable Implants

With nonbiodegradable implants, the implant must be removed once the steroid has been exhausted. Obviously, this could be avoided if a

Table 2. Alternative Steroids Incorporated in Silastic Subdermal Implants

Steroid	Loading dose—capsule or rod	Number of capsules or rods inserted	Average serum level in patients (μg/ml)	Average dose delivered to subject	References
Norgestrienone	47 mg	1–4	—[a]	—	42
R-2323	30–40 mg	2–5	~ 0.001	0.09 mg per day	43
Norethindrone (NET)	10 mg per 10 mm	60–200-mm lengths	—	10–12 μg per 10 mm per day	44
ST-1435	35 mg	3–5	—	—	45
Norgestrienone	47 mg	6	—	—	46
Norgestrienone	31.5 ± 1.1 mg per 34 mm	6	—	225 μg per day	47
Norethindrone acetate	40 mg	1	—	—	48
Levonorgestrel	—	6	(1.77 ± 0.82)–(2.21 ± 1.49)	—	49
ST-1435	40 mg	3	0.01–0.1	—	50
Levonorgestrel	34 mg	6	0.13–1.14	—	51
Levonorgestrel	30 mg	6	—	—	52
ST-1435	40 mg	1	0.05–0.22	—	53
Levonorgestrel	34 mg	6	—	29.9 μg per day	54

[a] Data not available in referenced article.

biodegradable carrier were used. The most natural biodegradable implant is one composed only of the steroid itself. Subdermal pellets of different steroids were studied during the 1940s[56,57] and release of the steroid was shown to be primarily dependent upon the surface area of the implant. It was also shown about that same time that the addition of cholesterol to the pellet reduced the rate of steroid release.[57] Both of these concepts, that is, solid implant rods of the steroid itself or pellets composed of the steroid plus an additive, are still being explored as delivery systems.

Investigators have implanted a norethindrone–cholesterol pellet subcutaneously in five patients and have studied its effects over a period of 229 days.[58] The pellets had uniform diameters of 2.4 mm, but varied in length from 5.8 to 6.2 mm. Each pellet contained between 23.9 and 25.6 mg of norethindrone and 4.2 and 4.5 mg of cholesterol. Preliminary results indicated that the NET was released too slowly, as evidenced by low NET plasma levels and anovulation during most of the treatment period. Thus, for this system to be effective, more pellets will have to be inserted.

Along these same lines, investigators are examining the incorporation of luteinizing hormone-releasing hormone (LHRH) in implants composed of cholesterol.[59] The implants so formulated contained 0.1 and 0.5% of the LHRH analogue and were subcutaneously implanted in rats. Results of these studies indicated the rats entered diestrus within three days. The duration of LHRH release from these implants was monitored by how long the constant diestrus phase was maintained. The 0.5% loading of the LHRH analogue (125 μg) maintained the diestrus condition in some of the rats for periods of up to 40 days. In addition, by placing a silicone sleeve over the pellet, diestrus was extended to 70–80 days in some of the rats.

The above implant also has been examined for synchronization and estrous suppression in cattle.[60] In addition, its utility for fertility control has been evaluated in the rhesus monkey and baboon.[61] Results of these monkey and baboon studies indicated that 1 mg per animal of D-Trp LHRH ethylamide delayed ovulation approximately 50 days and the same dose, repetitively implanted every 28 days for seven consecutive months, prevented ovulation in all but one of the 28 cycles of treatment.

Since polylactic and polyglycolic acid polymers have been used in the manufacturing and marketing of absorbable sutures,[62,63] and since they undergo a slow hydrolysis when implanted in tissues,[64] it was natural that they be investigated as a possible biodegradable subdermal implant material. One of the first studies in which these materials were used was that conducted by Jackanicz *et al.*[65] In these studies *d*-Norgestrel was incorporated into a solid matrix of poly-L-lactic acid. It was found that films containing 33% *d*-Norgestrel released the steroid at relatively constant rates when tested under *in vitro* conditions. However, when evaluated subcutaneously in rats, the initial release approximated 5.5 μg/day/cm^2. By 80 days, however, the release rate had declined to what had been observed under *in vitro* test conditions, that is, 3 μg/day/cm^2.

Perhaps the most promising subdermal polymer implant to date is one composed of a 90:10 ratio of lactide/glycolide with a molecular weight of 230,000.[66,67] Levonorgestrel (*d*-Norgestrel) has been incorporated at a 50 wt % level in the copolymer rods. These rods have been subcutaneously implanted in baboons. After 200 days the implant was still inhibiting ovulation as evidenced by the baboons' sex skin. More importantly, polymer release was similar to drug release and the lifetime of the system was calculated to be about eight years.

In addition to *d*-Norgestrel, norethisterone,[68] progesterone,[69] and β-estradiol[70] also have been incorporated in polylactic acid polymers and investigated for their effectiveness as subdermal implants. Other potentially biodegradable polymers, for example, poly-ξ-caprolactone,[71] polyglutamic acid,[68] glutamic acid/leucine copolymers,[72] pHEMA (polyhydroxyethyl methacrylate),[69] and Chronomer[TM],[73] a poly(ortho ester), have been explored as subdermal implants. The steroid release data from each of these systems to date pertain either to *in vitro* or animal studies. However, indications are that some of these systems may soon be evaluated clinically.

4. IUDs

The use of IUDs for contraceptive purposes is thought to date back as far as the eleventh century.[74] IUDs have evolved from pebbles placed in camel uteri, to exotic silver and diamond-studded pessaries, to the medicated devices in use today.[75] Richard Richter, a German physician, designed the first IUD to be used solely for human contraception.[74] This ring-shaped device, made of silkworm gut, was introduced in 1909. In the late 1920s, Ernest Grafenberg modified the silkworm gut ring with a spiral of silver wire, which became the first widely used IUD.[74] The use of metal intrauterine rings, designed by Tenrei Ota in Japan in 1933, marked the beginning of acceptance of IUDs in other countries.[76]

However, IUDs did not achieve widespread use for many years due to the risk of pelvic infection, side effects from poor insertion techniques, and a reluctance by the medical community to adopt their use. The development of antibiotics and biologically inert polyethylene are two major factors credited with advancing the development and reevaluation of modern IUDs.[74]

Although IUDs have been shown to be an effective means of contraception, there is not a single explanation as to the method by which they prevent pregnancy. The theory that appears to be most widely accepted is that IUDs inhibit implantation of the fertilized egg on the endometrium, due to a nonspecific inflammatory reaction.[75,77]

Since the resurgence of interest in IUDs for contraception in the 1950s and 1960s, several variations in sizes, shapes, and medications for IUDs have been researched. Presently, IUDs are generally divided into three groups: inert, copper bearing, and steroid releasing. The steroid-releasing IUDs were based upon data obtained from subdermal implants, that is, the sustained release of the active agent. However, the most important concept promoted by the steroid-releasing IUDs was that the steroid was being delivered to a local site, that is, the endometrium. The net effect was that low doses of the steroid were effective if delivered locally to its intended site of action.

4.1. Nonmedicated IUDs

The nonmedicated, or inert, IUDs do not release any substance within the uterus. Therefore, their mechanism of action is probably due primarily to an inflammatory reaction which inhibits implantation of the blastocyst on the endometrium.[75,77] The nonmedicated devices are generally fabricated from an inert plastic or copolymer, such as polyethylene, ethylene–vinyl acetate, or polypropylene. They vary in general configuration, stiffness, and surface area. However, these properties all appear to be related to the antifertility effect and expulsion rate of the IUD.[76]

Among the nonmedicated devices, the most popular IUD, and the one with which most comparisons are made, is the Lippes Loop.[75] The Loop comes in four sizes, to be accommodated by various-sized uteri, and is injection molded of alathon-20 polyethylene. Two monofilaments of nylon are attached to the loop to aid in detection and removal. The Loop was the first IUD to incorporate barium sulfate in order to render it opaque to X-rays for determining the position of the device.[74,77]

The Saf-T-Coil and the Dalkon Shield are two other IUDs which have been widely used in the United States.[76] The Saf-T-Coil is injection molded of an ethylene–vinyl acetate copolymer, has a marker tail, and comes in three sizes. The Dalkon Shield is a membrane-type device made of an ethylene–vinyl acetate copolymer. It has lateral fins and a cervical thread bound within a thin plastic sheath. This device, however, may be associated with a higher incidence of septic abortion and presently is being withdrawn from the market.[77]

4.2. Copper-Bearing IUDs

Zipper and co-workers first demonstrated the contraceptive effect of copper by placing a wire in the uterine horn of rabbits.[76,78,79] Zipper and Tatum then demonstrated an increased contraceptive effect of an inert polyethylene T-shaped IUD modified with copper wire.[78,79] Their preliminary data indicated a correlation between the surface area of the copper and the contraceptive effect of the IUD. Although the precise role of copper in the contraceptive effect is not known, copper has been shown to: (1) increase the inflammatory reaction; (2) interfere with enzymes in the uterus; (3) alter the biochemical composition of the cervical mucus, which in turn may affect sperm motility; and/or (4) increase contractions of the uterine muscle, thereby preventing implantation of the blastocyst.[77,78,80–82]

Three copper-bearing devices that have been subjected to extensive clinical evaluations are the copper-T (T-Cu 200), the copper 7 (Cu-7 or Gravigard), and the multiload (MLCu-250). The T-Cu 200 was the first medicated IUD tested for contraceptive use.[74] It is injection molded of polyethylene in a T-shape and the stem of the T is wound with copper wire having a surface area of 200 mm^2. The Cu-7 is injection molded of a polypropylene homopolymer and has copper wire wrapped around the lower arm of the 7, exposing a copper surface area of 200 mm^2. The multiload-Cu-250 is an injection-molded polyethylene vertical shaft with flexible arms that have serrated fins to hold the device in place. A 250-mm^2 surface area of copper wire is wrapped around the vertical stem.

The pregnancy rates of the above-mentioned copper-bearing IUDs compare favorably with other IUDs and differ only slightly among the various copper devices.[77,79,83–85] The copper metal on the devices gradually dissolves at a rate of 50 μg per day and lasts about two to three years.[77,78]

A second generation of copper-bearing IUDs has been developed in an effort to increase contraceptive effectiveness and lengthen the life span of the devices. These devices (T-Cu 380A, T-Cu 380Ag, T-Cu 220C, and Nova-T) incorporate the use of copper sleeves and/or silver cores. Increased surface area is obtained with the sleeves and the silver core is intended to reduce fragmentation of the copper wire.[78] Trials with these devices suggest the modifications may increase contraceptive efficacy while extending the time of effectiveness.[77,79,86]

4.3. Steroid-Releasing IUDs

The use of steroids in IUDs to increase the retention of the devices and to increase their contraceptive effectiveness was first investigated by Doyle *et al.*[87] However, these early trials with the medicated Silastic devices did not result in improved retention rates.[87] Release rates of steroids from the devices used in the early clinical studies by Doyle *et al.*[87] and Gibor *et al.*[88] suggested more of a first-order rate of steroid release rather than the projected zero-order rate. Pharriss *et al.*[89] adopted the use of the Tatum T-shaped IUD for release of progesterone in order to provide for contraceptive effectiveness superior to nonmedicated IUDs, but similar in effectiveness to oral contraceptives. This device was designed to eliminate harmful side effects caused by the systemic absorption of the steroid and to minimize pain and bleeding associated with inert IUDs.

The device fabricated, the Progestasert® (Alza Corporation, Palo Alto, California), delivers 65 μg of progesterone per day for one full year.[20,76,89,90] This was accomplished by developing a polymeric membrane that provided zero-order release of the drug. In addition, a device has now been fabricated which will release progesterone for up to three years.[91] Recent trials with the one-year device have shown that the pregnancy rates are comparable with those of the Nova T, the T-Cu 200, and the Cu-7.[77,81] An advantage of the Progestasert® device over other devices is that it reduces the amount of menstrual-blood loss.[20,77,92] In addition, the Progestasert® device uses the natural hormone progesterone. This hormone is released at a low dose that acts locally rather than systemically, thus reducing the potentials for side effects.

Other T-shaped devices containing levonorgestrel or dehydroretroprogesterone, more potent progestogens, have been investigated for extending the useful life of an IUD.[20,80,93,94] Release rates of 10, 20, and 30 μg per day of levonorgestrel were reported to exhibit results comparable to copper-releasing IUDs. In fact, some of these devices loaded with levonorgestrel release the steroid over a period of five years.[91] On the other hand, clinical trials with IUDs loaded with dehydroretroprogesterone, which released approximately 90 μg per day, resulted in pregnancy rates three times those of progesterone-loaded IUDs.[95] Thus, although this drug exhibited a potential useful life of two years, based upon the *in vitro* release data from Silastic rods, it was less effective than progesterone.[96]

In addition to synthetic progestogens, estrogens also have been evaluated as potential intrauterine contraceptive agents since they have been shown to inhibit blastocyst development and implantation in the uterus in rabbits.[97,98] Studies have documented that estriol-loaded IUDs, placed in the uterus of rabbits and baboons, have exhibited a contraceptive effect.[98] More recently, clinical trials have begun with IUDs releasing estriol at a rate of 11 μg per day.[99]

Another novel releasing medicated IUD, consisting of hollow fibers loaded with progesterone, is under investigation.[100] The advantages of this system are believed to include smaller size and greater flexibility, ease of insertion and less pressure on the uterine wall, increased surface area resulting in increased contact with the uterus, and possibly lower manufacturing costs.

The most significant finding when reviewing the various IUDs is that no single system is satisfactory for all women. The use of copper tends to allow for smaller devices than when using an inert IUD. However, the copper devices require recommended periodic replacement. The progesterone-releasing IUDs appear to reduce the volume of menstrual-blood loss, but tend to increase the number of days of bleeding and spotting. No one device appears to be profoundly superior in overall performance, since a change in design to improve one aspect often is detrimental to another.

5. Intravaginal Devices

Contraception by means of medicated intravaginal devices was first reported by Mishell *et al.* in 1970.[101,102] The devices were usually made of polydimethylsiloxane (Silastic) and were toroidal shaped. Various thicknesses and diameters of the ring were investigated to eliminate slippage and/or vaginal erosion. Several ring designs were incorporated to obtain constant release rates of the steroid.[103]

In the original study of Mishell *et al.*, silicone rubber rings ranging in size from 70 to 80 mm and containing 2 g of medroxyprogesterone acetate (MPA), were placed in the women's vagina on the first day of the menstrual cycle and remained in place for 28 days. Systemic absorption of MPA was evidenced by elevation of the BBT (basal body temperature), suppression of the midcycle LH peak, inhibition of ovulation, and endometrial changes.[101] Assay of the rings after removal indicated that two of the rings had lost 70 mg of MPA and the third ring had lost 220 mg.[104]

Mishell and Lumkin investigated the effectiveness of smaller rings containing lower amounts of MPA.[104] A drug loading as little as 50 mg of MPA in the smaller ring was sufficient to inhibit ovulation. Assay of the rings after removal indicated that between 11 and 27 mg of MPA had been lost from rings that had been in place for three weeks. Thus, the vaginal administration of medroxyprogesterone acetate, at a constant rate of 1 mg per day, appeared to be sufficient to inhibit ovulation.[102,105,106] However, MPA later was shown to induce breast nodules in beagles and work on these systems was discontinued.[20]

As a result of the MPA findings in dogs, studies similar to the above were conducted using vaginal rings containing levonorgestrel, a more potent progestogen. In these studies, ovulation was inhibited in 314 of 336 cycles and the release rates of these rings ranged from 120 to 830 μg per day.[102] However, the plasma levels of levonorgestrel varied widely indicating that constant release rates were not obtained with these rings.

Other progestogens which have been investigated in intravaginal devices have included chlormadinone acetate[107] and norethisterone.[102,108] Problems encountered with these systems have included too many ovulatory cycles, a large variation in plasma levels of the administered steroid, and breakthrough bleeding and bleeding irregularities.[20,102,108]

In an effort to control the bleeding irregularities, intravaginal rings containing a progestin and estrogen were investigated.[102,109] Systems combining progesterone and estradiol exhibited a high incidence of ovulatory cycles. However, favorable results (inhibition of ovulation and control of bleeding) were obtained when estradiol and *d*-Norgestrel were used.[102,109] The release rates from these intravaginal rings were 206–333 and 175–254 μg per day for *d*-Norgestrel and estradiol, respectively.

Vaginal rings have also been fabricated and designed to release low levels of progesterone (1400 μg per day), norethindrone (49.4 and 196 μg per day), and *d*-Norgestrel (21.6 μg per day).[110] Except for the high dose of norethindrone, ovulation usually occurred, but sperm penetration of the cervix was often inhibited, suggesting possible contraceptive properties.

Other intravaginal devices that have been investigated for contraceptive purposes include suppositories containing spermicides[111] or prostaglandins,[112,113] medicated collagen sponges,[114,115] tampons containing spermicides,[116] and oral-contraceptive tablets for vaginal administration.[117,118] All of these devices are designed to release the contraceptive agent at a sustained-controlled rate. However, additional work will be required to determine the release rates needed for contraception.

The advantages of the intravaginal contraceptive devices are: (1) the device can be inserted by the user; (2) patient and consort acceptance has been good[103,116]; (3) doses lower than oral-contraceptive doses can be used since these steroids are made directly available to their target organs before uptake by the portal system[109]; and (4) metabolic side effects are few.[103]

6. Intracervical Devices

The use of intracervical devices is based upon the idea that, following insertion, they can (1) cause a local inflammatory response which alters or changes the viscosity of the cervical mucus, the net effect being to modify sperm transport; or they can (2) release an agent at a sustained-controlled rate, for example, a spermicide or viscosity-altering agent, which again directly destroys or incapacitates the sperm or modifies cervical mucus viscosity.

Intravaginal devices, releasing progestogens, have been shown to inhibit sperm penetration through the cervix.[110] For this reason, several progestogens, that is, progesterone, chlormadinone acetate, norethindrone, and levonorgestrel, have been screened as contraceptive agents in intracervical devices placed in rabbits.[20,119] One such device, fabricated of silicone rubber, released chlormadinone acetate at a rate of 13.7 μg/kg/day and prevented fertilization of the ova. It was determined that a device which released 2.7–9.7 μg/kg/day of chlormadinone acetate lowered the number of sperm found in the rabbit uterine horn. However, a release rate of less than 7.0 μg/kg/day was not effective in preventing fertilization.(119) The effects of *d*-Norgestrel released from an intracervical device in the primate, *Erythrocebus patas*, were studied by Dagle *et al.*(120) Their results appeared to be similar to those when *d*-Norgestrel was administered systemically.

Intracervical devices containing "nonsteroids," that is, nonoxynol-9, quinine, and emetine, also have been investigated in rabbits.(119) Studies with these three spermicides indicate their potential usefulness as contraceptive agents.

Few clinical studies have been carried out using intracervical devices as a drug reservoir. One reason for this is that the original devices were not well retained in the cervix due to poor design.(20,119) However, since then, an extruded Silastic cervical device has been evaluated in women. This device was found to induce changes in the cervical mucus and release an estimated 150 μg per day of chlormadinone acetate.

A recent clinical study compared the effectiveness of an intracervical levonorgestrel-releasing mini-T device with an intrauterine levonorgestrel device and the Cu-T-200.(121) The mini-T device released about 10 μg per day of levonorgestrel. Its contraceptive efficacy was comparable to the Cu-T-200, but was lower than the levonorgestrel intrauterine device. The contraceptive effect was most likely the result of changes in cervical mucus as ovulation was assumed to have occurred.

The intracervical device is well suited to local delivery as long as the design is such that it is not expelled. Local delivery should offset the systemic problems that are often encountered with oral and parenteral administration of contraceptives. Further studies of effective compounds and release rates appear warranted.(20,119)

7. Oral Contraceptives

A very convenient route of administration for any medication is the oral route. Oral contraceptives have proven to be a very effective contraceptive method as witnessed by the 50 million women around the world taking the birth-control pill.(122) The first oral contraceptive was marketed in 1961. Since then, a wide variety of types and dosages of synthetic and natural steroids has been developed and marketed. However, since their inception, oral contraceptives have been associated with certain side effects which appear to be related to the steroid itself or to the total dose of the steroid.

Although the overall concentration of steroids has been reduced in many of the birth-control pills since their initial marketing, there remains a serum steroid concentration peak associated with ingestion of the pill on a daily basis. It is this daily peak which is believed to contribute to the side effects associated with oral contraceptives. Thus, one approach in minimizing this daily serum steroid peak was to develop an orally active time-released dosage form. The dosage form was designed to release the steroid over a 6–8 hour period as it traversed the gastrointestinal tract. It is believed that less total steroid might be required, yet remain effective, if absorbed slowly over the designated time.

In developing this dosage form, two approaches were sponsored by the National Institute of Child Health and Human Development. One approach involved the use of microcapsules containing a solution or suspension of ethynyl estradiol.[123] The second approach employed microspheres containing a homogeneous dispersion of ethynyl estradiol or norethindrone particles.[124] The microcapsules were prepared with a cellulose acetate butyrate polymer, while the microspheres were fabricated from hydrogenated mono-, di-, and triglycerides. The percentage of steroid incorporated in the microcapsules ranged from 3 to 11%, while the percentage of steroid incorporated in the microspheres ranged from 0.5 to 2%.

Both of these dosage forms were evaluated under *in vitro* test conditions. The microcapsules evaluated were of three size fractions: $< 125\,\mu$m, 125–210 μm, and 210–420 μm. The release of ethynyl estradiol from these microcapsules was designed to be zero order. However, the release from these capsules resulted in a linear relationship between the cumulative amount of steroid released and the square root of time.[123] The observed release rate was supported by the fact that the membrane could dissolve the total quantity of ethynyl estradiol initially loaded in the capsules. Thus, although microcapsules are considered a reservoir-type device, the microcapsules were acting as a thin planar membrane in this situation.

The microspheres, on the other hand, released the steroid over a period of eight or more hours.[124] In addition, only small differences were noted in the steroid release rate over the steroid concentrations examined, that is, some of the microspheres contained four times the steroid concentration yet released similarly to those that contained only one-quarter the steroid concentration. It was also noted that the steroid release rate was directly related to microsphere size and the glyceride composition of the microspheres. The microsphere sizes examined in these studies were 420–500, 600–710, and 850–1000 μm diameter.

Based upon these initial *in vitro* release data, an optimized microsphere formulation was prepared. This formulation contained an 80% mixture of mono- and diglycerides plus 20% triglyceride. Ethynyl estradiol or norethindrone was dispersed in this matrix at a 0.2% level. The microspheres

(600–700 μm) were evaluated in baboons and the results obtained were very encouraging. In fact, the results were so encouraging that limited clinical studies have been performed.[125]

8. Topical Application

One area which is receiving increased emphasis is the controlled release of drugs transdermally. This is particularly true for drugs such as nitroglycerin for treatment of angina pectoris and scopolamine for treatment of motion sickness. The topical application of steroids for fertility control purposes in animals[126,127] and for the treatment of menopausal disturbances in women has been investigated.[128] The initial emphasis of this approach in animals has centered around the administration of estrogens, which causes disturbances in ova transport.[126] The results of these early studies indicated that steroidal and nonsteroidal agents, possessing estrogenic activity and known to possess postcoital antifertility properties by the oral route, exerted the same effect when applied topically to the skin of rats—that is, these agents either prevented ova implantation in the uterus or caused an acceleration of ova transport through the fallopian tube.

Novel controlled-release devices may be developed to deliver selected antifertility agents transdermally. Such an initial delivery system might be best developed as a postcoital contraceptive, that is, a bandaid-like device which would deliver the antiferility agent when applied to the skin following intercourse. It is believed that this delivery system would have to deliver the agent over a three to four day period to provide the necessary protection. However, formidable problems in developing this concept include (1) that the application form be innocuous since skin can react to the treatment, for example, proliferation of the epidermal layer due to increased humidity levels; (2) physiological conditions, for example, sweating, which may affect the rate of drug absorption; and (3) individual variations in the pharmacokinetics of absorption.[128]

9. Intranasal Applications

Another potentially viable contraceptive method involves the intranasal administration of luteinizing hormone-releasing hormone (LHRH). The use of LHRH as a contraceptive was initially demonstrated in rats via a subcutaneous injection.[129] When LHRH was administered to rats, it inhibited ovulation. The use of these LHRH analogues now has been extended to women in whom the LHRH analogue is administered intranasally

(100 μg LHRH analogue per 100 μl solution).[130-132] The LHRH analogue was delivered at two doses, that is, 400 or 600 μg daily. The results from 51 female volunteers indicated no pregnancies over the 293 treatment months. In addition, there were no symptoms of estrogen deficiency, for example, vaginal discomfort, hot flashes, or other serious side effects. Of particular interest was that following cessation of treatment, ovulation and normal reproductive function returned rapidly. There were some slight bleeding abnormalities which may be a drawback to this method being used in developing countries. However, in many of the women, no menstrual bleeding occurred over periods exceeding six months.

The present method of delivering these LHRH analogues involves frequent inhaling of an aerosol mist. However, there is reason to believe that an inhaled particulate carrier might be developed to deliver these LHRH analogues at a sustained-controlled rate. Such a delivery system would have to be comprised of very small particulates which are innocuous and biodegradable. It is believed the sustained-controlled release of LHRH analogues from a small particulate carrier would reduce the frequency of administration now required.

Although the intranasal route of administration of LHRH analogues is a novel birth-control approach, other administration routes may be developed. In this regard, LHRH analogues have been administered to rats via the subcutaneous implantation of a minipump (Alza Corporation, Palo Alto, California).[133] Such a pump delivers the LHRH analogue at a rate of 1–2 μl/h over a three to seven day period.

10. Transcervical Device

Increased attention is being given to developing drug delivery systems which can be pretargeted. A contraceptive system which already satisfies this goal is an intrauterine device which releases progesterone locally to the uterus.[89] An alternative system which may have merit, but has yet to be developed, is a small particulate contraceptive delivery system which migrates transcervically. This system is based upon the use of microspheres or microcapsules, that is, small particulates ranging in size from 10 to 500 μm diameter.

In some cases, this delivery system could be used to deliver therapeutic medication, or as we envision, contraceptives. For instance, it has been demonstrated that inert carbon or ink particles migrate into the oviduct or peritoneal cavity after being placed in the posterior fornix of the human vagina.[134,135] Studies by Gardner *et al.*[136,137] have shown that microcapsules (containing an aqueous suspension of the radioactive tracer particles) migrate transcervically into the uterus or oviduct after placement in either the rabbit, stumptail monkey, or baboon vagina.

The results of the studies by Gardner *et al.*[137] in stumptail monkeys and baboons indicated that the percentage of transcervical migration was relatively low, that is, approximately 1% in 24 hours. However, this percentage is comparable to the percentage of sperm which migrate transcervically following deposition in the vagina. These low percentages of migration do not detract from the system's feasibility. Calculations by Gardner *et al.*[137] indicate that a drug 10 times more potent than progesterone will be required to make the system feasible. More recently, Beck and Cowsar[24] have investigated the transcervical migration of microspheres in baboons.

In considering this novel delivery system, however, many aspects of a particulate delivery system remain to be elucidated. These include: (1) determining the final destiny of the carrier particles following insertion; (2) controlling the amount of particulate leakage following vaginal insertion; (3) determining the local concentration of drug required for effectiveness at each site, for example, cervix, uterus, oviduct; and (4) defining the duration of contraceptive effectiveness, that is, is the contraceptive delivery system designed to last one month, only a few days surrounding ovulation, or could the system be administered just prior to coitus.

11. Immunological Response

Research during the past two to three decades suggests that fertility control might be accomplished by the production of an immunological response.[138,139] The primary advantage of an immunological response would be the infrequent administrations it would necessitate. Thus, if an antigen specific to a particular component of the reproductive system could be identified and isolated, then a specific antibody could be directed toward the antigen.

In particular, antigens associated with sperm, placenta, and ovum have been investigated for eliciting an immune response. In regard to sperm, it has been known that antifertility in women can be induced if antibodies are directed against sperm or the components of semen. In such cases, the antisperm may block fertility by: (1) immobilizing or causing the death of sperm; (2) modifying sperm transport; (3) enhancing phagocytosis within the female genital tract; or (4) interfering with the sperm–ovum contact.[138]

Isolation of placental proteins as potential antigens for antiferility regulation has concentrated on human chorionic gonadotropin (HCG). However, it has been found that when HCG is injected in patients, it affects the production of the human pituitary luteinizing hormone. This was overcome to a degree when only the β-subunit of HCG, attached to a tetanus-toxoid carrier, was injected. However, this still proved problematical since, when the β-tetanus-toxoid conjugate was alum precipitated for injection, much of the original antigenicity was lost. Efforts are now under way to develop synthetic adjuvants which will enhance the antigenicity of the β-subunit of HCG.[(140)]

The ovum is another potential immunological site where specific antigens might be produced and investigated. In particular, the zona pellucida (a gelatinous layer surrounding the ovum) is known to contain cell-specific antigens which may be involved in immunological reactions to infertility.[(141)] For instance, heterologous antibodies, specific to the zona pellucida, have been prepared which prevent sperm attachment and dissolution of the zona pellucida membrane. In addition, it has been shown that antibodies to porcine zona react with human zona and vice versa. Thus, isolated porcine zona now can be used as a test model to detect human antizona antibodies.

An alternative immunological approach is one in which a local immune response is generated in the vagina, cervix, uterus, or oviduct.[(142)] The production of a local concentration of antibodies, at any of these specific areas, would probably be a safer antiferility method since it would not involve a systemic response. Of these targets, the cervix is considered the most favorable region since it possesses columnar epithelium, associated with mucus-secreting cells. Furthermore, immunofluorescence has disclosed subepithelial secretory Ig and IgG plasma cells.[(142)]

From the above discussion it can be seen that an antifertility approach based upon an immune response is a real possibility. It is also believed that many of the controlled delivery systems discussed earlier in this chapter may be of value in delivering the immunogens. For instance, the use of small particulates (microspheres, microcapsules) which migrate transcervically might be used to deliver the immunogens to a specific site, for example, cervix or oviduct. In addition, intracervical devices might be ideally suited for delivering an immunogen. The major obstacle of utilizing these carrier systems at present is that large-molecular-weight materials must be released. Much of the prior work, involving controlled-release technology, has concentrated on low-molecular-weight materials. Release of macromolecules from controlled delivery devices is beginning to receive increased attention.[(143)] Thus, it is believed that delivery of large molecules, for example, antigens, will be an active area of future research utilizing controlled-release technology.

References

1. S. Coleman and P. T. Piotrow, *Population Reports*, Series H, No. 5, (September 1975).
2. W. Rinehart and P. T. Piotrow, *Population Reports*, Series A, No. 5, (January 1979).
3. W. N. Spellacy, W. C. Buhi, V. A. Dumbaugh, and S. A. Birk, *Fertil. Steril.* **30**(3), 289 (1978).
4. F. H. Min, in: *Recent Advances in Fertility Regulation* (C. C. Fen, D. Griffen, and A. Woolman, eds.), p. 378, Atar S. A., Geneva, Switzerland (1981).
5. G. S. Berger, L. G. Keith, and D. A. Edelman, in:*Proceedings Medicated IUDs and Polymeric Delivery Systems* (E. S. E. Hafez and W. A. A. van Os, eds.), Abstract 38, Amsterdam, Holland (1979).
6. W. Rinehart and J. Winter, *Population Reports*, Series K, No. 1 (March 1975).
7. E. T. Tyler, M. Levin, J. Elliot, and H. Dolman, *Fertil. Steril.* **21**(6), 469 (1970).
8. S. Jeppsson and E. D. B. Johansson, *Contraception* **14**, 461 (1976).
9. G. Howard, R. J. Warren, and K. Fotherby, *Contraception* **12**, 45 (1975).
10. E. Weiner and E. D. B. Johansson, *Contraception* **11**, 419 (1975).
11. V. Goebelsmann, F. Z. Stanczyk, P. F. Brenner, A. E. Goebelsmann, E. K. Gentzchein, and D. R. Mishell, *Contraception* **19**, 283 (1979).
12. H. K. Toppozada, S. Koetsawang, V. E. Aimakhv, T. Khan, A. Pretnar, T. K. Chatterjee, M. P. Molitor-Peffer, R. Apelo, R. Lichtenberg, P. G. Crosignani, J. C. de Souza, C. Gomez-Rogers, A. A. Haspels, R. H. Gray, P. Diethelm, G. Benagiano, and J. Annus, *Contraception* **25**(1), 1 (1982).
13. W. J. Bremmer and D. M. de Kretser, *N. Engl. J. Med.* **295**(20), 1111 (1976).
14. J. F. Melo and E. M. Coutinho, *Contraception* **15**(6), 627 (1977).
15. J. Frick, G. Bartsch, and W. H. Weiske, *Contraception* **15**(6), 649 (1977).
16. F. Alvarez-Sanchez, A. Faundes, V. Brache, and P. Leon, *Contraception* **15**(6), 635 (1977).
17. D. M. de Kretser, in: *Recent Advances in Fertility Regulation* (C. C. Fen, D. Griffen, and A. Wollman, eds.), p. 112, Atar S. A., Geneva, Switzerland (1981).
18. M. Sun, *Science* **217**, 424 (1982).
19. B. J. Culliton, *Science* **219**, 371 (1983).
20. L. R. Beck, D. R. Cowsar, and V. Z. Pope, *Research Frontiers in Fertility Regulation* (Program for Applied Research on Fertility Regulation), **1**(1), 1 (1980).
21. L. R. Beck, V. Z. Pope, D. R. Cowsar, D. H. Lewis, and T. R. Tice, *Contracept. Deliv. Syst.* **1**, 79 (1980).
22. T. K. Lee, T. D. Sokoloski, and G. P. Royer, *Science* **213**, 233 (1981).
23. L. R. Beck, D. R. Cowsar, D. H. Lewis, R. J. Cosgrove, C. T. Riddle, S. L. Lowry, and T. Epperly, *Fertil. Steril.* **31**(5), 545 (1979).
24. L. R. Beck and D. Cowsar, *Biomedical Application of Polymers*, Rubber and Plastics Association of Great Britain (RAPRA), **12**(5), 6 (1982).
25. L. R. Beck, R. A. Ramos, C. E. Flowers, Jr., G. Z. Lopez, D. H. Lewis, and D. R. Cowsar, *Am. J. Obstet. Gynecol.* **140**(7), 799 (1981).
26. V. Stella (T. Higuchi and V. Stella, eds.), p. 45, American Chemical Society Symposium Series No. 14, Washington, D.C. (1974).
27. J. Feijen, D. Gregonis, C. Anderson, R. V. Peterson, and J. Anderson, *J. Pharm. Sci.* **69**(7), 871 (1980).
28. M. A. Zupon, J. M. Christensen, R. V. Peterson, and S. M. Fang, *Polym. Prepr.* **20**, 604 (1979).
29. A. P. Gray and T. N. Yamauchi, *J. Med. Chem.* **21**(7), 712 (1978).
30. J. Folkman and D. M. Long, *J. Surg. Res.* **4**, 139 (1964).

31. H. W. Rudel, J. Martinez-Manautou, and M. Maqueo-Topete, *Fertil. Steril.* **16**, 158 (1965).
32. P. J. Dziuk and B. Cook, *Endocrinology* **78**, 208 (1966).
33. S. J. Segal and H. Croxatto, Paper Presented at the Twenty-Third Meeting of the American Fertility Society, Washington, D.C. (April 1967).
34. C. C. Chang and F. A. Kincl, *Steroids* **12**, 689 (1968).
35. H. Croxatto, S. Diaz, R. Vera, M. Etchart, and P. Atria, *Am. J. Obstet. Gynecol.* **105**(7), 1135 (1969).
36. H. J. Tatum, E. M. Coutinho, J. A. Filho, and A. R. S. Sant'anna, *Am. J. Obstet. Gynecol.* **105**(7), 1139 (1969).
37. S. Tejuja, *Am. J. Obstet. Gynecol.* **107**(6), 954 (1970).
38. G. Benagiano, M. Ermini, L. Carenza, and G. Rolfini, *Acta Endocrin.* **73**, 335 (1973).
39. G. Benagiano, M. Ermini, L. Carenzi, and P. Donini, *Acta Endocrin.* **73**, 347 (1973).
40. M. Ermini, F. Carpino, M. Russo, and G. Benagiano, *Acta Endocrin.* **73**, 360 (1973).
41. Editorial, *Brit. Med. J.* **5691**, 252 (1970).
42. E. M. Coutinho and A. R. Da Silva, *Fertil. Steril.* **25**(2), 170 (1974).
43. E. M. Coutinho, A. R. Da Silva, C. Carreira, M. Chaves, and M. C. Adeodata-Filho, *Contraception* **11**, 625 (1975).
44. J. Wiese, I. L. Marker, P. Holma, E. Vartiainen, M. Osler, T. Pyorala, E. Johannson, and T. Luukkainen, *Ann. Clin. Res.* **8**, 93 (1976).
45. E. M. Coutinho, A. R. Da Silva, and H-G. Kraft, *Int. J. Fertil.* **21**, 103 (1976).
46. A. R. Da Silva and E. M. Coutinho, *Int. J. Fertil.* **23**(3), 185 (1978).
47. S. Diaz, M. Pavez, E. Quinteros, J. Diaz, D. N. Robertson, and H. B. Croxatto, *Contraception* **18**(4), 429 (1978).
48. S. M. Shahani, P. P. Kulkarni, P. A. Bhate, and K. L. Patel, *Contraception* **19**(2), 135 (1979).
49. A. Faundes, V. B. de Mejias, P. Leon, D. Robertson, and F. Alvarez, *Contraception* **20**(2), 167 (1979).
50. P. Lahteenmaki, E. Weiner, P. Lahteenmaki, E. D. B. Johannson, and T. Luukkainen, *Contraception* **23**(1), 63 (1980).
51. H. B. Croxatto, S. Diaz, and P. Miranda, *Contraception* **22**(6), 583 (1980).
52. C. G. Nilsson and P. Holma, *Contraception* **35**(3), 304 (1981).
53. P. Lahteenmaki, E. Weiner, E. D. B. Johansson, and T. Luukkainen, *Contraception* **25**(3), 299 (1982).
54 S. Diaz, M. Pavez, P. Miranda, D. Robertson, I. Sivin, and H. B. Croxatto, *Contraception* **25**(5), 447 (1982).
55. H. A. Nash, D. N. Robertson, A. J. Moo Young, and L. Atkinson, *Contraception* **18**(4), 367 (1978).
56. T. R. Forbes, *Endocrinology* **32**, 282 (1943).
57. M. B. Shimkin and J. White, *Endocrinology* **29**, 1020 (1941).
58. V. Odlind, A. J. Moo Young, G. N. Gupta, E. Weiner, and E. D. B. Johansson, *Contraception* **19**(6), 639 (1979).
59. J. S. Kent, B. H. Vickery, and G. I. McRae, Proceedings of the Seventh International Symposium on Controlled Release of Bioactive Materials, Fort Lauderdale, Florida, (1980) p. 67.
60. R. C. Herschler and B. H. Vickery, *Am. J. Vet. Res.* **42**(8), 1405 (1981).
61. B. H. Vickery and G. I. McRae, *Int. J. Fertil.* **25**(3), 171 (1980).
62. A. R. Anscombe, N. Hira, and B. Hunt, *Brit. J. Surg.* **57**, 917 (1970).
63. J. B. Herrman, R. J. Kelly, and G. A. Higgins, *Arch. Surg.* **100**, 486 (1970).
64. D. E. Cutright, J. D. Beasley III, and B. Perez, *Oral Surg.* **32**, 165 (1971).

65 T. M. Jackanicz, H. A. Nash, D. L. Wise, and J. B. Gregory, *Contraception* **8**(3), 227 (1973).
66. D. L. Wise, J. D. Gresser, L. R. Beck, and J. F. Howes, in: *Biodegradables and Delivery Systems for Contraception* (E. S. E. Hafez and W. A. A. van Os, eds.), Vol. 1, p. 88, M.T.P. Press, Lancaster, England (1980).
67. J. D. Gresser, D. L. Wise, L. R. Beck, and J. F. Howes, in: *Biodegradables and Delivery Systems for Contraception* (E. S. E. Hafez and W. A. A. van Os, eds.), Vol. 1, p. 83, M.T.P. Press, Lancaster, England (1980).
68. L. C. Anderson, D. L. Wise, and J. F. Howes, *Contraception* **13**, 375 (1976).
69. J. R. Cardinal, S. W. Kim, S. Song, E. S. Lee, and S. H. Kim, *AIChE Symp. Ser.* **77**, 52 (1981).
70. S. Yolles, T. Leafe, M. Sartori, M. Torkelson, L. Ward, and F. Boettner (D. R. Paul and F. W. Harris, eds.), p. 123, American Chemical Society Symposium Series No. 33, Washington, D.C. (1976).
71. C. G. Pitt, D. Christensen, A. R. Jeffcoat, G. L. Kimmel, A. Schindler, M. E. Wall, and R. A. Zweidinger (H. L. Gabelnick, ed.), DHEW Publication No. (NIH) 77, (1976), p. 144.
72. K. R. Sidman, W. D. Steber, and A. W. Burg (H. L. Gabelnick, ed.), DHEW Publication No. (NHI) 77, (1976), p. 121.
73. G. Benagiano and H. L. Gabelnick, *J. Steroid Biochem.* **11**, 449 (1979).
74. P. T. Piotrow, W. Rinehart, and J. C. Schmidt, *Population Reports*, Series B, No. 3 (May 1979).
75. S. C. Huber, P. T. Piotrow, F. B. Orlans, and G. Kommer, *Population Reports*, Series B, No. 2 (January 1975).
76. B. T. Wagatsuma, in: *Recent Advances in Fertility Regulation* (C. Fen and D. Griffin, ed.) (September 1980), p. 228.
77. L. Liskin and G. Fox, *Population Reports*, Series B, No. 4 (July 1982).
78. G. Oster and M. Salgo, *New Engl. J. Med.* **293**, 432 (1975).
79 H. J. Tatum, *Clin. Obstet. Gynecol.* **17**(1), 93 (1974).
80. M. Heikkila, *Contraception* **25**(6), 561 (1982).
81. P. Fylling and M. Fagerhol, *Fertil. Steril.* **31**(2), 138 (1979).
82. T. Tamaya, Y. Nakata, Y. Ohno, S. Nioka, N. Furuta, and H. Okada, *Fertil. Steril.* **27**(7), 767 (1976).
83. N. Lauersen, L. Cederqvist, S. Donovan, and F. Fuchs, *Fertil. Steril.* **26**(7), 638 (1975).
84. E. Sadovsky and S. Yarkoni, *Fertil. Steril.* **30**(5), 519 (1978).
85. W. Nebel, J. Currie, and R. Lassiter, *Fertil. Steril.* **30**(5), 516 (1978).
86. J. Diaz, M. Diaz, L. Pastene, R. Araki, and A. Faundes, *Contraception* **26**(3), 221 (1982).
87. L. Doyle, T. Clew, and G. Chandler, *Fertil. Steril.* **26**(7), 649 (1975).
88. Y. Gibor, D. Seshadri, and A. Scommegna, *Fertil. Steril.* **22**(10), 671 (1971).
89. B. Pharriss, R. Erickson, J. Bashaw, S. Hoff, V. Place, and A. Jaffaroni, *Fertil. Steril.* **25**(11), 915 (1974).
90. G. Duncan, F. Burton, and W. Skiens, in: *Biodegradables and Delivery Systems for Contraception* (E. S. E. Hafez and W. A. A. van Os, eds.), Vol. 1, p. 109, M.T.P. Press, Lancaster, England (1980).
91. R. L. Dunn, D. H. Lewis, and L. R. Beck, in: *Controlled Release of Pesticides and Pharmaceuticals* (D. H. Lewis, ed.), Plenum Press, New York (1981), p. 125.
92. J. Martinez-Manautou, R. Aznar, M. Maqueo, and B. Pharriss, *Fertil. Steril.* **25**(11), 922 (1974).
93. C. Nilsson, T. Luukkainen, J. Diaz, and H. Allonen, *Lancet* **8220,1**, 557 (1981).
94. M. Heikkila, P. Lahteenmaki, and T. Luukkainen, *Contraception* **26**(3), 245 (1982).

95. A. Scommegna, R. Rao, and W. P. Dmowski, in: *Proceedings Medicated IUDs and Polymeric Delivery Systems* (E. S. E. Hafez and W. A. A. van Os, eds.), Abstract 6a, Holland (1979).
96. B. B. Pharriss and P. Rowe, in: *Proceedings Medicated IUDs and Polymeric Delivery Systems* (E. S. E. Hafez and W. A. A. van Os, eds.), Abstract 4, Holland (1979).
97. W. Dmowski, F. Aluletta, and A. Scommegna (H. L. Gabelnick, ed.), Proceedings Drug Delivery Systems, DHEW Publication No. (NIH) 77-1238, (1977), p. 12.
98. W. P. Dmowski, E. Kapetanakis, and A. Scommegna, in: *Proceedings Medicated IUDs and Polymer Delivery Systems* (E. S. E. Hafez and W. A. A. van Os, eds.), Abstract 90a, Holland (1979).
99. R. Baker, M. Tuttle, H. Lonsdale, and J. Ayres, *J. Pharm. Sci.* **68**(1), 20 (1979).
100. D. H. Lewis, W. E. Meyers, R. L. Dunn, and T. R. Tice, Proceedings of the Ninth International Symposium on Controlled Release of Bioactive Materials (Abstract), Fort Lauderdale, Florida, (1982), p. 61.
101. D. Mishell, M. Talas, A. Parlow, and D. Moyer, *Am. J. Obstet. Gynecol.* **107**(1), 100 (1970).
102. E. Diczfalusy and B. Landgren, in: *Recent Advances in Fertility Regulation* (C. Fen and D. Griffin, eds.), Atar S. A. publisher, Geneva, Switzerland (September 1980) p. 43.
103. S. Roy and D. Mishell, in: *Biodegradables and Delivery Systems for Contraception* (E. S. E. Hafez and W. A. A. van Os, eds.), Vol. 1, p. 163, M.T.P. Press, Lancaster, England (1980).
104. D. Mishell and M. Lumkin, *Fertil. Steril.* **21**(2), 99 (1970).
105. D. Mishell, M. Lumkin, and S. Stone, *Am. J. Obstet. Gynecol.* **113**(7), 927 (1972).
106. A. Vermeulen, M. Dhondt, M. Thiery, and D. Vandekerckhove, *Fertil. Steril.* **27**(7), 773 (1976).
107. M. Henzl, D. Mishell, J. Velazquez, and W. Leitch, *Am. J. Obstet. Gynecol.* **117**(1), 101 (1973).
108. B. Landgren, M. Oriowo, and E. Diczfalusy, *Contraception* **24**(1), 29 (1981).
109. D. Mishell, D. Moore, S. Roy, P. Brenner, and M. Page, *Am. J. Obstet. Gynecol.* **130**(1), 55 (1978).
110. F . Burton, W. Skiens, and G. Duncan, *Contraception* **19**(5), 507 (1979).
111. W. Masters, V. Johnson, R. Kolodny, and G. Tullman, *Fertil. Steril.* **32**(2), 161 (1979).
112. M. Salomy and P. Goldstein, *Fertil. Steril.* **29**(4), 456 (1978).
113. T. Roseman and C. Spilman, *AIChE Symp. Series* **206**(77), 69 (1981).
114. M. Chvapil, C. Kischer, J. Campbell, M. Kantor, J. Owen, and T. Chvapil. *Fertil. Steril.* **30**(4), 461 (1978).
115. R. Dorr, E. Surwit, W. Droegemueller, D. Alberts, F. Meyskens, and M. Chvapil, *J. Biomed. Mater. Res.* **16**, 839 (1982).
116. E. Page, *Contraception* **23**(1), 37 (1981).
117. E. Coutinho, E. J. Coutinho, M. Goncalves, and I. Barbosa, *Fertil. Steril.* **38**(3), 380 (1982).
118. U. Schwartz, E. Schneller, L. Moltz, and J. Hammerstein, *Contraception* **25**(3), 253 (1982).
119. F. Burton, W. Skiens, G. Duncan, and M. Sikov, in: *Biodegradables and Delivery Systems for Contraception* (E. S. E. Hafez and W. A. A. van Os, eds.), Vol. 1, p. 139, M.T.P. Press, Lancaster, England (1980).
120. G. Dagle, M. Sikov, S. Rowe, F. Burton, and W. Skiens, *Contraception* **22**(4), 409 (1980).
121. S. Mahgoub, *Contraception* **25**(4), 357 (1982).
122. A. Kols, W. Rinehart, P. T. Piotrow, L. Douchette, and W. F. Quillin, *Population Reports*, Series A, No. 6, A-198 (1982).
123. D. L. Gardner and D. J. Fink, in: *Biodegradables and Delivery Systems for Contraception* (E. S. E. Hafez and W. A. A. van Os, eds.), Vol. 1, p. 47, M.T.P. Press, Lancaster, England (1980).

124. D. J. Mangold and H. W. Schlamens, Proceedings of the Seventh International Symposium on Controlled Release of Bioactive Materials, Fort Lauderdale, Florida (1980), p. 62.
125. H. Gabelnick, personal communication.
126. A. B. Kar, B. S. Setty, and V. P. Kamboj, *Am. J. Obstet. Gynecol.* **102**(2), 306 (1968).
127. J. A. Copolla and J. L. Ball, *J. Reprod. Fertil.* **13**, 373 (1967).
128. H. Schaefer, G. Stuttgen, and W. Schalla, *Contraception* **20**(3), 225 (1979).
129. J. Humphries, Y. P. Wan, K. Folkers, and C. Y. Bowers, *Biochem. Biophys. Res. Commun.* **72**(3), 939 (1976).
130. C. Berquist, S. J. Nillius, and L. Wide, *Lancet* **2**, 215 (1979).
131. C. Berquist, S. J. Nillius, and L. Wide, *Fertil. Steril.* **38**(2), 190 (1982).
132. S. J. Nillius, C. Berquist, and L. Wide, *Contraception* **17**, 537 (1978).
133. C. Y. Bowers and K. Folkers, *Biochem. Biophys. Res. Commun.* **72**(3), 1003 (1976).
134. D. H. de Boer, *J. Reprod. Fertil.* **28**, 295 (1972).
135. G. E. Egli and M. Newton, *Fertil. Steril.* **12**, 151 (1961).
136. C. R. Hassler, D. L. Gardner, D. C. Emmerling, W. C. Baytos, and K. Williamson, *Proc. San Diego Biomed. Symp.* **13**, 1 (1974).
137. D. L. Gardner, D. J. Fink, and C. R. Hassler, in: *Biodegradables and Delivery Systems for Contraception* (E. S. E. Hafez and W. A. A. van Os, eds.), Vol. 1, p. 47, M.T.P. Press, Lancaster, England (1980).
138. V. C. Stevens, *Bull. WHO* **56**(2), 179 (1978).
139. W. R. Jones, *Fertil. Steril.* **33**(6), 577 (1980).
140. E. R. Gonzalez, *JAMA* **244**(13), 1414 (1980).
141. A. G. Sacco, *Obstet. Gynecol. Ann.* **10**, 1 (1981).
142. J. P. Vaerman and J. Ferin, Karolinska Symposium VII, Brussels, Belgium (1974), p. 281.
143. D. S. T. Hsieh, W. D. Rhine, and R. Langer, *J. Pharm. Sci.* **72**(1), 17 (1983).

5

Controlled Release and Plant-Growth Regulators

Jocelyn M. Miller and Amar Yahiaoui

Abstract. Efficient delivery of agricultural chemicals to crops has been the major focus of controlled release research. The resulting commercial applications have centered on traditional fertilizers, herbicides, insecticides, and fungicides. However, recent research suggests that significantly greater productivity can be achieved by using controlled release methods to deliver plant-growth regulators.

Traditional agricultural chemicals promote increased plant productivity by control of: disease, pest attack, weed competition, and soil components. Plant-growth regulators are substances which promote growth by modification of the plant characteristics. The mechanisms are not well understood and the technology has been hindered by lack of a suitable delivery system. Plant-growth regulators have to be applied in minute quantities and in a time-tailored manner. Those delivery system requirements are unique characteristics of controlled release systems.

Plant-growth regulators are generally classified by their dominant growth effect: cell elongation, cell division, dormancy induction, abscission, ripening acceleration, fruit and flower initiation, and bulb production. However, excess concentration or improperly timed application can produce undesired effects. For this reason, plant-growth regulators have received rather limited commercial application. They currently represent less than 1% of agricultural chemical sales.

Several research projects have proven the feasibility of delivering plant-growth regulators using controlled release systems. The next major step will be to perfect the technique for application to a commercial crop. Successful development will help solve the world food-supply problem.

1. Introduction

Plants provide an essential link between the earth and the sun. By the process of photosynthesis, solar energy is converted to a tangible form which man has utilized in many life-support systems. Obvious examples include food sources for mankind and livestock, energy supplies, clothing, and building materials. Less obvious but no less important to the sophistication

Jocelyn M. Miller and Amar Yahiaoui • College of Forest Resources, University of Washington, Seattle, WA 98040.

of man's existence is the use of plants for medicines and drugs, soil-erosion control, land reclamation, manufacturing, and beautification.

A major concern is that the supply of renewable plant resources is predicted to fall short of future accelerating demands.[1] World population growth predictions are subject to wide variations. However, even conservative estimates suggest the population will double in less than 50 years.[2] Presently, food-supply growth is approximately in balance with the world population growth (3% per annum). However, arable land and water are limited resources already subject to attack from population growth. For example, the United States is losing three million farmable acres per year to urban and industrial growth, which also demands a greater share of water supplies.[1]

Efficiency improvement will be necessary to balance supply with demand for plant resources.[3,4] Essential factors include increased plant yields of usable crops, elimination of waste products, and improved soil technology. The concept of efficiency improvement is not new.[5] Man has successfully achieved major improvements throughout the centuries with respect to crop diversification, quality, and yield. The study of plant genetics has provided disease and stress resistant varieties, fruit optimization, and extended growing seasons. Engineering sciences have brought improvements in irrigation, pesticide application, and mechanical harvesting. In recent decades chemical technology has provided new generations of herbicides, fungicides, insecticides, fertilizers, and plant-growth regulators. The technology of the first four in this group is well developed and although there will be future problems, changes, and improvements, man has a comprehensive understanding in this arena.[6] Plant-growth regulators are not well understood.

The plant-growth regulators (PGRs) are defined as natural or synthetic compounds which beneficially affect the physiological process of a plant. Man first became aware of these endogenous chemical messengers in plants in the late 1800s.[7] Subsequent knowledge has been slow to accumulate and only limited commercial use has been directed at increased yields, growth modification, and harvesting methods. The limitations are due to the rather complex interaction between the PGRs and the total physiology of the plant. For example, a regulator may produce diverse responses depending on minor variations in a plant's variety, size, age, and environment. Despite these difficulties, growth regulators offer plant-cultivation improvements beyond the present limits of man's imagination. Certainly, this technology may provide the type of quantum increase in plant production necessary for man's survival.

Current PGR applications include root promotion, control of fruit fall, storage deterioration, stimulating fruit set, sexual maturity, budding, and overcoming dormancy.[8] Yet these applications represent only 1% of all present agrochemical sales.[9] However, future PGR sales are forecast to overtake herbicides and become as important as fertilizers in world food production.[10]

Controlled release technology is considered to be the vital link in commercial application of plant-growth regulators. Laboratory research has already established the ability of controlled release systems to deliver minute quantities of PGRs to plants in patterns tailored to specific needs. The next major step will be to perfect the technique for application to a commercial crop. Further progress will be required to expand this technology to a wider range of crops, PGRs, and controlled release systems.

2. Plant-Growth Regulators

Growth coordination from internal chemical messengers (the growth regulators) was first conceived in the late 1800s by Charles Darwin and his son Francis[3] and Ciesielski.[11] Throughout the early 1900s this work was verified and developed.[12–16] A major success occurred in 1928 when F. W. Went[17] isolated the first growth substance from oat seedlings. The concept of altering plant growth was realized by the external applications of such growth materials.

Since the 1930s five separate classes of growth regulants have been discovered and accepted (Figure 1). These are auxins, gibberellins,

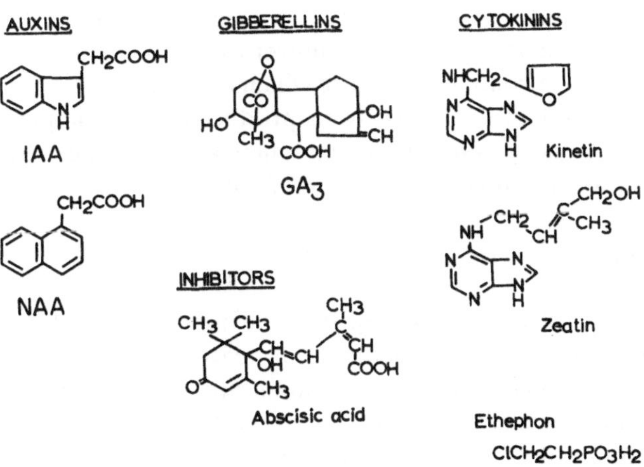

Figure 1. Representative examples of the different classes of plant-growth regulators.

cytokinins, inhibitors, and ethylene gas. Each is chemically distinct and creates characteristic growth responses. However, their total effect is extraordinarily diverse. Most aspects of plant growth are affected by each of the five regulators working alone or in concert with each other as their activities overlap and combine.

2.1. Auxins

Auxins were chemically identified and isolated as indol-3-ylacetic acid (IAA) and named auxins by Kögl and Haagen-Smit in 1934.[18] IAA is still considered to be the principal plant auxin, although a limited number of other naturally occurring indole compounds are recognized as auxins; all cause cell elongation.[19] By 1936 the auxins became available commercially and were quickly followed by a wide and diverse range of synthetic auxin-type compounds such as 1-naphthylacetic acid (NAA) and 4-(indol-3-yl)-butyric acid (IBA), which are commonly used for commercial rooting, fruit setting, and reducing fruit drop.[20]

2.2. Gibberellins

Gibberellins (GA) were discovered by Kurosawa[21] in 1926, and were successfully separated, crystallized, and named in 1935 by Yabuta.[22] However, it was not until 1955 that gibberellins received international recognition. More than 50 natural gibberellins have been isolated from a variety of plant materials. Gibberellins are diterpenoids, and their main effect is involved with stem elongation. Their overall effect on plant growth is impressive and has led to a great variety of commercial applications including overcoming dormancy,[23] inducing flowering,[24] and fruit development.[25] Gibberellins have a complex structure which has not yet been synthesized nor has any other unlike substance been produced with comparable activity. However, GA is produced in commercial quantities by large-scale fermentation of the fungus *Gibberella fujikuroi*.

2.3. Cytokinins

Cytokinins were first recognized as kinetin in 1955,[26] later isolated from maize, and chemically identified as zeatin.[27] This group was named cytokinins by Skoog in 1965. Cytokinins are specifically involved in cell division at the gene level. The list of naturally occurring cytokinins grow steadily[28] and, as with the auxins, a varied array of related compounds have been synthesized. Their range of effects dazzle: breaking dormancy in fruits and seeds, swelling roots, leaf enlargement, and delaying senescence in leaves and fruits. They have been shown successful in improving the shelf life of green vegetables,[29] but commercial application is slow due to problems in use and undesirable side effects. However, Cytese, a natural growth regulator containing a mixture of natural cytokinins, was recently registered for increasing fruit size and yield in tomatoes and oranges.[30]

2.4. Inhibitors

This group of PGRs was recognized in the late 1940s. Their physiology and effect was difficult to unravel and understanding came slowly. In 1965 abscisic acid (ABA),[31] the most important and widespread of the inhibitors, was discovered. Since then other diverse groups of inhibitors have been identified in a wide range of plants. Inhibitors induce dormancy,[32] abscission of leaves, and enhance a plant's response to environmental stresses.[33,34] There is also some correlation with maturity and senescence. Synthetic inhibitors were formed in 1949 and have become very important commercially. Maleic hydrazide (MH) prevents sprouting in stored crops. Plant-growth retardants, chlormequat (CCC) and daminozide, prevent lodging in cereal crops and morphactins suppress growth.

2.5. Ethylene Gas

Although the dramatic effects of ethylene gas were recognized in the early 1900s, it was not accepted as a natural plant-growth regulator until the late 1960s.[35] The development of gas chromatography allowed detailed experimental work which paved the way towards general acceptance of this unusual material as a plant hormone. Since then the importance of this PGR has magnified, and many think that other synthetic regulators exert their influence through an effect on the natural plant ethylene. Extremely small amounts of ethylene influence a wide variety of plant growth and development functions.[36] It is used commercially in controlled environments, and ethylene-generating compounds such as ethephon (Figure 1) and absorbent powders impregnated with ethylene[34] are used in the field to: promote uniform ripening, accelerate ripening, induce flowering, reduce stalk length, and increase bulb development.

Brassins[37] are a new group of steroidal substances with unique growth responses which suggest a separate grouping from the five established classes. With time the list of PGRs will continue to grow and perhaps even more classes will be defined. Each new PGR discovery increases the opportunity for plant-growth manipulation.

3. The Controlled Release Concept

The effectiveness of existing compounds can be improved by precise delivery with less environmental impact through the use of controlled release technology. Controlled release is defined as a combination of biologically active agent and polymeric material arranged to allow delivery of the agent to the target at controlled rates over a specified period of time. Controlled release compares favorably to conventional application methods. Controlled release provides greater effectiveness, economy, and safety, and minimum damage to target organisms and the environment. Other advantages have been reported.[38–53]

The principles of controlled release have been applied to several biologically active materials including pesticides, fertilizers, and pharmaceuticals; several formulations have been developed and tested. However, an area which has received little attention is plant-growth regulators; this is reflected by the number of publications reported.[59–67] The difficulties in predicting the optimum time of application and the very low concentrations of PGRs can be overcome by the use of controlled delivery systems. These formulations would regulate duration, maintain the required active amount in the plant, and minimize side effects from high concentrations. Such systems are classified into two categories: chemical and physical methods.

3.1. Physical Methods

These methods include the dissolving, dispersing, and encapsulating of the active agent which is released by diffusion through an inert matrix.

3.1.1. Dissolved Systems

The active agent is dissolved throughout the polymer matrix. The released amount increases with the square root of time, and the release rate initially falls inversely with the square root of time and eventually follows an exponential function.[54] Dissolved systems are easy to manipulate and are inexpensive; but low loading must be used[55] because permeability in the composite system is excessive with large amounts of active agent. Correspondingly, the desorption of the agent through the polymeric matrix is too fast.[54]

3.1.2. Dispersed Systems

Particles of the active agent are dispersed in a solid polymeric matrix which can be either biodegradable or nonbiodegradable. The matrix controls the release of the active agent by diffusion, erosion, or by a combination of both diffusion and erosion. Dispersed systems are inexpensive and easy to manipulate. The release depends on the geometry of the system and the polymer material used as the matrix.[55] The amount released in these systems is proportional to the square root of time as long as the concentration of the active agent present is higher than the solubility of the agent in the matrix.[56]

3.1.3. Encapsulated Systems

In this method, the active agent is totally encapsulated within a rate-controlling membrane device. The thermodynamic activity or driving force of the active agent from the device has to be maintained constant to achieve a constant level of active ingredient within the capsule over a prolonged period of time.[57,58] The release rate does not greatly depend on the shape of the system, but primarily depends on the wall thickness and the permeability of the polymeric film.

3.1.4. Laminated Systems

These consist of at least two polymeric films adhered or laminated together. The middle layer of the laminated structure contains a large amount of active ingredient and represents a reservoir layer. The release rate of these systems is similar to the encapsulated models.

3.2. Chemical Methods

Chemical combinations attach the active agent to the polymeric substrate by a chemical bond. The attachment can take any of the several

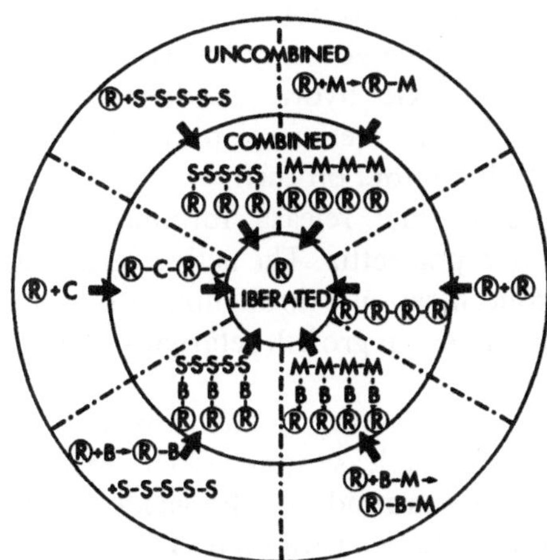

Figure 2. Synthesis of chemically bonded controlled release systems.

forms as illustrated in Figure 2, where R, S, B, M, and C, respectively, denote plant-growth regulator, polymeric substrate, bridging entity, monomer, and comonomer. The effectiveness of these bioactive polymer combinations depends on the release rate of the active agent which is controlled by:

1. The nature and stereochemistry of the bioactive-polymer bond.
2. The chemical characteristics of the monomers and comonomers.
3. The structure and dimension of the polymer substrate governed by the degree of polymerization and cross-linking.
4. Any other factors influencing the hydrophobicity of the system.

4. The Present State of the Art of Controlled Release Plant-Growth Regulators

The last 20 years have seen tremendous progress in the area of controlled release technology; notable products include time-release cold capsules, long-lasting flea collars, antifouling coatings, and time-release fertilizers. However, very few controlled release forms of plant-growth regulators have been developed.

Read *et al.*[(59)] reported that the microencapsulation of chlormequat stimulated and retarded the growth of tomatoes, petunias, snapdragons, and marigolds. At low rates, early blooming is stimulated; conversely, at higher levels of encapsulated chlormequat, plants are retarded by 50% when compared to controls.

Chemical linkages of cytokinins to starch and cellulose have been prepared by Bittner *et al.*[(60)] The hydrolysis of the carbamate bond is the rate-determining step, and the polymer plant-growth regulator combination releases the free cytokinin at a very low concentration over an extended period of time. This was monitored by using a soybean callus assay.

Frederick and Leonard[(61)] used a maleic hydrazide derivative in combination with methyl methacrylate ethyl methacrylate–acrylic acid copolymers as slow release plant-growth inhibitors on turf.

Einert *et al.*[(62)] applied a granular slow release formulation of ancymidol impregnated with clay to potted poinsettia. The active agent was available to the plant at very low concentrations throughout its entire growing period and thus eliminated unevenness in growth retardation of the plant stem.

Simulation of repeated treatments of active ingredients was used by Wilkins[(63)] to evaluate the viability of controlled release systems. This method suggested that a controlled release form of benzimidazole fungicides could control senescence in wheat. Sodium azide increased germination of wild-oat seed when delivered by impregnated clay granules coated with polyethylene glycol.[64] In most soils, leaching and dissipation of the sodium azide was reduced by the coating formulation.

The University of Washington Fiber and Polymer Science Group has been involved with controlled release technology for more than two decades.[38–43,49–51,65–72] The controlled release combinations which received particular attention are the auxin herbicides 2,4-dichlorophenoxyacetic acid (2,4-D) and 2,4-dichlorophenoxybutyric acid (2,4-DB) and their reaction products with natural and synthetic polymers.[50,72] The initial purpose of this research was to develop a controlled release herbicide delivery system for use in conifer regeneration with highly productive timber species such as Douglas fir and loblolly pine.

Selective herbicides can eliminate competitive weeds to ensure the seedling receives the necesary light, water, and nutrients for better growth. The herbicide 2,4-D and 2,4-DB combination with biodegradable polymer is toxic to weeds and nontoxic to the seedlings. Successful field trials were conducted on Douglas fir seedlings in the Pacific Northwest and loblolly pine

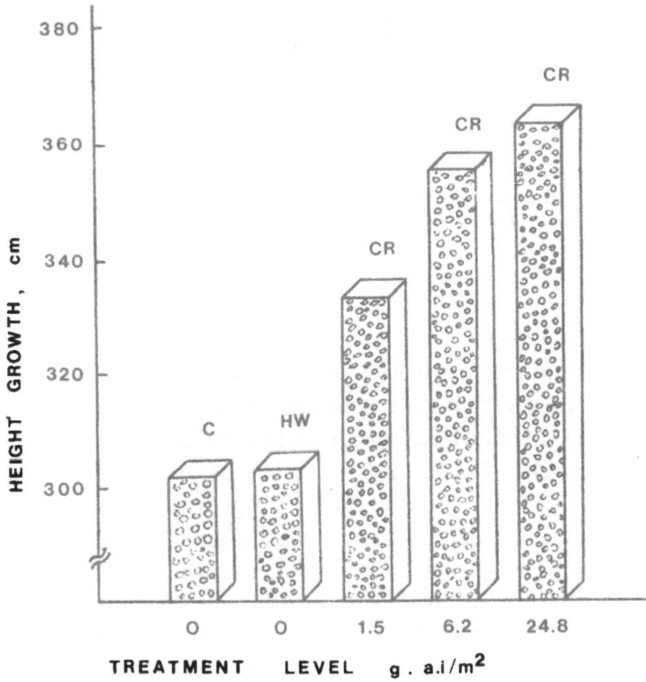

Figure 3. Height growth of loblolly pine seedlings treated with controlled release herbicide (C = control, HW = handweeded controls, CR = controlled-release formulations).

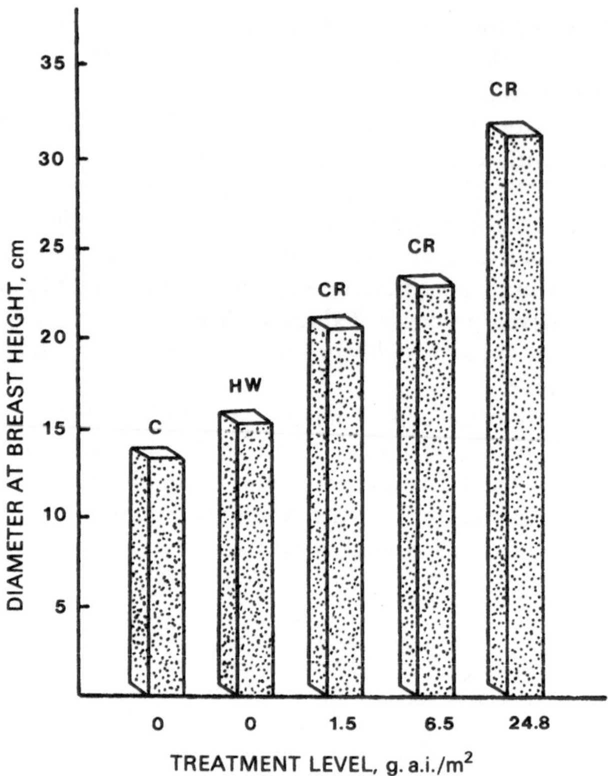

Figure 4. Diameter at breast height of loblolly pine seedlings treated with controlled release herbicide (C = control, HW = handweeded controls, CR = controlled release formulations).

seedlings in Mississippi. Trees treated with various levels of the controlled release 2,4-D combination were compared with trees which had no treatment at all and trees which were maintained weedfree. The weedfree controls were maintained by handweeding around the loblolly pine and by polyolefin film around the Douglas fir. For a period of six years the heights of all seedlings were recorded at the end of each growing season. The controlled release auxin–herbicide combination was found to keep vegetation away from the seedlings; moreover the apical stem of these treated trees was much elongated when compared to both controls. The seedlings kept weedfree by handweeding and polyolefin film were shorter and less bushy than the auxin–herbicide-treated seedlings. In addition to this height increase (Figure 3), the diameter at breast height was increased by 21% (Figure 4). This demonstrates that the slow release of the auxin herbicide provides seedlings with a metabolic stimulus over a period of years.

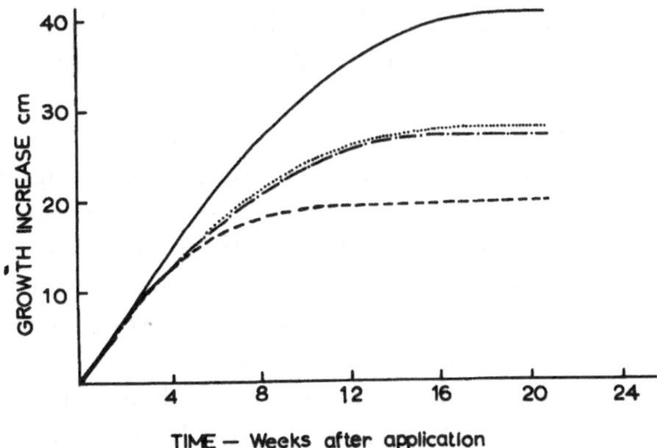

Figure 5. Growth-history curves of Douglas fir seedlings for 22 weeks after four different treatments where solid line (——) represents seedlings treated with controlled release—2,4-D triamido-*s*-triazine granules; dashed line (– – –) represents seedlings with no treatments; dot-dashed line (–·–·–) represents seedlings treated with controlled-release triamido-*s*-triazine hexazinone granules; and dotted line (· · ·) represents seedlings treated with an equivalent amount of free 2,4-D.

Further research led in 1982 to the development of a controlled release granule, 2,4-D triamido-*s*-triazine. This system was applied to the soil surface of potted Douglas fir seedlings.[73] The tree growth and the behavior of competing weeds were recorded and compared against three other systems:

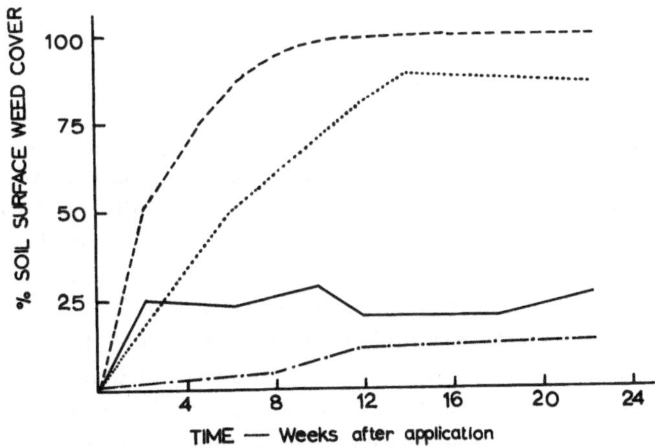

Figure 6. Growth development of weeds represented as a percentage of soil surface cover after treatments where dashed line (– – –) represents no treatment; solid line (——) represents soil surface treatment with controlled-release 2,4-D triamido-*s*-triazine granules; dot-dashed line (–·–·–) represents a soil treatment with controlled-release triamido-*s*-triazine hexazinone granules; and dotted line (· · ·) represents soil treatment with an equivalent amount of free 2,4-D.

(A) seedlings with no treatment, (B) seedlings treated with an equivalent amount of free 2,4-D, (C) another controlled release granule, triamido-*s*-triazine hexazinone (this herbicide has no auxinlike effect).

Figures 5 and 6 show growth and weed-control results of these four treatments. The controlled release 2,4-D system achieved significantly more growth than did the three other treatments. This result reflects the sustained regular delivery of the controlled release method combined with the auxin and herbicide effect of the 2,4-D. A notable result is that the controlled release hexazinone achieved the best herbicidal results, but the controlled release auxin (2,4-D) achieved much better tree growth. This suggests that plant-growth stimulation is not merely a matter of controlling competitive plants. Significant additional growth accompanies the use of plant-growth regulators.

5. The Future

Prediction of future technology requires recognition of the inherent socioeconomic pressures which lead to research investment. For example, grain shortages in the world, notably the USSR and India, can be expected to spur advances in grain production. Equally obvious is that the overall pressure of world resource demand will provide a broad stimulus to plant-growth research. The following sections identify those areas of PGR and controlled release systems which are likely candidates for research and development activity.

5.1. Long-life Crops

Crops grown on a continuing basis over many years include forests, orchards, rubber plantations, biomass farms, and grasslands. All have long productive lives. Growth-promoting PGRs are presently used to improve pasture tillerage and orchard productivity, and to promote latex flow in rubber trees. Repeated applications of GA_3 and $GA_{4/7}$ promote flowering in western red cedar, coastal cedar, and giant redwoods.[74,75] Anjou pears if treated for two years with a suitable PGR, mature two to three years earlier than normal and give an increase in fruit set and a better-quality fruit.[76] Two to three applications of GA_3 are used to set and accelerate fruit to maturity and also produce better berries in Delaware, Thompson seedless, and Perlette grapes.[77] Controlled release methods are ideally suited for use with these and other repetitive PGR doses in order to maintain effective applications over extended periods, up to several years for certain long-term crops.

Retardants are also used in permanent plantings, on lawns for reduced mowing,[78] in hedges, roadside plantings, and trees under power lines for limited maintenance. Inhibitors are used for sucker control in tobacco and orchards. To exend the continuity of these treatments would be advantageous both from the viewpoint of maintained control and by reducing energy and materials required for repeated applications.[79] The limited manner of delivery in controlled release systems should reduce some of the aesthetically undesirable foliar damage. It is often unattractive side effects which curtail the use of otherwise important aids.[3] Understanding and solving these unwanted side effects will be of primary research focus if PGRs are to be put to full use.

5.2. Seasonal Crops

Wherever an effect is sought which is longer than an immediate response to a one-shot application, then a controlled release technique is called for. These systems can be designed to fit the length of time involved, whether it be the whole or some part of the plant growing season. Gibberellins are used in sugar cane and cereal crops to increase tillering and thus the annual yield. A maximum response is produced by repeated applications.[80] Chlormequat (CCC) used repeatedly is effective in shortening the stem in wheat and oats and thus preventing lodging at harvest. Multiple sprays of chlorphonium and ancymidol[62,81,82] shorten the stem in potted ornamentals to produce desirable compact floral plants. Repeated spraying with an inhibitor as a pinching agent throughout the growing season controls the development of undesirable axillary buds in tomato plants. Cytokinins are used to lengthen the active life of a plant, delaying senescence and thus increasing the harvest period. These multiple applications could be accomplished in one efficient controlled release technique.

5.3. Environmental Vigor

One of the most exciting possibilities concerning growth regulators is that of adding vigor to a plant's metabolism, thus endowing the plant with an inner vitality to resist drought, salinity, temperature variations, disease, pest attack, and herbicide damage.[83] It would enable the modeling of a plant to fit an environment, growing plants beyond present natural restraints of latitude and season. Such growth regulators (abscisic acid is one possibility) would have to be reliably delivered throughout the life of the plant or during the principal periods of stress.

5.4. Improving Herbicide and Fertilizer Efficiency

It has been shown that the use of PGRs in conjunction with herbicides and fertilizers enhances the efficiency of these chemicals by improving the absorption and translocation. For these relationships to be most successful, the different active agents should be controlled to release throughout the same time period. A further development of this beneficial partnership could be expanded to involve a multiple of plant-growth enhancers. All the chemical requirements necessary for the plant (fertilizer, herbicide, PGRs, fungicide, insecticide) could be built into a total controlled release system thus saving on application energy and supplies. This concept is being approached at the University of Washington where fertilizer, herbicide, and PGR are delivered all in one granule.[73]

5.5. Promotion of Early Germination

PGRs can promote early seed germination. This applies to weed seed which could be treated in the fall immediately after harvest with PGRs in a controlled release system. Early the following year, this would initiate premature germination of the dormant weed seed which when established could be eradicated with a herbicide prior to planting the crop seed.

5.6. Protection from Degradation

Many PGRs are very unstable, subject to high water solubility, photodegradation,[84] heat sensitivity, and rapid decomposition from soil bacteria. As a result, the PGR material is often destroyed before it can be absorbed by the plant. In order to overcome these difficulties, large and repeated applications are made. An appropriate controlled release formulation will stabilize and improve delivery and so protect against degrading forces.

Potential growth regulators for use in controlled delivery systems are summarized in Table 1.

6. Conclusions

Plant-growth regulators help solve many crop-production problems. However, despite these promising advantages, their uses have been very limited, particularly when compared with other agricultural chemicals. The reasons are many; PGRs are expensive as are the costs for registration and marketing. Above all they are extremely difficult to evaluate and other alternate methods for plant management have been available. Now is the time for their ascendancy to an increasingly important role in plant cultivation. We have much better understanding of plant physiology plus a large and growing inventory of plant-growth regulator knowledge; add to this controlled release technology which has all the unique characteristics necessary for delivering PGRs in minute quantities over defined time periods. The partnership is ideal and growing pressures of world food demand will necessitate research and development of such systems which have the potential for satisfying man's increasing needs.

Table 1. Potential Use of Plant-Growth Regulators

Common name	Chemical name	PGR activity	References
Abscisic acid	5-(1-Hydroxy-2,6,6-trimethyl-4-oxo-2-cyclohexenyl) 3-methyl-2,4-pentadienoic acid	Growth inhibitor, defoliant	85, 86
Ancymidol	-Cyclopropyl-(4 methoxy-phenyl) 5-pyrimidine methanol	Growth retardant	62, 81, 82
Benzyladenine	6-Benzylaminopurine	Senescence delayer, fruit-shape changer, sex changer	60, 86, 87
BNOA	2-Naphthoxyacetic acid	Fruit set enhancer, disease controller	88, 89
Chlorflurecolmethyl	Methyl-2-chloro-9-hydroxy-fluorene-9-carboxylate	Growth retardant, fruit set enhancer, insect controller	91, 92
Chlormequat	2-Chloromethyl trimethyl-ammonium chloride	Germination inhibitor, senescence delayer, growth retardant, fruit set enhancer, stimulator of latex flow in rubber, antitransparent	59, 93–101
Chlorphonium	Tributyl-2,4-(dichlorobenzyl) phosphonium	Flower-bud stimulator, growth retardation	95, 102
2,4-D	2,4-Dichlorophenoxyacetic acid	Herbicide, latex stimulator, disease controller, insect controller, senescence delayer, fruit set enhancer, flowering inducer	71, 72, 103–108
Daminozide	Succinic acid-2,2-dimethyl hydrazine	Growth retardant, senescence delayer, multiflower stimulant,	109–121

(continued)

Table 1. (*Continued*)

Common name	Chemical name	PGR activity	References
		disease controller, flowering delayer, fruit set enhancer, antitranspirent	
DeKegulac	2,3,4,6-bis-(1-Methyl-ethylidene)-*O*-(L)-xylo-2-hexulo furanosonic acid, sodium salt	Growth retardant, chemical pruner	122–125
Dinoseb	2,4-Dinitro-6-*sec*-butylphenol	Herbicide, corn-yield enhancer	126–129
Ethephon	2-Chloroethyl phosphonic acid	Latex stimulator, growth regulator, color enhancer, flowering inducer, gametocide	130–145
Ethene	Ethylene	Flowering inducer, latex stimulator, defoliant	71, 72, 146, 147
Gibberellic acid	2,4a,7-Trihydroxy-1-methyl-8-methylene gibb-3-ene-1,10-carboxylic acid-1,4 lactone	Seedless grape, enlarger, senescence delayer, amylase stimulant, germination enhancer, shoot growth stimulant, insect and disease controller, ripening delayer	148–162
IAA	3-Indoleacetic acid	Plant-cell enlarger, disease controller fruit-ripening inhibitor	89, 163, 164
IBA	3-Indolebutyric acid	Fruit set stimulant, root inducer	165–168
Kinetin	6-Furfurylaminopurine	Germination stimulant, dormancy breaker, disease controller	60, 169, 170
Maleic hydrazide	1,2-Dihydro-3,6-pyridazinedione	Growth inhibitor, flowering inhibitor, insect controller, turf growth inhibitor	171–177
Mefluidide	*N*-2,4-dimethyl-5-(trifluoromethyl)-sulfonyl aminophenyl acetamide	Growth retardant, herbicide, turf growth inhibitor	178–182
NAA	1-Naphthaleneacetic acid	Latex stimulator, fruit inducer, root inducer, fruit set enhancer, disease controller, senescence delayer	88, 89, 145, 183–185
Triacontal	1-Hydroxytriacontane	Growth regulator	186–190

ACKNOWLEDGEMENTS

We thank Professor G. G. Allan in whose department the University of Washington research was carried out.

References

1. L. R. Brown, *Science* **214**, 995 (1981).
2. United States Departments of Agriculture, World Food Aid Needs and Availabilities, Economic Research Service (April 1982), p. 10.
3. R. P. Freeborg, *Weeds, Trees and Turf* **18**(8), 27 (1979).
4. R. E. Wagner, *Solutions* **26**(5), 16 (1982).
5. J. Bronowski, *The Ascent of Man*, Little, Brown, Boston (1973).
6. *Farm Chem.* **138**(3), 15 (1975).
7. C. Darwin, *The Power of Movement in Plants*, Murray, London (1880).
8. P. F. Wareing, *Outlook on Agric.* **9**(2), 42 (1976).
9. R. W. F. Hardy, in: *Plant Regulation and World Agriculture* (T. K. Scott, ed.), p. 165, Plenum Press, New York (1979).
10. L. G. Nickell, *Chem. Eng. News.* **56**(41), 18 (1978).
11. T. Ciesielski, *Beitr. Biol. Pflanz.* (*Cohn*) **1**(2), 1 (1872).
12. W. Pfeffer, *Planzenphysiologie: Ein Handbuch der Lehre vom Stoffwechsel und Kraftwechsel in der Pflanze. Bd II: Kraftwechsel*, Engelmann, Leipzig (1881, 2nd edition 1904).
13. W. Rolhert, *Beitr. Biol. Pflanz.* **7**, 1 (1894).
14. H. Fitting, *Jahrb. Wiss. Bot.* **44**, 177 (1907).
15. P. Boysen-Jensen, *Dan Vidensk. Selsk.* **3**, 1 (1911).
16. A. Paal, *Jahrb. Wiss. Bot.* **58**, 406 (1918).
17. F. W. Went, *Rec. Trav. Bot. Neerl.* **25**, 1 (1928).
18. F. Kögl, A. J. Haagen-Smit, and H. Erxleben, *Mitteilung. 2. Physiol. Chem.* **228**, 90 (1934).
19. M. Bevan and D. H. Northcote, *Planta* **152**, 32 (1981).
20. J. B. Zaerr, *Forest Sci.* **13**(3), 258 (1967).
21. E. Kurosawa, *Trans. Natl. Hist. Soc. Formosa* **16**, 213 (1926).
22. T. Yabuta, *Agric. Hortic.* **10**, 17 (1935).
23. R. P. Pharis and S. D. Ross, *Outlook on Agric.* **9**(2), 82 (1976).
24. N. C. Wheeler, R. L. Wample, and R. P. Pharis, *Physiol. Plant* **50**, 340 (1980).
25. F. G. Dennis, L. J. Edgerton, and K. G. Parker, *Hortscience* **5**, 158 (1970).
26. C. O. Miller, F. Skoog, F. S. Okhumura, M. H. Von Saltza, and F. M. Strong, *J. Am. Chem. Soc.* **78**, 1375 (1956).
27. D. S. Letham, *Life Sci.* **8**, 569 (1963b).
28. T. Yokoto and N. Takahashi, *Phytochemistry* **19**, 2367 (1980).
29. T. H. Thomas, *Outlook on Agric.* **9**(2), 62 (1967).
30. *Agrichem. Age* **22**(2), 38 (1979).
31. F. T. Addicott, in: *Handbuch der Pflanzenphysiologie* (W. Ruhland, ed.), Vol. XV, No. 1, p. 1094, Springer-Verlag, Berlin (1965).
32. A. A. Khan and C. M. Karssen, *Plant Physiol.* **66**, 175 (1980).
33. D. M. Fabijan, P. Dhindas, and D. Reid, *Planta* **152**(6), 481 (1981).
34. N. Jain and R. B. R. Yadava, *Agrochemica* **25**(1), 81 (1981).
35. H. K. Pratt and J. D. Goeschl, *Ann. Rev. Plant. Physiol.* **20**, 541 (1969).
36. P. B. Dickenson, *Outlook on Agric.* **9**(2), 88 (1979).

37. C. A. West, in: *Plant Growth Substances 1979* (F. Skoog, ed.), p. 289 Springer-Verlag, New York (1980).
38. G. G. Allan, Belgian Patent 705,509 (December 15, 1967); U.S. Patent 3,813,234 (May 28, 1974); French Patent 1,544,406 (1968).
39. G. G. Allan, C. S. Chopra, A. N. Neogi, and R. M. Wilkins, *Int. Pest Control.* **13**, 10 (1971).
40. G. G. Allan, C. S. Chopra, A. N. Neogi, and R. M. Wilkins, *TAPPI* **54**, 1293 (1971).
41. G. G. Allan, C. S. Chopra, A. N. Neogi, and R. M. Wilkins, *Nature* **234**, 349 (1971).
42. G. G. Allan, C. S. Chopra, and R. M. Russell, *Int. Pest Control* **14**(2), 15 (1972).
43. G. G. Allan, C. S. Chopra, J. F. Friedhoff, R. I. Gara, M. W. Maggi, A. N. Neogi, S. C. Roberts, and R. M. Wilkins, *CHEMTECH* **4**, 171 (1973).
44. N. F. Cardarelli, *CHEMTECH*, 482 (1975).
45. R. L. Goulding, ed., *Proceedings, 1977 International Controlled Release Pesticide Symposium*, Oregon State University, Corvallis (1977).
46. F. W. Harris, ed., *Proceedings, 1975 International Controlled Release Pesticide Symposium*, Wright State University, Dayton, Ohio (1975).
47. N. F. Cardarelli, ed., *Proceedings of the Controlled Release Pesticide Symposium*, University of Akron, Ohio (1974).
48. G. G. Allan, Canadian Patent 855,181 (November 3, 1970).
49. G. G. Allan, C. S. Chopra, M. W. Maggi, A. N. Neogi, and R. M. Wilkins, *Int. Pest Control* **15**(3), 8 (1973).
50. A. N. Neogi and G. G. Allan, in: *Controlled Release Pesticides* (A. C. Tanquary and R. E. Lacey, eds.), p. 195, Plenum Press, New York (1974).
51. G. G. Allan, J. W. Beer, M. J. Cousin, and J. C. Powell, *TAPPI* **61**, 33 (1978).
52. H. B. Scher, ed., *Controlled Release Pesticides*, American Chemical Society, ACS Symposium Series 53, Washington, D.C. (1977).
53. A. F. Kydonieus, ed. *Controlled Release Technologies: Methods, Theory and Applications*, Vols. 1 and 2, CRC Press, Fla. (1980).
54. R. W. Baker and H. K. Lonsdale, in: *Controlled Release of Bioactive Agents* (A. C. Tanquary and R. E. Lacey, eds.), p. 15, Plenum Press, New York (1974).
55. R. W. Baker and H. K. Lonsdale, in: *Proceedings, 1975 International Controlled Release Pesticide Symposium* (F. W. Harris, ed.), p. 9, Wright State University, Dayton, Ohio (1975).
56. A. F. Kydonieus, in: *Controlled Release Technologies: Methods, Theories and Applications* (A. F. Kydonieus, ed.), p. 1, CRC Press, Boca Raton, Fla. (1980).
57. J. E. Vandegaer, *Microencapsulation, Processes and Applications*, Plenum Press, New York (1973).
58. G. O. Fanger, *CHEMTECH* **6**, 397 (1974).
59. P. E. Read, V. L. Herman, and D. A. Heng, *Hortscience* **9**(1), 55 (1974).
60. S. Bittner, I. Perry, and Y. Knobler, *Phytochemistry* **16**, 305 (1977).
61. D. C. Frederick and E. G. Leonard, Belgian Patent 861,231 (March 16, 1978).
62. A. E. Einert, T. Pappas, and C. E. Williamson, *Arkansas Farm Res.* **27**, 3 (1978).
63. R. M. Wilkins, in: *Controlled Release of Bioactive Materials* (R. W. Baker, ed.), p. 343, Academic Press, New York (1980).
64. P. K. Fay, R. S. Gorecki, and P. M. Fuerst, *Weed Sci.* **29**(6), 674 (1980).
65. G. G. Allan, R. I. Gara, and R. M. Wilkins, *Int. Pest Control* **16**(4), 4 (1974).
66. G. G. Allan, J. F. Friedhoff, and J. C. Powell, *Int. Pest Control* **17**(2), 4 (1975).
67. G. G. Allan, J. W. Beer, and M. J. Cousin, *Int. Pest Control* **20**(2), 6 (1978).
68. G. G. Allan, M. J. Cousin, W. J. McConnell, J. C. Powell, and A. Yahiaoui, in Preprints, Chemical Marketing and Economics Division, American Chemical Society, San Francisco Meeting, p. 186 (1976).

69. G. G. Allan, M. J. Cousin, W. J. McConnell, J. C. Powell, and A. Yahiaoui, *Proceedings, Third International Controlled Release Pesticide Symposium*, University of Akron, Ohio (1976), p. 7.15.
70. G. G. Allan, J. W. Beer, M. J. Cousin, W. J. McConnell, J. C. Powell, and A. Yahiaoui, in: *Polymeric Drugs* (L. G. Donaruma and O. Vogl, eds.) p. 185, Academic Press, New York (1978).
71. G. G. Allan, R. A. Mikels, and M. J. Cousin, *ACS Polym. Prepr.* **20**, 341 (1979).
72. G. G. Allan, J. W. Beer, M. J. Cousin, and R. A. Mikels, in: *Controlled Release Technologies: Methods, Theories and Applications* (A. F. Kydonieus, ed.), Vol. 1, p. 7, CRC Press, Fla. (1980).
73. J. M. Miller (unpublished).
74. R. P. Pharis and C. G. Kuo, *Can. J. Forest Res.* **7**, 299 (1977).
75. R. P. Pharis and W. Morf, *Bioscience* **19**, 719 (1969).
76. E. A. Stahly and M. W. Williams, *Hortic. Science* **11**(5), 502 (1976).
77. L. Rappaport, *Plant Physiol.* **32**, 440 (1957).
78. T. L. Watschke, *Weeds, Trees and Turf* **18**(8), 32 (1979).
79. *Weeds, Trees and Turf* **18**(2), 60 (1979).
80. L. G. Nickell, in: *Plant Growth Substances 1979* (F. Skoog, ed.), p. 419, Springer-Verlag, New York (1980).
81. G. J. Wilfret, *Proc. of the Fl. State Hortic.* **91**, 220 (1978).
82. D. M. Elkins, J. W. Vandeventer, and M. A. Briskovich, *Agron. J.* **69**, 458 (1977).
83. S. D. Miller and J. D. Nalewaja, *Agron. J.* **72**(4), 662 (1980).
84. D. G. Crosby and C. S. Tang, *J. Agric. Food Chem.* **17**(6), 1291 (1969).
85. R. L. Wain, in: *Plant Regulation and World Agriculture* (T. K. Scott, ed.), p. 155, Plenum Press, New York (1979).
86. G. E. Tolla and C. E. Peterson, *Hortscience* **14**, 542 (1979).
87. R. H. Williams and P. M. Cartwright, *Ann. Botany* **46**, 445 (1980).
88. L. D. Nooden, G. M. Kahansk, and Y. Okatan, *Science* **206**, 841 (1979).
89. D. Davis and A. E. Dimond, *Phytopathology* **43**, 137 (1953).
90. G. M. Olympios, *Hortic. Res.* **16**, 65 (1976).
91. G. Schneider and G. Mohr, *Proc. Brit. Weed Control Conf.* **1**, 292 (1970).
92. L. M. Stahler and G. K. Harris, *Proc. Brit. Weed Control Conf.* **1**, 286 (1970).
93. J. Jung, in: *Plant Regulation and World Agriculture* (T. K. Scott, ed.), p. 273, Plenum Press, New York (1979).
94. Technical Brochure. "Cycocel™, Plant growth regulant," Cyanamid International, Wayne, New Jersey (1966).
95. A. M. M. Berrie and J. Robertson, *Plant Physiol.* **28**(2), 278 (1973).
96. N. E. Looney, *Can. J. Plant Sci.* **55**, 117 (1975).
97. E. C. Humphries, *Field Crop Abst.* **21**, 91 (1968).
98. E. C. Humphries, P. J. Welbank, and K. J. Witts, *Ann. Appl. Biol.* **56**, 351 (1965).
99. L. N. Dolgopolova and A. P. Lakhnov, *Khim Sel'sk Khoz* **17**(9), 27 (1979).
100. Y. Erner, R. Goren, and S. P. Monselise, *J. Hortscience* **51**, 367 (1976).
101. Y. Erner, R. Goren, and S. P. Monselise, *Am. Soc. Hortscience J.* **101**, 513 (1976).
102. P. C. Marth, *Proc. Am. Soc. Hortic. Sci.* **83**, 777 (1963).
103. F. B. Aboles, *Ethylene in Plant Biology*, Academic Press, New York (1973).
104. J. B. Adams, *Can. J. Zoology* **38**, 285 (1960).
105. R. C. Maxwell and R. F. Harwood, *Bull. Entomol. Soc. Am.* **4**, 100 (1958).
106. R. C. Maxwell and R. F. Harwood, *Ann. Entomol. Soc. Am.* **53**, 199 (1960).
107. C. W. Coggins, *Acta Hortic.* **34**, 469 (1972).
108. O. L. Hoffman, *Plant Physiol.* **28**, 622 (1953).
109. E. Baltazzi, U.S. Patent 3,343,941 (1967).

110. B. R. Roberts, *J. Arboriculture* **6**(3), 57 (1980).
111. D. Laycock and D. Tyson, *Proc. Brit. Weed Control Conf.* **1**, 244 (1970).
112. E. L. Bergman, *Hortscience* **1**, 53 (1966).
113. T. H. Thomas, I. E. Currah, and P. G. Salter, *Ann. Appl. Biol.* **75**, 63 (1973).
114. D. T. Sullivan and F. B. Widmoyer, *Hortscience* **5**, 91 (1970).
115. L. D. Tukey and H. K. Flemming, *Proceedings American Society of Hortscience*, Vol. **93**, p. 300 (1968).
116. D. Mishra and G. C. Pradhan, *Plant Physiol.* **50**, 271 (1972).
117. L. P. Batjer, M. W. Williams, and G. C. Martin, *Proceedings American Society of Hortscience*, Vol. **85**, p. 11 (1964).
118. N. E. Looney, *Acta Hortic.* **34**, 397 (1973).
119. N. E. Looney, *Can. J. Plant Sci.* **49**, 625 (1969).
120. R. C. Rom and K. R. Scott, *Hortscience* **6**, 134 (1971).
121. R. E. Byers and F. H. Emerson, *Hortscience* **8**, 48 (1973).
122. Dr. R. Maag Ltd., Technical Data Sheet, Atrinal Plant Growth Regulator (1974).
123. P. F. Bocion, W. H. De Silva, G. A. Huppi, and W. Szkrybalo, *Nature* **258**, 142 (1975).
124. R. M. Sachs, H. Hield, and J. De Bie, *Hortscience* **10**, 367 (1975).
125. W. H. De Silva, P. F. Bocion, and H. R. Walther, *Hortscience* **11**, 569 (1976).
126. *Agrichem. Age* **20**, 11 (1977).
127. *Agrichem. Age* **21**, 14 (1978).
128. *Agrichem. Age* **22**, 10 (1979).
129. O. E. Hathley, L. Herman, K. Collins, and A. J. Ohlrogge, *Down to Earth* **30**(1), 4 (1974).
130. W. L. Sims, R. E. Toss, and V. E. Rubatzky, *Study Guide for Plant Growth Regulators*, p. 47, University of California at Davis (1973).
131. D. E. Knavel and T. R. Kemp, *Hortscience* **8**, 403 (1973).
132. D. R. Tompkins and J. L. Bowers, *Hortscience* **5**, 84 (1970).
133. D. Levy, N. Kedar, and R. Karacinque, *Hortscience* **8**, 228 (1973).
134. A. Apelbaum and S. P. Burg, *Plant Physiol.* **50**, 117 (1972).
135. P. D. Abraham, J. B. Gomez, W. A. Southorn, and P. R. Wycherly, British Patent 1,281,524.
136. P. B. Dickanson, British Patent 1,315,131.
137. A. R. Cooke and D. I. Randall, *Nature* **218**, 974 (1968).
138. C. M. Brown and E. B. Early, *Agron. J.* **65**, 829 (1973).
139. R. L. Lower and C. H. Miller, *Nature* **222**, 1072 (1969).
140. R. W. Buescher, *Hortscience* **12**, 315 (1977).
141. R. W. Buescher and J. H. Doherthy, *J. Food Sci.* **43**, 1816 (1978).
142. Z. Worku and R. C. Harner, *Hortscience* **6**, 279 (1971).
143. R. M. Nakayama and F. B. Maha, *Hortscience* **8**, 252 (1973).
144. F. Jensens and H. Andris, *California Agric.* **31**(8), 18 (1977).
145. P. B. Dickenson, *Outlook on Agric.* **9**, 88 (1976).
146. A. B. Rodriquez, Department of Agriculture, Peurto Rico, **16**, 1 (1932).
147. A. E. Hitcock and P. W. Zimmerman, *Contrib. Boyce Thompson Inst.* **10**, 461 (1940).
148. J. N. Turner, *Acta Hortic.* **34**, 287 (1973).
149. D. R. Tompkins, *Abst. PGR Working Group* **6** (1975).
150. J. D. Fryer and R. J. Makepeace, *Weed Control Handbook*, 7th Edition, Vol. 2, Philadelphia, Lippincott (1972).
151. R. P. Pharis, *Modern Methods in Forest Genetics*, Springer, Berlin (1976).
152. H. Ikuma and K. V. Thimann, *Plant Physiol.* **35**, 557 (1960).
153. R. J. Weaver, *Nature* **183**, 1198 (1959).
154. P. W. Brian, J. H. P. Petty, and P. T. Richmond, *Nature* **184**, 69 (1959).
155. R. J. Henry, *Hortscience* **15**, 613 (1980).

156. S. D. Ross, *Acta Hortic.* **56**, 163 (1976).
157. P. B. Tompsett, *Ann. Botany* **41**, 1171 (1977).
158. P. M. Nelson and E. C. Rossman, *Science* **127**, 1500 (1958).
159. R. J. Cibulasky and G. M. Greene, *Proceedings Plant Growth Regulator Working Group*, Vol. 6, p. 180 (1979).
160. L. G. Nickell, *Chemical Growth Regulation in Sugar Cane, Outlook on Agriculture*, Vol. 9, p. 57 (1976).
161. J. C. Rogriguez and J. M. Campbell, *J. Econ. Entomol.* **54**, 984 (1960).
162. C. W. Coggins, *Acta Hortic.* **34**, 469 (1972).
163. Z. Erkan and F. Bangerth, *Angew. Bot.* **54**, 207 (1980).
164. R. M. Smock, L. J. Edgerton, and M. B. Hoffman, *Proceedings American Society of Horticultural Science*, Vol. 60, p. 184 (1952).
165. G. S. Avery, Jr. and E. F. Johnson, *Hormones and Horticulture*, McGraw-Hill, New York (1947).
166. H. T. Hartman and D. E. Kester, *Plant Propagation: Principles and Practices*, Prentice-Hall, Englewood Cliffs, New Jersey (1968).
167. A. C. Leopold, *Auxins and Plant Growth*, University of California Press, Berkeley and Los Angeles (1955).
168. R. J. Weaver, *Plant Growth Substances in Agriculture*, Freeman, San Francisco (1972).
169. C. O. Miller, *Plant Physiol.* **31**, 318 (1956).
170. T. H. Oswald and T. D. Wyllie, *Plant Disease Reporter* **57**, 789 (1973).
171. B. R. Roberts, *J. Arboriculture* **6**, 57 (1980).
172. S. H. Wittwer and R. C. Sharma, *Science* **112**, 597 (1950).
173. R. H. Moore, *Science* **112**, 52 (1950).
174. D. M. Elkins and D. L. Suttner, *Agron. J.* **66**, 487 (1974).
175. T. L. Watschke, F. W. Long, and J. M. Duich, *Weed Sci.* **27**, 224 (1979).
176. D. M. Elkins, *Weeds, Trees and Turf* **11**(5), 18 (1972).
177. A. G. Robinson, *Can. J. Plant Sci.* **41**, 413 (1961).
178. D. M. Elkins, J. A. Tweedy, and D. L. Suttner, *Agron. J.* **66**, 492 (1974).
179. G. J. Aageson and D. M. Elkins, *Agron. J.* **68**, 886 (1976).
180. D. M. Elkins, *Agron. J.* **66**, 426 (1974).
181. E. V. Parups and W. E. Cordukes, *Hortscience* **12**, 225 (1977).
182. T. L. Watschke, *Agron. J.* **68**, 787 (1976).
183. K. N. Murthy, P. M. Kumuran, and N. M. Nayar, *J. Plantation Crops* **3**, 81 (1975).
184. J. Tupy, *Planta* **88**, 144 (1969).
185. R. H. J. Corley, *J. Exp. Bot.* **27**, 533 (1976).
186. J. C. Bouwkamp and R. N. McArdle, *Hortscience* **15**, 69 (1980).
187. J. Jones, V. Wert, and S. Ries, *Planta* **144**, 277 (1979).
188. S. K. Ries, *Proceedings Plant Growth Regulator Working Group*, Vol. 6, p. 92 (1972).
189. J. L. Charlton, N. R. Hunter, N. A. Green, W. Fritz, B. M. Addison, and W. Woodbury, *Can. J. Plant Sci.* **60**, 795 (1980).
190. S. K. Ries, V. Wert, C. C. Sweeley, and R. A. Leavitt, *Science* **195**, 1339 (1977).

6

Hydrogels for Controlled Drug Release

Sung Wan Kim

Abstract. Hydrogels have been investigated for use in controlled-release drug delivery systems. Hydrogels show superior biocompatibility and permeation of both hydrophobic and hydrophilic drugs depending on the combination of copolymers. Drug diffusion through hydrogels depends on the hydrogel nature, mainly the percentage of cross-linking agents and hydrogel water content. Various devices designed with hydrogels are discussed in this chapter: reservoir systems, drug-dissolved monoliths, drug-dispersed monoliths, monolith devices with barrier layers, and novel delivery systems.

1. Introduction

Although cellulose materials which behave as hydrogels have long been used in hemodialysis applications. Wichterle and Lim first introduced a synthetic hydrogel poly(2-hydroxyethyl methacrylate) (P-HEMA) for biomedical applications.(1) Since that time, extensive hydrogel literature has been published. A comprehensive review published in 1976 included prior studies, synthesis, characterization, and the use of hydrogels utilizing various applications.(2) Recently, Ratner published a review comprising recent developments in hydrogel synthesis, structure, and medical applications.(3)

Hydrogels have in general the following unique properties: (1) water insolubility in physiological environments; (2) swellability to an equilibrium value of up to 90% H_2O content; and (3) superior biocompatibility.

Hydrogels have been investigated for use in controlled-release drug delivery systems. The primary advantages in using hydrogels are: (1) relatively easy extraction of polymerization initiators, decomposition products, and/or polymerization solvents prior to *in vivo* use; (2) the soft rubbery nature of hydrogels reduces mechanical irritation with *in vivo* implants; (3) the low hydrogel/water interfacial tension minimizes protein adsorption and cell adhesion (experiments with hydrogels and hydrogel grafted surfaces indicate low platelet adhesion); and (4) the permeation of

Sung Wan Kim • Department of Pharmaceutics, University of Utah, Salt Lake City, UT 84112.

both hydrophobic and hydrophilic drugs including charged solutes depending on the combination of hydrogel copolymers.

In the past decades, a number of hydrogels have been synthesized and characterized. Included among the relatively well-known hydrogels are: poly(hydroxyalkyl methacrylates), poly(vinyl alcohol), poly(acrylamide), poly(methacrylamide), and poly(*N*-vinyl pyrrolidone).

P-HEMA and its various copolymers with methoxyethoxyethyl methacrylate (MEEMA) or methoxyethyl methacrylate (MEMA), with and without ethylene glycol dimethacrylate (EGOMA), were synthesized for drug delivery systems. Basic studies have shown that drug diffusion through these hydrogels depends on the composition of copolymers, percentage of cross-linking agent, and water content of the hydrogels.[4]

Mechanically strong poly(vinyl alcohol) was prepared by cross-linking.[5] To improve mechanical strength, copolymer gels of poly(vinyl alcohol)–(vinyl pyrrolidone) were synthesized by the methanolysis of a radiation cross-linked vinyl acetate–vinyl pyrrolidone polymer.[6]

A cross-linked acrylonitrile was polymerized which can be hydrolyzed with HCl to form a hydrogel consisting of acrylonitrile, amide, and acid functional groups.[7]

As in other delivery systems, drug release from hydrogel devices shows zero-order or first-order kinetics. However, an additional factor in these polymers describes swelling which influences drug release. Several investigators have attempted to develop mathematical modeling for swellable polymeric systems. Lee provided approximate and exact analytical solutions, taking into consideration the moving boundary problems encountered during diffusional drug release from a polymeric matrix.[8]

Peppas *et al.*[9] developed pseudo-steady-state models for drug diffusion from swellable polymers exhibiting volume expansion with constant diffusion coefficients. Good published a mathematical model employing time-dependent drug diffusion coefficients with continuous swelling at a constant total volume of the hydrogel devices.[10] As swelling plays an important role in the hydrogels, solute permeation was investigated in an effort to develop models describing the transport phenomena in hydrogels with particular emphasis on the role of water in the process.[11] It was concluded that:

1. Hydrophilic drugs permeate P-HEMA and P-HEMA cross-linked with lower mole percent EGDMA via the "pore" mechanism. The diffusion coefficients of the solutes depend on the molecular size and may utilize the "bulklike" water in the hydrogels. As the water content of hydrogel increases, the solute permeability increases.
2. Hydrophobic drugs permeate P-HEMA and P-HEMA cross-linked with EGDMA via either the "pore" or "partition" mechanisms. Diffusion

coefficients are lower than those of hydrophilic drugs; however, steroids can permeate even in P-HEMA with 5.25 mol % EGDMA due to the predominant "partition" mechanism for hydrophobic drug permeation in this membrane. Hydrophilic drugs fail to permeate the highly cross-linked hydrogels.

3. Based on partition coefficient data, the hydrophilic drugs examined appear to permeate P-HEMA and P-HEMA with 1 mol % EGDMA via "bulklike" water regions. Partition coefficients of steroids in P-HEMA are dominated by the high solubility of the steroids in the hydrophobic regions of the hydrogels or "bulk water" domains, whereas, permeation is dominated by diffusion within "fluctuating pores."

2. Reservoir Devices

These devices can be obtained after hydrogel polymerization. A rod-shaped device is removed from the mold, then drilled out with a 1/8-in. bit. The device is filled with 100 mg of the progesterone mixture suspended in silicone oil and studied *in vitro*. The release of progesterone prepared from HEMA demonstrated constant release with an initial burst effect.(12)

A hydrophilic chamber based on HEMA was constructed by preparing sheets of hydrogel. This reservoir device in pouch form was filled with injections of freshly collected, viable neonatal rat or rabbit pancreas cells. This device released insulin for long periods, without rejection by the host's immunological systems and minimum adverse effects at implant sites.(13)

3. Monolithic Devices (Dissolved Systems)

Good developed a two-component hydrogel system described as macromer cross-linked graft copolymers(14) consisting of macromer component and P-HEMA. Variation of the macromer molecular weight allows a wide range for water swellability, drug loading, and release rates. To design oral delivery systes, disc-shaped copolymers were soaked in an antihyperglycemic drug, Phenformin–HCl–ethanol solution, for the drug-loading process. Networks were allowed to equilibrate in drug solution for 48 h at room temperature. The ethanol system utilized was 70% ethanol in water, a composition having solubility parameters near those of the networks. It was found that variation of cross-link density at a constant composition had a lesser effect on drug loadings. Release of Phenformin-HCl from the networks was typical for monolithic devices (first-order release kinetics). Tripelennamine HCl was also loaded in P-HEMA in various aqueous solutions of the drug and 36% (w/w) of load was obtained in 20% drug solution.[10]

P-HEMA and its copolymers show the greatest swelling in a 70% aqueous ethanol solution. As a cytoprotective agent, $PGF_{2\alpha}$ was loaded in hydrogels with maximum loading at 70% ethanol solution. *In vitro* release studies with $PGF_{2\alpha}$ dissolved gels (dried form) showed that $PGF_{2\alpha}$ release is dependent on drug-loading levels and hydrogel swelling. Hydrogel swelling was not affected by a medium pH and the devices were observed for their ability to maintain $PGF_{2\alpha}$ stability stored in the dried state as well as used in release experiments. $PGF_{2\alpha}$ released from these devices into a media of biological pH range can be achieved at levels required to produce cytoprotection with sufficient stability for the residence time of the device.[15]

Hydrogel oral dosage forms designed as monoliths (drug dissolved systems) have the ability to wash the devices clean of extractable impurities prior to loading the drug. The systems provide an attractive means of purification which not only renders the device safe for administration, but also eliminates unwanted additives and binders in conventional dosage forms which are absorbed in the gastrointestinal tract along with the active agent. Furthermore, diffusion mechanisms control the release of drug with a predetermined rate.

4. Monolithic Devices (Drug-Dispersed Systems)

These devices are relatively easy to formulate. Drug is embedded physically prior to the polymerization (drug mixed with monomer and initiator, and cross-linker if necessary) or during polymerization procedures of the hydrogel monoliths.

Monolithic hydrogel devices demonstrate drug release as a function of the hydrogel composition and initial drug load. If we plot the fraction of drug release vs. $t^{1/2}$ from matrix devices, we anticipate two straight lines with a breaking point due to changes in release characteristics during water absorption by the hydrogel devices.[16]

In earlier times, hydrogels were used for the controlled release of antibiotics,[17,18] narcotic antagonists,[19] and hydrocortisone succinate.[20] In the fabrication of soft contact lens, P-HEMA was combined with polymixin B, α-phenyl-ephrine, or pilocarpine for vision correction and ophthalmic therapeutics.[21,22]

An aprotinin delivery system using P-HEMA in IUD devices was studied for the reduction of menstrual hemorrhage.[23] Cervical hydrogel dilators were combined with prostaglandins for the dual purpose of studying mechanical dilation and drug action at the cervix.[24]

Heparin-releasing hydrogels have been developed. Heparin ionically bound to positively charged gel was coated on hydrophobic polymer and utilized for nonthrombogenic intravenous catheters. The *in vivo* heparin release rate from this device (H_2O content, 29–32%) into the patient's bloodstream at the beginning was about 10^{-2} units $cm^{-2} min^{-1}$, and gradually decreased 20 days after implantation. The results indicate that this device is markedly thromboresistant and especially effective in yielding low blood shear rates.[25]

A nonthrombogenic polymer was designed by providing simultaneously controlled release of heparin and prostaglandin. The matrix polymer must be capable of providing simultaneous controlled release of a low-molecular-weight solute (prostaglandin) and a macromolecule (heparin). The delivery systems selected were P-HEMA hydrogels because of the demonstrated capability of P-HEMA to provide a wide release range of solutes. Monolithic devices prepared by dispersing the mixture of prostaglandin (PGE_1) and heparin in P-HEMA matrices, were fabricated and evaluated for *in vitro* studies. Both agents released from the devices were shown to be biologically active.[26]

5. Monolithic Devices with Barrier Layer

The manufacturing of monolithic hydrogel devices is relatively uncomplicated; however, a constant release rate cannot be achieved. Methods are available to maintain a constant drug release rate; monolithic systems act as a core using a rate-controlling membrane which shows a constant release of the drug for long periods. For this system, the drug permeability through the outer layer has to be lower than permeability through the core. Progesterone-releasing devices were designed using this concept.[12] P-HEMA was chosen for the rate-controlling membrane and P-MEEMA or a MEEMA–HEMA copolymer was used as the core materials. The release rates were relatively constant after about day 20. The initial nonlinearity is due to the burst effect. It was shown that progesterone release increases as the water content of core material increases (MEEMA composition increases).

A continuous topical administration of fluoride to teeth using hydrogels was sought. The trilaminate devices are composed of a core of inorganic fluoride salt dispersed in a copolymer hydrogel of HEMA and MMA. A disk-shaped core made of 50/50 HEMA–MMA copolymer and sodium fluoride was coated with 30/70 HEMA–MMA copolymer. This device showed a nearly constant fluoride release rate of approximately 0.8 mg per day, for one month.[27]

Alternatively, it is possible to design constant release of a drug by cross-linking the hydrogel at the surface of a monolithic device. This was accomplished via soaking monolithic hydrogel devices in an ethanol solution of cross-linking agent, EDGMA, followed by exposure of the device to UV irradiation to initiate cross-linking reactions. Factors considered to control the release rate included soaking time in cross-linking solution, cross-linker concentration, and UV exposure time. Drug release from this device proved fairly constant.[28]

6. Novel Drug Delivery Systems

Recent progress has been made in the use of hydrogels as drug carriers targeted specifically to cell types where its action is required. *N*-(2-hydroxypropyl) methacrylamide copolymers have been synthesized which contain drugs attached to the polymer backbone via oligopeptide side chains. It is desired that such polymer–drug linkages be resistant to degradation during transport to the cell, but susceptible to hydrolysis by lysosomal enzymes. It was shown that designed oligopeptide side chains are readily hydrolyzed by lysosomal enzymes *in vitro* and also following pinocytic capture by rat visceral yolk sacs cultured *in vitro*. It was also demonstrated *in vivo* that more specific carbohydrate moieties can be used to target molecules to certain cell types and *N*-(2-hydroxypropyl) methacrylamide copolymers are directed efficiently to liver cells by inclusion of a small number of side chains terminating in galactase.[29,30]

Procainamide-coupled methacrylamide hydrogel was synthesized.[31] Experiments on mice indicated the monomer to be still active as an antiarrhythmic while having no enhanced toxicity. Dextrin and inulin were activated via periodate oxidation and the resulting polyaldehydes were coupled with procainamide or its glycine derivative. *In vivo* mice experiments with these drug-coupled hydrogels indicated significant pharmacological activity. The use of hydrogels as novel drug delivery systems is an important explorative area, notably in cancer chemotherapy.

As previously mentioned, most of the hydrogel drug delivery systems were designed as monoliths. Drug embedding is carried out during hydrogel polymerization. Therefore, drug-embedded hydrogels can be coated on medical devices for prophylactic purposes or for treatment with such drugs as antibiotics, corticosteroids, antithrombotic agents, and antitumor agents. Insertion of hydrogel devices dispersed with drugs into organs, as well as oral hydrogel drug delivery systems, are promising applications.

The hydrogels described above are nondegradable gels. Biodegradable hydrogels were synthesized to carry drugs as injectable systems.[32] However, drug release from these carriers is difficult to characterize due to the mixed process of drug release by diffusion and polymer degradation.

ACKNOWLEDGMENTS

The author wishes to thank Dr. J. D. Andrade, Dr. J. R. Cardinal, and Dr. D. G. Gregonis for their collaboration in the research of hydrogel drug delivery systems.

References

1. O. Wichterle and D. Lim, *Nature* **185**, 117 (1960).
2. J. D. Andrade, ed., *Hydrogels for Medical and Related Applications*, American Chemical Society Symposium Series No. 31, Washington, D.C. (1976).
3. B. D. Ratner, in: *Biocompatibility of Clinical Implant Materials* (D. Williams, ed.), Vol. 2, p. 145, CRC Press, Boca Raton, Fla. (1981).
4. S. Wisniewski and S. W. Kim, *J. Membr. Sci.* **6**, 299 (1980).
5. Y. Ikada, T. Mita, F. Horu, C. Sakurada, and M. Hatada, *Radiat. Phys. Chem.* **9**, 633 (1977).
6. N. A. Peppas and T. W. B. Gehr, *Trans. Am. Soc. Artif. Intern. Organs* **24**, 404 (1978).
7. J. Janacek, A. Stoy, and V. Stoy, *J. Polym. Sci. Symp.* **53**, 29 (1975).
8. P. I. Lee, *J. Membr. Sci.* **7**, 255 (1980).
9. N. A. Peppas, R. Gurny, E. Doelker, and P. Buir, *J. Membr. Sci.* **7**, 241 (1980).
10. W. R. Good, in: *Polymeric Delivery Systems* (R. Kostelnik, ed.), p. 39, Gordon and Breach, New York (1976).
11. S. W. Kim, J. R. Cardinal, S. Wisniewski, and G. M. Zentner, American Chemical Society Symposium Series **127**, 347 (1980).
12. J. R. Cardinal, S. B. Song, S. H. Kim, and S. W. Kim, *AIChE Symp. Ser.* **77**, 52 (1981).
13. S. H. Ronel, U.S. Patent 4,298,002 (November 3, 1981).
14. W. R. Good and K. F. Mueller, *AIChE Symp. Ser.* **77**, 42 (1981).
15. E. J. Mack, S. W. Kim, and W. R. Good, *A.Ph.A. Academy of Pharmaceutical Sciences Abstracts* **12**(2), 46 (1982).
16. S. Z. Song, S. H. Kim, J. R. Cardinal, and S. W. Kim, *J. Pharm. Sci.* **70**, 216 (1981).
17. M. Tollar, M. Stol, and K. Kliment, *J. Biomed. Mater. Res.* **3**, 305 (1969).
18. B. S. Levowitz, J. N. La Guerre, W. S. Calem, F. E. Gould, S. Joseph, and H. Scheonfeld, *Trans. Am. Soc. Artif. Intern. Organs* **14**, 82 (1968).
19. R. A. Abrahams and S. H. Ronel, *J. Biomed. Mater. Res.* **9**, 355 (1975).
20. J. M. Anderson, T. Konis, T. Nelson, M. Rorst, and D. S. Love, American Chemical Society Symposium Series **31**, 167 (1976).
21. H. E. Kaufman, M. H. Votila, A. R. Cassett, T. O. Wood, and E. D. Varnell, in: *Soft Contact Lens* (A. R. Cassett and H. E. Kaufman, eds.), Mosby, St. Louis, Mo. (1972).
22. S. M. Podes, B. Becker, C. Asseff, and J. Hartstein, *Am. J. Ophthal.* **73**, 336 (1972).
23. M. E. Tuttle, R. W. Baker, and L. W. Laufe, *J. Membr. Sci.* **7**, 351 (1980).
24. M. K. Akkapeddi, B. D. Halpern, R. H. Davis, and H. Balin, in: *Controlled Release of Bioactive Agents* (A. C. Tanquary and R. E. Lacey, eds.), p. 165, Plenum Press, New York (1974).
25. Y. Mori, S. Nagaoka, Y. Masubuchi, H. Tanzawa, M. Itoga, Y. Yamada, T. Yonaka, H. Watanobe, and Y. Idezuki, *Trans. Am. Soc. Artif. Intern. Organs* **24**, 736 (1978).

26. C. D. Ebert, J. C. McRea, and S. W. Kim, in: *Controlled Release of Bioactive Agents* (R. W. Baker, ed.), p. 107, Academic Press, New York (1980).
27. D. R. Cowsar, O. R. Tarwater, and A. C. Tanquary, American Chemical Society Symposium Series **31**, 150 (1976).
28. E. S. Lee, S. W. Kim, S. H. Kim, J. R. Cardinal, and H. Jacobs, *J. Membr. Sci.* **7**, 293 (1980).
29. R. Duncan, H. C. Cable, J. B. Lloyd, J. Kopecek, and P. Rejmanova, *Cell. Biol. Int. Reps. Suppl. A.* **5**, 14 (1981).
30. R. Duncan, P. Rejmanova, J. Kopecek, and J. B. Lloyd, *Biochim. Biophys. Acta* **678**, 143 (1981).
31. E. Schact, L. Ruys, J. Vermeersch, and P. Gyselinck, Symposium on Polymers in Medicine, Italian National Research Council, Porto Cervo, Italy, (1982), p. 12.
32. N. Tany, M. Van Dress, and J. M. Anderson, Seventh International Symposium on Controlled Release of Bioactive Materials, Fort Lauderdale, Florida (1980), p. 81.

7

Biodegradable Drug Delivery Systems Based on Polypeptides

Robert V. Petersen

Abstract. A new dimension to the use of polymeric materials as drug delivery devices is the incorporation of biodegrability into the system. A number of degradable polymers are potentially useful for this purpose, including synthetic and natural substances. A number of such substances are described in this chapter. However, major emphasis is placed on synthetic poly(α-amino acids). These polymers are useful for matrix and reservoir-type delivery systems and, in addition, when difunctional amino acids, such as glutamic acid, are utilized, polymer/drug conjugates represent another new dimension. In polymer/drug conjugates, the drug is covalently bonded directly to the polymeric "backbone" or to a spacer group between the drug and polymer. The bond must be of a type which hydrolyzes more rapidly than the amide bonds in the polymeric backbone. A number of such conjugates have been synthesized and evaluated for *in vitro* and *in vivo* hydrolysis and for potential toxicity. It appears that this concept can be utilized to advantage in the development of delivery systems for a wide variety of bioactive substances.

1. Introduction

Research on the use of synthetic polymers as carriers in drug delivery systems has increased rapidly in recent years. The polymers which have been investigated to the greatest extent are designed to be inert and nondegradable and depend upon controlled diffusion as the primary factor regulating drug release. Few such systems are currently available for general use.

The concept of utilizing biodegradable synthetic polymers as carriers for drugs in drug delivery systems is of more recent origin and is, necessarily, more restrictive in scope, due to the requirements of degradability and biocompatability of the polymer and its degradation products. In addition to the term biodegradable, these preparations have also been described as bioerodible and as bioabsorbable drug delivery systems. While there are shades of differences, precise distinctions have not been made and the terms frequently are used interchangeably.

Robert V. Petersen • Department of Pharmaceutics, College of Pharmacy, University of Utah, Salt Lake City, UT 84112.

1.1. Historical Background

Poly(lactic acid) (PLA) was first described in the literature in 1913 (French patent 456,824) and was reported by Kulkarni in 1966[1] as a material satisfactory for use in biodegradable surgical implants. In 1967 Schneider was issued a patent (French patent 1,478,694) on the use of PLA as an absorbable surgical suture material.

The first description of a biogradable implant for delivery of the narcotic antagonists cyclazocine, naltrexone, and naloxone, is attributed to Yolles *et al.*[2,3] The preparation used by Yolles was a composite of narcotic antagonists with poly(lactic acid), also referred to as polylactide. The delivery devices were either drug-laden films or drug-laden polymer powders suspended in a dispersing agent. In 1973, Jackanicz *et al.*[4] used PLA as a carrier for the contraceptive agent, *d*-norgestrel, in which the drug was incorporated into an implantable film.

According to investigations by Kulkarni,[1] PLA lost about 12–14% of its mass in three months. Schindler *et al.*[5] showed that after about 80 days postimplantation, the average molecular weight of PLA was reduced to about one-half the original molecular weight.

Following the early investigations on PLA, polymers of other hydroxy acids were synthesized and investigated for their suitability as implant materials and carriers in drug delivery systems. Poly(glycolic acid) (PGA) (polyglycolide) proved to be especially useful in surgical sutures. Copolymers of PLA and PGA were described by Schmitt and Polistina[6] in 1967 and their use in delivery systems for insecticides and fertilizers, as well as drugs, has been investigated. Various other copolymers of PLA and PGA, polyamides, and polyesters have also been synthesized and evaluated in delivery systems. Homopolymers and copolymers of ε-decalactone, and pivalolactone have been studied by Schindler *et al.*[5] and β-hydroxybutyric acid has been synthesized by Bissery and Puisieux.[7] Variations in biolifetimes from a few weeks to one year have been demonstrated.

The design of biodegradable implant devices based on PLA and PGA have been limited primarily to reservoir types. In such systems, a drug, in the crystalline state or as a suspension, is sealed within a polymer capsule such that the rate of release is regulated by the diffusion characteristics of the polymer membrane. While it is possible to form monolithic-type devices using PLA and PGA, little work has been reported on such studies. In the monolithic-type device, drug is dispersed throughout the polymer matrix. Since neither PLA nor PGA contain reactive functional groups, except in terminal positions on the polymer chains, it is not possible to form polymer/drug conjugates wherein covalent bonds exist between the polymer and a drug.

Another approach to the release of bioactive agents from degradable polymeric systems is the use of poly(α-amino acids) in which the drug is encapsulated in polymer shells, intermixed with polymer, or covalently bound to the polymer backbone. Most such systems are based on homopolymers of L-glutamic acid or L-aspartic acid or copolymers of these agents with other amino acids, such as L-leucine or L-valine, or with other substances. In preparations wherein the bioactive agent is covalently bound to the polymer, glutamic acid or aspartic acid is an essential component of the polymers because it provides reactive functional groups (carboxyls) to which the bioactive agents with appropriate functional groups can be chemically bound. Mitra *et al.*,[8] Feijen *et al.*,[9] Gregonis *et al.*,[10] Van Heeswijk *et al.*,[11] and Petersen *et al.*[12] have utilized poly(hydroxyalkyl-L-glutamines) as the polymer to which various drugs are bonded.

While most research has focused on polyesters of various types, it is beyond the scope of this chapter to provide detailed descriptions of these materials. Rather, emphasis will be placed upon poly(amino acids), which contain the amide or peptide bond and only brief descriptions will be given for polymeric systems which involve other bond types.

1.2. Advantages and Disadvantages

In the idealized situation, a biodegradable delivery system would provide a drug at a constant, controlled rate over a prescribed period of time following which the polymeric carrier would degrade into nontoxic, absorbable subunits which would subsequently be metabolized or eliminated from the body. Further, the system would be biocompatible, would not exhibit dose dumping at any time, and the polymer would retain its physical characteristics until after depletion of the drug. In practice, no system which exhibits all of these idealized characteristics has yet been described.

Many of the advantages of degradable systems are obvious. In general they embody the same advantages as nondegradable systems, such as uniform release of low levels of a drug, which eliminates the "sawtooth" effect of traditional unit dose forms, and assurance of patient compliance with dosage regimens. In addition, degradable systems eliminate the necessity for surgical removal of implanted devices following depletion of a drug. It must also be recognized that there are significant disadvantages in the use of degradable systems. In addition to several of the disadvantages exhibited by nondegradable systems, some types of degradable systems exhibit substantial dose dumping at some point or points following implantation or injection. The phenomenon of dose dumping by degradable systems is usually related to scission of polymer chains or cross-links to the extent that loss of integrity and sudden dispersion of polymer and release of substantial quantities of a drug results. A "burst effect," or high initial drug release soon after administration, is typical of most systems. In addition, degradable systems which are administered by injection of a particulate form are nonretrievable. This may be particularly troublesome if the patient exhibits a toxic reaction to the drug or carrier or the need for continued drug therapy is altered or eliminated.

In theory and in practice, biodegradable drug delivery systems can be designed to release a drug for periods of time ranging from a few days to several years. Under most circumstances, periods in excess of 3 to 12 months are impractical, particularly for the injectable systems which cannot be retrieved after injection.

1.3. Chemical Types

While a wide variety of polymer types may be used in nondegradable delivery systems, the types which are degradable in body fluids and which exhibit other requisites are quite limited. Chemical bonding types which are potentially biodegradable include amides, esters, ortho esters, acetals, glycosides, and related groups. Investigations have been conducted on a number of natural as well as synthetic polymers.

2. Poly(α-Amino Acids)

A variety of poly(amino acid) homopolymers, copolymers, terpolymers, and more complex combinations have been synthesized; however, those which have shown the greatest potential usefulness in drug delivery systems are homopolymers of L-glutamic acid or L-aspartic acid and copolymers of these amino acids with L-leucine or L-valine. These polymers have several advantages over other poly(amino acids). Since glutamic acid and aspartic acid are dicarboxylic acids, the polymer chains contain free carboxyl groups. The carboxyl groups influence aqueous solubility and provide functional groups which can subsequently be reacted with drugs, directly or via a variety of types of spacer groups. Copolymers of glutamic acid or aspartic acid with other amino acids exhibit a range of hydrophilic properties dependent on the relative proportions of the two amino acids in the chains. It must always be noted, however, that as the number of amino acids increases, the likelihood of immunogenic reactions also increases. Thus, it appears that the best candidates are the homopolymers of glutamic acid or aspartic acid.

Since poly(L-glutamic acid) and copolymers of L-glutamic acid with various α-amino acids or esters of dicarboxylic α-amino acids form films through which aqueous solutions of drugs may diffuse, most of the research on these polymers has involved drugs encapsulated in polymer shells or drugs intermixed with polymer in matrix-type devices. The incorporation of drugs via covalent bonding has only recently been described. Such systems are represented graphically as follows:

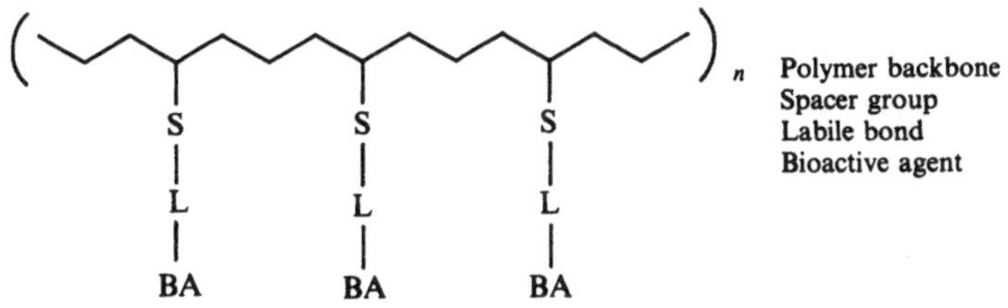

$$\left(\text{—}\underset{\substack{| \\ S \\ | \\ L \\ | \\ BA}}{\text{—}}\text{—}\underset{\substack{| \\ S \\ | \\ L \\ | \\ BA}}{\text{—}}\text{—}\underset{\substack{| \\ S \\ | \\ L \\ | \\ BA}}{\text{—}}\text{—} \right)_n$$

Polymer backbone
Spacer group
Labile bond
Bioactive agent

where:

1. The polymer backbone consists of a homopolymer of L-glutamic acid or L-aspartic acid, or copolymers of L-glutamic acid or L-aspartic acid with another amino acid, such as valine or leucine.
2. The spacer group consists of a straight chain, branched chain, aromatic, or other group having a terminal reactive functional group.
3. The labile bond consists of a functional group, such as ester, carbonate, oximino ester, amide, and so on, which is subject to enzymatic or chemical hydrolysis in body fluids.
4. The drug or bioactive agent contains a functional group capable of reacting with the terminal functional group on the spacer arm or directly with the carboxyl group on the polymer backbone.

3. Syntheses of Poly(α-Amino Acids) and Drug Conjugates

3.1. Polymer Backbone

The polymeric α-amino acid poly(L-glutamic acid) and poly(hydroxyalkylglutamines) have been utilized by Mitra *et al.*,[8] Feijen *et al.*,[9] and Gregonis *et al.*[10] for grafting of bioactive agents, such as norethindrone (a progestin), via several reactive functional groups. The synthesis of poly(L-glutamic acid) was carried out by the method of *N*-carboxyanhydride (NCA) polymerization as described by Lotan *et al.*,[13,14] according to Scheme 1.

The initial reaction involves esterification of the γ-carboxyl group of L-glutamic acid with a protecting group, such as benzyl alcohol, followed by reaction with phosgene to form the NCA of γ-benzyl-L-glutamate. The NCA is then polymerized using triethylamine as a catalyst followed by

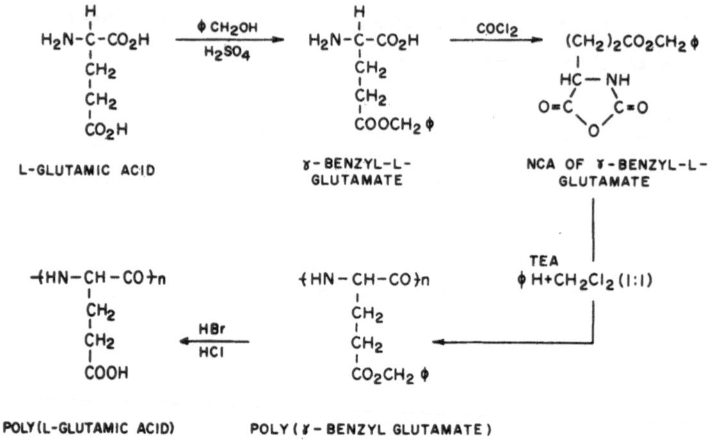

$H_2N-CH(CO_2H)-CH_2-CH_2-CO_2H$ $\xrightarrow[H_2SO_4]{\phi CH_2OH}$ $H_2N-CH(CO_2H)-CH_2-CH_2-COOCH_2\phi$ $\xrightarrow{COCl_2}$ NCA of γ-benzyl-L-glutamate

L-GLUTAMIC ACID — γ-BENZYL-L-GLUTAMATE — NCA OF γ-BENZYL-L-GLUTAMATE

NCA $\xrightarrow[\phi H + CH_2Cl_2\ (1:1)]{TEA}$ $\{HN-CH(CH_2CH_2CO_2CH_2\phi)-CO\}_n$ $\xrightarrow[HCl]{HBr}$ $\{HN-CH(CH_2CH_2COOH)-CO\}_n$

POLY(L-GLUTAMIC ACID) — POLY(γ-BENZYL GLUTAMATE)

SCHEME 1

debenzylation in HBr/HCl or by other methods. Van Heeswijk *et al.*[15] recently described improved methods of esterification by means of forming copper chelates. By this technique a wide variety of esters having a range of reactivities were synthesized. Most of the poly(L-glutamic acid) used for attachment of drugs had molecular weights in the range 15,000–100,000, although Feijen has succeeded in preparing these polymers with molecular weights in excess of one million.

3.2. Attachment of Norethindrone onto Poly(L-Glutamic Acid)

Poly(L-glutamic acid) has been used by Mitra *et al.*[8] as a backbone polymer for direct attachment of drugs, such as norethindrone, to the free carboxyl groups along the polymer chain. In this synthesis the oxime of norethindrone is first formed by reacting with hydroxylamine, then the oximino ester of poly(L-glutamic acid) is formed according to Scheme 2.

3.3. Attachment of Spacer Groups onto Poly(L-Glutamic Acid)

In most cases the drug component was attached via a spacer group between the drug and the polymer backbone. The procedure for these syntheses has been described by Gregonis *et al.*,[10] Feijen *et al.*,[9] and Mitra *et al.*[8] In these procedures, the spacer group was attached by reacting an alpha–omega hydroxyalkylamine with poly(γ-benzyl-L-glutamate) intermediate according to Scheme 3.

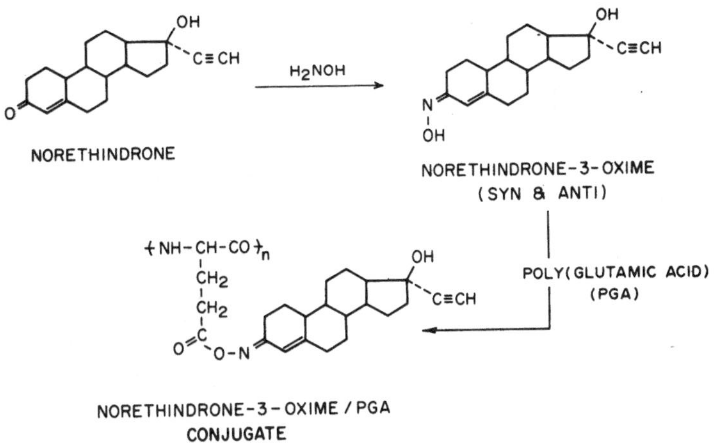

NORETHINDRONE

H_2NOH

NORETHINDRONE-3-OXIME (SYN & ANTI)

POLY(GLUTAMIC ACID) (PGA)

NORETHINDRONE-3-OXIME / PGA CONJUGATE

SCHEME 2

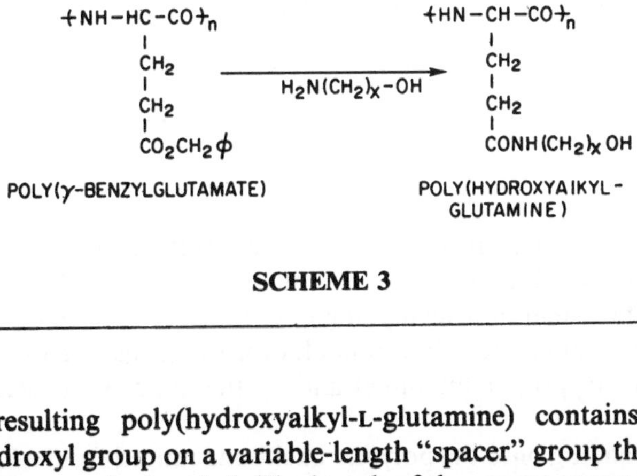

$H_2N(CH_2)_x$-OH

POLY(γ-BENZYLGLUTAMATE)

POLY(HYDROXYAIKYL-GLUTAMINE)

SCHEME 3

The resulting poly(hydroxyalkyl-L-glutamine) contains a reactive terminal hydroxyl group on a variable-length "spacer" group through which bioactive agents can be bonded. The length of the spacer group is dependent upon the chain length of the hydroxyalkylamine utilized. Increased chain length results in decreased hydrophilicity of the polymer.

3.4. Attachment of Bioactive Steroids onto Spacer Groups

An example of the method utilized to attach the steroid, norethindrone, via a carbonate linkage at the 17-position is shown by Scheme 4.

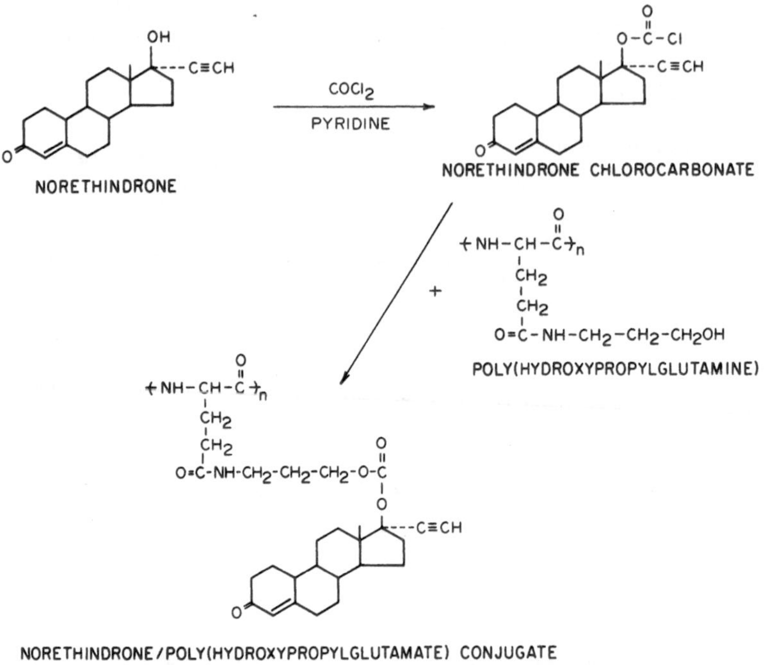

NORETHINDRONE

$COCl_2$

PYRIDINE

NORETHINDRONE CHLOROCARBONATE

+

POLY(HYDROXYPROPYLGLUTAMINE)

NORETHINDRONE/POLY(HYDROXYPROPYLGLUTAMATE) CONJUGATE

SCHEME 4

In this procedure it is necessary to convert the 17-hydroxy group of norethindrone to the more reactive chlorocarbonate by reaction with phosgene. The extent of drug loading onto the polymer can be controlled by varying the amount of norethindrone chloroformate used in the reaction with the poly(hydroxypropylglutamine) and by the appropriate selection of the catalyst.

Drug loading onto the polymer backbone ranged from a few percent to 70% based upon available spacer groups. The higher levels of drug loading resulted when reactions were catalyzed with dimethylaminopyridine. Drug loading above 70% was not achieved, probably due to steric hindrance produced by the bulky steroid. The higher the extent of drug loading, the lower was the hydrophilicity of the conjugate. This phenomenon seemed to work to an advantage in that increased hydrophilicity occurred as more and more drug was released, thus compensating in part for the lower total drug content.

3.4.1. Studies on Lability of the Carbonate Bond

Since chemical bonds in the polymer backbone, as well as the bond between polymer and drug, are subject to hydrolysis in body fluids, a question arises as to the composition of the released drug. A possibility exists that various fragments may remain attached to the drug. Such fragments may be a spacer group, spacer plus a glutamic acid, or spacer plus several glutamic acids. In order to study the lability of the carbonate ester bond and the possibility that fragments may remain attached to norethindrone, various ^{14}C-labeled derivatives were synthesized by Anderson *et al.*[12] and studied by *in vivo* methods by Fang *et al.*[12] The compounds designed to evaluate the lability of the carbonate ester bond were short-chain (methyl, propyl, and hexyl) carbonate esters of norethindrone:

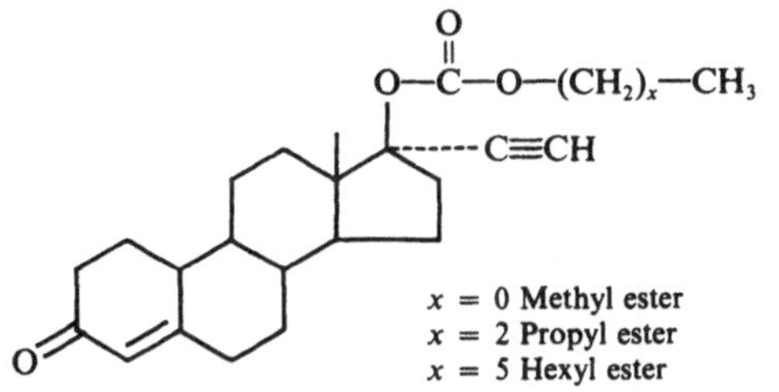

x = 0 Methyl ester
x = 2 Propyl ester
x = 5 Hexyl ester

When dissolved in aqueous solutions and subjected to nonspecific esterases, these carbonate esters exhibited half-lifes of 23.7 min or less, but were quite stable in pH 7.4 aqueous buffers as shown in Table 1. This suggests that solutions of polymer/drug conjugates such as poly(hydroxyalkylglutamine)/norethindrone-17-carbonates are quite rapidly hydrolyzed by esterases.

Anderson also attempted to attach 3-aminopropanol to norethindrone through a carbonate bond, which would represent an aminopropyl spacer group on norethindrone,

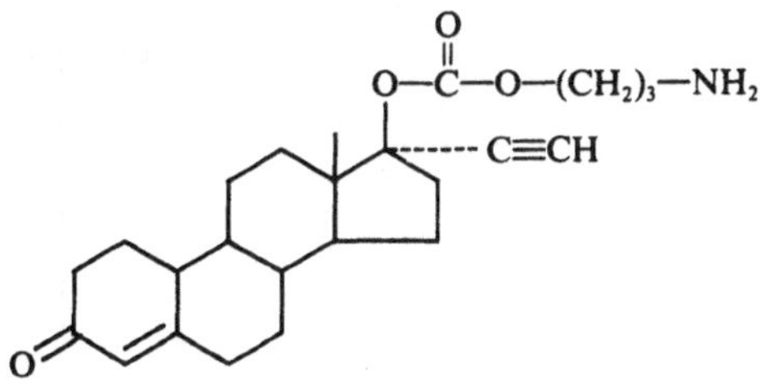

However, this compound was highly unstable and rearranged to the carbamate derivative,

Table 1. Hydrolysis of Norethindrone-17-Carbonate

NE-17-substituent	Concentration	Half-Life $t^{1/2}$ (min)		
		Rat plasma ln (NEAC)[a]	Rat plasma ln (NE-NE)[b]	Buffer (0.02 *M* tris; pH 7.4)
$-\overset{O}{\overset{\parallel}{C}}-O-CH_2$	$9.2 \times 10^{-6}\,M$	20.0	23.7	> 70[c]
$-\overset{O}{\overset{\parallel}{C}}-O-CH_2-CH_2-CH_3$	$1.12 \times 10^{-5}\,M$	4.2	4.7	> 70[c]
$-\overset{O}{\overset{\parallel}{C}}-O-(CH_2)_5-CH_3$	$9.6 \times 10^{-6}\,M$	2.7	3.9	> 70[c]
$-\overset{O}{\overset{\parallel}{C}}-O-(CH_2)_3-N(CH_3)_2$	$9.9 \times 10^{-6}\,M$	[d]	1.6	> 70[c]

[a] Determined by the disappearance rate of NEAC.
[b] Determined by the appearance of norethindrone (NE).
[c] No observable hydrolysis during a 70-min time period.
[d] Not determined.

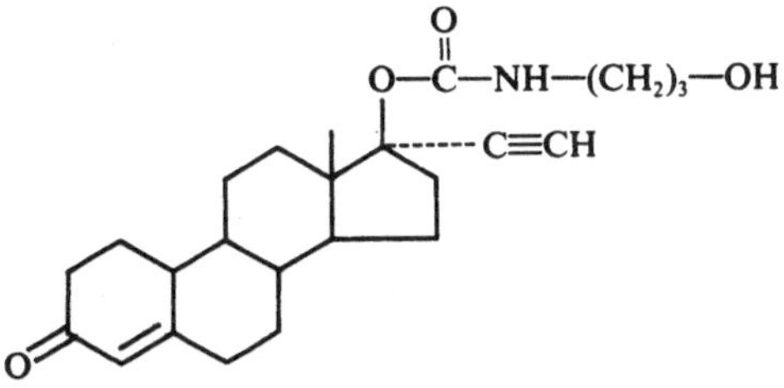

In order to block this rearrangement Anderson synthesized the dimethylaminopropyl derivative,

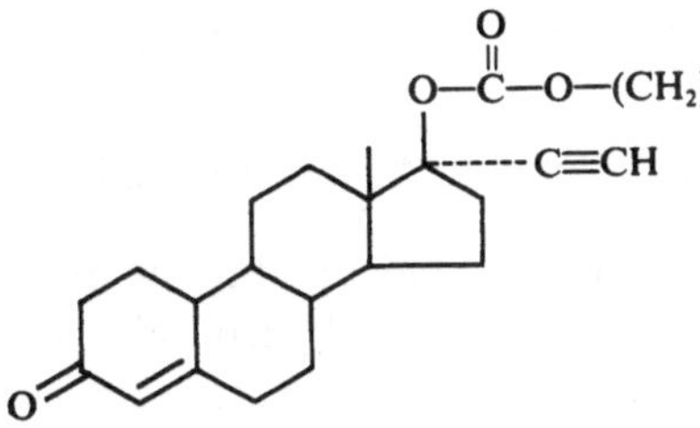

This compound exhibited an *in vitro* half-life in aqueous solution of nonspecific esterases of only 1.6 min as shown in Table 1. This was interpreted to indicate that the presence of fragments attached to norethindrone released from 17-carbonate esters into the bloodstream is unlikely.

The polymer/drug conjugates illustrated contain all of the components previously shown in the schematic diagram, namely, a biodegradable polymer backbone [poly(L-glutamic acid)], a spacer group (variable-length alkyl group), a labile bond (carbonate), and drug (norethindrone).

3.5. Other Polymer/Drug Conjugates Based on Poly(Hydroxyalkylglutamines)

Other compounds which conform to these models have been synthesized recently. For example, Van Heeswijk *et al.*[15] have described the synthesis and properties of the hormone, testosterone, bonded to poly(hydroxypropyl-L-glutamine) via a succinate diester spacer group as shown below:

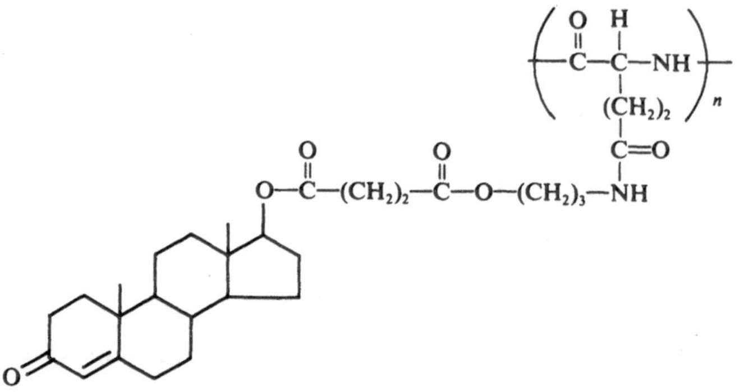

This molecule was designed to provide relatively rapid release of testosterone, unbound or as the hemisuccinate, followed by a more prolonged degradation of the polymer backbone.

3.6. Copolymerization of L-Glutamic Acid and L-Valine

Homopolymers of glutamic acid are highly hydrophilic due to the side-chain carboxyl group. However, the hydrophilicity can be regulated by copolymerization with a hydrophobic amino acid such as L-valine. Anderson *et al.*[(12)] have prepared several copolymers using glutamic acid and a second, less hydrophilic, amino acid. An example of such a reaction starting with L-glutamic acid and L-valine is shown in Scheme 5.

The side-chain carboxyl group of L-glutamic acid is initially prepared as the benzyl ester. The γ-benzyl-L-glutamate and valine are then converted separately to their respective *N*-carboxyanhydrides (NCAs) by reaction with phosgene. The copolymerizations were carried out using various feed ratios of γ-benzylglutamate and valine–NCAs in either dioxane or a 1 : 1 v/v mixture of benzene–methylene chloride. Triethylamine was used as the initiator for these copolymerizations. The benzyl protecting group was subsequently removed with acid catalysts.

Anderson found that the NCA of γ-benzyl-L-glutamate is more reactive than that of L-valine. Further, the variation of solvent was found to produce slightly different reactivity ratios—the NCA of γ-benzyl-L-glutamate was slightly more reactive in dioxane than in benzene–CH_2Cl_2. Table 2 shows the reactivity ratios of the two monomers.

After obtaining the reactivity ratios r_G and r_v of γ-benzylglutamate and valine, respectively, and knowing the monomer feed ratios F_G^0 and F_v^0, the instantaneous monomer feed (F_G and F_v), the average copolymer compositions, (f_G and f_v), and the incremental copolymer composition (ΔfG)

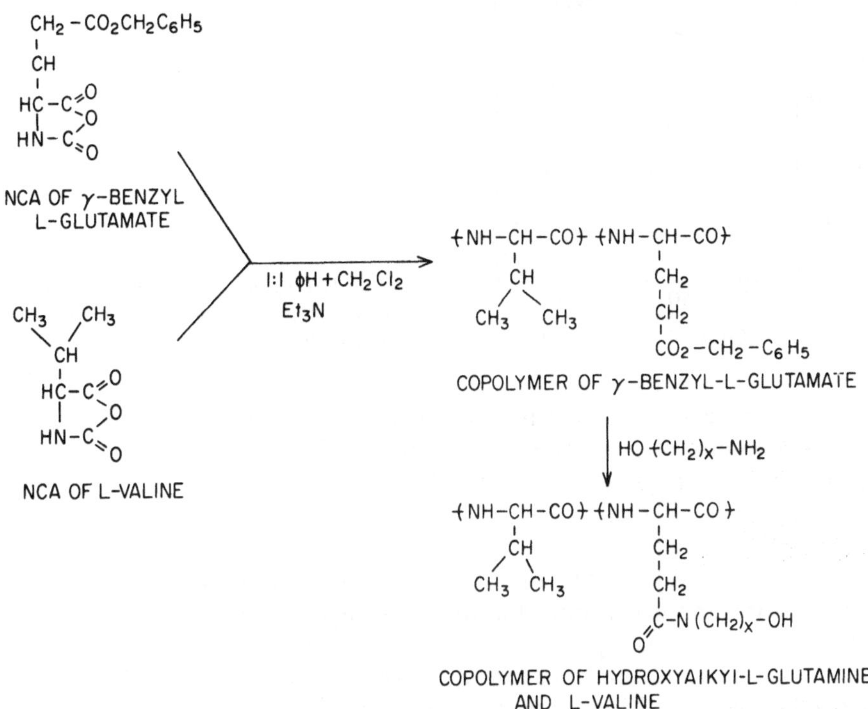

SCHEME 5

at any given conversion were determined. An example of these calculations is summarized in Figure 1 where the variations of f_G, Δf_G, and F_G, and F_G with percent conversion are given for $F_G^0 = 0.50$ for polymerizations carried out in dioxane. A similar pattern was found when reactions were carried out in a mixture of benzene and methylene chloride.

Table 2. Reactivity Ratios for γ-Benzylglutamate–Valine Copolymers

Dioxane	Benzene–Methylene chloride
$r_G = 2.7$	$r_G = 2.1$
$r_v = 0.3$	$r_v = 0.6$

where r_G is the reactivity ratio of glutamic acid and r_v is the reactivity ratio of valine.

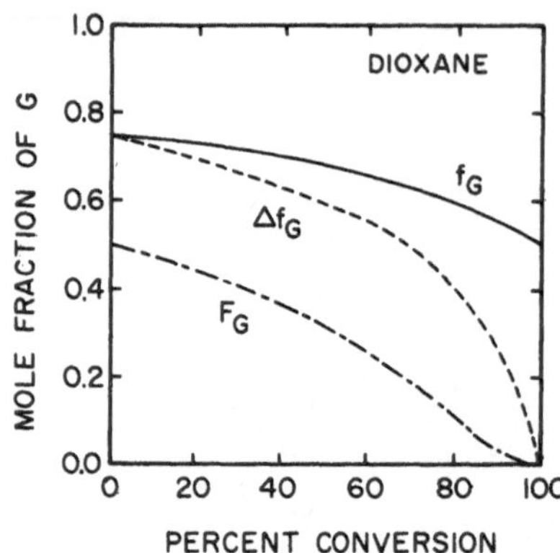

Figure 1. Polymer composition and feed ratios for copolymerization of L-glutamic acid and L-valine.

As expected from the high reactivity ratio of L-glutamic acid–NCA, larger amounts of glutamate are incorporated into the copolymer at low conversions. This greater inclusion of γ-benzyl-L-glutamate–NCA has been virtually totally converted into the copolymer and the remaining L-valine–NCA tends towards homopolymerization. The copolymerization, therefore, actually results in a family of copolymer chains of varying composition for which only the mean average copolymer composition at 100% conversion is equal to the initial or feed ratio on the NCA of the monomer. The instantaneous copolymer composition curves, f_G, from Figure 1 can be used to calculate the chain composition and heterogeneities which are formed over given conversions.

3.7. Copolymer/Drug Conjugates

Because of the heterogeneity of such copolymers, and because of the long duration of drug release from poly(L-glutamic acid) and poly(L-hydroxyalkylglutamines), Anderson did not pursue experiments in bonding of bioactive agents. However, Van Heeswijk *et al.*(16) recently prepared copolymers of L-glutamic acid and *N*-γ-glutamyl adriamycin onto which the antitumor antibiotic, adriamycin, was covalently bound. Bonding was achieved directly to the free carboxyls of glutamic acid via an amide linkage as shown below:

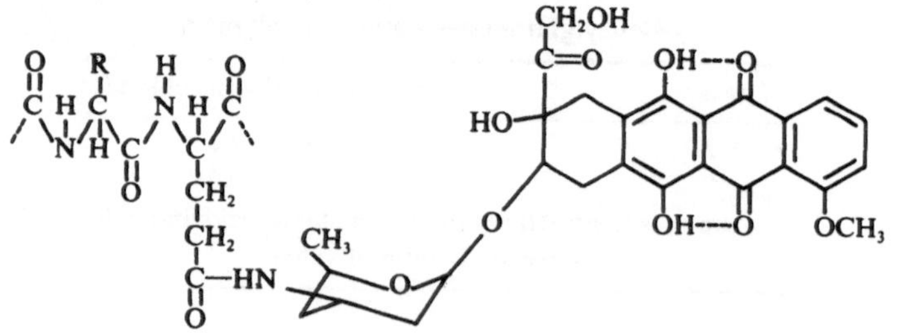

This drug/polymer conjugate was designed and synthesized for the purpose of achieving selective endocytosis into tumor cells with subsequent lysosomal digestion resulting in the release of adriamycin. This phenomenon (drug targeting) is discussed in greater detail in a later section of this chapter. Van Heeswijk *et al.*[15] have proposed a variety of spacer groups through which drugs containing various functional groups may be attached.

Another application of bioactive agents bound to biodegradable α-amino acid copolymers has been published by Ventner *et al.*[17] The approach taken by these investigators involves immobilization of catecholamine drugs onto copolypeptides of hydroxypropylglutamine and *p*-aminophenylalanine via an azo-linkage. Since the copolymer contains a random mixture of monomers, an exact structure cannot be assigned; however, an example structure involving D-isoproterenol bonded to the copolymer is shown below:

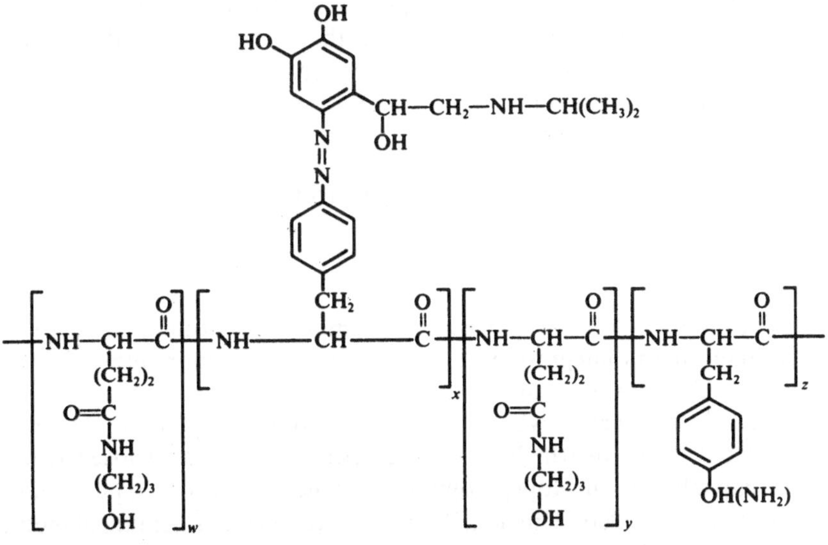

Such water-soluble drug/polymer conjugates retain their pharmacologic activity *per se* without being hydrolyzed. Thus, they are not considered to be prodrugs by classical definitions.

3.7.1. Copolymers as Reservoir Devices

Another method of utilizing copolymers of α-amino acids has been investigated by Sidman *et al.*[18,19] These investigators synthesized copolymers of L-glutamic acid and ethyl glutamate with average molecular weights in the range 140,000–250,000. The copolymers were prepared from the NCAs of the respective amino acids and had a structure as depicted on the following page:

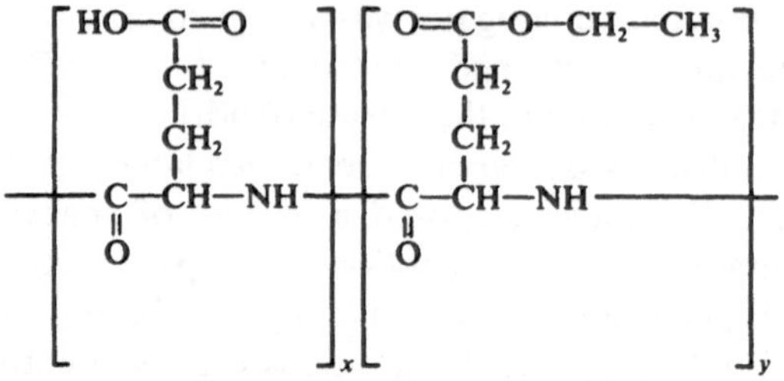

$$\left[-\underset{\underset{O}{\|}}{C}-\underset{\underset{\underset{\underset{HO-C=O}{|}}{CH_2}}{|}}{CH}\!-\!NH- \right]_x \left[-\underset{\underset{O}{\|}}{C}-\underset{\underset{\underset{\underset{O=C-O-CH_2-CH_3}{|}}{CH_2}}{|}}{CH}\!-\!NH- \right]_y$$

Instead of attaching bioactive agents to the polymer backbone, Sidman formed capsule shells of the polymer. The shells were then loaded with bioactive agents, such as norethindrone, naltrexone, leutenizing hormone-releasing hormone (LHRH), or other steroids or macromolecules having molecular weights up to 69,000. The loaded shells thus correspond to reservoir-type releasing devices. Sidman has indicated that release rates are relatively uniform and can be controlled with a range of a few days to a few years. These preparations exhibited a low order of toxicity. Matrix-type releasing devices can also be prepared from these copolymers.

4. Dosage-Form Formulation and Drug Release

4.1. Dosage Forms

The biodegradable drug delivery systems can be formulated into dosage forms much like nondegradable systems, namely: (1) monolithic or matrix forms in which the drug and polymer are intimately mixed together and fabricated into sheets, cylinders, rods, powders, or other shapes; (2) encapsulated forms in which solid or suspended drug is enclosed within a plastic capsule or envelope, commonly called reservoir types; and (3) suspensions of powders. Kim *et al.*[20] have developed mathematical methods for calculating release rates from these various forms.

4.2. *In vitro* Drug Release

The *in vitro* release of the progestin, norethindrone, from ^{14}C-labeled norethindrone bonded to poly(hydroxyalkylglutamines) via a carbonate linkage, has been studied by Kim *et al.*[12] and Fang *et al.*[12] A typical *in vitro* release curve is shown in Figure 2.

It can be seen that, following an initial burst likely due to entrapped norethindrone, release of the drug was relatively steady for about 30 days.

In polymeric delivery systems in which the bioactive agent is bound to

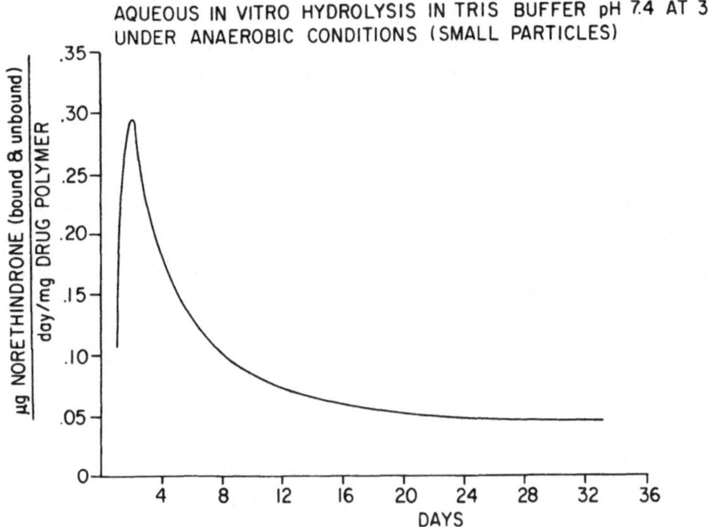

Figure 2. *In vitro* hydrolysis of poly-N^5-[3-hydroxypropyl)glutamine]/norethindrone-17-carbonate in aqueous buffer.

a polymer, drug release is effected through hydrolytic cleavage rather than diffusion through a membrane, although diffusion may be an additional rate-controlling factor in some instances. Systems in which a drug is covalently bound to a polymer backbone may be considered prodrugs since chemical or enzymatic biotransformation is necessary for liberation of the parent drug. Although it would appear that achievement of a zero-order release rate may be difficult or impossible, Kim has proposed a theoretical model which describes a mechanism for achievement of zero-order release kinetics. The model is

(i) Covalently bound drug $\xrightarrow{kh}$ free drug (hydrolysis).
(ii) Free drug $\xrightarrow{kd}$ release (diffusion through loose polymer matrix).
(iii) Constant release can be achieved by:
1. k_h (rate of hydrolysis) $\ll k_d$ (rate of diffusion).
2. $$\frac{-\mathrm{d}(C)_{bd}}{\mathrm{d}t} = k_h(C)_{bd} \qquad k = \text{const}$$
since activity of bound drug is nearly constant regardless of the remaining concentration $(C)_{bd}$.
3. Degradation of the backbone polymer should be slow.

There are five major parameters that can be varied to control the rate of drug release from the polymer: (1) hydrophilic character and molecular weight of the polymer; (2) length of the spacer group; (3) the lability of the covalent bond to the drug; (4) initial load percent of drug; and (5) particle size and geometry of the powdered or fabricated conjugate.

4.3. *In vivo* Drug Release

In vivo studies on the same compound as that shown in 4.2, injected subcutaneously as a suspended powder into rats, has been described by Fang *et al.*[12] Data showed that, following a moderate initial burst, drug release progressed at a relatively uniform rate with some reduction over a nine-month period. Figure 3 shows a typical release profile for norethindrone bound to a poly(hydroxyalkylglutamine). In this instance the polymer was poly(hydroxypropyl-L-glutamine) with norethindrone attached via a carbonate linkage.

A number of factors were shown to influence the rate of drug release in these systems, namely, particle size, extent of drug loading, and spacer-group length. Specifically, reduction of average particle sizes from 200 to 11 μm resulted in a tenfold increase in release rate. A drug load with 15–20% saturation of available spacer groups resulted in a drug release lifetime of a few months, whereas a conjugate with 50–60% drug load had a projected lifetime of several years. Conjugates with two carbon spacer groups released drug more rapidly than those with three carbon spacer groups. Thus, a polymer/drug conjugate having a low percent of drug load (5–10%), with drug attached to a long spacer group (three to six carbon units) via a carbonate linkage and injected subcutaneously in rats as a suspension of particles with a small average particle diameter (3–11 μm) would have a predicted lifetime of about 6–10 days. On the other hand, a drug/polymer conjugate having a high percent of drug load (60–70%), with drug attached to a short spacer group (one to two carbon units) via a

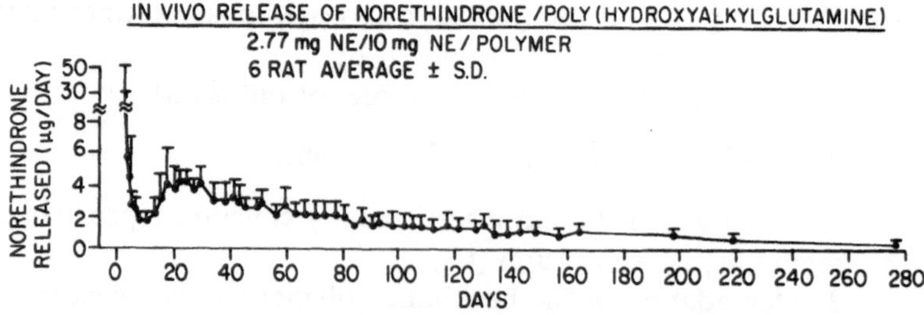

Figure 3. *In vivo* hydrolysis of poly-N^5-[(3-hydroxypropyl)glutamine]/norethindrone-17-carbonate injected as a suspension of small particles (11 μm) subcutaneously in rats.

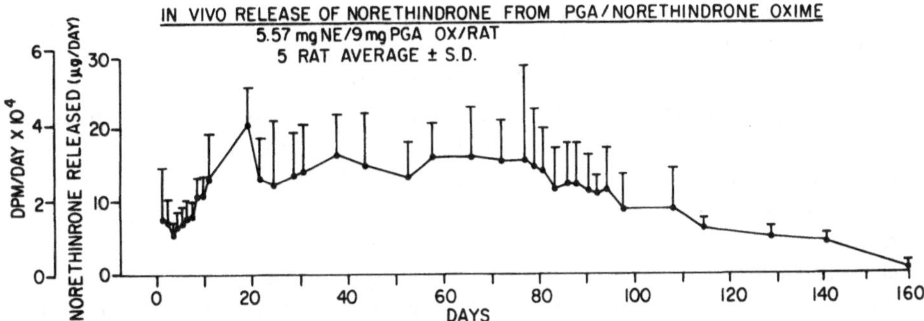

Figure 4. *In vivo* hydrolysis of poly(L-glutamic acid)/norethindrone-3-oximino ester injected as a suspension of medium-sized particles (50 μm) subcutaneously in rats.

carbonate linkage and injected subcutaneously in rats as a suspension of large particles (100–200 μm diameter) would have a predicted lifetime of several (2–10 years).

The conjugates synthesized by Mitra *et al.*,[8] namely, norethindrone oximino esters of poly(L-glutamic acid), were also studied by Fang *et al.*[12] for *in vivo* release rates in rats following subcutaneous injection of suspensions of ^{14}C-labeled conjugates in saline. Figure 4 shows the release profile of a conjugate having a 61% drug load and average particle diameter of about 50 μm.

It can be seen that, following an increase for about three weeks, drug release remained fairly constant for about two months before dropping to near zero over the subsequent three months.

5. Toxicity Studies

A major area of concern with nonbiodegradable systems is that of toxicity. Obviously, the device must be biocompatible, nonantigenic, and nonimmunogenic. The same concerns apply to biodegradable materials and, additionally, fragments resulting from chemical or enzymatic breakdown must not exhibit any of these toxic properties. Antigenicity and immunogenicity of polypeptides, polypeptides with attached spacer groups, and polypeptides with spacer group and bioactive agent attached, and breakdown products of these polymers are of major concern. It has been shown by Maurer[21] that the potential for immunogenic and antigenic reactions increases as the complexity of the polymer increases. For example, homopolymers of L-glutamic acid exhibited a low order of toxicity, whereas copolymers and terpolymers exhibited increasing toxicity. This has been confirmed by Nelson *et al.*[12] in studies using rats, rabbits, and guinea pigs. These investigators also studied the toxicity of poly(hydroxyalkylglutamines) and conjugates of this polymer with the progestin, norethindrone, in these same species. Experiments included skin-irritation tests in all three species, agglutination tests, precipitation tests, and gel diffusion tests in rabbits, and passive cutaneous anaphylaxis tests and complement fixation tests in rats and rabbits. In all cases test results were negative with one minor exception – the skin-irration test in guinea pigs using a drug/polymer conjugate showed a mild positive reaction. Histopathologic examination of excised rat tissue following various periods of time up to nine months' postimplantation with norethindrone/poly(hydroxyalkylglutamine) conjugates as the carbonate esters and norethindrone/poly(L-glutamic acid) conjugates as the oximino esters have also been performed by Coleman *et al.*[12] and by Woodward *et al.*[12] Although some mild to moderate histopathologic reactions (mild cellularity, occasional giant cells and macrophages, mild inflammation with amorphous eosinophilic material, lymphocytes, and plasma cells) were observed, none of these was considered severe or even moderately severe. Of the two drug/polymer bonding types, the oximino esters appeared to exhibit the greater toxicity.

Other biodegradable systems, such as the copolymers of L-glutamic acid and ethyl-L-glutamate used by Sidman *et al.*[18,19] and the polylactide/polyglycolide systems used by a number of researchers, appear to have good biocompatibility characteristics. In the case of polymers of L-glutamic acid, hydrolysis to the monomer merely means release of a natural amino acid, L-glutamic acid. Polylactides, if hydrolyzed to monomeric units, result in release of lactic acid, another natural metabolite. The products of hydrolysis to soluble fragments larger than monomers do not appear to have toxic effects. Likewise, glutamic acid esters or glycolates released from ethyl glutamate copolymers or polyglycolides, respectively, do not appear to exhibit toxicity from the quantities released.

6. Other Biodegradable Polymer/Drug Conjugates

The primary purpose of this chapter has been to describe various biodegradable drug delivery systems based on poly(α-amino acids), particularly those in which drugs are covalently bound to the polymers. However, there are a number of other degradable systems based on other types of polymers which have been evaluated. It is not possible to provide an exhaustive review of such systems within the confines of this chapter;

however, some representative examples will serve to illustrate the nature of some of the research currently being performed in this field.

A brief discussion of polymers of hydroxy acids such as lactic and glycolic acid or, more specifically, the cyclic dimers of these acids, the lactides and glycolides, was presented in the introduction. Such compounds have been studied rather extensively and show considerable promise. For example, Anderson *et al.*(22) showed that a near zero-order release of norethindrone for about 90 days could be achieved from matrix-type devices, while Beck *et al.*(23) obtained six months of release from microspheres of *dl*-poly(lactic acid) following intravenous injections in baboons. Almost all researchers have reported a terminal "dose-dumping" effect of substantial magnitude due to chain scission and loss of integrity prior to total drug release. There is also an autocatalytic effect produced by free carboxyls resulting in an accelerated rate of chain scission.

Wise *et al.*(24) also utilized special beads composed of various proportions of PLA/PGA copolymers with coatings made of pure polymers and implanted subcutaneously. By this means fairly constant levels were released for various periods of time as determined by relative proportions of the polymer components. Pitt and Schindler(25) have also studied these polymers and report results comparable with those of other investigators. A review on PLA and PGA has been published by Tanquary and Lacey.(26)

Another type of polyester delivery system involving polymerization of difunctional α-ketoacids and diols has been reported by Heller,(27) although this work was discontinued. In subsequent studies, Heller *et al.*(28) described a series of poly(ortho esters) to which drugs were covalently bonded. Structures of these polymers are shown below.

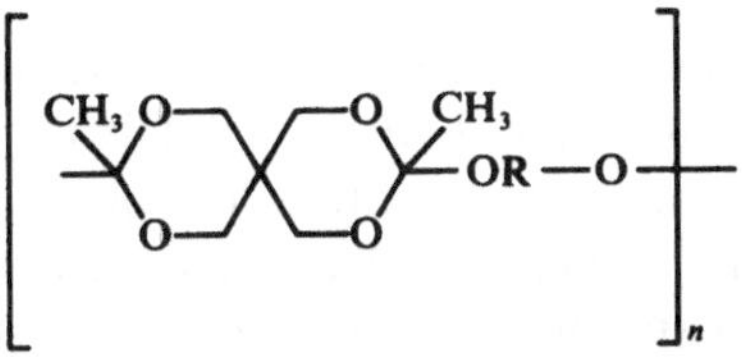

CH_3 O—O CH_3 —OR—O— (O—O) $]_n$

Discs of these polymer/drug conjugates, including soluble salts, exhibited near zero-order release rates for about 80–140 days in a buffered, aqueous environment. The salts, sodium chloride and sodium sulfate, served to imbibe water and produce swelling of the polymer. In the absence of salts, hydrolysis proceeded at a much slower rate, extending to more than one year.

Graham(29) has synthesized and evaluated a series of polyacetals prepared by polymerization of bisdihydropyrans and alkylene diols as shown on the following page:

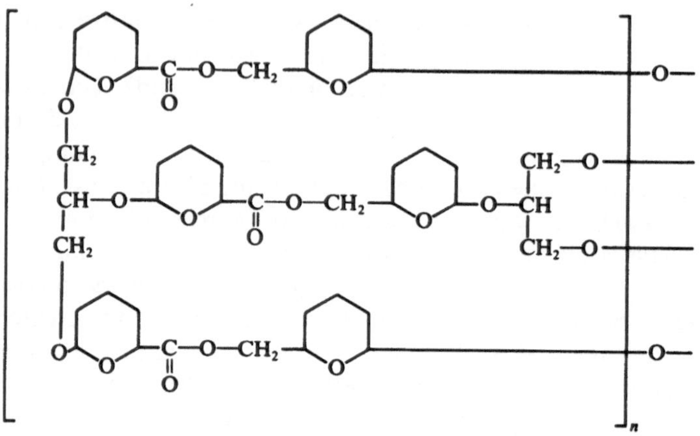

When drugs (e.g., antimalarials and steroids) were incorporated into matrix-type devices and administered as a powder to rats, bioabsorption was observed for periods ranging from 4 to 18 months.

Glycosaminoglycans, a class of polysaccharides, have been studied by Sparer *et al.*[30] as carriers for produgs. A typical structure for one of these carriers, condroitin-4-sulfate, is shown below:

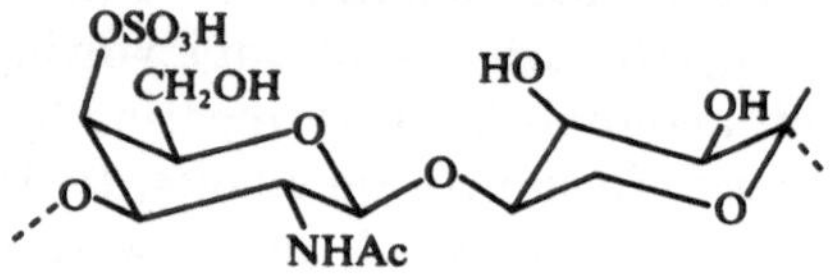

These compounds have been used as carriers for various bioactive agents, including antibiotics and anticancer drugs bonded directly or via small difunctional molecules such as amino acids. *In vivo* data in rats indicate that subcutaneous injections of conjugates of labeled cysteine provided slow release over a five-month period with a low order of toxicity.

McCormick and Anderson[31] have used other natural polymers, such as cellulose, chitin, and starch, as carriers for covalently bound herbicides.

In addition to their investigation on PLA/PGA, Yolles and Sartori[32] have used hydroxypropylcellulose as a backbone for covalent attachment of steroids via ester or amide bonds. Testosterone was released from these conjugates to the extent of 23% over 84 days.

7. Drug Targeting

Since Ehrlich first conceived the idea of the "magic bullet" in relation to specificity of action of the arsenicals, researchers have sought ways to

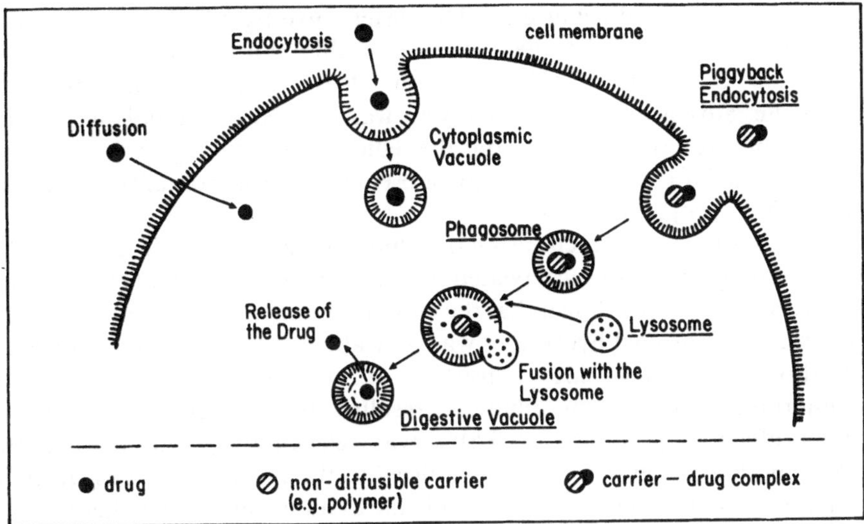

Figure 5. Graphic representation of cell uptake via diffusion, endocytosis, and "piggyback" endocytosis.

produce site-specific bioactive agents, that is, agents which have a high affinity for the diseased site or tissue and little or no affinity for normal tissues. Drug release at localized sites, such as that produced by ocular and uterine delivery systems, approaches that ideal, but residual systemic distribution has not been eliminated totally.

A number of novel concepts have been investigated in an attempt to produce a "targeted" drug action. Among these are the concepts of selective endocytosis and affinity labeling. Ringsdorf has used the term "homing device" when applied to polymeric delivery systems. A book on targeting of drugs, edited by Gregoriadis *et al.*,[33] has been published.

The phenomenon of selective endocytosis has been investigated by Van Heeswijk *et al.*,[16] Gregoriadis *et al.*,[34] Ringsdorf *et al.*,[35] Papahadjopolous *et al.*,[36] and several other researchers. Endocytosis or piggyback endocytosis involves cellular engulfment of a polymeric substance which serves as a carrier for a bioactive agent. The engulfed particle and surrounding cell membrane is termed a phagosome. In the idealized situation the phagosome fuses with a lysosome resulting in a digestive vacuole and subsequent release of the bioactive agent. These phenomena are shown graphically in Figure 5.

Most systems which are potentially selectively endocytized have been directed towards malignant cells. The system investigated by Van Heeswijk *et al.*,[16] which was described earlier in this chapter, utilizes polymer conjugates of L-glutamic acid which served as a carrier for the antitumor antibiotic, adriamycin. The adriamycin was bonded to the polymer via an amide linkage. Investigation on the efficacy of this agent in selective endocytosis into tumor cells and subsequent release of adriamycin are currently in progress.

In another experiment using drug/polymer conjugates, Trouet *et al.*[37] linked daunorubicin to succinylated serum albumin via an amide linkage. By *in vitro* studies, it was shown that the amount of daunorubicin released in the presence of lysosomal hydrolases was only about 2% in 10 hours. However, insertion of a spacer arm consisting of various amino acids between daunorubicin and serum albumin, resulted in an increased release of daunorubicin. Up to $(74.1 \pm 0.5)\%$ was released in 10 hours when the tetrapeptide Ala-Leu-Ala-Leu was utilized as the spacer arm. Subsequent *in vivo* experiments in mice and calves indicate that this substance holds promise as a useful antitumor agent. Other agents which have been investigated by Trouet *et al.*[37] include the antimalarial, primaquine, linked to asialofetiun via the Ala-Leu-Ala-Leu spacer arm. The resulting compound had a marked effect in extending the life span of mice challenged with plasmodium berghei sporozoites.

The concept of affinity labeling involves the binding of drugs onto molecules which have an affinity for the target cells. Little work has been done using biodegradable polymers, therefore, only a few representative examples will be given. An example of affinity labeling is the use of monoclonal and polyclonal antibodies as carriers for drugs to antigenic determinants. This has been investigated by a number of researchers particularly as it applies to cytotoxic drugs chemically bound to antibodies where the antigen is expressed on tumor cells. By this means, the drug-bearing antibody has a selective affinity for the tumor cell. A similar mechanism has been proposed for antitumor drugs linked to immunoglobulins by a dextran bridge or to fibrinogen or human serum albumin. Another approach to affinity labeling is the use of reconstituted membranous vesicles, for example, resealed erythrocytes, which may be phagocytized by other cells in culture, thereby introducing toxic substances onto or into secondary lysosomes.

Liposomes have been suggested as potential carriers in drug delivery systems by Gregoriadis *et al.*[33] A number of investigators have studied the effects of lipid composition, size, surface charge, and surface ligands on the ability of these vesicles to recognize and associate selectively with target cells. The combination of liposomes and antibodies also has potential for drug targeting.

8. Conclusions

The field of drug delivery systems is of recent origin, with most of the research having been done in the past 10 years. At this time only a very limited number of such systems are available for use in human medicine. However, as evidenced by the number of highly qualified researchers who have been attracted to the field, the number of novel ideas which have emerged in recent years, and expansion of the concept to include agricultural chemicals, pesticides, veterinary medicinals, pheromones, cosmetics, and related areas, it would appear that a number of highly beneficial products are near emergence. The concept of drug targeting, which offers some of the more exciting possibilities, is likely not to mature for several more years. Likewise, research on systems which release a drug in response to physiologic needs is still in the embryonic stages. The incorporation of biodegradable carriers into such products introduces new dimensions which further enhance the prospects for a bright future for drug delivery systems.

References

1. R. K. Kulkarni, K. Pani, C. Neuman, and F. Leonard, Poly(Lactic Acid) for Surgical Implants, Technical Report 6608, Walter Reed Medical Center, Washington, D.C. (1966).
2. D. A. Blake, S. Yolles, M. Helrich, H. F. Cascorbi, and M. J. Eagan, Release of Cyclazocine from Subcutaneously Implanted Polymeric Matrices, Abstract, Academy of Pharmaceutical Sciences, San Francisco (March 30, 1971).
3. J. H. R. Woodland, S. Yolles, D. A. Blake, M. Helrich, and F. J. Meyer, *J. Med. Chem.* **16**, 897–901 (1973).
4. T. M. Jackanicz, H. A. Nash, D. L. Wise, and J. B. Gregory, *Contraception* **8**, 227–234 (1973).
5. A. Schindler, R. Jeffcoat, G. L. Kimmel, C. G. Pitt, M. E. Wall, and R. Zweidinger, *Contemporary Topics in Polymer Science* (E. M. Pearce and J. R. Schaefgen, eds.), Vol. 2, pp. 251–286, Plenum Press, New York (1977).
6. E. E. Schmitt and R. A. Polistina, U.S. patent 3,297,033 (1967).
7. M. C. Bissery and F. Puisieux, Proceedings, Ninth International Symposium on Controlled Release of Bioactive Materials, The Controlled Release Society, Inc., Fort Lauderdale, Florida, (July 1982), pp. 30–34.
8. S. Mitra, M. Van Dress, J. M. Anderson, R. V. Petersen, D. Gregonis, and J. Feijen, *ACS Polym. Prepr.* **20**(2), 32–34 (1979).
9. J. Feijen, D. Gregonis, C. Anderson, R. V. Petersen, and J. Anderson, *J. Pharm. Sci.* **69**, 871–872 (1980).
10. D. Gregonis, J. Feijen, J. Anderson, and R. V. Petersen, *ACS Polym. Prepr.* **20**(1), 612–614 (1979).
11. W. A. R. Van Heeswijk, G. L. Brinks, and J. Feijen, Preprints, International Symposium on Polymers in Medicine, Porto Cervo, Sardinia, Italy, (1982), p. 6.
12. R. V. Petersen, S. W. Kim, J. M. Anderson, S. M. Fang, D. Gregonis, J. Nelson, D. Coleman, S. Woodward, and C. Anderson, Development and Testing of New

Biodegradable Drug Delivery Systems, Contract No. 1-HD-2824, Final Report submitted to NICHD, (December 14, 1980), pp. 1–125.
13. N. Lotan, A. Yaron, A. Berger, and M. Sela, *Biopolymers* **3**, 625–655 (1965).
14. N. Lotan, M. Bixon, and A. Berger, *Biopolymers* **8**, 247–257 (1969).
15. W. A. R. Van Heeswijk, M. J. D. Eenink, and J. Feijen, *Communications*, pp. 744–748, Georg Thieme Verlag, Stuttgart, (1982).
16. W. A. R. Van Heeswijk, M. E. Eenink, J. Feijen, H. M. Pinedo, J. Lankelma, and P. Leljeveld, Preprints, International Symposium on Polymers in Medicine, Porto Cervo, Sardinia, Italy, (1982), p. 23.
17. J. C. Ventner, M. S. Verlander, N. O. Kaplan, M. Goodman, J. Ross, Jr., and S. Sesayama, in;*Polymeric Delivery Systems* (R. L. Kostelnik, ed.) pp. 237–250, Gordon & Breach, New York (1978).
18. K. R. Sidman, W. D. Steber, A. D. Schwope, G. R. Schnaper, and R. Gayle, *Biopolymers* **22**(1), 547–556 (1983).
19. K. R. Sidman, A. D. Schwope, W. D. Steber, S. E. Rudolph, and S. B. Poulin, *J. Membr. Sci.* **7**, 277–291 (1980).
20. S. W. Kim, R. V. Petersen, and J. Feijen, in: *Drug Design*, Vol. X, pp. 193–250 (E. J. Ariens, ed.), Academic Press, New York (1980).
21. P. H. Maurer, *J. Immunol.* **88**, 330–338 (1961).
22. L. C. Anderson, D. L. Wise, and J. F. Howes, *Contraception* **13**(3), 375–384 (1976).
23. L. R. Beck, D. R. Cowsar, D. H. Lewis, J. W. Gibson, and C. E. Flowers, *Am. J. Obstet. Gynecol.* **135**(3), 419–426 (1979).
24. D. L. Wise, A. D. Schwope, S. E. Harrigan, D. A. McCarty, and J. F. Howes, in: *Polymeric Delivery Systems* (R. L. Kostelnik, ed.), pp. 75–89, Gordon & Breach, New York (1978).
25. C. G. Pitt and A. Schindler, in: *Biodegradables and Delivery Systems for Contraception, Progress in Contraceptive Delivery Systems* (E. S. Hafez and W. A. A. Van Os, eds.), Vol. I, pp. 17–46, MTP Press, Lancaster, England, (1980).
26. A. C. Tanquary and R. E. Lacey, eds., *Controlled Release of Biologically Active Agents*, Plenum Press, New York (1974).
27. J. Heller, personal communication (1979).
28. J. Heller, D. W. H. Penhale, R. F. Helwing, and B. K. Fritzinger, Proceedings, Eighth International Symposium on Controlled Release of Bioactive Materials, The Controlled Release Society, Inc., Fort Lauderdale, Florida (July 1981), pp. 90–96.
29. N. B. Graham, Proceedings, Ninth International Symposium on Controlled Release of Bioactive Materials, The Controlled Release Society, Inc., Fort Lauderdale, Florida, (July 1982), pp. 16–20.
30. R. V. Sparer, N. Ekwuribe, and A. G. Walton, Proceedings Eighth International Symposium on Controlled Release of Bioactive Materials, The Controlled Release Society, Inc., Fort Lauderdale, Florida, (July 1981), pp. 105–107.
31. C. L. McCormick and K. W. Anderson, Proceedings, Ninth International Symposium on Controlled Release of Bioactive Materials, The Controlled Release Society, Inc., Fort Lauderdale, Florida, (July 1982), pp. 32–34.
32. S. Yolles and M. F. Sartori, in: *Drug Delivery Systems* (R. L. Juliano, ed.), pp. 84–111, Oxford Press, New York (1980).
33. G. Gregoriadis, J. Senior, and A. Trouet, eds., *Targeting of Drugs*, Plenum Press, New York (1981).
34. G. Gregoriadis, C. Kirby, P. Lange, A. Meechan, and J. Senior, in: *Targeting of Drugs* (G. Gregoriadis, J. Senior, and A. Trouet, eds.), pp. 155–184, Plenum Press, New York (1981).

35. H. Ringsdorf, in: *Polymeric Delivery Systems* (R. L. Kostelnik, ed.), pp. 197–225, Gordon & Breach, New York (1978).
36. D. Papahadjopolous, G. Poste, W. J. Vail, and J. L. Biedler, *Cancer Res.* **36**, 2988–2993 (1976).
37. A. Trouet, R. Baurain, D. Deprez-De Campeneere, M. Masquelier, and P. Pierson, in: *Targeting of Drugs* (G. Gregoriadis, J. Senior, and A. Trouet, eds.), pp. 19–30, Plenum Press, New York (1981).

8

Controlled-Release Animal Repellents in Forestry

David I. Gustafson

Abstract. Reforestation efforts throughout the world are often hindered by animal browse damage. Numerous harmful natural phenomena affect seedling growth, including nutrient deficiencies, competing vegetation, insects, disease, and drought. However, animal damage is often found to have the most severely deleterious effects on the tree. Loss of time in reestablishing trees on harvested sites can necessitate costly replanting practices. The financial incentive for protecting seedlings has led to the development of several mechanical and chemical repellent systems.

The chemical deterrents generally must be applied in a sustained-release formulation in order to protect the tree for an extended length of time. Commercially available chemical repellents are of the contact variety: they are applied to the leaf surface, usually in conjunction with an adhesive. A systemic repellent is one which is absorbed by the tree, transported to the foliage, and thereafter expelled in a repugnant chemical form. Selenium, when applied in the selenite form, exerts such a systemic effect without causing the phytotoxic damage which has prevented the use of other systemic browse deterrents. Timed delivery forms for prolonging the effectiveness of systemic repellents are being developed.

Optimal application of controlled-release systems in both forestry and agricultural environments requires an *a priori* knowledge of both the release-rate behavior in the field and the efficiency of utilization of the active ingredient by the plant. Such insight requires extensive data collection, but may be predicted through the use of mass transfer models. Several complex phenomena combine to make the entire interaction between the plant and the slow release device highly unpredictable, but some progress has been made in defining the critical parameters that affect the efficiency of active ingredient utilization. A means for prescribing desirable release kinetics from soil applied sustained delivery systems should be available soon.

1. Introduction

Man's interaction with the forest ecosystem has resulted in a dilemma: he wishes to regenerate productive stands of timber, but in so doing he creates an optimal habitat for many natural enemies of trees.[1–8] As long as trees' natural enemies, such as deer and elk, remain desirable as game for sporting hunters, the only solution is to protect trees from these herbivorous antagonists of the forestry industry.

David I. Gustafson • Shell Development Company, P.O.Box 4248, Modesto, CA 95352.

The most susceptible of all trees are the young seedlings.[9] The amount of time and money invested in the planting of a tree warrants a certain effort to ensure that it reaches maturity as a lumber producer.[10] The systems which have been used to guarantee protected growth range from the mundane (paper sheets, rolled up, and stapled around the main seedling leader) to the bizarre (scattered human hair cuttings).[11] The mechanical protectants, such as the paper bud caps just described, often do physical harm to the seedling.[4] The chemical deterrents, such as human hair, often have very transient effectiveness, on the order of one to three months.[11,12] These two facts combine to make the animal damage dilemma an ideal target for controlled-release systems.[13]

The concept of controlled release, the prolonged delivery of active agents by incorporating them in an inert matrix, is a surprisingly new one given its many benefits. The advantages in the pharmaceutical field have already been exploited, and these include avoidance of patient compliance problems, minimization of subcutaneous and intramuscular injury from injection, and localization of drug delivery to the target organ.[14] In this forestry application, a sustained delivery system lowers the application cost, because only one, rather than multiple, treatments, is required. The efficiency of chemical deterrent utilization is greatly enhanced, because the rate of release is under more stringent control. Finally, the environmental contamination is reduced because there is no need for a wasteful excess application, as is sometimes done in conventional repellent treatments.

A major question arises in the use of such systems, however, regarding the rate of release that is desired. The situation is even more complex for systemic deterrents, which are applied to the soil, absorbed through the tree's root system, transported to the foliage, and thereafter eliminated in a repugnant, repelling form. Environmental and biochemical parameters combine to create an exceedingly complex mass transfer problem in such a case, but standard chemical engineering techniques may be used to identify the important dimensionless quantities that characterize the system.[15]

1.1. Animal Damage in Forestry

Although its severity varies considerably from site to site, there is virtually no forest plantation that is entirely free from animal damage.[16] The types of damage seen may be listed in approximate order of decreasing importance: browsing (generally defined as the consumption of the top 2 to 7 cm of the seedling by the browsing species), barking (removal of a strip of bark from a mature tree), root clipping, foliage clipping, and trampling.[13] The animals which are involved in such damage may be listed in order of decreasing importance: deer, porcupines, pocket gophers, hares, rabbits, elk, domesticated livestock, small rodents, mountain beavers, and bears.[13] Such generalizations have been arrived at through numerous surveys of foresters familiar with animal damage.[16]

1.1.1. Browse Damage to Seedlings

Many studies have identified browsing as the most prevalent type of animal damage encountered in forestry.[16] Both deer and elk are responsible for the majority of such activity. Browsing of seedlings takes place throughout the year, but it is often the heaviest during periods of rapid growth in the spring, when the most succulent foliage is present.[10] During particularly heavy winters, the lack of usual forage can lead to the browsing of tree seedlings, but this is much less commonly observed than spring browse.

1.1.2. Economic Impact of Browse

The total economic impact of animal browse damage is difficult to assess because of the uncertainty in assigning a monetary value to the seedling. The value of the lumber which it could produce can only be approximated, but the immediate cost of replanting an entire site because of extensive browsing is much easier to estimate. The cost due to deer browsing has been estimated to exceed several million dollars per year in Washington and Oregon alone.[10]

This rather high cost estimate is calculated by considering the value of lost time in reestablishing a viable, lumber-producing site on harvested land and the additional labor costs associated with replanting. Heavy browsing can sometimes slow the growth of the desired species to the extent that competitors gain a sufficient advantage to take over the entire site. In such a case, the site may be lost permanently unless very expensive mechanical preparation is carried out to remove the undesired species.

1.2. Repellent Systems

The number of protective schemes used in forestry nearly equals the number of foresters.[11] Each has a system with which the forester is most confident. The major criterion for a browse deterrent is that it prevents animal damage without harmfully affecting seedling growth. There are two classes of repellents commercially available at this time: mechanical barriers and chemical contact materials. A third class, systemic repellents, of which selenium is an example, should be available soon.

1.2.1. Mechanical Barriers

Commonly used mechanical barriers include the following: (1) a rigid, cylindrical plastic mesh (Vexar® tubing, E. I. Du Pont & de Nemours & Co.), which forms a fence around the seedling; (2) an elastic version of (1), available from the same source, which fits the tree much like a nylon stocking; (3) paper bud caps, water-resistant sheets, rolled up and stapled around the main seedling leader; and (4) nonwoven polyester fabric, sewn in the form of a sock and slipped over the main seedling leader (Reemay®, also from Du Pont). These mechanical systems represent the most popular means of protecting seedlings at the present time, even though there are many problems with each.[4,16]

The Vexar® tubes are quite effective, but expensive to deploy and maintain. Small pieces of wire are required to secure the tube in a vertical position, and at particularly windy sites workers must be sent into the plantations to adjust the tubing several times during the duration of browse susceptibility (usually three to five years). When the tube becomes tilted from the vertical, the seedling may contact the rigid mesh and alter its growth, often resulting in contorted, stunted development. The tubes have also not exhibited the biodegradability which was intended, and they may actually strangle the tree by retaining their initial rigidity long after the tree no longer requires browse protection. Some foresters have reported that the tube can "bake" seedlings at sunny, dry sites, apparently by reducing air flow around the tree. Finally, the tubes are generally no more than 50 cm in height, while browse can occur at greater seedling sizes, resulting in the fact that the exposed leader which protrudes from the top of the device may be clipped.

The Vexar® netting, because of its flexibility, does not contort the seedling to the extent that the rigid tubing can, but some constriction of growth is apparent from field observations. The netting is generally even shorter than the tubing, such that the main leader protrudes when the seedling is at a most vulnerable height.

Paper bud caps are quite simple to deploy and reasonably inexpensive (approximately $0.20 per tree), but their extremely small size (approximately 10 cm in length) renders their protective capabilities very temporary. The Reemay® sock is attractive in many ways since it remains on the seedling leader as the tree grows, but the fabric is only partially translucent, such that a reduction of photosynthetic input to the leader is effected, hampering the growth.

1.2.2. Chemical Contact Deterrents

The flaws which hinder the mechanical systems have induced researchers to seek out materials which, when applied to the foliage, dissuade browsing species from clipping the seedling. The commercially available materials have been selected through extensive trial and error studies.[17] Two products have found widespread use. BGR® (Big Game Repellent, available from McGlaughlin–Gormley King Co.) is a concoction containing 37% putrescent whole egg solids in a latex adhesive.[12] Thiram® is the common name for tetramethylthiuram disulfide, an odiferous material available from numerous sources throughout the world.[12]

Both materials have limited durations of effectiveness. Severe rainstorms can wash away the repellent the day after application.[12] Even under optimal meteorological conditions, the manufacturers of these substances recommend that they be applied every six to eight weeks. Foliage which appears after treatment is usually not protected by the presence of the material on old foliage, which is an important consideration since the critical seedling leader often produces fresh foliage.[11] A browsed seedling tip can induce the formation of multiple leaders, turning the tree into a shrub of limited utility to the forestry industry.

A second problem has severely limited the use of Thiram®. Its chemical structure is nearly identical to a compound used to treat alcoholics, which induces vomiting upon the ingestion of any alcohol. Certain planters have inadvertently poisoned themselves with trace of the chemical, thereafter experiencing severe nausea upon the consumption of alcoholic beverages.

1.2.3. Systemic Repellents

Problems with both mechanical barriers and chemical contact deterrents have led to the search for systemic repellents.[17] Such compounds should be able to exert browse reductive capability, while not significantly altering the normal growth of the seedling in a harmful way. Just as with the contact materials, hundreds of formulations have been tried and rejected. Selenium, when applied in the selenite form, has recently been shown to act as a systemic browse repellent.[13,18,19]

2. Selenium as a Timed-Release Systemic Repellent

The first indication of selenium's remarkable ability to affect browse behavior can be found in the travels of Marco Polo, who reported in the late 13th century the following observation from Tibet[20]:

> Throughout all the mountainous parts of it the most excellent kind of rhubarb is produced, in large quantities, and the merchants who come to buy it convey it to all parts of the world. It is a fact that when they take that road, they cannot venture amongst the mountains with any beasts of burden excepting those

accustomed to the country, on account of a poisonous plant growing there, which, if eaten by them, has the effect of causing the hoofs of the animal to drop off. Those of the country, however, being aware of its dangerous quality, take care to avoid it.

This hoof malady is called "alkali" disease, although it is now known to be caused by the consumption of foliage with large amounts of selenium, up to several thousand ppm.[20–25] Very few plant species are able to tolerate the high selenium foliar concentrations required to produce such behavior.[20,26]

2.1. Biochemistry of Selenium

Much of selenium's interaction with living tissue is caused by its chemical similarity to sulfur, which is directly above it in Group VIA of the periodic table. Selenium is thus often observed to enter into metabolic pathways by simply taking the place of sulfur, with slight differences in reactivity.[22–24] For instance, selenium analogues to the sulfur-containing amino acids such as methionine have been isolated from both plant and animal tissue.[27,28]

The toxic effects of large doses of selenium are well documented, and are thought to be caused by the altered biochemical behavior of molecules which have had their sulfur atoms replaced by selenium.[27] During the past 25 years the beneficial effects of selenium have been investigated.[29] These studies have shown that certain diseases are associated with low levels (less than 10 ppb Se) in the diet.[30–38] White muscle disease affects mainly lambs and calves, and its occurrence has been linked to selenium deficiencies.[34] Keshan disease is a cardiovascular myopathy which afflicts millions of humans deprived of selenium, mainly in China and Finland.[39]

The mechanism of selenium's metabolic role appears to be linked with both vitamin E and glutathione peroxidase activity.[22–24,35,40,41] Although once earmarked as a carcinogen, selenium is now being investigated as a cancer inhibitor.[42] The more recent data on the element's biochemistry have resulted in a partial lifting (as of August 1981) of the 1954 ban on selenium use in agriculture in the United States.[33]

2.1.1. Selenium in Plants

Plants capable of concentrating the high amounts of selenium necessary to cause alkali disease are known as accumulators.[20] Such species are referred to as primary accumulators if they can grow only in highly seleniferous soils. Secondary accumulators can tolerate high Se foliar levels, but are not restricted to seleniferous regions.[20] Most plants, including the

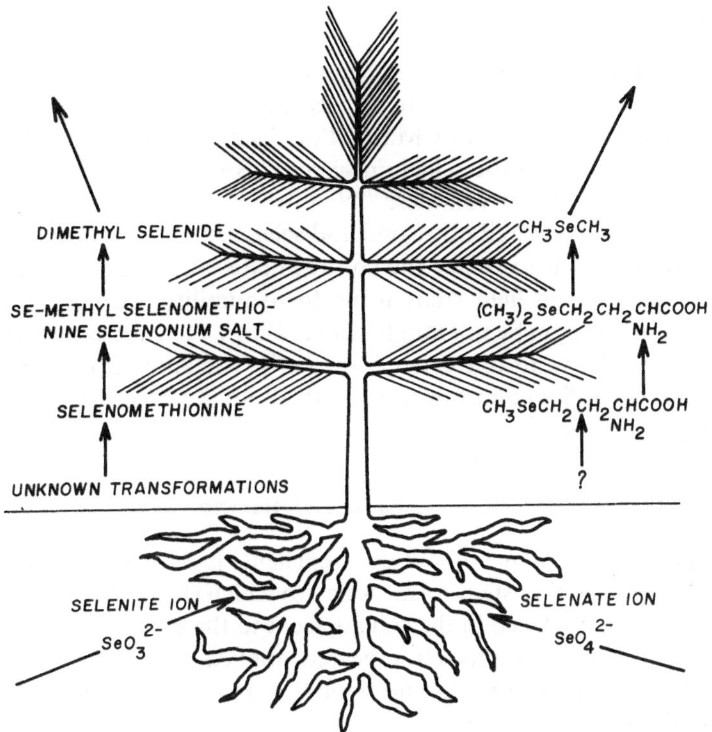

Figure 1. These final metabolic steps in the elimination of selenium from plants have been determined. Dimethyl selenide has a strong, garlicky odor and, being quite volatile, leaves the tree as it is formed.

commercially important trees, are not able to withstand foliar concentrations of the element above 100 ppm without suffering ill effects.[22–24]

As with many metals, selenium is eliminated from most living tissue by being methylated and thereafter expelled as a gas.[30,43–45] Some of the actual chemical species involved have been identified in nonaccumulating species, and these steps are shown in Figure 1. Selenium is absorbed through the root system as either selenite or selenate ion.[30] Other, more highly reduced forms of selenium, such as elemental selenium and the selenides, are not usually absorbed by plants.[30–32,37,46–49] Selenium which is absorbed goes through several unknown transformations which have the overall effect of reducing the atom, until selenomethionine is formed.[43] This amino acid is then methylated to yield a highly unstable selenonium salt which is enzymatically cleaved to give both dimethyl selenide and homoserine. Dimethyl selenide is quite volatile and leaves the plant as a gas. This compound has the same repugnant odor as most seleno-organics, and is thought to be responsible for bad breath in humans who have eaten garlic, which is rich in selenium.[50]

2.1.2. *Repellent Action of Selenium*

The strong odor of these selenium-containing volatiles renders the material an ideal systemic repellent. The learned behavior of the pack animals in Tibet, which was reported by Marco Polo, is most likely due to the ability of the foragers to detect the selenium scent.[20] Browsing animals are known to react strongly to plant volatiles.[51,52]

Caged rabbits offered untreated and selenized (10–20 ppm) Douglas fir seedlings were shown to prefer the control material.[13] Such repellency has also been observed in field trials using low-solubility selenium pellets. At foliar levels of 100 ppm Se in Douglas fir seedlings, 80% reduction of browse was observed relative to untreated controls.[19] The amount of protection is reduced at lower foliar concentrations, with about 50% reduction at 1–2 ppm Se in the tree needles.[19]

2.2. Formation of Timed-Release Browse Deterrents

The major advantage of systemic repellents is the ease with which they may be applied in a persistent form. A device which slowly releases the active ingredient is simply placed in the soil adjacent to the seedling.[19] Thus rather than being forced to administer repeated applications as with the use of contact deterrents, the forester need only apply the material once, most conveniently during planting.

The latex/adhesive formulations now used to deliver contact protectants, such as BGR®, may be regarded as a type of timed-release system (i.e., an eroding film), but this discussion will be limited to devices planted in the soil as described above. In particular, we will consider the production of a system for delivering selenium.

2.2.1. *Selection of A Selenite Salt*

Delivering the selenite, rather than the selenate ion, is desirable because the more highly oxidized species is more toxic and less readily eliminated from the seedling.[32,44,47] Selection of an appropriate salt of selenious acid is complicated by the fact that both selenium dioxide and its acid form (H_2SeO_3) are very powerful oxidation agents.[20,50] As noted earlier, the reduced forms of selenium are generally unavailable for absorption into the roots, thus one wishes to minimize the amount of selenite reduction which occurs before delivery to the soil solution. This necessitates selection of a highly stable cation, hopefully one with low solubility, in order to prolong the effectiveness of a device based on the salt.

A material which fits these specifications is melamine (triamino-*s*-triazine). It is highly stable to oxidation and forms salts with both organic and inorganic acids which are very sparingly soluble.[53–55] The dimelamine selenite salt is formed readily by mixing hot, stoichiometric aqueous solutions of melamine and selenium dioxide, followed by cooling and collection of the white crystals.[19] The salt has a solubility of about 1 g in 100 g H_2O, and an indefinitely long shelf life. Additionally, the melamine is responsible for some growth enhancement due to its high nitrogen content.[56]

2.2.2. *Production of Persistent Pellets*

Formation of devices which will deliver selenium for extended periods is relatively straightforward using the dimelamine selenite salt.[19] Melamine is commonly used, in conjunction with formaldehyde, to produce highly resistant resins through a polymeric, three-dimensional, cross-linking reaction between CH_2O and the three amino groups on each triazine center.[57] This reaction is catalyzed by both acid and base.[57] As the dimelamine selenite is inherently acidic (pH 5.05), cross-linking is catalyzed when an amount of aqueous formaldehyde is mixed with the salt. The pellets are formed by pressing such a mixture and heating briefly to accelerate the curing reaction.[19]

The pellets must then be coated to prevent dust formation during handling and to delay release of the selenium briefly in applications where immediate delivery is inappropriate. Any water-soluble polymer is adequate as a coating, so long as it is not toxic to the seedling and reasonably stable to oxidative attack. Chitosan, a natural derivative of arthropod skeletons, has been successfully used in such a manner.[19] Pellets were simply dipped in the polymer solution and allowed to dry.[19]

3. Characterizing the Performance of Controlled-Release Animal Repellents

The interaction of a soil-applied timed-release device with the target plant is highly complex. Superficially similar situations can lead to very different results, as indicated by the data in Figure 2.[18] The curves represent the selenium foliar concentrations observed in Douglas fir seedlings as a function of time following treatment with the slow-release pellets just described. In each case, the pellets were fabricated in exactly the same manner and were of the same dimensions (1 g; 1.27-cm diam cylinders, 0.7 cm in thickness). In addition, the (2-0) seedlings were healthy, containerized stock in both tests. The one difference was that at the Mutiny Bay field trial site the trees were treated with the pellets on the same day as they were planted, and

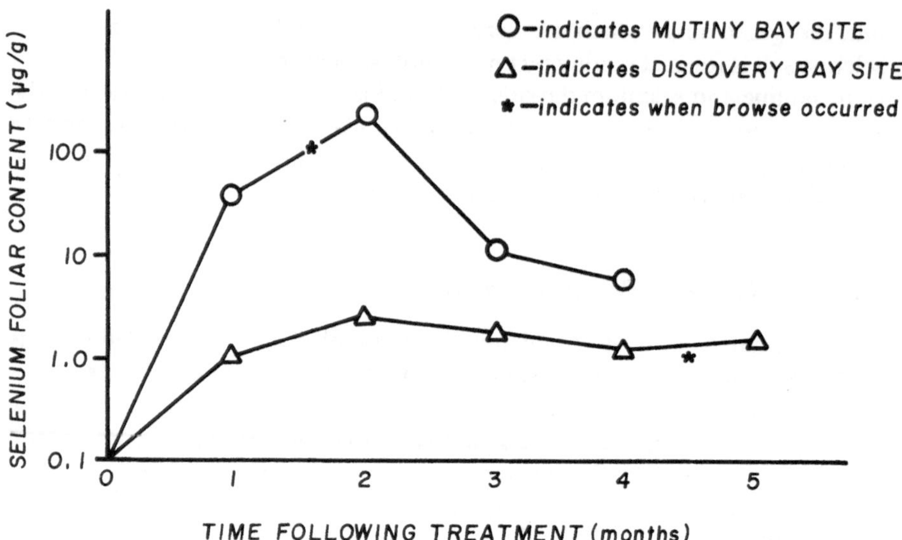

Figure 2. These two curves indicate the amount of selenium in the foliage as determined by neutron activation analysis. Browse occurred at the Mutiny Bay site two months following treatment, and 80% reduction of browse incidence was observed relative to trees not given pellets. The Discovery Bay site was browsed five months after treatment, and only 50% reduction of browse was obtained with the pellets.

the pellets were placed in the loose soil which had been disrupted by the planting tool, ensuring intimate root contact. At the Discovery Bay site, however, the pellets were placed in the soil adjacent to the stem two days after the tree had been planted. Important phenomena appear to have occurred immediately after planting at Mutiny Bay which caused the levels of selenium in the foliage to attain very high values.

The data which have been gathered at these sites provide a means of testing models of controlled-release animal-repellent performance. If developed with sufficient generality, such models could have application to other situations in which a sustained delivery system is placed in the soil, such as in the treatment of crops with timed-release pesticides.

3.1. Nature of the Problem

The situation we wish to describe is a sequential process involving the following steps: (1) delivery of active ingredient, in this case selenite ion, to the soil solution; (2) absorption of the active ingredient by the plant's root system; (3) metabolism and translocation of selenium to the foliage; and (4) volatilization of the repellent from the tree. In many agricultural applications, it is only steps (1) and (2) that we are interested in, but since this is a sequential, unidirectional process, no generality is lost by adding steps (3) and (4), which are crucial to the performance of the animal repellent. We will first describe in a phenomenological way the mass transfer process which escorts the selenite ion away from the timed-release device to the seedling.

3.1.1. *Mass Transfer from the Pellet to the Tree*

Selenite ion leaves the pellet through two independent mechanisms. The first is through simple dissolution of the dimelamine selenite salt. Since the solubility of the salt is approximately 0.02 *M*, this concentration of selenite ion should prevail in any aqueous solution contacting the solid, salt phase.

A complication arises, however, because of the method used to fabricate the pellets. An ion exchange matrix is formed when the melamine units are cross-linked through reaction with the formaldehyde.[18] This results in an additional transfer mode in which soil anions, such as carbonate and phosphate, exchange with selenite species ionically bound to the melamine/formaldehyde resin.

Both modes of selenite delivery depend in a direct way upon the moisture content of the soil. One would expect release to cease under arid conditions, and to accelerate during periods of extensive precipitation.

The mass transfer rate through the soil away from the pellet will also be a direct function of soil moisture content. One would expect molecular diffusion to be the primary mode of ion movement, but convective diffusion can also occur under conditions of complete soil saturation with water.[18]

Secondary effects on the mass transfer rate will include soil porosity, tortuosity, temperature, and pH. Dimelamine selenite solubility is a quite strong function of both temperature and pH, as indicated by data presented elsewhere.[18] The physical properties of the soil topography, as characterized by the simple porosity and tortuosity parameters, will have a distinct effect on the effective diffusivity of the selenite ion. Conductivity measurements indicate a quite low diffusion coefficient for the selenite ion of 3.0×10^{-6} cm^2/s at 25°.[118] Highly tortuous diffusional paths in the soil will lower this value even further, as will low porosities.

3.1.2. *Biokinetics of Selenium Metabolism*

The first biological process which one must model in assessing the performance of the sustained-release animal repellent is absorption through the root hairs. Studies of the absorption of selenium have indicated that the selenite and selenate ions are absorbed at different sites, and that sulfate ion competes with the selenate ion for its absorption site.[47] No analogous competitive inhibitor appears to exist for selenite absorption, thus one may assume simple Michaelis–Menten-type kinetics.[58] This kinetic model predicts linear dependence of absorption rate on the selenite concentration for small amounts of the ion in the soil. There is a saturation phenomenon, however, such that at high selenite concentration the absorption rate levels off to a constant value and becomes independent of the amount of the ion contacting the roots. This behavior is due to the fact that there are a finite number of absorption sites along the roots, and thus a limit to the pickup rate. Processes competing with absorption by root hairs include microbial volatilization[45,59,60] and formation of insoluble complexes with ferric hydroxides.[61,62] The latter is usually of the most importance.

Once in the plant, the selenite ion is subject to a dizzying number of different chemical transformations. In general, these reactions result in an overall reduction of the atom to selenides. Selenium is convected upward through the plant at a rate which is mainly determined by the evaporation rate of water from the foliage.[58] Capillarity draws water bearing the selenium up from the roots to replace evaporated fluids. By the time selenium has reached the tree needles it has probably undergone most of its transformations and can then be eliminated from the tree as the gas, dimethyl selenide. Indications of the data collected thus far are that this process is a simple, linear mechanism in which the amount of dimethyl selenide volatilized is directly proportional to the selenium foliar concentration.[18] It is this evolution rate of selenium which one is attempting to predict with the model, because there is both a lower limit for this parameter which must be exceeded for repellent action and an upper limit beyond which phytotoxicity is observed.[7,13,18,19]

A complication in modeling the biokinetic behavior of the tree is due to the fact that we are attempting to follow the mass transfer process over a period of several years. During this time the tree may grow four to five times its size at planting. It will also undergo fluctuations in metabolic activity with periodicities ranging from a day to a year. These two effects must also be taken into account in order to predict the repellent performance accurately.

3.2. Mathematical Analysis

Limits of space do not allow a complete description of the mathematical analysis of the mass transfer problem, thus this discussion will be confined to the foundations of the model. A more complete description of the model can be found elsewhere.[18] A few empirical correlations for describing the persistence of soil-applied materials have been proposed, but they have no generality, nor do they attempt to model the actual diffusional/kinetic interactions.[63–65]

Several definitions are required in order to begin, and we will first describe the rate of absorption of the active ingredient into the roots with the following expression:

$$r_a = \frac{v_{\max}^a C}{K_m^a + C} \tag{1}$$

In this expression we have used the standard Michaelis–Menten dependence of reaction rate on the molar concentration C of the active ingredient. This is a specific rate of root absorption, r_a, which has units of $\mathrm{mol\,cm^{-3}\,s^{-1}}$. The maximum rate of absorption, $v_{\max}^a$, is attained when C takes on very large values. The Michaelis constant, K_m^a, has units of molar concentration and corresponds to the concentration of active ingredient at which one-half the maximum rate of absorption is observed.

An assumption is implicit in the presentation of Eq. (1) that the root/soil system may be treated as a continuum with a certain uniform distribution of absorption sites. In actuality these sites will be concentrated on the nonuniformly dispersed root hairs. This assumption will be valid if the variations of the concentration C of active ingredient have a relatively long length scale compared to the length scale for variations in the absorption site number density distribution.

Using the same continuum hypothesis, and the same kinetic model, one can then obtain a similar expression for the specific rate of deactivation or insolubilization of the active ingredient:

$$r_d = \frac{v_{\max}^d C}{K_m^d + C} \tag{2}$$

The terms in this equation are entirely analogous to those in Eq. (1) with respect to both units and interpretation. This deactivation model is greatly simplified because, unlike absorption, the formation of the insoluble complexes which this represents is not a one-way process. When a selenite ion is absorbed by the root hair it is then transported through the tree and does not return to the soil solution. However, selenite ions which complex with ferric oxide groups in the soil can return to the soil solution as the equilibrium shifts. This simple model ignores the effect of desorption from colloidal particles, an approximation which becomes less accurate as time passes and as more selenite ions become adsorbed onto soil particles.

The concentration profile of the active ingredient will be heavily influenced by its effective diffusion coefficient in the soil, which we will denote as D_e. This diffusion coefficient will be much lower than the value which is found in pure water for the molecule. The major parameter affecting D_e will be the saturation S of the soil. S is generally defined as the volume of water actually present in a given representative amount of soil divided by the maximum volume of water which that same amount of soil could contain. As noted earlier, D_e will also be a function of porosity p, defined as the soil pore volume divided by the total soil volume, and of tortuosity t, defined as the diffusional path length divided by the actual displacement.

The sustained-release device will have the effect of maintaining the molar concentration of active ingredient at a certain value C^* in the soil solution immediately adjacent to its surface. This concentration will be a function of several parameters, but in this particular case the most important will be the temperature, the soil pH, and the ionic strength I, which has units of molar concentration and is defined by

$$I = \sum_{i=1}^{n} (C_i z_i^2) \tag{3}$$

In this definition the number of ionic species in solution is n, each having a molar concentration given by C_i and a valency z_i.

The tree's metabolism of the active ingredient is usually not known in detail, and this is certainly the case for selenium. We will assume, however, that the simple compartmental model given below is adequate to describe the overall effect of the process.

The model is illustrated in schematic form in Fig. 3. The active ingredient enters the root compartment at a rate R_a which has units of mol/s and is found through integration of the specific rate of absorption given by Eq. (1). This root compartment has a molar concentration of active ingredient given by C_r and a total volume V_r. The selenium then leaves this root compartment by virtue of a convective flow upward through the plant given by Q, a volumetric flow rate. Q will be a function of the evaporative loss of water from the tree, and will therefore be heavily influenced by weather conditions.

The destination of this flow is the foliage compartment which has a molar concentration of selenium, C_f, and a total volume V_f. Selenium will be modeled to leave the foliage at a rate governed by a simple first-order rate constant k_v, which will have units of inverse time. The total volatilization rate R_v at any given time will therefore be

$$R_v = k_v C_f V_f \tag{4}$$

Throughout the remainder of this analysis, we will consider a pseudo-steady-state situation, in which the tree has constant dimensions. It is not too difficult to include the effect of growth using numerical-solution

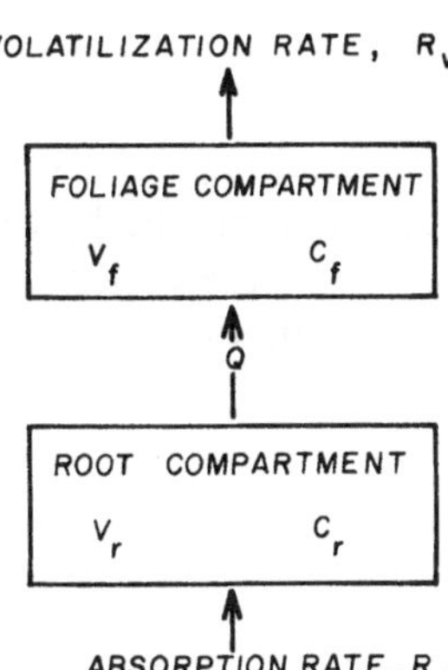

Figure 3. The seedling is modeled as a two-compartment system, with active ingredient absorbed through the root hairs to the root compartment, transported by a flow Q to the foliage compartment, and thereafter volatilized. Both the rate of flow to the foliage and the rate of elimination are first order in active-ingredient concentration, while the rate of absorption, R_a, follows Michaelis–Menten kinetics.

techniques, but we shall restrict the model to a static tree which has an extent of root development l_r. This length corresponds to the radius of a hemisphere in the soil, centered at the tree stem, in which there are root-hair absorption sites uniformly distributed. This characteristic length l_r is important to the phenomenon of leaching, in which active ingredient diffuses away from the root zone without being absorbed. The relatively high rate of absorption and insolubilization of selenite ion, combined with its low diffusivity, makes leaching somewhat unimportant to this particular example. However, this might not be the case for other materials.

3.2.1. Important Dimensionless Parameters

One standard method for identifying important dimensionless parameters is to write down the governing differential equations, nondimensionalize them and collect dimensionless groups.[15]

The first relationship we require is the convective diffusion equation for transfer through the soil. Assuming that there is no bulk flow through the soil we have

$$D_e\left(\frac{2}{r}\frac{\mathrm{d}C}{\mathrm{d}r} + \frac{\mathrm{d}^2C}{\mathrm{d}r^2}\right) = \frac{v^a_{\max}C}{K^a_m + C} + \frac{v^d_{\max}C}{K^d_m + C} \tag{5}$$

The radial coordinate r is measured from the center of the pellet, which we are approximating as a sphere of radius r_0. This radius is selected such that the surface area of the hypothetical sphere, $4\pi r_0^2$, is equal to the actual surface area of the slow-release device. The boundary conditions required to solve Eq. (5) are then

$$C(r_0) = C^* \tag{6}$$

$$C \rightarrow 0 \quad \text{as} \quad r \rightarrow \infty \tag{7}$$

The first boundary condition sets the concentration at the pellet surface at the value determined by the release kinetics, and the second states that far from the pellet the concentration of active ingredient must vanish.

Equation (5) is conveniently nondimensionalized by defining

$$\bar{C} = C/C^* \tag{8}$$

$$\bar{r} = r/r_0 \tag{9}$$

The differential equation then becomes

$$\frac{2}{\bar{r}}\frac{\mathrm{d}\bar{C}}{\mathrm{d}\bar{r}} + \frac{\mathrm{d}^2\bar{C}}{\mathrm{d}\bar{r}^2} = \frac{r_0^2 v_{\max}^a}{D_e K_m^a}\left[\frac{\bar{C}}{(C^*\bar{C}/K_m^a) + 1} + \frac{v_{\max}^d K_m^a}{v_{\max}^a K_m^d}\left(\frac{\bar{C}}{(C^*\bar{C}/K_m^d) + 1}\right)\right] \tag{10}$$

We see that four dimensionless groups have been derived through this method. The first is a Thiele modulus for absorption, which has the definition

$$h_T = r_0\left(\frac{v_{\max}^a}{D_e K_m^a}\right)^{1/2} \tag{11}$$

The Thiele modulus has the physical interpretation as the ratio of the rate of absorption to the rate of diffusion through the soil.

The second important group is defined as

$$f = \frac{v_{\max}^d K_m^a}{v_{\max}^a K_m^d} \tag{12}$$

This parameter is related to the efficiency of utilization of the active ingredient, and it has the physical interpretation as the ratio of the rate of deactivation to the rate of absorption at low concentrations.

The two other dimensionless groups are simple ratios of the delivery concentration to the Michaelis constants for either absorption or deactivation:

$$k_a = C^*/K_m^a \tag{13}$$

$$k_d = C^*/K_m^d \tag{14}$$

High values of these parameters would indicate maximum rates of the process near the pellet which are independent of concentration, while low values would indicate that the process is in the low-rate range where it is first order in concentration.

In terms of these newly defined groups the differential equation becomes

$$\frac{2}{\bar{r}}\frac{d\bar{C}}{d\bar{r}} + \frac{d^2\bar{C}}{d\bar{r}^2} = h_T^2\left(\frac{\bar{C}}{1 + \bar{C}k_a} + \frac{f\bar{C}}{1 + \bar{C}k_d}\right) \tag{15}$$

Numerical solution of this equation is required and has been reported for important ranges of the dimensionless groups.[18] The redimensionalized solution is used to calculate the total rate of absorption through the following integral:

$$R_a = 4\pi \int_{r_0}^{l_r} \left(r^2 \frac{v_{\max}^a C}{K_m^a + C}\right) dr \tag{16}$$

This overall rate of absorption feeds active ingredient to the root compartment, which has a selenium concentration governed by the following differential equation:

$$V_r \frac{dC_r}{dt} = R_a - QC_r \tag{17}$$

This equation may be nondimensionalized by defining

$$\bar{C}_r = C_r/K_m^a \tag{18}$$

$$\bar{t} = tv_{\max}^a/K_m^a \tag{19}$$

Then Eq. (18) becomes:

$$\frac{d\bar{C}_r}{d\bar{t}} = \frac{R_a}{V_r v_{\max}^a} - \left(\frac{QK_m^a}{V_r v_{\max}^a}\bar{C}_r\right) \tag{20}$$

The two dimensionless groups which arise can be defined as

$$d = \frac{R_a}{V_r v_{\max}^a} \tag{21}$$

$$q_r = \frac{QK_m^a}{V_r v_{\max}^a} \tag{22}$$

The first term quantitatively expresses the amount of dilution the active ingredient experiences upon entering the root compartment and is thus denoted by d. The second term, q_r, is a ratio of the time it takes for absorption to occur divided by the time it takes to empty the root compartment. The differential equation written in terms of these parameters is just

$$\frac{d\bar{C}_r}{d\bar{t}} = d - (q_r\bar{C}_r) \tag{23}$$

This linear, first-order equation is easily solved through the method of integrating factors, and the solution is

$$\bar{C}_r = \frac{d}{q_r} + \left[\bar{C}_r(0) - \frac{d}{q_r}\right] e^{-q_r\bar{t}} \tag{24}$$

This is for d and q_r independent of t, which can be made true by applying the equation for short time intervals.

The concentration of the systemic animal repellent in the foliage is the variable we are most interested in, and it is governed by

$$V_f \frac{dC_f}{dt} = QC_r - k_v C_f V_f \tag{25}$$

If we nondimensionalize this equation with the same characteristic concentration and time used to derive Eq. (23), we obtain

$$\frac{d\bar{C}_f}{d\bar{t}} = \left(\frac{QK_m^a}{V_f v_{\max}^a}\bar{C}_r\right) - \left(\frac{k_v K_m^a}{v_{\max}^a}\bar{C}_f\right) \tag{26}$$

The two groups which arise have the following definitions:

$$q_f = \frac{QK_m^a}{V_f v_{\max}^a} \tag{27}$$

$$v = \frac{k_v K_m^a}{v_{\max}^a} \tag{28}$$

The first group, q_f, gives the ratio of the time required for absorption divided by the time required to fill the foliage compartment. The second group, v, gives the ratio of the rate of volatilization to the rate of absorption.

Equation (26) may be solved through integrating factors to yield the solution:

$$\bar{C}_f = \frac{dq_f}{vq_r} + \left(\bar{C}_f(0) - \frac{dq_f}{vq_r}\right) e^{-v\bar{t}} + \left(\frac{[\bar{C}_r(0) - (d/q_r)]q_f}{q_f - v}\right)(e^{-q_r\bar{t}} - e^{-v\bar{t}}) \tag{29}$$

The volatilization rate from the tree is then calculated through Eq. (4) using the redimensionalized value of C_f found from Eq. (29).

3.2.2. *Results and Application in Practice*

Extensive data gathering is the major requirement for use of the model in any particular application. One must be able to evaluate the dimensionless groups given by Eqs. (11)–(14) in order to calculate the overall rate of absorption of the active ingredient. Models for predicting the effective diffusion coefficient D_e will generally require a means of converting weather data into values of the saturation S. Such conversions have been reported,[18] and are largely empirical in nature.

In applications where the distribution of the active ingredient in the plant tissue is important, as in the use of systemic repellents, additional dimensionless groups are required. These are given by Eqs. (21), (22), (27), and (28). Weather data implicitly affect the value of both Eqs. (22) and (27) by virtue of the water transport rate Q. Reasonable values for this parameter can be calculated from correlations reported elsewhere.[18]

Use of the model has been quite successful in predicting the selenium foliar concentration profiles at many sites, although the abnormal behavior observed at the Mutiny Bay site shown in Figure 2 has not been accounted for with this simple procedure. The results have been used, however, to design a new pellet for commercial applications with slightly different dimensions in order to "fine tune" the release rate.[18]

The assumptions concerning static tree dimensions can be easily removed by iterating the calculations over time intervals each with incrementally increased tree length and volume scales. Such iteration is also required to include variations in meteorological data.

The model presented in this section is widely applicable because the mathematical equations were solved in terms of dimensionless quantities. Generality is thus achieved which renders its accuracy universal for any soil-applied sustained-release device.

4. Future Impact of Bioactive Polymers in Forestry

The three-dimensional bioactive polymer which was formed during fabrication of the timed-release selenium pellets is among the first to be used in a forestry application. Other bioactive polymers have been used as controlled delivery systems for both herbicides and insecticides.[66]

4.1. Animal Repellents

The animal damage problem is one which can be expected to increase in intensity because of several factors. Deer populations are growing throughout the world because of the disappearance of their natural predators, and due to the optimal forage habitat produced by increasing lumbering activites. The use of both fertilizers and herbicides tends to create extremely healthy trees isolated within a forage-free circle, making the seedling all the more attractive to browsing species.

Foresters are becoming increasingly disillusioned with the currently available browse deterrents for the reasons outlined in Section 1.2; thus it is likely that controlled-release systemic repellents will make a strong impact on the science of reforestation in the very near future. The convenience of a one-time application is certainly a major consideration, and would justify the use of a slightly more expensive product. Market impact of timed-release systemic materials will depend heavily on their ability to reduce browse activity over the long term, however, and this sustained effectiveness has not yet been demonstrated conclusively.

4.2. Other Applications

There is little doubt that controlled-release materials will comprise the major use of bioactive polymers in forestry over the next 10–20 years. The convenience factors which improve the attractiveness of systemic repellents are equally valid for the application of other materials, such as fertilizers, herbicides, insecticides, and superabsorbents (used to increase water availability at dry sites).

In nearly all of these applications it will be important to know what delivery rate is required in order to maintain the active ingredient concentration within the desired limits. An important ratio which helps to determine how precise such delivery must be is the therapeutic index, or TI, defined as the lowest toxic concentration divided by the lowest effective concentration.[67] High values of this parameter are indicative of relatively safe materials for which the release rate need not be maintained at an extremely precise value. Low values, near one, would necessitate a very precise metering system to avoid poisoning the plant. Such dangerous materials could probably not be delivered with the simple type of device used to deliver selenium, since the rate of absorption would fluctuate too wildly with meteorological variations. The development of matrices to deliver such compounds is likely to be a topic of future research in bioactive polymeric systems.

ACKNOWLEDGMENTS

The author wishes to express his thanks to G. G. Allan for his constant advice and encouragement, and to INCO Research and Development for their continued support during the preparation of this manuscript.

References

1. A. Loudon, *New Scientist* **18**(3), 708 (1972).
2. G. L. Crouch, Spring-Season Deer Browsing of Douglas-Fir on the Capitol Forest in Western Washington, Pacific Northwest Forest and Range Experiment Station, Forest Service Research Notes, PNW-84 (August, 1968), p. 1.
3. G. L. Crouch and N. R. Paulson, Effects of Protection from Deer on Survival and Growth of Douglas-Fir Seedlings, Pacific Northwest Forest and Range Experiment Station, Forest Service Research Notes, PNW-94 (November, 1968), p. 1.
4. R. M. Anthony, Protecting Ponderosa Pine from Mule Deer and Plastic Tubes, *Tree Plant's Notes* (Summer, 1982), p. 22.
5. G. L. Crouch and M. A. Radwan, Effects of Nitrogen and Phosphorous Fertilizers on Deer Browsing and Growth of Young Douglas-Fir, Pacific Northwest Forest and Range Experiment Station, Forest Service Research Notes, PNW-368 (January, 1981), p. 1.
6. Deer Browse Control with a Repellent, *Forestation Notes*, USDA/Forest Service, No. 68 (March, 1982), p. 3.
7. J. H. Rediske and W. H. Lawrence, *Forest Sci.* **8**(2), 142 (1962).
8. G. A. Walters, Using Big Game Repellent to Protect *Acacia koa* Seedlings from Cattle, *Tree Planter's Notes* (Fall, 1981), p. 25.
9. W. E. Hines and C. E. Land, in: *Wildlife and Forest Management in the Pacific Northwest* (H. C. Black, ed.), pp. 121–132, School of Forestry, Oregon State University, Corvallis (1974).
10. E. J. Dimock and H. C. Black, in: *Symposium: Proceedings on Wildlife and Reforestation in the Pacific Northwest* (H. C. Black, ed.), pp. 80–81, School of Forestry, Oregon State University, Corvallis (1969).
11. M. Stockdale, Nipped in the Bud, *Tennessee Wildlife Magazine* (1981), p. 20.
12. Deer-Away Technical Report, International Reforestation Suppliers (July 8, 1981), p. 7.
13. S. Neogi, M.S. thesis, University of Washington (1980).
14. B. E. Ballard, in: *Sustained and Controlled Release Drug Delivery Systems* (J. R. Robinson, ed.), pp. 1–70, Marcel Dekker, New York (1978).
15. W. H. McAdams, *Heat Transmission*, McGraw-Hill, New York (1954).
16. E. J. Dimock, Cooperative Animal Damage Survey (CADS)—Biological Impacts and Highlights, Proceedings of the Western Forestry Conference, Sacramento, California (December 5, 1978).
17. W. R. Eadie, *Animal Control in Field, Farm and Forest*, Macmillan Co., New York (1954).
18. D. I. Gustafson, Ph.D. thesis, University of Washington (1983).
19. G. G. Allan, D. I. Gustafson, R. A. Mikels, J. M. Miller, and S. Neogi, The Reduction of Deer Browsing of Douglas-Fir (*Pseudotsuga menziesii*) Seedlings by Quadrivalent Selenium, *Forest Ecol. Manage.* (1983) (to be published).
20. I. Rosenfeld and O. A. Beath, *Selenium: Geobotany, Biochemistry, Toxicity and Nutrition*, Academic Press, New York (1964).
21. A. L. Moxon and M. Rhian, *Physiol. Rev.* **23**, 305 (1954).

22. A. Shrift, *Bot. Rev.* **24**(9), 550 (1958).
23. A. Shrift, *Ann. Rev. Plant Physiol.* **20**, 475 (1969).
24. A. Shrift, in: *Organic Selenium Compounds: Their Chemistry and Biology* (H. Guenther, ed.), pp. 763–814, Wiley, New York (1974).
25. S. F. Trelease and O. A. Beath, *Selenium*, The Champlain Printers, Burlington, Vt. (1949).
26. M. Sarquis and C. D. Mickey, *J. Chem. Ed.* **57**(12), 886 (1980).
27. D. C. Eustice, F. J. Kull, and A. Shrift, *Plant Physiol.* **67**, 1054 (1981).
28. B. H. Ng and J. W. Anderson, *Phytochemistry* **18**, 573 (1979).
29. K. Schwarz, Development and Status of Work on Factor-3 Selenium, *Fed. Proc.* **20**, 666–678 (1961).
30. W. H. Allaway, E. E. Cary, and C. F. Ehlig, in: *Symposium: Selenium in Biomedicine* (O. H. Muth, ed.), pp. 273–313, Avi, Westport, Conn. (1967).
31. B. Bisbjerg and G. Gissel-Nielsen, *Plant and Soil* **32**(2), 287 (1969).
32. G. W. Butler and P. J. Peterson, *Aust. J. Biol. Sci.* **20**, 77 (1966).
33. E. B. Gunby, *Bull. Selenium Tellurium Dev. Assoc.* **21**, 1 (1981).
34. O. H. Muth, *JAVMA* **142**(3), 272 (1963).
35. W. R. Meyre, D. G. Mahan, and A. L. Moxon, *J. Anim. Sci.* **52**(2), 302 (1981).
36. E. C. Segerson, W. R. Getz, and B. H. Johnson, *J. Anim. Sci.* **53**(5), 1360 (1981).
37. J. H. Watkinson and E. B. Davies, *New Zealand J. Agric. Res.* **10**, 116 (1967).
38. M. Yoshida, K. Iwami, K. Yasumoto, and K. Iwai, *Nippon Nogei Kagaku Kaishi* **55**(8), 680 (1981).
39. X. Chen, X. Chen, G. Q. Yang, Z. Wen, J. Chen, and K. Ge, in: *Selenium in Biology and Medicine* (J. E. Spallholz, J. L. Martin, and H. E. Gunther, eds.), pp. 171–175, Avi, Westport, Conn. (1981).
40. J. T. Rotruck, in: *Selenium in Biology and Medicine* (J. E. Spallholz, J. L. Martin, and H. E. Gunther, eds.), pp. 10–16, Avi, Westport, Conn. (1981).
41. A. T. Diplock, in: *Selenium in Biology and Medicine* (J. E. Spallholz, J. L. Martin, and H. E. Gunther, eds.), pp. 303–316, Avi, Westport, Conn. (1981).
42. D. V. Frost, *Chem. Eng. News* **59**(40), 4 (1981).
43. B. A. Gamboa-Lewis, in: *Environmental Biogeochemistry* (J. O. Nriagu, ed.), pp. 389–409, Ann Arbor Science Pub., Ann Arbor, Mich. (1976).
44. B. G. Lewis, C. M. Johnson, and T. C. Broyer, *Plant and Soil* **40**, 107 (1974).
45. D. C. Reamer and W. H. Zoller, *Science* **208**, 500 (1980).
46. G. Gissel-Nielsen, *Plant and Soil* **32**, 382 (1970).
47. A. M. Hurd-Karrer, *Am. J. Bot.* **24**, 720 (1937).
48. K. J. Jenkins and M. Hidiroglou, *Can. J. Biochem.* **45**, 1027 (1967).
49. O. E. Olson, E. E. Cary, and W. H. Allaway, *Agron. J.* **68**, 805 (1976).
50. H. Vokal-Borek, *Selenium*, University of Stockholm, Institute of Physics, Stockholm, Sweden (1979).
51. M. A. Radwan, Foliar Essential Oils and Deer Browsing Preference of Douglas-Fir Genotypes, Pacific Northwest Forest and Range Experiment Station, Forest Service Research Notes, PNW-324 (November, 1978), p. 5.
52. J. P. Scholl, R. G. Kelsey, and F. Shafizadeh, *Biochem. Sys. Ecol.* **5**, 291 (1977).
53. R. Chapman, P. Averell, and R. Harris, *Ind. Eng. Chem.* **35**, 137 (1943).
54. W. Scholl, R. Davis, B. Brown, and F. Reid, *Ind. Eng. Chem.* **29**, 202 (1937).
55. F. Knapp, *Ann. Chemie* **179**, 112 (1875).
56. R. A. Mikels, Ph.D. thesis, University of Washington (1982).
57. M. Okano and Y. Ogata, *J. Am. Chem. Soc.* **872**, 629 (1950).
58. J. F. Loneragan, in: *Trace Elements in Soil–Plant–Animal Systems* (D. J. D. Nicholas and A. R. Egan, eds.), pp. 109–134, Academic Press, New York (1975).
59. J. W. Doran and M. Alexander, *Soil Sci. Soc. Am. J.* **41**, 70 (1977).

60. R. Zieve and P. J. Peterson, *Sci. Total Environ.* **19**(3), 277 (1981).
61. F. J. Hingston, R. J. Atkinson, A. M. Posner and J. P. Quirk, *Int. Soil Sci.* **8**, 669 (1968).
62. J. P. Quirk and A. M. Posner, in: *Trace Elements in Soil–Plant–Animal Systems* (D. J. D. Nicholas and A. R. Egan, eds.), pp. 95–107, Academic Press, New York (1975).
63. M. G. Ford, R. Greenwood, and P. J. Thomas, *Pestic. Sci.* **12**, 175 (1981).
64. A. Walker and A. Barnes, *Pestic. Sci.* **12**, 123 (1981).
65. J. B. Passioura, *Plant and Soil* **18**(2), 225 (1963).
66. S. Roberts, Ph.D. thesis, University of Washington (1974).
67. V. H. Lee and J. R. Robinson, in: *Sustained and Controlled Release Drug Delivery Systems* (J. R. Robinson, ed.), pp. 71–122, Marcel Dekker, New York (1979).

9

Affinity Chromatography

C. R. Lowe and Y. D. Clonis

1. Introduction

The separation and purification of groups of biological macromolecules, such as enzymes, proteins, nucleic acids, and polysaccharides, by conventional procedures often requires considerable experience and expertise since individual members of the groups may differ only slightly in their physicochemical properties. However, one of the most characteristic properties of biological macromolecules is their ability to bind specifically and reversibly to other biomolecules. The technique of affinity chromatography exploits this formation of specific reversible complexes for the resolution of biological macromolecules.[1–3]

The concept of affinity chromatography is realized by covalently attaching one of the interacting species to a water-insoluble support and packing the support into a chromatographic column. In principle, when a solution containing the complementary species is passed through the column, the latter recognizes the immobilized partner and forms a specific and reversible complex; other species which do not recognize the bound molecule pass through the chromatographic bed unretarded and elute in the void volume. The specifically adsorbed biomolecule is subsequently eluted under conditions which favor dissolution of the complex (Figure 1).

The history of the affinity concept exploited in biomolecule separation commences as early as 1910 when Starkenstein[4] isolated α-amylase by adsorption on insoluble starch, while Willstätter *et al.*[5] purified lipase on powdered stearic acid. However, the modern concept of affinity chromatography as realized today using a ligand covalently bound to an insoluble matrix was first described only 30 years ago. Campbell *et al.*[6] isolated rabbit anti-bovine serum albumin antibodies on bovine serum albumin immobilized to diazotized *p*-aminobenzylcellulose. Lerman[7] subsequently isolated mushroom

C. R. Lowe and Y. D. Clonis • The Biotechnology Centre, University of Cambridge, Downing Street, Cambridge CB2 3EF, England.

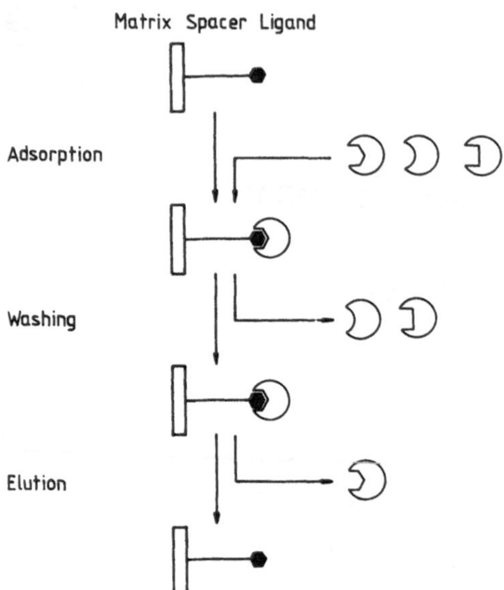

Figure 1. The principle of affinity chromatography.

tyrosinase on several *p*-azophenol-substituted cellulose adsorbents. Furthermore, Arsenis and McCormick[8,9] purified liver flavokinase on flavin-substituted cellulose and flavin mononucleotide-dependent enzymes on flavin phosphate-substituted cellulose. Until this time, affinity chromatography had found only limited application in biopolymer separations, primarily because of the poor qualities of available matrices and methods of activation and immobilization of ligands. However, the technique has witnessed a spectactular development in the last 15 years attributable to the introduction of agarose as a matrix,[10] the deployment of the cyanogen bromide activation procedure,[11] and the insertion of spacer molecules between the ligand and matrix backbone to alleviate steric effects.[10,12] These advances in the technology of affinity chromatography stimulated the extensive application of the technique in the separation and purification of various biological macromolecules and supramolecular structures.[13,14] Not surprisingly, therefore, a number of different terms have been proposed for both the technique and its individual components.[2,15] Thus, in order to distinguish the technique from nonspecific ion-exchange or hydrophobic chromatography, the terms "biospecific adsorption chromatography," "bioselective adsorption chromatography," "bioaffinity chromatography," "ligand specific chromatography," and "biospecific affinity chromatography" have been proposed. The polymer carrying the covalently attached species is invariably described as the "solid support," "carrier," or "matrix." The species bound to the matrix, often via a "spacer arm" is termed the "effector," "affinant," or simply the "ligand," whereas the substance to be isolated is called the "affiner partner," "ligate," or "complementary biomolecule."

2. The Constituents of the Ideal Affinity Adsorbent

The design of the ideal affinity adsorbent depends on how closely the experimental conditions chosen mimic those required for optimizing the ligand–macromolecule interaction. Thus, careful consideration must be given to the three main components of the adsorbent: the matrix, spacer arm, and ligand.

2.1. The Matrix

Four main classes of matrices have been used in affinity chromatography: natural polymers, semisynthetic polymers, synthetic polymers, and inorganic matrices. The ideal matrix should possess certain properties: (1) it must be insoluble, hydrophilic, and display a high degree of porosity; (2) the gel particles should be uniform, spherical, and rigid; and (3) the matrix should be physically and chemically stable and should have a large number of functional groups for coupling ligands.[1–3,10,12,13,15] In practice, however, a matrix possessing all these ideal properties is as yet unavailable, although a number of matrices possess some characteristics close to those of the ideal matrix.

Several natural polymers, such as agarose, cellulose, dextran, alginate, and carrageenan, are capable of forming gels. Among these, agarose is preeminent as a support matrix for affinity chromatography. Agarose is a linear polysaccharide comprising alternating 1,3-linked β-D-galactopyranose and 1,4-linked 3,6-anhydro-α-L-galactopyranose residues (Figure 2).[16]

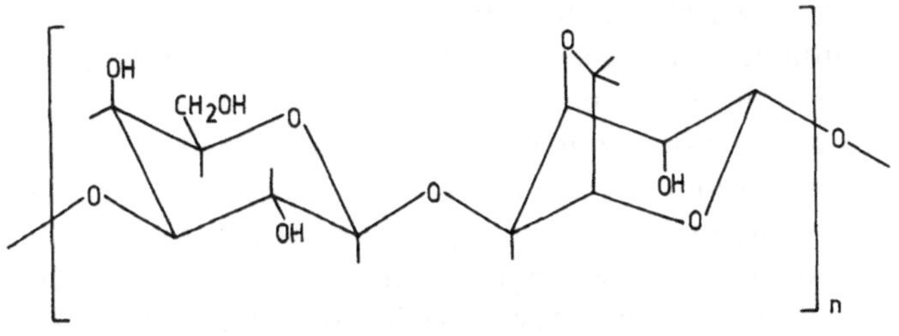

$$\left[\text{agarose repeating unit: OH, } CH_2OH \text{, O, OH, O, O, OH, O} \right]_n$$

Figure 2. The primary structure of agarose, an alternating copolymer of 3-linked β-D-galactopyranose and 4-linked 3,6-anhydro-α-L-galactopyranose residues.

Agarose gel comprises bundles of left-handed parallel double helices linked together with amorphous regions; the interior cavity of the helix is occupied by water molecules that participate in a hydrogen bonding system which stabilizes the overall gel structure.[17] Agarose displays low nonspecific adsorption of proteins, hydrophilic character, is very porous to large macromolecules, and can be prepared in beaded form.[18] The gel has good stability, especially after cross-linking with epichlorohydrin, 2,3-dibromopropanol, or divinylsulphone[19] and is therefore considered the best compromise as a matrix.[20]

Cellulose and dextran have been employed to a lesser extent, partly because cellulose frequently displays nonspecific adsorption, especially after activation,[21,22] and its fibrous and nonuniform structure impairs flow properties. Nevertheless, beaded cellulose derivatives are commercially available and cellulose derivatives have been exploited in affinity chromatography.[23] Dextran is a branched-chain glucose polysaccharide which when purified by fractional precipitation, epichlorohydrin cross-linked and beaded, is commercially available under the trade name Sephadex. However, the low porosity of Sephadex to large macromolecules seriously limits its general use in affinity chromatography. Nevertheless, for some types of ligand–macromolecule interaction, for example, when supramolecular structures such as cells, viruses, or organelles are to be separated and porosity is not a limiting factor[24,25] or where low molecular weight protens are being purified, dextran polymers may be used to great advantage.[26,27]

Besides the natural polysaccharides, synthetic polymers have also been used as matrices for affinity chromatography. Thus, polyacrylamide,[28–30] poly-*p*-aminostyrene,[31] and nylon [32,33] have been functionalized with a variety of ligands although their low porosity and hydrophobicity restricts their general use in affinity chromatography. On the other hand, poly(hydroxyalkylmethacrylate), commercially known as Spheron®, is available in a wide range of pore and bead sizes and contains a number of pendant hydroxyl functions which may be activated by the cyanogen bromide and other techniques.[13,34] Spheron appears to be an excellent adsorbent for affinity chromatography.

Semisynthetic matrices such as the agarose–polyacrylamide copolymer marketed under the trade name Ultrogel[36] and the allyl dextran-*N*,*N*′-methylenebis acrylamide copolymer available as Sephacryl[37] have received some attention although their applicability has yet to be tested more thoroughly. On the other hand, inorganic matrices offer some advantages over organic polymers. For example, controlled pore glass (cPG) is resistant to microbial attack and has a very rigid structure that is unaffected by changes in pH, pressure, ionic strength, or flow rate. Furthermore, its defined pore size offers sharp exclusion limits and good reproducibility.

Table 1. Some Adsorbents for Affinity Chromatography[1-3]

Agarose
Carrageenan
Cellulose
Cross-linked protein
Ethylene/maleic anhydride copolymer
Metal oxides
Metal oxide/agarose
Nylon
Plastics
Polyacrylamide/polyacrylic polymers
Polystyrene
Porous glass
Sephacryl
Sephadex
Silica
Titania
Trisacryl
Whole cells

Unfortunately, the ionized silanol groups of the surface of glass can create nonspecific adsorption effects. These problems can be circumvented either by coating the glass surface with dextran[38] or by silanizing with γ-glycidoxypropyltrimethoxysilane.[39-41] Microparticulate porous silica is a particularly effective adsorbent for high-performance liquid affinity chromatography.[40,41] Table 1 summarizes these and other adsorbents for application in affinity chromatography.

2.2. The Spacer Molecule

The requirement for a spacer molecule inserted between the ligand and the matrix backbone to minimize steric interference with the ligand–macromolecule interaction is now well established and will not be described in detail here.[42-47] It suffices to say that if a six-carbon chain is inserted between the matrix and ligand, dramatically enhanced adsorption of the complementary protein is observed. However, for ligands which display relatively high affinity for the complementary protein ($K_D \sim 10^{-6}\,M$) the effect of spacer molecules is not as profound as those with lower affinity; for example, staphylococcal nuclease binds to the strong competitive inhibitor 3′-(4-aminophenylphosphoryl)deoxythymidine 5′-phosphate immobilized directly to an agarose adsorbent.[43] Similarly, spacer molecules are not generally required where the ligand is a triazine dye,[48] lectin, antibody, or other high-affinity system.[1-3] In some cases, especially for low-affinity systems and/or high-molecular-weight complementary proteins, the deployment of macromolecular spacer molecules, such as denatured albumin,[49,50] poly(D,L-alanine),[51] polylysine,[52] and poly(lysylalanine)[53] simultaneously improves resolution and reduces ligand leakage due to solvolysis of the ligand–matrix junction.[2]

In most cases where low-molecular-weight ligands of average to moderate affinity are being immobilized, a 6-aminohexyl spacer molecule is now routinely employed. This is because it has become evident in recent years that true biospecific adsorption is a rare phenomenon[35] and is often supplemented by nonspecific interactions with the matrix–spacer arm assembly. In fact, in some cases, operational chromatography is only possible where nonspecific phenomena are present to reinforce the interaction between ligand and enzyme.[46,54,55] Thus, on a purely empirical basis and, in the absence of information to the contrary, a hexamethylene spacer molecule would normally be inserted between the ligand and matrix backbone to alleviate steric interference.

2.3. The Ligand

For effective purifications by affinity chromatography, a putative ligand must display the following characteristics: it must form specific and reversible complexes with dissociation constants ideally in the range 10^{-4}–$10^{-8}\,M$ in free solution with the biomolecules to be separated; it must possess at least one modifiable functional group not involved in the ligand–macromolecule interaction which would permit immobilization; and it must remain stable during the immobilization reactions. In addition, the selection of an appropriate ligand is contigent on the nature and mechanism of the ligand–macromolecule interaction, the affinity of the ligand for the macromolecule to be separated, and the position of attachment of the ligand to the spacer or matrix.[1–3] Typically, for enzyme purification the immobilized ligand may comprise a substrate, coenzyme, effector, or structurally related molecule displaying high affinity for the enzyme. However, for complex multisubstrate reactions the choice of appropriate ligand will depend more on the relative affinities for the complementary protein and their ease of immobilization.[56,57]

Careful consideration must also be given to the position of attachment of the ligand for immobilization; thus, it is important to avoid modifying groups vital for the ligand–macromolecule interaction.[58–60] In this context, a detailed knowledge of the three-dimensional structure of the macromolecule is useful in designing suitable affinity adsorbents.[61,62] In addition, the ligand–macromolecule interaction may be affected by other parameters such as immobilized ligand concentration,[59] bed geometry and dynamics,[63] temperature,[64] pH,[65] and dielectric constant.[66]

Table 2. Specificity in Affinity Adsorbents

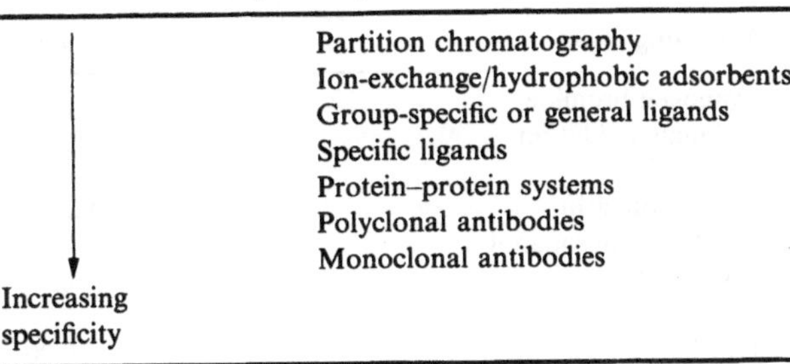

↓ Increasing specificity	Partition chromatography
	Ion-exchange/hydrophobic adsorbents
	Group-specific or general ligands
	Specific ligands
	Protein–protein systems
	Polyclonal antibodies
	Monoclonal antibodies

Ligands suitable for immobilization may be of two general types: "specific" ligands and "group-specific" or "general" ligands. The former class includes ligands which display narrow specificity for their complementary biomolecules but with the inherent drawback that a new adsorbent must be prepared for each individual substance to be purified. On the other hand, a group-specific or general ligand adsorbent once prepared may be used for whole groups of complementary molecules. Such adsorbents offer a compromise in terms of specificity, but this may often be overcome if specific elution techniques are employed. Table 2 illustrates these concepts of mono- and multispecificity as applied to affinity chromatography.

3. The Synthesis of Affinity Adsorbents

A recommended procedure for the synthesis of affinity adsorbents comprising immobilized low-molecular-weight ligands entails the synthesis of the spacer molecule–ligand conjugate by conventional organic chemistry and subsequent immobilization to an activated matrix material. The effort involved in isolating, purifying, and characterizing the intermediates of individual synthetic steps is rewarded by having precisely defined adsorbents that allow the experimental results to be interpreted with confidence. Furthermore, such a procedure avoids the introduction of superfluous unsubstituted spacer molecules and thus reduces the potentially damaging effects of nonspecific adsorption.[1–3,54,67]

Despite the advantages of defined adsorbents prepared in the above way, many workers continue to prepare bioselective adsorbents by prior activation of the matrix, coupling the spacer arm, and subsequent attachment of the ligand. The introduction of electrophilic groups into agarose by activation with cyanogen bromide[68–70] remains the most widely used procedure for coupling spacer molecules to polyhydroxylic matrices. Unfortunately, because of the formation of charged isourea functions,[71–74]

Table 3. Activation Procedures for Hydroxylic Supports

Activating agent	References
Cyanogen bromide	68–70
Periodate oxidation	81
Bisoxirane	75–77
Epichlorohydrin	82–84
2,3-Dibromopropanol	84
Divinylsulphone	85
p-Benzoquinone	86
Triazines	87–90
2,4,6-Trifluoro-5-chloropyrimidine	15
N,N'-carbonyldiimidazole	78
Tosyl chloride	92
Tresyl chloride	93
Metal chlorides	99, 100

the method invariably leads to a bioselective adsorbent displaying anion-exchange properties. Furthermore, isouronium functions are known to be susceptible to nucleophilic solvolysis and thus lead to ligand leakage.[(74)] Not surprisingly, therefore, a plethora of new activation procedures have been devised to produce stable, neutral, and leak-free linkages between the matrix and spacer molecule. Most of these procedures are listed in Table 3. In the authors' experience, activation procedures based on the use of bisoxiranes,[(75–77)] carbonyldiimidazole,[(78)] and various sulphonyl chlorides[(79)] are facile and produce leak-free neutral adsorbents. The relative advantages and disadvantages of the various activation procedures are given in several monographs.[(1–3,13)]

The attachment of bioligands to spacer arms involves a number of simple organic reactions[(1–3,13,43)] of which the most commonly used include acyl azides,[(94,95)] acid chlorides,[(96)] *N*-hydroxysuccinimide or *p*-nitrophenolate esters,[(97)] and carbodiimides.[(1,2,98)] Once the bioligand is covalently attached to the matrix the immobilized ligand concentration may be determined by difference analysis, spectroscopic methods, acid–base titration, elemental or chemical analysis, radioactive techniques, or hydrolysis of the ligand–matrix conjugate.[(1–3,13,43)]

4. The Chromatographic Properties of Affinity Adsorbents

A study of optimal conditions for the adsorption and elution of the complementary biomolecule from the affinity adsorbent may be undertaken once the design parameters for the adsorbent itself have been optimalized.

4.1. Adsorption of Complementary Biomolecules

As ionic or hydrogen bonds, hydrophobic interactions, and van der Waals forces may contribute to various extents in the binding of the complementary biomolecule to the immobilized ligand, the optimal conditions for the adsorption and desorption will depend on the system under investigation. In general, the choice of the column irrigating buffer is dictated largely by the conditions which favor optimal complex formation between the immobilized ligand and the isolatable substance. This will depend on a number of physical parameters, such as pH, ionic strength, temperature, and dielectric constant[(64–66)] and on the presence or absence of metal ions[(101)] and other specific factors. Careful juxtaposition of the four main physical parameters is a useful exercise in optimalizing the biospecific interaction while at the same time reducing nonspecific adsorption of macromolecules. Generally speaking, biospecific adsorption decreases with increasing temperature[(64)] although there are now a number of examples where the converse is true.[(102,103)] Furthermore, since both ionic and hydrophobic interactions are operative in protein binding sites, it follows that both pH and ionic strength are important factors in both adsorption and desorption.[(65,104)]

4.2. Bioselective Elution

Elution of a biomolecule from a chromatographic adsorbent by addition of low concentrations of competing substrate or inhibitor, such that there is no change in pH or ionic strength, is perhaps the best way of desorbing specifically bound proteins. Such a technique is effective for elution from ion exchange,[(105–109)] hydrophobic amphilyte,[(110,111)] and affinity adsorbents.[(1–3)] As a general rule, it is preferable to use an eluant ligand other than that which is immobilized to the matrix, since the system then entails dual specificity, for the immobilized and for the eluant ligand. For example, cAMP-dependent protamine kinase may be eluted from 8-(6-aminohexyl)-cAMP-agarose by 5 m*M* AMP. Furthermore, it is preferablc that the competing ligand displays higher affinity for the bound protein than the immobilized ligand; this ensures that low concentrations of eluant, ideally $< 10^{-3}\,M$, are required and thus minimizes nonspecific release of proteins. In cases where specific ternary complexes are required for adsorption of the desired protein, elution may be effected by withdrawing one of the interacting species from the column irrigants.[(112–114)] Elution specificity may also be enhanced by exploiting ternary complexes,[(115–117)] specific reduced NAD^+ adducts,[(118,119)] negative elution,[(114,120,129)] and the inhibition of ligand–macromolecule interactions.[(121,122)] In cases where the macromolecule to be purified and the immobilized ligand form extremely tight binary complexes, as for example in the case of hormone-receptor protein and immunological systems, elution may be effected with deforming buffers and chaotropes,[1–3] electrophoretic desorption,[123,124] or selective chemical,[43] or photolytic cleavage[125] of the ligand–matrix bond.

5. Applications of Affinity Chromatography

5.1. Protein Purification

In principle, the technique of affinity chromatography can be applied whenever a specific reversible interaction occurs between any two molecules. Not surprisingly, therefore, the literature abounds with innumerable examples of the purification of bioactive macromolecules and other complex structures by this technique.[1–3,13,126] Table 4 lists some examples of the use of group-specific ligands in affinity chromatography. This table does not include immobilized nucleotides and their analogues and the reactive triazine dyes, both of which have received much attention over the past decade or so. This is not surprising since of the 2000 or more enzymes, well over 30% participate in reactions involving nucleotides. Furthermore, the application of immobilized nucleotides and coenzymes in affinity chromatography has been the subject of a number of recent reviews.[60,62,127,128,131] The introduction of triazine dyes as immobilized group-specific media has made an enormous impact because of their low cost and ease of immobilization.[130]

The reactive anthraquinone dye, Cibacron Blue F3G-A, interacts with pyridine nucleotide-dependent dehydrogenases, kinases, coenzyme A-dependent enzymes, restriction endonucleases, synthetases, lyases, decarboxylases, and several blood proteins such as albumin, clotting factors, lipoproteins, complement factors, and interferon.[48,132,133] More recently, Procion Red HE-3B[132,134–136] and other dyes[48,101,114,129,130,137–140] have been extensively exploited in affinity chromatography. A number of studies have been performed, primarily with Cibacron Blue F3G-A, to establish the basis of these selective interactions.[133,136] It has been tentatively concluded that at least part of the structure of the dye mimics the naturally occurring bioactive heterocycles such as nucleoside phosphates, NAD^+, coenzyme A, and folic acid in terms of overall shape, charge, and aromaticity[141] and reflects coenzyme binding to the complementary protein.[61,142] Triazine dyes may thus be considered as nucleotide substitutes as used as pseudo-affinity ligands in protein purification. Such adsorbents are readily applied in large-scale purifications. For example, 3-hydroxybutyrate dehydrogenase was purified to homogeneity in two successive chromatographic steps on immobilized

Table 4. Some Immobilized Group-Specific Ligands for Application in Affinity Chromatography

Group-specific ligands	References
1. Coenzymes	
Biotin	155, 156
Cobalamine	157
Flavins	158, 159
Folates	160–165
Porphyrins	166, 167
Lipoic acid	168, 169
Pyridoxal coenzymes	170–172
2. Nucleotides	
Cytidine	173, 174
Guanine	175–182
Inosine	183–185
Thymidine	10, 186, 187
Uridine	188–190
3. Nucleic acids and polynucleotides	
DNA	191–194
poly(A)	195
poly(U)	196
poly(G)	197
4. Protein A	
Protein A	198–201
5. Lectins	
Concanavalin A	202–206
Soybean	207
Wheat germ	208
Pisum sativum	208
Lens culinaris	208, 209
Horseshoe crab	210
Bandeiraea simplicifolia	211
6. Boronates	
Boronates	212–214

Procion Red H-3B and Procion Blue MX-4GD with 70% yield.[48,140,143] This two-step affinity procedure replaced a conventional purification method involving eight steps with an overall yield of 9%. The same dye–ligand system may be used to purify malate dehydrogenase from the same source.[140,143] Furthermore, three tRNA synthetases, methionyl-, tryptophanyl-, and tyrosinyl-, were purified to homogeneity on Procion Green HE-4BD, Brown MX-5BR, and Orange MX-G adsorbents, respectively.[137] Conventional purification methods involved three further columns for tryptophanyl-tRNA synthetase and at least four for tyrosinyl- and methionyl-tRNA synthetases. Rapid large-scale purification was also reported for pig heart nucleoside diphosphate kinase on agarose-immobilized Cibacron Blue F3G-A.[144]

Affinity chromatography has also been used in the resolution and purification of antigen–antibody systems, hormone-binding proteins, and membrane-bound receptor proteins. For example, corticosteroid-binding globulin may be extracted from serum on agarose-bound cortisol, whereas oestradiol-binding proteins may be isolated on immobilized oestradiol.[145] Insulin receptors may be isolated on immobilized insulin[146,147] or various lectins.[148] Thyroid stimulating hormone (TSH) receptors were isolated on immobilized-TSH,[149] while nicotinic acetylcholine receptor protein was purified on agarose-bound pyridoxal phosphate-α-toxin.[150]

5.2. Purification of Supramolecular Structures

Affinity chromatography is also proving a potent technique for the resolution of viruses, phages, cells, organelles, and cell membranes.[151,14] Thus, agarose-bound D-mannose specific lectin from *Vicia ervilia* was effective in purifying influenza virus,[153] whereas other lectins were effective with bovine diarrhea viruses.[154]

5.3. Isoenzyme Resolution

All five isoenzymes of lactate dehydrogenase have been resolved on immobilized N^6-(6-aminohexyl)-AMP.[215] Similarly, specific elution with NAD^+ and cholic acid resolved the isoenzymes of horse liver alcohol dehydrogenase,[216] while a NADH gradient resolved cytoplasmic and mitochondrial malate dehydrogenase on the same adsorbent.[217] The isoenzymes of creatine kinase from the green sunfish were resolved on Blue Sepharose CL-6B.[218]

5.4. Removal of Contaminants

Contaminating lactate dehydrogenase activity may be removed from a crude commercial preparation of pyruvate kinase by chromatography on immobilized AMP.[219] A similar procedure has been used to remove an impurity of mitochondrial malate dehydrogenase from a commercial preparation of pig heart cytoplasmic enzyme,[217] to remove a number of dehydrogenases associated with a preparation of cytochrome C oxidase,[220] to remove albumin,[221] and other contaminants.[222]

5.5. Resolution of Mutant Proteins

Defective β-galactosidase forms produced from mutant strains of *E. coli* have been investigated by affinity chromatography,[223] whereas inactive mutant forms of *E. Coli* IMP dehydrogenase from guaB mutants have been purified on immobilized 8-(6-aminohexyl)-AMP following elution with a linear gradient of AMP.[224] Finally, the resolution of wild type from mutant enzymes and subunits of protein aggregates on immobilized Procion dyes has also been reported.[225]

5.6. Concentration of Dilute Solutions

Dilute solution of IgG may be concentrated on agarose-immobilized Protein A.[226]

5.7. Resolution of Chemically Modified Staphylococcal Nuclease from Native Proteins

Active site-modified staphylococcal nuclease may be resolved from active enzyme on agarose-bound deoxythymidine-3′-(*p*-aminophenylphosphate)-5′-phosphate,[227] whereas the resolution of functionally inactive $NADP^+$-isocitrate dehydrogenase from its active form has also been reported.[228]

5.8. Estimation of Dissociation Constants

Elution analysis and frontal analysis[230] may be used to determine the dissociation constants for the free and immobilized ligand although care should be exercised in the interpretation of such results in view of the complex time dependence observed in some cases.[231] Nevertheless, with some systems precise quantitative information may be obtained,[232] while in others a more qualitative approach may be preferable.[233]

5.9. Studies on Enzyme Kinetic Mechanisms

The compulsory ordered mechanism of lactate dehydrogenase in which the pyridine nucleotide binds first was elegantly confirmed by affinity chromatography.[234] Similarly, hexokinase only binds to immobilized ATP in the presence of the specific cosubstrate D-glucose.[55] In contrast, the equal effectiveness of all ADP and D-glucosamine adsorbents tested for hepatic glucokinase suggests that the kinetic mechanism for the binding of glucose and ATP to the enzyme is random and independent.[235] Immobilized coenzymes may also be utilized to distinguish between catalytic and effector sites, as with pyridine nucleotide transhydrogenase,[236] or to establish whether nucleotides may bind at the same or different sites on a particular enzyme, as with $NAD(P)^+$-dependent D-galactose dehydrogenase.[236]

Finally, information concerning the relative affinity of coenzyme fragments for the same enzyme binding site may be ascertained by affinity chromatography. Pig heart lactate dehydrogenase and horse liver alcohol dehydrogenase have been shown to bind strongly to all fragments greater than AMP.[236] Such studies in conjunction with investigations into the ionic, hydrophobic, and hydrogen bonding interactions between coenzyme fragments and enzyme, may make it possible to map the entire topography of the coenzyme binding site.

5.10. Clinical Applications

In vivo studies indicate that a number of substances, including bilirubin, digitoxin, cholic acid, salicylate, and immunoglobulins, may be removed from the blood by extracorporeal hemoperfusion through agarose-based affinity adsorbents.[237]

6. High-Performance Liquid Affinity Chromatography (HPLAC)

The resolving power and high speed of high-performance liquid chromatography (HPLC) may be combined with the biospecificity of affinity chromatography (AC) to yield a potent new technique, high-performance liquid affinity chromatography (HPLAC), capable of resolving proteins and other biomolecules with great ease. Thus, protein and other mixtures have been resolved on silica-immobilized nucleotides,[238] boronates,[239] dyes,[40,41] acriflavin,[240] lectins,[241] and immunoligands.[238]

7. Conclusions

The technique of affinity chromatography is a potent tool for the investigation and exploitation of specific and reversible interactions between biomolecules. In its widest sense the technique is also applicable to the construction of biosensors,[242,243] to the selective targeting of drugs,[244] and to the detoxification of human blood with appropriate extracorporeal adsorbents.[237]

References

1. C. R. Lowe and P. D. G. Dean, *Affinity Chromatography*, Wiley, Chicester (1974).
2. C. R. Lowe, in: *Laboratory Techniques in Biochemistry and Molecular Biology* (T. S. Work and E. Work, eds.), Vol. 7(II), Elsevier, Amsterdam (1979).

3. W. H. Scouten, *Affinity Chromatography: Bioselective Adsorption on Inert Matrices*, Wiley, New York (1981).
4. E. Starkenstein, *Biochem. Z.* **24**, 210 (1910).
5. R. Willstätter, E. Waldschmidt-Leitz, and J. Nemmed, *Z. Phys. Chem.* **123**, 93 (1923).
6. D. H. Campbell, E. R. Luescher, and L. S. Lerman, *Proc. Natl. Acad. Sci. U.S.A.* **37**, 575 (1951).
7. L. S. Lerman, *Proc. Natl. Acad. Sci. U.S.A.* **39**, 232 (1953).
8. C. Arsenis and D. M. McCormick, *J. Biol. Chem.* **239**, 3093 (1964).
9. C. Arsenis and D. M. McCormick, *J. Biol. Chem.* **241**, 330 (1966).
10. P. Cuatrecasas, M. Wilchek, and C. B. Anfinsen, *Proc. Natl. Acad. Sci. U.S.A.* **61**, 636 (1968).
11. R. Axén, J. Porath, and S. Ernbach, *Nature* **214**, 1302 (1967).
12. J. Porath, in: *Methods in Enzymology* (W. B. Jakoby and M. Wilchek, eds.), Vol. 34, pp. 13–30, Academic Press, New York (1974).
13. J. Turková, *Affinity Chromatography*, Journal of Chromatography Library, Vol. 12, pp. 246–318, Elsevier, Amsterdam (1978).
14. S. K. Sharma and P. P. Mahendroo, *J. Chromatogr.* **184**, 471 (1980)
15. T. C. J. Gribnau, *Coupling of Effector-Molecules to Solid Supports*, Drukkerij van Mameren B. V., Nijmegen (1977).
16. C. Araki, *Bull. Chem. Soc. Jpn.* **29**, 543 (1956).
17. S. Arnott, A. Fulmer, and W. E. Scott, *J. Mol. Biol.* **90**, 269 (1974).
18. S. Hjertén, *Biochim. Biophys. Acta* **79**, 393 (1964).
19. T. Kristiansen, *Biochim. Biophys. Acta* **362**, 567 (1974).
20. R. J. Boegman and M. J. Crumpton, *Biochem. J.* **120**, 373 (1970).
21. C. R. Lowe, M. J. Harvey, D. B. Craven, and P. D. G. Dean, *Biochem. J.* **133**, 499 (1973).
22. J. DeLarco and G. Guroff, *Biochem. Biophys. Res. Commun.* **50**, 486 (1973).
23. D. Mislovicová, P. Gemeiner, L. Kuniak, and Zemek, J., *J. Chromatogr.* **194**, 95 (1980).
24. C. F. Schlossman and L. Hudson, *J. Immunol.* **110**, 313 (1973).
25. D. B. Thomas and B. Phillips, *Eur. J. Immunol.* **3**, 740 (1973).
26. C. R. Lowe and P. D. G. Dean, *FEBS Lett.* **18**, 31 (1971).
27. J. Baird, R. Sherwood, R. F. G. Carr, and A. Atkinson, *FEBS Lett.* **70**, 61 (1976).
28. S. Hjertén and K. Mosbach, *Anal. Biochem.* **3**, 109 (1962).
29. J. K. Inman and H. M. Dintzis, *Biochemistry* **8**, 4074 (1969).
30. R. L. Schnaar and Y. -C. Lee, *Biochemistry* **14**, 1535 (1975).
31. I. H. Silman and E. Katchalski, *Ann. Rev. Biochem.* **35(II)**, 873 (1966).
32. W. E. Hornby and L. Goldstein, *Meth. Enzymol.* **44**, 118 (1976).
33. B. H. Yang, P. V. Sundaram, and A. Maelicke, *Biochem. J.* **199**, 317 (1981).
34. J. Turková, *J. Chromatogr.* **91**, 267 (1974).
35. Y. D. Clonis and M. J. Goldfinch, *J. Mol. Catal.* **16**, 1 (1982).
36. E. Boschetti, P. Girot, J. P. Secheresee, and J. S. Blancard, *J. Chromatogr.* **210**, 469 (1981).
37. D. W. Jacobsen, Y. D. Montejano, and F. M. Huennekens, *Anal. Biochem.* **113**, 164 (1981).
38. R. Regnier and W. H. Scouten, *Int. Lab. November/December 1974*, 13.
39. M. Glad, S. Ohlson, L. Hansson, M. -O. Månsson, and K. Mosbach, *J. Chromatogr.* **200**, 254 (1980).
40. C. R. Lowe, M. Glad, P. -O. Larsson, S. Ohlson, D. A. P. Small, A. Atkinson, and K. Mosbach, *J. Chromatogr.* **215**, 303 (1981).
41. D. A. P. Small, A. Atkinson, and C. R. Lowe, *J. Chromatogr.* **216**, 175 (1981).

42. P. Cuatrecasas, M. Wilchek, and C. B. Anfinsen, *Proc. Natl. Acad. Sci. U.S.A.* **61**, 636 (1968).
43. P. Cuatrecasas, *J. Biol. Chem.* **245**, 3059 (1970).
44. M. C. Hipwell, M. J. Harvey, and P. D. G. Dean, *FEBS Lett.* **42**, 355 (1974).
45. C. R. Lowe, M. J. Harvey, D. B. Craven, and P. D. G. Dean, *Biochem. J.* **133**, 499 (1973).
46. C. R. Lowe, *Eur. J. Biochem.* **73**, 265 (1977).
47. J. D. Aplin and L. D. Hall, *Eur. J. Biochem.* **110**, 295 (1980).
48. C. R. Lowe, D. A. P. Small, and A. Atkinson, *Int. J. Biochem.* **13**, 33 (1981).
49. S. A. Suleiman and R. Spector, *Arch. Biochem. Biophys.* **208**, 87 (1981).
50. B. H. Yang, P. V. Sundaram, and A. Maelicke, *Biochem. J.* **199**, 317 (1981).
51. R. Jost and A. Yaron, *Eur. J. Biochem.* **48**, 119 (1974).
52. M. Wilchek and T. Miron, *Meth. Enzymol.* **34**, 72 (1974).
53. G. Redenilh, R. Richard-Foy, C. Secco, Y. Torelli, R. Bucourt, E. E. Baulieu, and H. Richard-Foy, *Eur. J. Biochem.* **106**, 481 (1980).
54. P. O'Carra, S. Barry, and T. Griffin, *Biochem. Soc. Trans.* **1**, 289 (1973).
55. S. G. Doley, P. D. G. Dean, G. Dietz, M. J. Harvey, and P. J. Neame, in: *Chromatography of Synthetic and Biological Polymers* (R. Epton, ed.), Vol. 2, pp. 179–198, Ellis Horwood, Chichester (1978).
56. P. O'Carra and S. Barry, *FEBS Lett.* **21**, 281 (1972).
57. K. Slavik, W. Rode, and V. Slavikova, *Biochemistry* **15**, 4222 (1976).
58. C. R. Lowe, in: *Affinity Chromatography* (O. Hoffmann-Ostenhof *et al.* eds.), p. 39. Pergamon Press, London (1978).
59. M. J. Harvey, C. R. Lowe, D. B. Craven, and P. D. G. Dean, *Eur. J. Biochem.* **41**, 335 (1974).
60. C. R. Lowe, I. P. Trayer, and H. R. Trayer, *Meth. Enzymol.* **66**, 192 (1980).
61. J. F. Biellmann, J. -P. Samama, C. -I. Brändén, and H. Eklund, *Eur. J. Biochem.* **102**, 107 (1979).
62. C. R. Lowe, in: *Topics in Enzyme and Fermentation Biotechnology* (A. Wiseman, ed.), Vol. 5, pp. 13–146, Ellis Horwood, Chichester (1981).
63. C. R. Lowe, M. J. Harvey, and P. D. G. Dean, *Eur. J. Biochem.* **41**, 341 (1974).
64. M. J. Harvey, C. R. Lowe, and P. D. G. Dean, *Eur. J. Biochem.* **41**, 353 (1974).
65. C. R. Lowe, M. J. Harvey, and P. D. G. Dean, *Eur. J. Biochem.* **41**, 347 (1974).
66. C. R. Lowe and K. Mosbach, *Eur. J. Biochem.* **52**, 99 (1975).
67. P. O'Carra, S. Barry, and T. Griffin, *Meth. Enzymol.* **34**, 108 (1974).
68. R. Axén, J. Porath, and S. Ernbach, *Nature* **214**, 1302 (1967).
69. S. C. March, I. Parikh, and P. Cuatrecasas, *Anal. Biochem.* **60**, 149 (1974).
70. J. Porath, K. Aspberg, H. Drevin, and R. Axén, *J. Chromatogr.* **86**, 53 (1973).
71. R. Jost, T. Miron, and M. Wilchek, *Biochim. Biophys. Acta* **362**, 75 (1974).
72. B. Svensson, *FEBS Lett.* **29**, 167 (1973).
73. R. J. Yon and R. J. Simmonds, *Biochem. J.* **151**, 281 (1974).
74. G. I. Tesser, H. -U. Fisch, and R. Schwyzer, *Hew. Chim. Acta* **57**, 1218 (1974).
75. J. Porath and L. Sundberg, *Nature New Biol.* **238**, 261 (1972).
76. M. Caron, F. Fabia, A. Faure, and P. Corhillot, *J. Chromatogr.* **87**, 239 (1973).
77. R. G. Coombe and A. M. George, *Aust. J. Biol. Sci.* **29**, 305 (1976).
78. G. S. Bethel, J. S. Ayers, W. S. Hancock, and M. T. W. Hearn, *J. Biol. Chem.* **254**, 2572 (1979).
79. K. Nilsson and K. Mosbach, *Eur. J. Biochem.* **112**, 397 (1980).
80. R. Lamed, Y. Leven, and M. Wilchek, *Biochim. Biophys. Acta* **304**, 231 (1973).
81. M. Wilchek and R. Lamed, in: *Methods in Enzymology* (W. B. Jakoby and M. Wilchek, eds.), Vol. 34, p. 475 (1974).

82. J. Porath, J. -C. Janson, and T. Läas, *J. Chromatogr.* **60**, 167 (1971).
83. T. Kristiansen, *Biochim. Biophys. Acta* **362**, 567 (1974).
84. T. Låås, *J. Chromatogr.* **111**, 373 (1975).
85. J. Porath and R. Axén, in: *Methods in Enzymology* (K. Mosbach, ed.), **44**, 19 (1976).
86. J. Brandt, L. O. Andersson, and J. Porath, *Biochim. Biophys. Acta* **386**, 196 (1975).
87. G. Kay and M. D. Lilly, *Biochim. Biophys. Acta* **198**, 276 (1970).
88. J. R. Wykes, P. Dunnill, and M. D. Lilly, *Biochim. Biophys. Acta* **250**, 522 (1971).
89. K. Watanabe, K. Kimura, H. Marumo, and H. Samejima, *Agric. Biol. Chem.* **41**, 547 (1977).
90. Y. Morikawa, T. Tezuka, M. Teranishi, K. Kimura, Y. Fujimoto, and H. Samejima, *Agric. Biol. Chem.* **40**, 1137 (1976).
92. K. Nilsson, O. Norrlöw, and K. Mosbach, *Acta Chem. Scand.* **B35**, 19 (1981).
93. K. Nilsson and K. Mosbach, *Biochem. Biophys. Res. Commun.* **102**, 449 (1981).
94. E. E. Rickli and P. A. Cuendet, *Biochim. Biophys. Acta* **250**, 447 (1971).
95. C. J. Sanderson and D. V. Wilson, *Immunology* **20**, 1061 (1971).
96. H. H. Weetall and A. M. Filbert, in: *Methods in Enzymology* (W. K. Jakoby and M. Wilchek, eds.), Vol. 34, p. 59 (1974).
97. M. Robert-Gero and J. P. Waller, *Eur. J. Biochem.* **31**, 315 (1972).
98. C. R. Lowe, M. J. Harvey, D. B. Craven, and P. D. G. Dean, *Biochem. J.* **133**, 499 (1973).
99. A. N. Emery, J. S. Hough, J. M. Novais, and T. P. Lyons, *Chem. Eng.* **258**, 71 (1972).
100. J. F. Kennedy and V. W. Pike, *Enzyme Microb. Technol.* **1**, 31 (1979).
101. P. Hughes, C. R. Lowe, and R. F. Sherwood, *Biochim. Biophys. Acta* **700**, 90 (1982).
102. M. J. Comer, D. B. Craven, M. J. Harvey, A. Atkinson, and P. D. G. Dean, *Eur. J. Biochem.* **55**, 201 (1975).
103. F. Qadri and P. D. G. Dean, *Biochem. J.* **191**, 53 (1980).
104. K. Kasai and S. Ishii, *J. Biochem.* (*Tokyo*) **77**, 261 (1975).
105. B. M. Pogell, *Biochem. Biophys. Res. Commun.* **7**, 225 (1962).
106. J. A. Illingsworth and K. F. Tipton, *Biochem. J.* **118**, 253 (1970).
107. F. von der Haar, *Eur. J. Biochem.* **34**, 84 (1973).
108. H. G. Faulhammer and F. Cramer, *Eur. J. Biochem.* **75**, 561 (1977).
109. R. K. Scopes, *Biochem. J.* **161**, 253 (1977).
110. R. J. Yon, *Biochem. J.* **161**, 233 (1976).
111. R. J. Yon, *Biochem. J.* **185**, 211 (1980).
112. P. O'Carra and S. Barry, *FEBS Lett.* **21**, 281 (1972).
113. P. Andrews, *FEBS Lett.* **9**, 297 (1970).
114. Y. D. Clonis, M. J. Goldfinch, and C. R. Lowe, *Biochem. J.* **197**, 203 (1981).
115. R. Ohlsson, P. Brodelius, and K. Mosbach, *FEBS Lett.* **25**, 234 (1972).
116. L. Andersson, H. Jörnvall, and K. Mosbach, *Anal. Biochem.* **69**, 401 (1975).
117. C. R. Lowe, M. Glad, P. -O. Larsson, S. Ohlson, D. A. P. Small, A. Atkinson, and K. Mosbach, *J. Chromatogr.* **215**, 303 (1981).
118. N. O. Kaplan, J. Everse, J. E. Dixon, F. E. Stolzenbach, C. -Y. Lee, C. -L. Lee, S. S. Taylor, and K. Mosbach, *Proc. Natl. Acad. Sci. U.S.A.* **71**, 3450 (1974).
119. C. -Y. Lee, L. H. Lazarus, and N. O. Kaplan, in: *Enzyme Engineering* (E. K. Pye and H. H. Weetall, eds.), Vol. 3, p. 299, Plenum Press, New York (1978).
120. K. Slavik, W. Rode, and V. Slavikova, *Biochemistry* **15**, 4222 (1976).
121. R. Barker, K. W. Olsen, J. H. Shaper, and R. L. Hill, *J. Biol. Chem.* **247**, 7135 (1972).
122. I. P. Trayer and H. R. Trayer, *FEBS Lett.* **54**, 291 (1975).
123. M. J. Iqbal, L. Ford, and M. W. Johnson, *FEBS Lett.* **87**, 235 (1978).
124. M. R. A. Morgan, P. J. Brown, M. J. Leyland, and P. D. G. Dean, *FEBS Lett.* **87**, 239 (1978).

125. H. Sato, E. Hiei, S. Shimizu, and R. H. Abeles, *FEBS Lett.* **85**, 73 (1978).
126. C. M. Sturgeon and J. F. Kennedy, *Enzyme Microb. Technol.*, Literature surveys (1978–1985).
127. K. Mosbach, *Adv. Enzymol.* **46**, 205 (1978).
128. I. P. Trayer and M. A. Winstanley, *Int. J. Biochem.* **9**, 449 (1978).
129. Y. D. Clonis, *J. Chromatogr.* **236**, 69 (1982).
130. Y. D. Clonis, in *Affinity Chromatography of Nucleotide-Dependent Enzymes*, Ph.D. Thesis, University of Southampton, 1981.
131. Y. D. Clonis, *Chimika Chronika* (Athens), New Series **11**, 87 (1982).
132. P. D. G. Dean and D. H. Watson, *J. Chromatogr.* **165**, 301.
133. S. Fulton, in: *Dye–Ligand Chromatography* (M. Maroies, ed.), Amicon Corp., Lexington, Mass. (1980).
134. J. Baird, R. Sherwood, R. F. G. Carr, and A. Atkinson, *FEBS Lett.* **70**, 61 (1976).
135. A. J. Turner and J. Hryszko, *Biochim. Biophys. Acta* **613**, 256 (1980).
136. Y. D. Clonis and C. R. Lowe, *Biochim. Biophys. Acta* **659**, 86 (1981).
137. C. J. Bruton and A. Atkinson, *Nucl. Acid Res.* **7**, 1579 (1979).
138. C. R. Lowe, M. Hans, N. Spibey, and W. T. Drabble, *Anal. Biochem.* **104**, 23 (1980).
139. P. J. Neame and I. Parikh, *Appl. Biochem. Biotechnol.* **7**, 295 (1982).
140. M. D. Scawen, J. Darbyshire, M. J. Harvey, and A. Atkinson, *Biochem. J.* **203**, 699 (1982).
141. R. A. Edwards and R. W. Woody, *Biochemistry* **18**, 5197 (1979).
142. M. Land and P. G. H. Byfield, *Int. J. Biol. Macromol.* **1**, 223 (1979).
143. A. Atkinson, P. M. Hammond, R. D. Hartwell, P. Hughes, M. D. Scawen, R. F. Sherwood, D. A. P. Small, C. J. Bruton, M. J. Harvey, and C. R. Lowe, *Biochem. Soc. Trans.* **9**, 290 (1981).
144. J. Lascu, M. Duc, and A. Cristea, *Anal. Biochem.* **113**, 207 (1981).
145. W. Rosner and H. L. Bradlow, *J. Clin. Endocrinol. Metab.* **33**, 193 (1971).
146. P. Cuatrecasas and I. Parikh, *Meth. Enzymol.* **34**, 653 (1974).
147. T. W. Siegel, S. Ganguly, S. Jacobs, O. M. Rosen, and C. S. Rubin, *J. Biol. Chem.* **256**, 9266 (1981).
148. J. A. Hedo, L. C. Harrison, and J. Roth, *Biochemistry* **20**, 3385 (1981).
149. C. Rickards, P. Buckland, B. R. Smith, and R. Hall, *FEBS Lett.* **127**, 17 (1981).
150. F. A. Stephenson, R. Harrison, and G. G. Lunt, *Eur. J. Biochem.* **115**, 91 (1981).
151. N. Sharon, in: *Affinity Chromatography and Molecular Interactions* (J. M. Egly, ed.), Vol. 86, pp. 197–205, Inserm Colloque, Paris (1979).
153. T. Kristiansen, in: *Protides of Biological Fluids* (H. Peeters, ed.), p. 495, Pergamon Press, Oxford (1976).
154. T. Kristiansen, M. Sparrman, and J. Moreno-Lopez, in: *Affinity Chromatography and Molecular Interactions* (J. -M. Egly, ed.), Vol. 86, pp. 217–229, Inserm Colloque, Paris (1979).
155. P. Cuatrecasas and M. Wilchek, *Biochem. Biophys. Res. Commun.* **33**, 235–239 (1968).
156. G. Heney and G. A. Orr, *Anal. Biochem.* **114**, 92 (1981).
157. D. W. Jacobsen, Y. D. Montejano, and F. M. Huennekens, *Anal. Biochem.* **113**, 164 (1981).
158. A. H. Merrill and D. B. McCormick, *J. Biol. Chem.* **255**, 1335 (1980).
159. J. Sobhanaditya and N. Appaji-Rao, *Biochem. J.* **197**, 227 (1981).
160. H. L. Drake, S. I. Hu, and H. G. Wood, *J. Biol. Chem.* **225**, 7174 (1980).
161. R. L. Then, *Biochim. Biophys. Acta* **614**, 25 (1980).
162. F. A. Firgaira, R. G. H. Cotton, and D. M. Danks, *Biochem. J.* **197**, 31 (1981).
163. A. J. Wittwer and C. Wagner, *J. Biol. Chem.* **256**, 4102 (1981).
164. S. A. Suleiman and R. Spector, *Arch. Biochem. Biophys.* **208**, 87 (1981).

165. H. Nakata and H. Fujisawa, *J. Biochem.* (*Tokyo*) **90**, 567 (1981).
166. T. P. Conway and U. Muller-Eberhard, *Fed. Proc. Fed. Am. Soc. Exp. Biol.* **32**, 1386 (1973).
167. K. W. Olsen, *Anal. Biochem.* **109**, 250 (1980).
168. E. Schmincke-Ott and H. Bisswinger, *Eur. J. Biochem.* **114**, 413 (1981).
169. W. H. Scouten, F. Torok, and W. Gitomer, *Biochim. Biophys. Acta* **309**, 521 (1973).
170. E. Ryan and P. F. Fottrell, *FEBS Lett.* **23**, 73 (1972).
171. F. Kwok and J. E. Churchick, *J. Biol. Chem.* **254**, 6489 (1979).
172. C. D. Cash, M. Maitre, J. F. Rumigny, and P. Mandel, *Biochem. Biophys. Res. Commun.* **96**, 1755 (1980).
173. J. W. Baynes and F. Wold, *J. Biol. Chem.* **251**, 6016 (1976).
174. R. E. Scofield, R. P. Werner, and F. Wold, *Anal. Biochem.* **77**, 152 (1977).
175. P. T. Jackson, R. M. Wolcott, and T. Shiota, *Biochem. Biophys. Res. Commun.* **51**, 428 (1973).
176. H. -U. Siebeneick and B. R. Baker, *Meth. Enzymol.* **34**, 523 (1974).
177. D. L. Garbers, *J. Biol. Chem.* **254**, 240 (1979).
178. J. M. Wilson, B. W. Baugher, L. Landa, and W. N. Kelley, *J. Biol. Chem.* **256**, 10,306 (1981).
179. P. B. Iynedijan, *Enzyme* (*Basel*) **24**, 366 (1980).
180. L. Jervis, *Biochem. J.* **127**, 29P (1972).
181. S. Kanaya and J. Uchida, *J. Biochem.* (*Tokyo*) **89**, 591 (1981).
182. D. J. Ball and T. S. Nishimura, *J. Biol. Chem.* **255**, 10,805 (1980).
183. H. Rosemeyer and F. Seela, *Carbohydrate Res.* **62**, 155 (1978).
184. H. Rosemeyer and F. Seela, *Anal. Biochem.* **115**, 339 (1981).
185. Y. D. Clonis and C. R. Lowe, *Eur. J. Biochem.* **110**, 279 (1980).
186. W. Rohde and A. G. Lezius, *Hoppe-Seyler's Z. Physiol. Chem.* **352**, 1507 (1971).
187. E. P. Kowal and G. Markus, *Prep. Biochem.* **6**, 369 (1976).
188. R. Barker, K. W. Olsen, J. H. Shaper, and R. L. Hill, *J. Biol. Chem.* **247**, 7135 (1972).
189. R. K. Wierenga, J. D. Huizinga, W. Gaastra, G. W. Welling, and J. J. Beintema, *FEBS Lett.* **31**, 181 (1973).
190. K. W. Bock, D. Josting, W. Liliemblum, and J. Pfeil, *Eur. J. Biochem.* **98**, 19 (1979).
191. D. J. Arndt-Jovin, T. M. Jovin, W. Bähr, A. M. Frischauf, and M. Marquardt, *Eur. J. Biochem.* **54**, 411 (1975).
192. H. Schauer, C. Nüsslein, J. F. Bonhoefer, C. Kurz, and I. Nietzschman, *Eur. J. Biochem.* **26**, 474 (1972).
193. H. Tanaka, I. Sasaki, K. Tamashita, K. Miyazaki, Y. Matuo, J. Yamashita, and T. Hurio, *J. Biochem.* (Tokyo) **88**, 797 (1980).
194. P. Nehls and M. Renz, *Anal. Biochem.* **107**, 124 (1980).
195. H. C. Schröder, R. K. Zahn, K. Dose, and W. E. G. Müller, *J. Biol. Chem.* **255**, 4535 (1980).
196. J. DeMaeyer-Guignard, M. G. Tovey, and I. Gresser, *Nature* **271**, 622 (1978).
197. M. Pietrzak, H. Cundy, and M. Maluszynski, *Biochim. Biophys. Acta* **614**, 102 (1980).
198. P. L. Ey, S. J. Prowse, and C. R. Jenkin, *Immunochemistry* **15**, 429 (1978).
199. P. R. Field, S. Shankers, and A. M. Murphy, *J. Immunol. Meth.* **32**, 59 (1980).
200. B. W. Maidment, L. D. Papsidero, M. Gamarra, T. Nemoto, and M. Chut, *Anal. Biochem.* **111**, 336 (1981).
201. A. T. Yeung, K. J. Turner, N. F. Bascomb, and R. B. Schmidt, *Anal. Biochem.* **110**, 216 (1981).
202. M. C. Stuart, S. Ellis, L. Gowlland, and S. Tuff, *Clin. Chem.* **27**, 52 (1981).
203. K. Resch, S. Schneider, and M. Szamel, *Anal. Biochem.* **117**, 282 (1981).
204. J. Hermann and B. Keil, *Biochim. Biophys. Acta* **643**, 30 (1981).

205. E. Reinwald, P. Rautenberg, and H. J. Risse, *Biochim. Biophys. Acta* **668**, 119 (1981).
206. A. Jakobovits, Y. Eshdat, and N. Sharon, *Biochem. Biophys. Res. Commun.* **100**, 1484 (1981).
207. J. A. Hedo, L. C. Harrison, and J. Roth, *Biochemistry* **20**, 3385 (1981).
208. M. J. Hayman and M. J. Crumpton, *Biochem. Biophys. Res. Commun.* **47**, 923 (1972).
209. D. T. Dorai, B. K. Bachhawat, and S. Bishayee, *Anal. Biochem.* **115**, 130 (1981).
210. J. Viitala, K. K. Karhi, C. G. Gahmberg, J. Finne, A. Järnefelt, G. Myllydä, and T. Krusius, *Eur. J. Biochem.* **113**, 259 (1981).
211. R. P. Singhal, R. K. Bajaj, C. M. Buess, D. M. Smoll, and V. N. Vakharia, *Anal. Biochem.* **109**, 1 (1980).
212. V. Bouriotis, I. J. Galpin, and P. D. G. Dean, *J. Chromatogr.* **210**, 267 (1981).
213. T. A. Myöhänen, V. Bouriotis, and P. D. G. Dean, *Biochem. J.* **197**, 683 (1981).
214. P. Brodelius and K. Mosbach, *FEBS Lett.* **35**, 223 (1973).
215. L. Andersson, H. Jörnvall, and K. Mosbach, *Anal. Biochem.* **69**, 401 (1975).
216. R. A. Walk and B. Hock, *Eur. J. Biochem.* **71**, 25 (1976).
217. S. E. Fisher and G. S. Whitt, *Anal. Biochem.* **94** , 89 (1979).
218. K. Mosbach, H. Guilford, R. Ohlsson, and M. Scott, *Biochem. J.* **127**, 625 (1972).
219. J. J. Holbrook, J. Bucher, and R. Panniall, *Hoppe-Seyler's Z. Physiol. Chem.* **357**, 623 (1976).
220. J. Travis, J. Bowen, D. Tewksbury, D. Johnson, and R. Pannell, *Biochem. J.* **157**, 301 (1976).
221. M. Dao, J. Watson, R. Delaney, and B. Johnson, *J. Biol. Chem.* **254**, 9441 (1979).
222. M. R. Villarejo and I. Zabin, *Nature New Biol.* **242**, 50 (1973).
223. H. J. Gilbert, C. R. Lowe, and W. T. Drabble, *Biochem. J.* **183**, 481 (1979).
224. P. D. G. Dean and D. H. Watson, in: *Affinity Chromatography* (O. Hoffmann-Ostenhof *et al.*, eds.), pp. 25–38, Pergamon Press, London (1978).
225. J. W. Godding, *J. Immunol. Meth.* **13**, 215 (1976).
226. P. Cuatrecasas, *J. Biol. Chem.* **245**, 574 (1970).
227. P. Menter and W. Burke, *Fed. Proc., Fed. Am. Soc. Exp. Biol.* **38**, 673 (1979).
228. B. M. Dunn and I. M. Chaiken, *Biochemistry* **14**, 2343 (1975).
229. K. Kasai and S. Ishii, *J. Biochem. (Tokyo)* **72**, 629 (1976).
230. C. R. Lowe and M. G. Gore, *FEBS Lett.* **77**, 247 (1977).
231. S. Angal and I. M. Chaiken, *Biochemistry* **21**, 1574 (1982).
232. P. Brodelius and K. Mosbach, *Anal. Biochem.* **72**, 629 (1976).
233. P. O'Carra and S. Barry, *FEBS Lett.* **21**, 281 (1972).
234. I. P. Trayer, *Biochem. Soc. Trans.* **2**, 1302 (1974).
235. B. Höjeberg, P. Brodelius, J. Rydström, and K. Mosbach, *Eur. J. Biochem.* **66**, 467 (1976).
236. C. R. Lowe, in: *Theory and Practice in Affinity Techniques* (P. V. Sundaram and F. Eckstein, eds.), pp. 55–75,Academic Press, London (1978).
237. P. D. Berk, P. H. Plotz, and B. F. Scharschmidt, in: *Biomedical Applications of Immobilised Enzymes and Proteins* (T. M. S. Chang, ed.), Vol. 1, pp. 297–317, Plenum Press, New York (1977).
238. S. Ohlson, L. Hansson, P. O. Larsson, and K. Mosbach, *FEBS Lett.* **93**, 5 (1978).
239. M. Glad, S. Ohlson, L. Hansson, M. O. Månsson, and K. Mosbach, *J. Chromatogr.* **200**, 254 (1980).
240. D. A. P. Small, A. Atkinson, and C. R. Lowe, *J. Chromatogr.* **248**, 271 (1982).
241. A. Borchert, P. O. Larsson, and K. Mosbach, *J. Chromatogr.* **244**, 49 (1982).
242. C. R. Lowe, *FEBS Lett.* **106**, 405 (1979).
243. C. R. Lowe, *Trends in Biotechnology* **2**, 59 (1984).
244. G. Poste and R. Kirsh, *Biotechnology* **1**, 869 (1984).

10

Application of Radiation Grafting in Reagent Insolubilization

Chye H. Ang, John L. Garnett, Ronald G. Levot, and Mervyn A. Long

Abstract. Radiation grafting is shown to be a method with considerable research and industrial potential for the insolubilization of a wide range of organic reagents on polymer surfaces. The principle of the method is outlined in detail and involves radiation-induced copolymerization of a monomer containing an appropriate functional group to a polymer, then attachment of the reagent by subsequent chemical reactions. The relative merits of the two relevant grafting methods for this purpose, namely pre-irradiation and the mutual technique, are evaluated. Typical experimental procedures for each method are discussed. The mutual technique is shown to be more satisfactory for insolubilization reactions because of the lower radiation doses needed to achieve a particular percentage graft, resulting in less radiation damage to the backbone polymer. Variables influencing the efficiency of the mutual grafting method are reviewed, including solvent, dose rate, and dose. Additives for optimizing the grafting yield and properties are considered, including mineral acid and polyfunctional monomers. Methods for reducing competing homopolymerization are summarized. Three examples of the application of the mutual radiation grafting technique for insolubilization reactions are discussed in detail. These include immobilization of enzymes, heterogenization of catalytically active homogeneous metal complexes, and the anchoring of analytical reagents to form ion exchange resins.

1. Introduction

The concept of insolubilizing reagents on polymeric supports is a process with considerable commercial potential. Merrifield[1] originally introduced the technique of solid-phase synthesis of peptides by using an insoluble cross-linked macromolecule both as a protecting group and to simultaneously provide a facile method for isolating and purifying the product at each condensation step. Since this initial exposition of the solid-phase technique, the use of functionalized supports has expanded to include the anchoring of catalytically active homogeneous metal complexes and biologically active materials, as well as reagents for organic synthesis. The nature of the support and the means of attachment vary enormously both

Chye H. Ang, John L. Garnett, Ronald G. Levot, and Mervyn A. Long • School of Chemistry, The University of New South Wales, Kensington, New South Wales, Australia 2003.

within and between these different fields; but the one overriding requirement common to all methods is that the homogeneous reagent or catalyst must be permanently attached to a macromolecule in such a manner as to retain predominantly its original characteristics.

While individual supported reagents may have additional advantages over the homogeneous analogue, the major advantage of immobilization is in the ready isolation of the insolubilized reagent from the products of a reaction by energy-efficient physical means.

Numerous types of support have been used as the insoluble phase. These include polystyrene[2] and other vinyl polymers,[3] naturally occurring organic polymers such as cellulose,[4] and inorganic supports of silica[5] and alumina.[6] Methods for insolubilizing reagents on these supports have involved essentially conventional chemical reactions. Radiation-induced initiation[7,8] offers considerable potential for achieving the same process, particularly procedures involving radiation grafting. It is the purpose of this paper to discuss radiation grafting methods for insolubilization of reagents, particularly the advantages of such processes when compared with the chemical alternatives.

2. Principle of Radiation Grafting Insolubilization Method

The technique involves the use of ionizing radiation to graft a monomer containing an appropriate substituent to a backbone or trunk polymer. Attachment of a reagent to this system is then achieved by conventional chemical reactions involving the original substituent. In a typical example, *p*-nitrostyrene is radiation grafted to polypropylene powder, the nitro group in the resulting copolymer reduced to the amine [Eq. (1)] to which is attached the catalytically active species. The amine group can be chemically changed to any other functionality which may be required for the attachment reaction.

$$\sim\!\sim\!\sim + CH_2{=}CH{-}C_6H_4{-}NO_2 \longrightarrow \sim\!\sim\!\sim{-}CH_2{-}CH_2{-}C_6H_4{-}NO_2 \longrightarrow \sim\!\sim\!\sim{-}CH_2{-}CH_2{-}C_6H_4{-}NH_2 \qquad (1)$$

Many trunk polymers are radiation sensitive, especially naturally occurring macromolecules such as wool and cellulose. In such instances it is essential to use the lowest possible radiation dose for grafting to minimize degradation. Novel additives have been discovered to accelerate radiation grafting; thus inclusion of these materials in a radiation grafting system, particularly one involving naturally occurring backbone polymers, is particularly useful.

3. Radiation Grafting Procedures

When organic polymers are irradiated, simultaneous cross-linking and degradation of the chains occur due to the formation of free radicals on the polymer backbone. Irradiation in the presence of oxygen, an effective free-radical scavenger, produces peroxides and hydroperoxides within the polymer. The radicals formed have sufficiently long lifetimes such that when the irradiated polymer is placed in contact with an appropriate monomer, reaction occurs to yield a graft copolymer. This concept of radical formation in a monomer–polymer system results in two general methods for preparing radiation grafted materials.[9–11] The first of the procedures utilizes pre-irradiation of the backbone polymer before interaction with monomer, while the second procedure involves the simultaneous irradiation of monomer in contact with backbone polymer.

3.1. Pre-Irradiation Techniques for Grafting

In this method backbone polymer A_x is irradiated *in vacuo* or in the presence of an inert gas prior to exposure to the monomer which may be present as either vapor or liquid. On heating, the radicals formed during irradiation are mobilized and react with monomer giving graft copolymer in high yield compared to homopolymer produced.

A simple modification of the above procedure involves irradiation of backbone polymer A_x in the presence of oxygen to produce peroxide and hydroperoxide radicals which are then reacted with monomer B [Eqs. (2) and (3)]. This procedure offers the advantage of producing polymeric radicals with relatively long lifetimes, but introduces the problem of increased homopolymer due to homopolymerization initiated by hydroxy radicals generated by the decomposition of hydroperoxy radicals.

3.2. Mutual or Simultaneous Radiation Grafting Method

This technique utilizes the irradiation of trunk polymer A_x in contact with monomer B which may be present as vapor, liquid, or a solution. Irradiation may take place in air or *in vacuo*. Irradiation leads directly to the

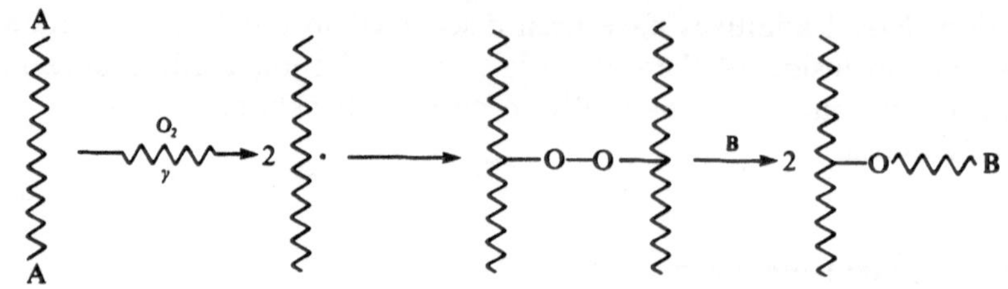

(2)

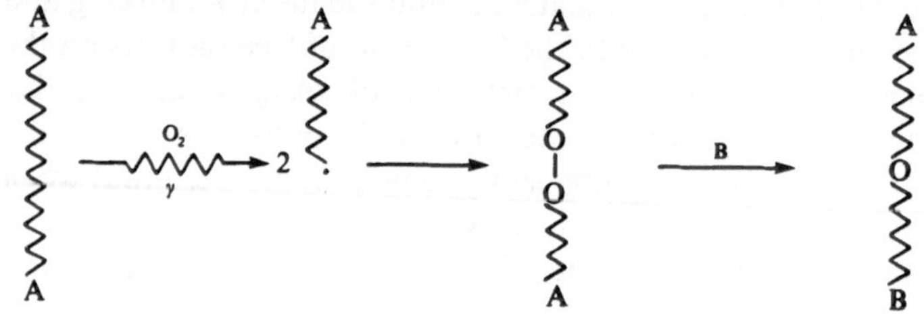

(3)

formation of active free radicals in both polymer A_x and monomer B resulting in graft copolymerization [Eqs. (4) and (5)]. This is the most efficient method of grafting since the radicals can react as fast as they are produced, however, homopolymer yields can be appreciable. Homopolymer formed can be removed by exhaustive soxhlet extraction.

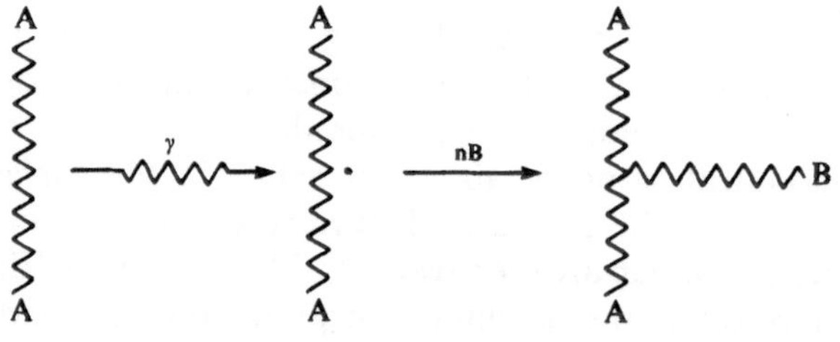

(4)

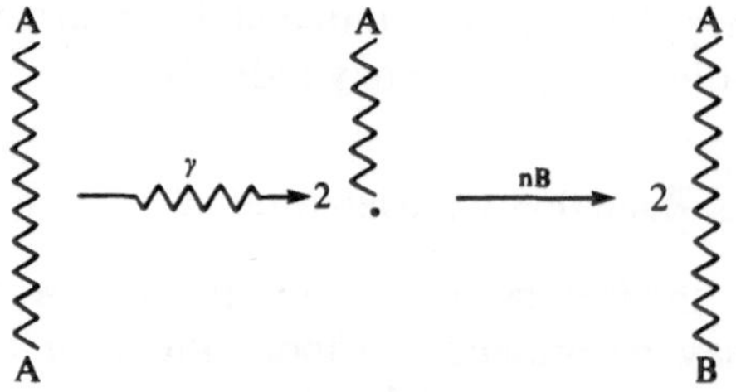

(5)

4. Typical Experimental Grafting Methods

For the grafting results reported in this paper purity of monomers and solvents utilized was carefully checked prior to use. Liquid monomers were freed from inhibitor and trace polymer by column chromatography on alumina, the grafting results being the same whether the monomers were purified by this method or by conventional distillation under nitrogen and reduced pressure.[12] After purification, monomers were reacted immediately or stored at $-2°$ C for, at most, one week prior to use. Solvents were high-purity grades and required no further treatment, except for methanol where care was taken to remove traces of ketones and aldehydes.[13]

4.1. Pre-Irradiation Grafting

For copolymerization in air, backbone polymer, either as film or powder, was placed in a stoppered glass vessel and irradiated in a pond cobalt-60 facility. After irradiation, the trunk polymer was reacted with solutions of monomer in glass tubes in an air oven at (60 ± 2)°C. The polymer was then removed from the tubes and washed with cold appropriate solvent prior to exhaustive extraction. For grafting *in vacuo*, the tubes were subjected to three freeze–thaw cycles and sealed off in ampoules at less than 1 Torr. Grafting yields are expressed as the percentage increase in weight of the original polymer which had been separated from homopolymer by exhaustive extraction with hot refluxing solvent for 100 h. For each particular grafting figure reported, up to eight determinations were made, this number being necessary because of the poor reproducibility, particularly with low grafts, by the pre-irradiation procedure.

4.2. Mutual or Simultaneous Grafting

Copolymerization experiments were performed in stoppered pyrex tubes, solvent being added first, then monomer to the required volume. The backbone polymer was fully immersed in the solution. After irradiation, grafted copolymer was removed quickly and treated as in the pre-irradiation method. Homopolymerization was determined by a modification[12] of the Kline procedure.[14]

5. Choice of Grafting Method for Insolubilization Reactions

Although considerable work has been reported using pre-irradiation grafting methods for the successful preparation of copolymers,[9–11,13] this pre-irradiation technique possesses a number of potential disadvantages for insolubilization processes. These problems originate from the basic concept of pre-irradiation since a minimum threshold radiation dose is required to rupture sufficient C-H bonds in the backbone polymer to give reasonable yields in the subsequent grafting step during reaction of these trapped radicals with monomer. By contrast, in the simultaneous method, the backbone polymers are irradiated in contact with monomer which structurally may be of a simple olefinic type (e.g., vinyl acetate) or a combination of aromatic and olefin bonding (e.g., styrene). Most of the useful backbone polymers required for insolubilization are predominantly hydrocarbon in nature (e.g., polyolefins, cellulose acetate). Olefinic compounds similar to the monomers normally protect aliphatic materials, such as these trunk polymers from radiolytic degradation, predominantly by scavenging reactions[(15)] particularly involving hydrogen atoms and electrons.[(16)] Thus, in the simultaneous method, there are additional mechanisms such as energy transfer which lead to grafting but do not involve prior rupture of C-C and C-H bonds in the backbone polymer. Hence grafting by the simultaneous method usually requires a much lower radiation dose to achieve a particular percentage degree of copolymerization and the resulting grafted material is inherently stronger than the corresponding copolymer from the pre-irradiation procedure.

Under certain irradiation conditions, particularly in air, the

Table 1. Pre-irradiation Grafting of Styrene and Vinylpyridines to Cellulose in Air[a]

Monomer	Time (h)[b]	Graft (%)[c] 20[d]	40[d]	60[d]	80[d]
Styrene	8	—	—	2	4
	15	—	—	8	22
	20	—	—	13	16
	25	—	—	5	17
	30	—	—	9	20
	41	—	—	19	31
	52	—	—	25	35
	65	—	22	—	—
	70	0	—	40	45
2-Vinylpyridine	386	0	0	0	6
4-Vinylpyridine	386	0	11	15	13

[a] Cellulose pre-irradiated in air at 1.5×10^5 rad/h to total dose of 2.5×10^6 rad, then reaction in monomer methanol solution.
[b] Time of reaction at 60°C for monomer in methanol.
[c] Reproducibility in quadruplicates poor for low yields (up to 30 h).
[d] Monomer concentration in methanol.

pre-irradiation technique can give poor reproducibility especially at low grafts. The data[13] in Table 1 demonstrate this property. Here styrene and the two vinylpyridines have been grafted to cellulose using the pre-irradiation technique. For the low grafting yields, that is, those samples where reaction was carried out for up to 30 h at 60°C, reproducibility in quadruplicates was poor and up to eight determinations had to be made to achieve meaningful results. This poor reproducibility is a distinct disadvantage of the pre-irradiation technique when compared with the simultaneous method. The latter method has therefore been chosen for most of the radiation work used in reagent insolubilization processes to date.

6. Variables Influencing Simultaneous Grafting

For insolubilization reactions a wide variety of backbone polymers are available for use, both polar and nonpolar types being of value for specific applications. For the purposes of discussing the variables which influence radiation grafting by the simultaneous method, two representative trunk polymers from each of the above groups will be considered. Thus polypropylene will be used as typical nonpolar synthetic trunk polymer and cellulose the corresponding, naturally occurring, polar material. The variables which predominantly influence grafting processes by the mutual irradiation technique include solvent, monomer structure, radiation dose, and dose rate.

6.1. Role of Solvent

The physical properties of a system markedly influence both the grafting yields and the type of copolymer that is obtained during radiation-initiated reactions. Thus grafting is enhanced with the appropriate backbone polymer if a swelling solvent is utilized. With cellulose, the low-molecular-weight alcohols act in this manner,[17–21] especially when reactive monomers such as styrene are being radiation grafted (Figure 1). The significant feature of the data in this figure is the peak in grafting which is observed at 30% monomer in methanol and ethanol. This is the Trommsdorff effect and is due essentially to diffusional properties of the radicals in the system. This gel effect is of practical significance since the absolute grafting rate is highest at the gel peak and chain lengths of graft copolymer are longer. Figure 1 also demonstrates the effect of wetting and swelling properties of the alcohol on grafting efficiency. Thus copolymerization drops markedly with *n*-butanol and even further with octanol, solvents that do not appreciably swell the trunk polymer. However, electron

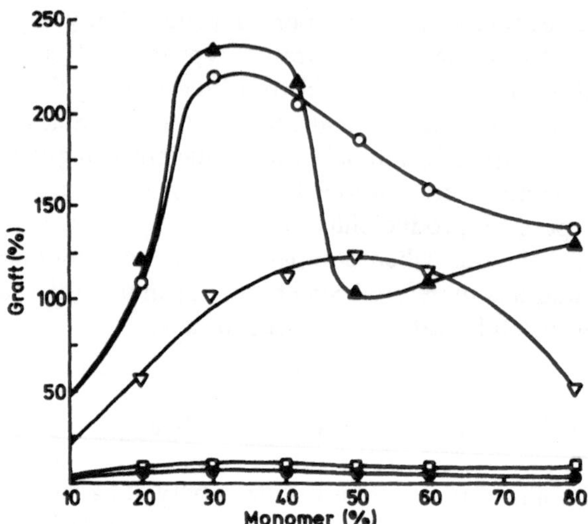

Figure 1. Gamma radiation grafting of styrene in alcohols to cellulose at 8.3 × 10^3 rad/h to total dose of 2.0 × 10^5 rad: (▲) methanol, (○) ethanol, (▽) *n*-butanol, and (●) *n*-octanol.

microscope studies show that the styrene graft in these two alcohols is essentially confined to the surface of the cellulose, whereas the corresponding copolymer in methanol consists of both surface and bulk graft. The choice of solvent can thus determine the specific nature of the copolymer.

When cellulose is replaced by polypropylene as backbone polymer, the analogous grafting data for styrene in the alcohols are different (Figure 2). Thus a Trommsdorff peak is now present in all solvents, even the higher-molecular-weight *n*-butanol and *n*-octanol. This enhancement observed in the polypropylene grafting is again attributed to swelling of the substrate facilitating diffusion of monomer to potential grafting sites. This is logical where the solvent has a greater affinity for the trunk polymer than does the monomer. The results for butanol and octanol in Figure 2 are consistent with this conclusion; however, enhancement in grafting is also observed with methanol which is precipitant of both the backbone polymer and the growing grafted chains. Odian and co-workers[22] have attributed this methanol behavior to the fact that methanol precipitates the growing grafted chains, leading to a reduction in the probability of bimolecular chain termination and hence increasing the grafting rate. However, these same data have been interpreted differently by the Silverman group[23] who proposed that methanol increased the viscosity of the grafting medium in the vicinity of the trunk polymer, especially near the grafting sites, thus reducing the mobility of the growing grafted chains leading to a corresponding decrease

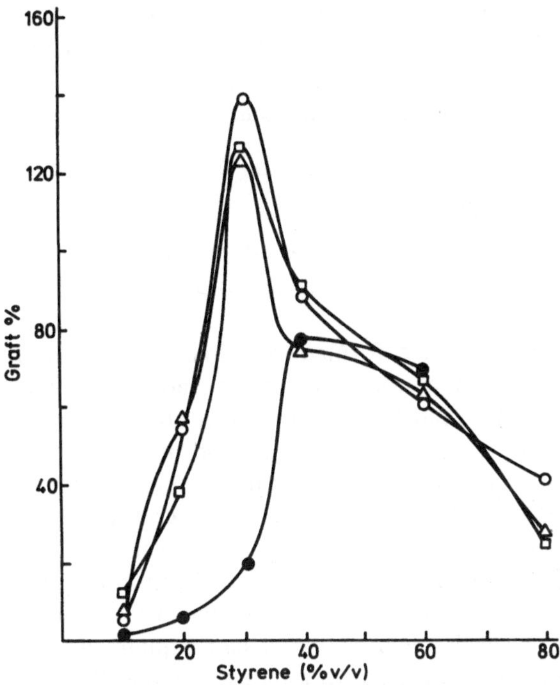

Figure 2. Effect of alcohols as solvents in styrene grafting to polypropylene film at dose rate of 4.5×10^5 rad/h to total dose of 3.0×10^5 rad; (○) methanol, (□) ethanol, (△) *n*-butanol, and (●) *n*-octanol.

in chain termination by the bimolecular process and a subsequent increase in grafting rate.

Both the Odian and Silverman models invoke essentially the physical properties of the grafting system to explain the observed copolymerization data. Swelling either from the solvent or monomer or both is also an important process in these reactions. However, if the alcohol results in Figures 1 and 2 are combined with additional solvent data for these reactions, involving benzene, pyridine, chloroform, acetone, and carbon tetrachloride,[(24)] a further theory would appear to be necessary to explain all solvent properties observed. Thus it is important to consider the radiation chemistry of the system and, in particular, the radiolysis products of the solvent in any complete analysis of the copolymerization process.[(25,26)] It has been suggested[(25)] that a contribution to the mechanism of the acceleration effect of methanol can be due to the radiolytic scavenging of styrene[(26,27)] and hence the relative numbers of styrene molecules and methanol radicals leading to charge transfer intermediates in the copolymerization reaction. In terms of the radiolytic theory, in a grafting system consisting of backbone polymer (PH), styrene monomer (M), and solvent (SH), the following primary initiating sequence of reactions will occur under irradiation:

$$PH \rightarrow P^{\cdot} + H^{\cdot} \quad (6)$$

$$SH \rightarrow S^{\cdot} + H^{\cdot} \quad (7)$$

$$PH + S^{\cdot} \text{ (or } H^{\cdot}) \rightarrow P^{\cdot} + SH \text{ (or } H_2) \quad (8)$$

$$M + S^{\cdot} \text{ (or } H^{\cdot}) \rightarrow MS^{\cdot} \text{ (or } MH^{\cdot}) \quad (9)$$

Grafting sites $P^{\cdot}$ are formed by direct bond rupture and also by hydrogen abstraction reactions with solvent radiolysis fragments $S^{\cdot}$ and $H^{\cdot}$, styrene monomer not readily forming radical products directly. Styrene can, however, scavenge radicals, the scavenged products also being capable of H atom abstraction reactions to give grafting sites. Following activation of these sites, chain initiation, growth, and termination occur either by bimolecular combination or disproportionation. As grafting proceeds, solvent radicals $S^{\cdot}$ and $H^{\cdot}$ are also scavenged by monomer to produce species $MS^{\cdot}$ or $MH^{\cdot}$ which may initiate homopolymerization. The important feature of this radiolytic mechanism is that it can explain aspects of the solvent grafting data that are difficult to rationalize by either the Odian or Silverman theories. In particular, all of the solvents that are mechanistically significant in grafting possess one common property, namely, that under radiolysis conditions they produce hydrogen atoms as a predominant species. From these studies, it does appear that the numbers of hydrogen atoms may well be a major contributing factor, in addition to the physical parameters defined by Odian and Silverman, in obtaining substantial copolymerization yields in general radiation grafting processes.

6.2. Significance of Dose Rate and Dose

As predicted by general grafting theory, copolymerization is inversely proportional to radiation dose rate; the data in Table 2 for the grafting of the styrene in methanol to the polyolefins confirm this trend. The results also clearly show the shift in the Trommsdorff peak from 50% monomer in solvent at 10,000 rad/h to 70% styrene at 75,000 rad/h. As the radiation dose rate increases, the magnitude of the gel peak progressively diminishes until it disappears at 112,000 rad/h. Although the dose rate and total dose reported for the polypropylene data in Table 2 are slightly different from those used to obtain the corresponding polyethylene results, a comparison of the data for the two polyolefins is useful and shows that grafting yields are comparable per radian of dose for both materials. However, the Trommsdorff peak

Table 2. Effect of Radiation Dose Rate on Grafting of Styrene in Methanol to the Polyolefins[a]

	Graft (%)					
	Polyethylene[b]					Polypropylene[c]
Styrene (% v/v)	10,000 rad/h	21,000 rad/h	41,000 rad/h	75,000 rad/h	112,000 rad/h	45,000 rad/h
10	—	—	—	—	—	6
20	24	24	14	9	7	54
30	61	48	37	18	14	140
40	51	92	76	27	23	89
50	409	216	109	39	25	—
60	—	196	89	43	28	61
70	223	159	89	53	35	—
80	—	130	68	51	35	41

[a] Total dose 2.3×10^6 rad, except polypropylene, which is 3.0×10^6 rad.
[b] Film thickness 0.12 mm (ex-Union-Carbide).
[c] Film thickness 0.06 mm (ex-Shell).

has shifted from 50% monomer for polyethylene to 30% monomer for the polypropylene, thus reflecting differences in structures of the two polymers, in particular, the presence of a tertiary carbon atom in polypropylene. The correction for dose is necessary when comparing polyethylene and polypropylene data, since grafting is proportional to total dose over the range of radiation conditions discussed in this paper.

7. Additive Effects in Grafting

Methods for increasing yields during radiation grafting are important, especially for those backbone polymers which are sensitive to ionizing radiation. In such instances, it is preferable to use the lowest total radiation dose to achieve a particular percentage graft. A number of type additives have been developed for giving increased copolymer yields. Some of these additives act by enhancing both graft and homopolymerization reactions, the former to a more significant extent while others are used to suppress homopolymerization preferentially.

7.1. Acid Effects in Grafting

The use of mineral acid as an additive to increase grafting yields has already been reported for a number of different backbone polymer systems.[(11,12,28,29)] Typical acid effects observed are shown in Table 3 where inclusion of 0.2 *M* sulfuric acid significantly enhances the grafting yields of styrene in methanol to polyethylene over a wide range of monomer concentrations studied, including that corresponding to the Trommsdorff peak. The mechanism of the acid effect is complicated.[29–32]

The presence of acid at the concentrations used should not markedly affect the precipitation of the grafted chains or the swelling of the backbone polymer, especially the polyolefins. Instead it has been proposed that the acid effect is due to a radiation chemistry phenomenon consistent with previous observations[12,26,30,31] and that in the radiolysis of methanol itself, addition of sulfuric acid increases $G(H_2)$ appreciably. The precursors of the extra hydrogen were thought to be hydrogen atoms ($H^\cdot$) and thermalized electrons [Eqs. (10) and (11)], both species being readily scavenged by styrene monomer[24,27]:

$$CH_3OH + H^+ \rightarrow CH_3OH_2^+ \quad (10)$$

$$CH_3OH_2^+ + e \rightarrow CH_3OH + H \quad (11)$$

Such processes can lead to the formation of increased grafting sites by abstraction reactions with the trunk polymer. Further work[33] on the acid effect indicates that acid leads to an increase in styrene–solvent intermediates ($MS^\cdot$) in the grafting solution, such species resulting in more grafting sites again by abstraction reactions. Finally, additional data from the analysis of the grafting solutions[32] indicate that in the presence of acid, chain length of oligomer in solution is shortened and also there are larger numbers of these shorter chains. Shorter oligomer chains could diffuse more readily into the

Table 3. Effect of Sulfuric Acid (0.2 *M*) on Grafting of Styrene in Methanol to Polyethylene Film at Different Dose Rates[a]

Styrene (% v/v)	Graft (%) 10,000 rad/h Neutral	10,000 rad/h 0.2 *M* H⁺	41,000 rad/h Neutral	41,000 rad/h 0.2 *M* H⁺	112,000 rad/h Neutral	112,000 rad/h 0.2 *M* H⁺
20	24	32	14	19	7	8
30	61	82	37	51	14	17
40	51	344	76	81	23	27
50	409	543	109	134	25	35
60	–	–	89	119	28	36
70	223	211	89	73	35	37
80	–	–	68	62	37	37

[a] Radians per hour with film 0.12 mm thickness (ex-Union-Carbide) to dose of 2.3×10^6 rad.

swollen backbone polymer to achieve more efficient termination at a grafting site. The increase in concentration of oligomer chains in the bulk of the solution would also result in an increase in the vicosity of both the grafting solution and the solution that is absorbed within the swollen backbone polymer, thus leading to the enhanced Trommsdorff peak as observed in the presence of acid.

7.2. Polyfunctional Monomers as Additives in Grafting

Inclusion of polyfunctional monomers, such as divinylbenzene (DVB) and trimethyl propane triacrylate (TMPTA), in additive amounts (≈ 1%), significantly enhances radiation grafting,[24,34] again especially at the Trommsdorff peak (Figure 3). The shapes of the acid and TMPTA graphs are different and suggest that the two enhancement reactions occur by different pathways. The polyfunctional monomers appear to have a dual function,[24,34] namely, to enhance the copolymerization and also cross-link the grafted polystyrene chains. In the grafting experiments, branching of the growing grafted polystyrene chains occurs when one end of the polyfunctional monomer (e.g., DVB) immobilized during grafting is bonded to the growing

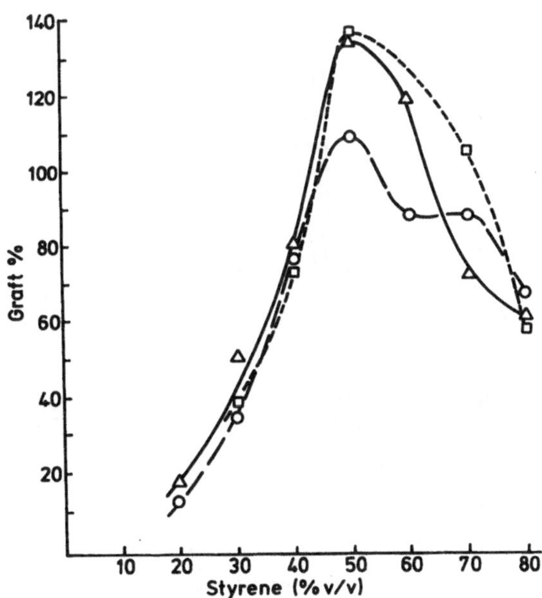

Figure 3. Effect of trimethylolpropane triacrylate on styrene grafting in methanol to polyethylene at dose rate of 4.1×10^4 rad/h to total dose of 2.4×10^5 rad: (○) styrene–methanol, (△) styrene–methanol–sulfuric acid (0.2 *M*), and (□) styrene–methanol–trimethylolpropane triacrylate (1% v/v).

chain. The other end is unsaturated and free to initiate new chain growth via scavenging reactions. The new branched polystyrene chain may eventually terminate, cross-linked by reacting with another polystyrene chain or an immobilized divinylbenzene radical. Grafting is thus enhanced mainly through branching of the grafted polystyrene chain.

7.3. Combined Effects of Acid and Polyfunctional Monomers

Although the yields of copolymer in the presence of TMPTA (or DVB) are similar in magnitude to those when acid is used, the shapes of the grafting

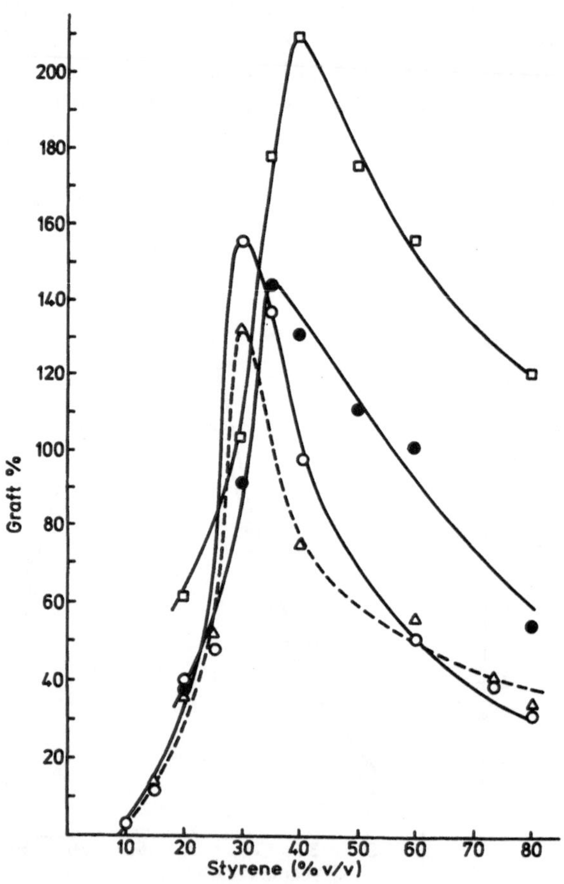

Figure 4. Effect of divinylbenzene and sulfuric acid on grafting of styrene on polypropylene film in methanol at dose rate of 4.1×10^4 rad/h to total dose of 2.4×10^5 rad: (△) styrene–methanol, (○) styrene–methanol–sulfuric acid (0.2 M), (●) styrene–methanol–divinylbenzene (1% v/v), and (□) styrene–methanol–divinylbenzene (1% v/v)–sulfuric acid (0.2 M).

curves with two type additives are different (Figure 3). This conclusion is consistent with the different mechanisms proposed for the mechanism of each type additive and suggests that both additives may be used simultaneously in solution to give a synergistic effect and increase the radiation grafting yield further.[34–36] The data in Figure 4 confirm this conclusion and show that the inclusion of both acid and DBC in a grafting solution in additive amounts gives large increases in copolymerization yield, a result which is of potential value in a preparative context, especially for the insolubilization of reagents.

7.4. Miscellaneous Additives to Reduce Homopolymerization

A problem with radiation grafting generally, and the mutual technique specifically, is that homopolymerization is always a potential detrimental competing reaction. The acid and polyfunctional monomer additives already discussed not only increase grafting yields, but also homopolymerization; however, grafting efficiency is also improved in the presence of both acid and polyfunctional monomers and thus the copolymerization process is preferentially favored under these conditions at certain monomer concentrations (Table 4). By contrast, other methods for minimizing homopolymer formation have involved inclusion of metal cations such as Fe^{2+} and Cu^{2+}; however, by this technique both grafting and homopolymerization are suppressed, the latter preferentially.[37] Thus potentially higher doses of radiation are required in the presence of these additives to achieve a particular percentage graft and the technique needs to be used with care for radiation-sensitive backbone polymers.

This limitation does not exist with a third method developed for reducing homopolymerization, namely, the use of a comonomer technique[38] where small percentages ($\approx$ 20%) of styrene are used with radiation-sensitive monomers to achieve uniform efficient grafting. Thus when ethyl acrylate is

Table 4. Grafting Efficiency of Styrene to Polyethylene Film in Methanol Containing Acid and Divinylbenzene Additives[a]

	Grafting efficiency (%)[b]		
Styrene (% v/v)	Neutral	H_2SO_4 (0.2 *M*)	DVB (1% v/v) and H_2SO_4 (0.2 *M*)
20	51.7	49.6	54.5
30	56.2	59.4	58.8
40	73.9	83.0	74.2
50	75.1	79.4	85.2
70	45.7	41.2	65.4

[a] Dose rate of 10,000 rad/h to total dose of 2.4 × 10^5 rad.
[b] Ratio of graft/graft and homopolymer × 100.

Table 5. Radiation-Induced Grafting of Ethyl Acrylate (EA) and Styrene (S) to Belmerino Wool Cloth[a]

Grafting solution				S/EA in grafted polymer	
EA	S	Dose ($\times 10^6$ rad)	Graft (% wt)	Found	Calculated
25.0[b]	50.0	0.10	1.2	0.91	2.24
25.0	50.0	0.20	2.4	1.12	2.24
37.5	37.5	0.20	2.7	0.40	1.44
37.5	37.5	0.40	9.5	0.87	1.44
62.5	12.5	0.10	2.1	0.26	0.56

[a] Radiation conditions as in Reference 38.

[b] Grafting of ethyl acrylate without styrene leads to excessive homopolymer and graft is difficult to extract.

grafted to wool (Table 5), it is observed that using the monomer alone, at 2×10^5 rad, copolymerization is inefficient due to homopolymer formation. Inclusion of styrene not only leads to satisfactory grafting of ethyl acrylate with little homopolymer, but at high ethyl acrylate/styrene ratios (5:1), the graft consists predominantly of ethyl acrylate (76%) and the copolymer exhibits essentially the properties of an ethyl acrylate graft with little contribution from the styrene. The technique has also been used for a variety of other grafting systems.(11)

8. Application of Radiation Grafting to Enzyme Insolubilization

Enzymes are highly efficient, specific biological catalysts which usually function in soluble form. However, since they are difficult and costly to isolate in a reasonably pure form, their use in industrial processing is restricted, because once used they cannot be conveniently recovered. To overcome these limitations, various techniques, which are summarized in Figure 5, have been developed to render enzymes water insoluble.(39) Of these procedures, the most widely used involve cross-linked polystyrene beads which incorporate, by chemical means, a functional group suitable for covalent attachment of a particular enzyme.

All of these preceding methods suffer from a number of disadvantages(39) which can be essentially overcome if a radiation grafting technique is used with a relatively inert polymer such as polypropylene. A suitable monomer for this purpose is *p*-nitrostyrene which can be radiation grafted to the surface of polypropylene powder. This nitro group can then undergo conventional chemical reactions such as reduction to the amine

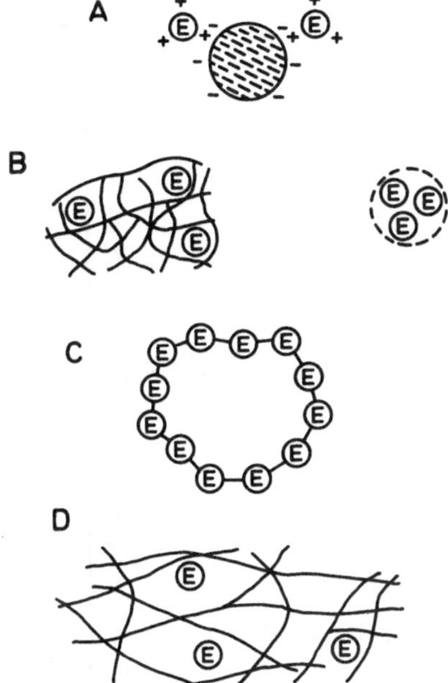

Figure 5. Methods for the preparation of water-soluble enzymes: (A) physical absorption – to clays, silica, ion exchange, and so on; (B) entrapment–within a cross-linked polymer lattice (gel) or within microcapsules; (C) intermolecular cross-linking – using difunctional reagents to form a water-insoluble "superpolymer"; (D) covalent attachment – - to water-insoluble polymers of natural or synthetic origin.

followed by conversion to the isothiocyanato or diazo derivative to which enzymes may be bound to produce water-insoluble conjugates [Eq. (12)]. In Table 6 three enzymes have been successfully immobilized onto a copolymer of poly(*p*-isothiocyanatostyrene-g-propylene). The percentage copolymerized for the first three runs in Table 6 was kept low (2%) to evaluate the performance of the copolymer with the higher graft material (30%) in run 4. For the percentage graft utilized, the 2% copolymer appears to be more efficient (run 3 *vs*. 4). Compared with previous chemical methods for enzyme immobilization, radiation grafting involves a one-step copolymerization operation, gives only a surface graft so that enzyme is readily accessible to soluble reactant, and is applicable to a wide range of polymeric supports of differing chemical structures.

$$\sim CH_2-CH_2-C_6H_4-NO_2 \xrightarrow[HCl]{Sn} \sim CH_2-CH_2-C_6H_4-NH_2 \xrightarrow{SCCl_2} \sim CH_2-CH_2-C_6H_4-NCS \xrightarrow{NH_2\text{-protein}} \sim CH_2-CH_2-C_6H_4-NH-CS-NH\text{-protein} \qquad (12)$$

Table 6. Reactivity of Polymer–Enzyme Conjugates Using Radiation Grafted Polymer Supports

Run	Polymer support	Enzyme	Substrate	Activity (%)[c]
1	pNO_2ST/PP[a]	Peroxidase	H_2O_2 with *o*-dianisidine	10
2	pNO_2ST/PP[a]	*β*-Galactosidase	*o*-Nitrophenyl-D-galactopyranoside	14
3	pNO_2ST/PP[a]	Trypsin	*N*-α-Benzoyl-L-arginine ethyl ester	10
4	pNO_2ST/PP[b]	Trypsin	*N*-α-Benzoyl-L-arginine ethyl ester	31

[a] Prepared by grafting[11] *p*-nitrostyrene (12.6 g) in methanol (142.0 g) containing 0.1 *M* H_2SO_4 to polypropylene powder (PP) (20.0 g) at dose rate of 0.04×10^6 rad/h to dose of 4.0×10^6 rad. Graft (2%) converted to poly(*p*-isothiocyanatostyrene-g-propylene) for enzyme attachment.

[b] Graft (30%).

[c] Relative to equal weight of soluble enzyme.

This latter aspect is important since specific polymers can be used for particular purposes. Thus polypropylene is not subject to chemical or microbial attack, as are natural polymers which have also been employed for radiation grafting but have not previously been used to immobilize enzymes. Even with the natural polymers, grafting reactions can be used to modify the materials and give them the required chemical properties.

For this enzyme work, *p*-nitrostyrene was the primary monomer chosen for grafting. The nitro group is deactivating in grafting and the total dose required for sufficient copolymerization of this monomer for the attachment reaction to the backbone polymer powders was considerable (3.0×10^6 rad). Recent work using the additives reported in this paper (acid and polyfunctional monomer) has shown that this dose can be reduced to approximately 1.0×10^6 rad to achieve the same level of grafting. Such reductions in total dose are extremely beneficial since doses above 2.0×10^6 rad can lead to significant degradation in certain trunk polymers. With polypropylene, depolymerization may be observed, while with PVC, dehydrohalogenation can occur, both reactions reducing the stabilities of the copolymers for insolubilization reactions. The other advantage in grafting *p*-nitrostyrene directly to the trunk polymer, rather than grafting styrene then nitrating with fuming nitric acid, is that depolymerization and oxidation of the trunk polymer during nitration is avoided. Also, only one nitro group is present in the ring rather than a distribution of non-, mono-, and polynitrated rings. Furthermore, in *p*-nitrostyrene grafts, all nitro groups in the radiation copolymer are readily accessible to chemical reagents for subsequent reactions required to complete the immobilization process, that is, no functional groups are sterically hindered to reaction as would be the situation with ortho- and, to a lesser extent, meta-, substituted nitro groups.

9. Application of Grafting to Heterogenization of Complexes

Homogeneous metal complexes, of Group VIII transition metals particularly, are active for catalytic processes such as hydrogenation, hydroformylation, isomerization, and related reactions. At present such reactions are performed in a bulk reactor which limits the efficiency of the process. Conversion of the homogeneous complex to its heterogeneous analogue by attachment of the complex to a polymer surface has a number of advantages including high reaction rates, acceptable thermal and mechanical stability, and capacity for use in packed and fluidized beds, as well as being relatively easy to reclaim and reactivate after the process has been completed.

Chemical methods have been extensively used for the heterogenization

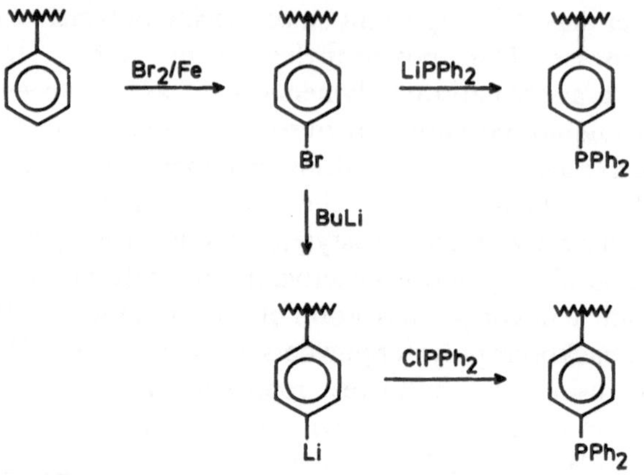

Figure 6. Typical chemical methods for phosphination of polystyrene for heterogenization reactions.

reactions[40] with cross-linked polystyrene as the support. Attachment of the complex to the polymer can be via an appropriate functional group introduced by a conventional chemical reaction. Since many catalytically active homogeneous metal complexes are phosphine derivatives, the phosphine ligand is used for attachment in many instances. Typical phosphination processes with polystyrene are shown in Figure 6 where bromination of polystyrene is followed by phosphination. Heterogenization can then be achieved by either direct complexation on the surface [Eq. (13)] or by ligand exchange [Eq. (14)]:

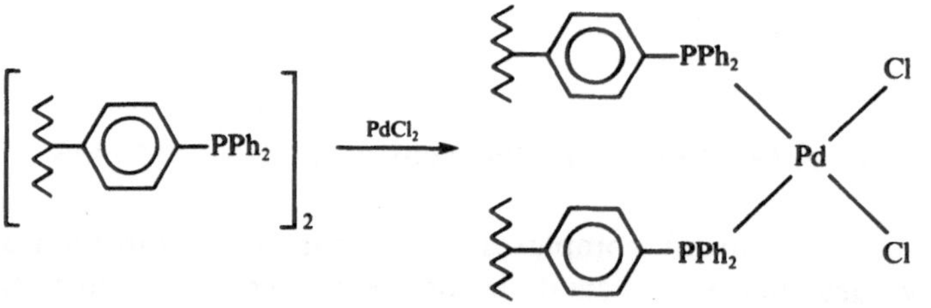

$$[\text{P}\text{—C}_6\text{H}_4\text{—PPh}_2]_2 \xrightarrow{PdCl_2} (\text{P}\text{—C}_6\text{H}_4\text{—PPh}_2)_2PdCl_2 \tag{13}$$

$$\text{P}\text{—C}_6\text{H}_4\text{—PPh}_2 \xrightarrow{RhCl(PPh_3)_3} \text{P}\text{—C}_6\text{H}_4\text{—PPh}_2RhCl(PPh_3)_2 + PPh_3 \tag{14}$$

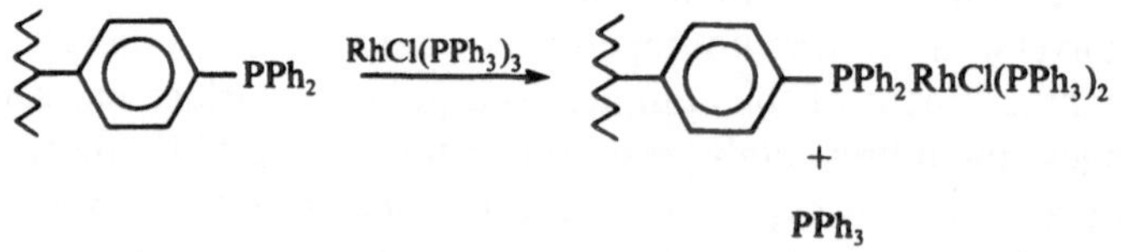

It is generally agreed that desirable support characteristics include: (1) good mechanical properties such as attrition resistance; (2) readily accessible sites for anchoring potential active centers; and (3) limited solubility in the reaction medium. For polystyrene resins most of these properties are controlled by the degree of cross-linking in the polymer. Unfortunately, the high degree of cross-linking required for insolubility and thermal and mechanical strength reduces the extent of swelling in the resin and thus the accessibility of sites within the polymer matrix. Therefore, highly cross-linked supports may present problems in achieving reaction within the bead during functionalization and later, when the bead is used as a catalyst, diffusional problems may also occur. In practice a compromise is reached. Where the degree of cross-linking is $< 1\%$ swollen resins generally have low mechanical stability and readily fragment even under careful handling. In contrast, commercially available microporous resins with $> 8\%$ cross-linking are mechanically very stable, but give rise to acute diffusional limitations resulting in slow and incomplete reactions. Resins of approximately 2% cross-link ratio provide a satisfactory compromise generally allowing adequate penetration by most reagents and yet retaining sufficient mechanical stability for easy handling.

A major problem with the above chemical techniques for such heterogenization is that metal is gradually leached from the surface of the polymer during reaction. The difficulty appears to be due to the nature of the bonding, particularly to the chemical attachment system, and the use of styrene–divinylbenzene copolymers for this purpose. Potentially improved heterogenization can be achieved by a radiation grafting process[11,29] since a novel type of surface copolymer can be prepared from a wide range of backbone polymers of differing chemical structures.

Table 7. Hydrogenation of Olefins with Homogeneous Metal Complexes Heterogenized on Polymer Supports[a]

Run	Support[b]	Graft (%)	Catalyst	Conversion[c] of olefin (%)
1	SDP/PST	45	Chlorocarbonyltris(triphenylphosphine) iridium	29
2	SDP/PST	45	Chlorotris(triphenylphosphine) rhodium (I)	30
3	SDP/PVC	23	Chlorotris(triphenylphosphine) rhodium (I)	3
4	PhST/PP	50	Chlorotris(triphenylphosphine) rhodium (I)	88
5	SDP/PST	16	Chlorotris(triphenylphosphine) rhodium (I)	25
6	*d*	0	Nil	0

[a] Benzene used to heterogenize catalysts.

[b] Runs 1 and 2, styryldiphenylphosphine (PST) grafted to polystyrene in a manner analogous to grafting conditions[11] in Table 6; Run 3, poly(vinyl chloride) (PVC) as backbone polymer; Run 4, styrene grafted to polypropylene (PP), then copolymer phosphinated; Run 5, polystyrene 2% cross-linked with DVB used as backbone polymer.

[c] Hydrogenation with cyclohexene for 65 h at 80°C for runs 1–3, while hexene-1 used in runs 4 (16 h at 45°C) and 5 (16 h at 17°C).

[d] Polystyrene used as blank.

In Table 7 representative data are shown for the hydrogenation of olefins with a number of catalytically active homogeneous metal complexes which have been heterogenized on a variety of polymer surfaces. When styryldiphenylphosphine is radiation grafted to polystyrene, the conversion of cyclohexene is appreciable, whereas the corresponding reaction with PVC is not efficient. In run 3, the support was charred after reaction, indicating that the PVC had partially decomposed during hydrogenation. The results in runs 4 and 5 are significant since the hydrogenations were performed under relatively mild conditions and the conversions are appreciable. There was also no apparent color in the supernatant benzene at the completion of these reactions, indicating no significant leaching of the complex from the surface under the experimental conditions used. This conclusion is to be verified by instrumental analytical techniques. From these studies it would appear that the radiation grafting method may possess significant advantages over the alternative chemical methods for heterogenizing catalytically active homogeneous metal complexes, especially if leaching rates of the anchored complexes from the radiation grafted copolymers for a wide variety of catalytic systems prove to be as consistently low as those reported in Table 7.

10. Application of Grafting to Insolubilization of Analytical Reagents

The use of analytical reagents chemically bound to polymer surfaces is of potential value for the formation of novel ion exchange resins containing high specificity for a particular metal. Styrene–divinylbenzene copolymer beads, produced by suspension polymerization, have previously been utilized for this purpose. The copolymer is nitrated, reduced to the amine, diazotized, and the polymeric diazotate coupled to a suitable analytical reagent in alkaline solution. Reagents such as 8-hydroxyquinoline have already been immobilized in this manner[41–44]; however, the technique yields an ion exchange resin of low capacity[43] as the impurities from the side reactions occurring at each stage of the synthesis build up in the resin. In addition, ion exchange resins prepared from styrene polymers generally possess low ion exchange rates[43]; thus when 8-hydroxyquinoline is attached to a styrene–divinylbenzene copolymer, the resulting ion exchange properties are poor.[42] Although this resin has the necessary physical characteristics and chemical stability with high selectivity for metal ions, rates of sorption of metal ions are too low for the resin to be of practical value.

Radiation graft copolymers possess potential advantages in this application, particularly the fact that the basic polymer structure can be

Table 8. Ion Exchange Properties of 8-Hydroxyquinoline (HQ) When Insolubilized on Styrene Radiation Graft Copolymers

Polyaminostyrene copolymer	Run	Diazotization conditions[c]	HQ attached to copolymer (g)	Cu capacity of copolymer (meq/g)[d]	$t^{1/2}$ (min)[e]
1[a]	1	KNO_2(0.22 g) HCl (1 N, 20 ml)	0.05	0.26	35–40
	2	KNO_2(0.22 g) HCl (3 N, 20 ml)	0.42	0.61	30–33
2[b]	3	KNO_2(0.22 g) HCl (3 N, 20 ml)	0.69	0.67	~ 90

[a] Copolymer (3.00 g) of polypropylene powder containing styrene radiation grafted (38%), then nitrated and reduced.[41]

[b] Similar to copolymer 1, except styrene graft (42%) prepared in presence of DVB (1% v/v with respect to styrene).

[c] Amino group on copolymer diazotized using pH conditions shown, then HQ attached in 10% NaOH (450 ml).[41]

[d] Determined at pH 5.5.

[e] Time for half the maximum absorption of copper to occur.

readily varied to give different properties, whereas in the preceding technique, the use of the system is somewhat limited by the properties of the one resin used, namely, the DVB chemically cross-linked polystyrene.

In the work reported here, styrene is radiation grafted to an inert backbone polymer, polypropylene and the resulting copolymer is nitrated, reduced, and coupled to 8-hydroxyquinoline by diazotization.[45] The data with these copolymers (Table 8) show that they contain moderately high levels of immobilized 8-hydroxyquinoline and as potential ion exchange resins they exhibit a large capacity for copper ions. The capacities are affected by the pH of the solution during diazotization, an increase in strength of acid from 1 to 3 N leading to a large enhancement in copper capacity of the resin (runs 1 and 2). The exchange rates of the resins are still relatively slow, this speed being further reduced when the copolymer is cross-linked with DVB (run 3). Such cross-linking usually increases the physical strength of the polymer; however, it also appears to increase diffusion time of copper to the coordination site in the resin thus lowering the capacity.

These preliminary studies show that radiation grafted copolymers can be used as ion exchange resins. Because of the versatility of the copolymers and the large number of resins available for consideration in this application, the potential of the method is large. Thus by commencing with an inert backbone polymer such as polypropylene, polarity can be built into the resin by direct copolymerization of monomers containing the appropriate functionality (e.g., —OH, —COOH, —NO_2, etc). The degree of cross-linking can also be easily controlled; hence the technique is promising as a viable alternative to the chemical methods currently used.

11. General Conclusions

Radiation grafting is a procedure which offers great potential as an essentially one-step method for converting chemical reagents to insolubilized form. The mutual radiation technique has been discussed in detail in this current treatment; however, the alternative process of pre-irradiation is still of use, particularly for those backbone polymers which are not radiation sensitive yet form sufficient trapped radicals under irradiation to give reasonable grafting yields. Future work in this area will involve an evaluation of the comparative use of ionizing radiation with UV[46] as initiators for these copolymerization reactions. Sensitized UV is a useful developing tool for copolymerization[46] and its value in reagent insolubilization reactions needs to be evaluated, particularly for two of the most urgent applications, namely, enzyme insolubilization and heterogenization of catalytically active homogeneous metal complexes.

ACKNOWLEDGMENTS

The authors wish to thank the Australian Institute of Nuclear Science and Engineering, the Australian Research Grants Committee, and the Australian Atomic Energy Commission for the support of this research.

References

1. G. R. Marshall and R. B. Merrifield, in: *Biochemical Aspects of Reactions on Solid Supports* (G. R. Stark, ed.), pp. 111–169, Academic Press, New York (1971).
2. K. G. Allum, R. D. Hancock, I. V. Howell, R. C. Pitkethly, and P. J. Robinson, Supported transition metal complexes. IV. Rhodium catalysts for the liquid phase hydroformylation of hexene-1, *J. Catal.* **43**, 322–330 (1976).
3. K. G. Allum, R. D. Hancock, I. V. Howell, R. C. Pitkethly, and P. J. Robinson, Supported transition metal complexes. V. Liquid phase catalytic hydrogenation of hexene-1, *J. Catal.* **43**, 331–338 (1976).
4. Y. Kawabata, M. Tanaka, and I. Ogata, Asymmetric hydrogenation by a rhodium catalyst complexed with a phosphinite derived from cellulose, *Chem. Lett. 1976*, 1213–1214.
5. T. J. Pinnavaia and P. K. Welty, Catalytic hydrogenation of 1-hexene by rhodium complexes in the intracrystal space of a swelling layer lattice silicate, *J. Am. Chem. Soc.* **97**, 3819–3820 (1976).
6. M. Capka and V. Kavan, Catalytic activity of rhodium (I) complexes containing poly(siloxy)alkyl diphenyl phosphines, *Collect. Czech. Chem. Commun.* **45**, 2100–2107 (1980).

7. J. L. Garnett, R. S. Kenyon, and M. J. Liddy, Enzyme immobilization by covalent attachment to novel polymer matrices prepared by a radiation grafting technique, *J. Chem. Soc. Chem. Commun. 1974*, 735–736.
8. I. Kaetsu, M. Kumakura, M. Yoshida, M. Asano, M. Himei, M. Tamura, K. Hayashi, Immobilization of enzymes by radiation, *Rad. Phys. Chem.* **14**, 595–602 (1979).
9. A. Charlesby, *Atomic Radiation and Polymers*, Pergamon Press, Oxford (1960).
10. A. Chapiro, *Radiation Chemistry of Polymeric Systems*, Interscience, New York and London (1962).
11. J. L. Garnett, Grafting, *Rad. Phys. Chem.* **14**, 79–99 (1979).
12. J. L. Garnett and N. T. Yen, Acid effects in the radiation grafting of monomers to polymers, particularly polyethylene, *ACS Symp. Ser.* **121**, 243–261 (1980).
13. S. Dilli and J. L. Garnett, Radiation-induced reactions with cellulose. IX. Copolymerization with styrene using a pre-irradiation procedure and the effect of additives on the grafting reaction, *Aust. J. Chem.* **24**, 981–987 (1971).
14. G. M. Kline, *Analytical Chemistry of Polymers Part 1*, 3rd Edition, Interscience, New York (1966).
15. A. Ekstrom and J. L. Garnett, Radiolysis of binary mixtures. Part IV. The effect of polycyclic aromatic additives in methanol, *J. Chem. Soc. A 1968*, 2416–2418.
16. A. Chapiro, A. M. Jendrychowska-Bonamour, and G. Lelievre, Molecular products in the radiolysis of vinyl monomers, *Faraday Discuss Chem. Soc.* **63**, 134–140 (1977).
17. S. Dilli and J. L. Garnett, Radiation-induced reactions with cellulose. III. Kinetics of styrene copolymerization in methanol, *J. Appl. Polym. Sci.* **11**, 859–870 (1967).
18. J. L. Garnett, Grafting to cellulose using UV and gamma radiation as initiators, *ACS Symp. Ser.* **48**, 334–360 (1977).
19. R. B. Phillips, J. Quere, G. Guiroy, and V. T. Stannett, Modification of pulp and paper by graft copolymerization, Tappi **55**, 858–867 (1972).
20. R. J. Demint, J. C. Arthur, Jr., A. R. Markezich, and W. F. McSherry, Radiation-induced interactions of styrene with cotton, *Text. Res. J.* **32**, 918 (1962).
21. A. Hebeish and J. T. Guthrie, *The Chemistry and Technology of Cellulosic Copolymers*, Springer-Verlag, Berlin and Heidelberg (1981).
22. G. Odian, T. Acker, and M. Sobel, Accelerated effects in radiation induced graft polymerization, *J. Appl. Polym. Sci.* **7**, 245–250 (1963).
23. S. Machi, I. Kamel, and J. Silverman, Effect of swelling on radiation-induced grafting of styrene to polyethylene, *J. Polym. Sci. A-1* **8**, 3329–3337 (1970).
24. C. H. Ang, J. L. Garnett, and R. Levot, Use of polyfunctional monomers as additives in accelerating the radiation grafting of styrene to polyolefins, Proceedings of the American Chemical Society, 183rd National Meeting, Las Vegas, Nevada (March 1982).
25. S. Dilli and J. L. Garnett, A charge-transfer theory for the interpretation and radiation-induced grafting of monomers to cellulose, *J. Polym. Sci. A-1***4**, 2323–2324 (1966).
26. A. Ekstrom and J. L. Garnett, Radiolysis of binary mixtures. I. Liquid phase studies with benzene–methanol, *J. Phys. Chem.* **70**, 324–330 (1966).
27. D. F. Sangster and A. Davison, Pulse radiolysis of styrene and acrylate monomers, *J. Polym. Sci. Symp. Polym. 49*, 191–210 (1975).
28. J. L. Garnett and J. D. Leeder, Recent developments in grafting of monomers to wool keratin using U.V. and γ-radiation, *ACS Symp. Ser.* **49**, 197–220 (1977).
29. H. Barker, J. L. Garnett, R. Levot, and M. A. Long, Use of additives to enhance radiation grafting of monomers to poly(vinyl chloride) and application of these PVC copolymers to immobilization of enzymes and heterogenization of homogeneous metal complexes, *J. Macromol. Sci., Chem. A12(2)*, 261–273 (1978).
30. J. H. Baxendale and F. W. Mellows, The gamma radiolysis of methanol and methanol solution, *J. Am. Chem. Soc.* **83**, 4720–4726 (1961).

31. W. J. Chappas and J. Silverman, The effect of acid on the radiation-induced grafting of styrene to polyethylene, *Rad. Phys. Chem.* **14**, 847–852 (1979).
32. J. L. Garnett, S. V. Jankiewicz, and D. F. Sangster, Effect of mineral acid on polymer produced during radiation-induced grafting of styrene monomer, *J. Polym. Sci., Polym. Lett. Ed.* **20**, 171–175 (1982).
33. G..Fletcher and J. L. Garnett (unpublished work).
34. C. H. Ang, J. L. Garnett, R. Levot, and M. A. Long, Polyfunctional monomers as additives for enhancing the radiation copolymerization of styrene to polyethylene, polypropylene and PVC, *J. Appl. Polym. Sci.* **27**, 4893–4897 (1982).
35. C. H. Ang, J. L. Garnett, R. Levot, and M. A. Long, Accelerated radiation-induced grafting of styrene to polyolefins in the presence of acid and polyfunctional monomers, *J. Polym. Sci., Polym. Lett. Ed.* **21**, 257–261 (1983).
36. C. H. Ang, J. L. Garnett, R. Levot, and M. A. Long, Novel additives for accelerating radiation grafting of monomers to polymers in acid media, Proceedings of the Fourth International Meetings on Radiation Proceeding, Yugoslavia, 1982 (in press).
37. M. B. Huglin and B. L. Johnson, Role of cations in radiation grafting and homopolymerization, *J. Polym. Sci. A-1*, **7**, 1379–1384 (1969).
38. J. L. Garnett and R. S. Kenyon, Acid effects in the styrene comonomer technique for radiation grafting to wool, *J. Polym. Sco., Polym. Lett. Ed.* **15**, 421–425 (1977).
39. M. J. Liddy, J. L. Garnett, and R. S. Kenyon, The insolubilization of trypsin by attachment to radiation graft copolymers of polypropylene, *J. Polym. Sci., Symp. Polym. 49*, 109–116 (1975).
40. R. H. Grubbs and E. M. Sweet, Polymer attached catalysts. A comparison between polystyrene attached and homogeneous Rh(I) hydrogenation catalysts, *J. Mol. Catal.* **3**, 259–270 (1977/8).
41. R. V. Davies, J. Kennedy, E. S. Lane, and J. L. Williams, Synthesis of metal complexing polymers, IV. Polymers containing miscellaneous functional groups, *J. Appl. Chem.* **9**, 368–371 (1959).
42. J. R. Parrish and R. Stevenson, Chelating resins from 8-hydroxyquinoline, *Anal. Chim. Acta* **70**, 189–198 (1974).
43. F. Vernon and H. Eccles, Chelating ion-exchangers containing 8-hydroxyquinoline as the functional group, *Anal. Chim. Acta* **63**, 403–414 (1973).
44. F. Vernon and H. Eccles, Chelating ion-exchangers containing salicyclic acid, *Anal. Chim. Acta* **72**, 331–338 (1974).
45. C. H. Ang and J. L. Garnett (unpublished work).
46. N. P. Davis, J. L. Garnett, and R. G. Urquhart, Comparison of photosensitized and γ-ray-induced graft copolymerization of monomers to cellulose, *J. Polym. Sci., Symp. Polym. 55*, 287–301 (1976).

11

Immobilized Enzymes

Melvin H. Keyes and Seshaiyer Saraswathi

Abstract. The literature abounds with procedures to immobilize enzymes which generally fit into one of six main categories. The earliest methods developed were enzyme adsorption and cross-linking. Adsorption–cross-linking, the combination of these methods, results in a more utilitarian catalyst. The most widely used method is covalent bonding of the enzyme to a support material; but entrapment and microencapsulation are valuable techniques for immobilization of enzymes.

The properties of immobilized enzymes are affected by the support material. In essence, the physical properties are those of the support material. Inorganic supports result in rigid catalysts with moderate enzyme loading. Very high enzyme loading can be achieved with organic supports which are easily molded into desired shapes. The support used for immobilization of enzymes also affects the chemical properties of the catalyst. The optimum pH is often different for an immobilized enzyme than for the soluble enzyme and changes in substrate specificity have also been observed.

Immobilized enzymes find applications both in industrial processing and analytical testing. In addition, immobilized enzymes have been used for analysis of the structure–function relationships of enzymes. Although showing considerable promise, therapeutic applications generally remain in the research stage.

1. Introducton

Natural catalysts, the enzymes, have two advantages over the traditional chemical catalysts, namely, specificity and efficiency. However, the high cost and low stability of many enzymes have limited the application of these efficient catalysts for commercial purposes. Enzymes are expensive because they are present in very low concentrations *in vivo*, and therefore must be extracted and purified. Moreover, in general, enzymes are unstable in solution, losing their catalytic activity in a few hours.

In spite of their limitations enzymes are used for industrial processing, analytical testing, and medical treatment.[1–3] In the food industry, glycoamylase and α-amylase are used for converting cornstarch to corn syrup. Hydrolysis of pectic substances in fruit juices by pectic enzymes and

Melvin H. Keyes and Seshaiyer Saraswathi • Owens–Illinois, Inc., One SeaGate, Toledo, OH 43666.

the use of proteases in meat tenderizing are further examples of the industrial use of enzymes.

Analytical testing with enzymes is illustrated by the enzymatic assays of glucose and urea, as well as enzyme immunoassays. For therapeutic applications, many enzymes have been derivatized to reduce the immune response.[3,4] A particular example is the application of enzymes derivatized with polyethylene glycols.[5] Soluble enzyme derivatives have also been studied for many years in the analysis of the structure and function of enzymes.[6,7]

The study of soluble enzyme derivatives probably led to the extensive research of enzymes immobilized on solid supports.[8] Surprisingly, the immobilized enzymes generally retained considerable activity. Even more important, from a practical point of view, the activity that was retained often appeared to be quite stable in an aqueous environment. Thus, by immobilization, one obtained a catalyst that was specific, efficient, and stable, and one that could be reused easily. The advantages which result from immobilization often make the use of enzymes more practical.

This review will discuss the nature and methods of immobilization. In addition, applications of immobilized enzymes will be considered in relationship to the method of immobilization. The reader may also wish to study the other numerous reviews that have appeared in the literature over the last decade.[9–18]

The methods commonly used for immobilization of enzymes fall into one of the following categories: adsorption, cross-linking, adsorption

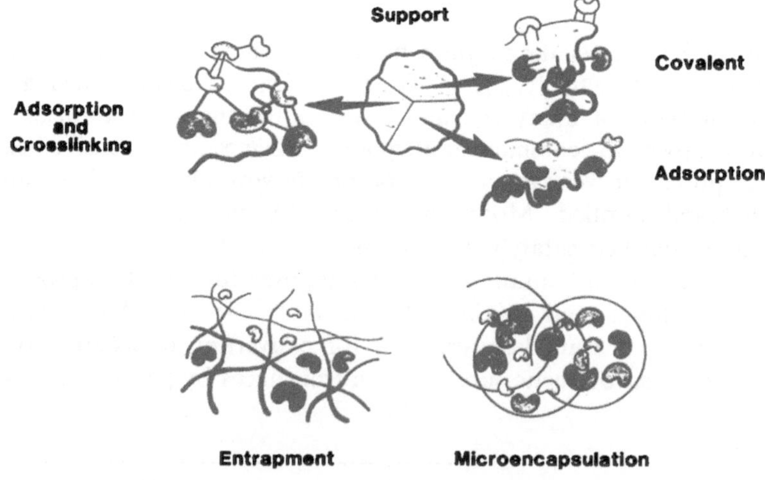

Figure 1. General methods of immobilization.

followed by cross-linking, covalent bonding, entrapment, and microencapsulation. Figure 1 illustrates some of these methods. Before discussing each method, it is helpful to become familiar with the types of support materials, as described in the next section.

2. Supports

Immobilized enzymes consist of two components—the enzyme which provides the catalytic functions, and the support which imparts the physical properties as well as modifies the catalytic properties. The support or carrier can be almost any insoluble material, from stainless-steel spheres to feathers. Support materials are usually divided into two main classes: inorganic and organic. Table 1 lists the typical support materials. The support material should have several characteristics depending on the application of the immobilized enzyme:

1. Chemical functionality.
2. Mechanical strength.
3. Thermal stability.
4. Microbial resistance.
5. Chemical durability.
6. Cost.
7. Hydrophilicity.
8. Loading capacity.
9. Reusability.
10. Accessibility.

Organic and inorganic supports will be discussed in light of these characteristics. Both porous and nonporous forms of particulate inorganic supports can be used for immobilization. Generally the porous form is

Table 1. Common Supports for Enzyme Immobilization

Organic supports		
Natural	Synthetic	Inorganic supports
Cellulose	Polyacrylamide	Controlled pore glass
Starch	Polystyrenes	Ceramics
Lipids	Nylons	Sand
Erythrocytes	Methacrylates	Silica
Collagen		Alumina
Dextrans		Titania
Agarose		Stainless steel

favored since it increases enzyme loading. A porous support having a particle size from 50 to 300 μm will have sufficient enzyme loading and yet be easy to handle. It also possesses good flow properties for use in a column rector. Indeed, these inorganic supports are most often used in packed columns, whether the application is industrial processing or analytical testing. The most commonly used inorganic supports are alumina and silica,[19,20] and immobilization is usually carried out by one of two methods. In the covalent attachment method, the surface is first derivatized to create a chemically reactive site, and then reacted with side chains of the enzyme. The second method requires the enzyme to first be adsorbed onto the surface and then cross-linked with a bifunctional reagent. These methods are discussed in detail below.

Organic supports can be in either a particulate or membrane configuration. As with inorganic supports, these materials are usually in a porous form. One main advantage is the ease with which one can form a covalent bond between enzyme and support. Since their organic nature allows the selection from many reactive side chains, it is relatively easy to obtain high enzyme loading. Hydrophilic, organic supports can also be designed which lack specific interactions with protein. Generally these supports are inferior to inorganic materials in microbial resistance, thermal stability, mechanical strength, and ease of regeneration.

3. Methods of Immobilization

3.1. Adsorption

The earliest examples of immobilization recorded in the literature describe adsorption to solid supports.[21] For immobilization by adsorption, the surface area and chemical composition of the support are highly important. Generally a larger surface area allows greater enzyme loading. Moreover, hydrophilic groups usually increase the enzyme loading.

Unfortunately, this very simple method has the disadvantage that enzyme is easily lost through desorption from the support. Not only does this lead to the rapid loss of activity of the immobilized catalyst with time, but also it will release enzyme molecules into the reaction medium.

Enzymes are adsorbed onto supports that are charged or uncharged. Charged supports usually possess carboxyl acid or amino side chains. With uncharged or slightly charged supports, the adsorption is mainly due to the surface effects. These types of supports include glass, alumina, clays, carbon, and calcium phosphate gel. Examples of immobilization by adsorption are given in the literature.[22–24]

3.2. Cross-linking

Other than adsorption, the oldest method of immobilization is probably cross-linking in which the enzyme is reacted with a bifunctional reagent. During cross-linking at moderate to high protein concentration, intermolecular bonds are formed causing the formation of large aggregates which are insoluble. Cross-linking reagents which change the charge on the protein surface also favor precipitation. Numerous cross-linking reagents have been investigated[25] and many of them are available commercially. Almost all of these reagents react predominantly with the amino side chains of the protein. Some reagents will also react with arginine, histidine, and tyrosine side chains.

Glutaraldehyde has been used more frequently for immobilization by cross-linking than any other reagent, since it is expensive, readily available, and frequently precipitates protein from solution. The mechanism of immobilization is complex, but it is recognized that glutaraldehyde reacts with amines and is generally thought to form uncharged bonds which cause precipitation.[026,27] When purchased commercially, the reagent is actually a mixture of glutaraldehyde and condensation products known as "polyglutaraldehyde." From the studies that have been carried out, it is almost certain that both reagents participate in enzyme cross-linking. In spite of numerous studies, the mechanism of reaction and the nature of the bonds formed remain unclear.[26,27] Commercial glutaraldehyde solutions react with amines over a wide range of pH. When reacted in the acid region, the bonds formed are unstable and can be reversed. In the neutral or basic region, the bonds formed are quite stable. Other bifunctional reagents that have been used include isocyanate derivatives, *N,N′*-ethylene bismaleimide, bis-diazobenzidine, and *N,N′*-polymethylene bisiodoacetamide.[11]

3.3. Adsorption–Cross-linking

The adsorption–cross-linking method consists of two parts. First, the enzyme is adsorbed onto a support and then a bifunctional reagent is added to cross-link the enzyme. Since many of the bonds formed are intermolecular, a large sheet of enzyme is formed on the surface of the support. This large aggregate of enzyme is tightly bound to the support since all the interactions between the support and aggregate must be broken for desorption.

The best supports for this type of immobilization are those which interact strongly with enzymes. Many inorganic supports, such as porous glass or alumina, work very well. Organic supports having ionic groups also are very effective in adsorbing proteins. However, when the enzyme is adsorbed onto a support containing amino side chains, this method is difficult to distinguish from immobilization by covalent bonding. The cross-linking reagents added to the adsorbed enzyme can form covalent bonds between support and enzyme or between enzyme molecules. In most instances the enzyme is probably immobilized by a combination of both methods. In other words, a sheet of enzyme molecules form which is covalently bonded to the support.

As with immobilizing an enzyme by cross-linking in the absence of a support, the most common cross-linking reagent in the presence of a support is glutaraldehyde. One of the earliest examples of this method is that of Haynes and Walsh. They describe the immobilization of trypsin onto collodial silica, in which trypsin was adsorbed onto silica followed by the addition of a 5 M excess of glutaraldehyde.[28] Later, it was shown to be more effective to first coat the silica with polyimine.[29]

A particularly useful support for this method is porous alumina. Not only do many enzymes readily adsorb on this material, but it is stable over a wide range of pH. One method developed uses aliphatic amines in combination with aliphatic dihalides.[30] Another method immobilizes the enzyme by means of disulfide rearrangement while the enzyme is adsorbed onto alumina.[31] If the enzyme which one wishes to immobilize has no or very few disulfide bonds, an additional protein can be adsorbed onto the support with the desired enzyme.[32] Thus the desired enzyme is immobilized with the aid of the disulfide bond rearrangement of a "filler" protein.

A process was developed by Messing which involves papain adsorbed onto a porous glass.[33] The support containing adsorbed enzyme is removed and placed in contact with a solution of 1,6-diisocyanathohexane in an organic solvent.

In addition to particulate supports, membranes have been prepared by this method. An early paper reports the use of cellophane and glutaraldehyde.[34] In similar experiments many investigators report the preparation of immobilized enzymes onto and within protein films and membranes.

The adsorption–cross-linking process results in a more tightly bound immobilized enzyme than simple adsorption. Unlike covalent bonding, the support material need not be activated before use. Moreover, when enzymes are immobilized simply by cross-linking, the catalyst obtained is a "gummy" precipitate. By adsorbing the enzyme onto a support, the physical properties of the support are imparted to the resulting catalyst.

3.4. Covalent Bonding

The most widely used method for enzyme immobilization is covalent attachment to a support which yields a strongly bound enzyme. Table 2 lists

Table 2. Methods of Covalent Bonding

Support	Activation	References	Coupling method	Reaction conditions
Organic	CNBr activation	45,46	Activated support	Agarose derivatives in acid pH; dextran derivatives in alkaline pH; coupling in neutral or alkaline pH
	Bisoxiranes	57	Epoxide–polymer	Activation and coupling reaction at alkaline pH
	Tosyl and tresyl chloride	58–60	Sulfonate ester derivatives	Activation at alkaline pH in organic solvents; coupling at alkaline pH
Inorganic	Silanation	82–85	Carbodiimide	Silanation in aqueous or organic medium; coupling at acid pH in cold
	Glutaraldehyde treatment of alkylamine support	83–85	Carbonyl derivative	Derivatization and coupling in neutral buffers

some of the common procedures for covalent attachment of an enzyme to a support. Several functional groups of proteins are available for covalent linkage: the $—NH_2$, —COOH, —SH, —OH groups of serine and threonine, the phenolic group of tyrosine, and the imidazole group of histidine. Support materials are first activated by introduction of an electrophilic group on the matrix. In cases where the activated matrix contains amino or carboxyl groups, a coupling agent is employed to bind the enzyme to the carrier. A third method is to activate the enzyme so that it can be linked with the support. This method is less frequently used since enzyme activity may be lost due to the harsh conditions of activation.

The conditions such as pH, temperature, or reaction medium for the activation and immobilization steps will depend largely on the nature of the protein, as well as the support. Optimization of these parameters is necessary since immobilization can lead to considerable loss of enzyme activity and changes in the properties of the enzyme. The optimization of immobilization conditions is generally accomplished by trial and error.

3.4.1. Organic Supports

Enzymes have been covalently linked to a wide variety of natural as well as synthetic organic polymers. Many support materials are activated with reagents which react with the hydroxyl groups on the support and thus introduce an electrophilic group. Alternately, reactive groups such as $—NH_2$, diazo, and —COOH are introduced into the polymer during synthesis. These reactions have been reviewed extensively in other articles and textbooks.[35–38]

One of the most popular techniques for covalent coupling of proteins to polysaccharides and other hydroxyl supports involves the reaction of the support with cyanogen bromide.[39,40] The activation reaction was originally attributed to the formation of imidocarbonate derivatives of the polysaccharides,[39–41] and in view of the stability of these derivatives in alkaline solutions, the reaction was generally done in a highly alkaline medium. However, several reports subsequently indicated that the CNBr-activated Sepharose (Pharmacia, Inc.) was less stable in alkaline solution than in the acid region, while for CNBr-activated Sephadex (Pharmacia, Inc.) derivatives, the stability was in the alkaline region, as expected.[42–45] Recent studies by Kohn and Wilchek[45,46] demonstrate that the polygalactose supports such as Sepharose are indeed activated differently by CNBr than a polyglucose support such as Sephadex. Whereas imidocarbonates, which are stable in alkaline condition, are the reactive intermediates of Sephadex, the enzyme coupling by activated Sepharose takes place through the cyanate ester derivatives, which are better preserved under acidic conditions. These findings indicate that the proper pH of the reagents and washing buffers is critical in preserving the activated derivatives of the CNBr-activated polysaccharides.

Since the CNBr reaction is easy to perform and is done under relatively mild conditions, this procedure has been employed for the immobilization of a large number of enzymes on a variety of hydroxyl polymers.[38–49] However, the method has many drawbacks. First, degradation of the reagent, resulting in toxic fumes, is an undesirable side reaction during the process, which necessitates extreme care during the experiment. Moreover, the linkage between the protein and the CNBr-activated supports can be somewhat unstable, and slow detachment of the enzyme during storage has been reported.[50–53] Finally, the traditional procedure uses an excess quantity of CNBr which is an expensive reagent and difficult to store.

Epoxide-containing support materials have been frequently used for enzyme immobilization,[54–56] since the method of introducing oxirane groups on Sepharose gels was first developed by Sundberg and Porath.[57] The technique involves the reaction of the polymeric support with bisoxiranes such as 1,3- or 1,4-diglycidyl ether, thus introducing a long spacer arm with a reactive group onto the matrix. Both activation of the support and coupling of the enzyme is optimal in the alkaline region.

A method for activating the polysaccharide supports for enzyme linkage has been described by Nilsson *et al.*[58] and involves the reaction of the support with *p*-toluenesulfonyl chloride to form a toluene sulfonic ester. Alternately, 2,2,2-trifluoroethanesulfonyl chloride has been introduced as a more reactive sulfonating agent to yield highly reactive ester derivatives of agarose, cellulose, and inorganic hydroxyl supports.[59,60] Activation with these reagents is done under nonaqueous solvents to prevent the hydrolysis of the reagents, and an alkaline pH is maintained with pyridine. Coupling of enzymes to these sulfonic esters is most effective at or above neutral pH.

Many enzymes have been covalently immobilized using dichlorotriazinyl derivatives of polysaccharide supports. The triazinyl derivatives are prepared by reacting the polymer with trichlorotriazine, in highly alkaline solutions. The reaction is rapid and the derivatization is usually complete in about 30 min at room temperature.[61–63] The triazinyl support reacts rapidly with proteins to form covalent bonds between the $—NH_2$ group of the protein and the free chloro group of the support over a wide range of pH.[63] Use of 2-amino-4,6-dichlorotriazine in place of trichlorotriazine generates derivatives which react at a slower rate and thus react more selectively than the dichlorotriazine derivatives.[18]

Several other methods to activate hydroxy supports for covalent attachment of enzymes have been demonstrated. *N*-hydroxysuccinimide esters of agarose,[64] available as Affigel-15 (Biorad Laboratories), form stable amide bonds with proteins.[65] Cellulose supports have been reacted with ethyl chloroformate to obtain cellulose-*trans*-2,3-carbonate derivatives for enzyme coupling.[66–68] Agarose supports have been derivatized with benzoquinone to yield reactive quinone-linked polymers,[69] and hydroxy polymers have been converted to the vinylsulfonyl derivatives by treating with divinyl sulfone.[70] Alternately, periodate oxidation of polysacchrides yields a dialdehyde support, for covalent cross-linking of proteins.[71,72]

Besides the activation reactions involving the hydroxyl groups described above, organic supports have been made more versatile for enzyme coupling by introducing groups such as diazo, thiol, or amino groups. Diazonium derivatives of organic supports have been used for covalent coupling primarily to the phenolic group of tyrosine or the imidazole of histidine.[72–74] The diazo supports are generally prepared by a multistep reaction. Initially, the polymer containing amino side chains is reacted with nitrobenzoyl chloride solution in pyridine. The resulting derivative is reduced with reagents such as sodium dithionate to yield an arylamine support, which is then diazotized by reacting with $NaNO_2$ and HCl at 0–5°C. The details of this immobilization technique have been discussed by Cohen.[74]

Interaction of the sulfhydryl groups of the protein with the disulfide groups in thiolated carriers has been adopted for the covalent immobilization.[75–78] The method offers a way to immobilize the enzymes reversibly, since the bound enzyme can be eluted out by disulfide reagents. Agarose-glutathione-2-pyridyl disulfide gel is offered by Pharmacia, Inc. under the name Activated Thiol-Sepharose.

Another method which does not involve hydroxyl activation has been shown to be effective. For example, a four-component system, involving the condensation of amine, carboxyl, carbonyl, and isocyanide groups for enzyme immobilization, has been developed.[79–81] By this process, coupling can be via the amino or carboxy function of the protein, by providing the reagents containing the other three groups.

3.4.2. *Inorganic Supports*

A wide variety of inorganic supports (see Table 1), including minerals as well as prefabricated materials, have been used for covalent immobilization of enzymes.[82,83] Linkage of proteins generally involves the prior activation of the support materials by means of a coupling agent, the most popular of which is the silane reagent, aminopropyl-triethoxy-silane (APTS).

This reagent was first introduced by Weetall[84,85] for derivatization of controlled pore glass (CPG) for enzyme coupling, and has since found wide application for linkage of proteins to many inorganic supports, including ZrO_2, TiO_2, Al_2O_3, and NiO. Reaction of APTS with CPG involves the multiple displacement of the ethoxy residues on APTS by the hydroxyl groups on the surface of the glass, resulting in a covalently bonded silane polymer. Silane reagents with other pendant groups have also been used as coupling agents.[86]

Support material is prepared for silanation by prior washing with nitric acid at 80° and subsequently with water.[82] Reaction of CPG with the silane reagent has been performed in organic medium such as toluene or acetone, as well as in aqueous medium. Organic silanations have been reported to give derivatives with a higher amine loading, while aqueous silanation has been found to yield supports possessing lower enzyme loading but greater stability.[82]

Enzymes can be immobilized onto the alkylamine inorganic support either by direct coupling or after further derivatization of the support. Direct linkage to the protein–COOH groups is accomplished by means of carbodiimide coupling reagent.[84–86] Alternately, carboxyl derivatives of the silanated supports are prepared by reaction with succinic anhydride, and coupled to the amino group of the proteins by means of carbodiimide reagent.[87] Activated derivatives of the aminoalkyl carriers have been generated by reacting with excess glutaraldehyde at pH 7.0. The derivatized support is used directly for coupling with the $—NH_2$ group of the protein, after removal of the excess glutaraldehyde. This method has been used to immobilize a variety of enzymes to the inorganic supports.[88–95] The alkylamino supports have also been converted to the arylamine materials by reaction with nitrobenzoyl chloride plus dithionite and further diazotized for enzyme coupling involving tyrosine.[96,97]

Hydroxyl groups on coated glass surfaces can be activated for enzyme coupling by procedures discussed for polysaccharide supports. For example, periodate oxidation of glycophase-G (Pierce Chemical Co.) yields an aldehyde derivative, which facilitates enzyme coupling.[98,99] Treatment of glucophase-G with cyanogen bromide also allows covalent attachment of enzymes.[98–100] Sulfonate esters of glucophase-G have been prepared by reacting with 2,2,2-trifluoroethanesulfonyl chloride, and these esters are good supports for enzyme immobilization.[59]

3.5. Entrapment

An enzyme can be entrapped or occluded in a semipermeable constraining structure, which will prevent the diffusion of the enzyme into the surrounding medium, while allowing the entry of low-molecular-weight substrates into the structure. Vinyl monomers such as acrylamide, 2-hydroxyethyl methacrylate, and *n*-vinyl pyrrolidone, as well as polymers such as starch, agar, collagen, silicon rubber, silica gel, polyvinyl alcohols, and seaweed polysaccharides have been effectively used to entrap enzymes.

Among the monomers used for enzyme entrapment, acrylamide has found the widest application.[100–103] Entrapment is generally performed by polymerizing an aqueous solution containing the acrylamide monomer and the enzyme, in the presence of methylene-bis-acrylamide as the cross-linking agent. Polymerization takes place in presence of a catalyst such as potassium persulfate, riboflavin, tetraethyl methylene diamine, or diaminopropionitrile, under conditions similar to those for the preparation of acrylamide gels for electrophoresis, which is generally at neutral pH.[104] Gels of different shapes and sizes can be prepared, by using appropriate containers. However, acrylamide gels are highly hydrophilic, and hence mechanically weak, which may result in enzyme leakage through the pores. Moreover, polymerization of acrylamide is exothermic, and necessitates careful temperature control to preserve the enzyme activity. Alternatively, enzymes have been entrapped in 2-hydroxyethyl methacrylate gels cross-linked with ethylene glycol dimethacrylate in the presence of polymerization initiators such as di(*sec*)-butylperoxycarbonate or benzoin ethyl ester.[105–107]

A method of polymerizing acrylamide by X-ray irradiation was introduced by Dobo.[108] This method offers the advantage that the immobilization can be performed at low temperatures, even in the frozen state, thus minimizing the loss of enzyme activity. Moreover, the resulting polymer will be free of any chemical catalyst. A wide variety of enzymes has been entrapped by irradiation–polymerization[109–111] of acrylamide gels; however, irradiation-induced immobilization requires costly equipment which limits its applicability.

Inorganic materials such as silica gel, as well as silicon rubber, and biological polymers such as alginates, seaweed polysaccharides, and collagen have been used for entrapment of enzymes.[11] Trypsin has been entrapped in silica gel, by adding the enzyme to silicic acid solution of ionic strength of 0.1 *M*. The gel is set in 30 min and allowed to age for 12 h. Enzymes have been entrapped in the hydrophobic matrix of silicon rubber by polymerizing the Silastic™ resin and a vulcanizing agent in the presence of enzyme. Entrapment of enzymes in polyvinyl alcohol (PVA) gels has been performed by exposing an aqueous solution of the reagent containing the enzyme to gamma irradiation, or electron beam.[112,113] PVA films with entrapped enzymes have been prepared by *in vacuo* drying of the PVA reagent containing the desired enzymes.[114] Jancsik *et al.*[115] have reported several methods of preparing PVA-entrapped enzymes using lactase and aldolase.

Several procedures have been developed for the entrapment of enzymes in polysaccharides. For carrageenan, the method consists of mixing a solution of enzyme and a solution of the polysaccharide at 37–60°C and cooling the mixture to 10°C to facilitate the gel formation.[116] A suspension of enzyme in cellulose solubilized in a mixture of ethyl pyridinium chloride and dimethyl formamide forms cellulose fibers upon mixing with water at room temperature.[117] Starch gels containing enzymes have been prepared by dispersing a mixture of enzyme in a warm starch solution and covering the mixture to obtain the starch gel with entrapped enzymes, and expressing the water from the gel over a polyurethane pad.[118]

Condensation copolymerization involves the simultaneous reaction of three components, namely, a water-soluble preformed polymer, a low-molecular-weight diamine, and the enzyme. The method has been described by Pollak *et al.*[119] who have immobilized nearly 60 enzymes using a water-soluble copolymer of acrylamide-*N*-acryloxysuccinamide and triethylenetetramine.

In another method of immobilization, the enzymes are first grafted with a long spacer arm, by reacting with reagents such as glycidyl methacrylate. The grafted enzyme is then added to the suspension of the support, and the immobilization is initiated with redox systems such as Fe^{+2}–H_2O_2. Polysaccharide supports such as Sephadex, Sepharose, and cellulose have been used for this method. The enzymes immobilized in this manner are not necessarily diffusion limited, since they are located on the outer surface of the polymer by virtue of the long spacer arm. The graft–copolymerized enzymes are hence shown to have kinetics similar to those of the free enzymes.[120,121]

3.6. Microencapsulation

By microencapsulation the enzyme is enclosed within a semipermeable membrane which prevents the enzyme from leaking out while allowing the free flow of substrates and small molecules. Ultrathin spherical membranes of cellulose acetate and nylon[122–124] have been created around aqueous emulsions of enzyme droplets by the method of interfacial polymerization. First, an aqueous solution of the enzyme containing a hydrophobic monomer dissolved in the same organic solvent is emulsified. During this process, monomers are polymerized in the interphase between the aqueous and organic phase, and the enzyme in the aqueous phase is enclosed within the membrane. The size of the microcapsules can be adjusted by changing the concentration of the emulsifying agent. Furthermore, suitable combinations of the monomers must be chosen according to the nature of the enzyme and the application for which it is intended.

In the liquid-drying method of microencapsulation, an enzyme solution is dispersed in a water-immiscible organic solvent containing dissolved polymer, so as to form a water-in-oil–type emulsion. This emulsion containing the aqueous microdroplets is then dispersed in an aqueous, protective solution of colloidal substances, such as polyvinyl alcohol, gelatin, and surfactants. This secondary emulsion is dried *in vacuo*, resulting in a polymer membrane containing enzymes.[125]

Instead of forming spherical membranes, fiber-entrapped enzymes have been prepared by extruding the emulsion of the enzyme–polymer solution to form fibers containing microdroplets of the enzymes.[126] These types of immobilized enzymes are valuable in industrial reactors, where large flow rates are generally required.

Biological materials such as lipids and erythrocytes have been used to encapsulate enzymes, which may have potential application as therapeutic agents. When water-insoluble polar lipids are introduced into water, a highly ordered system on concentric closed membranes is formed. Each membrane represents an unbroken bimolecular sheet of lipid molecules, called liposomes.[127,128] During the process of membrane formation, water-soluble substances are entrapped in the aqueous compartment in the lipid layer, provided they are of the size to fit into the aqueous space between the planes of the hydrophilic head groups. For immobilization, the lipids are dissolved in an organic solvent, which is then evaporated under vacuum. leaving a thin film. This film is dispersed in an aqueous solution of the enzyme, to form the liposomes. A wide variety of enzymes have been encapsulated in liposomes by this method.[127]

When red blood cells are placed in a hypotonic solution, they swell and become permeable to proteins. As a result, the cell proteins leach out while allowing the entry of proteins or enzymes from the outside solution. Next, the blood cells are placed in an isotonic solution and regain their normal shape, simultaneously enclosing the foreign enzymes. The resulting immobilized enzyme is biological and hence biocompatible; however, because of the very low permeability of the red blood cells to many substrates, the activity of the immobilized enzyme may be quite low.[129]

3.7. Electrochemical Methods

Enzymes can be immobilized on porous alumina discs by electrodeposition. A solution of the enzyme is taken in the electrolytic apparatus, and fitted with the anode and cathode terminals. The porous disc is interposed between the terminals, adjacent to the anode. Upon passing the current, the enzyme migrating towards the anode is deposited on the disc.[130] An electrolytic codeposition method has been used to prepare an enzyme–collagen membrane.[131,132] Upon electrolysis of a mixture of collagen fibrils and an enzyme solution, the enzyme–collagen membrane is formed on the surface of the electrode.

4. Properties of Immobilized Enzymes

4.1. Physical Properties

The physical properties of an immobilized enzyme are determined by the support material (see Section 2). Inorganic support materials are rigid and generally particulate, resulting in excellent flow properties when packed into a column reactor. They are usually prepared in porous form to increase enzyme loading, but even in a porous form the quantity of enzyme which can be bound per unit volume is lower than for most organic supports.

Organic supports afford the advantage of a much wider selection of reactive, functional groups. A high density of the proper functional group on the support can result in very high enzyme loading. In addition, enzymes that have proven difficult to immobilize successfully by typical methods, can often be immobilized on organic supports by using less common reactive sites.

4.2. Chemical Properties

The chemical properties of immobilized enzymes are determined not only by the enzyme, but also by the method of immobilization and the support. Changes in catalytic properties upon immobilization result from two causes. First, the microenvironment of the enzyme has changed. It is no longer simply in an aqueous solvent. In essence the enzyme is now located in a mixed "solvent" made up of the support material and the original aqueous solvent. This new "solvent" can alter the catalytic activity by changing the pH of the enzyme environment which changes the observed pH of optimum activity. Since the solvent surrounding the enzyme is different, the concentration of the substrate and/or product may be different around the enzyme than in the bulk solution which will affect the apparent catalytic rate.

Second, the catalytic activity could also be affected by the method of immobilization. Derivatization of side chains of the enzymes can cause changes in the tertiary and secondary structure. Such changes can have an effect on the enzyme activity, but are difficult to detect since the enzyme is a small part of the support-enzyme composite. Moreover, immobilized enzymes are difficult to study by optical methods so commonly employed by enzymologists.

4.3. Stability

There are two basic reasons to immobilize an enzyme: (1) when an enzyme is immobilized it can be easily separated from the products of the reaction for reuse; and (2) an immobilized enzyme is often more stable than its soluble counterpart. Thus an enzyme, properly immobilized, can be easily used over and over again for an extended period of time.

Enzymes do not always become more stable in the process of immobilization.[133] In many instances they may appear more stable even though this is not the case. For example, when an enzyme is immobilized with high enzyme loading, the measured rate of reaction may be diffusion limited. Under these circumstances, a constant rate of reaction may be observed for weeks or even months, until the rate is no longer diffusion limited. When the reaction finally becomes limited by the enzymatic rate of catalysis, the rate of conversion of substrate to product may drop dramatically. On the surface, it would appear that the enzyme had suddenly lost activity or been "poisoned," when in fact the enzyme activity had decreased gradually with time.

The stability of an immobilized enzyme is usually dependent upon the conditions of storage and use. Immobilized enzymes stored wet at 0–5°C often remain stable over several months or even years.[134] Obviously, stability in storage is important so that immobilized enzymes can be readily available for replacement. The stability of an immobilized enzyme in use is a very important parameter to determine if a process is commercially feasible. Generally stability of several months can be achieved as long as a temperature well below the unfolding transition of the enzyme is maintained. The reason for loss in activity is sometimes difficult to identify.[135] In some instances it is clearly due to attrition at the surface of the support material,[136] while in other cases it can be traced to "poisoning" of the enzyme with an inhibitor.[137] Probably it is most often due to loss of the native conformation of the enzyme, but this cause is difficult to establish.

Increased stability at high temperature can be accomplished through immobilization. It has been shown that the increase in temperature stability is most likely due to increased "rigidity" caused by the bonds formed with the side chains in the process of immobilization.[133] Thus when enzymes are immobilized by covalent bonding, the thermal stability increases with the number of bonds between the enzyme molecule and the support.[133]

4.4. Specificity

The specificity of an enzyme can be affected by the process of immobilization. When an immobilized enzyme has a macromolecular substrate as well as low-molecular-weight substrates, the specificity will nearly always change in favour of the low-molecular-weight substrates.[11] The predominant reason is the diffusional limit of the large substrate both to and from the immobilized enzyme. The microenvironment discussed above also affects the specificity.[11] The concentration of substrate in the vicinity of the enzyme is affected by the nature of the support material. If the support material is hydrophobic, the concentration of a hydrophobic substrate will be higher in the vicinity of the enzyme than in the bulk solution. This higher concentration will result in a higher rate of reaction than would otherwise be expected. If the support is charged, similar phenomena will be observed. If the substrate possesses the opposite charge, the rate will be enhanced. On the other hand, a substrate of the same charge would exhibit a lower rate than otherwise expected. Charged support material also can result in a different pH optimum for the immobilized enzyme compared to the free soluble enzyme. These phenomena are observed at low ionic strength when the distribution of hydrogen ions results in a different pH in the vicinity of the enzyme than in the bulk solution.

It is also possible that a conformational change upon immobilization could cause a change in specificity. As mentioned above, the structure of the immobilized enzyme is difficult to study. One example in the literature relates to the immobilization of chymotrypsin in which a change in conformation during immobilization was established.[138] No change in specificity was observed in this case.

5. Uses of Immobilized Enzymes

Immobilized enzymes are currently used commercially in two types of applications. Most large-scale industrial applications are in the food industry and use immobilized enzymes in column reactors. By contrast, analytical applications require small quantities of immobilized enzymes.

5.1. Industrial Applications

The first commercial application of immobilized enzymes was developed by the Tanabe Pharmaceutical Company in 1969 to produce L-amino acids.[139] The procedure involved organic synthesis of the acyl derivative of the desired amino acid, followed by treatment with immobilized L-aminoacylase. Separation of the L-amino acid from the acyl derivative of the D-amino acid is quite easy. After considerable research, the method of immobilization selected was ionic adsorption of the L-aminoacylase onto DEAE-Sephadex.

Of the industrial applications, the use of immobilized glucose isomerase for the production of high fructose corn syrup (HFCS) is the largest. The enzyme used commercially, which is available from several sources, is actually D-xylose isomerase. Fortunately, this enzyme happens to catalyze the conversion of glucose to fructose, as well as the isomerization of pentoses.

Since its introduction approximately seven years ago, HFCS has increased its share of the sweetener market dramatically. Presently, this sweetener accounts for one-third of the $6 billion market.[140] Although growth has slowed, if soft-drink manufacturers switch to HFCS the market share could increase considerably.

The traditional method of making glucose syrup from cornstarch was by acid hydrolysis. This process had two disadvantages: (1) the many side reactions that occurred in the acid environment resulted in a syrup of poor quality and (2) the syrup lacked the sweetness required for many applications. Today's typical process overcomes these problems. First, α-amylase is used to liquify the starch, followed by glucoamylase to hydrolyze the starch fragments to glucose. This high-quality glucose syrup is then converted to HFCS by passing it through a column containing immobilized glucose isomerase which increases the sweetness.

Another enzyme of interest to the food industry is invertase which hydrolyzes sucrose to a mixture of glucose and fructose. Raffinose, which interferes with the crystallization of sucrose, can be hydrolyzed with an α-galactosidase. Commercial processes involving these enzymes have been developed by Hokkaido Sugar Co. and are discussed in detail by Brodelius.[141]

Immobilized lactase (β-galactosidase) is of commercial interest in the treatment of milk and whey.[13] Since many people are intolerant to lactose found in milk, it would be beneficial to treat milk with immobilized lactase. Whey also contains lactose and is a major byproduct of cheese making. Immobilized lactase can be used to hydrolyse lactose to glucose and galactose and the resulting whey product could be used as a sweetener.

5.2. Analytical Applications

Analytical tests using immobilized enzymes can be divided into two groups. The first group consists of those instruments in which the immobilized enzyme is in contact with the sensor. With other types of instruments the immobilized enzyme is in a column or reactor separated from the sensor.

5.2.1. Biochemical Sensors

The term "biochemical sensor" might be defined as any sensor which measures a biochemical substance, but in this review we shall use the term in a somewhat more restricted sense. *"Biochemical sensors" are those instruments which incorporate immobilized enzymes in contact with the detector.* Thus "enzyme electrodes" are biochemical sensors. The term "enzyme electrode," although frequently used, is misleading. It is used to describe an immobilized enzyme in contact with an electrochemical detector, but it is not an electrode to measure enzyme activity as the name implies. Biochemical sensors as defined here, include those instruments traditionally considered enzyme electrodes, as well as those that do not involve electrochemistry as a means of detection.

Although most biochemical sensors use a form of electrochemical detector,[142] several other means of detection are presently being investigated. Several studies are now in the literature illustrating the use of thermometric probes.[143–146] The advantage of this form of detection lies in its universality since almost any reaction that is catalyzed by the immobilized enzyme either absorbs or gives off heat. Unfortunately, any side reaction that might occur upon addition of a sample would likewise be detected, and the high sensitivity that is required in many applications has proven difficult to achieve. Biochemical probes have also been demonstrated by attachment of the immobilized enzyme to the tip of a fiber optic device. Table 3 illustrates the scope of biochemical sensors that have been reported in the literature.

Biochemical sensors have been well studied because of their many potential advantages:

1. Ease of measurement.
2. Little pretreatment of sample.
3. Rapid response.
4. Small sample volume.
5. Inexpensive to construct.
6. Many potential applications.
7. Microprobes can be designed.

Many problems have been discovered which have slowed the utilization of biochemical sensors. The lack of long-term stability is sometimes a problem and can be effected by many factors:

1. Stability of the base sensor.
2. Thickness of the enzyme layer.
3. Physical strength of the enzyme layer.
4. Method of immobilization.
5. Initial enzyme activity.
6. Stability of enzyme activity.
7. pH and temperature of use.
8. Accumulation of inhibitor.
9. Storage conditions.

Table 3. Biochemical Sensors

Detected substance	Reference	Enzyme(s)	Support	Immobilization method	Detector type
Glucose	147	Glucose oxidase	Polycarbonate	Cross-linking glutaraldehyde collagen	Amperometric Pt, Ag
Glucose	148	Glucose oxidase	Polyacrylamide	Entrapment	Amperometric Pt, Ag
Sugars	149	Sugar oxidases	Controlled pore glass	Silanation glutaraldehyde	Thermistor
Urea	150	Urease	Controlled pore glass	Silanation glutaraldehyde	Thermistor
Lactate	151	Lactate dehydrogenase	Cellulosic dialysis membrane	Cross-linking glutaraldehyde	Amperometric glass carbon
Amino acids	152	Amino acid oxidase	Polyacrylamide	Entrapment	Potentiometric
Ethanol	153	Alcohol oxidase	Albumin	Cross-linking glutaraldehyde	Amperometric Pt
Penicillin	154	Penicillinase	Polyacrylamide	Entrapment	pH Electrode

The temperature and pH of operation of the biochemical sensor is very important. The optimum pH of the enzyme layer is often different from that of the sensor. Usually the pH chosen for operation will be that of the enzyme rather than the sensor. If the enzyme is opertated at a pH other than its optimum, not only will the activity be greatly reduced, but the stability may be very low. The temperature of operation is likewise determined mainly by the enzyme. The highest enzyme activity is usually obtained near the temperature of thermal inactivation, but to operate at this temperature would greatly reduce the life of the immobilized enzyme. Thus the temperature should be at least 10° below the highest temperature, which results in maximum activity.

In the authors' opinions, the most serious problem with biochemical sensors is interferences. Interferences may arise from any material which interacts with the enzyme, measured species, or base sensor. Most interferences involve the sensor. Since the species to be measured is rarely in pure form, many other substances present may be detected by the sensor. An example is the glass ammonium-ion sensor married to enzymes which catalyze reactions that release ammonia. Unfortunately, these electrodes are responsive to Li^+, K^+, Na^+, and H^+, as well as the desired ammonium ions. An amperometric sensor is less likely to have serious interference problems. When covered with a membrane (Clark electrode) and used to detect oxygen, and reducible gases interfere.

Yellow Springs Instrument Co. has developed an instrument to measure glucose which uses an electrode covered with a series of membranes. One membrane contains immobilized glucose oxidase to convert glucose and O_2 to H_2O_2 and gluconic acid, while the other membranes are present to reduce interferences. H_2O_2 is detected by the platinum electrode.

5.2.2. Partitioned-Enzyme Sensors

Partitioned-enzyme sensors differ from biochemical sensors in that the immobilized enzyme is separated from the sensor. The primary advantage lies in the fact that by separating the enzyme from the sensor it is possible to use both enzyme and sensor at or near their optimum operating conditions.

When designing the immobilized enzyme for partitioned-enzyme sensors, several parameters must be considered. This type of instrument requires relatively rapid and very constant flow rate. Moreover, the amount of catalyst must be sufficient for high initial activity as well as sustained activity. Since the immobilized enzyme will probably remain in the instrument during its lifetime, stability under operating conditions is paramount.

Two types of supports have been used extensively for these instruments:

porous inorganic and tubular organic supports. Both of these supports have excellent flow properties, but the quantity of enzyme which can be loaded per unit volume is in the moderate range. Porous, particulate, organic supports do allow high enzyme loading, but are much more likely to generate flow irregularities. When using tubular supports, the enzyme is usually immobilized by partial hydrolysis of the interior surface of the tube followed by covalent attachment. For porous inorganic supports, the method of choice is either covalent attachment or adsorption–cross-linking.

Many of the early systems used spectrophotometric detectors. More than 15 years ago, a system was reported using polyacrylamide gel-entrapped lactate dehydrogenase and glucose oxidase which detected either lactate or glucose.[155] Immobilized enzymes on the interior of nylon tubes used in conjunction with spectrophotometric detectors were investigated by Hornby and co-workers.[161] Some of the enzymes used in analysis were urease, glucose oxidase and invertase. Table 4 lists examples from the literature which describe partitioned-enzyme sensors.

Several commercial instruments have been developed using the partitioned-enzyme sensor concept. Technicon Instruments Corp. developed tubing immobilized enzymes required in the hexokinase method of glucose analysis. These immobilized enzymes could be inserted into the SMA or SMAC clinical instruments. The product, NADH, is measured spectrophotometrically.

Instruments to measure the concentration of sugars in industrial processes were developed by the Leeds & Northrup Company. The enzymes were immobilized on porous glass which resulted in excellent flow characteristics and sufficient enzyme loading. One instrument measured glucose with the aid of immobilized glucose oxidase, while the more complex instruments measured glucose and sucrose or glucose and lactose.[166] All the instruments depend upon a three-electrode amperometric detector.

A clinical instrument to measure urea was developed by Owens–Illinois, Inc.[156,157,167] The enzyme urease was immobilized on porous alumina. This irregular-shaped inorganic support gave excellent flow characteristics and good enzyme loading. The injected sample containing urea passes through the column of immobilized urease. The ammonium ions generated are converted to dissolved ammonia gas when the flowing stream mixes with a stream containing a solution of sodium hydroxide. The dissolved ammonia gas is then sensed as it passes by an ammonia detector which consists of a pH electrode covered with a gas-permeable membrane.

5.3. Structure–Function Studies

A method has been developed by Chan[168] to study the subunit interactions in oligomeric enzymes by techniques of immobilization. First,

Table 4. Partitioned-Enzyme Sensors

Detected substance	Reference	Enzyme(s)	Support	Immobilization method	Detector type
Urea	156,157	Urease	Porous alumina	Cross-linking by disulfide rearrangement	Ammonia electrode
Glucose	158	Glucose oxidase	Porous alumina	Cross-linking alkane dihalide and diamine	3-Amperometric electrodes
Lactose	159	Lactase Glucose oxidase	Phenol formaldehyde resin	Cross-linking glutaraldehyde	Colorimetric Potentiometric
Estriol	160	β-Glucuronidase	Controlled pore glass	3-Amino propyl-methyl-dimethyl-silane	Fluorescence
Sucrose Maltose Lactose	161	Glucose oxidase Amyloglucosidase Invertase β-Galactosidase	Tubular (nylon)	Glutaraldehyde	Spectrophotometer
Nitrate	162	Nitrate reductase	Controlled pore glass	Silanation–diazonium salt	Spectrophotometer
Amino acids	163	Amino acid oxidase	Controlled pore glass	Silanation Glutaraldehyde	Ammonia electrode
α-Amylase	164	Glucose oxidase Catalase Glucoamylase	Porous alumina	Cross-linking alkane dihalide and diamine	3-Amperometric electrodes
Mercury (II)	165	Urease	Controlled pore glass	Silanation Glutaraldehyde	Ammonia electrode

the oligomeric enzyme is covalently attached to the matrix through one subunit per enzyme molecule.[169,170] The immobilized enzyme is then washed extensively with a protein-denaturing agent to dissociate and remove the subunits not linked to the support. The support-linked subunit is then renatured to obtain a single subunit of the enzyme covalently immobilized to the matrix, which is then tested for enzyme activity, in the presence and absence of a soluble subunit prepared from the soluble form of the enzyme. The relationship of subunit association to activity and conformational stability can thus be investigated.

5.4. Therapeutic Applications

Even though the potential use of enzymes as therapeutic agents in many disease conditions has been recognized,[4] their application has been very restricted. Not only do enzymes initiate an immune response, but they are also rapidly inactivated by proteases *in vivo*. For many diseases, it is difficult to target the enzyme to the pathological organs in the body.[171] The application of immobilized enzyme offers an alternative whereby many of the above problems can be circumvented.[172] Microencapsulated enzymes have the advantage of protection from inactivation by external proteases, as well as reduced immune response compared to the free enzyme. Liposomes have elicited considerable interest as potential enzyme carriers in the treatment of lysosomal diseases.[172,173] Liposomes administered *in vivo* are taken up into the cell by endocytosis, and the endocytic vacuoles fuse to the lysosomes.[174] Within the lysosome, the liposomes are disrupted, releasing the entrapped enzyme. The possibility of targeting liposomes by incorporating anticell antibodies,[175] as well as tissue-specific glycoprotein,[176] has been reported. However, liposome-entrapped enzymes, when administered *in vivo*, still elicit some immune response.[177] Encapsulation in erythrocyte membranes offers another technique to target enzymes to organs, such as the liver and spleen, without the problem of immune reaction.[178]

In addition to the *in vivo* application, immobilized enzymes have potential use in the extracorporeal treatment, as well as in local application.[178] In the local application, the possible absorption of soluble enzymes through the skin, and consequent immune response, can be avoided by using immobilized preparations.

6. Future Directions

The field of catalysis will no doubt grow in the 1980s and beyond because of the need to conserve energy and reduce environmental pollution.

Hundreds of methods to immobilize enzymes have been investigated and are reported in the literature. In addition, over the past two decades, considerable experience has been gained in the use of immobilized enzymes. Certainly, the basic information is available to develop applications as opportunities arise. The pharmaceutical field should be a fertile area for the application of immobilized enzymes. This industry has used fermentation for the production of drugs, but this method is inherently inefficient since the drug desired is only a very small fraction of the total biomass produced. With immobilized enzymes the drug could be produced "cleanly" reducing the need to dispose of unwanted biomass.

Genetic engineering will make the use of many enzymes economically feasible that would be too expensive otherwise. Even so, the enzymes produced by means of genetic engineering will have the same characteristics as all soluble enzymes. Thus immobilization of genetically engineered enzymes will be required.

Potential application of immobilized enzymes in the area of solar-energy conversion has gathered considerable attention. Systems incorporating the photosynthetic electron transport with the bacterial ferredoxin–hydrogenase system have been studied in many laboratories, with limited success because of the inherent instability and oxygen sensitivity of the hydrogenase. Use of immobilized hydrogenase has improved the outlook for the long-term usefulness of this system to generate power.[177–181]

Modification of enzyme properties during the process of immobilization will no doubt be an important consideration in future applications. It has long been known that immobilization can be used to increase the stability of an enzyme, as well as to change the pH optimum. Furthermore, it is sometimes possible to use an immobilized enzyme under conditions in which the soluble enzyme is unstable.

Acknowledgments

The authors wish to thank Dr. Barry Watson for helpful discussions during preparation of this manuscript. Thanks are also due to Mr. Mark Bohnett for the design of the figure on immobilization. The preparation of this manuscript was made possible by the financial support of Owens–Illinois, Inc.

References

1. A. Wiseman, in: *Handbook of Enzyme Biotechnology* (A. Wiseman, ed.), pp. 111–124, Wiley, New York (1975).
2. M. Keyes, *Encyclopedia of Chemical Technology*, Vol. 9, 148–172, Wiley, New York (1980).

3. T. Everse, C. L. Ginsburg, and N. O. Kaplan, *Methods Biochem. Anal.* **25**, 135–201 (1977).
4. J. S., Holcenberg and J. Roberts, ed., *Enzymes as Drugs*, pp. 1–353, Wiley, New York (1981).
5. P. J. Lisi, T. van Es, A. Abuchowski, N. C. Palczuk, and F. F. Davis, *J. Appl. Biochem.* **4**, 19–33 (1982).
6. C. H. W. Hirs, ed., *Methods in Enzymology*, Vol. 11, pp. 481–640, Academic Press, New York (1967).
7. C. H. W. Hirs and S. N. Timasheff, ed., *Mewthods in Enzymology*, Vol. 25, pp. 339–651, Academic Press, New York (1972).
8. M. A. Mitz and L. J. Summaria, *Nature (London)* **169**, 576–577 (1961).
9. P. W. Carr and L. D. Bowers, *Immobilized Enzymes in Analytical and Clinical Chemistry*, Wiley, New York (1980).
10. L. B. Wingard, E. Katchalski-Katzir and L. Goldstein, eds., *Applied Biochemical Bioengineering*, Vol. 1, Academic Press, New York (1976).
11. I. Chibata, *Immobilized Enzymes*, Wiley, New York (1978).
12. T. K. Ghose, A. Fiechter, and N. Blakebrough, eds., *Advanced Biochemical Engineering*, Vol. 10, Springer-Verlag, New York (1978).
13. R. A. Messing, ed., *Immobilized Enmzymes for Industrial Reactors*, Academic Press, New York (1975).
14. K. Mosbach, ed., *Methods in Enzymology*, Vol. 44, Academic Press, New York (1976).
15. L. B. Wingard, Jr., E. Katchalski-Katzir, and L. Goldstein, eds. *Applied Biochemical Bioengineering*, Vol. 3, Academic Press, New York (1981).
16. A. C. Olson and C. L. Cooney, eds., *Immobilized Enzymes Food Microbiology Processes*, (Proceedings of Symposium), Plenum Press, New York (1974).
17. M. Salmona, C. Saronio, and S. Garottini, eds., *Insolubilized Enzymes*, Raven Press, New York (1974).
18. T. M. S. Chang, ed., *Biomedical Applied Immobilized Enzymes Proteins*, Vol. 1, Plenum Press, New York (1977).
19. R. A. Messing, in: *Advanced Biochemical Engineering* (T. K. Ghose, A. Fiechter, and N. Blakebrough, eds.), Vol. 10, pp. 51–73, Springer-Verlag, New York (1978).
20. G. P. Royer, G. M. Green, and B. K. Sinha, in: *Polymer Grafts in Biochemistry* (H. F. Hixson, Jr. and E. P. Goldberg, eds.), pp. 289-307, Marcel Dekker, New York (1976).
21. O. Zaborsky, in: *Biomedical Applications of Immobilized Enzymes and Proteins* (T. M. S. Chang, ed.), p. 37, Plenum Press, New York (1977).
22. R. A. Messing, *Biotechnol. Bioeng.* **16**, 897–908 (1974).
23. P. Grunwald, W. Gunsser, F. R. Heiker, and W. Roy, *Anal. Biochem.* **100**, 54–57 (1979).
24. G. A. Kovalenki, N. B. Shitova, and U. D. Sokolowski, *Biotechnol. Bioeng.* **23**, 1721–1734 (1981).
25. R. Uy and F. Wold, in: *Advanced Experimental Medical Biology* (M. Friedman, ed.), Vol. 86A, pp. 169–186, Plenum Press, New York (1977).
26. F. M. Richards and J. R. Knowles, *J. Mol. Biol.* **37**, 231–233 (1968).
27. P. Monsan, G. Puzo, and H. Mozargui, *Biochimie* **57**, 1281–1292 (1975).
28. R. Haynes and K. A. Walsh, *Biochem. Biophys. Res. Commun.* **36**, 235–242 (1969).
29. R. Haynes and K. A. Walsh, U.S. Patent No. 3,796,634 (1974).
30. M. H. Keyes, U.S. Patent No. 3,933,589 (1976).
31. M. H. Keyes, U.S. Patent No. 4,008,126 (1977).
32. M. H. Keyes, U.S. Patent No. 4,204,040 (1980).
33. R. A. Messing, U.S. Patent No. 3,804,719 (1974).
34. G. Brown, E. Selegny, S. Avrameas, and D. Thomas, *Biochim. Biophys. Acta* **185**, 260–262 (1969).

35. J. Porath, in: *Methods in Enzymology* (W. B. Jacoby and M. Wilchek, eds.), Vol. 34, pp. 13–30, Academic Press, New York (1974).
36. J. Porath and R. Axen, in: *Methods in Enzymology* (K..Mosbnach, ed.), Vol. 44, pp. 19–42, Academic Press, New York (1976).
37. M. D. Lilly, in: *Methods in Enzymology* (K. Mosbach, ed.), Vol. 44, pp. 46–53, Academic Press, New York (1976).
38. I. Chibata, *Immobilized Enzymes*, pp. 15–46, Wiley, New York (1978).
39. R. Axen and S. Ernback, *Eur. J. Biochem.* **1971**, 351–360.
40. R. Axen, J. Porath, and S. Ernback, *Nature(London)* **214**, 1302–1304 (1967).
41. R. Axen and P. Vretblad, *Acta Chem. Scand.* **25**, 2711–2716 (1971).
42. R. L. Schnaar, T. F. Sparks, and S. Roseman, *Anal. Biochem.* **79**, 513–525 (1977).
43. M. Wilchek, T. Oka, and Y. J. Trooper, *Proc. Natl. Acad. Sci. U.S.A.* **72**, 1055–1058 (1975).
44. K. Kohn and M. Wilchek, *Biochem. Biophys. Res. Commun.* **84**, 7–14 (1978).
45. J. Kohn and M. Wilchek, *Anal. Biochem.* **115**, 375–382 (1981).
46. J. Kohn and M. Wilchek, *Enzyme Microb. Technol.* **4**, 161–164 (1982).
47. H. Rosemeyer, E. Kornig, and F. Seela, *Eur. J. Biochem.* **122**, 375–380 (1982).
48. L. R. Benkova, M. Mrackova, and K. Baber, *Collect. Czech. Chem. Commun.* **45**, 160–167 (1980).
49. M. Iwaki and M. Nozaki, *J. Biochem.(Tokyo)* **91**, 1549–1553 (1982).
50. M. Wilchek, in: *Enzyme Engineering* (E. K. Pye and H. H. Weetall, eds.), Vol. 3, pp. 283–289, Plenum Press, New York (1978).
51. J. F. Kennedy and J. A. Barnes, *Int. J. Biol. Macromol.* **2**, 289–296 (1980).
52. T. Oka and Y. J. Trooper, *Proc. Natl. Acad. Sci. U.S.A.* **71**, 1630–1636 (1974).
53. M. Wilchek and T. Miron, in: *Methods in Enzymology* (W. B. Jakoby and M. Wilchek, eds.), Vol. 34, pp. 72–76, Academic Press, New York (1974).
54. O. Hannibal-Friedrich, M. Chun, and M. Sernetz, *Biotechnol. Bioeng.* **22**, 157–175 (1980).
55. I. Matsumoto, H. Kitagaki, Y. Akai, Y. Ito, and N. Seno, *Anal. Biochem.* **116**, 103–110 (1981).
56. I. Matsumoto, Y. Mizuno, and H. Seno, *J. Biochem.* (*Tokyo*) **85**, 1091–1098 (1979).
57. L. Sundberg and J. Porath, *J. Chromatogr.* **90**, 97–98 (1974).
58. K. Nilsson, O. Norrlow, and K. Mosbach, *Acta Chem. Scand., Ser, B* **35**, 19–27 (1981).
59. K. Nilsson and K. Mosbach, *Biochem. Biophys. Res. Commun.* **102**, 449–457 (1981).
60. L. Bulow and K. Mosbach, *Biochem. Biophys. Res. Commun.* **107**, 456–464 (1982).
61. T. H. Finlay, V. Troll, M. Levey, A. J. Johnson, and L. T. Hodgins, *Anal. Biochem.* **87**, 77–90 (1978).
62. G. Kay and M. D. Lilly, *Biochim. Biophys. Acta* **198**, 276–285 (1970).
63. N. C. Smith and H. M. Lenhoff, *Anal. Biochem.* **61**, 302–305 (1974).
64. P. Cuatrecasas and I. Parikh, *Biochemistry* **11**, 2291–2298 (1972).
65. R. G. Frost, J. F. Monthony, S. C. Engelhorn, and C. J. Siebert, *Biochim. Biophys. Acta* **670**, 163–169 (1981).
66. S. A. Berker, H. C. Tun, D. H. Doss, C. J. Gray, and J. F. Kennedy, *Carbohydr. Res.* **17**, 471–474 (1971).
67. J. F. Kennedy and A. Zamir, *Carbohydr. Res.* **29**, 497–501 (1973).
68. C. J. Grey and T. H. Yeo, *Carbohydr. Res.* **27**, 2325–238 (1973).
69. J. Porath, T. Laas, and J. C. Janson, *J. Chromatogr.* **103**, 49–62 (1975).
70. J. Brandt, L. Andersson, and J. Porath, *Biochim. Biophys. Acta* **386**, 196–202 (1967).
71. M. Singh, A. R. Ray, P. Vasudevan, P. Verma, and S. K. Guha, *Biomater. Med. Dev. Artif. Organs* **7**, 495–512 (1979).
72. F. B. Weakley and C. L. Mehltretter, *Biotechnol. Bioeng.* **15**, 1189–1192 (1973).
73. S. A. Barker and J. F. Kennedy, in: *Handbook of Enzyme Biotechnology* (A. Wiseman, ed.), pp. 203–209, Wiley, New York (1975).

74. L. A. Cohen, in: *Methods in Enzymology* (W. B. Jakoby and M. Wilchek, eds.), Vol. 34, pp. 103–108, Academic Press, New York (1974).
75. K. Brocklehurst, J. Carlsson, M. P. J. Kiersten, and E. M. Crook, *Biochem. J.* **133**, 573–584 (1973).
76. J. S. Lin and J. F. Foster, *Anal. Biochem.* **63**, 485–490 (1975)
77. J. Carlsson, R. Axen, and T. Unge, *Eur. J. Biochem.* **59**, 567–572 (1975).
78. J. Carlsson, R. Axen, K. Brocklehurst, and E. M. Crook, *Eur. J. Biochem.* **44**, 189–194 (1974).
79. R. Axen, O. Vretblad, and J. Porath, *Acta Chem. Scand.* **25**, 1129–1132 (1971).
80. P. Vretblad and R. Axen, *Acta Chem. Scand.* **27**, 2769–2780 (1973).
81. L. Goldstein, *J. Chromatogr.* **215**, 31–43 (1981).
82. H. H. Weetall, in: *Methods in Enzymology* (K. Mosbach, ed.), Vol. 44, pp. 134–148, Academic Press, New York (1976).
83. A. Wiseman, in: *Topics in Enzyme Fermentation Biotechnology* (A. Wiseman, ed), Vol. 2, pp. 49–57, Halsted Press, New York (1978).
84. H. H. Weetall, *Nature* (*London*) **223**, 959–960 (1969).
85. H. H. Weetall, in: *Advanced Experimental Medical Biology* (R. B. Dunlap, ed.), Vol. 42, pp. 191–212, Plenum Press, New York (1974).
86. W. F. Line, H. Wong, and H. H. Weetall, *Biochim. Biophys. Acta* **242**, 194–202 (1971).
87. H. H. Weetall and A. M. Filbert, in: *Methods in Enzymology* (W. B. Jakoby and M. Wilchek, eds.), Vol. 34, pp. 295–297, Academic Press, New York (1974).
88. M. V. Wondolowski and T. H. Woychik, *Biotechnol. Bioeng.* **16**, 1633–1654 (1974).
89. B. E. Dale and D. H. White, *Biotechnol. Bioeng.* **21**, 1639–1648 (1979).
90. K. Parkin and H. O. Hultin, *Biotechnol. Bioeng.* **21**, 939–953 (1979).
91. V. Ramesh and C. Singh, *J. Appl. Biochem.* **4**, 81–85 (1982).
92. B. Danielsson, B. Mattiason, R. Karlsson, and F. Winqvist, *Biotechnol. Bioeng.* **21**, 1749–1766 (1979).
93. C. C. Hon and P. J. Reilly, *Biotechnol. Bioeng.* **21**, 505–511 (1979).
94. O. V. Lomako, I. I. Menyailova, C. A. Nakhapetyan, Y. Nikitin, and A. V. Kiselev, *Enzyme Microb. Technol.* **4**, 89–92 (1982).
95. C. D. Bowers and P. B. Johnson, *Anal. Biochem.* **116**, 111–115 (1981).
96. G. P. Royer, F. A. Liberatore, and G. M. Green, *Biochem. Biophys. Res. Commun.* **64**, 478–484 (1975).
97. L. D. Bowers and P. W. Carr, *Anal. Chem.* **48**, 549–558 (1976).
98. J. M. Cabral, J. M. Novais, and J. P. Cardoso, *Biotechnol. Bioeng.* **23**, 2083–2092 (1981).
99. R. D. Mason and H. H. Weetall, *Biotechnol. Bioeng.* **14**, 637–645 (1972).
100. B. Weiss, M. Hui, and A. Lajtha, *Biochem. Med.* **18**, 330–343 (1977).
101. T. Mori, F. Sato, T. Tosa, and I. Chibata, *Enzymologia* **43**, 217–226 (1972).
102. S. E. Brolin, A. Agren, B. Ekman, and S. Joholm, *Anal. Biochem.* **78**, 577–581 (1977).
103. A. Szewczuk, A. Ziomek, M. Mordarski, M. Siewinski, and J. Wieczorek, *Biotechnol. Bioeng.* **21**, 1543–1552 (1979).
104. K. F. O'Driscoll, in: *Methods in Enzoymology* (K. Mosbach, ed.), Vol. 44, pp. 169–175, Academic Press, New York (1976).
105. K. F. O'Driscoll, M. Izu, and R. Korus, *Biotechnol. Bioeng.* **14**, 847–850 (1972).
106. S. Fukui, A. Tanaka, T. Iida, and E. Hasegawa, *FEBS Lett.* **66**, 179–182 (1976).
107. I. Kaetsu, K. Minoru, and Y. Yoshida, *Biotechnol. Bioeng.* **21**, 847–861 (1979).
108. J. Dobo, *Acta Chim. Acad. Sci. Hung.* **63**, 453–456 (1970).
109. K. Kawashima and K. Umeda, *Biotechnol. Bioeng.* **16**, 609–621 (1979).
110. K. Kawashima and K. Umeda, *Agric. Biol. Chem.* **40**, 1151–1157 (1979).
111. I. Kaetsu, M. Kumakura, and M. Yoshida, *Biotechnol. Bioeng.* **21**, 847–861 (1979).

112. H. Maeda, H. Suzuki, and A. Yamauchi, *Biotechnol. Bioeng.* **15**, 607–610 (1973).
113. H. Maeda, H. Suzuki, and A. Yamauchi, *Biotechnol. Bioeng.* **15**, 827–829 (1973).
114. T. Yagi, *Appl. Biochem.* **1**, 448–454 (1979).
115. V. Jancsik, Z. Belezani, and T. Keleti, *J. Mol. Catal.* 297–306 (1982).
116. T. Tosa, T. Sato, K. Mori, I. Yamamoto, Y. Takata, Y. Nishida, and I. Chibata, *Biotechnol. Bioeng.* **21**, 1697–1707 (1979).
117. Y. Y. Linko, L. Pohtola, R. Viskari, and M. Linko, *FEBS Lett.* **62**, 77–80 (1976).
118. E. K. Bauman, L. H. Goodson, G. G. Guilbault, and D. N. Kramer, *Anal. Chem.* **37**, 1378–1381 (1965).
119. A. Pollak, H. Blumenfeld, M. Wax, R. Bughn, and G. M. Whiteside, *J. Am. Chem. Soc.* **102**, 6326–6336 (1980).
120. L. D'Angiuro, P. Cremonesi, G. Mazzola, B. Fochev, and G. Vecchio, *Biotechnol. Bioeng.* **22**, 2251–2272 (1980).
121. L. D'Angiuro, G. Mazzola, G. Vecchio, B. Fochev, and P. Cremonesi, *J. Appl. Biochem.* **2**, 208–217 (1980).
122. T. M. S. Chang, *Nature* (*London*) **229**, 117–118 (1971).
123. T. Mori, T. Tosa, and I. Chibata, *Biochim. Biophys. Acta* **321**, 653–661 (1973).
124. R. D. Aisina and G. B. Nakdarni, *Biotechnol. Bioeng.* **23**, 431–436 (1981).
125. I. Chibata, *Immobilized Enzymes*, pp. 57–59, Wiley, New York (1978).
126. D. Dinelli, W. Marconi, and F. Morisi, in: *Methods in Enzymology* (K. Mosbach, ed.), Vol. 44, pp. 227–231, Academic Press, New York (1978).
127. G. Gregoriadis, *N. Engl. J. Med.* **295**, 706–710 (1976).
128. G. Gregoriadis, in: *Methods in Enzymology* (K. Mosbach, ed.), Vol. 44, pp. 218–227, Academic Press, New York (1976).
129. M. B. Fiddler, C. D. S. Hudson, and R. J. Desnick, *Biochim. J.* **168**, 191–195 (1977).
130. M. H. Keyes, U.S. Patent No. 3,839,175 (1974).
131. I. Karube and S. Suzuki, *Biochem. Biophys. Res. Commun.* **47**, 51–54 (1972).
132. W. R. Vieth and K. Venkatasubramanian, in: *Methods in Enzymology* (K. Mosbach, ed.), Vol. 44, pp. 248–250, Academic Press, New York (1976).
133. A. M. Klibanov, *Anal. Biochem.* **93**, 1–25 (1979).
134. H. H. Weetall, *Biochim. Biophys. Acta* **212**, 1–7 (1970).
135. W. H. Pitcher, Jr., in: *Advanced Biochemical Engineering* (T. K. Ghose, A. Fiechter, and N. Blakebrough, eds.), Vol. 10, pp. 1–26, Springer-Verlag, New York (1978).
136. D. L. Regan, P. Dunnill, and M. D. Lilly, *Biotechnol. Bioeng.* **16**, 333–343 (1974).
137. P. W. Carr and L. D. Bowers, *Immobilized Enzymes in Analytical and Clinical Chemistry*, pp. 249–251, Wiley, New York (1980).
138. D. Gabel, I. Z. Steinberg, and E. Katchalski, *Biochemistry* **10**, 4661–4669 (1971).
139. I. Chibata, *Immobilized Enzymes*, pp. 168–178, Wiley, New York (1978).
140. K. A. Hughes, *Wall Street Journal* (Three Star), East Ed., November 9, 1982, p. 31.
141. P. Brodelius, in: *Industrial Applications of Immobilized Biocatalysts in Advanced Biochemical Engineering* (T. K. Ghose, A. Fiechter, and N. Blakebrough, ed.), Vol. 10, pp. 75–109, Springer-Verlag, New York (1978).
142. P. W. Carr and L. D. Bowers, *Immobilized Enzymes in Analytical and Clinical Chemistry*, pp. 197–299, Wiley, New York (1980).
143. B. Danielsson, I. Lundstrom, K. Mosbach, and L. Stiblert, *Anal. Lett.* **12**, 1189–1199 (1979).
144. K. Mosbach, U.S. Patent No. 4,021,307 (1977).
145. L. D. Bowers and P. W. Carr, *Thermochim. Acta* **10**, 129–142 (1974).
146. J. C. Weaver, C. L. Cooney, S. P. Fulton, P. Schuler, and S. R. Tannenbaum, *Biochim. Biophys. Acta* **452**, 258–291 (1976).

147. L. C. Clark, Jr., U.S. Patent No. 3,539,455 (1970).
148. S. J. Updike and G. P. Hicks, *Nature (London)* **214**, 986–988 (1967).
149. B. Mattiasson and B. Danielsson, *Carbohydr. Res.* **102**, 273–282 (1982).
150. B. Danielsson, K. Gaold, B. Mattiasson, and K. Mosbach, *Anal. Lett.* **9**, 987–1001 (1976).
151. W. J. Blaedel and R. A. Jenkins, *Anal. Chem.* **48**, 1240–1247 (1976).
152. G. G. Guilbault and E. Hrabankova, *Anal. Chem.* **42**, 1779–1783 (1970).
153. M. Nanjo and G. G. Guilbault, *Anal. Chim. Acta* **75**, 167–180 (1975).
154. G. J. Papariello, A. K. Mukherji, and C. M. Shearer, *Anal. Chem.* **45**, 790–792 (1973).
155. G. P. Hicks and J. J. Updike, *Anal. Chem.* **38**, 726–730 (1966).
156. M. H. Keyes and R. C. Barabino, in: *Enzyme Engineering* (E. K. Pye and H. H. Weetall, eds.), Vol. 3, pp. 51–56, Plenum Press, New York (1976).
157. B. Watson and M. H. Keyes, *Anal. Lett.* **9**, 713–725 (1976).
158. B. Watson, D. N. Stifel, and F. E. Semersky, *Anal. Chim. Acta* **106**, 233–242 (1979).
159. B. Volesky and C. Emond, *Biotechnol. Bioeng.* **21**, 1251–1276 (1979).
160. P. R. Johnson and L. D. Bowers, *Anal. Chem.* **54**, 2247–2250 (1982).
161. D. J. Inman and W. E. Hornby, *Biochem. J.* **137**, 25–32 (1974).
162. D. R. Senn, P. W. Carr, and L. N. Klatt, *Anal. Chem.* **48**, 954–963 (1976).
163. G. Johansson, K. Edstrom, and L. Ogren, *Anal. Chim. Acta* **85**, 55–60 (1976).
164. R. C. Barabino, D. N. Gray, and M. H. Keyes, *Clin. Chem.* (Winston-Salem, N.C.), **24**, 1393–1398 (1978).
165. L. Ogren and G. Johansson, *Anal. Chim. Acta* **96**, 1–11 (1978).
166. D. N. Gray, M. H. Keyes, and B. Watson, *Anal. Chem.* **49**, 1067A–1072A (1977).
167. D. N. Gray, and M. H. Keyes, *CHEMTECH* **7**, 642–648 (1977).
168. I. Chan, in: *Methods in Enzymology* (K. Mosbach, ed.), Vol. 44, pp. 491–503, Academic Press, New York (1976).
169. G. F. Bickerstaff, *Int. J. Biochem.* **11**, 201–206 (1980).
170. S. McCraken and E. Meighen, *Can. J. Biochem.* **57**, 834–842 (1979).
171. J. S. Holcenberg, *Ann. Rev. Biochem.* **57**, 795–812 (1982).
172. G. Gregoriadis, in: *Methods in Enzymology* (K. Mosbach, ed.), Vol. 44, pp. 698–709, Academic Press, New York (1976).
173. G. Gregoriadis, *N. Engl. J. Med.* **295**, 765–770 (1976).
174. G. Gregoriadis and E. D. Neerunjun, *Eur. J. Biochem.* **7**, 179–185 (1974).
175. R. L. Juliano and D. Stamp, *Biochem. Biophys. Res. Commun.* **63**, 651–658 (1975).
176. P. Gosh, P. K. Das, and B. K. Bachhawat, *Arch. Biochem. Biophys.* **213**, 266–270 (1982).
177. L. D. S. Hudson, M. B. Fiddler, and R. J. Desnick, *J. Pharmacol. Exp. Ther.* **208**, 507–514 (1979).
178. G. A. Grabowski and R. J. Desnick, in: *Enzymes as Drugs* (J. C. Holcenberg and J. Roberts, eds.), pp. 167–208, Wiley–Interscience, New York (1981).
179. I. V. Berezin and S. D. Varfolomeev, *Appl. Biochem. Bioeng.* **2**, 259–290 (1979).
180. E . C. Hatchikian and P. Monsan, *Biochim. Biophys. Res. Commun.* **92**, 1091–1096 (1980).
181. D. A. Lappi, F. E. Stolzenbach, N. O. Kaplan, and M. D. Kamen, *Biochem. Biophys. Res. Commun.* **69**, 878–884 (1976).

12

Biomedical Polypeptides—A Wellspring of Pharmaceuticals

James Samanen

Abstract. Decades of peptide research have created a wide variety of biomedical polypeptides: hormones, neurotransmitters, antigenic protein fragments, synthetic peptides, enzyme substrates and inhibitors, chemotactic agents, and ionophoric agents. A major benefit arising from this research is the creation of a rich source of new pharmocophores from which drugs can be designed.

Regardless of their origin, all peptides address a common group of drug development problems. This chapter reviews some of these problems and the possibilities for development of new drugs from peptides.

The diverse activities of peptides arise from the types of amino acids, from their sequence in the peptide chain, and from the conformations of the peptide backbone. Sequences can be assigned different roles: message sequences, signal sequences, prohormone sequences, and stability sequences. Flexibility of the peptide chain gives rise to many conformations in solution, only a small number of which allow for favorable receptor interaction. Such bioactive conformations can be stabilized in constrained synthetic analogues of the peptide, improving potency, stability, and selectivity.

Although some natural peptides display sufficiently selective action and proteolytic stability to produce therapeutic actions in humans, many more peptides are degraded too rapidly or elicit a multitude of responses on pharmacologic administration. Chemists have discovered a variety of synthetic modifications which can dramatically improve proteolytic stability, potency, and selectivity. A number of synthetic peptides have now been developed which display useful therapeutic actions in humans, many of which can be administered by oral and intranasal routes.

Past research with native peptides developed a strong bias against the potential for either adequate peptide intestinal transport or adequate peptide blood–brain barrier permeability. Recent research suggests that both barriers may be surmountable.

The classic approach to pharmacologic control of endogenous peptide action would be with agonists and antagonists to the peptides. A variety of alternate approaches to control of endogenous peptide levels are offered by the peptide biosynthetic pathway: (1) blockage of posttranslational processing with selective enzyme inhibitors; (2) modulation of peptide release with agonists and antagonists to the natural peptide release factors and to the natural peptide release-inhibiting factors, and with selective enzyme inhibitors to posttranslational processing of these factors; and (3) enhancement of endogenous peptide levels by blocking degradation with selective enzyme inhibitors.

James Samanen • Peptide Chemistry Department, Smithkline Beckman, Inc., 1050 Page Mill Road, Palo Alto, CA 94304.

Peptides can also be developed to serve as proteinase inhibitors, synthetic vaccines, and as ionophoric antibiotics.

A wide variety of methods are available for synthesizing peptides of any size or structure: chemical synthesis, biological synthesis, and semisynthesis. In essence, all of the tools are in hand to develop novel therapeutic agents from peptides for both human and animal health maintenance.

1. Introduction

In a limited sense polypeptides can be considered as ordinary polymers. Polypeptides are molecular chains formed from α-amino acids linked together by amide (peptide) bonds. Polymeric chains can be formed from α-amino acids in the presence of powerful condensation reagents. Such poly-amino acids have been important in developing an understanding of peptide conformation,[1] but poly-amino acids are of little biomedical interest.

Of greater interest are the polypeptides constructed in nonrandom, nonrepetitive, specific sequences of amino acids. Such peptides assume specific three-dimensional structures that can interact stereospecifically with cellular membrane-bound receptors to trigger powerful physiological responses in living organisms. These peptides can be found in living organisms serving as modulators of many important life processes.

Peptides have been demonstrated to function in most tissues of the body: skeletal and visceral muscle, the heart, the sensory organs, digestive tissues and glands, endocrine glands, the genitourinary system, reproductive organs, the placenta, the immune system, and both the peripheral and central nervous systems.

With the development of reliable techniques for synthesis, peptides have been synthesized to meet a wide variety of biomedical interests: peptide hormones and neurotransmitters, synthetic peptide analogues, antigenic protein fragments, enzyme substrates and inhibitors, and ionophoric antibiotic agents. Research with these biomedical polypeptides has increased our understanding of their physiological roles and has revised conceptions of brain, gut, and endocrine physiology. The biomedical polypeptides will also be important in the development of new clinical diagnostic assays.

A major benefit arising from the synthesis of biomedical polypeptides is the creation of a rich source of new pharmacophores from which drugs can be designed. In the last 10 years clinical interest in peptides has soared. Over 170 clinical trials with natural peptides and analogues have been conducted. This chapter will focus on the possibilities and problems of pharmaceutical development of biomedical polypeptides.

This report does not present an exhaustive review of the field. The report is intended to represent the field's major trends. As a consequence, the valuable work of many scientists cannot be included. For a more detailed review, the reader is referred to the series: *Amino Acids, Peptides and Proteins*, Royal Society of Chemistry, Burlington House, London, Vols. 1 (1969)–15 (1984).

2. Methods of Peptide Synthesis

A number of peptide synthesis methods have evolved over the years. They may be grouped into three categories: chemical synthesis, biological synthesis, and semisynthesis. The virtues of each method complement the others giving the medicinal chemist a broad range of capabilities, from synthesis of the simplest dipeptide up to the synthesis of peptides containing over 100 amino acids. These methods may be performed on any scale from research to industrial. Since the technologies are rather involved, only the comparative virtues of each method will be outlined here.

2.1. Chemical Synthesis[2–6]

The chemical synthesis of peptides falls into two strategic categories: solution synthesis and solid-phase synthesis. As the name implies, each reaction in a solution synthesis occurs in a chemical solution followed by product purification involving extraction, purification, crystallization, or even chromatography. In solid-phase synthesis the first amino acid (and hence the growing peptide chain) is attached to a resin bead, allowing for purification by simple filtration of all reagents from the product peptide attached to the resin. This method has allowed for automation of the synthesis[7] and manufacture of instruments for automated solid-phase synthesis.[8–11]

The solution method is best employed for synthesis of small peptides containing less than about six amino acids. The solid-phase method is best employed for synthesis of medium-sized peptides of up to about 40 amino acids, even though large peptides can be made by either method, for example, ribonuclease A (124 amino acids) by solution synthesis[12] and β-lipotropin (91 amino acids) by solid phase.[13] By either method as the size of peptide increases, final product yield decreases, synthetic and purification times lengthen, and expenditures for chemicals increase. Industrial-scale synthesis is practicable by either method, but only if the peptide chain length falls within the mentioned length limits. For large peptides the biosynthetic methods are best employed.

The chemical synthesis methods permit the synthesis of molecules not found in nature. Consequently, these methods have been invaluable to peptide pharmaceutical development.

2.2. Biological Synthesis

All biological synthesis methods employ natural ribosomal translation processes to assemble peptides and proteins. At least four main methods have been employed:

1. *Extraction from Natural Sources.* Animal and human cadaver organs can be processed to obtain many important peptides, for example, insulin; growth hormone; and pituitary, thymic, and thyroid hormones.
2. *Cell Culture—Hybridoma Technique.* This method has been applied primarily for the production of monoclonal antibodies.[14] Ideally, hybridomas of peptide-secreting cells might be useful sources of peptides.
3. *Fermentation Processes.* Industrial fermentation processes produce antimicrobial peptides, for example, gramicidin and bacitracin.[15] Through recombinant DNA techniques, fermentable microorganisms can be altered to produce new peptide and protein products, such as human insulin,[16] growth hormone,[17] and interferon.[18]
4. *Cell-Free Translation.* The ribosomal apparatus can be removed from cells and in the presence of mRNA and a supply of amino acyl-tRNA's peptides can be synthesized.[19] Such processes might some day reach a preparative scale and be modified to allow for assembly of unnatural peptides, for example, peptides containing D-amino acids. The possibility for biological synthesis of stabilized potent and selective analogues of large peptides may then come into existence.

The following are advantages of biological synthesis methods:

1. They are the only feasible way to obtain larger peptides and proteins.
2. These methods are probably less expensive than chemical methods for industrial-scale synthesis of midrange peptides.
3. Once the method is established, actual synthesis and purification times are faster for mid- to large-sized peptides.[16]

The disadvantages of the biological synthesis methods are:

1. The high startup costs and development time. Thus the biological methods are not suitable for use in an analogue development program.
2. Limited to natural biological structures. Unnatural amino acids cannot be readily incorporated into a peptide synthesized in a cell. Cell-free systems, as mentioned earlier, might ultimately be developed that allow for synthesis of peptides containing unnatural amino acids.

For development and production of large peptide analogues bearing unnatural amino acids, one may wish to consider semisynthesis.

2.3. Semisynthesis

The semisynthesis method[(20)] is a blend of chemical and biological approaches to produce modified structures of large peptides and proteins. Semisynthesis is a sort of molecular surgery, whereby a piece of peptide chain is cut away and replaced with a new section. This form of peptide modification is useful for peptides above approximately 30 amino acids. Total synthesis by solid phase is more practical for smaller peptides. The conversion of porcine insulin to human insulin[(21)] serves as an excellent example of the value of semisynthesis.

A potentially viable way to produce stabilized analogues of large peptides would be to cleave them enzymatically at their most sensitive peptide bonds and chemically couple segments back into the peptide chain that render proteolytic resistance to the peptide.

With biological synthesis as a method for generating starting materials, and semisynthesis as a method for modification, earnest drug development efforts with large peptides are now possible.

3. Structural and Conformational Specificity*

3.1. Peptide Structure: Polyamide Backbone with Various Side Chains

Peptides are small proteins (containing less than 100 amino acids). Their functions are determined by:

1. The kinds of amino acids (there are 20 different amino acids found in mammalian peptides and proteins).

*The words "polypeptide" and "oligopeptide" are used interchangeably with the word "peptide," which will be used here for any chain of amino acids.

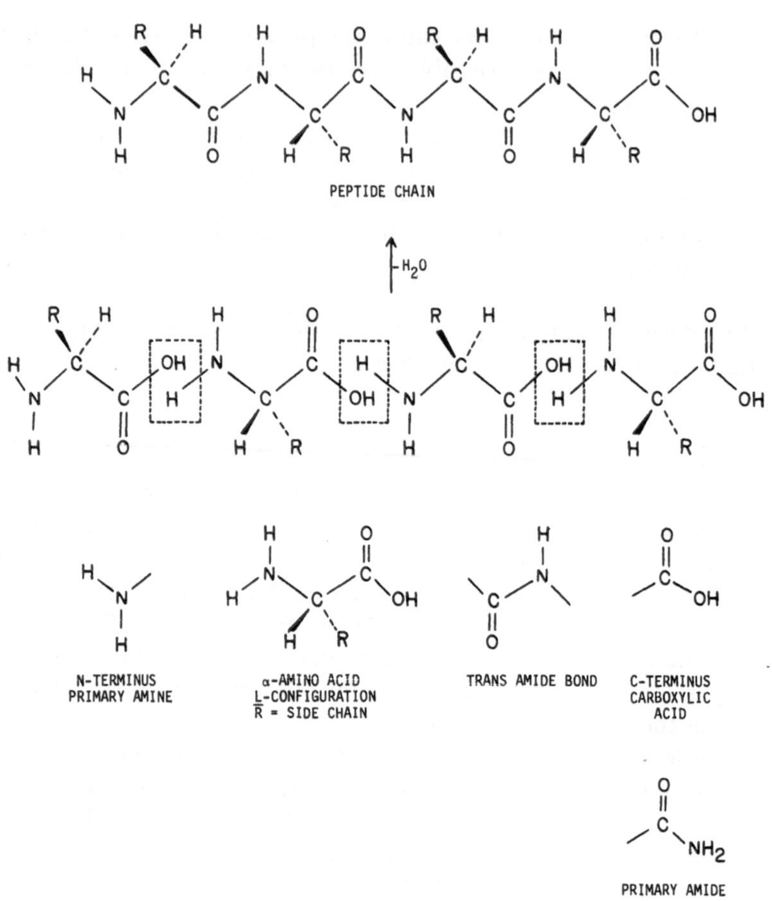

Figure 1. Elements of peptide structure.

2. The length of the peptide molecule and the sequence or ordering of the amino acids in the peptide chain.
3. The conformation or three-dimensional shape of the peptide chain.

Figure 1 displays the elements of peptide structure. Peptides are composed of polyamide (polypeptide) chains with tetrahedral carbon atoms positioned between each amide group. Various groupings of atoms, called "side chains," can branch off each tetrahedral carbon atom along the peptide backbone.

A peptide chain is derived from α-amino acids by chemical or biological combination of the amine group from each amino acid with the carboxylic acid group from each preceding amino acid. A linear peptide contains an

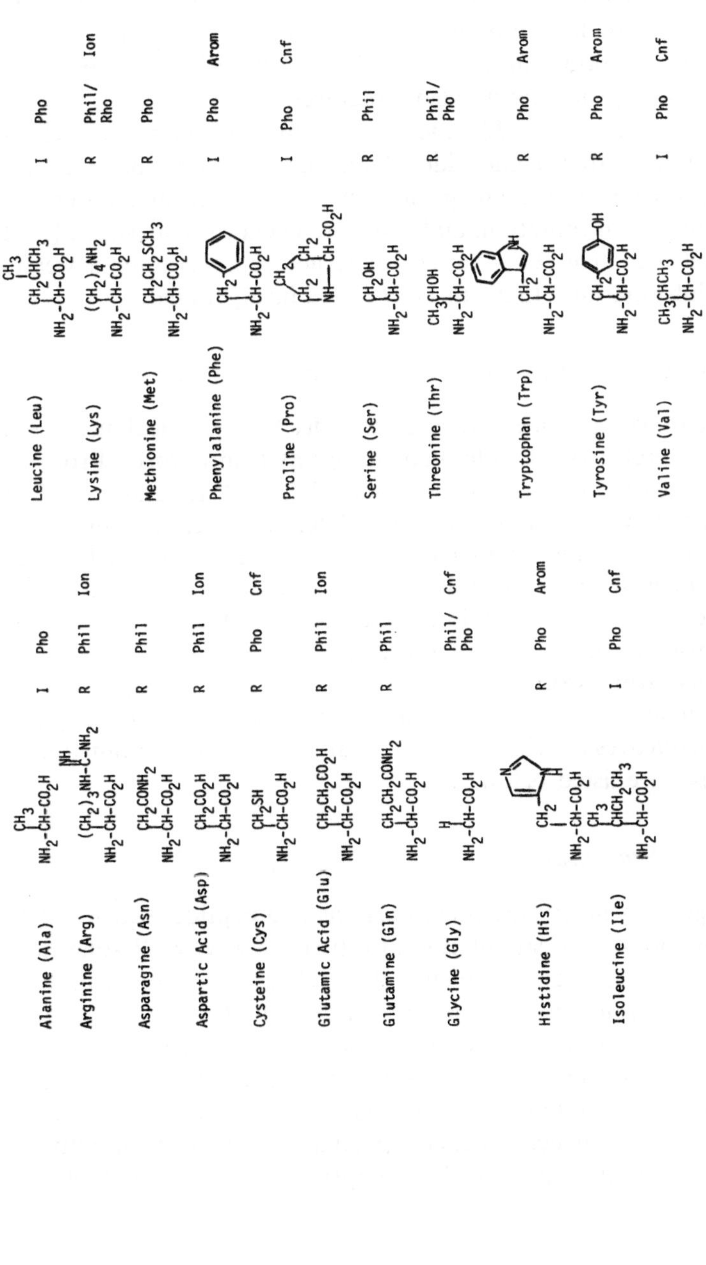

Amino acid	Structure			
Alanine (Ala)	CH_3 $NH_2\text{-}CH\text{-}CO_2H$	I	Pho	
Arginine (Arg)	$(CH_2)_3NH\text{-}C(=NH)\text{-}NH_2$ $NH_2\text{-}CH\text{-}CO_2H$	R	Phil	Ion
Asparagine (Asn)	CH_2CONH_2 $NH_2\text{-}CH\text{-}CO_2H$	R	Phil	
Aspartic Acid (Asp)	CH_2CO_2H $NH_2\text{-}CH\text{-}CO_2H$	R	Phil	Ion
Cysteine (Cys)	CH_2SH $NH_2\text{-}CH\text{-}CO_2H$	R	Pho	Cnf
Glutamic Acid (Glu)	$CH_2CH_2CO_2H$ $NH_2\text{-}CH\text{-}CO_2H$	R	Phil	Ion
Glutamine (Gln)	$CH_2CH_2CONH_2$ $NH_2\text{-}CH\text{-}CO_2H$	R	Phil	
Glycine (Gly)	H $NH_2\text{-}CH\text{-}CO_2H$		Phil/ Pho	Cnf
Histidine (His)	CH_2-[imidazole] $NH_2\text{-}CH\text{-}CO_2H$	R	Pho	Arom
Isoleucine (Ile)	CH_3 $CHCH_2CH_3$ $NH_2\text{-}CH\text{-}CO_2H$	I	Pho	Cnf
Leucine (Leu)	CH_3 CH_2CHCH_3 $NH_2\text{-}CH\text{-}CO_2H$	I	Pho	
Lysine (Lys)	$(CH_2)_4NH_2$ $NH_2\text{-}CH\text{-}CO_2H$	R	Phil/ Rho	Ion
Methionine (Met)	$CH_2CH_2SCH_3$ $NH_2\text{-}CH\text{-}CO_2H$	R	Pho	
Phenylalanine (Phe)	CH_2-[phenyl] $NH_2\text{-}CH\text{-}CO_2H$	I	Pho	Arom
Proline (Pro)	CH_2 CH_2 CH_2 $NH\text{—}CH\text{-}CO_2H$	I	Pho	Cnf
Serine (Ser)	CH_2OH $NH_2\text{-}CH\text{-}CO_2H$	R	Phil	
Threonine (Thr)	CH_3CHOH $NH_2\text{-}CH\text{-}CO_2H$	R	Phil/ Pho	
Tryptophan (Trp)	CH_2-[indole] $NH_2\text{-}CH\text{-}CO_2H$	R	Pho	Arom
Tyrosine (Tyr)	CH_2-[phenyl]-OH $NH_2\text{-}CH\text{-}CO_2H$	R	Pho	Arom
Valine (Val)	CH_3CHCH_3 $NH_2\text{-}CH\text{-}CO_2H$	I	Pho	Cnf

I = inert, R = reactive, Pho = hydrophobic, Phil = hydrophilic, Arom = aromatic, Ion = ionic, Cnf = conformationally signficant

Figure 2. Structure of 20 mammalian amino acids.

"*N*-terminus" at which the amine group of the *N*-terminal amino acid is free (not linked to another amino acid). At the other end of the linear peptide is the "*C*-terminus" at which the carboxylic acid group of the *C*-terminal amino acid is free. Mammalian peptides often bear a primary carboxamide ($—CONH_2$) at the *C*-terminus.

The alpha-carbon atoms of 19 of the 20 amino acids found in mammals bear side chains. Such amino acids are asymmetric and of the L absolute configuration.[22] Peptides in microorganisms contain both L- and D-amino acids and many unusual amino acid structures.

Figure 2 displays the 20 different mammalian amino acid chemical structures. Notice that some side chains are hydrophobic and others hydrophilic; some bear aromatic groups and others bear ionic groups; some are important to conformation; and some are chemically inert, while others are chemically reactive. This variety of structure constitutes a wealth of structural "information" that can be encoded into peptides.

3.2. The Contribution of Sequence to Biological Specificity

Nature utilizes peptide sequences as codes for biological messages in a manner comparable to the way humans string letters into words. From a pool of 26 letters one can arrange one "H," one "S," one "A," and one "C" into the sequence "C–A–S–H" to form the word "cash," which has an important meaning to many people. Similarly, the sequence Tyr-Gly-Gly-Phe-Leu of five amino acids forms a peptide which has analgesic properties in the brain.[23] There are over five million unique pentapeptide sequences that can be formed from a pool of 20 amino acids.

Humans string words into sentences to derive messages of greater meaning. Similarly, nature often builds peptide chains containing several component sequences to achieve a greater variety of biological functions. The sequence types are listed below.

3.2.1. Message Sequences

Message sequences[24] are those portions of a peptide sequence that are specifically involved with stimulation of a tissue. Tyr-Gly-Gly-Phe-Leu has already been described as a sequence that delivers the biological message "analgesia" to appropriate areas of the brain. The partial sequences Gly-Gly-Phe-Leu and Tyr-Gly-Gly do not cause analgesia.[25] Just as two different words can bear similar meanings, two different sequences can bear similar messages: Tyr-Gly-Gly-Phe-Met is another analgesic peptide.

The message sequence in the enkephalins encompasses the entire chain length. Message sequences can also comprise a small portion of a large sequence, for example, the analgesic message is encoded in just the first five amino acids of the 31-amino acid peptide β-endorphin. Table 1* lists some of the peptides found in nature that contain the analgesic message.[23,26–33]

It is possible for a peptide sequence to contain several messages. Parathyroid hormone (84 amino acids) produces both hypotension and hypercalcemia. Hypotensive action is triggered by residues 24–34, while hypercalcemic action is triggered by the *N*-terminal portion.[34]

Words like "fine" and "swell" have different meanings for different people. Peptide messages can similarly trigger different biological responses in different tissues. In the gut, Tyr-Gly-Gly-Phe-Leu causes constipation;[35] in the brain, it promotes the release of growth hormone.[35]

The different actions of a peptide on different tissues are controlled in the body by local utilization of the peptide. If synthesis, release, and rapid degradation of the peptide all occur in the vicinity of the site of action, then the peptide is not likely to stimulate other tissues.

* The abbreviations employed here follow the "IUPAC–IUB Commission on Biochemical Nomenclature Symbols for Amino-Acid Derivatives and Peptides Recommendations 1971," *J. Biol. Chem.* **247**, 977–983 (1972).

The following abbreviations are used to represent modified and unnatural amino acids:

Aib = 2-aminoisobutyric acid
Aha = 7-aminoheptanoic acid
Ala(CH_2Cl) = Alanine chloromethylketone, $NH_2CH(CH_3)COCH_2Cl$
Arg(CHO) = Arginine aldehyde, $NH_2CH[(CH_2)_3NH{-}(C{=}NH){-}NH_2]CHO$
Boc = *t*-Butyloxycarbonyl
Carba Cys = CH_2— (methylene substituted for S)

```
    CH2—
    |
    CH2
    |
—NH—CH—CO—
```

CHO = formyl
Dmp = 3,3-dimethylproline
αMePhe = L-2-amino-3-phenylisobutyric acid
Met(O) = Methionine sulfoxide
Met(O_2) = Methionine sulfone
Met(O)ol = Methioninol sulfoxide, $NH_2CH(CH_2CH_2SOCH_3)CH_2OH$
Nal(2) = β-naphylalanine
Pen = Penicillamine
Phe(ol) = Phenylalaninol, $NH_2CH(CH_2C_6H_5)CH_2OH$
Suc = Succinic acid
Thz = Thiazolidine-5-carboxylic acid
Thr(ol) = Threoninol, $NH_2CH[CH(CH_3)CH_2OH]CH_2OH$
Val(CH_2Cl) = Valine chloromethylketone, $NH_2CH[CH(CH_3)_2]COCH_2Cl$

Table 1. Some Opiate Peptides Containing Tyr-Gly-Gly-Phe-(Met or Leu)

Peptide	References	Structure
[Met[5]]-Enkephalin	23	*Tyr-Gly-Gly-Phe-Met*
[Leu[5]]-Enkephalin	23	*Tyr-Gly-Gly-Phe-Leu*
α-Neo-endorphin	26	*Tyr-Gly-Gly-Phe-Leu*-Arg-Lys-Tyr-Pro-Lys
β-Neo-endorphin	27	*Tyr-Gly-Gly-Phe-Leu*-Arg-Lys-Tyr-Pro
Bovine adrenal medulla dodecapeptide	28	*Tyr-Gly-Gly-Phe-Met*-Arg-Arg-Val-Gly-Arg-Pro-Glu
Bovine adrenal peptide E	29	*Tyr-Gly-Gly-Phe-Met*-Arg-Arg-Val-Gly-Arg-Pro-Glu-Trp-Trp-Met-Asp-Tyr-Gln-Lys-Arg-*Tyr-Gly-Gly-Phe-Leu*
Bovine adrenal peptide F	30	*Tyr-Gly-Gly-Phe-Met*-Lys-Lys-Met-Asp-Glu-Leu-Tyr-Pro-Leu-Glu-Val-Glu-Glu-Glu-Ala-Asn-Gly-Gly-Glu-Val-Leu-Gly-Lys-Arg-*Tyr-Gly-Gly-Phe-Met*
Porcine prodynorphin (209–240)	31	*Tyr-Gly-Gly-Phe-Leu*-Arg-Arg-Ile-Arg-Pro-Lys-Leu-Lys-Trp-Asp-Asn-Gln-Lys-Arg-*Tyr-Gly-Gly-Phe-Leu*-Arg-Arg-Gln-Phe-Lys-Val-Val-Thr
Porcine dynorphin A	32	*Tyr-Gly-Gly-Phe-Leu*-Arg-Arg-Ile-Arg-Pro-Lys-Leu-Lys-Trp-Asp-Asn-Gln
Human β-endorphin	33	*Tyr-Gly-Gly-Phe-Met*-Thr-Ser-Glu-Lys-Ser-Gln-Thr-Pro-Leu-Val-Thr-Leu-Phe-Lys-Asn-Ala-Ile-Ile-Lys-Asn-Ala-Tyr-Lys-Lys-Gly-Glu

3.2.2. Binding and Stimulation Sections of the Message Sequence

The term "message sequence" is a useful way to describe the specific information relayed from one cell to another cell via a peptide hormone or neurotransmitter. On a molecular level the interaction of the message sequence with the biological receptor involves binding to the receptor and stimulation of the receptor.

In many peptides, receptor binding and receptor stimulation are relegated to separate sections of the message sequence. From these peptides, analogues can be made in which the stimulation section is altered to prevent receptor stimulation, while the binding section remains unaltered to allow the receptor binding. Such an analogue acts as an "antagonist" as it may block the interaction of endogenous peptide with the receptor. The peptide hormone angiotensin II can be turned into a potent antagonist by merely removing the side chain from the *C*-terminal amino acid.[36] The first 34 amino acids in parathyroid hormone (PTH) (containing 84 amino acids) are the shortest sequence that retains full biological activity.[37] The fragment PTH(3–34) completely loses activity yet binds with equal avidity to the PTH receptor. It is an antagonist to PTH(1–34) and PTH(1–84).[38] Examination of smaller PTH fragments has revealed that most of the binding activity of parathyroid hormone (3–34) resides in the *C*-terminal decapeptide PTH(25–34).[38]

In both examples receptor stimulation was relegated to small sections of the molecule distinct from sections involved with binding. In other peptides receptor binding and receptor stimulation involve the same sections of the message sequence. For these peptides, analogues must be found which can adopt conformations that bind well but do not stimulate the receptor. In Section 4.5, examples of antagonists created by conformational alterations will be described.

Potent, selective peptide antagonists can have important therapeutic properties. Antagonists to angiotensin II can lower blood pressure,[39] antagonists to LHRH (leuteinizing hormone-releasing hormone) are known to block ovulation in women,[40] and antagonists to substance P may be useful as analgesics.[41]

3.2.3. *Signal Sequences*

Proteins and peptides are biosynthesized in the nuclei of cells as larger precursor molecules that transiently contain *N*-terminal extensions of 16–30 amino acids called "leader" or "signal" sequences.[42] These sequences are thought to be involved with binding of the nascent peptide–polysome complex to the membrane bilayer of the rough endoplasmic reticulum (RER), and to promote insertion of the growing peptide chain into the RER during peptide translation. The signal sequences are cleaved during translation by specific RER membrane-bound enzymes. The structures of 22 signal peptides have been identified by cell-free translation of peptides. All such peptides display a double amphipathic structure: a hydrophobic core of 9–24 amino acid residues and two charged polar ends.[43]

Recently demonstrated[44] inhibition of translation of placental, parathyroid, and pituitary hormones by a synthetic signal sequence to parathyroid hormone suggests nonselective participation of the signal sequence in the insertion process. Thus, signal sequences are probably not useful as therapeutic agents to modulate overproduction of peptide hormones, for example, growth hormone in acromegaly. Signal sequences may be useful as prodrug sequences to increase transport across the gut or the blood–brain barrier. Such a function has yet to be demonstrated.

3.2.4. *Prohormone Sequences*

For successful translation and insertion of a peptide from the polyribosome–mRNA complex to the RER, it has been proposed that the peptide must traverse a distance equal to the length of a 70-amino and peptide chain.[45] This would necessitate the biosynthesis of all peptide hormones and neurotransmitters from large precursor peptides. Following translation, the propeptide (prohormone or proneurotransmitter) would be cleaved by selective proteolysis into the peptide of desired chain length. The proposed[46] functions of prohormone sequences are:

1. Intracellular signaling, enabling the transport of various proteins and peptides to their proper site of action or release.
2. The potential generation of multiple biological actions from a common gene product, enabling cells to selectively express or depress certain biological actions by cell-specific co- and posttranslational processing of the common precursor polypeptide into different active peptide products.

The propeptide sequences of many peptides have been delineated. In many cases the active peptides are connected through a pair of basic amino acids (Lys or Arg).[47] A propeptide can contain more than one active peptide sequence. Preproglucagon contains the sequences for glucagon and a peptide homologous to gastric inhibitory peptide;[46] pro-opiomelanocortin contains the sequences of α-MSH, ACTH (adrenocorticotropic hormone), and β-endorphin.[48]

Pairs of basic amino acids in a new propeptide sequence may serve as markers to reveal new peptide chains that bear unique activities, for example, the insulin "C" peptide connecting the A and B chains,[49] CLIP, and γ-MSH from the pro-opiomelanocortin precursor to ACTH and β-endorphin,[48] an *N*-terminal 12-amino acid peptide extension of somatostatin,[46] precursor chains in preproglucagon,[46] and preprocalcitonin.[46]

A protease specific for consecutive basic residues has been isolated recently from pituitary granules.[50] As more of these enzymes are discovered and their specifications delineated, the prospect will increase for the development of specific enzyme inhibitors to therapeutically block the activation of endogenous peptides. The mode of action of the antihypertensive agent, captopril (Figure 6), will illustrate this approach to drug development.

3.2.5. *Stability Sequences*

Stability sequences can serve the following functions:

1. Increase plasma half-life by sterically blocking proteolytic degradation. This is best illustrated by comparing the half-life of

Met^5-enkephalin in rat plasma (2.0 min)[51] with the half-life of β-endorphin in human plasma (37 min).[52]

2. Stabilize the bioactive conformation of the message sequence.

Peptides that occur naturally as cyclic structures typically require the cyclic structure for activity. Both somatostatin and vasopressin are cyclic by virtue of the disulfide bridge between two cysteine residues in the peptides. Analogues of either peptide that lack cyclic structures are inactive.

In larger peptides spatially proximal sections of peptide chain can stabilize bioactive conformation by hydrogen bonding and hydrophobic interactions, for example, between Asn^5 and Thr^{12}, and between Phe^6 and Phe^{11} in somatostatin.[53]

3.3. Contribution of Peptide Conformation to Biologic Specificity

Peptide chains may adopt many conformational shapes in solution. Only a small number of these conformations, the so-called bioactive conformations, will place the side chains into the proper three-dimensional array that allows for binding and stimulation of the biological receptor. A peptide analogue that holds a bioactive conformation in solution may be more potent than one that does not. Hence, peptide chemists have found it profitable to develop conformationally restricted analogues of peptides. Rational design of conformationally restricted peptide analogues necessitates an understanding of peptide conformation.

For a peptide containing all L-amino acids, the potential number of conformations is reduced by the following structural features:

1. Amide bonds generally adopt a trans configuration which is more stable than the cis configuration by at least 2 kcal/mol,[54] with a rotational barrier of about 20 kcal/mol,[55] due to the partial double bond character of the N^α—C' bond. Peptide bonds that include the secondary amino group of the cyclic amino acid proline can adopt either cis or trans isomers since the trans is no longer favored energetically.[56] Trans amide groups predominate, however, in all but cyclic peptides.

2. The first side chain angles X′ prefer angles of 60°, − 60°, or 180°,[57] with − 60° the predominate X′ angle for L-amino acids.

3. The possible values for the amide backbone angles, ϕ and ψ, are limited by steric interactions between the side chain β-methylene or methyl groups and the amide groups to the regions of ϕ, ψ space shown in Figure 3 for acetyl-L-alanyl methylamide. This map was generated by a computer program called ECEPP, which calculates the conformational energy of a peptide as the sum of all energies of interaction between the atoms in a

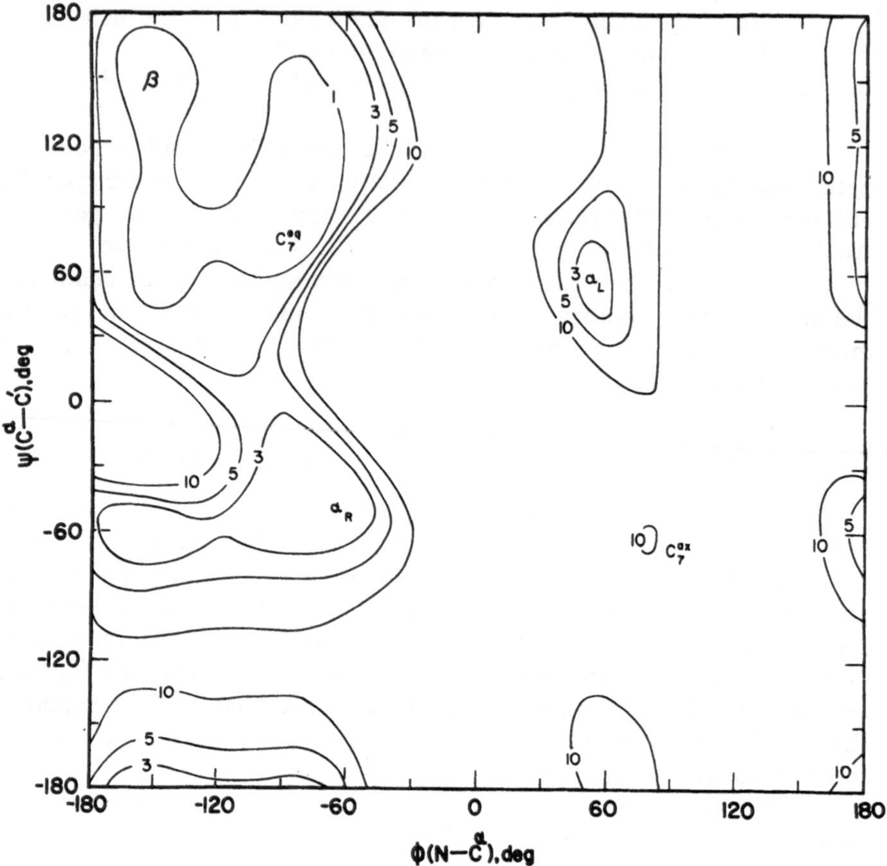

Figure 3. ϕ, ψ Map of acetyl-L-alanyl-methylamide (supplied by Frank Momany, Memphis State University).

peptide chain (see Figure 4.)[58] The solid lines connect all points in the map of equal energy establishing isoenergetic contours, similar to a geographic contour map. The contour map outlines regions of low-energy ϕ, ψ combinations. The actual conformation of acetyl-L-alanyl methylamide would oscillate among the various low-energy sets of ϕ and ψ, with certain conformations more prevalent in a given solvent. All natural L-amino acids give ϕ, ψ maps similar to Figure 3 except that the low-energy regions may be more narrowly defined.

Using the low-energy combinations of ϕ and ψ as starting points for

A COMPUTER PROGRAM THAT:

1) CONSTRUCTS PEPTIDE CHAIN WITH BOND LENGTHS AND FIXED BOND ANGLES FROM X-RAY AND NEUTRON DIFFRACTION DATA.

2) CALCULATES ALL DIHEDRAL ANGLES ϕ, ψ, ω, and χ (see below).

3) CALCULATES THE TOTAL CONFORMATIONAL ENERGY, E: THE SUM OF ALL ENERGIES OF INTERACTION BETWEEN ATOMS IN PEPTIDE CHAIN.

$$E = E_{ES} + E_{NB} + E_{HB} + E_{TOR}$$

E_{ES} = Electrostatic Energy

E_{NB} = Non-bonded Energy

E_{HB} = Hydrogen Bonded Energy

E_{TOR} = Torsional Energy

4) VARIES DIHEDRAL ANGLES, RECALCULATES TOTAL CONFORMATIONAL ENERGY, DRIVING TOWARD CONFORMATIONS OF LOW TOTAL ENERGY.

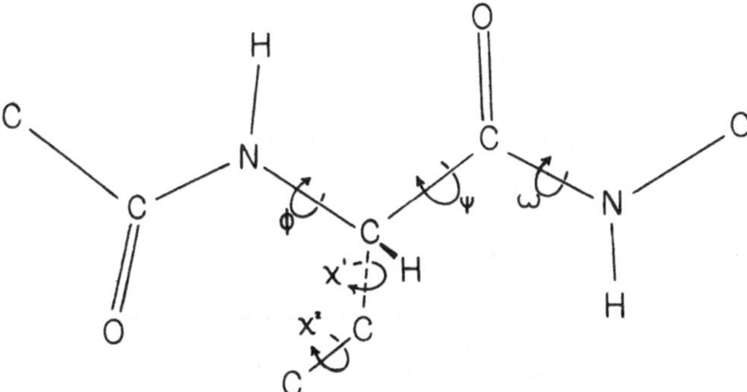

Figure 4. Minimum conformational energy calculations.

each amino acid in a peptide, one can search for low-energy conformations of peptides with the aid of a computer.[59–61]

The conformations of natural peptides are stabilized by intramolecular hydrogen bonds, intramolecular nonbonded interactions between side chains, covalent disulfide bridges between cysteine residues, and cyclizing *N*-terminal to *C*-terminal amide bonds.

One cannot attempt to determine bioactive conformation of small linear peptides by the techniques of conformational analysis that observe peptides in solution—ORD (optical rotatory dispersion), CD (circular dichroism), IR, and NMR spectrometry—since the conformations of small linear peptides in solution are numerous.[62] For small linear peptides X-ray crystal structures, if solvable, need not represent the bioactive conformations. The crystal-lattice forces, being different from biological receptor interactions, can favor low-energy conformations that differ from bioactive conformations. For larger peptides the conformations may be sufficiently stabilized that the larger peptides would tend to maintain shapes in solution that are similar to those at the receptor.

For all peptides that lack conformational rigidity the determination of bioactive conformation is only approachable by synthesis and bioassay of peptide analogues in which conformational restrictions have been added. As will be seen in Section 4.5, the synthesis and bioassay of conformationally restricted analogues of peptides can lead to dramatic increases in potency and selectivity of action.

4. Modifications of Peptides to Enhance Therapeutic Utility

4.1. Motives for Peptide Modification

Although peptides of significant potency and selectivity have evolved to serve various functions in the body, nature has not employed peptides of optimal potency or selectivity. The peptide chemist has found considerable latitude between the activities of natural peptides and synthetic analogues.

Table 2. Unmodified Peptides Displaying Therapeutically Useful Actions in Humans[a]

Adrenocorticotropic hormone (ACTH)
- √ (iv) Corticosteroid release in treatment of severe inflammatory conditions, for example, lupus erythmatosus, arthritis, severe allergic conditions, respiratory diseases.[66]

Bacitracin
- √ (topical) Treatment of superficial bacterial infections.[67]

Bestatin
- (oral) Increased effectiveness of cancer irradiation therapy and chemotherapy.[68]

Calcitonin (CT)
- √ (sc or im) Treatment of Paget's Disease.[69]
- Treatment of osteoporosis.[70]
- √ (iv) Control of hypercalcemia in metastatic breast cancer.[71,72]
- (oral) Inhibition of gastric secretion.[73]
- (sa) Analgesia in patients with intractable pain.[74]

Cholecystokinin (CCK-8 or CCK-32)
- (in) Treatment of chronic pancreatitis.[75]
- (iv) Induction of satiety in obese[76] and normal humans.[77]
- (iv) Treatment of postoperative paralytic ileus.[78]

Delta sleep-inducing peptide (DSIP)
- (iv) Improvement of disturbed sleep (insomnia).[79]

β-Endorphin
- (iv, ith) Relief from pain of cancer.[80,81]
- (ith) Pain relief during delivery of obstetric patients.[82]
- (iv) Suppression of narcotic abstinence syndrome.[83]

Gramicidin
- √ (topical) Treatment of superficial bacterial infections.[87]

Table 2. (*Continued*)

Growth hormone (GH)
- √ (im) Increased height in growth-hormone deficient children.[84–86]
- (im) Increased height in children with normal variant short stature.[86]

Insulin
- √ (iv) Treatment of diabetes mellitus.[88]

Luteinizing hormone-releasing hormone (LHRH)
- (iv) Induction of ovulation in women with hypothalamic amenorrhea.[89]
- (in) Descent of testicle in boys with cryptorchidism.[90,91]

Melanocyte inhibiting factor-I (MIF-I)
- (oral) Mood improvement in depressed patients.[92]
- (oral) At low doses decline in severity and frequency of episodes in patients with tardive dyskinesia.[93]

Neurotensin (NT)
- (iv) Inhibition of gastric juice secretion.[94]

Oxytocin (OT)
- √ (iv) Induction of labour; control of postpartum bleeding.[95]
- √ (in) Stimulation of milk letdown during nursing.[96]

Parathyroid hormone (PTH 1–34)
- (sc) Increased vertebral bone volume in patients with osteoporosis.[97]

Somatostatin (SS)
- (iv) Reduced bleeding of gastric ulcers.[98–100]

Teprotide
- (par) Lowers blood pressure in humans with hypertension.[101]

The following constitute thymic hormones:

- Serum thymic factor (FTS)
  - (iv) Reduced frequency and severity of infection in children with immune deficiencies.[102]
- Crude thymosin
  - (par) Improved clinical status of patients with autoimmune disorders.[103]
  - (im) Improved clinical status of patients with collagen vascular disease.[104]
  - (iv) Improved survival rate of cancer patients with initial low T-cell count undergoing chemotherapy.[105]
  - (iv) Improved clinical status of patients with rheumatoid arthritis.[106]
- Thyrotropin releasing hormone (TRH)
  - (oral) Prolonged infertility and lactation period in women who are breastfeeding.[107]
  - (iv) Decreased ataxia in patients with spinocerebellar degeneration.[108]

Arginine^{8}–vasopressin and lysine^{8}–vasopressin
- √ (iv) Treatment of diabetes insipidus.[109]

[a] The following abbreviations represent routes of administration: im, intramuscular; in, intranasal; ith, intrathecal; ip, intraperitoneal; iv, intravenous; par, parenteral; sc, subcutaneous; sa, subarachnoid; sl sublingual. The √ = application approved for humans by the U.S. Food and Drug Administration.

For example, [D-Ser(tBu)^{6}, des-Gly^{10}]-LHRH ethylamide is 50–70 times as potent as natural LHRH in stimulating release of LH and FSH (follicle stimulating hormone).[63] The enkephalin analogue Tyr-D-Met(O)-Gly-MePhe(ol) is 14,562 times as active as [Met^{5}]-enkephalin in contracting guinea pig ileum.[64] An analogue of α-MSH, which could be made by biological synthesis, [Cys^4, Cys^{10}]-α-MSH, is over 10,000 times as active as α-MSH in a frog skin bioassay.[65]

Not all peptides require structural optimization. Table 2 lists the therapeutic actions displayed by unmodified peptides in humans.[66–109] The applications indicated with a check (√) have been approved for humans by the U.S. Food and Drug Administration. The therapeutic value of many of these peptides may be increased, nevertheless, through structural modification or controlled delivery. Table 3 lists potentially useful activities of peptides in humans or mammals that need the development of analogues with increased potency, selectivity, or stability before these actions would constitute a useful human therapy.[110–138]

Table 3. Potentially Useful Activities of Peptides That Need Development of Analogues with Greater Stability or Selectivity

ACTH
- Corticosteroid release in treatment of inflammation.[66]
- Improvement of mood and attention.[110]
- Treatment of seizures in children with epilepsy.[111]

Angiotensin II (AII)
- Increased effectiveness of cancer chemotherapy.[112]

Angiotensin II antagonists
- Lowering of blood pressure in hypertensives.[113]
- Improving heart function after cardiac failure.[114]
- Controlling thirst.[115]

Bombesin
- Induction of hypothermia.[116]

Bradykinin (BK)
- Lowering blood pressure.[117]
- Improving peripheral circulation.[117]
- Immunostimulant.[117]

Calcitonin (CT)
- Induction of analgesia.[74]
- Control of hypercalcemia.[71,72]
- Control of osteoporesis.[70]
- Inhibition of gastric secretion.[73]
- Treatment of Paget's Disease.[69]

Cholecystokinin (CCK)
- Control of hunger in obese humans.[76]
- Treatment of chronic pancreatitis.[75]

Enkephalins (EK)
- Induction of analgesia.[35]
- Antidiarrheal agent.[35]

Gastrin antagonists
- Reduction of gastric acid secretion in gastric ulcers.[118]

Growth-hormone releasing agents
- Stimulation of GH release in GH-deficient children.[119]

Table 3. (*Continued*)

LHRH
- Inhibition of ovulation.[120]
- Induction of ovulation in women with secondary amenorrhea.[121]
- Improve sperm count in men with idiopathic oligoasthenozoospermia.[122]
- Control prostatic carcinoma.[123]
- Induction of puberty in males with idiopathic hypogonadotrophic hypogonadism.[124]
- Promote testicular descent in boys with cryptorchidism.[90]
- Treat precocious puberty.[125]
- Inhibit spermatogenesis.[126]

LHRH antagonists
- Inhibit ovulation.[127]
- Inhibit spermatogenesis.[128]

Muramyl dipeptide (*N*-acetylmuramyl-L-alanyl-D-isoglutamine)
- Display the adjuvant effect of complete Freund's adjuvant.[129]
- Stimulate nonspecific resistance to bacterial infections.[129]
- Mitogenic for splenocytes.[129]
- Untoward reactions include pyrogenicity, leukopenia, endotoxin sensitization.[129]

Neurotensin (NT)
- Lower blood pressure.[130]
- Lower body temperature.[130]
- Induce analgesia.[130]
- Reduce gastric acid secretion.[94]

Parathyroid hormone (PTH)
- Treatment of osteoporosis.[97]

Somatostatin (SS)
- Control bleeding of gastric ulcers.[98–100]
- Inhibit GH release in diabetics receiving insulin.[131]
- Treatment of acute pancreatitis.[131]
- Control endocrine tumors.[132]

Thymic hormones (FTS, Thymopoietin, Thymosin, etc.)
- Treatment of immunodeficiency diseases.[102]
- Treatment of autoimmune disease.[103]
- Treatment of cancer.[105]
- Treatment of arthritis.[106]

Thyrotropin releasing hormone (TRH)
- Prolonged infertility and period of lactation in women who are breastfeeding.[107]
- Decreased ataxia in patients with spinocerebellar degeneration.[108]

Tuftsin
- Antileukemia activity.[133,134]
- Immunostimulant.[133]

Vasopressin (VP)
- Improve cognitive functioning of patients with progressive idiopathic dementia,[135] affective illness.[136]
- Cessation of hemorrhagic episodes in hemophiliacs.[137]
- Control of central diabetes insipidus.[109]

Vasopressin antagonists
- Diuretic agent.[138]

Table 4. Peptide Analogues with Enhanced Potency[a]

ACTH

  [D-Ser[1], Lys[17,18]]-ACTH (1–18) amide

    10-Fold increased potency in corticosteroid release, 14-hour duration of action by intranasal administration in man.[139]

Met(O_2)-Glu-His-Phe-D-Lys-Phe

    1000-Fold increase over ACTH(4–10) in rate of extinction of rat pole jumping avoidance response.[140]

Bradykinin (BK)

  [Aib[7]]-BK

    4 × BK in guinea pig ileum contractions.[141]

    5.7 × BK in blood-pressure lowering.[141]

Chemotactic peptides

  CHO-Met-Leu-Phe

    Extremely potent chemoattractant ($ED_{50} = 7 \times 10^{-11} M$).[142]

Enkephalins (EK)

  Tyr-D-Met(O)-Gly-MePhe(ol) SD-26

    23,300 times morphine in blocking electrically stimulated guinea pig ileum.[64]

    9.1 times morphine in the rattail flick test (sc).[64]

    6.6 times morphine in the rat writhing test (sc).[64]

  [D-Thr[2], Thz[5]]-EK-NH_2

    1.7 (oral) or 4.8 (iv) times morphine in the rattail flick test.[143]

  Tyr-D-Ala-Gly-MePhe-Met(O)ol (FK 33-824)

    Four times (iv) or two times (sc) greater activity than morphine in the rattail flick test.[144]

    Postoperative analgesia (epidural) in four of eight patients.[145]

    Suppression of ACTH release.[146]

    Elevation of serum prolactin, growth hormone, lowering of serum LH, FSH, ACTH, and cortisol levels.[147]

Tyr-D-Ala-Gly-Phe-MeMet-NH_2 (metkephamid)

    Increased but nonselective affinity for both μ and δ receptors.[148]

    Greater analgesic effect in humans (parenteral) than meperidine.[149]

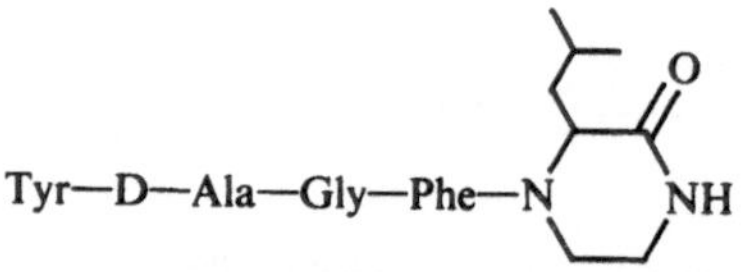

O

Tyr—D—Ala—Gly—Phe—N   NH

    Analogue with prolonged duration and potency comparable to morphine.[150]

LHRH

  [D-Nal(2)[6]]-LHRH

    200 times the potency of LHRH in estrus suppression.[151]

  [D-Ser(tBu)[6], des-Gly[10]]-LHRH ethylamide (buserelin or HOE 766)

    50–70 times the potency of LHRH.[63]

    (in) Inhibition of ovulation in 50 women for 147 or 150 treatment months.[120]

    (in) Normal menstrual cycle in 7 of 10 women with secondary amenorrhea.[121]

  [D-Trp[6], des-Gly[10]]-LHRH ethylamide

    140 times the potency of LHRH.[152]

    (sc) Inhibition of spermatogenesis in five of eight males.[126]

    (in) Induction of ovulation in three of six sterile females.[153]

    (iv) Inhibition of LH/FSH in five girls with precocious puberty.[125]

Table 4. (*Continued*)

α-MSH
  [Cys^4, Cys^{10}]-α-MSH
    Over 10,000 times α-MSH activity in frog skin bioassay.[65]
Somatostatin (SS)

Cyclo (-Aha-Cys-Phe-D-Trp-Lys-Thr-Cys-) [S—S bridge between the two Cys residues]
    2.7 Times somatostatin in glucagon release inhibition.[154]
    3.5 Times somatostatin in insulin release inhibition.[154]
    2.6 Times somatostatin in growth-hormone release inhibition.[154]
  Cyclo(-Pro-Phe-Phe-D-Trp-Lys-Thr-)
    1.74 Times somatostatin in growth-hormone release inhibition.[155]
    5.2 Times somatostatin in insulin release inhibition.[155]
    6.0 Times somatostatin in glucagon release inhibition.[155]
Thyrotropin releasing hormone (TRH)
  Analogues with increased potency and resistance to enzymatic degradation[156]:

pGlu-His-3,3′-DiMePro-NH_2 (RX 77368)

[6-oxopiperidine-2-carbonyl]-CO-His-Thz-NH_2 (MK 771)

and

R-His-Pro-NH_2

where

R = [orotyl: 2,6-dioxo-tetrahydropyrimidine-4-yl]-CO— (CG 3509)

[5-methyl-3-oxothiomorpholine]-CO— (CH_3, S, O, N–H) (CG 3703)

[5-oxotetrahydrofuran-2-yl]-CO— (DN 1417)

[a] iv = intravenous; in = intranasal; sc = subcutaneous.

Table 5. Peptide Analogues with Enhanced Selectivity[a]

ACTH

[(*S*-carboxamidomethyl)Cys^{25}]-$ACTH^{1-26}$

Increased aldosterone over corticosterone release.[157]

ACTH(4–10)

Increased behavioral activities over adrenal activity.[158]

ACTH(4–9) Analogue Met(O_2)-Glu-His-Phe-D-Lys-Phe

More potent, long-acting analogue lacking steroidogenic properties displaying (oral):

1. Beneficial effect on goal-directed motivation.[159]
2. Improved mood and performance without affecting sleep.[160]
3. Improved attentional processes in mentally retarded adults.[161]
4. Reduced anxiety and depression and increased feelings of competence in geriatric patients and patients with senile dementia.[110]

Angiotensin II (AII)

[Sar^1, Ala^8]-AII

(iv) Antagonist analogue lowers blood pressure in human hypertensives with high renin levels.[113]

Bradykinin (BK)

[L-α$MePhe^8$]-Bradykinin

Selective toward constriction of rat uterus.[162]

γ-Endorphin (β-lipotropin 61–77)

[des-Tyr^1]γ-Endorphin (β-lipotropin 62–77)

Improved conditions of schizophrenics.[163]

Analogue displaying behavioral activities, devoid of analgesic activity.[164]

No effect on schizophrenics.[164]

Enkephalins (EK)

Tyr-D-Ser-Gly-Phe-Leu-Thr

Potent enkephalin selective for δ receptors.[165]

MeTyr-D-Ala-Gly-Phe-$NHCH_2CH_2C_6H_5$

Potent enkephalin selective for μ receptors.[166]

```
     S————————S
     |        |
Tyr-D-Cys-Gly-Phe-Cys-NH2
```

Potent enkephalin selective for μ receptors.[168]

```
     S————————S
     |        |
Tyr-D-Pen-Gly-Phe-Cys-NH2
```

Potent enkephalin selective for δ receptors.[168]

```
     ε-NH——————————┐
     |             |
Tyr-D-Lys-Gly-Phe-Leu-CO
```

Potent enkephalin selective for μ receptors.[169]

Tyr-D-Ala-Gly-MePhe-N($CH_2CH_2C_6H_5$)$CH_2CH_2SCH_3$

Potent antagonist to opioid receptor interaction.[170]

Growth-hormone releasing peptides

Tyr-D-Trp-Ala-Trp-D-Phe-NH_2

Over 1000-fold greater GH releasing activity than enkephalin analogues.[171]

Human chorionic gonadotropin (HCG)

Table 5. (*Continued*)

Deglycosylated HCG
HCG deglycosylated with hydrogen fluoride gives an analogue with superior binding to gonadal receptors and HCG specific antibodies, and acts as a potent HCG antagonist both *in vitro* and *in vivo*.[172]

LHRH
[DPhe^2, DTrp^{3,6}]-LHRH
Inhibition of ovulation in 6 of 10 normal women through one menstrual cycle.[40]
Ac-[D-Phe^1, D-pClPhe^2, D-Trp^{3,6}]-LHRH
100% Blockade of ovulation in rats at 62 μg.[173]
Ac-[Thr^1, DPhe^2, D-Trp^3, D-Ser^4, DTyr^5, DTrp^6, D-Arg^8]-LHRH
70% Inhibition of ovulation in rats at 25 μg.[174]
Ac-[dehydro Pro^1, D-pClPhe^2, DPhe^2, DTrp^{3,6}, MeLeu^7]-LHRH
Greater than 90% inhibition of ovulation in rats at 10 μg.[175]
Ac-[D-pClPhe^{1,2}, DTrp^3, DPhe^6, DAla^{10}]-LHRH
88% Inhibition of ovulation in rats at 7.5 μg.[176]

S——————S (disulfide bridge between DCys^1 and Cys^{10})
[DCys^1, DPhe^2, DTrp^3, DTrp^6, Cys^{10}]-LHRH
Antagonist with greater receptor affinity than LHRH.[175]

Muramyl dipeptide
MurNAc-L-Ala-D-Gln-O-*n*-C_4H_9
Adjuvant active without pyrogenic activity.[177]

Neurotensin
[D-Trp^{11}]-Neurotensin
A selective antagonist for neurotensin-induced coronary vessel constriction.[178]
Novel increase in motor activity in rats, opposite to neurotensin.[179]

Oxytocin (OT)
[Thr^4, Gly^7]-OT
High oxytocic with negligible antidiuretic and pressor activities.[180]
des amino-[Carba Cys^1, Tyr(Me)^2]-OT
Induction of porcine[181] and bovine[182] uterine contractions without vasoconstriction lasting two hours.
[Pna^1, Leu^2]-OT
Potent antagonist to *in vitro* uterine contractions (pA_2 7.1) and *in vivo* milk ejection induced by OT.[183]
[Pna^1, Phe^2, Thr^4]-OT
Potent antagonist to *in vitro* (pA_2 7.7) and *in vivo* uterine contractions induced by OT.[184]
des-amino-[Pna^1, Tyr(Me)^2, Thr^4]-OT
[(β-mercapto-β,β-cyclopentamethylene propionic acid)^1-Thr^4]-OT
des-amino-[Pna^1, Tyr(Me)^2, Thr^4]-OT
des-amino-[Pna^1, Thr^4]-OT
Potent antagonists to *in vivo* uterine contractions[185] and to *in vivo* milk ejection[186] induced by OT.

Parathyroid hormone (PTH)
bPTH(1–34)
Minimum sequence for full biological activity.[34]
bPTH(24–28)
bPTH(24–34)

(*continued*)

Table 5. (*Continued*)

Selective hypotensive activity in rats and dogs with no hypercalcemic action.[34]

[Nle^8, Nle^{18}, Tyr^{34}]-bPTH(3–34) amide

A PTH antagonist with enhanced binding and resistance to oxidation.[37]

Somatostatin (SS)

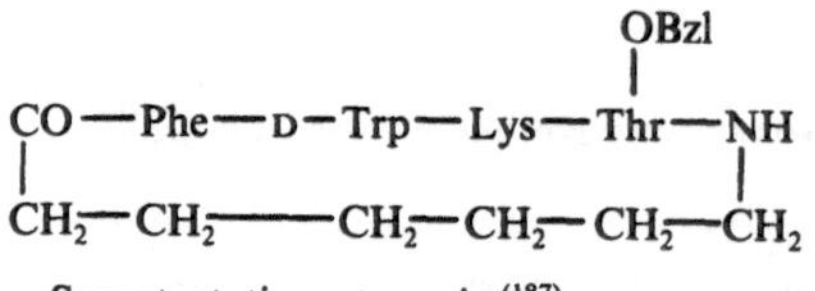

```
                              OBzl
                               |
CO—Phe—D—Trp—Lys—Thr—NH
|                               |
CH2—CH2——CH2—CH2—CH2—CH2
```

Somatostatin antagonist[187]

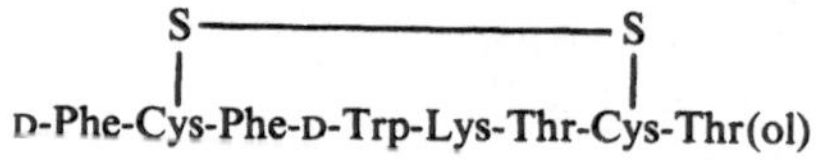

```
       S———————————S
       |           |
D-Phe-Cys-Phe-D-Trp-Lys-Thr-Cys-Thr(ol)
```

Potent analogue selective for inhibition of growth-hormone release over insulin and glucagon release.[188]

Opiate antagonist selective to μ receptors.[189]

[D-Trp^{5,8}]-Somatostatin

Potent selective inhibition of growth hormone and insulin over glucagon release.[190]

[Phe^4]-Somatostatin

Increased selectivity toward inhibition of growth hormone over insulin and glucagon release.[191]

[des-(Ala^1, Gly^2), His^{4,5}, D-Trp^8]-Somatostatin

Increased inhibition of growth hormone over glucagon and insulin release.[192]

[des-(Ala^1, Gly^2), D-Trp^8, D-Asu^{3,14}]-Somatostatin

Increased inhibition of glucagon over insulin and growth-hormone release.[193]

Substance P

[D-Pro^2, D-Trp^{7,9}]-Substance P

Potent antagonist to Substance P contraction of guinea pig ileum with very low agonist activity.[194]

[D-Pro^9, D-Trp^{7,9}]-SP(4–11)

pA_2 5.65 in inhibiting Substance P contraction of guinea pig ileum with no agonist activity.[195]

Tuftsin

[Ser^1]-Tuftsin

Antagonist.[196]

Thyrotropin releasing hormone (TRH)

γ-Butyrolactone-γ carbonyl-His-Pro-NH_2

Potent CNS activities with nominal TSH-releasing activity.[197]

Pyro-L-α-aminoadipyl-His-L-thiazolidine-5-carboxylic acid

Increased CNS over hormonal actions.[198]

pGlu-His-L-3,3-dimethylproline amide

pGlu-His-L-*trans*-3-methylproline amide

Increased CNS over hormonal effects due to increased enzymatic stability.[199]

Vasopressin (VP)

[des-amino, D-Arg^8]-Vasopressin (DDAVP)

Increased antidiuretic over vascular effects (2000:1).[200]

(sl, in) Control of central diabetes insipidus.[201]

(in) Improved cognitive functioning of patients with progressive idiopathic dementia.[135]

Table 5. (*Continued*)

(in) Improved long-term memory in patients with affective illness.[136]
(in) Controlled hemorrhagic episodes in hemophiliacs.[137]
[(β-mercaptopropionyl)^1, Phe^2, Arg^8]-VP
Increased ADH over pressor and uterotonic activity.[202]
[(β-mercapto propionic acid)^1, Dpr^7, Arg^8]-VP
Selective antidiuretic over pressor activity.[203]
des-amino-[Phe^2, Dpr^7, Arg^8]-VP
Increased ADH over pressor activity.[204]
Ac-[Tyr(Me)^2, Arg^8]-VP
Antagonist to vasopressor response of [Arg^8]-VP and lacks vasopressor and antidiuretic activity; also antagonist to uterotonic activity of OT.[205]
[(β-mercapto-β,β-diethylpropionyl)^1, D-Arg^8]-Vasopressin
Potent antivasopressor analogue with weak antidiuretic and antioxytoxic activity.[206]
[(β-mercapto-β-β-cyclopentamethylenepropionyl)^1, D-Leu^2, Val^4, Arg^8]-Vasopressin
Potent anti-antidiuretic agents (AVP antagonists) with weak antivasopressor activity.[207]
Arginine Vasotocin
[(3-mercaptopropionyl)^1, Orn^8]-Vasotocin
Potent vasotocin analogue selective toward anovulatory activity.[208]

[a] iv = intravenous; sl = sublingual; in = intranasal.

Peptide analogues in which structural modifications have resulted in significantly enhanced potency are shown in Table 4,[139–156] and peptides in which structural modifications have resulted in greater selective action are shown in Table 5.[157–208] Many of these improved analogues have been tested in humans, as shown in these two tables.

The following are potential goals of structural modification:

1. Develop analogues of a peptide that display increased resistance to enzymatic degradation.
2. Develop analogues with increased selectivity of action to reduce the number of side effects by:
   a. Shortening the peptide chain to remove conflicting message sequences.
   b. Refining a message sequence that stimulates many tissues into a message sequence that stimulates only one target tissue.
3. Develop analogues that are more potent.
4. Develop analogues with increased duration of action.
5. Develop analogues with increased ability to diffuse across the intestinal epithelium, the kidney tubules, or the blood–brain barrier.

Methods for approaching these goals are discussed in the following sections.

4.2. Enhancement of Resistance to Proteolytic Degradation

The major mode of peptide degradation in the body involves enzymatic cleavage of amide bonds in the peptide chain. Peptidases are found in many areas of the body: the intestinal wall, blood plasma, the kidney, the liver, and other tissues.[209] The most important sites of enzymatic degradation are located in the intestine and kidney.[210] From port of entry into the body in transit to tissue receptors, all of the amide bonds in a peptide are potentially susceptible to enzymatic attack. In reality, however, only certain distinct amide groups are critical for maintenance of an active peptide. Protection of these amide groups is most important.

4.2.1. Synthetic Modifications Which Reduce Peptide Bond Cleavage

Peptide chemists have developed modified amino acid residues and amide groups[211,212] that resist enzymatic attack and can be substituted in the region of the peptide containing susceptible amide bonds. Some substitutions are shown in Figure 5. Olefin substitution is the most satisfactory amide group substitution in terms of bond angles and rigidity.[213,214] It is less hydrophilic than an amide group and is synthetically challenging. Carbonyl reduction[215,216] is less of a synthetic problem, but it may introduce a cationic ammonium group in the chain that may be unfavorable to receptor interaction. Alternatively, —CH_2—S— and —CH_2—O— substitutions[217–219] have been developed which lack a cationic group. D-amino acid substitution[220] is perhaps the simplest substitution, given the commercial availability of D-amino acid derivatives. This substitution can alter conformation, however.[221] N^{α}-methyl substitution,[144] C^{α}-methyl substitution,[222] C^{α}–C' methylene insertion,[223] and dehydro amino acid substitution[224] provide moderate synthetic challenges and can alter conformation.[225] Some of these amino acid derivatives are commercially available. Retroinverso alteration[226] constitutes a sophisticated modification that can also alter receptor binding.

Not shown in Figure 5 are the following modifications:

1. Removal of the *N*-terminal amine or reduction of the *C*-terminal carboxylic acid group.[225] These are modifications which may reduce proteolytic susceptibility but also reduce hydrophilicity.
2. *N*-terminal to *C*-terminal amide bond formation gives cyclic analogues which would resist amino peptidase and carboxy peptidase action;[227] cyclization imposes a conformational restriction that may be undesirable.

The consequent pharmacologic changes that can be observed are illustrated with the series of enkephalin analogues[228] in Table 6. Analogue I

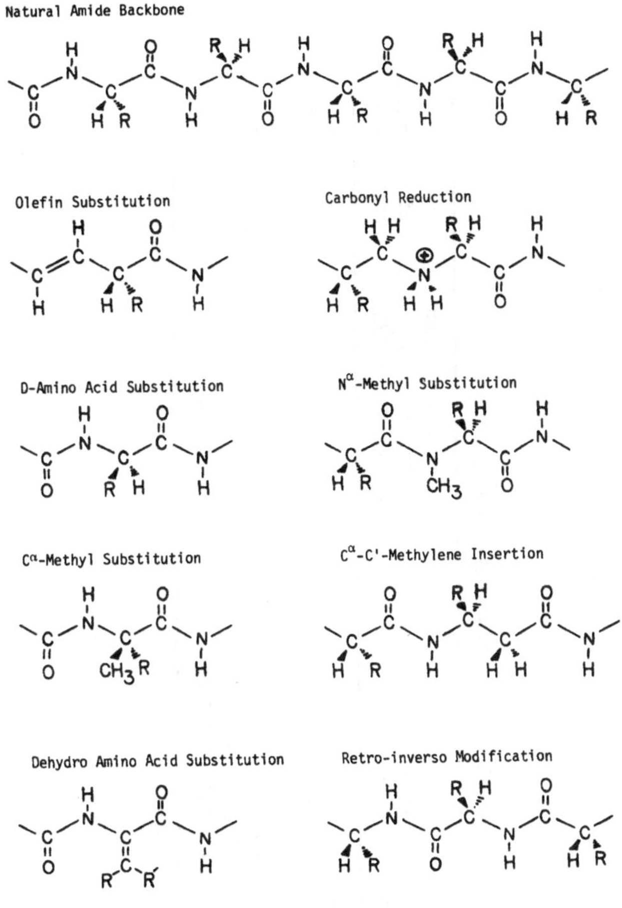

Natural Amide Backbone

Olefin Substitution

Carbonyl Reduction

D-Amino Acid Substitution

N^{α}-Methyl Substitution

C^{α}-Methyl Substitution

C^{α}-C'-Methylene Insertion

Dehydro Amino Acid Substitution

Retro-inverso Modification

Not Shown: N-terminal to C-terminal cyclization.

Figure 5. Structural modifications which reduce proteolysis.

contains a D-alanine instead of glycine in the second position, and a *C*-terminal leucine residue lacking a carboxyl group. This analogue is over nine times more potent *in vitro* than Met^5-enkephalin. This increased activity may be due either to increased stabilization to membrane-bound enzymes in the target tissue or to a profitable stabilization of bioactive conformation. Analogue I is still a poor analgesic *in vivo*, perhaps due to its rapid degradation by kidney enzymes.

Either N^{α}-terminal methylation in analogue II or N^{α}-methylation of phenylalanine in analogue III significantly improves stability to kidney

Table 6. N-Methylated Enkephalin Analogues[228][a]

Compound No.	Tyr-D-Ala-Gly-Phe-Leu-[Leu^5]-enkephalin analogue	Kidney $T^{1/2}$ (min)	Guinea pig ileum (potency relative to Met^5 EK = 1.0)	Mouse writhing test ED_{50}(mg/kg) (iv)[b]
I	Tyr-D-Ala-Gly-Phe-$NHCH_2CH_2CH(CH_3)_2$	2	9.4	100
II	MeTyr-D-Ala-Gly-Phe-$NHCH_2CH_2CH(CH_3)_2$	5	5.2	0.2
III	Tyr-D-Ala-Gly-MePhe-$NHCH_2CH_2CH(CH_3)_2$	6	5.0	0.2
IV	MeTyr-D-Ala-Gly-MePhe-$NHCH_2CH_2CH(CH_3)_2$	120	2.35	0.3
V	MeTyr-D-Ala-Gly-Phe-$N(CH_3)CH_2CH_2CH(CH_3)_2$	60	10.00	0.4
VI	MeTyr-D-Ala-Gly-MePhe-$N(CH_3)CH_2CH_2CH(CH_3)_2$	60	0.710	3.0

[a] MePhe = N^{α}-methylphenylalanine and MeTyr = N^{α}-methyltyrosine.
[b] iv = intravenous.

Table 7. Stabilized Enkephalin Analogues[a]

Compound	References	Analogue	Relative potency: Vas deferens	Relative potency: Guinea pig ileum	Relative potency: ^3H Naloxone binding	Analgesia hot plate (morphine = 1.0)
VI		Tyr-Gly-Gly-Phe-Met-OH	1.0	1.0	1.0	0.012 (iv)
VII	229	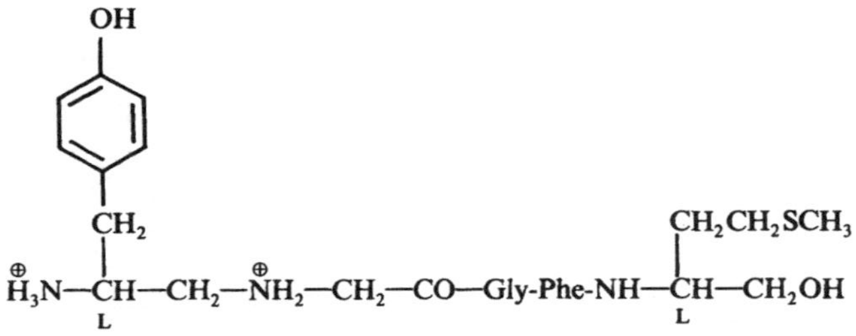 $\overset{\oplus}{H_3N}-\underset{L}{CH}(CH_2-C_6H_4-OH)-CH_2-\overset{\oplus}{N}H_2-CH_2-CO-$Gly-Phe-NH$-\underset{L}{CH}(CH_2CH_2SCH_3)-CH_2OH$	–	0.83	2.0	–
VIII	230	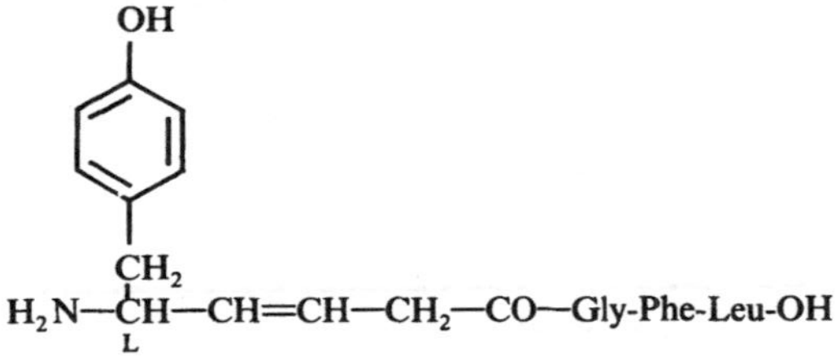 $H_2N-\underset{L}{CH}(CH_2-C_6H_4-OH)-CH{=}CH-CH_2-CO-$Gly-Phe-Leu-OH	–	–	0.67	–
IX	35, 161	Tyr-D-Ala-Gly-Phe-Met-$CONH_2$	3.81	7.75	1.5	0.1 (iv)

(*Continued*)

Table 7. (*Continued*)

Compound	References	Analogue	Relative potency: Vas deferens	Relative potency: Guinea pig ileum	Relative potency: ^{3}H Naloxone binding	Analgesia hot plate (morphine = 1.0)
X	231, 232	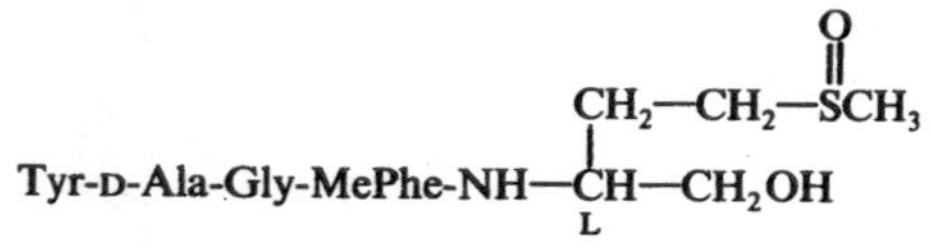 Tyr-D-Ala-Gly-MePhe-NH—CH(L)(CH_2—CH_2—S(=O)CH_3)—CH_2OH	0.94	2.0	9.86	4.0 (iv) 2.0 (sc) 0.2 (oral)
XI	233	Tyr-D-Ala-Gly-Phe-MeMet-NH_2	2.6	—	—	4.0 (sc)
XII	234	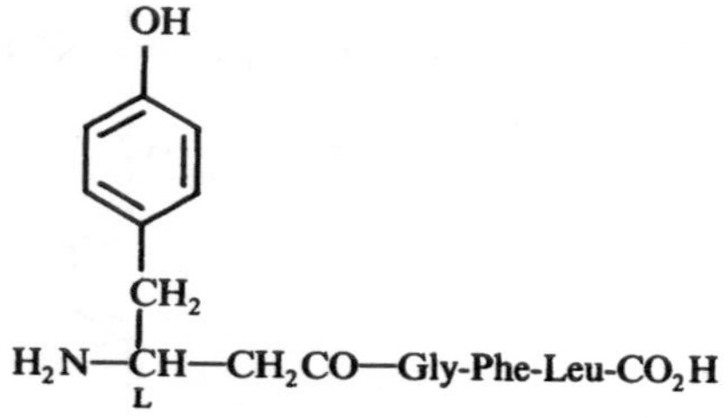 H_2N—CH(L)(CH_2—C_6H_4—OH)—CH_2CO—Gly-Phe-Leu-CO_2H	0.037	—	< 0.0023	—
XIII	235	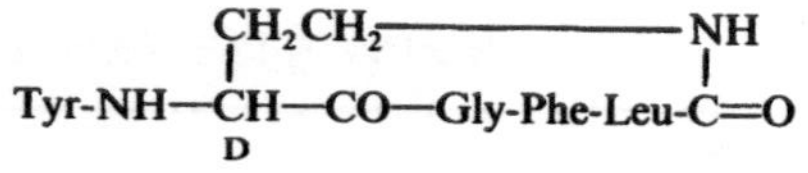 Tyr-NH—CH(D)—CO—Gly-Phe-Leu-C=O, with CH_2CH_2 on the CH bridged to NH on the C=O (cyclic)	—	17.5	2.0	—
XIV	236	[des-amino[1], Met[5]]-enkephalin	< 0.0002			
XV	234	Tyr-Gly-Gly-Phe-$NHCH_2CH_2CH(CH_3)_2$	1.19	0.36	1.82	

[a] iv = intravenous; sc = subcutaneous.

enzymes and both analogues display greater activities *in vivo*. Notice that the receptor affinities of analogues II and III were actually diminished in the guinea pig ileum. Thus, stabilized peptide analogues can achieve greater plasma concentrations and consequently appear to be more potent *in vivo* than the natural peptide, even though receptor affinity may not have increased.

Double methylations in analogues IV and V increase stability to kidney tissue dramatically, but do not enhance *in vivo* potency further. Triple methylations in analogue VI actually decrease potency both *in vivo* and *in vitro*, presumably by restricting the analogue to conformations that are unfavorable for receptor interaction. No data were given on duration of action in the series of analogues.

Table 7[229–236] displays enkephalin analogues that bear many of the substitutions in Figure 5. Only potency data are listed. Many of the substitutions maintained or increased potency. Only C^{α}–C' methylene insertion in analogue XI and *N*-terminal amine group removal in analogue XIV resulted in lower potency. Analogue X is sufficiently stable and potent to display oral analgesic activity in rats. Both analogues X and XI display analgesia in humans.[237–239] These examples of stabilized enkephalins demonstrate the wide variety of methods available to stabilize a peptide against enzymatic attack without loss of biologic potency.

Stabilized analogues of peptides are usually prepared by total synthesis. The peptide chemist should not overlook the possibility of stabilizing a peptide to enzymatic attack by selective synthetic modification of the whole peptide. For example, reductive methylation of β-endorphin gave a single homogeneous product that showed greater resistance to degradation in the presence of pituitary homogenate.[240] For larger peptides semisynthetic modification may be useful (see Section 9 and Reference 241).

4.2.2. *Natural Structural Influences on Proteolytic Susceptibility*

In Section 3.2.5 portions of peptide sequences called stability sequences were described that naturally protect the message sequence from enzymatic degradation. Knowledge of the three-dimensional structure of a peptide can be utilized to build peptide analogues in which the protective properties of the stability sequence are enhanced.

The analgesic peptides enkephalin and β-endorphin are both rapidly inactivated by proteolytic cleavage in the message sequence Tyr^{1}-Gly^{2}-Gly^{3}-Phe^{4}-Met^{5} at the amide bonds between residues 1–2 and 3–4.[242] The good *in vivo* analgesic potency of β-endorphin is attributed to the stabilizing influence of the helical tail, residues 6–31, attached to the message sequence, 1–5.[243] Two analogues of β-endorphin were synthesized with increased helical character of the tail sequence,[243] as observed by circular dichroism measurement. Both analogues displayed greater resistance to proteolysis in rat brain homogenate than β-endorphin.

4.2.3. The Altered Selectivity of Stabilized Analogues

A peptide analogue that can achieve greater plasma concentration may also reach certain tissues more effectively than the native peptide. Such an analogue would appear to have a different selectivity than the native peptide. For example, a stabilized TRH (thyrotrypin releasing hormone) analogue was shown to be selective toward CNS actions over peripheral actions due to increased plasma concentrations and increased ability to penetrate the brain.[244]

4.3. Enhancement of Selectivity by Chain Shortening to Remove Undesirable Message Sequences

Several examples of peptide analogues, developed to have increased selective action by chain shortening, may be found in Table 5. The 4–10 and 4–9 sequences of ACTH display important behavioral activities without stimulating adrenal secretion.[158,159]

The minimal sequence for complete activity in bovine parathyroid hormone (bPTH) is the sequence bPTH(1–34).[34] Removal of the first two amino acids from the amine terminus gives rise to antagonists, for example, [Nle^8, Nle^{18}, Tyr^{34}]-bPTH(3–34) amide.[37] The sequence bPTH(24–28) and bPTH(24–34) continue to display hypotensive action but lack hypercalcemic activity.[34]

The central and peripheral actions of substance P appear to be divided between the amine and carboxyl terminal portions of the peptide: SP(1–7) and SP(1–8) display the pain-modulating activities of the full molecule but display no smooth muscle or blood-pressure activities,[245] while pGlu-SP(7–11) is fully active in smooth muscle and blood-pressure bioassays.[245]

There are undoubtedly many other peptides which contain separate message sequences with selective actions. Not all peptides contain separate message sequences for each action. In spite of the multiplicity of actions contained in the 14-amino acid peptide somatostatin, all of its activities are expressed by the internal tetrapeptide sequence -Phe-Trp-Lys-Thr.[154] Selectively acting somatostatin analogues required development by side-chain and peptide-backbone modifications (Table 5).

4.4. Enhancement of Potency and Selectivity by Side-Chain Modification

In several cases modifications of amino acid side chains have given rise to peptide analogues that act as antagonists to the parent peptide; for example: [Sar^1, Ala^8]-angiotensin II,[36] [D-Ala^2, MePhe^4, (*N*-phenethyl-2-methylthioethylamine)^5]-enkephalin,[170] [Ser^1]-tuftsin,[196] Ac-[Tyr(CH_3)^2, Arg^8]-vasopressin,[205] and an analogue of somatostatin, cyclo(-Ala-Phe-D-Trp-Lys-Thr(Bzl)-Thr-).[187] Side-chain modifications were important in the enhancement of potency in ACTH, enkephalin, LHRH, and TRH analogues in Table 4. Side-chain modifications also increased selectivity of action in many analogues listed in Table 5.

Peptides and proteins can bear carbohydrate residues off of certain side chains (Asn, Ser, Thr, and hydroxylysine).[246] Removal of these carbohydrate groups can alter the activities of glycosylated peptides and proteins. A potent antagonist to HCG (human chorionic gonadotropin) was created by deglycosylation of HCG with anhydrous hydrogen fluoride.[172] Since maintenance of the corpus luteum in the pregnant state is dependent on HCG, an antagonist may constitute a novel means of fertility control. The methods of biological synthesis prepare peptides and proteins without glycosidic side chains. Thus antagonists and other selective analogues of glycopeptides, for example, HCG, LH, and FSH can be prepared by these methods.

4.5. Enhancement of Potency and Selectivity by Peptide-Backbone Modification to Promote Bioactive Conformation

A peptide analogue that maintains a bioactive conformation in solution is likely to be more potent than one that does not. Hence, peptide chemists have found it profitable to develop conformationally restricted analogues of peptides.[221]

The synthetic options for conformational restriction include:

1. Cyclization through *N*-terminal to *C*-terminal amide bond formation.
2. Cyclization through a disulfide bridge between a pair of cysteines.
3. Cyclization through other side chains, for example, amide bond formation between Glu and Lys side chains.
4. Restriction of individual amino acid conformations by incorporation of *N*-methyl amino acids, α-methyl amino acids, D-amino acids, β-disubstituted amino acids, and cyclic amino acids that involve the amide backbone in a ring (e.g., proline).
5. Restriction of individual side-chain conformations, for example, via dehydroamino acid analogues.[224]

Analogues containing one of the above conformational restrictions are bioassayed. If biologic potency is maintained, then one can assume that the bioactive conformations can include that particular conformation. The study of a series of conformationally restricted analogues gives rise to a "conformational structure–activity relationship."[62] This information serves as a set of criteria by which models of bioactive conformation can be judged. It also narrows the number of possible conformations that need to be considered as candidate bioactive conformations. With sufficient information the search for bioactive conformations can be limited to a range of possible conformations that can be surveyed by computer.[59–61]

Small, highly restricted peptide analogues that are biologically potent may assume solution conformations that are much closer to the bioactive conformation, allowing for conformation determination by NMR spectrometry[62] and future analogue design from highly refined structural models.

In practice, chemists may be led to potent, constrained analogues by empirical design without having developed a detailed three-dimensional model. They will have, nevertheless, profited by orienting their analogue design towards analogues bearing conformational restrictions.

It should be noted that many of the conformational restrictions listed at the beginning of this section are also listed as modifications that block enzymatic degradation in Figure 5. Potency increases observed *in vivo* for analogues containing conformational restrictions may be due to increased stability as well as to better receptor fit.

All of the peptide analogues with enhanced potency in Table 4 contain conformational restrictions. Many of the selectively acting analogues in Table 5 bear conformational restrictions, including the antagonists to LHRH,[173–175] neurotensin,[178] and substance P.[194,195]

The importance of conformational restrictions toward enhancing peptide selectivity is particularly well demonstrated by the cyclic enkephalins. Opiate receptors have been classified by their relative affinities for different ligands: μ (morphine), δ (enkephalins), σ (SKF 10,047), and κ (ketocyclazocine).[247] [Met^5]- and [Leu^5]-enkephalin show a somewhat greater preference for δ-type over μ-type receptors. The disulfide-bridged analogue [D-Cys^2, Cys^5]-enkephalin amide, however, shows a much greater preference for μ receptors.[168] The disulfide-bridged analogue [D-Pen^2, D-Cys^5]-enkephalin amide containing penicillamine (Pen), which differs from cysteine (Cys) by two methyl groups off the β-carbon, displays the opposite preference for δ receptors.[169]

Peptides that can adopt a β-bend conformation may be stabilized by a number of bend-stabilizing groups, including 3-amino-2-piperidone-6-carboxylic acid,[248] *trans*-3-oxo-5β-formamidomethyl-8-a-phenylmethyl perhydronaphthalene,[(249)] 5H-6-oxo-2,3,4,4a,7,7a-hexahydropyrano [2,3,b] pyrrole,[(250)] and dipeptide-γ-lactams.[(251)] The latter two units have been successfully incorporated into active enkephalin[(250)] and LHRH[(251)] analogues.

4.6. Enhancement of Potency and Selectivity with Peptide Oligomers

Concurrent with peptide–receptor binding, receptor microaggregation may be an important step in the activation of a tissue by a peptide hormone.[(252)] Peptide analogues which facilitate receptor aggregation might therefore be more potent. This consideration led to the conjugation of polymercaptosuccinyl tobacco mosaic virus with bromoacetyl and maleimide derivatives of α-MSH, [D-Ala^2, Lys^5]-enkephalinamide, and ACTH.[(253)] These conjugates displayed dramatically superior activities to the parent hormones, as well as greater receptor affinities and prolonged actions.

Both [D-pGlu^1, D-Phe^2, D-Trp^3, D-Lys^6]-LHRH amide, A, and the lysyl–lysyl cross-linked dimer, A-EGS-A, resulting from treatment of A with ethylene glycol bis(succinimidyl succinate), EGS, are antagonists to endogenous LHRH.[(252)] Reaction of antibody to [D-Lys^6]-LHRH, (Ab), with A-EGS-A gave a divalent complex Ab-$(A\text{-}EGS\text{-}A)_2$ which, in contrast to the parent compounds, is an *agonist*:—it stimulates the release of LH from pituitary cultures, presumably by promoting receptor aggregation.

Perhaps by a similar mechanism, attachment of the inactive stereoisomer *N*-acetylmuramyl-D-Ala-D-Glu-NH_2 of muramyl dipeptide to multipoly(D,L-Ala)–poly(L-Lys) gives a conjugate capable of increasing nonspecific immunity, yet remains devoid of other muramyl dipeptide activities.[(254)]

Ethylene diamine dimerization of [D-Ala^2, Leu^5]-enkephalin via the *C*-terminal carboxyl group gives a dimer with greater δ receptor selectivity than the parent monomer.[(255)] In this case the dimer is presumed to selectively aggregate δ receptors over μ receptors.

Until receptor microaggregation is demonstrated with these peptide oligomers, an alternate explanation may account for their superior activities. An oligomeric peptide derivative brings several peptide molecules to the receptor at once, increasing the local concentration of peptide ligand in the vicinity of the receptor, providing the appearance of increased binding and potency. The larger oligomers should also diffuse more slowly from the receptor area increasing duration of action.

These examples suggest nevertheless that larger oligomeric derivatives of peptides may lead to greater potency and selectivity than smaller analogues.

4.7. Enhancement of Potency with Peptide Analogues Displaying Superior Membrane Permeability

There are three potential barriers toward delivery of a peptide to its target tissue:

1. The intestinal epithelium, which must be crossed by orally administered peptides.
2. The proximal kidney tubule epithelium, which must be crossed by a peptide to remain in plasma at suitable concentrations for periods greater than several minutes.
3. The capillary walls at the blood–brain barrier, which should be crossed by centrally acting peptides.

All three "barriers" are composed of a layer of epithelial cells, with cell walls consisting of lipid bilayer membranes. "Active" carrier systems may promote membrane transport for certain peptides, for example, active transport of amino acids, dipeptides, tripeptides, and tetrapeptides into the intestinal epithelial cell.[(256)] Diffusion-controlled passive transport is a more probable process for most peptides,[(257)] especially for synthetic peptide analogues.

Peptides inherently contain an array of polar amide and side-chain groups. Peptides containing nonpolar side chains, however, can be sufficiently lipophilic to penetrate membranes. Schwyzer[(24)] has shown that ACTH (1–24) can at least poke through a lipid bilayer membrane in a capacitance minimization experiment. Presumably, the signal peptides described in Section 3.2.3 can also penetrate lipid bilayer membranes, although this remains to be demonstrated.

The antibacterial ionophoric peptide gramicidin A penetrates membranes[(258)] to serve as an ion channel,[(259)] as do the peptides alamethicin,[(260)] antamanide,[(261)] and cyclo(-Glu-Sar-Gly-decylGly-)$_2$.[(262)] A ^{13}C-NMR paramagnetic shift reagent study demonstrated that the lipophylic peptide Boc-Pro-Leu-Val-OMe resides completely within the hydrophobic environment of a phospholipid vesicle.[(263)]

It should thus be possible for sufficiently lipophilic peptides to penetrate and pass through lipid bilayer membranes. The number of polar groups that would have to be replaced by nonpolar groups to afford sufficient membrane permeability may be rather small. It may not be necessary to replace the amide backbone of a suitably potent and stable peptide with a hydrocarbon skeleton. Specific challenges presented by each membrane barrier will be discussed individually.

4.7.1. Peptide Transport across the Gut: The Key to Oral Activity

The digestive system evolved to break peptides and proteins down to their constituent amino acids to serve as nutrients to the body. In the interest of nutrition, decades of research have accumulated a wealth of knowledge about the metabolic decomposition of peptides and proteins in transit from mouth to blood vessels. A consequence of these efforts is that "peptide absorption" and "peptide transport" have come to mean the degradative processing from peptides in the lumen to amino acids in the blood. These terms *rarely* refer to the active or passive transport processes by which *intact* peptides cross the gut into the blood.[264]

Experiments with radiolabeled protein have shown that small amounts of intact proteins (or peptides) can cross the intestinal epithelium under physiologic conditions.[265] The requirements for optimal intact peptide transport across the gut remain to be delineated. A speculative list would include:

1. Structural deterrents to aminopeptidase, carboxypeptidase, and selective endopeptidase action.
2. A sufficiently large ratio of lipophilic to hydrophilic amino acids to optimize lipid membrane permeability.
3. Inclusion of sequences for which carrier-mediated active transport processes may exist.[264]
4. Inclusion of lipophylic sequences, for example, signal peptides, to assist membrane penetration.

Since a peptide could be absorbed through the oral epithelium, this list would apply to the sublingual route of administration as well.

A number of peptides have now been shown to display activity in humans by oral administration (see Table 10). Given the potential utility of peptides as pharmaceuticals, it is important that the optimal requirements for intact peptide gut transport be delineated.

4.7.2. Peptide Transport across the Kidney Tubule: The Key to Long Plasma Lifetime

The kidney is perhaps the most active site of peptide clearance from plasma.[209] The epithelial cells at the luminal brush border of the proximal tubule are similar in structure and function to their sister cells in the small intestine.[210] The high capacity of the reabsorption process in the proximal tubule epithelium[210] suggests that more stable and lipophylic peptides should reside longer in plasma than native peptide hormones.[266]

4.7.3. Pathways for CNS Stimulation by Peripherally Administered CNS Peptides

The physiological roles of candidate peptide neurotransmitters are initially determined by direct injection into the brain. Isolation and localization experiments further delineate the brain neurotransmitter roles of such peptides.[267] A number of brain peptides and analogues have now been shown to elicit central effects by peripheral administration[268] (see Table 8).[269–274] The possible therapeutic application of CNS peptide analogues is raised by such experiments, as is the consideration of the routes by which peripherally administered CNS peptides stimulate the brain.

4.7.3a. Blood–Brain Barrier Transport. There normally are many avenues for passage of substances from the interior of a blood capillary to the extracapillary environment:[275]

1. Passage through intracellular clefts.
2. Passive diffusion through thin membranous fenestrae in capillary cells.
3. Transport by pinocytosis.
4. Transcellular passive diffusion of lipid soluble materials.

Brain capillaries lack intracelluar clefts, fenestrae, and pinocytotic processes.

Table 8. CNS Peptides Displaying Central Effects upon Peripheral Administration

Peptide	Effect
ACTH (39 amino acids)	Delays extinction of avoidance conditioning in adrenalectomized rats.[169]
CCK (33 amino acids)	Induces satiety in lean or obese humans.[77]
DSIP (9 amino acids)	Improves disturbed sleep (insomnia) without sedation.[79]
β-Endorphin (31 amino acids)	Suppresses narcotic abstinence syndrome.[83] Relieves pain in cancer patients.[81] Increases alpha activity in EEG.[270] Improves mood of depressed patients.[271]
LHRH (10 amino acids)	Stimulates mating behavior in male and female rats.[272]
MIF-I (3 amino acids)	Improves mood in depressed patients.[92] Reduces symptoms in Parkinsonism patients.[93]
α-MSH (13 amino acids)	Alters EEG.[273]
TRH (3 amino acids)	Inhibits opiate withdrawal hypothermia.[244]
1-Des-amino-8-D-arginine–vasopressin (DDAVP) (9 amino acids)	Improves learning and memory in cognitively impaired and cognitively unimpaired adults.[274]

Table 9. Fractional Extraction (E) of Radiolabeled Peptides by Rat Brain[276]

Compound	E(%)
^{125}I cholecystokinin	0.005
^{125}I somatostatin	0.01
^{3}H leu^{5}-enkephalin	2.4
^{3}H met^{5}-enkephalin	2.4
^{3}H TRH	1.0
^{3}H glutathione	0.6

Some substances, such as glucose, cross by protein carrier systems, but most blood-borne substances presumably pass into the brain by passive transcellular diffusion only.[275] With lipophilicity as the key structural feature to passive diffusion, peptides with their abundance of amide and other polar groupings should be unable to diffuse across this "blood–brain barrier."

The classic experiment[275] for measuring the percentage of brain uptake of a substance in blood is performed as follows: A radiolabeled peptide is injected into the brain by way of the common carotid artery along with a second diffusable radiolabeled compound, for example, ^{3}HOH. By measuring the differential amounts of radiolabeled materials in brain tissue upon sacrifice of the animal, the percentage of peptide absorbed can be calculated. The experiment is fraught with many experimental problems,[275,276] but is still the primary tool for measuring amounts of materials crossing the blood–brain barrier.

Table 9[276] lists percentages of radiolabeled peptides absorbed by rat brain tissue. All were absorbed in small amounts, less than 3%. Kastin *et al.*[277] have argued that these small amounts may be sufficient for CNS stimulation by peripheral injection, for example, following injection of 200 μg of delta sleep-inducing peptide (DSIP) a doubling of DSIP brain levels was observed by RIA.

4.7.3b. Blood–CSF Transport. Another route for peptide entry into the brain might be by diffusion into the cerebrospinal fluid (CSF). Although the blood–CSF barrier is only 0.02% of the surface area of the blood–brain barrier,[278] the capillaries within the cerebroventricular organs lack the tight junctions of the blood–brain barrier and may comprise a route for peptide entry into the brain. Gerner *et al.*[279] recently measured CSF β-endorphin levels in three depressed patients following peripheral injection of the peptide. In all three subjects, dramatic increases (100–400%) in CSF β-endorphin were measured. Intravenous β-endorphin is known to improve the mood of depressed patients.[280] It is also known that β-endorphin relieves pain in humans, either by spinal injection[281] or intravenous injection.[282] Thus, for

β-endorphin the route of passage from plasma to CSF may be sufficient to induce analgesia and mood elevation.

Measurements of CSF/plasma concentration ratios indicate that other peptides (insulin and prolactin) are transported across the blood–CSF barrier.[283] Once in the CSF a peripherally administered peptide must still diffuse across brain tissue to the site of stimulation. This may be a rather circuitous journey unless specific receptors for the peptide exist on the CSF surface of the brain organs.

4.7.3c. Binding to Receptors at the Blood–Brain Barrier. A third route for CNS stimulation by peripherally administered peptides may be by direct stimulation of specific receptors on the blood side of the blood–brain barrier. The existence of specific saturable binding sites in the brain capillary bed has been demonstrated for both insulin[283] and angiotensin II.[284] The physiological function of such binding sites remains to be elaborated. Almost all peptides found in the brain as neurotransmitters also function in the periphery. It seems likely that many of these peptides will be shown to bind to specific saturable receptors at the blood–brain barrier.

4.7.3d. Transport of Peptides to the Hypothalamus from the Pituitary. Pituitary peptide stimulation of the brain at the hypothalamus, by retrograde blood flow from the pituitary to the hypothalamus, has been suggested from the following experiments:

1. Significant increases of tritiated ACTH(4–9) were detected in the brain and hypothalamus following intrapituitary injection of the labeled peptide into sheep.[285]
2. A significant drop in colonic temperature followed intrapituitary, but not intravenous, injection of neurotensin into rats.[286]

Whether pituitary peptides gain access to interior brain tissue receptors via blood–brain barrier diffusion or whether the peptides interact with receptors at the blood–brain barrier has yet to be determined. Regardless of the pathway by which blood-borne peptides stimulate the brain, the number of peptides and analogues that have been shown to elicit central effects by peripheral administration (see Table 8) suggests that analogues of CNS peptides may be of significant therapeutic value.

Among the many approaches to enhancement of *in vivo* potency and selectivity described in Section 4, no one approach is predictably superior to the other. In a peptide development program the peptide chemist is wise to consider all approaches. Most of the enhanced peptides in Tables 4 and 5 resulted from at least two types of modifications. Compared to most molecules that become candidates for drug development, peptides are quite large. Upon first consideration, a medicinal chemist's response might be,

"How can a peptide be made smaller?" The numerous examples of potent, stable peptide analogues presented here suggest that peptides do not necessarily need to be made smaller.

Table 10[287–295] lists a number of natural and modified peptides which are sufficiently potent and stable to evoke important pharmacologic responses in humans and mammals by oral or intranasal administration. As stabilized analogues of other peptides are developed, we should expect to see this list grow dramatically through the 1980s.

5. Physicochemical Vehicles to Modulate Proteolysis and Absorption

In Section 4.2 modifications to peptide structures that block proteolysis were discussed. A number of pharmaceutical techniques show promise in reducing proteolysis and increasing duration.

Liposome encapsulation has been promoted as a vehicle for a variety of drugs.[296] Liposome-entrapped insulin proved to be effective by oral administration in diabetic animals,[297] but its stability and effectiveness were variable.[298] Liposome-entrapped angiotensin II effectively elevated blood pressure intravenously, but produced no effects intragastrically.[299] The lack of response was attributed to the short plasma lifetime of angiotensin II. Stabilized analogues would presumably be more effective. Liposome-encapsulated TRH caused significant increases in body temperature over free TRH upon intravenous injection.[300] Since intraventricularly injected TRH also increases body temperature, the liposome encapsulation seemed to facilitate blood–brain barrier passage. Liposome-encapsulated muramyl dipeptide increased the degree and duration of activation of macrophages as tumoricidal agents.[301] Subcutaneous LHRH-impregnated silicone rubber implants effectively released LHRH for at least 10 days, as did a TRH implant.[302] A zinc-gelatin vehicle provided markedly prolonged duration of an ACTH analogue by intramuscular injection.[303]

With the rapid development of novel pharmaceutical techniques such as the ones described here, as well as erodible polymers and transdermal and intraocular devices,[304] the natural potency but instability of peptides qualifies them as important candidates for deployment of new delivery devices.

6. Modes of Pharmacologic Control of Endogenous Peptidergic Processes

A peptidergic process that has been targeted for pharmacologic control need not only be controlled by agents that mimic or block peptide–receptor

Table 10. Peptides Demonstrating Activity by Oral and Intranasal Administration

Peptide	Administration	Activity (in humans, unless noted otherwise)
[Suc^1, Val^5, Phg^8]-AII	Intranasal (rat)	Blood-pressure reduction.[287]
ACTH analogue (Org 2766): Met(O_2)-Glu-His-Phe-D-Lys-Phe	Oral (rat, human)	Antiamnesic effects.[288]
		Moderate reduction in seizure frequency in children with intractable seizures.[289]
		Beneficial effect on goal-directed motivation.[159]
		Improved mood and performance without affecting sleep.[160]
		Improved attentional processes in mentally retarded adults.[161]
		Reduced anxiety and depression; increased feelings of competence in geriatric patients and patients with senile dementia.[111]
Aspartame	Oral	Beverage sweetener.[290,291]
Bestatin	Oral	Increased effectiveness of cancer irradiation and chemotherapy.[68]
Calcitonin	Oral	Inhibition of gastric secretion[73]
Captopril	Oral	Blood-pressure reduction in hypertension.[292]
		Improved exercise tolerance in severe heart failure cases.[293]
CCK (26–33)	Intranasal	Treatment of chronic pancreatitis.[75]
Enalapril	Oral	Blood-pressure reduction in hypertension.[294]
Enkephalin analogues: Tyr-D-Ala-Gly-MePhe-Met(O)ol	Oral (rat)	Analgesia.[144]
D-Thr^2-Thz^5-enkephalins-NH_2	Oral (rat)	Analgesia.[143]

LHRH analogue: D-Ser(tBu)^{6}-des-Gly^{10}-LHRH ethyl amide	Intranasal	Inhibition of ovulation.[120]
MIF-I	Oral	Mood improvement in depressed patients.[92]
		Decline in severity and frequency of episodes in patients with tardive dyskinesia.[93]
Muramyl dipeptide	Oral (mouse)	Increased humoral immune response.[295]
Oxytocin	Intranasal	Stimulation of milk letdown in lactating women.[96]
Somatostatin analogue: D-Phe-Cys-D-Trp-Lys-Thr-Cys-Thr(ol) (S—S bridge between the two Cys)	Oral (rat)	Growth-hormone suppression.[188]
TRH	Oral	Prolonged infertility and lactation period in women who are breastfeeding.[108]
[Dmp^{3}]-TRH	Oral (mouse)	Reversed hypothermia induced by reserpine.[244]
Des-amino-[D-Arg^{8}]–Vasopressin (DDAVP)	Sublingual/intranasal	Antidiuresis in diabetes insipidus.[201]
		Enhanced recall ability in patients with depression and normal humans.[136]
		Control of hemorrhagic episodes in hemophiliacs.[137]

interaction. Potentially, there are many points in the scheme of a peptidergic process where pharmacologic control can be established.

6.1. Inhibition of Ribosomal Translation

This should be an ideal control point, especially in disease processes that involve overproduction of a protein or peptide, for example, growth hormone in acromegaly,[305] gastrin in Zollinger–Ellison syndrome,[306] renin in hypertension.[307]

In Section 3.2.3., signal peptides were described as having too little specificity to be useful as competitive inhibitors. As the molecules involved in the specific recognition and binding of the nascent peptide chain–ribosomal-mRNA complex to the RER membrane become identified, it is possible that competitive inhibitors could be developed.

6.2. Modulation of Posttranslational Processing

Following ribosomal translation a propeptide is processed by a series of enzymatic reactions into the active form of the peptide. Once a selective processing enzyme has been isolated and characterized, the medicinal chemist has the opportunity to develop competitive inhibitors to the processing enzyme of the prohormone sequence.

The processing enzymes arise in a similar manner by posttranslational processing from proenzymes.[308] The medicinal chemist might therefore be able to control the processing of a peptide by developing a competitive inhibitor to the posttranslational processing of the appropriate processing enzyme.

Most peptides are released from the cell of origin in the active form. Some peptides, however, are released into plasma as an inactive precursor, for example, angiotensin[309] and bradykinin.[117] Their extracellular activation is dependent upon the release of the critical processing enzyme renin for activation of angiotensin[309] and kallikrein for bradykinin.[117]

Besides renin, activation of angiotensin involves a second processing enzyme: angiotensin-converting enzyme (ACE). Blood pressure can be lowered by blocking the action of angiotensin II. Figure 6[310–316] displays various agents, both enzyme inhibitors and antagonists, that have been developed to modulate angiotensin II and effectively lower blood pressure in mammals and humans.

6.3. Modulation of Peptide Release: Peptide Release Factors, Release Inhibitors, and Their Antagonists

Many physiological processes are controlled by hormones released

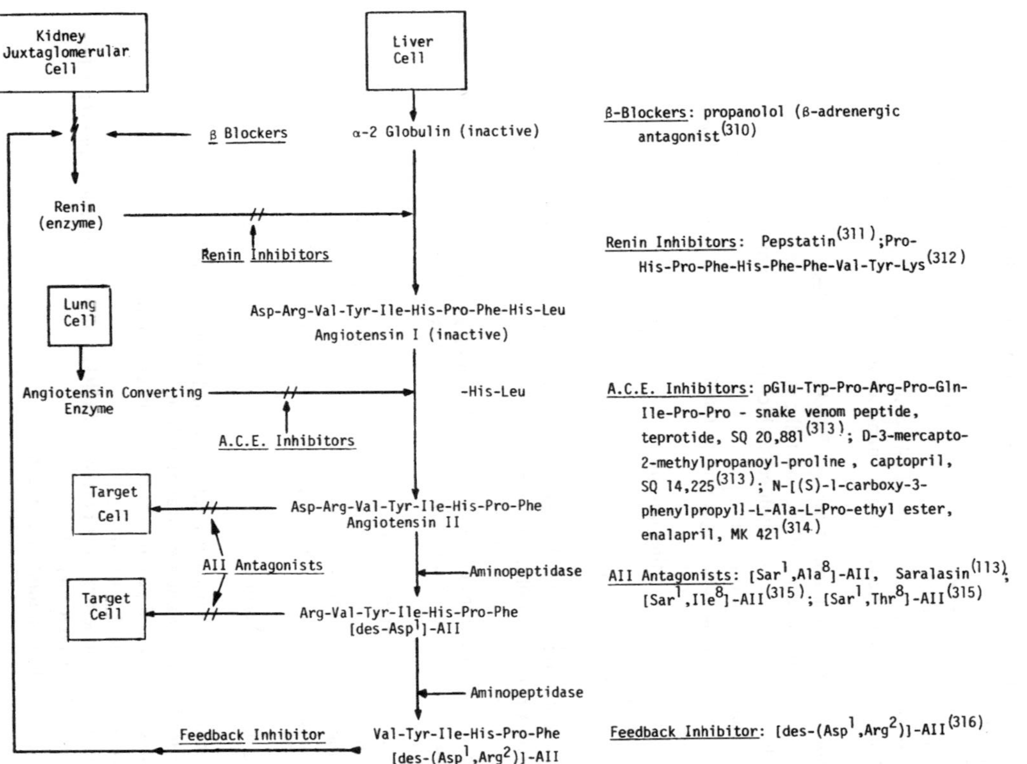

Figure 6. Scheme depicting plasma activation of angiotensin II (AII) and pharmacologic modulators of AII.

Table 11. Pituitary Hormones and Their Hypothalamic Releasing and Release-Inhibiting Factors[317]

Adrenocorticotropic hormone (ACTH, corticotropin; 44 amino acids)
- Actions:
  - Stimulates adrenal cells to produce glucocorticoids (aldosterone, cortisol, testosterone, estradiol) from cholesterol.
  - ACTH(4–10) common to α-MSH induces potent behavioral actions.[158]
- Hypothalamic releasing hormone:
  - Corticotropin releasing factor (CRF; 41 amino acids).

Follicle stimulating hormone (FSH; 210 amino acids)
- Actions:
  - Promotes early maturation of ovarian follicles in females.
  - Promotes final maturation of ovarian follicles.
  - Stimulates development of Sertoli cells which produce spermatozoa in males.
- Hypothalamic releasing hormone:
  - Luteinizing hormone–releasing hormone (LHRH, luliberin, gonadotropin releasing hormone; 10 amino acids).
- Hypothalamic release-inhibiting hormone:
  - Somatostatin-28.[319]

Growth hormone (GH, somatotropin; 190 amino acids)
- Actions:
  - Promotes statural growth via long bone development up to puberty.
  - Stimulates red blood cell production.
  - Promotes tissue protein synthesis.
  - Diabetogenic (promotes glucose conversion from glycogen).
  - Stimulates the release of other growth factors (somatomedins).
  - Promotes acromegaly.
- Hypothalamic releasing hormone:
  - Growth-hormone releasing factor (GHRF, somatocrinin; 44 amino acids).[318]
- Hypothalamic Release-inhibiting hormone:
  - Growth-hormone release inhibiting factor (GHRIF, GIH, somatostatin; 14 or 28 amino acids).[319]

Luteinizing hormone (LH, interstitial cell-stimulating hormone; 204 amino acids)
- Actions:
  - Triggers ovulation in females at about second week after menses.
  - Promotes final maturation of ovarian follicle.
  - Stimulates Leydig cell production of testosterone in males.
- Hypothalamic releasing hormone:
  - Luteinizing hormone-releasing hormone (LHRH, luliberin, gonadotropin releasing hormone; 10 amino acids).
- Hypothalamic release-inhibiting hormone:
  - Somatostatin-28.[319]

Leuteotropic hormone (LTH, leuteotropin, prolactin; 198 amino acids)
- Actions:
  - Promotes full lubuloalveolar development of breasts during pregnancy.
  - Inhibits LHRH secretion, delaying ovulation to six weeks after end of lactation.
  - Inhibits dopamine secretion from the hypothalamus.
  - Actions in male unknown.
- Hypothalamic releasing hormone:
  - Thyrotropin releasing hormone (TRH; 3 amino acids).

Table 11. (*Continued*)

Hypothalamic release-inhibiting hormone:
Dopamine.
Somatostatin.[319]
Thyroid stimulating hormone (TSH, thyrotropin; 204 amino acids)
Actions:
Stimulates thyroid hormone synthesis.
Promotes growth of thyroid gland.
Stimulates release of prolactin.
Hypothalamic releasing hormone:
Thyrotropin releasing hormone (TRH; 3 amino acids).
Hypothalamic release-inhibiting hormone:
Somatostatin.[320]

from the pituitary. The release of anterior pituitary hormones is controlled in turn by peptide releasing factors and release-inhibiting factors from the hypothalamus, as shown in Table II.[317–320]

Peptide hormones released from other tissues of the body, for example, the pancreas and other organs of the digestive system, are modulated by other peptides (see Table 12).[321,322] A dipeptide Tyr-Arg (kyotorphin) induces analgesia in the rat by acting as an enkephalin releasing factor.[323] Release of other brain peptides may be modulated by peptide factors as well.

For each of these release-modulating peptides, pharmacologic control of a disease process may be achieved by developing selective agonists or antagonists to the release-modulating peptide. Tables 4 and 5 list many peptides that have enhanced activities as release modulators. The reader should recognize that some of the natural peptides have displayed useful activities in humans (Table 2) as release modulators, for example, LHRH, TRH, and somatostatin.

6.4. Enhancement of Endogenous Peptide Levels by Blocking Degradation

Therapeutic benefit may be achieved in some instances by raising the levels of an endogenous peptide through blockage of its major degrading enzyme. Enkephalinase is a membrane-bound dipeptidyl carboxypeptidase that cleaves enkephalines at the Gly^3-Phe^4 amide bond,[324] with localization patterns that mimic opiate receptor localizations. The enzyme is a prime candidate for being an important enkephalin-degrading enzyme. Roques *et al.*[325] found that a potent inhibitor to enkephalinase, D,L-(3-mercapto-2-benzylpropanoyl)-glycine blocks enkephalin degradation *in vitro* in nanomolar concentrations and displays analgesic activity in rats lasting over 2 h.

Part of the blood-pressure lowering action of ACE inhibitors (see Figure 6), which block the release of the pressor peptide angiotensin II, has

Table 12. Gut Hormones and Their Modulators[321,322]

Gut hormones	Release stimulated by	Release inhibited by
Cholecystokinin (CCK; 8 or 33 amino acids)		Somatostatin
Gastrin (14, 17, or 34 amino acids)	GRP	GIP Somatostatin Substance P
Gastric inhibitory peptide (enterogastrone, GIP; 42 or 43 amino acids)	—	Somatostatin
Glucagon (glicentin, enteroglucagon; 29 and 101 amino acids)	CCK Enkephalin Gastrin GIP GRP Glucagon Substance P VIP (vasoactive intestinal peptide)	Secretin Somatostatin
Insulin (30 amino acids)	CCK Enkephalin Gastrin GIP GRP Glucagon Secretin VIP	Somatostatin Substance P —
Molilin (23 amino acids)	—	Insulin Somatostatin
Pancreatic polypeptide (36 amino acids)	CCK GRP	Somatostatin TRH
Secretin (27 amino acids)	—	Enkephalin Somatostatin
Somatostatin (14 and 28 amino acids)	VIP	CCK Gastrin GIP Secretin
Vasoactive intestinal peptide (VIP) (28 amino acids)	—	Somatostatin

been ascribed to its retardation of bradykinin degradation, a hormone which has hypotensive properties.[326]

6.5. Feedback Inhibitors

The process of feedback inhibition is well established for multienzyme systems. It is thus unremarkable that a peptide–enzyme cascade system such as the renin–angiotensin system is inhibited by its end product (see Figure 6).[316] It may be important, however, to exploit such inhibition, for example, to lower blood pressure by selective inhibition of renin release with analogues of [des-(Asp^1, Arg^2)]-angiotensin II.[316]

6.6. Common Dosage Problems with Endogenous Peptide Modulation

Accurate therapeutic modulation of endogeneous peptide levels may be complicated by several factors:

1. *Deviations from Sigmoidal Dose-Response Curves.* Some peptides display bell-shaped dose-response curves (increased effect at low doses; decreased effect at high doses). For example, MIF-I (melanocyte inhibiting factor-I)[92,93] and DSIP[210] show reduced effects at high doses. Peptides can also display sine-wave-shaped dose-response curves (negative effects at high doses). For example, hPTH(1–34) shows losses of both calcium and phosphate at high doses,[97] and SP(1–7) decreases latency times to painful stimuli at high doses.[245]

2. *Variations in Response with Dosage Schedule.* The type of response may depend on dosage schedule. For example, continuous infusion of LHRH drastically reduced LH secretion in rhesus monkeys, whereas pulsatile delivery (6 min per hour) repeatedly stimulated LH secretion.[175] An initial agonist response to bolus injection of [Sar^1, Ala^8]-angiotensin II is not observed on infusion of the antagonist.[113]

3. *Variations in Physiological States of the Target Tissues.* A tissue targeted for peptidergic control may be regulated by many factors other than the endogenous peptide. Therapeutic dosage may vary with the physiological state of the tissue. Gonadotropin response to LHRH varies widely during the menstrual cycle.[175] The effectiveness of angiotensin II antagonists varies with renin levels.[113] The effectiveness of gut hormones varies widely with digestive state.[321]

4. *Accumulation of Effect.* Peptides not only stimulate and inhibit the cellular release of hormones or neurotransmitters; they can also modulate the overall activity of cells. Hence, the response of tissue to subsequent doses of a peptide may vary dramatically from the initial dose. For example, cumulative effects have been observed with ACTH analogues,[139] DDAVP (-desamino-8-D-anginine-vasopressin),[200] LHRH superagonists,[175] and ACE inhibitors.[314] A novel peptide should be studied over a broad dosage range, by several modes of administration, by different dosage schedules, and under different physiological conditions.

7. Peptide Analogues with Increased Duration of Action

A number of peptide analogues display increased duration of action over the native peptide. The enkaphalin analogue Tyr-D-Ala-Gly-MePhe-Met(O)ol displayed a five-hour duration of analgesic action in rhesus monkeys following intravenous injection of the minimum effective dose.[(144)] Elevated LH levels are observed for five hours after LHRH superagonist intravenous injection, followed by an insensitivity to subsequent injections for at least 72 hours.[(175)] In this case receptor down-regulation is responsible for the long duration of effect.

[D-Ser^1, Lys^{17,18}]-ACTH(1–18) elevated human cortisol levels for 7–14 h at 1 mg intranasally.[(139)] [Suc^1, Val^5, Phg^8]-angiotensin II induced a 22% reduction in rat blood pressure for 120 min at 1 mg/kg intranasally during renin infusion.[(287)] No pressor effect was seen via intranasal administration. [D-Thr^2, Thz^5]-enkephalinamide induced analgesia in rats lasting 30 (9 μmol/kg oral) to 90 min (9 μmol/kg sc).[(143)] Disulfide-bridged Cys-Phe-Thr-Lys-D-Trp-Phe-Phe-Cys, a somatostatin analogue, suppressed plasma levels of many hormones by greater than 50% for up to 11 h.[(132)] DDVAP (30 μg sublingual) reduced diuresis for 12 h with a maximal effect at 4 h, where Arg^8-VP (vasopressin) showed almost no effect.[(201)] In these examples slow absorption via subcutaneous, intranasal, and sublingual routes were cited as the cause for increased duration over intravenous administration. Even though larger doses are required for intranasal and sublingual administration, the increased duration of action observed by these routes suggests a definite advantage over parenteral routes of administration.

8. Types of Therapeutic Biomedical Polypeptides

8.1. Peptide Hormones and Neurotransmitters

Over 30 peptides have been characterized as hormones and neurotransmitters. Like resistors in a television set, peptides are utilized in many areas of the body. Most peptides are not limited to function in one organ of the body, but are utilized in many different locations, for example, both brain and gut.[(321)] Most of these peptides are not limited to neurotransmitter or hormonal roles, but can serve either role in different areas of the body.[(267,321)] Methods of therapeutic modulation of peptide hormones and neurotransmitters have been discussed in previous sections.

Table 13. Proteinases Targeted for Therapeutic Control and Their Proteinase Inhibitors

1. *Acrosin*—a serine protease present in the head of spermatazoa that may play a role in egg penetration.
   Proteinase Inhibitor: Tosyl-L-lysine chloromethyl ketone (TLCK).
   TLCK treated spermatozoa fail to fertilize female rabbits.[328]
2. *Cathepsin B and D*—enzymes capable of degrading myosin and actin. Proteinase inhibitors: leupeptin, pepstatin.
   Potent inhibitors of cathepsin D and B.
   Inhibits protein degradation in cultured muscle cells and the muscle wasting in dystrophic chickens.[329,330]
3. *Elastase*—a leucocyte proteinase that degrades lung elastin and collagen in emphysema. Emphysema may develop in people with an inherited abnormality in α_1-proteinase inhibitor (α_1 PI), which normally blocks proteolytic activity of elastase released by leucocytes.
   Proteinase inhibitors: Ac-Ala-Ala-Pro-Ala(CH_2Cl), Suc-Ala-Pro-Val(CH_2Cl).
   Diminished experimentally induced emphysema in rats.[331]
   Proteinase inhibitor: MeO-Suc-Ala-Ala-Pro-Val(CH_2Cl).
   (oral) Prevented induction of experimental emphysema in rats.[332]
4. *Thrombin*—catalyzes the polymerization of fibrinogen into fibrin.
   Proteinase inhibitors: D-Phe-Pro-Arg(CHO), Boc-D-Phe-Pro-Arg(CHO).
   In vitro inhibition of blood clotting.[333]
5. *Aminopeptidase B and Leucine Aminopeptidase*—enzymes located on the surfaces of macrophages and lymphocytes.
   Proteinase inhibitors: [(2*S*,3*R*)-3-amino-2-hydroxyl-4-phenylbutanoyl]-L-leucine, bestatin
   (oral) Retarded tumor growth in mice.[334]
   (oral) Increased effectiveness of cancer irradiation therapy and chemotherapy in humans.[68]

8.2. Peptidase and Proteinase Inhibitors

In Section 6 control of peptide posttranslational processing and peptide inactivating enzymes with peptidase inhibitors was described. Another major use for peptide enzyme inhibitors is in the control of proteinases,[327] enzymes which act on proteins. Table 13[328–334] lists some proteinase inhibitors that display potentially useful actions.

8.3. Antigenic Peptides as Synthetic Vaccines

As useful as vaccines have been as therapeutic agents, the development and production of vaccines is fraught with many problems.[335] Peptides can be synthesized that are identical to portions of the exterior regions of proteins specific to the cell membrane of a virus, bacterium, or cancer cell. These peptides can be employed as antigens to raise antibodies against the virus, bacterium, or cancer cell. Such peptides can then serve as synthetic vaccines.

Lerner[335] described the synthesis of surface peptides from a coat protein of foot-and-mouth virus. Immunization with one of the synthetic peptide vaccines protected a group of guinea pigs against infection by large doses of foot-and-mouth virus. Cloning and biological synthesis of whole viral proteins has also been accomplished for foot-and-mouth disease virus, hepatitis B virus, and influenza virus.[336]

Selection of the appropriate cell-surface protein and identification of the exterior regions of the selected protein are crucial and time-consuming steps in the development of a synthetic peptide vaccine. The prospects[337] of targeting previously uncontrollable diseases via the immune system are exciting nevertheless.

8.4. Peptide Ionophoric Antibiotics

A number of peptides, isolated from microorganisms, are cytotoxic to bacteria, for example, alamethicin,[338,339] gramicidin,[340] tyrocidin,[340] and valinomycin.[341,342] These peptides contain cyclic structures that endow them with the remarkable ability to alter cation permeability of membranes by acting as ionophores. The structures of these peptides complexed to cations have been examined in some fascinating studies.[338–342] Their modes of action may not involve membrane cation permeability, but may involve binding to DNA and inhibition of RNA polymerase.[340,343] Synthesis by microorganisms[344] constitutes the best approach to these peptides.

8.5. Peptide Chemoattractant Immunostimulants

The process which draws leukocytes to invading bacterial cells is called chemotaxis—the directional migration of cells via a chemical gradient. A group of *N*-formylated peptides derived from bacteria has been identified as natural leukocyte chemoattractants, and have demonstrated immunostimulant activity.[345] Structure–activity studies have demonstrated a stereospecific sensitivity of leukocytes for these formylated peptides. A number of superpotent chemoattractants, for example, CHO-Met-Leu-Phe,[142] and antagonists, for example, Boc-Phe-D-Phe-Leu-Phe-D-Leu-Phe,[142] have been developed. Future physiological studies with these peptides shall be important in unraveling the mysteries of the immune system.

8.6. Flavorful and Poisonous Peptides

The vast range of physiological properties of peptides is bounded by flavorful peptides on the one hand and poisonous peptides on the other. Among the flavors available one finds peptides that are:

1. *Sweet*. Aspartame, or Asp-Phe-OMe, being marketed as an over-the-counter sweetener.[290,291]
2. *Delicious*. Lys-Gly-Asp-Glu-Glu-Ser-Leu-Ala.[346]
3. *Bitter*. Cyclo(-Leu-Trp-).[347]

In the continued search for new pharmaceuticals from poisonous organisms several peptides have been identified as toxic principles, including erubatoxin B,[348] actinomycin D,[349] and bleomycin,[350] the latter two having tumoristatic properties.

9. Conclusion

Decades of peptide research have generated scores of polypeptides that attract a broad range of biomedical interests. A large number of synthetic peptides displaying potentially therapeutic actions have stimulated a great deal of interest in peptides as pharmaceuticals. Well over 170 clinical trials have been performed with peptides in the last 10 years. Nineteen natural peptides and 19 peptide analogues have demonstrated therapeutic actions in human trials. Many of these peptides can be administered by oral and intranasal routes. Twelve peptides are currently available as pharmaceuticals. Employing the concepts and tools described in this chapter, medicinal and peptide chemists are actively developing peptide agents to control a multitude of ailments in both animals and humans. The revolution in physiology that peptides ignited in the 1970s has inflamed medicinal chemistry in the 1980s.

Acknowledgments

The author extends his inestimable gratitude to Wilma Miller for her expert typing of this manuscript. He thanks Betty Henderson for her assistance with the tables and figures and Pradip Bhatnagar and Jeannette Poumadere for their suggestions. He appreciates past discussions with Frank Momany, Domenico Regoli, Bob Samuels, and John Hughes, and the encouragement provided by Tobias Yellin.

References

1. M. Goodman and R. P. Saltman, *Biopolymers* **20**, 1929–1948 (1981).
2. M. Bodanszky, Y. S. Klausner, and M. A. Ondetti, *Peptide Synthesis*, 2nd Edition, Wiley, New York (1976).
3. E. Gross and J. Meienhofer, eds., *The Peptides, Analysis Synthesis Biology*, Vol. 1, *Major Methods of Peptide Bond Formation*, Academic Press, New York (1979).

4. E. Gross and J. Meienhofer, eds., *The Peptides, Analysis Synthesis Biology*, Vol. 2, *Special Methods in Peptide Synthesis, Part A*, Academic Press, New York (1980).
5. E. Gross and J. Meienhofer, eds., *The Peptides, Analysis Synthesis Biology*, Vol. 3, *Protection of Functional Groups in Peptide Synthesis*, Academic Press, New York (1981).
6. M. Bodanszky, *Synthesis* **5**, 333–356 (1981).
7. R. B. Merrifield and J. M. Stewart, *Nature* **207**, 522–523 (1965).
8. Spinco Division, Beckman Instruments, Inc., 1050 Page Mill Road, Palo Alto, CA 94304.
9. Vega Biochemicals, 420 East Columbia, Tucson, AZ 85714.
10. Biosearch, Inc., 2980 Kerner Blvd., San Rafael, CA 94901.
11. Peninsula Laboratories, Inc., 611 Taylor Way, Belmont, CA 94002.
12. H. Yajima and N. Fujii, *J. Chem. Soc., Chem. Commun.* **1980**, 115–116.
13. D. Yamashiro and C. H. Li, *J. Am. Chem. Soc.* **100**, 5174–5179 (1978).
14. C. Milstein, *Sci. Am., October 1980*, 66–74.
15. C. L. Cooney, *Science* **219**, 728–733 (1983).
16. I. S. Johnson, *Science* **219**, 632–637 (1983).
17. J. A. Martial, R. A. Mallewell, J. D. Baxter, and H. M. Goodman, *Science* **205**, 602 (1979).
18. R. A. Hitzeman, D. W. Leung, L. J. Perry, W. J. Kohr, H. L. Levine, and D. V. Goeddel, *Science* **219**, 620–625 (1983).
19. H. R. B. Pellham and R. J. Jackson, *Eur. J. Biochem.* **67**, 247–256 (1976).
20. R. E. Offord, in: *Semisynthetic Peptides and Proteins* (R. E. Offord and C. Di Bello, eds.), pp. 3–19, Academic Press, New York (1978).
21. K. Inouye, K. Watanabe, K. Morihara, Y. Tochino, T. Kanaya, J. Emura, and S. Sakakibara, *J. Am. Chem. Soc.* **101**, 151–152 (1979).
22. J. P. Greenstein and M. Winitz, in: *Chemistry of the Amino Acids*, Chapters 1 and 2, Wiley, New York (1961).
23. J. Hughes, T. W. Smith, H. W. Kosterlitz, L. A. Fothergill, B. A. Morgan, and H. R. Norris, *Nature* **258**, 577–579 (1975).
24. R. Schwyzer, *Trends Pharm. Sci. August 1980*, 329–331.
25. H. H. Buscher, R. C. Hill, D. Roemer, F. Cardinaux, A. Closse, D. Hauser, and J. Pless, *Nature* **261**, 423–425 (1976).
26. K. Kangawa, N. Minamino, N. Chino, S. Sakakibara, and H. Matsuo, *Biochem. Biophys. Res. Commun.* **99**, 871–878 (1981).
27. N. Nimano, K. Kangawa, N. Chino, S. Sakakibara, and H. Matsuo, *Biochem. Biophys. Res. Commun.* **99**, 864–870 (1981).
28. K. Mizuno, N. Minamino, K. Kangawa, and H. Matsuo, *Biochem. Biophys. Res. Commun.* **97**, 1283–1290 (1980).
29. D. L. Kirkpatrick, T. Taniguchi, B. N. Jones, A. S. Stern, J. E. Shively, J. Hullihan, S. Kimura, S. Stein, and S. Udenfriend, *Proc. Natl. Acad. Sci. U.S.A.* **78**, 3265–3268 (1981).
30. B. N. Jones, A. S. Stern, R. V. Lewis, S. Kimura, S. Stein, S. Udenfriend, and J. E. Shively, *Arch. Biochem. Biophys.* **204**, 392–395 (1980).
31. H. Kakidini, Y. Furutani, H. Takahashi, M. Noda, Y. Morimoto, T. Hirose, M. Asai, S. Inayama, S. Nakanishi, and S. Numa, *Nature* **298**, 245–249 (1982).
32. A. Goldstein, W. Fischli, L. I. Lowney, M. Hunkapiller, and L. Hood, *Proc. Natl. Acad. Sci. U.S.A.* **78**, 7219–7223 (1981).
33. C. H. Li, D. Yamashiro, L. -F. Tseng, and H. H. Loh, *J. Med. Chem.* **20**, 325–328 (1977).
34. P. K. T. Pang, M. C. M. Yang, H. T. Keutmann, and A. D. Kenny, *Endocrinology* **112**, 284–289 (1983).
35. J. S. Morley, *Ann. Rev. Pharmacol. Toxicol.* **20**, 81–110 (1980).

36. M. C. Khosla, R. R. Smeby, and F. M. Bumpus, *Handbook Exp. Pharmacol.* **37**, 126–161 (1974).
37. J. E. Mahaffey, M. Rosenblatt, G. L. Shepard, and J. T. Potts, Jr., *J. Biol. Chem.* **254**, 6496–6498 (1979).
38. S. S. Nussbaum, M. Rosenblatt, and J. J. Potts, Jr., *J. Biol. Chem.* **255**, 10,183–10,187 (1980).
39. D. B. Case, H. J. Keim, J. M. Wallace, and J. H. Laragh, *Kidney Int.* **15**, S104–S107 (1979).
40. A. Zarati, E. S. Canales, I. Sthorey, D. M. Coy, A. M. Comaru-Schally, and A. V. Schally, *Contraception* **24**, 315–320 (1981).
41. S. Caranikas, J. Mizrahi, E. Escher, and D. Regoli, *J. Med. Chem.* **25**, 1313–1316 (1982).
42. G. Bloebel, *Proc. Natl. Acad. Sci. U.S.A.* **77**, 1496–1500 (1980).
43. J. Gainer, P. Gaze, J. -C. Mercier, and B. Robson, *Biochemie* **62**, 231–239 (1982).
44. J. A. Majzoub, M. Rosenblatt, B. Fennick, R. Maunus, H. M. Kronenberg, J. T. Potts, Jr., and J. F. Habener, *J. Biol. Chem.* **255**, 11,478–11,483 (1980).
45. D. Shields, T. G. Warren, R. F. Green, S. E. Roth, and M. J. Brenner, The Primary Events in the Biosynthesis and Post-Translational Processing of Different Precursors to Somatostatin, in: *Peptides: Synthesis, Structure and Function; Proceedings of the Seventh American Peptide Symposium* (D. H. Rich and E. Gross, eds.), pp. 471–479, Pierce Chemical Co., Rockford, Ill. (1981).
46. J. F. Habener, P. K. Lund, J. W. Jacobs, P. C. Dee, and R. H. Goodman, Polyprotein Precursors of Regulatory Peptides, in: *Peptides: Synthesis, Structure and Function; Proceedings of the Seventh American Peptide Symposium* (D. H. Rich and E. Gross, eds.), pp. 452–467, Pierce Chemical Co., Rockford, Ill. (1981).
47. B. A. Eipper and R. E. Mains, Production of α-N-Acetyl-β-Endorphin (1–26) from α-N-Acetyl-β-Endorphin (1–27), in: *Peptides: Synthesis, Structure and Function; Proceedings of the Seventh American Peptide Symposium* (D. H. Rich and E. Gross, eds.), pp. 451–456, Pierce Chemical Co., Rockford, Ill. (1981).
48. T. L. O'Donohue and D. M. Dorsa, *Peptides* **3**, 353–395 (1982).
49. B. H. Frank, J. M. Petter, R. E. Zimmerman, and P. J. Burk, in: *Peptides: Synthesis, Structure and Function; Proceedings of the Seventh American Peptide Symposium* (D. H. Rich and E. Gross, eds.), pp. 729–738, Pierce Chemical Co., Rockford, Ill. (1981).
50. B. M. Austin, in: *Peptides: Synthesis, Structure and Function; Proceedings of the Seventh American Peptide Symposium* (D. H. Rich and E. Gross, eds.), pp. 493–496, Pierce Chemical Co., Rockford, Ill. (1981).
51. J. M. Hambrock, B. A. Morgan, M. J. Rance, and C. F. C. Smith, *Nature* **262**, 782–783 (1976).
52. K. M. Foley, I. A. Kourides, C. E. Inturrisi, R. F. Kaiko, C. G. Zàroulis, J. B. Posner, R. W. Houde, and C. H. Li, *Proc. Natl. Acad. Sci. U.S.A.* **76**, 5377–5381 (1979).
53. D. F. Veber, F. W. Holly, W. J. Palaveda, R. F. Nutt, S. J. Bergstrand, M. Torchiana, M. S. Glitzer, R. Saperstein, and R. F. Hirschman, *Proc. Natl. Acad. Sci. U.S.A.* **75**, 2636–2640 (1978).
54. L. A. La Planche and M. T. Rogers, *J. Am. Chem. Soc.* **86**, 337 (1964).
55. B. Pullman and A. Pullman, *Adv. Protein Chem.* **28**, 347–526 (1974).
56. G. N. Ramachandran and V. Sasisekharan, *Adv. Protein Chem.* **23**, 283–437 (1968).
57. T. N. Bhat, V. Sasisekharan, and M. Vijayan, *Int. J. Peptide Protein Res.* **13**, 170–184 (1979).
58. F. A. Momany, R. F. McGuire, A. W. Burgess, and H. A. Scheraga, *J. Phys. Chem.* **79**, 2361–2381 (1975).
59. F. A. Momany, *J. Am. Chem. Soc.* **98**, 2990–2996 (1976).

60. F. A. Momany, *J. Am. Chem. Soc.* **98**, 2996–3000 (1976).
61. F. A. Momany, *J. Med. Chem. Soc.* **21**, 63–68 (1978).
62. V. J. Hruby and H. I. Mosberg, *Peptides* **3**, 329–336 (1982).
63. H. Koch, *Pharm. Int. May 1981*, 99–100.
64. Y. Keso, M. Yamaguchi, T. Akita, H. Moritoki, M. Takei, and H. Nakamura, *Naturwissenschaften* **68**, 210–211 (1981).
65. J. J. Knittel, T. K. Sawyer, V. J. Hruby, and M. E. Hadley, *J. Med. Chem.* **26**, 125–129 (1983).
66. Chas. R. Baker, *Physicians Desk Reference*, 36th Edition, Medical Economics Co., Oradell, N.J. (1982), p. 589.
67. Chas. R. Baker, *Physicians Desk Reference*, 36th Edition, Medical Economics Co., Oradell, N.J. (1982), pp. 799–800.
68. *Drugs of the Future* **6**, 604–606 (1981).
69. I. M. A. Evans, *Lancet I*, 1232–1233 (1979).
70. L. A. Austin and H. Heath, *N. Eng. J. Med.* **304**, 269–278 (1978).
71. T. D. Koelmeyer and E. J. W. Stephens, *N. Zeal. Med. J.* **87**, 434–435 (1978).
72. L. A. Wisneski, W. P. Groom, O. L. Silva, and K. L. Becker, *Clin. Pharmacol. Ther.* **24**, 219–222 (1978).
73. R. Ziegler, H. Minne, J. Hotz, and H. Goebell, *Digestion* **11**, 157–160 (1974).
74. F. Fracoli, A. Fabri, L. Gnessi, C. Moretti, C. Santoro, and M. Felici, *Eur. J. Pharmacol.* **78**, 381–382 (1982).
75. A. Papp, Z. Berger, and V. Vacro, *Digestion* **21**, 163–168 (1980).
76. F. X. Pi-Sunyer, H. R. Kissileff, J. Thornton, and G. P. Smith, *Clin. Res.* **29**, A631 (1981).
77. G. Stacher, H. Bauer, and H. Steinringer, *Physiol. Behav.* **23**, 325–331 (1979).
78. I. Magnusson and T. Ihre, *Lancet I*, 792 (1982).
79. D. Schneider-Helmut, F. Gnirss, M. Monnier, J. Schenker, and G. A. Schonenberger, *Int. J. Clin. Pharmacol. Ther. Toxicol.* **19**, 341–348 (1981).
80. D. H. Catlin, K. K. Hui, H. H. Loh, and C. H. Li, *Commun. Psychol. Pharmacol.* **1**, 493–500 (1977).
81. T. Oyama, S. Fukushi, and T. Jin, *Can. Anesth. Soc. J.* **29**, 24–26 (1982).
82. T. Oyama, A. Matsuki, T. Taneichi, N. Ling, and R. Guilleman, *Am. J. Obstet. Gynecol.* **137**, 613–616 (1980).
83. D. H. Catlin, K. K. Hui, H. H. Loh, and C. H. Li, *Adv. Biochem. Psychopharmacol.* **18**, 341–350 (1978).
84. Chas. R. Baker, *Physicians Desk Reference*, 36th Edition, Medical Economics Co., Oradell, N.J. (1982), p. 1001.
85. D. Frasier, G. Castin, B. M. Lippe, T. Aceto, and P. F. Bunger, *J. Clin. Endocrinol. Metab.* **53**, 1213–1217 (1981).
86. D. Rudman, M. H. Kutner, R. D. Blackstone, R. A. Cushman, R. P. Bain, and J. H. Patterson, *N. Engl. J. Med.* **305**, 123–131 (1981).
87. Chas. R. Baker, *Physicians Desk Reference*, 36th Edition, Medical Economics Co., Oradell, N.J. (1982), p. 1842.
88. Chas. R. Baker, *Physicians Desk Reference*, 36th Edition, Medical Economics Co., Oradell, N.J. (1982), p. 1111.
89. G. Leyendecker, L. Wildt, and M. Hansmann, *J. Clin. Endocrinol. Metab.* **51**, 1214–1216 (1980).
90. P. Pirazolli, F. Zapparella, F. Bernardi, M. P. Villa, D. Aleksandrowicz, A. Scandola, P. Stancaci, A. Cicognani, and E. Caiccari, *Arch. Dis. Childhood* **53**, 235–238 (1978).
91. J. Happ, F. Kollmann, C. Kawehl, M. Neubauer, U. Krause, K. Demisch, J. Sandow, W. von Rechenberg, and J. Beyer, *Fertil. and Steril.* **29**, 546–551 (1978).

92. R. H. Ehrensing and A. J. Kastin, *Am. J. Psych.* **135**, 562–566 (1978).
93. R. H. Ehrensing, A. J. Kastin, P. F. Larsens, and G. A. Bishop, *Dis. Nervous Syst.* **38**, 303–307 (1977).
94. A. M. Blackburn, S. R. Bloom, R. G. Cory, D. R. Fletcher, N. D. Christofides, M. L. Fitzpatrick, and J. H. Baron, *Lancet I*, 987–989 (1980).
95. Chas. R. Baker, *Physicians Desk Reference*, 36th Edition, Medical Economics Co., Oradell, N.J. (1982), p. 1689.
96. Chas. R. Baker, *Physicians Desk Reference*, 36th Edition, Medical Economics Co., Oradell, N.J. (1982), p. 1690.
97. J. Reeve, P. J. Meunier, J. A. Parsons, M. Bernat, O. L. M. Bijovet, P. Courpron, C. Edouard, L. Klenerman, R. M. Neer, J. C. Renier, D. Slovik, F. J. F. E. Vismans, and J. T. Potts, Jr., *Brit. Med. J.* **1980**, 1340–1344.
98. L. Kayasseh, U. Keller, K. Gyr, G. A. Stalder, and M. Wall, *Lancet* **I**, 844–845 (1980).
99. J. P. Galmiche, M. V. Veyroe, and R. Colin, *Lancet I*, 1306–1307 (1980).
100. B. Limberg and B. Kommerall, *Lancet II*, 916–917 (1980).
101. D. B. Case, J. M. Wallace, H. J. Keim, M. A. Weber, J. E. Sealey, and J. H. Laragh, *N. Engl. J. Med.* **296**, 641–646 (1977).
102. P. Bordigoni, G. Faure, M. C. Bene, M. Dardenne, J. F. Bach, J. Duheille, and D. Olive, *Lancet I*, 293–297 (1982).
103. M. T. Lavastida and J. C. Daniels, *Fed. Proc.* **37**, 1669 (1978).
104. J. Costanzi, J. Daniels, G. Thurman, A. Goldstein, and J. Hokanson, *Ann. N.Y. Acad. Sci.* **332**, 148–159 (1979).
105. P. B. Chretien, S. D. Lipson, R. W. Makuch, D. E. Kenady, and M. H. Cohen, *Ann. N.Y. Acad. Sci.* **332**, 135–147 (1979).
106. E. M. Veys, E. C. Huskisson, M. Rosenthal, T. L. Vischer, H. Mielants, P. A. Thrower, J. Scott, H. Ott, M. Scheijgord, and J. Symoens, *Ann. Rheum. Dis.* **41**, 441–443 (1982).
107. J. E. Tyson, U.S. Patent No. 4,125,605 (November 14, 1978).
108. I. Sobue, H. Yamamoto, M. Konagaya, M. Iida, and T. Takayanagi, *Lancet I*, 418–419 (1980).
109. D. R. Brown and D. L. Uden, *Minn. Med.* **62**, 427–432 (1979).
110. *Drugs of the Future* **7**, 319–322 (1982).
111. K. Pentella, D. S. Bachman, and C. A. Sandman, *Neuropediatric* **13**, 59–62 (1982).
112. M. Suzuki, K. Hori, I. Abe, S. Saito, and H. Saito, *J. Natl. Cancer Inst.* **67**, 663–669 (1981).
113. D. B. Case, J. M. Wallace, H. J. Keim, J. E. Sealey, and J. M. Laragh, *Am. J. Med.* **60**, 825–836 (1976).
114. A. J. Cowley, J. M. Rowley, K. L. Stainer, and J. R. Hampton, *Lancet I*, 730–732 (1982).
115. R. Di Nicolantorio, F. A. O. Mendelsohn, J. S. Hutchinson, Y. Takata, and A. E. Doyle, *Can. J. Physiol. Rev.* **242**, R498–R504 (1982).
116. J. E. Morley, A. S. Levine, M. M. Oker, M. Grace, and J. Kneip, *Peptides* **3**, 1–6 (1982).
117. D. Regoli and J. Barabe, *Pharmacol. Rev.* **32**, 1–46 (1980).
118. J. H. Walsh, in: *Gut Hormones* (S. R. Bloom and J. M. Polak, eds.), pp. 163–170, Churchill Livingstone, New York (1981).
119. M. O. Thorner, J. Rivier, J. Spiess, J. L. Borges, M. L. Vance, S. R. Bloom, A. D. Rogol, M. J. Cronin, D. L. Kaiser, W. S. Evans, J. D. Webster, R. M. MacLeod, and W. Vale, *Lancet I*, 24–28 (1983).
120. C. Berquist, S. J. Nillius, and L. Wide, *Clin. Endocrinol.* **17**, 91–98 (1982).
121. T. Katzorke, D. Propping, M. Vander Ohe, and P. F. Tauber, *Fertil. and Steril.* **33**, 35–42 (1980).
122. O. Levalle, L. Tropea, N. Aparicio, A. Guitelman, A. Manieri, M. Comaru-Schally, and A. V. Schally, *Andrologia* **13**, 207–211 (1981).

123. G. Tolis, D. Ackman, A. Stellos, A. Mehta, F. Labrie, A. T. A. Fazekas, A. M. Comaru-Schally, and A. V. Schally, *Proc. Natl. Acad. Sci. U.S.A.* **79**, 1658–1662 (1982).
124. A. R. Hoffman and W. F. Crowley, *N. Engl. J. Med.* **307**, 1237–1241 (1982).
125. F. Comite, G. B. Cutler, J. Rivier, W. W. Vale, D. L. Loriaux, and W. F. Crowley, *N. Engl. J. Med.* **305**, 1546–1550 (1981).
126. R. Linde, G. C. Doelle, N. Alexander, F. Kirchner, W. W. Vale, J. Rivier, and D. Rabin, *N. Engl. J. Med.* **305**, 663–667 (1981).
127. A. Zarati, E. S. Canales, I. Sthorey, D. H. Coy, A. M. Comaru-Schally, and A. V. Schally, *Contraception* **24**, 315–320 (1981).
128. C. Rivier, J. Rivier, and W. Vale, *Science* **210**, 93–94 (1980).
129. E. Lederer, *J. Med. Chem.* **23**, 819–825 (1980).
130. M. Fernstrom, R. E. Carraway, and S. E. Leeman, *Frontiers Neuroendocrin.* **6**, 103–127 (1980).
131. N. E. Gyr, L. Kayasseh, and U. Keller, in: *Gut Hormones* (S. R. Bloom and J. M. Polak, eds.), pp. 581–585, Churchill Livingstone, New York (1981).
132. A. J. Barnes, R. G. Long, T. E. Adrian, W. Vale, M. R. Brown, J. E. Rivier, J. Hanley, M. A. Ghatei, D. L. Sarson, and S. R. Bloom, *Clin. Sci.* **61**, 653–656 (1981).
133. K. Nishioka, G. F. Bobcock, J. M. Phillips, and R. D. Noyes, *Mol. Cell Biochem.* **41**, 13–18 (1981).
134. K. Nishioka, *Brit. J. Cancer* **39**, 342–345 (1979).
135. H. Weingartner, W. Kaye, P. Gold, S. Smallby, R. Petersen, J. C. Gillin, and M. Ebert, *Life Sci.* **29**, 2721–2726 (1981).
136. P. W. Gold, H. Weingartner, J. C. Bollinger, F. K. Goodwin, and R. M. Post, *Lancet I*, 992–994 (1979).
137. I. Kobayashi, *Thromb. Res.* **16**, 775–779 (1979).
138. F. L. Stassen, R. W. Erickson, W. F. Huffman, J. Stefankiewicz, L. Sulat, and V. D. Wiebelhaus, *J. Pharmacol. Exp. Ther.* **223**, 50–54 (1982).
139. G. Baumann, A. Walser, P. A. Desaulles, F. J. A. Paesi, and L. Geller, *J. Clin. Endocrinol. Metab.* **42**, 60–63 (1976).
140. D. de Weid, A. Witter, and H. M. Greven, *Biochim. Pharmacol.* **24**, 1463–1468 (1975).
141. R. J. Vavrek and J. M. Stewart, *Peptides* **1**, 231–235 (1980).
142. H. J. Showell, R. J. Freer, S. H. Zigmond, E. Schiffmann, S. Aswanikumar, B. A. Corcoran, and E. L. Becker, *J. Exp. Med.* **143**, 1154–1169 (1976).
143. L. F. Tseng, H. M. Loh, and C. H. Li, *Life Sci.* **23**, 2053 (1978).
144. D. Roemer, H. H. Buescher, R. C. Hill, J. Pless, W. Bauer, F. Cardinaux, A. Closse, D. Hauser, and R. Huguenin, *Nature* **268**, 547–549 (1977).
145. H. B. Anderson, B. C. Jorgenson, and A. Engquist, *Acta Anaesth. Scand.* **26**, 69–71 (1982).
146. E. del Pozo, J. Martin-Perez, A. Stadelmann, J. Girard, and J. Brownell, *J. Clin. Invest.* **65**, 1531–1534 (1980).
147. W. A. Stubbs, G. Delitata, A. Jones, W. J. Jeffcoate, C. R. W. Edwards, S. J. Rotter, G. M. Besser, S. R. Bloom, and K. G. M. M. Alberti, *Lancet I*, 1225–1227 (1978).
148. C. Burkhardt, R. C. A. Frederickson, and G. W. Pasternak, *Peptides* **3**, 869–871 (1982).
149. J. F. Calimlim, W. M. Wardell, K. Sriwatanakul, L. Lasagna, and C. Cox, *Lancet I*, 1374–1375 (1982).
150. M. W. Moon, R. A. Lahti, P. F. von Voigtlander, and J. Samanen, in: *Peptides: Synthesis, Structure and Function; Proceedings of the Seventh American Peptide Symposium* (D. H. Rich and E. Gross, eds.), pp. 641–644, Pierce Chemical Co., Rockford, Ill. (1981).
151. J. J. Nestor, Jr., T. L. Ho, R. A. Simpson, B. L. Horner, G. H. Jones, G. I. McRae, and B. H. Vickary, *J. Med. Chem.* **25**, 795–801 (1982).

152. W. Vale, C. Rivier, and M. Brown, *Ann. Rev. Physiol.* **39**, 473–527 (1977).
153. J. Botella-Llusia, L. Jaramillo, and A. Charro-Salgado, *Acta Obstet. Gynecol. Scand.* **56**, 337–341 (1977).
154. D. F. Veber, F. W. Holly, R. F. Nutt, S. J. Bergstrand, S. F. Brady, R. Hirschmann, M. S. Glitzer, and R. Saperstein, *Nature* (*London*) **280**, 512–514 (1979).
155. D. F. Veber, in: *Peptides: Synthesis, Structure and Function; Proceedings of the Seventh American Peptide Symposium* (D. H. Rich and E. Gross, eds.), pp. 685–693, Pierce Chemical Co., Rockford, Ill. (1981).
156. G. Metcalf, *Brain. Res. Rev.* **4**, 389–408 (1982).
157. C. H. Li, J. Blake, and C. H. K. Cheng, *Biochem. Biophys. Res. Commun.* **102**, 697–702 (1981).
158. D. de Wied, A. Witter, and H. M. Greven, *Biochem. Pharmacol.* **24**, 1463–1468 (1975).
159. A. W. K. Gaillard and C. A. Varey, *Physiol. Behav.* **23**, 79–84 (1979).
160. A. N. Nicholson and B. M. Stone, *Neuropharmacology* **19**, 1245–1246 (1980).
161. B. B. Walker and C. A. Sandman, *Am. J. Mental Def.* **83**, 346–352 (1979).
162. J. Turk, P. Needleman, and G. R. Marshall, *J. Med. Chem.* **18**, 1139–1142 (1975).
163. J. M. van Ree, D. de Wied, W. M. A. Verhoeven, and H. M. van Praag, *Lancet I*, 1363–1364 (1980).
164. H. M. Emrich, M. Zaudig, W. Kissling, G. Dirlich, D. von Zerssen, and A. Herz, *Pharmakopsychiat. Neuropsychopharm.* **13**, 290–298 (1980).
165. G. Gacel, M. -C. Fournie-Zaluski, and B. P. Roques, *FEBS Lett.* **118**, 245–247 (1980).
166. B. A. Morgan, J. D. Bower, K. P. Guest, B. K. Handa, G. Metcalf, and C. F. C. Smith, in: *Peptides, Proceedings of the Fifth American Peptide Symposium* (M. Goodman and J. Meienhofer, eds.), pp. 111–113, Wiley, New York (1977).
167. P. W. Schiller, B. Eggimann, J. Di Maio, C. Lemieux, and J. M. -D. Nguyen, *Biochem. Biophys. Res. Commun.* **101**, 337–343 (1981).
168. M. I. Mosberg, R. Hurst, V. J. Hruby, J. J. Galligan, T. F. Burks, K. Gee, and H. I. Yamamura, *Biochem. Biophys. Res. Commun.* **106**, 506–512 (1982).
169. J. DiMaio, T. M. -D. Nguyen, C. Lemieux, and P. W. Schiller, *J. Med. Chem.* **25**, 1432–1438 (1982).
170. J. D. Bower, B. K. Handa, A. C. Lane, B. A. Morgan, M. J. Rance, C. F. C. Smith, and A. N. A. Wilson, in: *Peptides: Synthesis, Structure and Function; Proceedings of the Seventh American Peptide Symposium* (D. H. Rich and E. Gross, eds.), pp. 607–611, Pierce Chemical Co., Rockford, Ill. (1981).
171. F. A. Momany, C. Y. Bowers, G. A. Reynolds, D. Chang, A. Hong, and K. Newlander, *Endocrinology* **108**, 31–39 (1981).
172. M. R. Sairan and P. Manurath, *J. Biol. Chem.* **258**, 445–449 (1982).
173. D. H. Coy, I. Mezo, E. Pedroza, M. V. Nekola, J. Vilchey, P. Piyachaturawat, and A. V. Schally, in: *Peptides, Structure and Biological Function; Proceedings of the Sixth American Peptide Symposium* (E. Gross and J. Meienhofer, eds.), pp. 775–779, Pierce Chemical Co., Rockford, Ill. (1979).
174. K. Folkers, C. Y. Bowers, F. Momany, and K. J. Friebel, *Z. Naturforsch. B* **37**, 872–876 (1982).
175. W. W. Vale, C. Rivier, M. Perrin, M. Smith, and J. Rivier, in: *Neurosecretion and Brain Peptides* (J. B. Martin, S. Reichlin, and K. L. Bick, eds.), pp. 609–625, Raven Press, New York (1982).
176. J. Erchegyi, D. H. Coy, M. V. Nekola, E. Pedroza, E. J. Coy, I. Mezo, and A. V. Schally, *Peptides* **2**, 251–253 (1981).
177. P. Lefancier, M. Derrien, X. Jamet, J. Choay, E. Lederer, F. Audibert, M. Parant, F. Parant, and L. Chedid, *J. Med. Chem.* **25**, 85–87 (1982).

178. R. Quirion, F. Rioux, D. Regoli, and S. St. Pierre, *Eur. J. Pharmacol.* **61**, 309–312 (1980).
179. F. B. Jolicoeur, A. Barbeau, F. Rioux, R. Quirion, and S. St. Pierre, *Peptides* **2**, 171–175 (1981).
180. G. L. Stahl and R. Walter, *J. Med. Chem.* **20**, 492–495 (1977).
181. N. Cort, S. Einarsson, and S. Viring, *Am. J. Vet. Res.* **40**, 430–432 (1979).
182. Z. Veznik, A. Holub, Z. Zraly, V. Kummer, V. Holcak, K. Jost, and J. H. Cort, *Am. J. Vet. Res.* **40**, 425–429 (1979).
183. V. J. Hruby, K. K. Deb, D. M. Yamamoto, M. E. Hadley, and W. Y. Chan, *J. Med. Chem.* **22**, 7–12 (1979).
184. V. J. Hruby, H. I. Mosberg, M. E. Hadley, W. Y. Chan, and A. M. Powell, *Int. J. Peptide Protein Res.* **16**, 372–381 (1980).
185. J. Lowbridge, M. Manning, J. Seto, J. Haldar, and W. H. Sawyer, *J. Med. Chem.* **22**, 565–569 (1979).
186. W. H. Sawyer, J. Haldar, D. Gazis, J. Seto, K. Bankowski, J. Lowbridge, A. Turan, and M. Manning, *Endocrinology* **106**, 80–91 (1980).
187. J. L. Fries, W. A. Murphy, J. Suieras-Dias, and D. H. Coy, *Peptides* **3**, 811–814 (1982).
188. W. Bauer, U. Briner, W. Doepfner, R. Haller, R. Hugeinin, P. Marbach, T. J. Petcher, and J. Pless, *Life Sci.* **31** , 1130–1140 (1982).
189. R. Maurer, B. H. Gaehwiler, H. H. Buscher, R. C. Hill, and D. Roemer, *Proc. Natl. Acad. Sci. U.S.A.* **79**, 4815–4817 (1982).
190. V. M. Garsky, J. Bicksler, R. L. Fenichel, and E. L. Lien, in: *Peptides, Proceedings of the Fifth American Peptide Symposium* (M. Goodman and J. Meienhofer, eds.), pp. 547–550, Wiley, New York (1977).
191. C. A. Meyers, D. H. Coy, W. A. Murphy, J. W. Redding, A. Arimura, and A. V. Schally, *Proc. Natl. Acad. Sci. U.S.A.* **77**, 577–579 (1980).
192. F. J. Bex, A. Corben, D. Sarantakis, and E. Lien, *Nature* **284**, 342–343 (1980).
193. Y. Harano, H. Hidaka, T. Nakano, J. Emura, T. Kimura, S. Sakakibara, and Y. Shigeta, *Biomed. Res.* **1**, 560–564 (1980).
194. K. Folkers, J. Horig, G. Rampold, P. Lane, S. Rosell, and U. Bjorkroth, *Acta Chem. Scand. B* **36**, 389–395 (1982).
195. S. Caranikas, J. Mizrahi, E. Escher, and D. Regoli, *J. Med. Chem.* **25**, 1313–1316 (1982).
196. M. Fridkin, Y. Stabinsky, V. Zakuth, and Z. Spirer, *Biochim. Biophys. Acta* **496**, 203–211 (1977).
197. N. Fukuda, O. Nishimura, M. Shikato, C. Hatanaka, M. Miyamoto, Y. Saji, R. Nakayama, M. Fujino, and Y. Nagawa, *Chem. Pharm. Bull.* **28**, 1667–1672 (1980).
198. R. F. Nutt, F. W. Holly, C. Homnick, R. Hirschmann, D. F. Veber, and B. H. Arison, *J. Med. Chem.* **24**, 692–698 (1981).
199. G. Metcalf, D. W. Dettmar, A. G. Lynn, D. Brewster, and M. E. Havler, *Regul. Peptides* **2**, 277–284 (1981).
200. D. R. Brown and D. L. Uden, *Minn. Med.* **62**, 427–432 (1979).
201. F. Laczi, G. Mezei, J. Julesz, and F. A. Faszlo, *Int. J. Clin. Pharmacol. Ther. Toxicol.* **18**, 63–68 (1980).
202. C. R. Botos, C. W. Smith, Y. L. Chen, and R. Walter, *J. Med. Chem.* **22**, 926–935 (1979).
203. C. W. Smith, C. R. Botos, and R. Walter, in: *Peptides, Proceedings of the Fifth American Peptide Symposium* (M. Goodman and J. Meienhofer, eds.), pp. 161–164, Wiley, New York (1977).
204. C. W. Smith and R. Walter, *Science* **199**, 297–298 (1978).
205. D. A. Jones, Jr. and W. H. Sawyer, *J. Med. Chem.* **23**, 696–698 (1980).
206. M. Manning, B. Lammek, M. Kruszynski, J. Seto, and W. H. Sawyer, *J. Med. Chem.* **25**, 408–414 (1982).

207. M. Manning, W. A. Klis, A. Olma, J. Seto, and W. H. Sawyer, *J. Med. Chem.* **25**, 414–419 (1982).
208. D. W. Cheesman, R. Schlegel, A. M. Sugasay, and P. H. Forsham, *Endocrinology* **112**, 269–276 (1983).
209. H. P. J. Bennett and C. McMartin, *Pharmacol. Rev.* **30**, 347–392 (1978).
210. F. A. Carone, P. R. Peterson, and G. Flouret, *J. Lab. Chem. Med.* **100**, 1–14 (1982).
211. J. S. Morley, *Trends in Pharm. Sci. December 1980*, 463–468.
212. A. F. Spatola, in: *Chemistry and Biochemistry of Amino Acids, Peptides and Proteins* (B. Weinstein, ed.), Marcel Dekker, New York (1983) (in press).
213. M. M. Hahn, P. O. Sammes, P. D. Kennewell, and J. B. Taylor, *J. Chem. Soc., Chem. Commun.* **1980**, 234–235.
214. M. T. Cox, J. J. Gormley, C. F. Hayward, and N. N. Petter, *J. Chem. Soc., Chem. Commun.* **1980**, 800–802.
215. M. Szelke, B. Leckie, A. Hallett, D. M. Jones, J. Suciras, B. Atrash, and A. F. Lever, *Nature* **299**, 255 (1982).
216. G. Van Lommen, F. Al-Obeidi, E. Delock, E. Destrijker, G. van Binst, and V. J. Hruby, in: *Peptides 1980, Proceedings of the 16th European Peptide Symposium* (K. Brunfeldt, ed.), pp. 248–252, Scriptor, Copenhagen (1981).
217. J. T. Shaw, L. Miller, M. J. Turnbull, J. T. Gormley, and J. S. Morley, *Life Sci.* **31**, 1259 (1982).
218. A. F. Spatola and A. L. Betto, *J. Org. Chem.* **46**, 2393–2394 (1981).
219. S. L. Baxter and J. S. Bradshaw, *J. Org. Chem.* **46**, 831–832 (1981).
220. C. B. Pert, D. L. Bowie, B. T. W. Fong, and J. -K. Chang, *Opiates and Endogenous Opioid Peptides*, Elsevier/North-Holland Biomedical Press, Amsterdam (1976), pp. 79–86.
221. V. J. Hruby, *Life Sci.* **31**, 189–199 (1982).
222. M. J. O'Donnell, B. Le Cleff, D. B. Rusterholtz, L. Ghosez, J. -P. Antoine, and M. Navarro, *Tetrahedron Lett.* **23**, 4259–4262 (1982).
223. K. Stachowiak, M. C. Khosla, K. Placinska, P. A. Khairallah, and F. M. Bumpus, *J. Med. Chem.* **22**, 1128-1130 (1979).
224. Y. Shimohigushi, M. L. English, C. H. Stammer, and R. Costa, *Biochem. Biophys. Res. Commun.* **104**, 583–590 (1982).
225. R. Kalir and A. Patchornik, *Polym. Lett. Edit.* **16**, 35–39 (1978).
226. M. Chorev, R. Shavitz, M. Goodman, S. Minick, and R. Guillemin, *Science* **204**, 1210 (1979).
227. N. Izumiya, T. Kato, H. Aoyagi, M. Waki, and M. Kondo, Synthetic Aspects of Biologically Active Cyclic Peptides – Gramicidin S and Tyrocidines, Halsted Press/Wiley, New York (1979).
228. G. Metcalf, *Pharm. J. 1979*, 356–358.
229. D. Hudson, R. Sharpe, and M. Szelke, *Int. J. Peptide Protein Res.* **15**, 122–129 (1980).
230. M. M. Hann, P. G. Sammes, P. D. Kennewell, and J. B. Taylor, *J. Chem. Soc., Chem. Commun. 1980*, 234–235.
231. D. H. Coy and A. J. Kastin, *Pharmacol. Ther.* **10**, 657–668 (1980).
232. J. Pless, W. Bauer, F. Cardinaux, A. Closse, D. Hauser, R. Huguenin, D. Roemer, H. H. Buescher, and R. C. Hill, *Helv. Chim. Acta* **62**, 398–411 (1979).
233. R. C. A. Frederickson, E. L. Smithwick, and R. Shuman, in: *Characteristics and Functions of Opioids* (J. M. Van Ree and L. Terinius, eds.), pp. 215–216, Elsevier North-Holland Biomedical Press, Amsterdam (1978).
234. C. R. Beddell, R. B. Clark, G. W. Hardy, C. A. Lowe, F. B. Ubatuba, J. R. Vane, F. R. S. Wilkinson, S. Wilkinson, K. -J. Chang, P. Cuatrecasas, and R. J. Miller, *Proc. R. Soc. London, Ser B* **198**, 249–265 (1977).

235. J. Di Maio and P. W. Schiller, *Proc. Natl. Acad. Sci. U.S.A.* **77**, 7162–7166 (1980).
236. A. S. Dutta, J. J. Gormley, C. F. Hayward, J. J. Morley, J. S. Shaw, G. J. Stacey, and M. J. Turnbull, *Acta Pharm. Suec. Suppl.* **14**, 14–15 (1977).
237. J. F. Calimlim, W. M. Wardell, K. Sriwatanakul, L. Lasagna, and C. Cox, *Lancet* **I**, 1374–1375 (1982).
238. R. C. A. Frederickson, E. L. Smithwick, and D. P. Henry, in: *Neuropeptides and Neural Transmission* (C. A. Marson and W. Z. Traczyk, eds.), pp. 227–235, Raven Press, New York (1980).
239. G. Stacher, P. Bauer, H. Steinringer, E. Schrieber, and G. Schmierer, *Pain* **7**, 159–172 (1979).
240. M. R. Boarder, E. Erdelyi, and J. D. Barchas, *Biochem. Pharmacol.* **30**, 1289–1293 (1981).
241. K. Inouye, K. Watanabe, Y. Tochino, M. Kobayashi, and Y. Shigeta, *Biopolymers* **20**, 1845–1850 (1981).
242. K. Bauer, Degradation of Neuropeptides, in: *Brain and Pituitary Peptides, Ferring Symposium, Munich, 1979* (W. Wuttke, A. Weindl, K. H. Voigt, and R. R. Dries, eds.), pp. 213–222, Karger, Basel (1980).
243. J. W. Taylor, R. J. Miller, and E. T. Kaiser, *Mol. Pharmacol.* **22**, 657–666 (1982).
244. D. Brewster, P. W. Dettmar, and G. Metcalf, *Neuropharmacology* **20**, 497–503 (1981).
245. J. M. Stewart, M. E. Hall, J. Harkins, R. C. A. Frederickson, L. Terenius, T. Hokfeldt, and W. A. Krivoy, *Peptides***3**, 851–857 (1982).
246. A. B. Zinn, J. J. Plantner, and D. M. Carlson, in: *The Glycoconjugates, Vol. 1, Mammalian Glycoproteins and Glycolipids* (M. I. Horowitz and W. Pigman, eds.), pp. 69–85, Academic Press, New York (1977).
247. K. -J. Chang and P. Cuatrecasas, *Fed. Proc.* **40**, 2729–2734 (1981).
248. D. S. Kemp and E. T. Sun, *Tetrahedron Lett.* **23**, 3759–3760 (1982).
249. P. C. Belanger, C. Dufresne, J. Scheigetz, R. N. Young, J. P. Springer, and G. I. Dmitrienko, *Can. J. Chem.* **60**, 1019–1029 (1982).
250. J. L. Krstenansky, R. L. Baranowski, and B. L. Currie, *Biochem. Biophys. Res. Commun.* **109**, 1368–1374 (1982).
251. R. M. Freidinger, in: *Peptides: Synthesis, Structure and Function; Proceedings of the Seventh American Peptide Symposium* (D. H. Rich and E. Gross, eds.), pp. 673–683, Pierce Chemical Co., Rockford, Ill. (1981).
252. P. M. Conn, D. C. Rogers, J. M. Stewart, J. Niedel, and T. Sheffield, *Nature* **296**, 653–656 (1982).
253. R. Schwyzer and V. M. Kriwaczek, *Biopolymers* **20**, 2011–2020 (1981).
254. L. Chedid, M. Parant, F. Parant, F. Audibert, F. Lefrancier, J. Choay, and M. Sela, *Proc. Natl. Acad. Sci. U.S.A.* **76**, 6557–6561 (1979).
255. T. Costa, Y. Shimohigashi, S. A. Krumins, P. J. Munson, and D. Rodbard, *Life Sci.* **31**, 1625–1632 (1982).
256. D. B. A. Silk and A. M. Dawson, *Int. Rev. Physiol. Gastroint. Physiol. III* **19**, 153–204 (1979).
257. D. M. Matthews and S. A. Adibi, *Gastroenterology* **71**, 151–161 (1976).
258. W. Veatch, S. Weinstein, B. A. Wallace, and E. R. Blout, in: *Peptides, Structure and Biological Function; Proceedings of the Sixth American Peptide Symposium* (E. Gross and J. Meienhofer, eds.), pp. 635–638, Pierce Chemical Co., Rockford, Ill. (1979).
259. E. Bamberg, H. -J. Appell, H. Alpes, and P. Lauger, in: *Peptides, Structure and Biological Function; Proceedings of the Sixth American Peptide Symposium* (E. Gross and J. Meienhofer, eds.), pp. 629–638, Pierce Chemical Co., Rockford, Ill. (1979).
260. G. Boheim and H. A. Kolb, *J. Membr. Biol.* **38**, 99–150 (1978).
261. I. L. Karle, in: *Peptides, Structure and Biological Function; Proceedings of the Sixth American Peptide Symposium* (E. Gross and J. Meienhofer, eds.), pp. 681–690, Pierce Chemical Co., Rockford, Ill. (1979).

262. C. M. Deber, P. D. Adawadkar, M. E. M. Young, and J. Tom-Kun, in: *Peptides, Structure and Biological Function; Proceedings of the Sixth American Peptide Symposium* (E. Gross and J. Meienhofer, eds.), pp. 691–694, Pierce Chemical Co., Rockford, Ill. (1979).
263. B. A. Wallace and E. R. Blout, in: *Peptides, Structure and Biological Function; Proceedings of the Sixth American Peptide Symposium* (E. Gross and J. Meienhofer, eds.), pp. 697–698, Pierce Chemical Co., Rockford, Ill. (1979).
264. D. B. A. Silk, *Clin. Sci.* **60**, 607–615 (1981).
265. A. L. Warshaw, W. A. Walker, and K. J. Isselbacher, *Gastroenterology* **66**, 987–992 (1974).
266. M. A. Stelzter-Stevenson, F. Carone, G. Flouret, and D. Peterson, in: *Peptides: Synthesis, Structure and Function; Proceedings of the Seventh American peptide Symposium* (D. H. Rich and E. Gross, eds.), pp. 821–824, Pierce Chemical Co., Rockford, Ill. (1981).
267. H. Gainer, ed., *Peptides in Neurobiology*, Plenum Press, New York (1977).
268. A. J. Kastin, L. A. Wade, D. H. Coy, A. V. Schally, and R. D. Olson, in: *Brain and Pituitary Peptides, Ferring Symposium, Munich, 1979* (W. Wuttke, A. Weindl, K. H. Voight, and R. -R. Dries, eds.), pp. 71–78, Karger, Basel (1980).
269. R. E. Miller and N. Ogawa, *J. Comp. Physiol. Psychol.* **55**, 211–213 (1962).
270. A. Pfefferbaum, P. A. Berger, G. R. Elliott, J. R. Tinklenberg, B. S. Kopell, J. D. Barchas, and C. H. Li, *Psych. Res.* **1**, 83–88 (1979).
271. D. A. Gorelich, D. M. Catlin, and R. H. Gerner, *Mod. Problems: Pharmacopsych.* **17**, 236–245 (1981).
272. R. L. Moss and S. M. McCann, *Science* **181**, 177–179 (1973).
273. A. J. Kastin, L. H. Miller, D. Gonzalez-Barcena, W. D. Hawley, K. Dyster-Aas, A. V. Schally, M. L. Velasco-Parra, and M. Velasco, *Physiol. Behav.* **7**, 893–896 (1971).
274. H. Weingartner, P. Gold, J. L. Ballinger, S. A. Smallberg, R. Summers, D. R. Rubinow, R. M. Post, and F. K. Goodwin, *Science* **211**, 601–603 (1981).
275. W. H. Oldendorf, *Exp. Eye Res. 1977 Suppl.*, 177–190.
276. W. H. Oldendorf, *Peptides* **2**, 109–111 (1981).
277. A. J. Kastin, C. Nissen, and D. H. Coy, *Pharmacol. Biochem. Behav.* **15**, 955–959 (1981).
278. C. Crane, in: *Ion Homeostasis of the Brain*, (B. K. Siesjo and S. C. Sorensen, eds.), pp. 52–62, Munksgaard, Copenhagen (1971).
279. R. H. Gerner, B. Sharp, and D. H. Catlin, *J. Clin. Endocrinol. Metab.* **55**, 358–360 (1982).
280. D. A. Gorelich, D. H. Catlin, and R. M. Greene, *Mod. Problems: Pharmacopsych.* **17**, 236–245 (1981).
281. T. Oyama, S. Fukushi, and T. Jin, *Can. Anaesth. Soc. J.* **29**, 24–26 (1982).
282. D. H. Catlin, K. K. Hui, H. H. Loh, and C. H. Li, *Commun. Psycho. Pharmacol.* **1**, 493–500 (1977).
283. W. M. Pardridge, H. J. L. Frank, E. M. Cornford, L. D. Braun, P. D. Crane, and W. H. Oldendorf, in: *Neurosecretion and Brain Peptides* (J. B. Martin, S. Reichlin, and K. L. Bick, eds.), Raven Press, New York (1981).
284. M. van Houten, E. L. Schiffrin, J. F. E. Mann, B. I. Posner, and R. Boucher, *Brain Res.* **186**, 480–485 (1980).
285. R. Bergland, H. Blume, A. Hamilton, P. Monica, and R. Peterson, *Science* **31**, 541–543 (1980).
286. D. M. Dorsa, E. R. de Kloet, E. Mezey, and D. de Wied, *Endocrinology* **104**, 1663–1666 (1979).
287. B. A. Shoelkens, R. Lux, and R. Steinbach, *Arch. Int. Pharmacodyn.* **229**, 244–250 (1977).
288. *Drugs of the Future* **7**, 319–322 (1982).
289. K. Pentella, D. S. Bachman, and C. A. Sandman, *Neuropediatrics* **13**, 59–62 (1982).
290. J. Fox, *Chem. Eng. News, April 21, 1980*, 27–28.
291. G. Bylinsky, The Battle for America's Sweet Tooth, *Fortune* (July 26, 1982), pp. 28–32.

292. H. R. Brunner, H. Gavras, B. Waeber, G. A. Turini, and J. P. Wauters, in: *Captopril and Hypertension* (D. B. Case, E. H. Sonnenblick, and J. H. Laragh, eds.), pp. 149–170, Plenum Medical Book Co., New York, 1980.
293. A. J. Cowley, K. L. Stainer, J. M. Rowley, and J. R. Hampton, *Lancet II*, 730–732 (1982).
294. C. S. Sweet, D. M. Gross, P. T. Arbegast, S. L. Gaul, P. M. Britt, C. T. Ludden, D. Weitz, and C. A. Stone, *J. Pharmcol. Exp. Ther.* **216**, 558–566 (1981).
295. L. Chedid, F. Audibert, P. Lefrancier, J. Choay, and E. Lederer, *Proc. Natl. Acad. Sci. U.S.A.* **73**, 2472–2475 (1976).
296. G. Gregoriadis, *Drug* **29**, 261–266 (1982).
297. G. Dapergolas and G. Gregoriadis, *Lancet I*, 824–827 (1976).
298. J. F. Arieta-Molero, K. Aleck, M. K. Sinha, C. M. Brownscheidle, L. J. Shapiro, and M. A. Sperling, *Hormone Res.* **16**, 249–256 (1982).
299. S. Papaioannou, P. -C. Yang, and R. Novoteny, *Clin. Exp. Hypertension* **1**, 407–422 (1978).
300. T. J. Postmes, M. Hukkelhoven, A. E. J. M. Vanden Bogaard, S. G. Halders, and J. Loenegracht, *J. Pharm. Pharmacol.* **32**, 722–724 (1980).
301. A. J. Schroit and I. J. Fidler, *Cancer Res.* **42**, 161–167 (1982).
302. W. Lotz and B. Syllwasschy, *J. Pharm. Pharmacol.* **31**, 649–650 (1979).
303. S. Futaguchi, K. Odaguchi, A. Tanaka, and M. Hirata, *J. Pharm. Pharmacol.* **34**, 343–344 (1982).
304. A. Zaffaroni, Delivering drugs, CHEMTECH, *February 1980*, 82–88.
305. W. F. Ganong, *Review of Medical Physiology*, 10th Edition, Lange Med. Pub., Los Altos, Calif. (1981), pp. 329–330.
306. J. H. Walsh, in: *Gut Hormones* (S. R. Bloom and J. M. Pollak, eds.), pp. 163–170, Churchill Livingstone, New York (1981).
307. J. H. Laragh, in: *Captopril and Hypertension* (D. B. Case, E. H. Sonnenblick, and J. H. Laragh, eds.), pp. 73–184, Plenum Medical Book Col., New York (1980).
308. B. Kiel, in: *Proteases and Hormones* (M. K. Agarwall, ed.), pp. 1–18, Elsevier/North-Holland Biomedical Press, New York (1979).
309 L. T. Skeggs, F. E. Dorer, M. Levine, K. E. Lenty, and J. R. Kahn, *Adv. Exp. Med. Biol.* **130**, 1–23 (1978).
310. W. S. Peart, in: *Captopril and Hypertension* (D. B. Case, E. H. Sonnenblick, and J. H. Laragh, eds.), pp. 39–56, Plenum Medical Book Co., New York (1980).
311. E. Haber and J. Burton, *Fed. Proc.* **38**, 2768–2773 (1979).
312. J. Burton, R. J. Cody, Jr., J. A. Herd, and E. Haber, *Proc. Natl. Acad. Sci. U.S.A.* **77**, 5476–5479 (1980).
313. D. W. Cushman, H. S. Cheung, E. F. Sabo, B. Rubin, and M. A. Ondetti, *Fed. Proc.* **38**, 2778–2782 (1979).
314. J. Biollaz, M. Burnier, G. A. Turin, D. B. Brunner, M. Porchet, H. J. Gomez, F. H. Jones, F. Ferber, W. B. Abrams, H. Gavras, and H. R. Brunner, *Clin. Pharmacol. Ther.* **29**, 665–670 (1981).
315. T. Hata, T. Ogihara, M. Nakamura, S. Gotoh, K. Masuo, S. Saeki, A. Kumagai, and Y. Kumahara, *Eur. J. Clin. Pharmacol.* **23**, 7–10 (1982).
316. T. Kono, F. Ikeda, F. Oseko, Y. Ohmori, R. Nakano, H. Muranaka, A. Taniguchi, H. Imura, M. C. Khosla, and F. M. Bumpus, *Acta Endocrinol.* **99**, 577–584 (1982).
317. W. F. Ganong, *Review of Medical Physiology*, 10th Edition, Lange Med. Pub., Los Altos, Calif. (1981).
318. W. B. Wehrenberg, N. Ling, P. Brazeau, F. Esch, P. Bohlen, A. Baird, S. Ying, and R. Guillemin, *Biochem. Biophys. Res. Commun.* **109**, 382–387 (1982).

319. R. P. Millar, L. J. Klaff, J. Barron, N. S. Levitt, and N. Ling, *Clin. Endocrinol.* **17**, 103–107 (1982).
320. W. Vale, P. Brazeau, C. Rivier, J. Rivier, G. Grant, R. Burgus, and R. Guillemin, *Fed. Proc.* **32**, 211–214 (1973).
321. *Gut Hormones*, (S. R. Bloom and J. M. Polak, eds.), Churchill Livingstone, New York (1981).
322. N. Bethege, F. Diel, and K. H. Usadel, *J. Clin. Chem. Clin. Biochem.* **20**, 603–613 (1982).
323. H. Yajima, H. Ogawa, H. Ueda, and H. Takagi, *Chem. Pharm. Bull.* **28**, 1935–1938 (1980).
324. B. Malfroy, J. P. Swartz, A. Guyon, B. P. Roques, and J. -C. Schwartz, *Nature* **276**, 523–526 (1978).
325. B. P. Roques, M. C. Fournie-Zaluski, E. Soroca, J. M. Lecomte, B. Malfroy, C. Llorens, and J. -C. Schwartz, *Nature* **288**, 286–288 (1980).
326. B. Rubin, M. J. Antonaccio, and Z. P. Horovitz, in: *Captopril and Hypertension* (D. B. Case, E. H. Sonnenblick, and J. H. Laragh, eds.), pp. 115–135, Plenum Medical Book Co., New York (1980).
327. A. J. Barrett, in: *Enzyme Inhibitors as Drugs* (M. Sandler, ed.), pp. 219–229, MacMillan Press, London (1980).
328. L. Zaneveld, R. Robertson, and W. Williams, *FEBS Lett.* **11**, 345–347 (1970).
329. P. Libby and A. L. Goldberg, *Science N.Y.* **199**, 534–536 (1978).
330. A. Stracher, E. B. McGowan, and S. A. Shafig, *Science N.Y.* **200**, 50–51 (1978).
331. J. Kleinerman, V. Ranga, D. Rynbrandt, M. P. C. Ip, J. Sorensen, and J. C. Powers, *Am. Rev. Resp. Dis.* **121**, 381–387 (1980).
332. A. Janoff and R. Dearing, *Am. Rev. Resp. Dis.* **121**, 1024–1025 (1980).
333. S. Bajusz, E. Barabas, P. Tolnay, E. Szell, and D. Bagdy, *Int. J. Peptide Protein Res.* **12**, 217–221 (1978).
334. H. Umezawa, *J. Antibiot. 30* (*Suppl.*), 138–163 (1977).
335. R. A. Lerner, Synthetic vaccines, *Sci. Am.*, *February 1983*, 66–74.
336. R. F. Peterson, *Ann. Clin. Res.* **14**, 245–252 (1982).
337. J. Beale, *Nature* **298**, 14–15 (1982).
338. G. R. Marshall and T. M. Balasubramanian, in: *Peptides, Structure and Biological Function; Proceedings of the Sixth American Peptide Symposium* (E. Gross and J. Meienhofer, eds.), pp. 639–646, Pierce Chemical Co., Rockford, Ill. (1979).
339. G. Jung, H. Bruckner, R. Oekonomopulos, G. Boheim, E. Breitmaier, and W. K. Konig, in: *Peptides, Structure and Biological Function; Proceedings of the Sixth American Peptide Symposium* (E. Gross and J. Meienhofer, eds.), pp. 647–654, Pierce Chemical Co., Rockford, Ill. (1979).
340. N. Izumiya, T. Kato, H. Aoyagi, M. Waki, and M. Kondo, in: *Synthetic Aspects of Biologically Active Cyclic Peptides—Gramicidin S and Tyrocidines*, pp. 99–108, Halsted Press/Wiley, New York (1979).
341. B. F. Gisin, D. G. Davis, J. A. Hamilton, M. N. Sabeson, and L. K. Steinrauf, in: *Peptides, Structure and Biological Function; Proceedings of the Sixth American Peptide Symposium* (E. Gross and J. Meienhofer, eds.), pp. 711–714, Pierce Chemical Co., Rockford, Ill. (1979).
342. D. G. Davis and B. F. Gisin, in: *Peptides, Structure and Biological Function; Proceedings of the Sixth American Peptide Symposium* (E. Gross and J. Meienhofer, eds.), pp. 719–722, Pierce Chemical Co., Rockford, Ill. (1979).
343. W. Veatch, N. Sarker, P. K. Mukherjee, D. Langley, H. Paulus, V. T. Ivanov, and E. N. Shepel, in: *Peptides, Structure and Biological Function; Proceedings of the Sixth American Peptides Symposium* (E. Gross and J. Meienhofer, eds.), pp. 707–710, Pierce Chemical Co., Rockford, Ill. (1979).

344. S. G. Laland, T. -L. Zimmer, and O. Froyshou, in: *Bioactive Peptides Produced by Microorganisms* (M. Umezawa, T. Takita, T. Shiba, eds.), pp. 7–34, Halsted Press/Wiley, New York (1978).
345. E. Schiffman, H. J. Showell, B. A. Corcoran, P. A. Ward, E. Smith, and E. L. Becker, *J. Immunol.* **144**, 1831–1837 (1975).
346. Y. Yamasaki and K. Maekawa, *Agric. Biol. Chem.* (*Tokyo*) **42**, 1761–1765 (1978).
347. T. Shiba, H. Uratani, I. Kubota, and Y. Sumi, *Biopolymers* **20**, 1985–1987 (1981).
348. L. P. Taylor, R. A. Hudson, and M. S. Dorscher, in: *Peptides: Synthesis, Structure and Function; Proceedings of the Seventh American Peptide Symposium* (D. H. Rich and E. Gross, eds.), pp. 241–244, Pierce Chemical Co., Rockford, Ill. (1981).
349. I. H. Goldberg, M. Rabinowitz, and E. Reich, *Proc. Natl. Acad. Sci. U.S.A.* **48**, 2094–2101 (1962).
350. T. Takita, Y. Umezawa, S. Saito, H. Morishima, H. Umezawa, Y. Muraoka, M. Suzuki, M. Otsuka, S. Kobayashi, M. Ohno, T. Tsuchiya, T. Miyake, and S. Umezawa, in: *Peptides: Synthesis, Structure and Function; Proceedings of the Seventh American Peptide Symposium* (D. H. Rich and E. Gross, eds.), pp. 29–39, Pierce Chemical Co., Rockford, Ill. (1981).

13

Drug Delivery with Protein and Peptide Carriers

John M. Whiteley

Abstract. Drugs or toxins such as methotrexate, adriamycin, daunomycin, fluorodeoxyuridine, neocarzinostatin, and ricin can be bound, either covalently or by occlusion, to proteins, synthetic polypeptides, and antibodies. Common reactions for coupling these drugs to carriers include the use of carbodiimides, the generation of Schiff bases followed by reduction, and the formation of disulfide linkages. Often a spacer group such as a dextran or polypeptide is included to facilitate the interaction. The resultant conjugates usually contain from approximately 5 to 25 mol of drug per mole of carrier and most retain attenuated antimetabolic properties typical of the free drug. The complexes, however, also possess the properties of the carrier, and in this way delivery and uptake of a bound drug by cells can be significantly different from that of the free drug. For example, *in vivo*, the higher molecular weight of the conjugate can lead to a larger retention time for the drug prior to excretion, with the attendant greater opportunity for interaction with target cells. Additionally, the mechanism by which the drug is taken up into a cellular target may be altered for the drug-carrier complex and, in the case of an antibody carrier, tissue specificity may also be added to the properties of the drug. These and other characteristics of the complexes are discussed in detail in the text and their possible relevance to chemotherapy is outlined.

1. Introduction

Specificity is one of the most important factors governing the successful use of a pharmacologically active agent in the treatment of disease. Unfortunately, most agents do not discriminate exclusively between the desired target and other sites which should remain viable features of thc host. This is evident in both antiviral and anticancer chemotherapy, where the slow rate of successful cures can be attributed to the failure of cytotoxic molecules to interact unequivocally with the invasive species. Specificity, until recently, has usually been achieved by the discovery or design of molecules which carry out limited and selective destructive functions, as is exemplified by the infinite variety of antibiotics which exploit the distinctive metabolic differences separating the prokaryotic and eukaryotic worlds. Such an

John M. Whiteley • Division of Biochemistry, Department of Basic and Clinical Research, Scripps Clinic and Research Foundation, La Jolla, CA 92037.

approach has so far met only limited success in the cases of malignant and viral-induced lesions; therefore it has been necessary to search for more discriminating techniques to direct a drug to its target site. Unfortunately, most anticancer drugs are small molecules which gain entry into cells by carrier-mediated mechanisms that are little different for either the tumor or healthy cell. In particular, toxicity to the rapidly dividing marrow and intestinal mucosal cells is often a limiting factor in continued and successful therapy.

The problems of tumor destruction are also enhanced by the many natural cellular defense mechanisms by which the invasive cells derive a selective advantage because of their increased proliferative powers. For example, target cells may elevate the concentration of key enzymes with which a drug interacts, or they may use alternate biosynthetic routes to a required metabolite, which bypass those blocked by the drug, or they may develop transport blocks which prevent entry of the drug into the cell. Moreover, in whole animals the drug may be rapidly excreted, metabolized, or, because of the drug's intrinsic physical properties, be blocked by physiological features, such as the blood–brain barrier.

To overcome some of these problems a wide range of structural variations of a known specific antimetabolite have been synthesized and used to secure continuous successful therapy in the patient, for example, the many tetracyclines, antihistaminics, prostaglandins, alkylating agents, and nucleoside analogues. In these cases simple variations in molecular structure, such as altered alkyl, aryl, or halogen substituents, isomeric displacement of hydroxyl groups, and so on, have often led to greater drug specificity. However, in the case of antitumor agents, limitations inherent to this approach have led recently to renewed interest in a concept first proposed in 1906 by Ehrlich, which suggested that a drug may be directed to its target tissue by a carrier possessing a specific affinity for the tissue.[1]

Proteins have been considered as promising carrier molecules for a variety of reasons. For example, it has been shown that their transport into cells, which takes place by endocytosis,[2,3] occurs more readily *in vivo* in some types of tumor cells than in bone marrow cells[4] and that radioactively labeled proteins injected into tumor-bearing rats *in vivo* show greater accumulation in the tumor than normal tissue.[5–7] These observations of selective uptake, when coupled with the known enhanced lysosomal activity of many tumor cells,[8] which is necessary for breakdown of the drug–polymer complex and release of free drug, suggests that such complexes might enhance the ability of a drug to hit its target. Additionally, the restricted mobility of a drug-carrier complex because of its large molecular size can lead to a high localized drug concentration in a body cavity containing a malignant mass, where slow uptake of the complex may give prolonged release of the free drug at a rate commensurate with malignant cell kill. The carrier, by virtue of its much greater molecular size, can also effectively mask the biochemical properties of the attached low-molecular-weight drug until the agent is released at its site of action, thus minimizing undesired side effects to the drug. The carrier, of course, may initiate its own responses by the host, but these will usually be of a different type to those created towards the small-molecular-weight species. Whatever the effect, there is now sufficient evidence to suggest that the host response to the carrier-bound complex will be different from that of the free drug.

It is the aim of this review to concentrate primarily on the polypeptide, protein, and glycoprotein carriers and to describe their various methods of preparation and their physical and biological properties. Wherever possible, examples of their effects on biological systems will be described and an attempt will be made to emphasize or clarify any significant features unique to the carrier-bound drug concept.

2. Preparation of Drug-Carrier Complexes

During the past decade a wide range of cytotoxic drugs have been coupled to a variety of high-molecular-weight molecules by many investigative groups. Examples of such carriers include immunoglobulins,[9,10] particularly monoclonal antibodies,[11-13] albumins,[14,15] dextrans,[16] DNA,[17,18] erythrocyte ghosts,[19] hepatocytes,[20] nonbiodegradable synthetic systems such as nylon semipermeable microcapsules,[21] glass beads,[22] magnetic microspheres,[23] biodegradable systems such as lactic acid polymers,[24] liposomes,[25,26] and synthetic polymers such as those from divinyl ether maleic anhydride (DIVEMA).[27]

The ideal carrier should be readily available, have multiple sites for drug coupling, be sufficiently stable to reach the target site yet be readily excretable, be nontoxic, and preferably have properties which might aid the direction of the drug to its target site. Polypeptides, proteins, and glycoproteins meet many of these carrier requirements, although the possibilities of immune response by the host exist. The synthetic polymers such as DIVEMA, or homopolymers such as polylysine, may minimize the immune response, but in the former case problems in metabolism and excretion may occur. Of special interest is the possibility that tumor-specific antibodies may be isolated which can act as carriers and target bound drugs directly to the tumor site. Early attempts to realize this potential produced evidence to support enhanced delivery of the various agents by carriers, but the magnitude of enhanced selectivity was not compelling,[16] Most approaches used covalent coupling reactions or bifunctional cross-linking reagents to attach the drug to the carrier. Unfortunately, such reactions often inactivated the antibody combining site or disturbed the integrity of the cytotoxic agent. A resurgence of interest in the antibody carrier concept has occurred recently, stimulated both by biochemical and structural studies of plant and bacterial toxins[28,29] and by the rapid development of techniques for generating monoclonal antibodies.[30] Structural similarities between the identical disulfide-linked halves of antibody $F(ab')_2$ molecules and the disulfide-connected toxic A and lectin B chains of, for example, ricin, has afforded the possibility of a mild coupling procedure[31] via the generation of a disulfide bond between one-half of a specific antibody molecule and one of the toxic chains.

Table 1 contains a selection of the more commonly used procedures for coupling drugs to proteins, polypeptides, and glycoproteins. Early experiments from this laboratory showed that methotrexate (MTX) could be readily bound via its distal carboxyl functions in a carbodiimide-promoted reaction to aminoalkyl polysaccharides.[42] This reaction was initially extended to bovine serum albumin (BSA) ($M_r \sim 150{,}000$) and poly-L-lysine ($M_r \sim 35{,}000$). In every case the water-soluble carbodiimide 1-ethyl-3(3′-dimethylaminopropyl)carbodiimide (EDC) hydrochloride was used as the condensing agent, and usually from 5 to 25 mol of MTX were bound covalently per mole of carrier probably via a γ-carboxamide linkage. Alternate methods reported for linking MTX to carrier include diazotization[35] and the use of activated esters generated by *N*-hydroxysuccinimide.[34] Unfortunately, the amino groups of MTX do not readily diazotize[43] and therefore reported couplings by this procedure should be viewed with caution. The activated esters are attractive as they permit substitution into antibody molecules with little disruption of the carrier structure, a situation which contrasts with the use of carbodiimides where cross-linking and aggregation of the carrier can occur. A common coupling agent is glutaraldehyde, which can cross-link amino groups, contained in a drug, to free amino groups of a carrier.[40] Unfortunately again, protein–protein cross-linking can occur with resultant aggregation.[41]

In the case of drugs which contain a carbohydrate moiety such as daunomycin, periodate oxidation may be employed to generate free carbonyl groups that can be coupled to protein carriers via Schiff-base formation and stabilized by borohydride reduction.[40] Sometimes coupling is better achieved by a bridging group, such as a dextran, or by a peptide fragment of polylysine or polyglutamate. In the case of the dextran, periodate converts it to a polyaldehydic form, which is then reactive with both drug and carrier; with the peptide spacer, carbodiimide provides the linking stimulus. A bridging group can offer two advantages; (1) it can provide more reactive centers for drug coupling so that a greater number of drug molecules can be

Table 1. Methods for the Preparation of Drug-Carrier Complexes

Drug	Spacer	Carrier	Reaction	References
Methotrexate	—	BSA[a]	Carbodiimide	14
Methotrexate	—	Poly-L-lysine	Carbodiimide	32,33
Methotrexate	—	Ig[b]	NHS[a]	34
Methotrexate	—	Ig	Diazotization	35
DDS[a]	—	Poly-L-lysine	Occlusion	36
Chlorambucil	—	Ig	Occlusion	37,38
NCS[a]	—	Ig hybrid	Occlusion	39
Daunomycin	—	Ig	Glutaraldehyde	40
Daunomycin	Dextran	Ig	Periodate	41
Ricin, diphtheria toxin	—	Ig	DTNB or SPDP[a]	11,31
PDM[a]	Polyglutamic acid	Ig	Carbodiimide	10

[a] Abbreviations: BSA, bovine serum albumin; NHS, *N*-hydroxysuccinimide; DDS, 4,4′-diaminodiphenylsulfone; NCS, neocarzinostatin; DTNB, 5,5′-dithio-bis(2-nitrobenzoic acid); SPDP, 3-(2-pyridyldithio)propionic acid *N*-hydroxy-succinimide ester; PDM, *N,N*-bis(2-chloroethyl)-*p*-phenylenediamine.

[b] Ig is used to represent antibody from a mono- or polyclonal source.

bound to a single carrier molecule; and (2) it may allow the bound drug to be more readily recognized once the carrier has reached its target site.

A procedure particularly suitable for coupling antibodies to plant lectins or bacterial toxins is based on their similarities in structure, namely, that they are composite molecules comprised of more than one polypeptide chain held together by disulfide bonds. Thus the antibody $F(ab')_2$ unit can be split and the newly available thiol group, preferably blocked by DTNB or SPDP (Table 1), to prevent self-annealing, can then be reacted with the free thiol groups contained in the toxic fragments of molecules such as ricin or diphtheria toxin.[11,13] Variations of this coupling procedure have recently been reported from many laboratories.[13,44,45]

Other methods of attachment may include the use of chlortriazines[46] and even noncovalent interactions.[47] One particularly interesting example of this latter situation is afforded by DDS (Table 1) occluded with polylysine.[36] This antileprosy drug is unable to penetrate intact *Mycobacterium leprae in vitro*, as determined by its effect on *o*-diphenoloxidase in the bacilli, but when combined with polylysine, the sulfone drug readily passes through the cell membranes giving 100% inhibition of enzyme activity. Another interesting variant of the occlusion procedure is the concept of developing hybrid antibodies in which one unit of the antibody hybrid contains a specific site for drug binding, while the second portion retains specificity for a particular cell surface determinant. A typical example is offered by the antineocarzinostatin (NCS)/antihuman IgG hybrids. NCS is a polypeptide antibiotic ($M_r \sim 10{,}700$), which inhibits cellular DNA synthesis and replication.[48,49] A rabbit antiserum against NCS,[50] and antibodies to human IgG $F(ab')_2$ fragments, which react strongly with immunoglobulin molecules expressed by human lymphoid cell lines,[51] were prepared. Both purified antibody preparations were digested with pepsin and the isolated $F(ab')_2$ fragments were mixed in equimolar portions. Mild reduction with thiol followed by reoxidation produced the required hybrid antibody. The product showed the expected dual functions of cell specificity and NCS binding.

Although interest in the use of carriers has tended to concentrate on the development of antitumor agents, other areas have also been explored. For example, Verlander *et al.* have coupled catecholamines to a series of synthetic polymers[52] and have established that the agents retain their biological activity. Shen and Ryser have demonstrated that poly-L-lysine can be used to enhance cellular uptake of proteins such as albumin and horseradish peroxidase.[53] In addition, Balboni *et al.*[54] have used albumin conjugates of 5-fluorodeoxyuridine and cytosine arabinoside to enhance their antiviral activities. In this instance, the conjugates are rapidly absorbed by macrophages, and pox viruses, whose infectious cycle begins in these cells, are then particularly susceptible to the drugs.

3. Chemical and Biological Properties

Very few drug-carrier systems have been rigorously examined with regard to their physical and chemical properties. In many instances the compounds have not been isolated to homogeneity, and therefore it has been difficult to assign specific properties to the complex. For example, when MTX was coupled to BSA using carbodiimide it was noted that the covalent linkages between albumin chains increased when the concentration of the carbodiimide greatly exceeded that of the MTX; even with equivalent concentrations, some aggregation occurred. However, chromatography on Bio-Gel P-100 separated the aggregated species, which could be identified by polyacrylamide electrophoresis.[14] The desired monomeric product had an electrophoretic profile similar to that of the starting material. The uptake of MTX could be measured spectrophotometrically, λ_{max} (0.1 N NaOH) = 370 nm, ε = 7100,[55] or by using the radioactively labeled drug. Surprisingly, despite the high content of lysine in BSA [59 residues[56]], the ratio of MTX bound to albumin was never greater than ~12. The albumin-bound derivative was still an inhibitor of dihydrofolate reductase, the target enzyme for MTX, although its inhibitory powers were an order of magnitude less than those exhibited by the free drug (Table 2). Similar observations were made when MTX was bound to other proteins and polypeptides. It should be noted, however, that MTX is somewhat unique in that the principal source of its interaction with the reductase is via the 2,4-diaminopteridine portion of the molecule, a region which probably remains accessible after coupling since the reaction is accomplished through the minimally interacting distal carboxyl entity.

Table 2. Inhibition of L1210 R6 Dihydrofolate Reductase with MTX and MTX–BSA

Inhibitor	I_{50}[a] (μM)
MTX	1.25
MTX–BSA	7.5

[a]The concentration of drug necessary to cause 50% inhibition of activity of dihydrofolate reductase (2 μM, specific activity 185 μmol min^{-1} mg^{-1} protein), I_{50}, was measured using the enzyme isolated from the MTX-resistant L1210 R6 strain of ascites tumor cells.[57] The assays were performed in a system that contained, in 1.0 ml, 56 μM dihydrofolate, 150 μM NADPH, 0.1 M K-phosphate, 0.1 M K-HEPES, and 0.5 M KCl, pH 7. Initial velocities (at 37°C) were determined via absorbance changes at 340 nm after a 10-min, 25°C preincubation period of the enzyme with MTX or the MTX derivative.[58]

The pharmacological activity of daunomycin and adriamycin both free and carrier bound can be assessed by inhibition of cellular RNA synthesis, as measured by [^{3}H] uridine incorporation into test cells.[40] Results indicate that a substantial amount of drug activity is preserved when covalently bound to immunoglobulin. Again the activity of the conjugates is lower than that of the free drug at low concentrations, but at high concentrations both produce the same maximal inhibition, particularly if allowed to interact with cells for a longer time. The kinetics of the free and the protein-bound drug also indicate that the free form is more active in the short term, but that after longer interaction the conjugate exerts almost the same effect.

The properties of the drug-carrier complex are strongly influenced by the coupling method used to produce the linkage. This is apparent even in the case of two very similar drugs such as daunomycin and adriamycin, which when coupled to the carrier via the dextran-bridge procedure, show very different properties—daunomycin is active, whereas adriamycin is either inactive or shows very low activity. However, when adriamycin is linked in an alternate way, through its C13 keto group to a hydrazine derivative of dextran, which in turn is coupled to antibody via glutaraldehyde, the derivative now shows high activity.[59] Further evidence, which indicates that the method chosen for coupling drug to carrier is important, is contained in a report from Kulkarni and co-workers[34] where different methods of binding MTX to antibody were concerned. These included carbodiimide, mixed anhydride, and active esters with NHS (Table 1). The results showed that the active esters gave the product with best retention of drug activity and antibody specificity.

In experiments carried out by Raso and Griffin, the mild coupling procedure of disulfide bond formation used to combine the toxic ricin A chain to Fab′ fragments led to a Fab′—A conjugate which could discriminate between Ig-bearing and Ig-nonbearing lymphoid cells.[31] The product was cytotoxic or cytostatic for all the Ig positive lines tested and was inert for Ig negative lines. Furthermore, the species specificity of the antibody was also preserved. Thus the ricin A chain bound to an antibody fragment specific for human antigenic determinants was nontoxic in mice.

4. Mechanisms of Action and Pharmacological Properties

Early studies from this laboratory showed that MTX could be readily bound via its carboxyl groups to aminoalkyl polysaccharides in a carbodiimide-promoted reaction.[42] This method was extended to the use of albumin as the polymeric carrier and the products were tested in a model animal system,[14] where it was shown that female BDF_1 mice treated with

MTX covalently bound to bovine or murine serum albumin (BSA, MSA), exhibited a higher, more prolonged serum concentration and a decreased rate of excretion of MTX compared with a similar group of mice treated with free MTX. Additionally, tritiated MTX-albumin derivatives circulated to the tissues as the covalent complexes but were cleaved prior to excretion and exited primarily as unbound MTX. Moreover, when the tritiated derivatives were injected intraperitoneally into L1210 tumor-bearing mice there was prolonged localization of drug in the ascitic fluid and elevated intracellular MTX levels even after 24 h. In contrast to the rapid excretion normally observed with free MTX, approximately 90% of the tritium label found in the L1210 cells was located in the cell lysate as free, unmetabolized MTX, whereas when the albumin carrier was labeled with ^{125}I, 80% of the radioactivity was found associated with the cell membrane. A single dose of MTX–BSA (equivalent to 15 mg of MTX per kilogram) injected into BDF_1 mice 24 h after inoculation of 10^5 L1210 cells was as effective as MTX in prolonging survival time from 8 (control) to 15 days. The observations suggested at this time that the high-molecular-weight MTX-albumin derivatives were retained in the serum and extracellular compartments until hydrolysis occurred and thus markedly increased the lifetime of MTX within the animal.

The nature of these results was sufficiently promising to suggest that an alternate type of malignancy should be examined, and a solid tumor, the Lewis lung carcinoma, was chosen.[60] Here, it was found that ip (intraperitoneal) injected MTX–BSA was more effective than an equivalent dose of free MTX in reducing the number of metastases observed in female BDF_1 mice bearing the sc (subcutaneous) implanted tumor. Additionally, treatment with the MTX derivative caused a decreased rate of growth of the primary tumor and a modest increase in the life span of the tumor-bearing animal. When tumor-bearing mice were killed after receiving injections of [^{3}H]MTX or [^{3}H]MTX–BSA, no difference in the amount of drug was found at the tumor site after 1 h; however, after 8 or 24 h twice as much radioactivity was found in the tumors of mice treated with the carrier-bound drug. Clearly, the carrier-bound drug was showing altered biological properties to the free drug.

It was therefore decided that a closer examination should be made of the interaction occurring between the carrier-bound MTX and L1210 cells.[61] For these experiments MTX was covalently linked to two further proteins, α-chymotrypsinogen and bovine IgG. Each complex was as effective as free MTX and MTX–BSA in prolonging the life span of BDF_1 mice bearing the L1210 tumor, suggesting that MTX can act as an effective chemotherapeutic agent when covalently bound to a variety of proteins. In *in vitro* experiments MTX was more toxic than MTX–BSA to L1210 cells at 37°C, an observation which correlated with the relative internal drug concentrations recorded after incubation for 1 h with free or derivatized MTX. Further *in vitro* experiments employing [^{3}H]MTX–BSA as the prototype of a drug–protein carrier indicated that MTX–BSA was only partially dependent on the same transport mechanism for cellular uptake as MTX. For example, MTX transport into cells was almost zero at 0°C, whereas MTX–BSA entry was only inhibited 50%. Additionally, MTX–BSA uptake was much less inhibited by *p*-chloromercuribenzenesulfonate and 5-formyltetrahydrofolate than the parent compound.[62,63] Experiments using [^{3}H]MTX–BSA or MTX–^{125}I–BSA indicated that the MTX–BSA complex was not degraded on the external surface of the cell membrane, as 60–65% of both radioactive labels was found in the cell lysate fraction, with much of the iodinated material degraded to low-molecular-weight fragments. In contrast, the ^{125}I marker which remained bound to the cell membrane was still present as a nondialyzable high-molecular-weight molecule.

Because of the therapeutic significance implied in the observations that carrier structure could influence drug uptake and that the mechanism of uptake might be different for carrier-bound and free drug, it was decided to examine the interaction of MTX–BSA with a nonsystemic cellular source such as the MTX transport-resistant Reuber H35 hepatoma cells, which had been developed in the laboratory of Dr. J. Galivan, N.Y. State Health Laboratories, Albany, New York.[64] However, MTX–BSA was ineffective in suppressing growth of these cells ($I_{50(MTX)} \sim 3.5\,\mu M$). Therefore, because carrier structure can determine cellular interaction, and because it was known that poly-L-lysine possessed exceptional cellular-membrane-penetrating properties,[53] MTX was bound to this alternate carrier and the product was tested with the hepatoma cell line.[32] Both the parent H35 line and transport-resistant sublines showed a similar response to treatment with MTX poly(L-lysine) ($I_{50} \sim 80$ nM). Depressed cell growth after drug treatment of the H35 cells and the resistant sublines could be reversed by treatment with thymidine/hypoxanthine. Additionally, folinic acid was effective in preventing MTX poly(L-lysine) toxicity in H35 cells but could not do so for the MTX transport-deficient sublines, presumably because of its inability to enter the cells. These results suggested that despite an alternate uptake route MTX poly(L-lysine) was toxic to both cell lines via blockage of the one-carbon metabolic pathway. A similar conclusion was reached independently by Shen and Ryser when observing the uptake of MTX poly(L-lysine) into L929 mouse fibroblasts.[65]

A more definitive examination of the mechanism of cellular uptake and cleavage of MTX poly(L-lysine) was clearly of interest to confirm these earlier results, which suggested a mechanism of cellular entry requiring endocytosis followed by lysosomal degradation. Support for this hypothesis was obtained

Table 3. Effect of Folinic Acid, Leupeptin, and NH_4Cl on MTX and MTX Poly(L-Lysine) Disposition in H35 and $H35R_{0.3}$[a] Cells[b]

		Cell MTX (dpm/mg cell protein extract)[c]	
Cell type	Inhibitors	MTX poly(L-lysine) (1 μM)	MTX (0.5 μM)
H35	None	195	15,120
	Folinic acid (5 μM)	187	3,103
	Leupeptin (100 μM)	41	17,170
	NH_4Cl (10 mM)	133	14,516
	NH_4Cl (25 mM)	59	11,794
$H35R_{0.3}$	None	258	1,538
	Folinic acid (5 μM)	288	1,460
	Leupeptin (100 μM)	104	1,278
	NH_4Cl (10 mM)	210	1,972
	NH_4Cl (25 mM)	130	1,952

[a] A hepatoma cell line which is resistant to $0.3 \times 10^{-6} M$ MTX.

[b] Ten 100-mm plates at confluence were incubated with 0.5 μM [^{3}H]MTX (5.75×10^5 dpm/nmol) or 1 μM [^{3}H]MTX poly(L-lysine) (2.4×10^5 dpm/nmol) and the indicated additions for 20 h. Cell extracts were made and the cellular contents resolved by a previously described procedure.[66]

[c] Cell MTX was calculated from the radioactivity eluted from fractions 53–78 (Figure 1), which includes MTX and polyglutamates. Where resistant cells were treated with 0.5 μM MTX, both reductase-bound and free species are included because as much as 80% of intracellular MTX is reductase bound under these conditions. All results are normalized as dpm/mg cell protein extract.

(Table 3) when it was shown that leupeptin, an inhibitor of lysosomal cathepsins in hepatic cells, inhibited the liberation of MTX from MTX poly(L-lysine). Studies with folinic acid also illustrated that MTX from the poly(L-lysine) conjugate did not enter the cell after hydrolysis in the medium, as the uptake from this source was unaffected by this competitive transport inhibitor, whereas the uptake of free MTX by H35 cells was severely depressed. Additional support for lysosomal cleavage was offered by the influence on uptake of the lysosomotropic inhibitor NH_4Cl. Incorporation of NH_4Cl into the cultures significantly reduced free cellular MTX in those cells treated with the carrier-bound agent when compared to those exposed to MTX alone. It was also noted that there was increased uptake of the conjugate by cells as a function of their age in culture and that the addition of dexamethasone and tocopherol had no effect. The opposite was true for free MTX.[67] The results of Table 3 could be explained by inhibition of uptake of MTX poly(L-lysine), however, leupeptin did not alter the uptake into H35 cells of either free MTX or the poly(L-lysine) conjugate. As expected, folinic acid merely interfered with the uptake of free MTX.

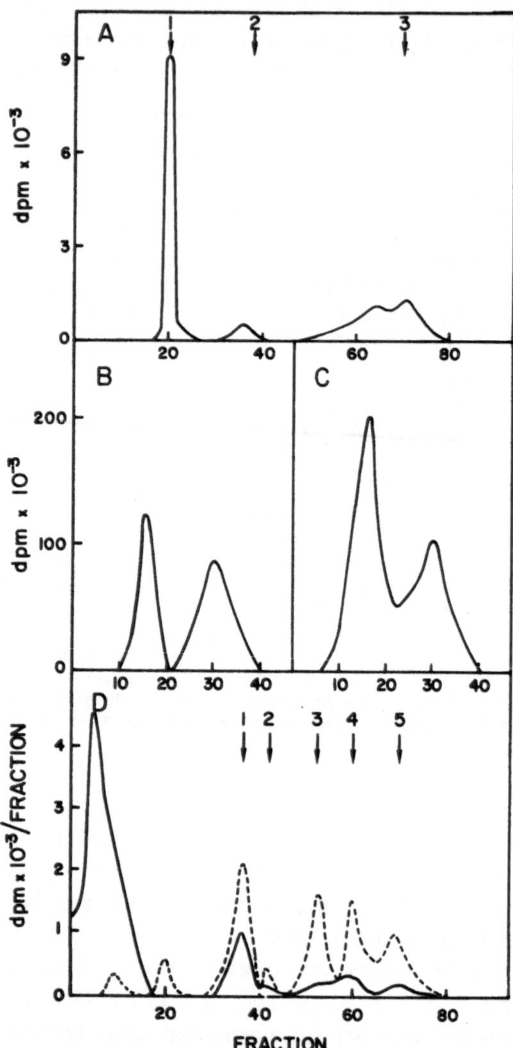

Figure 1. (A). Sephadex G-75 chromatography of extracts of H35 cells exposed to [^{3}H]MTX poly(L-lysine) (0.5 μM). Thirty 60-mm plates were exposed to the reagent after 72 h in culture. Cell extracts were then made and chromatographed as outlined in Reference 66. The numbers above each elution profile designate: (1) Blue Dextran void volume; (2) H35 cell dihydrofolate reductase; and (3) MTX. (B). The contents of peak 2 (fractions 33–41) were placed in a boiling-water bath for 10 min and an aliquot was chromatographed on Sephadex G-25. MTX polyglutamates eluted between fractions 6 and 20 and MTX between 22 and 40. (C). An aliquot of pooled fractions 53–78 chromatographed as in (B). (D). DEAE-cellulose chromatography of the dihydrofolate reductase-bound (-----) and free (——) MTX species according to the procedures outlined in Reference 68. The arrows indicate MTX standards; the numbers 1–5 correspond to the number of glutamate residues in the molecule (1, MTX; 2, MTX–Glu, etc.).

The fate of tritium labeled MTX poly(L-lysine) after *nonresistant* H35 cell entry was examined by Sephadex G-75 gel filtration of the cell extracts (Figure 1). The derivative exhibited extensive hydrolysis. Radioactivity occurred in three regions of the eluate: a high-molecular-weight component (unhydrolyzed material) associated with a Blue Dextran 2000 void volume marker; a second component with retention comparable to that of dihydrofolate reductase; and a third in the elution region of MTX, MTX polyglutamates, or MTX bound to lysine oligopeptides. The various species were identified by further Sephadex G-25 gel filtration and DEAE-cellulose chromatography. By using known standards it was demonstrated that MTX hydrolyzed from the conjugate was free to either interact directly with dihydrofolate reductase or be glutamylated with up to four glutamate residues in a manner similar to that already established for MTX alone.[69] Identical behavior was observed when the *resistant* cell line was treated with MTX poly(L-lysine). In contrast when cells were treated with MTX poly(D-lysine) uptake was again observed but cell toxicity was no greater than that observed with poly(D-lysine) alone, nor were any metabolic products observed in the cell extracts.

In contrast to the detailed investigation caried out with MTX-carrier complexes, the mode of action of daunomycin conjugates has not yet received similar analysis. However, in one study, the drug was attached to an antibody carrier by the nonhydrolyzable covalent linkage resulting from the reduced Schiff-base intermediate, and experiments were then designed to measure the capacity of tumor cells to take up the radioactively labeled conjugate into which the label had been introduced by using [^{3}H] borohydride in the reduction. Thus the label was present exclusively at the conjugation point of the drug and macromolecule. The *in vitro* uptake of the conjugate into either YAC Moloney virus-induced lymphoma cells or normal rat lymphocytes was measured.[70] The results showed clearly that both free daunomycin and its conjugate were taken up by the YAC cells; however, the uptake of the free drug was more rapid than that of the conjugate. Nevertheless, accumulation of the drug, free or conjugated, in the nuclei paralleled its uptake by intact cells, and corresponded roughly to two-thirds of that which entered. Daunomycin bound directly to antibody fragments entered the cells at higher levels than when bound through dextran or when bound to normal Ig. In all cases the label was present in the nuclear pellet, but it was unknown whether penetration of the nucleus had occurred.

It is conceivable that the conjugate with daunomycin linked directly to antibody could enter the nucleus as such, or alternatively, might be digested by lysosomic enzymes prior to its penetration of the nucleus. The labeling method used would not differentiate between these two alternatives; however, when daunomycin was linked to antibody through dextran and labeled at the reduction step, most of the radioactive label was in the dextran bridges. The finding of the label in or on the nucleus, even though there are no dextranases in lysosomes, gives weight to the possibility that a large conjugate is capable of entering the cell and approaching the nucleus. These results do not, however, prove a particular mechanism for the cytotoxic activity of the large-molecular-weight covalent complexes of daunomycin. It is also possible that the free drug, released from the complexes *in situ*, is the compound that possesses the cytotoxic activity. However, the results demonstrate that complexes are able to reach the nucleus, provided that they show some affinity for the cell surface.

When hybrid antibodies containing dual specificity for both cell surface antigen and toxin have been generated, the mechanism of cellular uptake is of interest, particularly because of current speculation regarding the mode of entry chosen by native ricin.[71] One possibility is that the ricin A chain might be held in close proximity to the cell surface as a result of the hybrid antibody being bound to its cellular receptor site and that when released it might penetrate directly through the membrane. Alternatively, the hybrid-antibody ricin A chain complex could be internalized as a unit, with dissociation occurring within the cell, perhaps following degradation of antibody. Since previous experiments have shown that anti-A chain antibodies can block A chain activity *in vitro*,[72] intracellular destruction of hybrid may be essential for the observed cytotoxic effects. Hybrid-bound whole ricin has additional entry possibilities since now the galactose-specific B component is also present. Experiments conducted in the presence of lactose, partially blocked hybrid-mediated toxicity,[73] suggesting that some of the toxic molecules were released from the antibody prior to entry via the natural B-chain-mediated process. However, toxicity remaining refractory to the lactose block indicated that a portion still entered complexed to the hybrid. Hybrid-mediated delivery of the free ricin A chain to cells is of course more attractive since it circumvents the undesirable toxicity incurred when using whole ricin.

Similar mechanistic possibilities also exist for the cellular uptake of antigen-specific poly- or monoclonal antibodies acting as the carrier and in those instances where a single F(ab′) fragment is bound directly to the toxic component.

The generation of the disulfide linkage between antibody and toxin is attractive because of the mild requirements for bond formation, which lead to little loss of antibody specificity. However, the disulfide bond is fragile and susceptible to dissociation; therefore, breakup of the conjugate may often occur before the target site is reached. Clearly, alternate linkages between toxin and antibody, which still allow retention of antibody specificity, yet offer a more stable interaction between the two components, could be of great utility.

5. Conclusion

Naturally occurring proteins, glycoproteins, or synthetic polypeptides can be readily combined with a wide selection of drugs to generate derivatives, whose properties are often sufficiently different from the free drug that they may initiate novel host cell responses. Experiments conducted *in vitro* indicate that the conjugates can retain the specific antimetabolic actions usually associated with the free drug, although the activity is usually somewhat attenuated. Generally, the carrier-bound derivatives are endocytosed, then undergo intracellular hydrolysis to liberate the free drug, which is then able to carry out its antimetabolic function. In the case of MTX poly(L-lysine), the mechanism of cell kill has been shown to be identical to that of free MTX. Furthermore, in this instance, endocytic uptake of the carrier-bound complex enables the drug to bypass transport resistance which can develop in cells exposed to high concentrations of free MTX. To date, there are little data on this point regarding MTX therapy clinically. However, an interesting series of experiments in a model system has been conducted by Sirotnak and co-workers.[74] Leukemic mice were treated with MTX and a total of 14 different resistant L1210 sublines were isolated. Of these, nine were transport resistant and the remainder were resistant via an elevation in dihydrofolate reductase—the target enzyme for MTX. Clearly, in this system, transport-dependent resistance is a frequent mechanism by which transformed cells become insensitive to MTX treatment. If this model is indicative of a general trend in other mammalian species, there may be a definite role for carrier-bound drugs in cancer chemotherapy.

The carrier-bound derivatives may also alter the pharmacokinetic distribution of the drug within a whole animal. If this leads to an increased concentration of drug in restricted regions of the host containing a malignant lesion, the opportunities for drug interaction will be increased. However, the overall retention of high levels of an antimetabolite will also be deleterious to tissues with high turnover, such as the mucosal membranes. Therefore, a balance between these two factors must be determined in order to design a useful therapeutic protocol.

The use of antibodies having binding affinities for selected cellular determinants appears to surmount some of the problems associated with nontargeted carriers and is a very attractive concept for drug delivery. However, the sensitivity of antibody structures to chemical manipulation with regard to the retention of specificity minimizes the number of chemical coupling reactions which can be applied. Furthermore, the generation of truly tumor-selective antibodies can be a laborious and unrewarding pastime and without such selectivity, the transport of toxic agents, such as ricin or diphtheria toxin, can be a very hazardous mode of drug therapy.

It should be noted that so far the most successful experiments have been conducted with selected *in vitro* cellular systems. The usual caution must therefore be exercised in extrapolating experimental results to situations employing whole animals. When drug-carrier derivatives have been applied to treatment of tumors within a host, results have been equivocal suggesting that other mechanisms, yet unresolved, may provide obstacles to this form of therapy. Although *in vitro* systems are invaluable for the preliminary analysis of the efficacy of a carrier-bound drug, the achievement of selectivity and total target-cell destruction in the whole animal still remains the elusive goal.

ACKNOWLEDGMENTS

The work described from the author's laboratory was supported by U.S. PHS Grant No. CA11778. The author is indebted to Dr. Stephanie Webber for her critical review of this manuscript.

References

1. P. Ehrlich, *Collected Studies on Immunity*, Vol. II, Wiley, New York (1906), pp. 442–447.
2. S. C. Silverstein, R. M. Steinman, and Z. A. Cohn, Endocytosis, *Ann. Rev. Biochem.* **46**, 669–722 (1977).
3. H. J.-P. Ryser, Uptake of protein by mammalian cells: An underdeveloped area, *Science* **159**, 390–396 (1968).
4. G. C. Easty, The uptake of fluorescent labelled proteins by normal and tumour cells, *Brit. J. Cancer* **18**, 368–377 (1964).
5. J . L. Mego and J. D. McQueen, The uptake of labeled proteins by particulate fractions of tumor and normal tissues after injection into mice, *Cancer Res.* **25**, 865–869 (1965).
6. T. Ghose, R. C. Nairn, and J. E. Fothergill, Uptake of proteins by malignant cells, *Nature* **196**, 1108–1109 (1962).
7. R. G. Long, J. G. McAfee, and J. Winkelman, Evaluation of radioactive compounds for the external detection of cerebral tumors, *Cancer Res.* **23**, 98–108 (1963).
8. C. deDuve, in: *Biological Approaches to Cancer Chemotherapy* (R. J. C. Harris, ed.), p. 101, Academic Press, London (1961).
9. T. Ghose and A. H. Blair, Antibody-linked cytotoxic agents in the treatment of cancer: Current status and future prospects, *J. Natl. Cancer Inst.* **61**, 657–676 (1978).
10. G. F. Rowland, G. J. O'Neill, and D. A. L. Davies, Suppression of tumour growth in mice by a drug-antibody conjugate using a novel approach to linkage, *Nature* **255**, 487–488 (1975).
11. D. G. Gilliland, Z. Steplewski, R. J. Collier, K. F. Mitchell, T. H. Chang, and H. Koprowski, Antibody-directed cytotoxic agents: Use of monoclonal antibody to direct the action of toxin A chains to colorectal carcinoma cells, *Proc. Natl. Acad. Sci. U.S.A.*, **77**, 4539–4543 (1980).
12. A. Huang, L. Huang, and S. J. Kennel, Monoclonal antibody covalently coupled with fatty acid, *J. Biol. Chem.* **255**, 8015–8018 (1980).
13. P. E. Thorpe, D. W. Mason, A. N. F. Brown, S. J. Simmonds, W. C. J. Ross, A. J. Cumber, and J. A. Forrester, Selective killing of malignant cells in a leukaemic rat bone marrow using an antibody-ricin conjugate, *Nature* **297**, 594–596 (1982).

14. B. C. F. Chu and J. M. Whiteley, High molecular weight derivatives of methotrexate as chemotherapeutic agents, *Mol. Pharmacol.* **13**, 80–88 (1977).
15. G. Barbanti-Brodano and L. Fiume, *In vitro* effect of a 5-fluorodeoxyuridine albumin conjugate on tumour cells and on peritoneal macrophages, *Experientia* **30**, 1180–1182 (1974).
16. N. G. L. Harding, Amethopterin linked covalently to water-soluble macromolecules, *Ann. N.Y. Acad. Sci.* **186**, 270–283 (1971).
17. A. Trouet, D. D. Campeneere, and C. de Duve, Chemotherapy through lysosomes with a DNA-daunorubicin complex, *Nature* **239**, 110–112 (1972).
18. G. Atassi, M. Duarte-Karim, and H. J. Tagnan, Comparison of adriamycin with DNA-adriamycin complex in chemotherapy of experimental tumors and metastases, *Eur. J. Cancer* **11**, 309–316 (1975).
19. R. Green, J. Miller, and W. Crosby, Enhancement of iron chelation by desferrioxamine entrapped in red blood cell ghosts, *Blood* **57**, 866–872 (1981).
20. A. J. Matas, D. E. R. Sutherland, M. W. Steffes, S. M. Mauer, A. Lowe, R. L. Simmons, and J. S. Najarian, Hepatocellular transplantation for metabolic deficiencies: Decrease of plasma bilirubin in gunn rats, *Science* **192**, 892–894 (1976).
21. T. M. S. Chang, *Artificial Cells*, Charles C Thomas, Springfield, Ill. (1972).
22. J. C. Venter, B. R. Venter, J. E. Dixon, and N. O. Kaplan, A possible role for glass bead immobilized enzymes as therapeutic agents (immobilized uricase as enzyme therapy for hyperuricemia), *Biochem. Med.* **12**, 79–91 (1975).
23. A. Senyei, K. Widder, and G. Czerlinski, Magnetic guidance of drug-carrying microspheres, *J. Appl. Phys.* **49**, 3578–3583 (1978).
24. N. Mason, C. Thies, and T. J. Cicero, *in vivo* and *in vitro* evaluation of a microencapsulated narcotic antagonist, *J. Pharm. Sci.* **65**, 847–850 (1976).
25. G. Gregoriadis, The carrier potential of liposomes in biology and medicine, *New Engl. J. Med.* **295**, 707–710 and 765–770 (1976).
26. H. K. Kimelberg, T. F. Tracy, S. M. Biddlecome, and R. S. Bourke, The effect of entrapment in liposomes on the *in vivo* distribution of [^{3}H] methotrexate in a primate, *Cancer Res.* **36**, 2949–2957 (1976).
27. W.-P. Fung, M. Przybylski, H. Ringsdorf, and D. S. Zaharko, *in vitro* inhibitory effects of polymer-linked methotrexate derivatives on tetrahydrofolate dehydrogenase and murine L5178Y cells, *J. Natl. Cancer Inst.* **62**, 1261–1264 (1979).
28. S. Olsnes and A. Pihl, Different biological properties of the two constituent peptide chains of ricin. A toxic protein inhibiting protein synthesis, *Biochemistry* **12**, 3121–3126 (1973).
29. S. Olsnes, K. Refsnes, and A. Phil, Mechanism of action of the toxic lectins abrin and ricin, *Nature* **249**, 627–631 (1974).
30. J. W. Goding, Antibody production by hybridomas, *J. Immunol. Meth.* **39**, 285–308 (1980).
31. V. Raso and T. Griffin, Specific cytotoxicity of human immunoglobulin-directed Fab′–ricin A chain conjugate, *J. Immunol.* **125**, 2610–2616 (1980).
32. J. M. Whiteley, Z. Nimec, and J. Galivan, Treatment of Reuber H35 hepatoma cells with carrier-bound methotrexate, *Mol. Pharmacol.* **19**, 505–508 (1981).
33. H.J.-P. Ryser and W.-C. Shen, Conjugation of methotrexate to poly(L-lysine) increases drug transport and overcomes drug resistance in cultured cells, *Proc. Natl. Acad. Sci. U.S.A.* **75**, 3867–3870 (1978).
34. P. N. Kulkarni, A. H. Blair, and T. I. Ghose, Covalent-binding of methotrexate to immunoglobulins and the effect of antibody-linked drug on tumor growth *in vivo*, *Cancer Res.* **41**, 2700–2706 (1981).
35. E. Calendi, G. Constanzi, F. Indiveri, G. Lotti, and C. Zini, Histoimmunologic specificity of an anti-lymphoid tissue sarcoma γ-globulin bound to methotrexate, *Boll. Chim. Farm.* **108**, 25–28 (1969).

36. K. Prabhakaran, E. B. Harris, and W. F. Kirchheimer, A possible method for improving the efficacy of dapsone, *Experientia* **36**, 1350–1351 (1980).
37. T. Ghose and S. P. Nigam, Antibody as carrier of chlorambucil, *Cancer* **29**, 1398–1400 (1972).
38. D. A. Davies and G. J. O'Neill, *In vivo* and *in vitro* effects of tumor-specific antibodies with chlorambucil, *Brit. J. Cancer* **28** (Supp. 1), 285–298 (1973).
39. V. Raso, Antibody mediated delivery of toxic molecules to antigen bearing target cells, *Immunolog. Rev.* **62**, 93–117 (1982).
40. E. Hurwitz, R. Levy, R. Maron, M. Wilchek, R. Arnon, and M. Sela, The covalent binding of daunomycin and adriamycin to antibodies with retention of both drug and antibody activities, *Cancer Res.* **35**, 1175–1181 (1975).
41. R. Arnon and M. Sela, *In vitro* and *in vivo* efficacy of conjugates of daunomycin with anti-tumor antibodies, *Immunolog. Rev.* **2**, 5–27 (1982).
42. G. P. Mell, J. M. Whiteley, and F. M. Huennekens, Purification of dihydrofolate reductase via amethopterin–aminoethyl starch, *J. Biol. Chem.* **243**, 6074–6075 (1968).
43. R. B. Angier, J. H. Boothe, J. H. Mowat, C. W. Waller, and J. Semb, Pteridine chemistry. II. The action of excess nitrous acid upon pteroylglutamic acid and derivatives, *J. Am. Chem. Soc.* **74**, 408–411 (1952).
44. Y. Masuho, K. Kishida, M. Saito, N. Umemoto, and T. Hara, Importance of the antigen-binding valency and the nature of the cross-linking bond in ricin A chain conjugates with antibody, *J. Biochem.* **91**, 1583–1591 (1982).
45. T. F. Bumol, Q. C. Wang, R. A. Reisfeld, and N. O. Kaplan, Monoclonal antibody and an antibody-toxin conjugate to a cell surface proteoglycan of melanoma cells suppress *in vivo* tumor growth, *Proc. Natl. Acad. Sci. U.S.A.* **80**, 529–533 (1983).
46. T. Lang, C. J. Suckling, and H. C. S. Wood, Affinity chromatography using agarose–triazine derivatives, *J. Chem. Soc.* 2189–2194 (1977).
47. M. Szekerke and J. S. Driscoll, The use of macromolecules as carriers of antitumor drugs, *Eur. J. Cancer* **13**, 529–537 (1977).
48. H. Sawada, K. Tatsumi, S. Masataka, T. Makumuka, and W. Gyoichi, Effects of neocarzinostatin on DNA synthesis in L1210 cells, *Cancer Res.* **34**, 3341–3346 (1974).
49. T. A. Beerman and J. H. Goldberg, DNA strand scission by the antitumor protein neocarzinostatin, *Biochem. Biophys. Res. Commun.* **59**, 1254–1261 (1974).
50. T. S. A. Samy and V. Raso, Radioimmunoassay of neocarzinostatin, on antitumor protein, *Cancer Res.* **36**, 4378–4381 (1976).
51. L. Chess, R. P. MacDermott, and S. F. Schlossman, Immunologic functions of isolated human lymphocyte subpopulations. I. Quantitative isolation of human T and B cells and response to mitogens, *J. Immunol.* **113**, 1113–1121 (1974).
52. M. S. Verlander, J. C. Venter, M. Goodman, N. O. Kaplan, and B. Saks, Biological activity of catecholamines covalently linked to synthetic polymers: Proof of immobilized drug theory, *Proc. Natl. Acad. Sci. U.S.A.* **73**, 1009–1013 (1976).
53. W. C. Shen and H.J.-P. Ryser, Conjugation of poly(L-lysine) to albumin and horseradish peroxidase: A novel method of enhancing the cellular uptake of proteins, *Proc. Natl. Acad. Sci. U.S.A.* **75**, 1872–1876 (1978).
54. P. G. Balboni, A. Minia, M. P. Grossi, G. Barbanti-Brodano, A. Mattioli, and L. Fiume, Activity of albumin conjugates of 5-fluorodeoxyuridine and cytosine arabinoside on poxviruses as a lysosomotropic approach to antiviral chemotherapy, *Nature* **264**, 181–183 (1976).
55. R. L. Blakley, in: *The Biochemistry of Folic Acid and Related Pteridines*, (A. Neuberger and E. L. Tatum, eds.), p. 93, North-Holland, Amsterdam and London (1969).
56. T. Peters, Serum albumin, in: *The Plasma Proteins* (F. W. Putnam, ed.), pp. 133–181, Academic Press, New York, San Francisco, and London, (1975).

57. R. C. Jackson, D. Niethammer, and F. M. Huennekens, Enzymic and transport mechanisms of amethopterin resistance in L1210 mouse leukemia cells, *Cancer Biochem. Biophys.* **1**, 151–155 (1975).
58. C. Fan, G. Henderson, K. Vitols, and F. M. Huennekens, Molecular targets for methotrexate, in: *Antimetabolites in Biochemistry, Biology and Medicine* (J. Skoda and P. Langen, eds.), pp. 313–326, Pergamon Press, Oxford, England and Elmsford, New York (1979).
59. E. Hurwitz, M. Wilchek, and J. Pitha, Soluble molecules as carriers for daunorubicin, *Appl. Biochem.* **2**, 25 (1980).
60. B. C. F. Chu and J. M. Whiteley, Control of solid tumor metastases with a high-molecular-weight derivative of methotrexate, *J. Natl. Cancer Inst.* **62**, 79–82 (1979).
61. B. C. F. Chu and J. M. Whiteley, The interaction of carrier-bound methotrexate with L1210 cells, *Mol. Pharmacol.* **17**, 382–387 (1980).
62. J. L. Rader, D. Niethammer, and F. M. Huennekens, Effect of sulfhydryl inhibitors upon transport of folate compounds into L1210 cells, *Biochem. Pharmacol.* **23**, 2057–2059 (1974).
63. A. Nahas, P. F. Nixon, and J. R. Bertino, Uptake and metabolism of N^5-formyl-tetrahydrofolate by L1210 leukemia cells, *Cancer Res.* **32**, 1416–1421 (172).
64. J. H. Galivan, Transport and metabolism of methotrexate in normal and resistant cultured rat hepatoma cells, *Cancer Res.* **39**, 735–743 (1979).
65. W.-C. Shen and H. J. P. Ryser, Selective protection against the cytotoxicity of methotrexate and methotrexate-polylysine by thiamine pyrophosphate, heparin and leucovorin, *Life Sci.* **28**, 1209–1214 (1981).
66. J. H. Galivan, Evidence for the cytotoxic activity of polyglutamate derivatives of methotrexate, *Mol. Pharmacol.* **17**, 105–110 (1980).
67. J. H. Galivan, Transport of methotrexate by primary cultures of rat hepatocytes: Stimulation of uptake *in vitro* by the presence of hormones in the medium, *Arch. Biochem. Biophys.* **206**, 113–121 (1981).
68. J. Galivan, M. Balinska, and J. M. Whiteley, Interaction of methotrexate poly(L-lysine) with transformed hepatic cells in culture, *Arch. Biochem. Biophys.* **216**, 544–550 (1982).
69. J. H. Galivan, Transport and metabolism of methotrexate in normal and resistant cultured rat hepatoma cells, *Cancer Res.* **39**, 735–743 (1979).
70. E. Hurwitz, R. Maron, A. Bernstein, M. Wilchek, M. Sela, and R. Arnon, The effect *in vivo* of chemotherapeutic drug–antibody conjugates in two murine experimental tumor systems, *Int. J. Cancer* **21**, 747–755 (1978).
71. R. C. Hughes, How do toxins penetrate cells? *Nature* **281**, 526–527 (1979).
72. S. Olsnes, C. Fernandez-Puentes, L. Carrasco, and D. Vazquez, Ribosome inactivation by the toxic lectins abrin and ricin. Kinetics of enzymic activities of the toxin A chains, *Eur. J. Biochem.* **60**, 281–288 (1975).
73. V. Raso and T. Griffin, Hybrid antibodies with dual specificity for the delivery of ricin to immunoglobulin-bearing target cells, *Cancer Res.* **41**, 2073–2078 (1981).
74. F. M. Sirotnak, D. M. Moccio, L. E. Kelleher, and L. J. Goutas, Relative frequency and kinetic properties of transport-defective phenotypes among methotrexate-resistant L1210 clonal cell lines derived *in vivo*, *Cancer Res.* **41**, 4447–4452 (1981).

14

Biomedical Applications of Polysaccharides

Conrad Schuerch

Abstract. Progress has been made in correlating physical properties of soluble, insoluble, gel-, film-, and fiber-forming polysaccharides with their molecular and supermolecular structure and conformation.

Products derived from polysaccharides are available for ion exchange, for gel permeation or gel filtration, or affinity and conventional chromatographies, as gel media for microbial cultures, for electrophoresis, and so on. Polysaccharides are used as demulcents, in drug formulations, for dental-impression materials, dusting powders, hemostatics, and for treatment of mild intestinal disorders. They are used for plasma replacement, and as anticoagulants both in solution and as surface treatment on artificial organs. Polysaccharides are converted into bioactive textiles and are formed into membranes and hollow fibers for hemofiltration and hemodialysis. They can control the release of drugs as polymeric carriers or by means of microencapsulation. They have been used to enhance the rate of healing in surgery and burn therapy.

Polysaccharides are important components of interstitial fluids and connective tissue, providing mechanical strength and lubrication. Shorter sequences of saccharides on soluble proteins and on cell surfaces maintain conformation and act as important antigens interacting with soluble and membrane-bound proteins. Similar interactions control inter- and intra-cellular transport of proteins and their removal from blood serum; are involved in control of tissue growth by contact inhibition; and are also involved in bloodtyping. They can also be used to distinguish normal from malignant cells, in cell surface investigations, radioimmunoassay, and in targeting drugs to specific tissues. Similarly, the extracellular polysaccharides and surface carbohydrates of bacteria activate the immune system and are used as vaccines and adjuvants.

1. Structure and Physical Properties

Polysaccharides have been used in nearly all the applications discussed elsewhere in this volume and have other biomedical applications specific to themselves as well. Polysaccharides are an extremely diverse set of materials and include insoluble structural elements of plants and animals; water-soluble and insoluble reserve foodstuffs; gel formers; physiological information-carrying species; acidic, basic, and neutral entities; and so on.[1,2]

Conrad Schuerch • Department of Chemistry, State University of New York, College of Environmental Science and Forestry, Syracuse, NY 13210.

It should, therefore, be helpful to discuss first how their structures and their sources relate to their properties and potential uses.

Polysaccharides are composed of simple sugar units linked by oxygen (acetal) bonds between the aldehyde or ketone (carbonyl) function of one monosaccharide and an alcoholic hydroxyl of another monosaccharide unit. Since there are one carbonyl function and several hydroxyl functions on each monosaccharide, polysaccharides can in principle be linear, branched, or cyclic. All three structural types are known. They are also in some cases linked to other substances, such as phosphate esters, proteins, lipids, and polypeptides.[3] These may form insoluble three-dimensional network structures or materials with complex solubility, liquid-crystalline, or surface-active properties.

In addition to homopolymers, composed of a single kind of monosaccharide, there are a variety of copolymer classes. These may be regular in structure with two to eight different sugars in a repeating sequence. They may be copolymers with a backbone consisting of two different sugars distributed randomly, or a linear backbone of a single sugar with randomly distributed branches of another sugar. The architecture of branching in both homopolymers and copolymers can be enormously varied—comblike or treelike. Interrupted and blocky linear copolymers are also known, and aperiodic sequences of sugars on proteins, lipids, and cell surfaces have also been found to have important physiological functions in living organisms.

Although the commonest class of monomer unit is that of six-carbon aldehydo sugars in a six-membered ring (aldohexopyranoses), many others are also found. These include sugars with ketone rather than aldehyde functions (ketoses), sugars with five-membered rings (furanoses), and sugars with hydrogen or one of many other functional groups replacing the usual hydroxyl groups. Sugars with five carbon atoms are not uncommon and eight- and nine-carbon sugars are also known. Some of these are of great physiological importance.

The properties of polysaccharides are profoundly affected by their molecular structure and resultant supermolecular architecture, and by the configurations of the individual sugars and the functional groups present.[4] These features affect especially polymer–polymer and solvent–polymer interactions and a number of broad generalizations have recently been made that are helpful in correlating structure–physical property relationships. Let us consider first the structural variations possible in linear polysaccharide chains. The individual pyranose rings are generally present in rigid chair conformations and the ring can, therefore, be treated as an extended bond between the two oxygen atoms linking it into the chain. The angles between this "virtual bond" and the bonds to adjacent rings differ with each different linkage and thus cause wide variations in the preferred conformation of the chains both in solution and in the solid state. The angles tend to produce molecular helices of a preferred shape that may be on the one hand very thin, steep, and extended or on the other, very flat and open.

Typically, the trans-diequatorial β-(1 $\rightarrow$ 4) linkages result in a ribbonlike helix with each unit nearly parallel to the helix axis. Polysaccharides of this configuration pack well, are good film and fiber formers, and are found as structural elements in plants and arthropods, for example, cellulose, the hemicelluloses glucomannan and xylan, and chitin.

Open helices are formed by polysaccharides with α- or β-(1 $\rightarrow$ 3) or α-(1 $\rightarrow$ 4) (the latter cis axial–equatorial) linkages. Depending upon the kind of linkage and preferred conformation, the helices may be steep or flat, and the number of sugars in a repeat unit of the crystal lattice can vary widely. Frequently, the helices interpenetrate and form double or triple helices.

Polysaccharides of the open helix family have a variety of biological functions, perhaps the most prominent of which is that of reserve foodstuff—starch and glycogen. A number are found in molds, algae, fungi, and higher plants and some may have some structural functions. Some are generated as extracellular polysaccharides by microorganisms and are used as gel formers in industrial, food, and medical applications.

The interior of the open helix is relatively hydrophobic and can often form canal complexes with hydrophobic species. The butanol and iodine complexes of the linear portion of starch (amylose) are well known. Similarly, the derivatized (1 $\rightarrow$ 4)-α-D-glucopyranan and 3-*O*-methyl-(1 $\rightarrow$ 4)-D-mannopyranan of *Mycobacterium smegmatis* have been shown to complex with fatty acids.[5,6] The complexation is involved in physiological control mechanisms.[7]

The equatorial–equatorial β-(1 $\rightarrow$ 2) linkage tends to form a crumpled structure, and long helical segments are difficult to form. Probably because of these steric restraints, this linkage is not common in polymeric backbones of natural polysaccharides. However, cyclic β-(1 $\rightarrow$ 2) glucans containing 17 to 23 glucose residues are formed by *Agrobacter* and *Rhizobium* species.[8]

In the preceding cases, an oxygen atom links the anomeric (carbonyl) center on one ring to a secondary position on the adjacent monosaccharide. Rotation can only occur around these two carbon–oxygen bonds and the preferred conformation of the chain is determined by dipole interactions, nonbonded interactions, and hydrogen bonding between adjacent rings. If the anomeric center is, instead, linked through oxygen to the primary C-6 carbon, rotation can also occur around the C-5, C-6 bond. This provides a relatively loosely jointed structure with an additional degree of freedom. The 1 $\rightarrow$ 6 linkage is found in many water-soluble gums, extracellular microbial polysaccharides, and as a branch point or blocky sequence in gel formers.[2]

The introduction of points of disorder into a polysaccharide has a profound effect on its properties. The disorder may be by random partial substitution, the product becomes soluble first in alcohol, and then in esters and ketones. Solubility and compatibility then decrease at higher DS, near becomes less regular, and solubility usually increases or is greatly modified.

The influence of disorder in a polysaccharide structure on solubility is especially well-defined in cellulose derivatives. Cellulose itself swells but does not dissolve in alkali. At very low degrees of substitution, DS = 0.1–0.5 (methoxyl groups per anhydro glucose unit), methyl cellulose is alkali soluble. At DS = 0.8–1.7, it is water soluble. At higher degrees of substitution, the product becomes soluble first in alcohol, and then in esters and ketones. Solubility and compatibility then decrease at higher DS, near 3, as the polymer again approaches an ordered, fully substituted structure. The same pattern of solubility is followed with ethyl cellulose and cellulose acetate with variations in the range of solubility in a solvent class. The influence of disorder is also apparent in starch. Amylose, the linear portion of starch, reprecipitates or "retrogrades" from aqueous solution on standing, while branched starch, amylopectin, remains in solution although its molecular weight is 10–30 times larger.

Similarly, ivory nut mannan, $(1 \rightarrow 4)$-β-D-mannopyranan, is a hard horny polymer, formerly used to make buttons. The same backbone is also found in water-soluble gums from *Leguminosae* like locust bean and guar, but in these polysaccharides the backbone is modified with α-D-galactosyl units linked to the 6-position of mannose (ratio 1:1.8 in guar; 1:4 in locust bean gum). For industrial purposes these polysaccharides are modified further by hydroxyethylation. The modified products are less readily flocculated, are more compatible with other additives, and serve better as thickeners in various formulations. Xanthan gum is also a comb-shaped polysaccharide. It has a cellulose backbone with partially substituted trisaccharide units attached to alternate glucose units. It forms pseudoplastic aqueous solutions that gel when mixed with locust bean gum.

The properties of polysaccharides are influenced not only by the kind of linkage between saccharide units and the degree of branching and disorder in the structure, but also depend on the configurations on the sugar moieties and the functional groups present. A glucan composed of chains of glucose units linked α-$(1 \rightarrow 6)$ and corresponding to the backbone of clinical dextran is water soluble, as might be expected from the flexibility of the backbone. However, the corresponding mannan forms cloudy solutions in water and requires some dimethylsulfoxide present for molecular solution. The corresponding galactan is insoluble in virtually all solvents effective with other hydrogen-bonding polymers except for a limited number of powerful complexing solvents. Presumably the axial hydroxyl groups on mannose and especially galactose enhance interchain bonding and decrease solubility.

Acidic functions, carboxylate or sulfate groups, are common components of polysaccharides. These tend to enhance water solubility or swelling in the presence of sodium or other alkali-metal cations, but can serve as cross-links in the presence of calcium or other di- and trivalent ions.

Polysaccharides and polysaccharide derivatives are used in a variety of physical forms. As solids they are used as coatings and as adhesives, films, fibers, and powders. In solution they are used as viscosity-control aids and colloidal stabilizers. They also can form partially swollen solids and gels of varying degrees of stiffness. In all of these forms the products can be used for industrial, food, or biomedical applications.[9]

A vast technology has been developed for the production of useful products by modifying the most available polysaccharides – cellulose,[10] and to a lesser extent, starch. The methods have also been applied to other polysaccharides as they have become important. Usually the hydroxyl functions are partially derivatized to give a simple ether or ester linkage or a functionalized ether such as hydroxyethyl, hydroxypropyl, *N*,*N*-diethylaminoethyl, *N*,*N*-diethanol-aminoethyl, sodium carboxymethyl, sulfate, or phosphate. The reactions involved in these syntheses are identical to those used on small molecules. The only complication relates to finding conditions under which partial substitution at the desired level occurs and the substituents are distributed randomly (or for special purposes, selectively) throughout the polymer matrix.

Very low degrees of substitution can cause profound changes in the behavior of the polymer. A slightly cationic starch, for example, associates strongly with solids in water suspension and flocculates with them because of its positive surface charge. Functionalized celluloses at low degrees of substitution are commonly used as insoluble ion-exchange substrates for chromatographic separations. At higher degrees of substitution, cross-linking with epichlorohydrin or other reagents is necessary to avoid excessive swelling. At the other extreme, highly sorptive products have value in personal products such as diapers and sanitary napkins. For these applications, a water-soluble carboxymethyl cellulose can be lightly cross-linked by gentle heating at acidic pHs. This product will imbibe and hold large quantities of aqueous fluids. Graft copolymers of starch and partially hydrolyzed polyacrylonitrile behave similarly.[11]

The structure of natural gels is under active investigation. In general, gelation is caused by regular sequences of identical units interacting to form cross-links.[4] They may associate as individual chains or as clusters of double or triple helices. The regular sequences are usually interrupted by disordered or more complex sequences that act as unassociated bridges between the regular junction points. This network structure can immobilize large volumes of solvents. Rigid elastic gels like agar are believed to have ropelike structures at the cross-links that are difficult to separate. Viscoelastic gels are believed to have "nested" helices that can be disrupted by shear. Alginic acid has regular sequences of α-L-guluronate units that can form "eggbox" structures around calcium ions and these structures act as the network cross-links,[4] while sequences of β-D-mannuronic acid units act as bridges, between the cross-links.

Although the relationship between some physical properties and polysaccharide structure is known to follow generally the broad outline discussed above, the near-infinite variety of materials that are found in this class of natural products ensures unexpected effects and necessitates investigation of each individual system.

2. Applications in Biochemistry

The use of polysaccharides and polysaccharide derivatives is commonplace in biochemical experimentation, and a wide variety of products developed for specific applications is available from laboratory supply houses and product manufacturers. The most common products are derived from cellulose, agarose, and dextran. These are: (1) a partially crystalline fibrous glucan with (1 → 4)-β-D-linkages, from plant cell walls; (2) a gel-forming agent from the marine algae, Gelidium, composed of an alternating sequence of 4-0-β-D-galactopyranosyl and 3-*O*-(3,6-anhydro)-α-L-galactopyranosyl units lightly sulfated (ca. 0.4%); and (3) a water-soluble exocellular polysaccharide produced by the microorganism *Leuconostoc mesenteroides*.[2] It is composed of a backbone of (1 → 6)-α-D-glucopyranosyl units with about 5% short branches and a few long branches of the same structure. Related dextrans are produced by other genera and species.[12]

Specially prepared cellulose is available for thin-layer, paper and column chromatography, as a zone-stabilizing medium in electrophoresis, and as hollow fibers and membranes for dialysis, concentration, fractionation, and osmometry of biopolymers. Standard derivatization reactions are used (usually at DS less than 0.15 and ion-exchange capacity less than 1.5 meq/g) to produce cationic and anionic ion-exchange resins. These include aminoethyl, diethylaminoethyl, triethylaminoethyl, and hydroxyalkylaminoethyl celluloses as anion-exchange media, and in some cases these products are etherified or esterified for special purposes. Cationic products include oxidized cellulose, carboxymethyl cellulose, and cellulose phosphate. Some products are cross-linked with epichlorohydrin to prevent excessive swelling. Ion-exchange media are used for partition chromatography of a series of natural oligomers as well as for ion exchange *per se*.

Agarose, the linear purified galactan from agar, forms a stiff gel at

1.5–2% concentration that is useful for electrophoresis, immunoelectrophoresis, and immunodiffusion. A beaded form is used for gel filtration and molecular-weight estimation of proteins. Derivatized to cationic (carboxymethyl) and anionic (diethylamino) forms, the ion-exchange gels have been used for protein separations and other purposes. The incorporation of chitin into microbiological media has been recommended to enhance selective growth of *Actinomycetes*.

Comparable products are produced by controlled cross-linking of dextran (Sephadex). The unsubstituted beads are used for gel filtration by column and thin-layer techniques, and ion-exchange forms are also available. In addition the soluble ionic forms – dextran sulfate and diethylaminoethyl dextran – have been introduced for use in protein precipitation, cell agglutination, and as polymers in two-phase aqueous solutions for the separation and purification of biological substances, such as viruses, bacteria, proteins, and nucleic acids, by countercurrent distribution. Very high-molecular-weight blue dextran is used for calibrating gel filtration columns.

Some of these products and a range of related derivatives have been applied to affinity chromatography for immobilization of cells, enzymes, glycoproteins, lectins, and for the preparation of immunoabsorbents. Product listings and pertinent literature can be obtained from suppliers.

A most extensive listing of experimental systems of these types can be found in the annual specialist periodical report *Carbohydrate Chemistry* of the Royal Society of Chemistry.[13] The lists include matrix, the mode of linkage, the ligand, and the specific use for which the product was made. Insoluble and soluble matrices are given and also examples of microencapsulation and the use of hydrogels in immobilizing enzymes. Agarose is by far the most common matrix material but cellulose, cellulose derivatives, dextran, and other organic polymeric and inorganic materials are also reported. The use of cycloimidocarbonate as linkage directly to the ligand and through spacer units appears to be the preferred method of immobilizing most substrates.

3. Uses as Surface-Acting Drugs and in Medicinal Formulations

Polysaccharides have long been used in pharmaceuticals and as surface-acting drugs. Their utility in medicinal formulations depends upon the extraordinary range and combination of useful properties that can be found among them and their simple derivatives. Among the materials approved for pharmaceutical use are U.S.P. grades of acacia (gum arabic); tragacanth (gum tragacanth); agar; glycyrrhiza; sodium and calcium alginate; starch; and methyl; carboxymethyl, and hydroxypropyl cellulose. Typically, these are used as suspending and emulsifying agents, tabletting aids, viscosity-control aids, and coating materials.[14]

Essentially all can act as protective colloids in formulations[9]; methyl cellulose is favored, for example, for use with quarternary ammonium and heavy metal salts because of its nonionic nature. In tablet formation, amylose can act as a binder, filler, and disintegrant and can also complex with and encapsulate water-insoluble materials, such as vitamin A palmitate; methyl cellulose is used as a granulating and disintegrating agent. Sodium and calcium alginate combinations act as both binder and disintegrating agent. Carboxymethylcellulose has been used as a tablet-coating material for enteric treatments since it is insoluble in stomach acid and soluble at alkaline pHs. The water-soluble cellulose derivatives are also used in contact-lens solutions and ophthalmic preparations.[14] Alginates are used as dental-impression materials with di- or trivalent metal salts as gelling agents. A special cornstarch is the preferred dusting powder for surgical gloves because of its ready bioabsorption and good lubricating properties (Absorbable Dusting Powder, U.S.P.). Specific applications are dealt with in more detail elsewhere.

The aqueous solutions of these materials also have demulcent properties, that is they have the ability to soothe and alleviate irritation of mucous membranes and abraded skin surfaces.[14] They are, therefore, useful components of lotions, ointments, wet dressings, demulcent drinks, enemas, lozenges, and gargles. Their bland flavor masks the flavor of unpleasant drugs. These properties have obvious value in many formulations.

Polysaccharides have application in the treatment of mild intestinal disorders.[14] Pectin, a polygalacturonic acid, is useful in the treatment of diarrhea and a number of natural gums and synthetic cellulose derivatives are common bulk-forming laxatives, as are bran and vegetable fiber. Cellulose phosphate has been used in Europe since 1963 to bind calcium ion and prevent excessive absorption in some kidney-stone sufferers.[15] It is presently nearing final marketing approval by the U.S. Food and Drug Administration.

4. Uses in Blood, Body Fluids, and Biomaterials, and in Trauma

In the applications discussed above, the polymers cannot readily cross the skin or mucous-membrane barrier and enter the body proper, so many synthetic and natural polymers may be safely used. For the following uses, polysaccharides are introduced into the organism and the requirements for safe and effective use are far more stringent. In general, the substance should remain within the tissue or the general circulation for a period of time adequate to perform its desired function and yet should be eventually excreted or metabolized. This requirement is more readily met by many polysaccharides than by most synthetic polymers so far tested. It should not affect adversely any physiological function or tests, such as blood typing, nor should it have an antigenic, allergenic, or pyretic effect. As natural products, the ability of polysaccharides to meet these latter criteria is very much a function of their structure, since individual carbohydrates play widely diverse roles in natural systems. Some of the pertinent factors will be discussed in later sections. Finally, polysaccharides should be able to be easily sterilized, have reasonable shelf life, and in liquid form have a viscosity suitable for parenteral administration over a reasonable temperature range.[14]

Polysaccharides have been used as pharmacologically inert polymers in the body fluids, as well as for specific physiological interactions. Inulin has, for example, in the past been used for the measurement of extracellular water. The plasma volume corresponds to about 7.5% and interstitial water 27.5% of body fluids. The molecular size of inulin is appropriate for the measurement of these, since it is too large to enter intracellular areas.[14]

A more important function is that of plasma expander, and dextran is considered to possess nearly ideal attributes and has been used successfully in the treatment of problems associated with loss of whole blood or plasma. Clinical dextran is a partially hydrolyzed and fractionated extracellular polysaccharide produced by *Leuconostoc mesenteroides* strain B 512. A number of molecular-weight ranges are available for experimental purposes, but those in use clinically have weight-average molecular weights around 40,000 and 70,000. Dextran has little effect on the blood, does not interfere with typing, crossmatching, or Rh determinations. It appears to have no deleterious effects on kidney or liver functions. Its main role is to provide a volume of fluid of an oncotic (osmotic) pressure comparable to that of plasma. The lower-molecular-weight fractions are excreted by the kidney over a period of 24 hours, but the remainder is slowly degraded over a period of weeks. Dextran has an inhibitory action on the clotting mechanism since it coats platelets and interferes with their aggregation. It also coats erythrocytes and plasma-clotting proteins and decreases blood viscosity. It can be given concurrently with anticoagulant drugs in cases of thromboembolic disorders, and the treatment has produced encouraging results.[14]

Dextran is antigenic. However, in the massive doses given in infusions, it is tolerogenic and the sensitivity reactions that have occasionally been observed are usually mild. As manufacturing controls have led to less branched products with narrower molecular-weight distribution, these reactions have been minimized. Recently, it has been established that the prior injection of dextran oligosaccharides results in hapten inhibition and prevents serious effects (anaphylactic shock) in a very small minority of highly sensitized individuals that are at risk from dextran injection.[16]

As an alternative to dextran as a plasma extender, hydroxyethyl starch has attracted clinical attention in Europe and Japan and carboxymethyl starch has been developed by the Academy of Science of the People's Republic of China and is in clinical use in that country.[17] Pullulan has been suggested as another possibility.[18] Hydroxyethyl starch has been coupled to hemoglobin to increase its oxygen-carrying capacity as a blood expander and to increase the dwell time of stroma free hemoglobin.[19] Dextran has been coupled to 3-amino-2,4,6-triiodobenzoic acid to act as a macromolecular medium for direct lymphography.[20]

One of the most generally recognized physiologically active roles for a polysaccharide is the inhibition of blood coagulation by heparin. Heparin is a mucopolysaccharide or proteoglycan composed of alternating sulfated units of D-glucosamine and D-glucuronic acid or L-iduronic acid, with more than two sulfate residues per disaccharide sequence.[2] It is formed in mammalian lung and liver tissues and is associated with mast cells. Heparin is isolated commercially from beef lung and liver and used in cases of embolism, thrombosis, and related disorders.[1] As a polymer, it requires parenteral administration and is effective after intravenous, intramuscular, or subcutaneous injection. Its action after intravenous injection is immediate, in contrast to the action of anticoagulants of small molecular weight administered orally.[14] The coagulation of blood involves a cascade of interactions of 12 identified blood factors.[14] Heparin appears to act on those steps in the mechanism that involve calcium ion. Heparin also activates the enzyme lipoprotein lipase that catalyzes hydrolysis of triglycerides, clearing the turbidity of plasma caused by eating fats.

Perhaps the only significant use of polysaccharides in biomaterials and artificial organs is the surface treatment of polymers with heparin to enhance their biocompatibility and to prevent thrombosis. Tridodecylmethylammonium chloride, a cationic hydrophobic species, adsorbs readily on a wide variety of synthetic polymers and complexes with and retains heparin hydrogel on the surface of implants.[21] Heparin has also been linked covalently to polymers,[22] and combined with polyvinyl alcohol for the same purpose.[23]

A number of less expensive sulfated polysaccharides have been tested for anticoagulant properties, among them indophycan, fucoidin, carrageenan,[9] and most recently sulfated chitin.[24a] In general, they are active but less effective and in some cases show significant toxicity.

An oxidized cellulose (Oxycel Surgical) containing a substantial proportion of carboxyl functions on C-6 is available in the form of sterile-gauze pads and strips for use as an absorbable hemostatic.[14] This is one of a limited number of materials that can be used in trauma or surgery to form an artificial blood clot and to control oozing from minute vessels. Absorption of moderate amounts occurs within a few days. Other acidic cellulose derivatives have been proposed for this application in the patent literature.[25]

Acceleration of wound healing by application of cartilage powders and extracts has been described in the medical literature and attributed largely to the presence of *N*-acetyl-D-glucosamine.[24b] The application of the monomeric sugar is not effective because of rapid elimination. However, chitin powder, of less than 10-μ particle size, is more effective than cartilage. Inflammatory cells transport the enzyme lysozyme to wound sites, and it gradually degrades chitin to *N*-acetylglucosamine dimer that is an active agent in the healing process. Regenerated chitin in various forms is also effective.[24b]

An artificial skin has been developed that acts as a polymeric template for the regeneration of new skin over extensive burned areas.[26] The material is a bilayer membrane consisting of an external silicone film that controls moisture loss and a biodegradable layer of collagen cross-linked with about 8% of chondroitin 6-sulfate. Autologous epidermal cells are seeded into the protein–polysaccharide layer and the membrane grafted over the burnt areas. The collagen–chondroitin layer undergoes biodegradation at a controlled rate and is replaced by neodermal tissue. The silicone film is spontaneously ejected following formation of a neoepidermal layer. The polysaccharide eliminates a minor allergic reaction to collagen and probably also serves as a source of *N*-acetylglucosamine for healing. Patients with burns covering over 50% of the body surface have been successfully treated with this remarkable material.

5. Physiological Functions, Immunological Relationships, and Vaccines

An appreciation of the many physiological functions of polysaccharides and those of complex carbohydrates is necessary for an intelligent application of these materials in medicine. Saccharides take part in intraorganism functions and in interspecies interactions. A knowledge of both can be applied in designing medicines and in medical treatments and in recognizing structures that must be avoided. For the purpose of brevity, the important classes of complex carbohydrates found in higher organisms and their functions will be discussed first, then those of microorganisms, and practical implications of both.

Generally the most important saccharides in mammalian tissues are linked as side chains to protein molecules, and conversely there is reason to believe that most proteins are combined with carbohydrates. These complex molecules can vary from predominantly carbohydrate to predominantly protein, and the chain lengths of the carbohydrate moieties can vary from one to a hundred or thousands of monosaccharide units.[3] Those that are predominantly carbohydrate are known as proteoglycans;[27] those predominantly protein we will call glycoproteins, although this term is frequently used to cover both categories.

The proteoglycans that have traditionally been called mucopolysaccharides consist of a protein core covalently linked to many long-chain heteropolysaccharides. The saccharide chains consist of D-glucosamine or D-galactosamine alternating with another sugar, usually acidic D-glucuronic acid or L-iduronic acid. *N*- and *O*-sulfate and *N*-acetyl groups are attached to the sugar units. The common proteoglycans are the imporant anticoagulant heparin, hyaluronic acid, the chondroitin sulfates and dermatan, and heparan and keratan sulfates.[2] The carbohydrate chains of hyaluronic acid have molecular weights up to 3 million, the others up to 50,000. Proteoglycans are found in synovial and other fluids and in connective tissue such as skin, bone, ligaments, and cartilage. Together with collagen or elastin fibers, they form a matrix in which connective-tissue cells are imbedded. They have mechanical functions, stabilizing and strengthening tissues and acting as lubricants. As polyelectrolytes they can form highly viscous and elastic solutions even at high dilution and can serve to maintain the salt and water balance in the body. A number of serious genetic disorders have been identified as enzyme deficiencies related to the metabolism of these substances. Heparan sulfate and dermatan sulfate from pig mucosa are available for the clinical diagnosis and study of connective-tissue diseases. Medical applications for heparin and chondroitin-6-sulfate have already been mentioned.

Glycoproteins, that generally contain less carbohydrate in shorter chains, are ubiquitous in the body. One group, known as mucins, covers the lining of the digestive and respiratory tracts, the cervix, and the eyesocket, and is found in saliva.[3] These mucins have hundreds of short saccharide sequences along the chain, many of which are disaccharide units terminated with *N*-acetylneuraminic acid or related sialic acids. The negatively charged units immobilize large quantities of water and the repulsive forces cause extension of the protein chain. The high-viscosity solutions resulting act as lubricants and as protective agents for the tissues against mechanical or chemical stress and against bacterial attack.

There are many other classes of glycoproteins with varying amounts of saccharides and lengths of saccharide side chains. They include intestinal and intracellular enzymes, immunoglobulins, and indeed most serum proteins except serum albumin, membrane-bound proteins, and so on.[3] The carbohydrate sequences in these glycoproteins also have mechanical functions, enhancing solubility and stabilizing particular conformations. Of far more interest for our purposes is the fact that in many cases these carbohydrate sequences have an information-carrying role that mediates many physiological processes.

Information-carrying functions are usually associated with protein or polypeptide hormones and with nucleic acids. However, the amount of information that can be carried per monomer unit is far less with them than with carbohydrates. Only one dimer can be formed from two identical amino acids. Twenty different reducing disaccharides can be formed from two identical hexose molecules. The linear arrays in proteins and nucleic acids permit easy "reading" of long complex directions, whereas the multiple shapes possible with only a few sugar units make them ideal labels for identification, addresses, or simple commands. This seems to be the kind of information carried by carbohydrate sequences. The information is often in terminal units and probably never in sequences longer than seven or eight successive monosaccharide units of a polymer chain.

Information transfer in many physiological processes typically involves a protein–carbohydrate reaction, for which lectin–carbohydrate interactions can serve as a model. Lectins are carbohydrate-binding proteins, originally found in plants, that react selectively with one or a limited number of carbohydrate structures.[28] Usually the lectin has more than one binding site so it can cause aggregation and precipitation of polysaccharides that have several sequences of complementary structure, or they can cause aggregation of leucocytes, erythrocytes, or other cells with membrane-bound carbohydrates of appropriate structure.

Protein–carbohydrate interactions of this type have important practical applications. They can be the basis of separations of specific polysaccharides, proteins, or glycoproteins from complex mixtures—physiological fluids or aqueous extracts—by affinity chromatography using either protein or saccharide as column ligand. Lectins can be used in bloodtyping, separating cell types, distinguishing normal from malignant cells, investigating the distribution and mobility of cell-surface glycoproteins, and so on. Lectin interactions with cell-bound carbohydrate can also induce physiological responses, such as lymphocyte division, and D-glucose transport and have been used to study the biochemical events accompanying growth, differentiation, and division which in some cases they initiate.[28]

The preceding discussion describes interactions of soluble proteins reacting with soluble polysaccharides or cell-bound carbohydrate sequences. Cell-bound proteins also have receptors for carbohydrates and these interactions are of profound physiological importance.[28,29] One well-characterized example is the elimination of glycoprotein from blood serum. Typically, certain glycoproteins have carbohydrate sequences of *N*-acetylneuraminic acid (NANA) linked to galactose linked to *N*-acetylglucosamine.[30] Removal of the terminal NANA and exposure of the penultimate galactose unit results in interaction of the galactose with a glycoprotein on receptor sites of the plasma membrane of the liver. The serum glycoprotein is promptly removed from circulation, transported into the cell, and catabolized. This is a normal physiological process and its implication for the targeting of polymer–drug complexes has not been lost on the polymer science community.

Another well-defined system is that of a lysosomal enzyme, a glycoprotein acid hydrolase that is labeled with terminal mannose-6-phosphate. This label acts as a recognition symbol for receptor-mediated uptake of the enzyme from a culture medium and also to segregate the enzyme at the appropriate locus for action within the cell.[31–33] Other protein–saccharide control systems include the following: The uptake of certain glycoproteins are mediated by terminal mannose and *N*-acetyl-glucosamine.[31,34] Lymphocytes with L-fucose on their surfaces migrate to the spleen, but when this sugar is removed they migrate to the liver.[3] Influenza virus has been shown to bind to sialic acid residues on the surface of erythrocytes and to cause hemagglutination.[3] The limitation of organ growth during the development of embryos is almost certainly the result of a contact inhibition probably mediated by similar protein–carbohydrate interaction on the surfaces of adjacent tissues. It is obvious that this field of biochemistry will be of great interest to those interested in the selective uptake of polymer–drug complexes by specific tissues, and may suggest other therapeutic uses of carbohydrates.

The basis of the ABO blood group system is the interaction of carbohydrates on the surface of erythrocytes with immunoglobulins in blood serum of another type.[3] Blood type A has A antigen, a carbohydrate determinant, on its red cells and antibodies to type B in its serum. The converse is true for the B group. The O group blood has neither antigen on its red cells, but has antibodies to both A and B group in its serum. Mixing of different blood groups results in agglutination, a multivalent protein–carbohydrate reaction.

The same kind of phenomenon is generally involved in immune or allergic reactions.[35] If a foreign body, usually a macromolecule, is introduced into the body, it becomes attached to the surface of an antibody-producing cell. The latter is caused to proliferate and the daughter cells introduce into the blood serum quantities of antibody complementary to the original antigen. If more antigens of the same type are present, they react with the antibodies and initiate a complex defense mechanism against the invader. The antigen need not be polysaccharide, but because of the presence of saccharides on and around the surface of microoganisms, polysaccharides are of especial importance in immune reactions.[36]

Many pathogenic bacteria produce a specific polysaccharide outside their cell walls.[36] This constitutes a slime layer, or capsule or matrix within which the organism lives. The polysaccharide in some cases may allow the bacterium to remain at a particular site, as a branched α-(1 $\rightarrow$ 3)-linked glucan allows *Streptococcus* strains to cling to the surfaces of the teeth or heart muscle. In general, however, it acts as a protective layer that inhibits attachment and ingestion by phagocytes. In the case of *Cryptococcus neoformans* the capsule permits the organism to cross the blood–brain barrier with a life-threatening meningitis the result.[37] In general, capsular polysaccharides are heteropolysaccharides with a repeating sequence of two to eight sugar units. The particular sequence is characteristic of a particular *strain* of a microorganism. There are, for example, about 80 strains of *Streptococcus pneumoniae* that cause pneumonia in man. Each has a specific capsular polysaccharide.[38] The structure of some of them have been elucidated. Although these capsular polysaccharides act to protect the bacteria, they do act as antigens and as such can serve as vaccines. A polysaccharide vaccine in active clinical use against pneumonia consists of capsular polysaccharides of 14 strains and protects against 80% of the cases of pneumonia in North America.[39] Successful meningitis vaccines of the same type have been developed and licensed. Vaccine therapy is increasing in importance with the development of bacterial strains resistant to antibiotics, and polysaccharide vaccines against a number of other diseases are under investigation.

Many bacteria are acapsular, and therefore the host animal's defense mechanisms interact with the external features of the cell wall. These differ between the Gram-negative and Gram-positive classes.[40] In the case of Gram-negative bacteria, the cell wall consists of a peptidoglycan sac for structural strength and a mobile liquid-crystalline array of lipoprotein, lipopolysaccharide, and protein. The hydrophilic end of the lipopolysaccharide extends far out into the medium and is the major antigenic determinant of the cell wall. It consists of many repeating tetra- or pentasaccharide units and is called the O-antigen of this group of bacteria. The structure of O-antigens of *Klebsiella, Salmonella* and others are under active investigation, and in principle can be used as vaccines as the capsular polysaccharides have been. Alternatively, haptens – single repeat units – of both capsular polysaccharides and O-antigen have been synthesized and coupled to protein (15–30 per protein molecule) to form synthetic antigens which can be used to induce immunity against the parent organism or for use in monoclonal antibody technology.[41]

Gram-positive bacteria also have hydrophilic chains projecting out from the cell wall that serve as their major antigenic determinants. These are the teichoic acids, the tail of which consists of 30 or more repeat units of glycerol phosphate or ribitol phosphate often linked to various sugars and to D-alanine.[3,40]

A few high-molecular-weight stereoregular homo- and heteropolysaccharides with random sequences of two sugars have been synthesized by purely chemical methods and used to investigate enzymic, lectin, immunological, and allergic mechanisms.[42] The more difficult problem of the chemical and chemical-enzymic synthesis of high polymers with regular repeat units of different saccharides is under investigation at the Zelinsky Institute of Organic Chemistry in Moscow.[43]

Carbohydrate polymers can induce body defensive reactions by other routes than by causing the development of antibodies and complexing with them.[44,45] They can also interact with a component of the complement system, known as properdin, which is part of an alternate immune pathway.[46] Polysaccharides are also, in some cases, interferon inducers. Interferon is involved in protection against virus attack. A number of polysaccharides have been used as immunological adjuvants.[44] When a particular antigen is of excessive toxicity, it cannot be used alone in sufficient quantity to induce immunity. It is then administered in much smaller quantity together with a relatively innocuous material, an adjuvant, that enhances the body's sensitivity, causing antibody production at the lower level of antigen concentration.

Finely divided chitin suspended in physiological saline and injected into dogs induced the formation of antibodies that affected ticks and fleas. In a few days the dogs were reported to have been freed of these parasites.[24c]

A number of fungal, yeast, and bacterial polysaccharides have been reported to have antitumor activity.[47] Prominent among them are $(1 \rightarrow 3)$-β-D-glucans.[48–50] A polysaccharide preparation (PSK) containing 38% protein is under clinical investigation in Japan.[51,52] In general, the polysaccharides appear to have their greatest effect against implanted tumors in experimental animals, but some positive results have been recognized against spontaneous human tumors. Their effect is clearly host mediated and operates in a variety of ways on the immune system. At present they are being tested in combination therapy with surgery, radiation, and chemotherapy.[51,52]

Monoclonal antibodies against a carbohydrate antigen 19-9, characteristic of colorectal cancer, are available for radioimmunoassay. They are reported to have low false positive rates and to be useful for prognostic management of colorectal cancer.[53]

6. Textiles, Membranes, Microencapsulation, Controlled-Release Agents, and Targeted Drugs

Modification of textile materials to introduce antibacterial, antifungal, and antiviral properties is an active area of research that has led to a number of products marketed for medical, hygienic, and aesthetic applications, Generally they are useful to control pathogenic organisms in hospital environments, to prevent skin infections, and to suppress odor-forming bacteria on clothing and skin. Usually an antibacterial agent is bonded to the fiber surface by covalent, ionic, or chelate bonds and is slowly released by hydrolysis or related reaction. Alternatively, a permanent antibacterial silicone-based barrier can be fixed to the fiber, or for special purposes, encapsulation techniques are used. The modification of cotton or rayon generally follows treatments similar to those used on other textile materials, although the functionality of cellulose permits a greater range of options.[25,54,55]

Membrane technology received a major impetus in the early 1960s from the development of an anisotropic cellulose acetate membrane consisting of a very dense ultrathin skin that provided permselectivity and a thick microporous backing that provided mechanical strength while permitting high transport rates.[56] Since that time a variety of polymers have been used to construct similar anisotropic membranes for specific separation problems.[57] Nevertheless, cellulose and cellulose acetate have continued as important membrane-forming materials for biomedical purposes.[58] Specifically, cellulose membranes are spun in the form of anisotropic hollow fibers from cuprammonium solution or regenerated by deacetylating cellulose acetate fibers. These are assembled into cartridges for hemodialysis.[58] The relatively low cost and good biocompatibility of cellulose are two factors that favor the continued use of this product. A cellulose acetate unit is also marketed for hemofiltration. Deacetylated chitin, or chitosan, has also been proposed as a membrane material for artificial kidneys, and may have some advantages over cellulose acetate.[24d]

Oxygen transport rate is one of the properties limiting the use of some polymers in contact lenses. Cellulose acetate butyrate is under clinical evaluation as an alternative material for hard contact lenses since its oxygen permeability is about twice that of polymethylmethacrylate, the most commonly used material.[57]

The general principles and problems involved in drug treatment have been well summarized in a standard reference on therapeutics.[14] Traditional drug therapy has serious limitations in that medicines are generally not delivered to the desired locus at constant concentration over an extended period. Instead, high concentrations are usually present and widely diffused immediately after administration and then decrease gradually through the dosing interval. The dosing interval is determined by physiological reactions to the drug and may be too short for patient convenience and compliance. Resulting concentration fluctuations may also be hazardous with drugs of low therapeutic index like those used in cancer therapy. The distribution of a drug is determined by its solubility characteristics, its ability to cross cell membranes, and its tendency to associate with all body constituents rather than with its necessary site of action. The ability to control the delivery of a drug to a specific location and to maintain its concentration over time, therefore, would provide many obvious advantages. A number of new methods and devices for this purpose are under development and some are in clinical use.[(59)]

One of the early methods of controlled delivery was the use of acidic derivatives of cellulose—carboxymethylcellulose or cellulose phthalate—to encapsulate drugs for enteric administration. These derivatives are insoluble in the acidic stomach environment but soluble at the higher pHs characteristic of the intestine.[(60)] The variable nature of that environment, however, limits this method for drug administration.

The high permeability of the mucous membrane permits the transport of drugs into the body, and placement of an appropriately designed device on mucous membrane surfaces can control the rate of entry. This method is advantageous for treatment especially at or near those sites, where a relatively high concentration of drug can be maintained.[(59)] For example, treatment of eye disorders by placement of devices under the eyelid is being actively pursued. Lacrisert (Merck, Sharp and Dohme) is one device that releases a hydrophilic cellulosic polymer into the tear film to treat dry-eye syndrome. Related products are expected as carriers to the eye for a number of medications. Related systems for topical application and controlled release through the skin are also becoming available.[(59)]

In these systems, control is provided only by the location and transport properties of the device. There is some inherent advantage in biodegradability of the carrier but less than there is in cases of parenteral administration. Parenteral administration may be via an implant of an insoluble polymer, a slowly swelling or slowly dissolving polymer, or a biodegradable polymer.[(60)] Alternatively, the implant may be of a drug depot or reservoir inside a semipermeable membrane, or osmotic pump. The kinetics of drug release is of great importance in the design of these systems.[(59)] The technique of microencapsulation and development of microspheres allows the injection of drugs or encapsulated enzymes or cells by means of hypodermic syringes. The biocompatibility and biodegradability of polysaccharides suggest their application for these purposes.[(61)] Calcium alginate has been used for the encapsulation of islets of Langerhans,[(62)] chitin has been used in an eye implant,[24] and techniques for microencapsulation with starch and cellulose derivatives are well developed.[63]

Soluble polymer–drug complexes are useful for retention of a drug within the circulatory system or wherever there is a slow rate of fluid exchange. Bonding to polysaccharides by covalent link, association, or chelation is simple. A number of polysaccharide–iron complexes are in clinical application,[14] but a combination of oxytocin and carboxymethyl dextran reported earlier[45] is not available.[14,64]

Factors influencing the effectiveness of polymeric drugs have been evaluated by Pitha[65] in part by coupling dextran to beta-adrenergic agents through connecting chains. He found that although dextran itself was accessible to antibodies, drugs attached to it even by quite long side chains were bound only weakly to cell-membrane receptors. However, the persistence of action of drug–polymer complexes was enhanced when several drug molecules were linked to one molecule of carrier. Vert[66] has established that very minor changes in structure altering the hydrophilic–hydrophibic ratio of a drug–polymer complex can dramatically alter its conformation, enclose the drug in a polymeric coil, and render it inaccessible. In this respect, dextran with its relatively flexible chain and random coil conformation may well be a poorer choice for drug–polymer complexes than hydroxyethyl amylose would be. The helical conformation of amylose should ensure a more extended and accessible structure for drugs attached to it. The extended conformation of water-soluble cellulosic derivatives should also be preferable (if FDA approved).

Ringsdorf[67] has discussed in detail the difficulties and advantages in targeting drug–polymer complexes to a specific site, and there is now much effort in this direction. Preliminary results using protein–carbohydrate interactions for targeting have been reported. Rembaum[68] has coupled the lectin conconavalin A to methacrylate microspheres to map lectin-receptor sites on cell surfaces. Injection of mitomycin linked to the same lectin has been used to bind the drug to tumor tissues.[69] The clearing of a synthetic polymer by liver cells has been shown to be much more rapid when a terminal galactose unit is linked to side chains on the polymer.[70]

The mode of entry of polymer–drug complexes into cells has also been evaluated.[67] This is generally by endocytosis, an engulfing action that results in formation of a vesicle that fuses with lysosomes. If the drug is not linked by an acid-sensitive[71] or enzyme-hydrolyzable bond, the drug can be retained in the lysosome away from the protoplasm and may not be effective.[65] An alternative mode of encapsulating drugs for delivery to cells is within liposomes, cell-like aggregates of phospholipids or glycolipids.[59] These may deliver the drug to cells by endocytosis or, alternatively, by fusion of the lipid with the plasma membrane. In the latter case, the drug is delivered directly to the cell cytoplasm.[59] Modified liposomes derived from synthetic glycolipids are being investigated by Ringsdorf.[72]

It is to be expected that the role of carbohydrates and polysaccharides in drug delivery systems will continue to grow in importance.

References

1. W. Pigman and D. Horton (eds.), *The Carbohydrates*, Vol. IIA, IIB, Academic Press, New York (1970).
2. G. O. Aspinall, *Polysaccharides*, Pergamon Press, Oxford (1970).
3. N. Sharon, *Complex Carbohydrates*, Addison-Wesley, Reading, Mass. (1975).
4. D. A. Rees, *Polysaccharide Shapes*, Chapman and Hall, London (1977).
5. W. L. Smith and C. E. Ballou, *J. Biol. Chem.* **248**, 7118 (1973).
6. H. Yamada, R. E. Cohen, and C. E. Ballou, *J. Biol. Chem.* **254**, 1972 (1979).
7. K. K. Yabusaki and C. E. Ballou, *Proc. Natl. Acad. Sci. U.S.A.* **75**, 691 (1978).
8. H. Hisamatsu, A. Amemura, and T. Hasada, Paper 361, *Abstracts, XIth International Carbohydrate Symposium*, Vancouver, Canada, August 22–28, 1982, National Research Council of Canada, Ottawa, Ontario.
9. R. L. Whistler and J. N. BeMiller, *Industrial Gums*, 2nd Edition, Academic Press, New York (1973).
10. E. Ott, H. M. Spurlin, and M. W. Grafflin (eds.), *Cellulose and Cellulose Derivatives*, Second Edition, Parts I–III; Interscience, New York (1954–1955); N. M. Bikales and L. Segal (eds.), Parts IV and V, Wiley, New York (1971).
11. M. O. Weaver, R. R. Montgomery, L. D. Miller, V. E. Sohns, G. F. Fanta and W. M. Doane, *Staerke* **29**, 413 (1977).
12. R. L. Sidebotham, in: *Advances in Carbohydrate Chemistry and Biochemistry* (R. S. Tipson and D. Horton, eds.), Vol. 30, p. 371, Academic Press, New York (1974).
13. J. F. Kennedy and N. R. Williams (eds.), *Carbohydrate Chemistry, A Review of the Literature*, Royal Society of Chemistry, London.
14. L. S. Goodman and A. Gilman (eds), *The Pharmacological Basis of Therapeutics*, 4th, 5th, and 6th Editions, Macmillan Co., New York (1970, 1975, 1980).
15. *Chem. Eng. News 60* (49), p. 8 (December 6, 1982).
16. W. Richter, H. Hedin, K. Messmer, and K. -G. Ljungstrom, *Int. Arch. Allergy Appl. Immun.* **66** (Suppl. 1), 288 (1981).
17. Personal communication, Institute of Organic Chemistry, Chinese Academy of Science, Shanghai, People's Republic of China
18. J. E. Zajic and A. LeDuy, in: *Encyclopedia of Polymer Science and Technology*, (H. F. Mark *et al.*, ed.), Supplementary Volume 2, p. 651, Wiley, New York (1977).
19. L. C. Cerny, D. M. Stasiw, E. L. Cerny, and M. Cerny, *Proceedings, International Union of Pure and Applied Chemistry, 28th Macromolecular Symposium*, Amherst, Massachusetts, July 12–16, 1982, p. 343.
20. J. Kalab, *Proceedings, International Union of Pure and Applied Chemistry, 28th Macromolecular Symposium*, Amherst, Massachusetts, July 12–16, 1982, p. 391.
21. B. D. Halpern and W. Karo, in: *Encyclopedia of Polymer Science and Technology*, (H. F. Mark *et al.*, ed.), Supplementary Volume 2, p. 368, Wiley, New York (1977).
22. N. A. Plate, *Proceedings, International Union of Pure and Applied Chemistry, 28th Macromolecular Symposium*, Amherst, Massachusetts, July 12–16, 1982, p. 385.
23. K. Kataoka, *Proceedings, International Union of Pure and Applied Chemistry, 28th Macromolecular Symposium*, Amherst, Massachusetts, July 12–16, 1982, p. 387.

24. R. A. A. Muzzarelli, *Chitin*, Pergamon Press, Oxford (1977) [(a) p. 260, (b) p. 293, (c) p. 258, (d) p. 257].
25. Z. A. Rogovin and A. D. Virnik, in: *Cellulose and Cellulose Derivatives* (N. M. Bikales and L. Segal, eds.), Part V, p. 1334, Wiley, New York (1971).
26. I. V. Yannas, J. F. Burke, E. Chen, D. P. Orgill, and E. M. Skrabut, *Proceedings, International Union of Pure and Applied Chemistry, 28th Macromolecular Symposium*, Amherst, Massachusetts, July 12–16, 1982, p. 336.
27. J. F. Kennedy, *Proteoglycans, Biological and Chemical Aspects in Human Life*, Elsevier, New York (1979).
28. I. J. Goldstein and C. E. Hayes, in: *Advances in Carbohydrate Chemistry and Biochemistry* (R. S. Tipson and D. Horton, eds.), Vol. 35, p. 127, Academic Press, New York (1977).
29. I. J. Goldstein (ed.), *Carbohydrate–Protein Interaction*, American Chemical Society Symposium Series, No. 88, American Chemical Society, Washington, D.C. (1979).
30. G. Ashwell and A. G. Morell, *Advances in Enzymology*, Vol. 41 (Atton Meister, ed.), p. 99, Krieger, Huntington, N.Y. (1974).
31. W. S. Sly and P. Stahl, in: *Transport of Macromolecules in Cellular Systems* (S. C. Silverstein, ed.), p. 229, Dahlem Konferenzen, Berlin 1978.
32. M. R. Natowicz, M. M. -Y. Chi, O. H. Lowry, and W. S. Sly, *Proc. Natl. Acad. Sci. U.S.A.* **76** (9), 4322 (1979).
33. H. D. Fischer, M. Natowicz, W. S. Sly, and R. K. Bretthauer, *J. Cell Biology* **84**, 77 (1980).
34. J. Distler, V. Hieber, R. Schmickel, and G. W. Jourdian, in: *Advances in Carbohydrate Chemistry and Biochemistry* (R. S. Tipson and D. Horton, eds.), Vol. 28, p. 163, Academic Press, New York (1977).
35. M. Sela, ed., *The Antigens*, Vols. 1–3, Academic Press, New York (1973–1975).
36. K. Jann and O. Westphal, in *The Antigens* (M. Sela, ed.), Vols. 1–3, Academic Press, New York (1973–1975).
37. R. Cherniak, E. Reiss, and S. H. Turner, *Carbohydr. Res.* **103**, 239 (1982).
38. O. Larm and B. Lindberg, in: *Advances in Carbohydrate Chemistry and Biochemistry* (R. S. Tipson and D. Horton, eds.), Vol. 33, p. 295, Academic Press, New York (1976).
39. J. B. Robbins, *Immunochemistry* **15**, 839 (1978).
40. R. Y. Stanier, E. A. Adelberg, and J. L. Ingraham, *The Microbial World*, 4th Edition, Prentice-Hall, Englewood Cliffs, N.J. (1976).
41. C. P. Stowell and Y. C. Lee, in: *Advances in Carbohydrate Chemistry and Biochemistry* (R. S. Tipson and D. Horton, eds.), Vol. 37, p. 225. Academic Press, New York (1980).
42. C. Schuerch, in: *Advances in Carbohydrate Chemistry and Biochemistry* (R. S. Tipson and D. Horton, eds.), Vol. 39, p. 157, Academic Press, New York (1981).
43. N. K. Kochetkov, in: *Bacterial Lipopolysaccharides, Structure, Synthesis, and Biological Activities* (F. M. Unger and L. Anderson, eds.), American Chemical Society Symposium Series No. 231, American Chemical Society, Washington, D.C. (1983).
44. S. Raffel, *Immunity*, Appleton-Century-Crofts, New York (1961).
45. C. Schuerch, in: *Advances in Polymer Science* (H. J. Cantow, ed.), Vol. 10, Springer-Verlag, Berlin (1972), p. 173.
46. H. N. Eisen, *Immunology*, Harper & Row, Hagerstown, Md. (1974).
47. R. L. Whistler, A. A. Bushway, P. P. Singh, W. Nakahara, and R. Tokuzen, in: *Advances in Carbohydrate Chemistry and Biochemistry* (R. S. Tipson and D. Horton, eds.), Vol. 32, p. 235, Academic Press, New York (1976).
48. W. Browder, E. Jones, R. McNamee, and N. R. Di Luzio, *Surg. Forum 27* (62), 134 (1976).
49. T. Sasaki, N. Abiko, Y. Sugino, and K. Nitta, *Cancer Res.* **38**, 379 (1978).
50. R. Bomford and C. Moreno, *Brit. J. Cancer 36(1)*, 41 (1977).
51. K. Taguchi, Recent results, *Cancer Res.* **68**, 174 (1978).
52. K. Taguchi, Recent results, *Cancer Res.* **68**, 236 (1978).
53. Commercial literature, Centocor Inc., Malvern, PA 19335 (1982).

54. T. L. Vigo and M. A. Benjaminson, *Textile Res. J.* **51**, 454 (198).
55. T. L. Vigo, in: *Modified Cellulosics* (R. Rowell and R. Young, eds.), Academic Press, New York (1978).
56. S. Loeb and S. Sourinajan, *U.C.L.A. Eng. Rept. 1960*; 60–60 *Adv. Chem.* **38** 117 (1963).
57. D. R. Paul and G. Morel, in: *Encyclopedia of Chemical Technology*, 3rd edition (Kirk & Othmer, eds.), Vol. 15, p. 92, Wiley, Interscience, New York (1981).
58. I. Cabasso, in: *Encyclopedia of Chemical Technology*, 3rd edition (Kirk & Othmer, eds.), Vol. 12, p. 492, Wiley, Interscience, New York (1980).
59. H. Benson, B. Harley, and E. E. Schmitt, in: *Encyclopedia of Chemical Technology*, 3rd edition (Kirk & Othmer, eds.) Vol. 17, p. 290, Wiley Interscience, New York (1982).
60. B. D. Halpern and W. Karo, in: *Encyclopedia of Polymer Science and Technology* (H. F. Mark and N. M. Bikales, eds.), Supplementary Vol. 2, p. 368, Wiley, New York (1977).
61. R. L. Kronenthal, Z. Oser, and E. Martin (eds.), *Polymers in Medicine and Surgery*, Plenum Press, New York (1975); quoted in Reference 7.
62. F. Lim and A. M. Sun, *Science* **210**, 908 (1980)
63. J. A. Herbig, in: *Encyclopedia of Polymer Science and Technology* (H. F. Mark, N. G. Gaylord, and N. M. Bikales, eds.) Vol. 8, p. 719, Wiley, New York (1968).
64. *Physicians Desk Reference*, 32nd Edition, Medical Economics Co., Oradell, N.J. (1978).
65. J. Pitha, *Proceedings, International Union of Pure and Applied Chemistry, 28th Macromolecular Symposium*, Amherst, Massachusetts, p. 380, July 12–16, 1982.
66. M. Vert, *Proceedings, International Union of Pure and Applied Chemistry, 28th Macromolecular Symposium*, Amherst, Massachusetts, p. 377, July 12–16, 1982.
67. H. Ringsdorff, *J. Polym. Sci., Polym. Symp.* **51**, 135 (1975).
68. A. Rembaun, *Proceedings, International Union of Pure and Applied Chemistry, 28th Macromolecular Symposium*, Amherst, Massachusetts, p. 374, July 12–16, 1982.
69. E. P. Goldberg, W. E. Longo, and H. Iwata, *Proceedings, International Union of Pure and Applied Chemistry, 28th Macromolecular Symposium*, Amherst, Massachusetts, p. 337, July 12–16, 1982.
70. J. B. Lloyd and R. Duncan, *Proceedings, International Union of Pure and Applied Chemistry, 28th Macromolecular Symposium*, Amherst, Massachusetts, p. 400, July 12–16, 1982.
71. W. -C. Shen and H. J.-P. Ryser, *Proceedings, International Union of Pure and Applied Chemistry, 28th Macromolecular Symposium*, Amherst, Massachusetts, p. 368, July 12–16, 1982.
72. H. Bader and H. Ringsdorf, *Proceedings, International Union of Pure and Applied Chemistry, 28th Macromolecular Symposium*, Amherst, Massachusetts, p. 341, July 12–16, 1982.

15

Interferon Induction by Polymers

Hilton B. Levy and Thomas Quinn

1. Introduction

This chapter will deal primarily with the induction of interferon by polymers. A section dealing with the biological role of interferon and the molecular biology of its action will be presented first. More complete reviews are given in References 1–4.

There is a phenomenon in virology called interference, in which the presence of one virus in a group of cells inhibits the growth of a second infecting virus. During research on this subject in 1957, Isaacs and Lindenman,[5] in England, found that cells infected with the first virus produced a protein, which the investigators called interferon. When this protein was incubated with uninfected cells, these cells were modified so that they produced less virus when subsequently infected than did cells that had not been pretreated with the interferon.

Interferon is produced by most types of cells upon infection with most viruses, although the amounts produced vary tremendously with both the cell type and the inducing virus. In many instances the inducing virus does not have to replicate in the cells in order to induce the synthesis of interferon. As a matter of fact, sometimes inactivated viruses are better inducers than live viruses. Interferon is more or less species specific, in that interferon induced in chicken cells is not active in mouse cells and vice versa. However, there is some crossing between closely related species, like monkey and human. Occasionally, there are peculiar examples of cross protection, as for example between rabbit and human.

Viruses differ greatly in their sensitivity to interferon, some being exquisitely sensitive and some relatively resistant to the inhibitory action. There is no immunologic or genetic relationship between the inducing virus and the interferon produced. That is, interferon induced in a given cell type by virus A, B, or C, unit for unit, would be equally effective against virus D.

There are three major types of interferon, depending primarily on the

Hilton B. Levy and Thomas Quinn • National Institute of Allergy and Infectious Diseases, Bethesda, Maryland, and Johns Hopkins Hospital, Baltimore, Maryland.

type of cell that has produced it. Interferon produced by leucocytes, formerly called Le interferon, is now called α interferon; interferon produced in fibroblasts, formerly called F interferon, is now β interferon; and interferon produced by lymphocytes sensitized to a specific antigen when reexposed to that antigen is called γ interferon. These interferons differ in immunologic characteristics. α and β have slightly different physical properties from γ interferons. By the use of genetic engineering techniques 20 different subtypes of α interferon and two of β interferon have been isolated.

A major problem with interferon research is the imprecision of the assay. Interferon is assayed by biological procedures. Serial dilutions of the material to be assayed are incubated with separate plates of tissue-culture cells. The interferon is then removed, the treated cells are infected with a known amount of an interferon-sensitive virus, and the amount of virus growth is measured. That dilution of interferon that causes a 50% reduction in virus yield is defined as one unit of interferon. The reciprocal of that dilution thus defines the concentration, in units, of the original preparation. Reproducibility of $\pm$ twofold is considered satisfactory.

Until recently, interferon had never been isolated in pure form. One of the problems in purifying interferon is associated with its very high specific biological activity. Preparation of interferon containing 10^8 units per milligram of protein has been prepared, but these preparations are not homogeneous. The preparation of this partially purified 10^8 units would require 10^9 units of crude starting material. 10^9 units of interferon represents a good deal of crude material. Successful efforts have been made to prepare large quantities of human interferon in order to purify it. This has led to the determination of the amino acid sequence in the protein. It is hoped to seek a fragment that would have the biological action of the interferon, thus making synthesis possible.

The body has several defenses against the invasion of viruses. Interferon is one of them. Others include antibody production, cell-mediated immunity, and the various aspects of the inflammatory response.

The kinetics of interferon production, in relationship to virus growth in an infected host, are indicated schematically in Figure 1 (courtesy of Dr. S. Baron, University of Texas, Galveston). The virus enters the body and after a short time it starts to replicate, usually with a period of viremia. Shortly thereafter, interferon can first be detected in the blood. The clinical manifestations of disease also appear shortly after the initiation of virus growth. The amount of virus declines because of the interferon and other factors. Then the amount of interferon declines. Finally, evidence of disease decreases. During the later part of these events, neutralizing antibodies specific to that virus begin to appear, and the disease is over. All these statements apply to the first infection of a host with a given virus. If the same

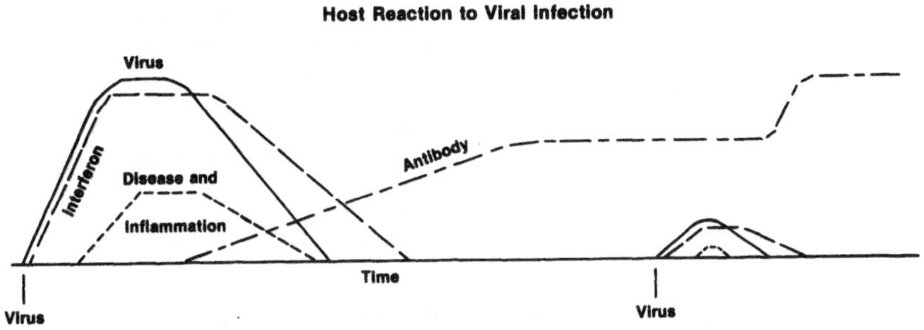

Host Reaction to Viral Infection

Figure 1. Schematic representation of host response to virus infection.

virus infects that host a second time virtually the only response is an increase in antibody to that virus, with an abortion of the disease.

There is frequent reference to "the interferon system." The term "system" is appropriate because the antiviral effect of interferon involves the interaction of two components. The first is the group of cells that are infected with the attacking virus. These cells, in addition to producing new virus, also synthesize a new messenger RNA which encodes for the new protein interferon. The second component consists of those cells that are protected by the interferon. These second cells also undergo several DNA derepressions, leading to the synthesis of several new proteins, which enable the second cells to selectively reject viral RNA for translation, while continuing to translate cells *m*-RNA more or less normally.

It was soon realized that interferon might represent a broad-spectrum antiviral agent, useful for human therapy of a number of viral diseases. Realization of this hope has been painfully slow. It is not too difficult to understand why. A mouse synthesizes up to 10^6 units of interferon in response to a virus infection. Dose-response curves show that perhaps 2–6 $\times$ 10^6 units of exogenous interferon would have to be given to a mouse to have a significant enhancing effect over that interferon which the mouse had synthesized. During the early days of interferon research, a preparation of mouse interferon of 10^3 units/ml was about the best one could do. Therefore, one would have to give a mouse 2–6 liters of that interferon solution to hope for a therapeutic effect. With larger animals the situation was even worse. Recently, due to work by many investigators, the supply of interferon has been greatly increased and limited, very costly experiments with humans are under way.

Attention was turned, therefore, to finding nonviral inducers that would cause the host to synthesize large quantities of its own interferon, and a number of such materials were found. The types are tabulated in Table 1. Some are active both in tissue culture and *in vivo*, and others are effective

Table 1. Chemical (Nonreplicating) Inducers of Interferons

Endotoxins	Polyacrylic acid
Nucleic acids	Polyacrylamide
Mitogens	Maleic acid/vinyl pyrollidine
Polysaccharides	Crotonic acid/vinyl pyrollidine
Antibiotics	Polyvinyl sulfate
Pyran copolymer (maleic anhydride/divinyl ether)	

primarily *in vivo*. We shall limit our discussion to just a few of these types of polymers.

2. Types of Polymers

2.1. Polycarboxylic Acids

The most widely studied polycarboxylic acid interferon inducer is pyran copolymer. Pyran copolymer is an anionic material prepared from 1:2 divinyl ether (DIVE) and maleic anhydride (MA). It is sometimes referred to as Divema. During the polymerization a six-membered pyran ring is presumed to be formed; hence the name pyran copolymer. However, evidence for the pyran ring is equivocal, and perhaps a five-membered ring is actually present.[6,7] There is in the polymer a high density of carboxylic acids, apparently a requirement for interferon induction. Other copolymers with high carboxylic acid density include an acrylic acid–maleic anhydride copolymer and a maleic anhydride homopolymer. These materials are generally not biodegradable, a fact that contributes to both their prolonged action and their toxicity. The molecular weight of the original pyran copolymer was very heterogeneous. The biological activities of the original pyran include the ability to induce good levels of serum interferon, to activate macrophages for antitumor activity *in vitro*, and to inhibit the growth of tumors *in vivo* in rodents.[8] Unfortunately, pyrans proved to be toxic.[9] They can cause anemia, leucocytosis, hepatosplenomegaly, liver enzyme release, and can sensitize to the lethal action of endotoxins. Recent work would suggest that higher-molecular-weight (mw) fractions may be responsible for much of the toxicity.

Fractionation of the original polydisperse pyran into low-mw and high-mw fractions revealed the following.[8,9] While the parent material showed the toxicities mentioned above, the low-mw material was much less toxic. The very-low-mw fractions had antitumor activity but no antiviral activity, while the high-mw fractions had both. The low-mw fraction stimulated phagocytosis, while the high-mw material inhibited it. In particular, one narrow mw range fraction, MVE 2, with a mean molecular weight of 15,500 daltons retained both antitumor and antiviral activity.

The mode of action of pyran against tumors is probably different from the mode of action against viruses. The latter effect involves both the induction of interferon and in some instances, an inhibition of viral polymerases,[10–15] while the action *vs.* tumors probably depends strongly on activation of macrophages and natural killer cells. Human clinical experience with pyran has been restricted to the original unfractionated material. Only very low levels of interferon were induced,[16–18] with no evidence of therapeutic effect. Chills, fever, malaise, and thrombocytopenia, leucopenia, hypertension, and acute nervous-system symptoms were noted.

2.2. Polysaccharides

A number of synthetic and natural polysaccharides, both neutral and anionic, have been found to induce very small amounts of interferon and to exert some very modest antitumor and antiviral activity. These include mannan, galactomannan, and glucan among the neutral components, and polycarboxylic acids of amylose, cellulose, and dextran, polysulfates of dextran, and seaweed polysaccharides and polyphosphates of dextran and mannan among the anionic components.[19–22]

2.3. Nucleic Acids

By far the most effective nonviral interferon inducers have been the double-stranded ribonucleic acids. Two series of experiments led to their discovery. Isaacs[5] postulated that interferon production is the cell's response to the presence of foreign nucleic acid. He presented some data that indeed indicated that treatment of tissue-culture cells with nucleic acids extracted from heterologous cells induced the formation of interferon. Even chemical modification of homologous nucleic acids with nitrous acid was sufficient to make the nucleic acid an interferon inducer. However, the amount of interferon induced was very small, the experiments were hard to reproduce, and the question of nucleic acid induction of interferon was held in abeyance. More direct evidence leading to the development of the double-stranded RNAs came from work with Helenine,[2,3] a crude material, found in cultures of *Penicillium funiculosum*, which shows antiviral activity. Helenine contains a ribonucleoprotein that stimulates tissue-culture cells and mice to make interferon. When Helenine was extracted with phenol and the resulting product partially purified, a double-stranded RNA was obtained that was able to induce interferon. It was later shown that the double-stranded RNA was derived from a virus that infected the *P. funiculosum*.

A series of papers from workers at Merck, Sharp and Dohme[25-29] showed that a variety of both natural and synthetic double-stranded RNAs were effective interferon inducers in tissue cultures and in rodents. The homopolymer pair, polyriboinosinic–polyribocytidylic acid was the most effective of the synthetics, with polyriboadenylic–polyribouridylic acid being significantly less so. As little as 0.5 μg of poly(I)–poly(C) given intravenously to a rabbit produces detectable interferon. In some tissue-culture cells, even less of the compound is effective. In general, single-stranded RNAs are less active as interferon inducers although, under some conditions, they can cause the production of significant amounts.[30,31]

The observations regarding the capacity of poly(I)–poly(C) to induce interferon in rodents triggered a search for compounds that would be even more effective. From these studies there has emerged the realization that a number of structural requirements must be combined in a compound in order for it to be a good interferon inducer.

1. There needs to be a secondary structure that is stable at the temperature of the test. The two strands of each double-stranded RNA disassociate from each other at a specific transition temperature T_m, which is a measure of the stability of double-stranded RNA. If one attempts to correlate the T_m of a group of double-stranded RNAs with their ability to induce interferon, one can discern a trend toward correlation. However, there are so many exceptions that it is obvious that other factors in addition to the degree of thermal stability are important. Single-stranded DNAs that have secondary structures can induce interferon both *in vivo* and in tissue culture although, in general, not nearly so well as double-stranded RNA.[30,31] It would appear that secondary structure, not necessarily double strandedness, is a requirement for activity.

2. Another factor that is important, though not dominant in determining the degree of activity of a double-stranded RNA, is its resistance to the action of ribonucleases. The single-stranded RNA, poly AU, is a poor inducer of interferon. When thiophosphates are substituted for the phosphate groups, the resultant poly (As US) is much more resistant to nuclease action and is a better inducer than poly AU. However, there are exceptions to this generality. The effect of differences in the amount of nuclease action in the sera of different animal species will be mentioned later.

3. In the one case where it has been possible to study a series of chemically identical double-stranded RNAs with differing molecular weights, it appears that a certain minimum molecular weight is necessary for action.[32-34] (P34). Poly(I)–poly(C) with molecular weight 2.7×10^5 daltons is inactive, but the compounds with molecular weight higher than 2.7×10^5 daltons are active.

4. A ribose backbone is needed. Single- or double-stranded DNAs induce little or no interferon ( ) in spite of generally high T_m. If the 2′ hydroxy groups on the ribose are esterified with a methyl group, the double-stranded RNA loses its interferon-inducing activity.

The list of antiviral drugs of any type that are effective *in vivo* is small indeed. Poly(I)–poly(C) is the most successful. Its effect has been largely limited to rodents and rabbits. This fact will be referred to again later. Two examples of the antiviral action are as follows. When rabbit eyes are abraded and infected with herpes virus, they develop a keratoconjunctivitis resembling the human disease. Figure 2 shows the data obtained by Park and Boron in treating this disease with poly(I)–poly(C), in the form of eye drops.[35] The abscissa is the time in days after infection; the ordinate is a number obtained by combining the evaluation of several clinical parameters. The higher the number, the more severe the disease. It can be seen that if treatment is begun on the same day as infection, no disease develops. One can wait as long as three days after infection to begin treatment and still have a significant curative effect. However, if treatment begins on day four, the drug is without therapeutic value.

In other experiments by Worthington and Baron,[36] mice were infected with semiliki forest virus, a virus that causes fatal encephalitis. Figure 3 plots the percentage of the animals that die as a function of days after infection, without the treatment by poly(I)–poly(C). It can be seen that untreated animals are dying by day five, and that virus is replicating in the brain by day

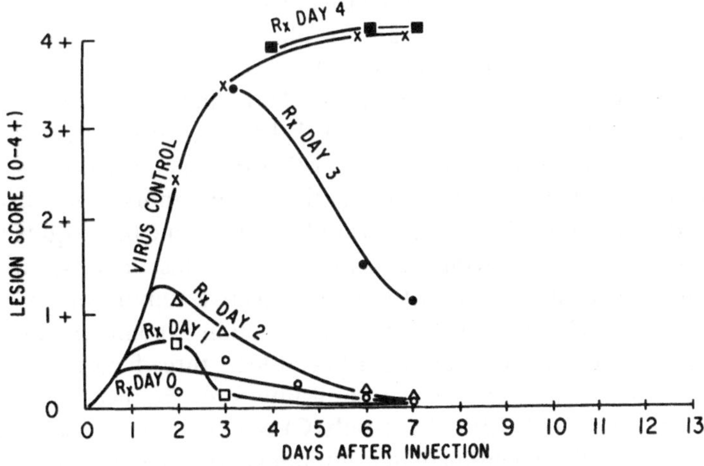

Figure 2. Response of herpetic keratoconjunctivitis to topical treatment with poly(I)–poly(C) acid; R_x indicates treatment.

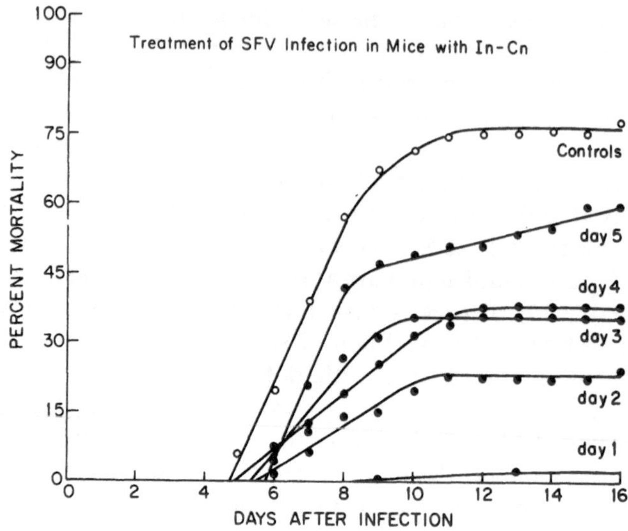

Figure 3. Treatment of semliki forest virus infection in mice with poly(I)–poly(C).

four. However, initiation of treatment even after the virus was in the brain resulted in a significant decrease in the mortality rate.

These laboratories studied the effect of poly(I)–poly(C) on tumors.[37] Together with Law and Rabson, we looked at a tumor that had originally been induced with an adenovirus, and then was carried by transplantation. There was no detectable infectious virus at that time. About 12 days after transplantation, the tumor was quite visible. After five daily interperitoneal treatments with 150 μg per day of poly(I)–poly(C), about 40% of the tumors were rejected, with hemorrhagic necrosis. These tumors grew back in spite of continued treatment and ultimately killed the mice, although survival time was longer than in untreated mice. With other tumors there was a significant number of cures. Table 2 gives a partial list of the tumors. The sensitivity of different tumors to the drug is quite variable, with some being quite sensitive and some, the fast-growing leukemias, being just barely statistically significantly affected.[37–40] With the more sensitive tumors up to one-third survived. The mechanism of this antitumor action is complex, and will not be discussed in detail. There appear to be at least three factors involved: (1) poly(I)–poly(C) induces the formation of interferon in mice, and interferon has antitumor action[41]; (2) poly(I)–poly(C) is a potent enhancer of immune reactivity, particularly cell-mediated immunity, the type that is thought to play an important role in natural host defense mechanisms against tumors[42,43]; (3) there is a more or less specific inhibition of tumor

Table 2. Effect of Poly(I)–Poly(C) on Animal Tumors[a]

Tumor	Increase in median survival over control (%)
J96132-Reticulum cell sarcoma (subcutaneous)	130[b]
J96132-Reticulum cell sarcoma (ascites)	96[b]
Carcinosarcoma Walker 256	100
Reticulum cell sarcoma RCSL	89
Ehrlich ascites tumor	70
S91 Melanoma	55
Fibrosarcoma	52
B1237-Lymphoma (ascites)	45
L1210 Leukemia	42
Plasma cell YPC-1	39
B1237-Lymphoma (subcutaneous)	28
MT-1 tumor (subcutaneous)	26
Reticulum cell sarcoma ovarian	20
Leukemia P388	16
Leukemia L1964	12

[a] Treatment, in most cases, was 150 μg per mouse daily, or three times weekly, by intraperitoneal route. With the exception of the J96132-Reticulum cell sarcoma, some Ehrlich ascites tumors, and a few Walker carcinosarcoma, all animals ultimately died.
[b] Mean day of death of the animals that died. About 30% of all the animals treated have survived, although treatment had been stopped at about day 50.

macromolecule synthesis in some of the animal tumor systems.[44] These three elements may interplay in different quantitative degrees in the different tumors.

On the basis of these findings in animals we embarked on a cooperative program with the National Cancer Institute to test poly(I)–poly(C) against tumors in humans. The results were very disappointing. Unfortunately, the human response to poly(I)–poly(C) is very weak, even with high doses of the drug.[45–47] Levels of 50 units of interferon per milliliter of serum were found, but usually not much more, as contrasted with perhaps 50,000 units/ml in mice. Rhesus monkeys and chimpanzees were totally nonresponsive to poly(I)–poly(C).

Some experiments by Nordlund *et al.* shed light on the poor primate response.[48] Poly(I)–poly(C) is a pyrogen. In rabbits the fever response to poly(I)–poly(C) is shown in Figure 4. If poly(I)–poly(C) is incubated with primate serum and then it is injected into rabbits, the pyrogenic effect is gone, as shown in Figuire 5. Further testing revealed that the poly(I)–poly(C) had been hydrolyzed to acid-soluble products and was no longer able to induce interferon. A comparison of the ability of human and rabbit sera to hydrolyze poly(I)–poly(C) is shown in Table 3. It can be seen that rabbit serum hydrolyzes much less than does human serum. By and large those

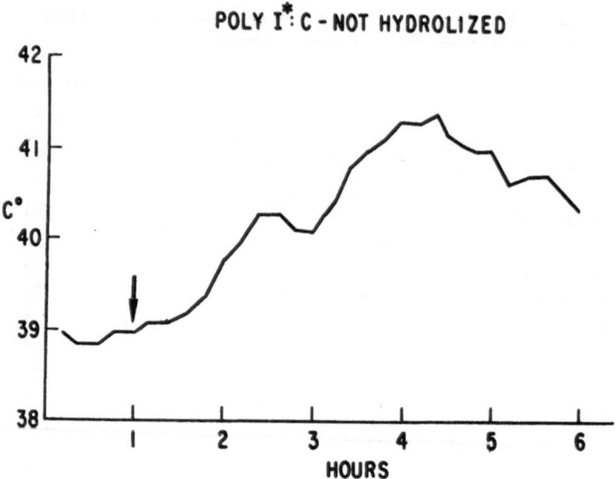

Figure 4. Fever response in rabbits (mean of 3) after iv injection of 30 μg poly(I)–poly(C) in 0.17 ml of 0.15 *M* pyrogen-free saline.

species of animals that showed a large capacity to hydrolyze were poor responders, and good responders had low hydrolytic capacity. This correlation does not establish a cause-and-effect relationship between hydrolysis capacity and response to the drug, but it is suggestive. An attempt was made to prepare a complex of poly(I)–poly(C) that would be more resistant to hydrolysis than the parent compound, that would not be very

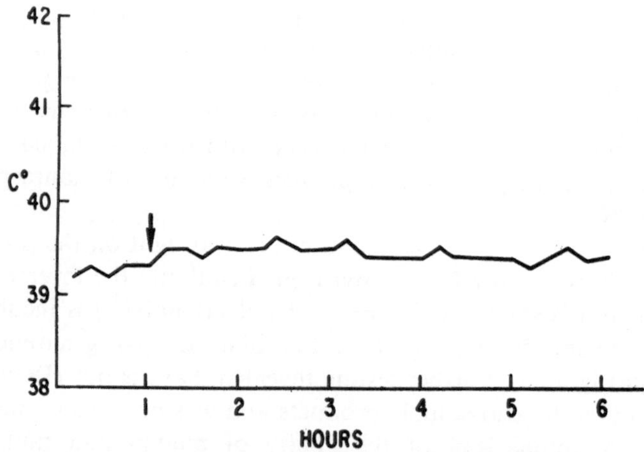

Figure 5. Fever response in rabbits (mean 3) after iv injection of 30 μg poly(I)–poly(C) in 0.15 ml of 0.15 *M* pyrogen-free saline that had been treated with human serum.

Table 3. Acid-Soluble Labeled Poly(I)–Poly(C) (%) after Incubation with Human and Rabbit Sera for Different Periods at 37°C

Serum	Isotope	Hours of incubation			
		0	1	4	16
Human	^{3}H (Cytid)	4.5	48	85	91
	^{14}C (Inos.)	2.4	2.8	4.2	7.9
Rabbit	^{3}H (Cytid)	0.4	0.5	1.5	4.5
	^{14}C (Inos.)	2.1	2.3	2.8	3.9

toxic, and that would induce interferon. A complex of poly(I)–poly(C) that is partially resistant to this hydrolysis was prepared in the following way. A hydrophilic complex was made between poly-L-lysine of molecular weight of 27,000 daltons and carboxymethylcellulose. This was further complexed with poly(I)–poly(C) to give a colloidal solution containing complexed poly(I)–poly(C), a complex called poly(ICLC).[49] A comparison of resistance to hydrolysis of two different preparations of poly(ICLC) and poly(I)–poly(C) is shown in Figure 6. Poly(ICLC) is about 10 times more resistant to hydrolysis than poly(I)–poly(C).

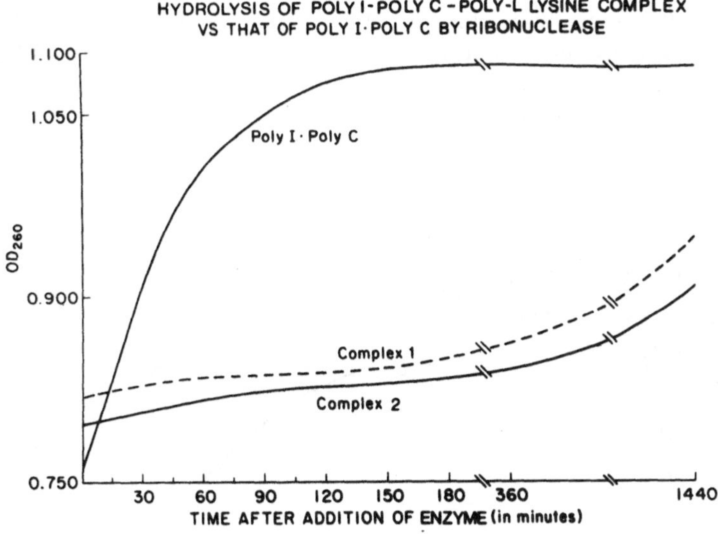

Figure 6. Hydrolysis of poly(I)–poly(C) and two different lots of the poly-L-lysine complex of poly(I)–poly(C) by pancreatic RNase. The complexes, at a concentration of 50 μg poly(I)–poly(C) per milliliter in 0.15 *M* NaCl–0.001 *M* phosphate buffer (pH 7.2), were exposed to 5 μg pancreatic RNase per milliliter at room temperature (about 24°C). Optical density (OD) readings at 260 nm were taken at 10-min intervals.

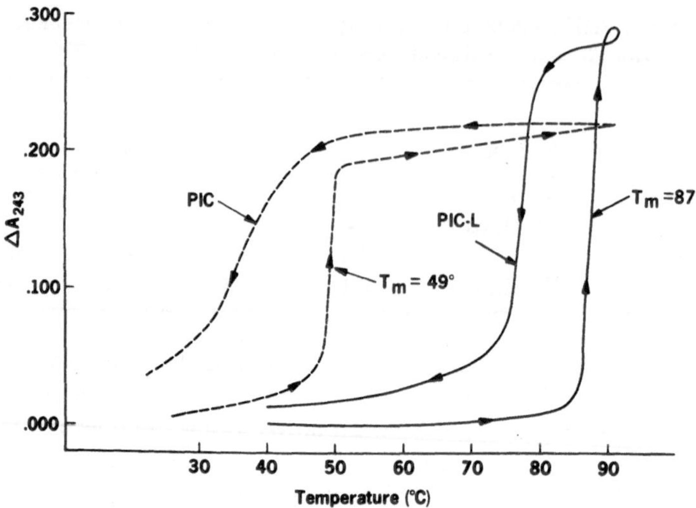

Figure 7. Thermal denaturation of poly(I)–poly(C) and the poly-L-lysine complex of poly(I)–poly(C) (PIC-L). The compounds, at a concentration of 50 μg poly(I)–poly(C) per milliliter in 0.1 standard saline citrate were heated to the indicated temperatures in a recording spectrophotometer set at 245 nm. T_m represents melting temperature.

Poly(ICLC) is a more stable structure to thermal denaturation than poly(I)–poly(C) is and this is shown in Figure 7. Poly(ICLC), in 0.15 *M* NaCl, does not denature below 100°, while poly(I)–poly(C) had a T_m of 62.5°. It was necessary to dilute the salt to 0.015 *M* to obtain a T_m of 87° for the complex, while plain poly(I)–poly(C) melted at about 49°C.

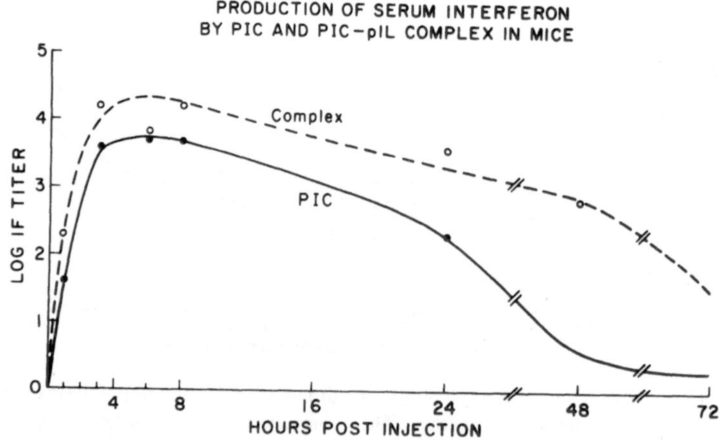

Figure 8. Kinetics of induction of serum interferon (IF) in mice after iv administration of 5 mg/kg poly(I)–poly(C) or poly(ICLC).

The compound was slightly more effective in mice as an interferon inducer than poly(I)–poly(C), as shown in Figure 8. Serum interferon was detectable slightly earlier, rose to a somewhat higher titer, and lasted longer. Not surprisingly, poly(ICLC) is a somewhat better antiviral agent in mice than is a poly(I)–poly(C) (L. Glasgow, unpublished observation).

Of greater interest was the fact that poly(ICLC) was an effective inducer in monkeys and chimpanzees (Figure 9). Interferon levels as high as 15,000 units/ml of serum have been found in cynomolgus and rhesus monkeys, under conditions where no interferon was induced by poly(I)–poly(C). However, levels of 200–2000 units are more regularly seen.

The new compound is an effective antiviral agent in monkeys. Monkeys injected with street rabies virus could be completely protected by poly(ICLC) together with antirabies vaccine, as shown in Table 4. Vaccine alone had no protective effect.[50]

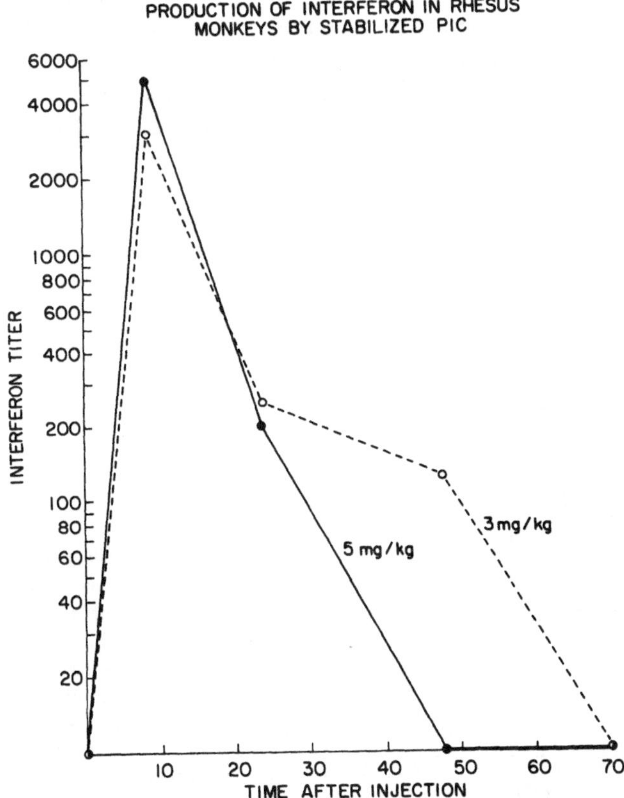

Figure 9. Kinetics of induction of serum interferon in rhesus monkeys by iv administration of 3 or 5 mg/kg poly(ICLC) (one dose per monkey).

Table 4. Mortality of Rabies-Infected Rhesus Monkeys Treated Six Hours after Challenge, with Rabies Vaccine Plus Interferon or Rabies Vaccine Plus Inducer

Treatment	Mortality
None (control)	7/8
Vaccine + poly(ICLC)	0/8
Vaccine + interferon	1/8

Comparable results were obtained with yellow fever virus, and with simian hemorrhagic fever virus, both of which are fatal viruses for monkeys.[51,52]

There is an animal model of chronic hepatitis in young chimpanzees. Figure 10 shows the effect of treatment of such chimps with poly(ICLC).[53] It can be seen that when the animals are injected with poly(ICLC) they produce interferon. A marker of the progress of the infection is the level of DNA-dependent polymerase in the blood. During the course of the treatment the polymerase activity fell to background level. Unfortunately, when treatment stopped, evidence of the disease returned. Whether really prolonged treatment would have a more permanent effect has yet to be determined.

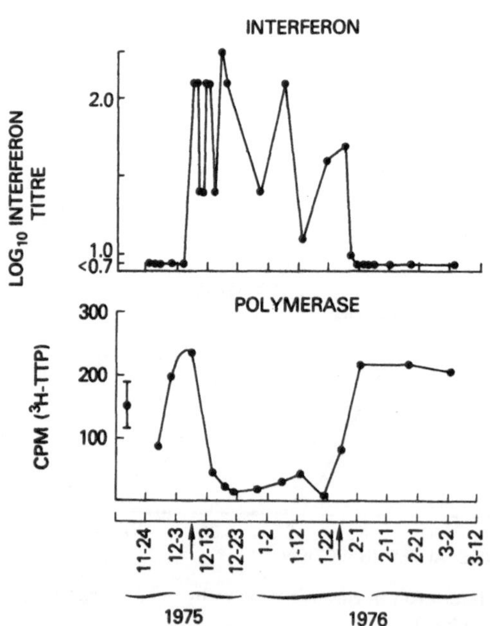

Figure 10. Effect of treatment with poly(ICLC) on markers of chronic hepatitis-B virus infection in chimps. The beginning and end of the treatment period are indicated by arrows on the abscissa. The dot and bar on the graph of DNA polymerase response indicate the mean (± SD) of polymerase activity detected in six serum samples obtained during the five weeks immediately preceding the experiment.

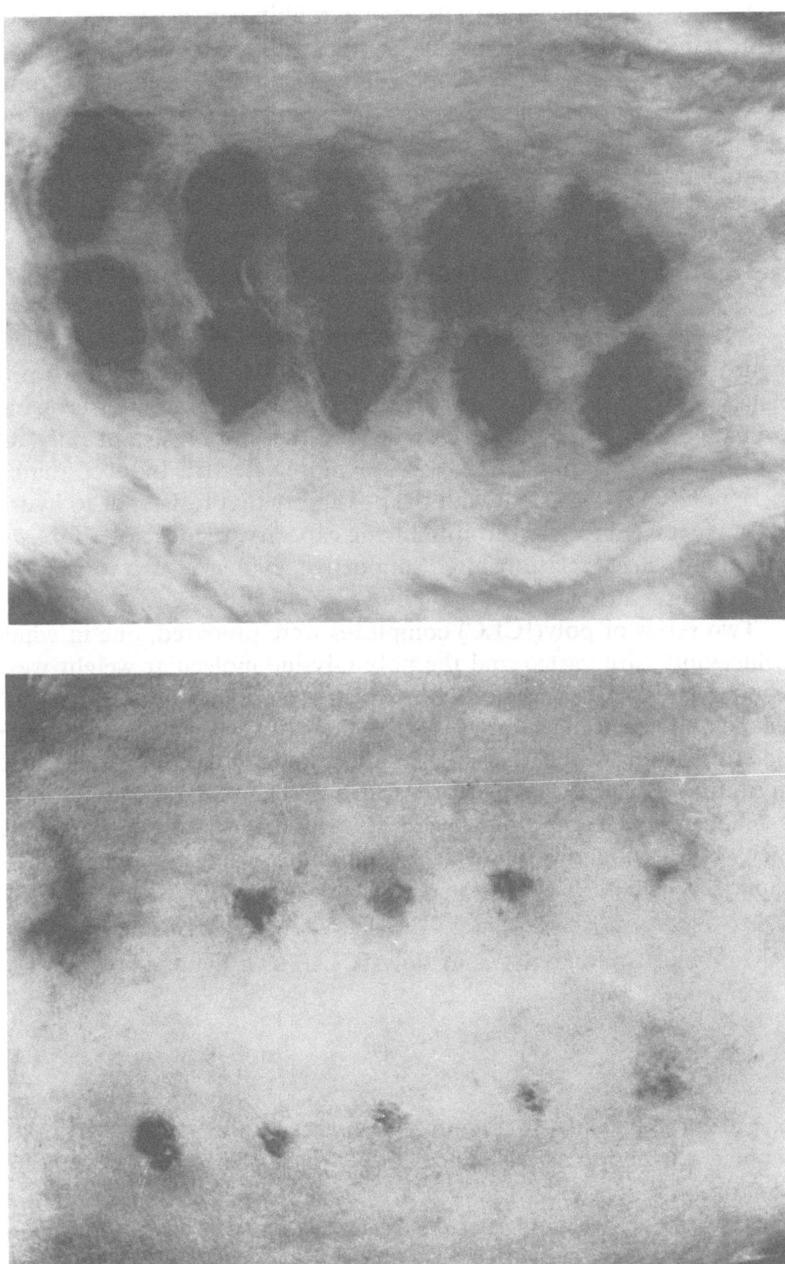

Figure 11. Effect of poly(ICLC) ointment on already-established vaccinia lesions on rabbit skin.

Vaccinia virus infection in the skin in rabbits can be a severe, sometimes fatal, infection. Rabbits were injected intradermally with vaccinia virus. After the lesions became visible, some rabbits were treated with an ointment containing the drug. A comparison of the treated and the nontreated rabbits is seen in Figure 11. The infection essentially did not progress after initiation of treatment.[54]

3. Effect of Size of Components on Effectiveness of Poly(ICLC)

As indicated above, the presence in primate sera of high concentrations of hydrolytic activity against poly(I) · poly(C) was advanced as a possible explanation of the ineffectiveness of poly(I) · poly(C) as an interferon inducer in primates, and led to the development of the more resistant poly(ICLC). While the fact that poly(ICLC) was a more effective inducer than plain poly(I), poly(C) is consistent with the possibility that resistance to hydrolysis by serum is related to interferon-inducing capacity, it does not establish such a cause-and-effect relationship. Supportive evidence was given by the following observations.

Two series of poly(ICLC) complexes were prepared, one in which the polynucleotide size varied and the poly-L-lysine molecular weight was kept constant (series A), and the other in which the polynucleotide size was kept constant and the molecular weight of the poly-L-lysine varied (series B). A determination of the thermal denaturation (T_m) profile, sedimentation pattern, and resistance to hydrolysis at different concentrations of RNase was made.[55] All compounds were studied for their interferon-inducing capacity in nonhuman primates (rhesus monkeys) and in mice. The data are summarized below.

3.1. Molecular Characteristics of Poly(ICLC) as a Function of the Component Size

3.1.1. T_m

Formation of a stable complex between double-stranded ribonucleic acid and a polyamino acid is indicated by a rise in T_m.[56] In our experiments, formation of 4S, 6S and 9S poly(ICLC) complexes was also associated with an increase in T_m (Figure 12). All complexes appeared homogeneous, as judged by the shape of the thermal transition curve, with a single-step sharp increase in A_{260} near the T_m point. There were reproducible increases in T_m of about 2–2.5°C, as one went from 4S poly(ICLC) to 6S poly(ICLC) to 9S

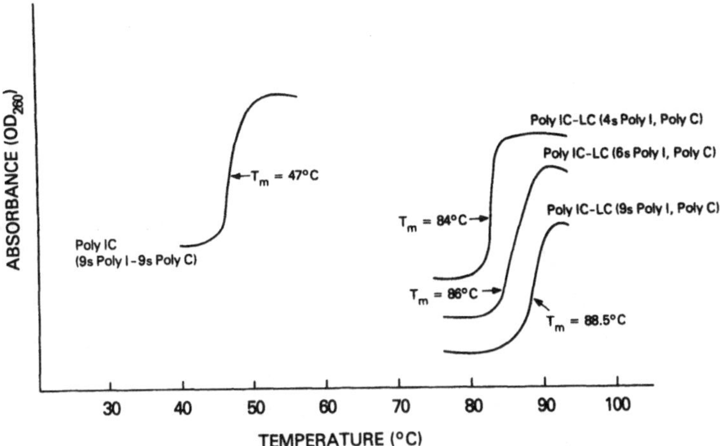

Figure 12. T_m profiles of 9S poly(I) · poly(C) and modified poly(ICLC) complexes in 0.1X standard saline citrate: 4S poly(ICLC); 6S poly(ICLC); and 9S poly(ICLC). The initial A_{260} was about 1.0.

poly(ICLC), with the latter having a T_m of 88.5.* In the series B complexes, changing the molecular weight of polylysine had only a very slight effect on the T_m.

3.1.2. Hydrolysis

It was shown previously that 9S poly(ICLC) is more resistant to hydrolysis by RNase than is poly(I) · poly(C).[49] Figure 13 shows the percent increase in optical density of 4S, 6S, and 9S poly(ICLC) upon incubation with different concentrations of RNase. Poly(I) · poly(C) was readily hydrolyzed by low levels of RNase. The complexes with poly-L-lysine and carboxymethylcellulose were more resistant to hydrolysis. At any given concentration of RNase, the rate of hydrolysis with time was most rapid with poly(I) · poly(C) and was slowest with 9S poly(ICLC) (data not shown). Also for 9S poly(I) · poly(C), a maximum hydrolysis of 56% was achieved at an RNase concentration of 5 μg/ml. For 9S poly(ICLC), the maximum level of hydrolysis was 15% at 35 μg of RNase per milliliter. Any further increase in RNase concentration did not result in higher levels of hydrolysis for this complex. A higher level of hydrolysis (33%) at the same concentration of RNase (35 μg/ml) was seen with 6S poly(ICLC). Maximum hydrolysis for 4S

* The difference between the T_m's of Figures 12 and 7 is attributed to the fact that two different machines were used to obtain the data. The data of Figure 12 are considered more authentic.

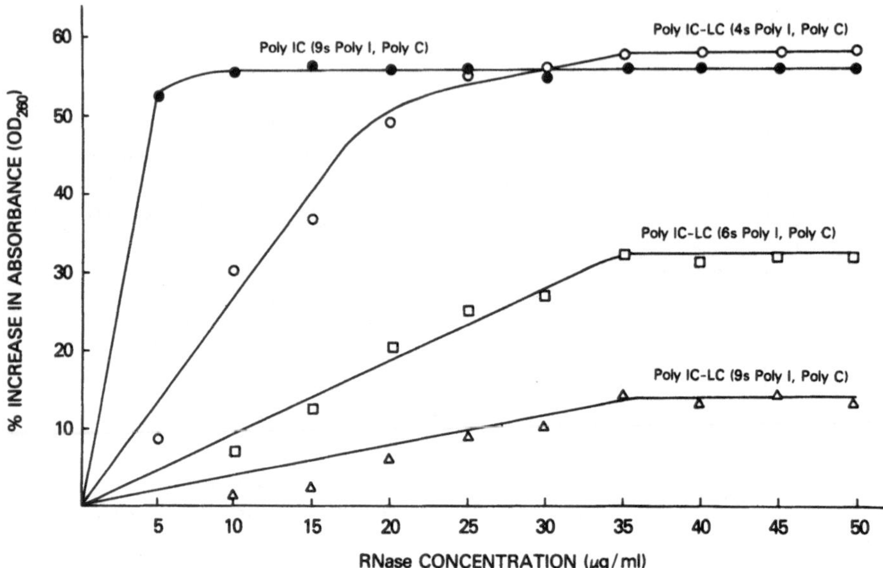

Figure 13. Hydrolysis by RNase of poly(I) · poly(C) and poly(ICLC) complexes in 0.01 M PO_4 (pH 7.2)–0.15 M NaCl (percent increase in A_{260} for 1 h at room temperature by the indicated concentration of RNase):(●) 9S poly(I) · poly(C); (○) 4S poly(ICLC); (□) 6S poly(ICLC); (△) 9S poly(ICLC). Poly-L-lysine was 27,000 daltons.

poly(ICLC) by 35 μg of RNase per milliliter was similar to that of 9S poly(I) · poly(C) (56%). Resistance to hydrolysis also increased when the molecular weight of the polylysine increased.

The larger the polynucleotide or polylysine strand, the smaller the maximum rise in absorbance on hydrolysis. One can only conjecture why this should be so, but it may reflect the possibility that, with the shorter components, there are more unprotected linkages than with the longer strands. This concept would be consistent with the relative T_m values of the complexes, at least for those complexes with different polynucleotides.

Although the complexes of poly-L-lysine with poly(I) · poly(C) made with any of the homopolymers tested resulted in compounds with a much higher T_m than that of the uncomplexed poly(I) · poly(C), not all of the poly-L-lysine complexes induced interferon in monkeys. The ability to resist hydrolysis by RNase (and presumably primate serum) appears to correspond most closely with the interferon-inducing capacity in primates (Table V). This was so whether one measured the final extent of hydrolysis or the rate of hydrolysis.

These data lend additional support to the idea that the degree of resistance to hydrolysis of the complexes is an important determinant in the ability to induce interferon in primates.

4. Effect of Inducers on the Immune System—Antibody Production

The realization that interferon inducers can modify antibody production developed early, with studies on synthetic polynucleotides playing a pivotal role.[57] Three assay systems have been used in studying the effects of interferon inducers on antibody production: (1) a totally *in vivo* system in which animals are immunized, and, after a suitable time, the amount of circulating antibody is measured; (2) an *in vivo–vitro* system, where animals are immunized, and, after a shorter period of time, spleens are removed. The number of spleen cells (lymphocytes) capable of engaging in producing antibody to the immunogen is determined by use of a plaque-forming technique in tissue culture (Jerne Plaque number); and (3) a totally *in vitro* system, the Mishell–Dutton technique which is described later.

It had been shown that natural nucleic acids were active in augmenting antibody formation to sheep red blood cells (SRBC) in mice, but single-stranded synthetic polynucleotides were inactive. When double-stranded synthetic polynucleotides, such as poly(A) · poly(U), poly(I) · poly(C), or poly(G) · poly(C) were used, pronounced augmentation was obtained. The single-stranded polynucleotides did prove active in tissue-culture test systems. Braun and co-workers suggested that the difference between the results of the *in vivo* and *in vitro* systems might be attributable to a rapid enzymatic degradation of the single-stranded polynucleotides in an *in vivo* situation, while the breakdown is decreased when the polynucleotides are double stranded.[58,59] The antibody augmenting effect was demonstrable earliest (within 48 hours) in the *in vivo–in vitro* test where the number of plaque-forming cells was measured, but could be seen later when circulating precipitating antibody levels were determined. Double-stranded polynucleotides such as poly(A) · poly(U) or poly(I) · poly(C), can restore impaired antibody response or elicit antibody response under conditions where such response is not normally made. As an example of the latter, four-day old mice do not ordinarily form plaque-forming cells to sheep red blood cells, but when SRBC are given along with poly(A) · poly(U), poly(I) · poly(C), or endotoxin, the mice make a good response.[60] C57 B1 mice, as they age, develop a decreased antibody response to sheep red blood cells. This decreased response is more than overcome by immunizing together with a double-stranded RNA.

In connection with the question of whether the immune modulation brought about by the inducers is attributable solely to the interferon induced, it is important to note that at least insofar as antibody response to SRBC in mice is concerned, poly(A) · poly(U) is as good as poly(I) · poly(C) ; but poly(A) · poly(U) is a poor interferon inducer, while poly(I) · poly(C) is a good interferon inducer.

The augmenting effects of double-stranded RNA, specifically poly(I) · poly(C) and poly(ICLC), on antibody production have been seen with other antigens, including bovine gamma globulin,[61] influenza virus vaccine,[62] a polysaccharide vaccine made for *Hemophilus influenzae*, a vaccine against Rift Valley fever virus,[63] and an envelope antigen from herpes virus.[64] However, poly(ICLC) is not a universal adjuvant. When used with strong antigens, such as serum albumin or pneumococcal polysaccharide type III, there was seen an inhibition of antibody production.[65]

5. Effects on Cell-Mediated Immunity

In addition to the effects on antibody production, poly(ICLC) has effects on aspects of cell-mediated immunity. It enhances delayed-type hypersensitivity in mice as measured by foot-pad swelling, an effect opposite to that brought about by exogenous interferon. Comparable differences between interferon and poly(ICLC) have been found with regard to graft *vs.* host reaction, with interferon being inhibitory and poly(ICLC) being stimulatory. These differences also are found in the effects on the formation of glycoprotein cell-stimulating factor (CS) and on the growth of a pluripotential stem cell colonies (CFC).[66,67]

Two natural defense mechanisms that are currently considered important in biological defenses against tumors are natural killer (NK) cells, and macrophages. Both of these cell types are activated *in vivo* and *in vitro*, in mouse and humans by poly(ICLC). The effect of exogeneous interferon can be inhibitory or augmentatory, depending on several factors.

So even though poly(ICLC) leads to the formation of high concentrations of serum interferon which is inhibitory by itself, the net effect of poly(ICLC) is generally stimulatory. A reasonable explanation is that poly(ICLC) induces the formation of stimulatory factors, in addition to induction of interferon.

6. Interferon Inducers and Human Disease

In this section, we will discuss the efficacy and toxicity of interferon inducers as therapeutic agents in human disease. The reader is referred to several other reviews for further in-depth discussion of clinical trials with interferon and interferon inducers.[1–4]

6.1. Polyinosinic–Polycytidylic Acid [Poly(I)–Poly(C)]

As discussed earlier, the interferon inducer poly(I)–poly(C) has been shown to be a good inducer of interferon in rodents and has been shown to be effective in several experimental animal viral infections and neoplasias.[49,50,53] In 1971, Field *et al.* treated 20 patients with advanced cancer with poly(I)–poly(C), 14 of whom developed serum interferon levels of 3–32 units/ml after a single intravenous adminstration of the drug.[45] Interferon appeared in the serum of these patients in two hours, with peak levels generally between the 12th- and 48th-hour postinoculation. There was no apparent antitumor effect in these 20 patients with advanced carcinoma. Side effects included nausea, vomiting, and fever. The toxicity appeared to be dose related and the drug was relatively well tolerated at lower dosages. Development of a hyporesponsive period to poly(I)–poly(C) therapy was observed in several subsequent studies; capacity for induction was renewed six to seven days after the first admistration. Perhaps the most disappointing finding in comparison to the animal studies, was that poly(I)–poly(C) was a poor interferon inducer in man and had no detectable antitumor or antiviral effect. This was confirmed in 1976 by Robinson *et al.*[46] who examined poly(I)–poly(C) in a phase I clinical trial of patients with advanced cancer. *In vitro* studies showed that poly(I)–poly(C) was rapidly hydrolyzed by primate sera, suggesting that such hydrolysis could explain its relative ineffectiveness in primates *in vivo*.[47] As discussed earlier in the chapter, this eventually led to the development of the hydrolysis-resistant compound poly(ICLC).[49]

Despite the low induction of interferon by poly(I)–poly(C), the efficacy of the drug both as an antiviral agent and immunostimulant was assessed in several clinical trials over the past decade. Hill *et al.*[40] administered poly(I)–poly(C) intranasally to volunteers over a seven-day period prior to inoculation of either rhinovirus 13 or type A2 influenza virus. In these studies, toxic effects were not detected and there was a small, but definite reduction in symptoms of upper respiratory tract illness associated with drug treatment as compared to placebo. However, in only one of the two rhinovirus studies was there a reduction in viral shedding. Treatment had no effect on the shedding of A2 influenza virus. The authors' impression of the results was that the minimal amounts of nasal interferon stimulated by the poly(I)–poly(C) were responsible for the less than optimal results obtained.

In a study by Feldman *et al.*[69] 24 children with advanced cancer and localized herpes zoster infection were randomized to treatment with topical poly(I)–poly(C) in patients receiving frequent topical treatment. The median duration of lesions in the treated group was 3.1 days, as opposed to 3.5 days in the placebo-treated patients; this difference was not statistically significant. Moreover, the total median days of lesions and percentage of dermatome involvement were similar in both groups, as were complications. Similarly, 15 children with viral neurological disease, including subacute panencephalitis, congenital rubella, and Reye's syndrome, were also treated with poly(I)–poly(C) with relatively low peak levels of interferon induction ranging from 8 to 500 units/ml in response to 0.1–1.0 mg/kg.[70] While there were few side effects when given intravenously, there did not appear to be any significant clinical response to the drug. Other negative studies involved with the clinical application of poly(I)–poly(C) included the lack of efficacy in treatment of either herpes simplex virus infection or varicella zoster virus infection in 24 patients with cancer, in a male infant with herpes simplex encephalitis, and in a six-year-old girl with terminal acute leukemia with varicella infection.[71,72]

In one recent study by Kemeny *et al.*[73] poly(I)–poly(C) was utilized as adjunctive therapy in 32 patients with superficial urinary bladder tumors. Poly(I)–poly(C) was administered intravenously (25 ug/kg every two weeks for one year) after complete transurethral resection and/or fulguration of tumor. Cystoscopies were performed at three to four month intervals for several years following initial therapy. While there were no statistical differences in the tumor recurrence rates in the two groups during the first year, a significant increase in survival at four years was observed between the poly(I)–poly(C) group as compared to the control group. While studies such as the above need confirmation, this particular study did indicate a favorable effect of poly(I)–poly(C) after an initial decrease in the overall tumor mass. These effects did not appear to be directly related to interferon production since levels in these patients of interferon were as low as in previous studies, suggesting that perhaps the immunostimulant effect of this drug may have been effective. Certainly, the concept of adjunctive therapy with interferon inducers in patients whose tumor burden was decreased by prior surgery or chemotherapy should be considered.

6.2. Polyinosinic–Polycytidylic Acid Poly-L-Lysine [Poly(ICLC)]

The failure of poly(I)–poly(C) to induce significant levels of interferon in monkeys or in patients with cancer and viral infections eventually led to the understanding of the *in vivo* hydrolysis of poly(I)–poly(C). A modification of poly(I)–poly(C) with the addition of poly-L-lysine and carboxymethylcellulose led to a compound poly(ICLC) which was relatively stable in *in vivo* hydrolysis.[49] This substance induced much higher levels of interferon in primates than did poly(I)–poly(C), and decreased the hyporesponsive period that had been observed with poly(I)–poly(C).[74] Overall, dose-limiting side effects were reported, but tolerable dosages, which were associated with nausea, vomiting, and fever, still resulted in acceptable levels of interferon production. As commented earlier, high levels of protection against lethal simian hemorrhagic fever, rabies, yellow fever, and other viruses have been obtained in primates.[50,52–54,75] Therefore, in 1978 a phase I clinical human trial was initiated at the National Cancer Institute. In this study by Levine *et al.*, poly(ICLC) was given intravenously, in 15 daily doses of 0.5–27.0 mg/m^2, to 19 patients with various solid tumors and to 6 patients with acute leukemia.[76,77] The drug was administered over one hour each day for a total of 14 days. Three complete trials were conducted at each of six dose levels. The maximum tolerated dose for all patients was 12 mg/m^2, at this dose the mean peak interferon titer was 1940 reference units/ml. At 18 mg/m^2 the mean peak interferon titer was 4473 reference units/ml, but toxic side effects, such as myalgia, arthralgia, and significant hypotension, limited the administration of this dosage. Peak levels of interferon were obtained eight hours after each dose and were dose dependent. At high doses, intravenous poly(ICLC) also induced interferon in the cerebrospinal fluid. Great variability in peak interferon levels was seen from patient to patient, but the correlation between dose and peak interferon titer was linear.

In regards to antitumor efficacy, most of the patients who received the drug in this initial study had advanced cancer, including acute leukemia, sarcoma, nonHodgkins lymphoma, melanoma, and carcinoma of lung, kidney, liver and prostate. Over a six-week period of observation, 23 of 25 patients demonstrated progressive tumor involvement. One adult with acute undifferentiated leukemia demonstrated a brief period of stabilization. One child with acute lymphoblastic leukemia appeared to enter a complete remission after treatment, but evaluation of the child's response was complicated by prior exposure to chemotherapy immediately before administration of poly(ICLC). In this initial study, toxic drug reactions included fever (100%), nausea (44%), hypotension (28%), thrombocytopenia and leukopenia (68%), erythema (12%), and polyarthralgia/myalgia (16%). Hypotension and arthralgia–myalgia related specifically to the administered dose and/or the magnitude of interferon induction, whereas other toxic manifestations were not dose related. Since this study clearly demonstrated that poly(ICLC) was an effective interferon inducer, although the toxicities and lack of clinical response were disappointments, additional clinical trials were instituted over the following several years.

Lampkin *et al.*[78] analyzed the effect of poly(ICLC) on childhood acute leukemia. This study encompassed 17 terminally ill children, 14 of whom had null-type lymphoblastic leukemia, the remainder having acute myelocytic leukemia. Unfortunately, there have been no clinical responses in any of the trials, at varying dose levels. There were a number of patients whose marrow was cleared of tumor cells, but they did not survive. However, this study added to further information regarding toxicities of poly(ICLC). Unlike adult patients, the dose of 12 mg/m^2 was too large in these children, with most experiencing unacceptable fever, chills, hypotension, and seizures. Only the 3–6 mg/m^2 dose was well tolerated. Consequently, subsequent studies have used lower dosages of poly(ICLC) to avoid the more severe side effects.

Other studies have investigated the efficacy of poly(ICLC) in a variety of different tumor states, with variable results and toxicity. Krown *et al.*[79] administered poly(ICLC) to 14 patients with advanced cancer including chronic lymphocytic leukemia, renal cell carcinoma, multiple myeloma, and salivary gland tumors. These patients were first treated with low doses (0.01–1.0 mg/m^2) with the dose being gradually increased thereafter until limiting toxicity was reached. Using this escalating-dose schedule, the mean maximum tolerated dose was 4.8 mg/m^2, but individual patients varied considerably in their ability to tolerate the side effects of the drug. Side effects included fever, chills, diaphoresis, leukopenia, thrombocytopenia, and hypotension, clinically significant in 5 of 14 trials. Common to most studies, poly(ICLC) induced significant titers of serum interferon. Levels of interferon were directly related to the administered dose. While hyporesponsiveness to interferon induction was observed with repeated daily administration of the same dose, this effect could be altered by administration every third day or once a week. Despite interferon induction, there was no clinical response in any of the treated patients. Somewhat more encouraging results were demonstrated in patients with refractory multiple myeloma. In nine patients with refractory disease, two showed greater than a 50% reduction in myeloma protein, while four showed subjective improvement with lesser degree of decrease of myeloma protein.[80]

Encouraging results with poly(ICLC) treatment have recently been reported in patients with recurrent laryngeal papillomatosis. Leventhal *et al.* treated three children with recurrent laryngeal papillomatosis with poly(ICLC) following surgical therapy.[81] There was clearly a dramatic decrease in the rate of recurrence of endobronchial disease in patients treated with poly(ICLC) after surgery, but continued administration of the drug was required to maintain the effect. There was no regression of established pulmonary lesions in these cases. Four additional cases also showed improvement in their disease (Leventhal and Whiznant, personal communication).

Poly(ICLC) has also demonstrated clinical efficacy in patients with chronic dyssimmune neuromuscular disease. These patients develop a severe, crippling, sometimes fatal disease of nerves characterized by often profound muscle weakness, including respiratory muscles. They are usually refractory to high, prolonged doses of prednisone plus cytotoxic agents (azathioprine or cyclophosphamide). Engle *et al.*[82] have reported on the response of one patient with dysimmune dysschwannian motor neuropathy who has been treated with poly(ICLC) for four years. Several months after therapy with poly(ICLC) this patient, who had been previously confined to a wheelchair, was able to walk eight miles per day and perform routine muscular tasks. He continues to do well with monthly administration of poly(ICLC). Similar results have been documented in several other patients with this neuromuscular disease. One patient with chronic dysimmune motor-sensory dysschwannian neuropathy and a woman with chronic dysimmune myopathy (polymyositis) both demonstrated marked improvement after several months of intravenous poly(ICLC). These diseases demonstrated the most dramatic response to poly(ICLC), but the mechanism of action is unknown. We favor an antidysimmune mechanism, namely, an anti-lymphocytic effect rather than an antiviral effect. Toxicity of poly(ICLC) in these patients consisted of a transient flu-like syndrome (fever, chills, headache, malaise, and myalgia) for 12–24 hours following therapy. There was an associated leukopenia to 2–20% of baseline values with return to normal by three to five days; polymorphonuclear leukocyte response was biphasic, with a twofold to threefold rise at 6–24 hours, a 30% fall below baseline at two to three days, and return to normal by five days. Interferon levels have ranged widely, varying with both patient and dose, from 20–1000 units with each poly(ICLC) dose. No persistent side effects have occurred in any patients.

A few small clinical trials of poly(ICLC) in patients with viral infections have been instituted. In the first study by Champney *et al.*[83] fourteen patients with several viral illnesses, including St. Louis encephalitis, disseminated varicella–zoster infection, and herpes simplex virus encephalitis were treated with poly(ICLC) intravenously at varying doses. No evident benefit resulted from the use of poly(ICLC) in these patients. Topical poly(ICLC) did not shorten the duration of genital herpetic recurrences and the pain experienced was unaltered.[84] The progression of healing and size of the involved area were similar in both the drug and placebo cohorts. In addition, the shedding of the virus was the same in each sexual stratum, but with a significant fall in virus titer within 60 hours after the onset of clinical symptoms in women given poly(ICLC) ($p = 0.05$). Unfortunately, poly(ICLC) was not assessed in patients with initial genital herpes which has shown a more dramatic response to other antiviral agents which have failed in recurrent herpes.

In summary, over 182 reported patients with either advanced cancer or severe viral infections have been treated with poly(ICLC) with varied results. From the dramatic responses to therapy with poly(ICLC) demonstrated in patients with neuromuscular diseases and in laryngeal papillomatosis, further studies should be instituted to study the effect of this drug in other similar diseases. The efficacy of poly(ICLC) is unproven in other viral infections in man, and hence, should be studied. While it is clear that the drug has significant associated toxicities, administration at lower doses (3–6 mg/m^2) with intermittent dosage schedules has less associated clinically severe side effects. The most common immediate drug side effects include fever, chills, nausea, vomiting, myalgias, and arthralgias. Less common side effects seen occasionally at lower doses include transient hypotension, weakness or malaise, abdominal discomfort, diarrhea, dizziness, and seizure. Transient elevation of hepatic enzymes and altered hematologic values have been demonstrated but are readily reversible. Rare side effects include elevation of blood urea nitrogen, acute renal failure, gastrointestinal bleeding, rash and edema, but these were associated with high doses, greater than 12 mg/m^2. From these earlier studies, a great deal has been learned about the pharmacokinetics of poly(ICLC) and interferon induction. Most of the studies clearly demonstrated a dose-related response to interferon levels that peaked 8–24 hours following intravenous adminstration. Similarly, a period of hyporesponsivness to administration of the drug was demonstrated in several of the studies with a recovery period of several days. Spacing of injections at intervals greater than three days avoids the hyporesponsiveness.

Other investigations are now in progress to assess the efficacy of poly(ICLC) in a wide range of illnesses. These include patients with metastatic renal carcinoma, neuroblastoma, melanoma, hepatoma with associated hepatitis B surface antigen, metastic breast cancer, multiple sclerosis, and Landry Guillian–Barre syndrome. Although interferon production is induced by poly(ICLC) and this effect may be beneficial to patients with viral infections, it also has a strong immunomodulating effect which may eventually be proven to be of importance in patients with certain types of cancer. A combination of both effects may indeed be important in certain diseases such as neuromuscular disease. Poly(ICLC) may be preferred to interferon for the treatment of these diseases because: (1) it is readily available; (2) it is far less expensive than synthetic interferons; (3) interferon production may not be the mechanism of the poly(ICLC) benefit; and (4) it has proven efficacy in certain neuromuscular diseases based on immunologic abnormalities. In conclusion, poly(ICLC) is a drug that exerts strong biological effects in primates and in man. Within the following years we will hopefully be able to determine its effectiveness in a variety of clinical diseases.

References

1. H. B. Levy, S. Baron, and C. E. Buckler, in: *The Biochemistry of Viruses* (H. B. Levy, ed.), Dekker, New York, (1969).
2. S. E. Grossberg *New Engl. J. Med.* **287**, 13–19, 79–85, 1213–1220, (1972).
3. N. B. Finter (ed.), *Interferons and Interferon Inducers*, North-Holland, Amsterdam (1973).
4. H. B. Levy, F. L. Riley, and C. E. Buckler, in: *Molecular Biology of Viruses* (Nayak, ed.), Dekker, New York (1977).

5. A. Isaacs and J. Lindenmann, *Proc. R. Soc. London, Ser. B.* **147**, 258 (1957).
6. R. J. Samuels, *Polymer* **18**, 452 (1977).
7. A. E. Munson, K. L. White, and P. C., Klykken, in: *Pharmacology of MVE Polymers in Augmenting Agents in Cancer Therapy* (E. M. Hersh, M. A. Chirigos, and M. D. Mastrangelo, ed.), Raven Press, New York (1981).
8. A. E. Munson, W. Regelson, W. Lawrence, and W. R. Wooles, *J. Reticuloendothel. Soc.* **7**, 375 (1970).
9. W. Regelson, P. Morahan, and A. Kaplan, in: *Polyelectrolytes and Their Applications*, (A. Rembaum and E. Selegny, ed.), Vol. II, Reidel, Holland (1975).
10. A. K. Field, A. A. Tytell, E. Piperno, G. P. Lampson, M. M. Nemes, and M. P. Hilleman, *Medicine* **51**, 169–174 (1972).
11. S. Feldman,, W. T. Hughes, R. W. Darlington, and H. K. Kim, *Antimicrob. Agents Chemother.* **8**, 289–294 (1975).
12. M. A. Guggenheim and S. Baron, *J. Infect. Dis.* **136**, 50–58 (1977).
13. A. I. Freeman, S. Bogger-Goren, M. L. Brecher, and J. A. O'Malley, *J. Interferon Res.* **1**, 457–462 (1981).
14. A. I. Freeman, N. Al-Bussam, J. A. O'Malley, L. Stutzman, S. Gjornsson, and W. A. Carter, *J. Med. Virol.* **1**, 79–93 (1977).
15. N. Kemeny, A. Yagoda, Y. Wang, K. Field, H. Wrobleski, and W. Whitmore, *Cancer* **48**, 2154–2157 (1981).
16. A. Lindenbaum *et al.*, ANL-7970, *Annual Report* (1972), pp. 121–125.
17. L. Aschoff and Keymo, *Folia Haematol.* **15**, 383 (1913).
18. F. F. Pindak, *Infect. Immun.* **1**, 217 (1970).
19. P. W. A. Mansell, N. R. DiLuzio, R. McNamee, G. Rowden, and J. W. Proctor, *Ann. N.Y. Acad. SDci.* **277**, 20–28 (1976).
20. T. C. Merigan, in: *Interferons and Interferon Inducers* (N. B. Finter, ed.), pp. 45–71, Elsevier/North-Holland, New York, (1973).
21. P. P. Singh, R. L. Whistler, R. Tokuzen, and W. Nakahara, *Carbohydr. Res.* **37**, 245–247 (1974).
22. H. D. Suit, A. Elman, R. Sedlacek, and V. Silobrcic, in: *Immune Modulation and Control of Neoplasia by Adjuvant Therapy* (M. A. Chirigos, ed.), pp. 235–240, Raven Press, New York (1978).
23. R. E. Shope, *J. Exp. Med.* **97**, 627 (1953).
24. G. T. Banks, K. W. Buck, E. B. Chain, F. Himmelwelt, J. E. Marks, J. M. Tyler, M. Hollings, F. T. Lost, and O. M. Stone, *Nature* (*London*) **218**, 542 (1968).
25. A. K. Field, A. A. Tytell, G. P. Lampson , and M. R. Hilleman, *Proc. Natl. Acad. Sci. U.S.A.* **58**, 1004 (1967).
26. A. K. Field, G. P. Lampson, A. A. Tytell, M. M. Nemes, and M. R. Hilleman, *Proc. Natl. Acad. Sci. U.S.A.* **58**, 2102 (1967).
27 A. K. Field, A. A. Tytell, G. P. Lampson, and M. R. Hilleman, *Proc. Natl. Acad. Sci. U.S.A.* **61**, 340 (1968).
28. A. A. Tytell, G. P. Lampson, A. K. Field, and M. R. Hilleman, *Proc. Natl. Acad. Sci. U.S.A.* **58**, 1719 (1967).
29. G. P. Lampson, A. A. Tytell, A. K. Field, M. M. Nemes, and M. R. Hilleman, *Proc. Natl. Acad. Sci. U.S.A.* **58**, 782 (1967).
30. S. Baron, N. N. Bogomolova, A. Billiau, H. B. Levy, C. E. Buckler, R. Stern, and R. Naylor, *Proc. Natl. Acad. Sci. U.S.A.* **64**, 67 (1969).
31. A. Billiau, C. E. Buckler, F. Dianzani, C. Uhlendorf, and S. Baron, *Proc. Soc. Exp. Biol. Med.* **132**, 790 (1969).
32. A. K. Field, A. A. Tytell, G. P. Lampson, M. M. Nemes, and M. R. Hilleman, *J. Gen. Physiol.* **56**, 905 (1970).
33. P. Jameson and S. E. Grossberg, *Bacteriol. Proc.* **1970**, 155.

34. J. Vilcek, M. H. Ng, A. E. Friedman-Kien, and T. Krauciu, *J. Virol.* **2**, 648 (1968).
35. J. H. Park and S. Baron, *Science* **162**, 811 (1968).
36. M. Worthington and S. Baron, *Proc. Soc. Exp. Biol. Med.* **136**, 323 (1971).
37. H. B. Levy, L. W. Law and A. S. Rabson, *Proc. Natl. Acad. Sci. USA.* **62**, 357 (1969).
38. H. V. Gelboin and H. B. Levy, *Science* **167**, 205 (1970).
39. P. S. Sarma, S. Baron, R. J. Huebner, and G. Shiu, *Nature* **224**, 604 (1969).
40. L. D. Zeleznick and B. K. Bhuyan, *Proc. Soc. Exp. Biol. Med.* **130**, 126 (1969).
41. I. Gresser, L. Berman, G. DeThe, D. Brouty-Boye, J. Coppey, and E. Falcoff, *J. Natl. Cancer Inst.* **41**, 505 (1968).
42. H. Cantor, R. Asofsky, and H. B. Levy, *J. Immunol.* **104**, 1035 (1970).
43. W. Turner, S. P. Chan, and M. A. Chirigos, *Proc. Soc. Exp. Biol. Med.* **133**, 334 (1970).
44. H. B. Levy and F. Riley, *Proc. Soc. Exp. Biol. Med.* **135**, 141–145 (1970).
45. C. W. Young, *Med. Clin. North Am.* **55**, 720–728 (May 1971).
46. R. A. Robinson, V. T. DeVita, H. B. Levy, S. Baron, S. P. Hubbard, and A. S. Levine, *J. Natl. Cancer Inst.* **57**, 599 (1975).
47. D. A. Hill, S. Baron, H. B. Levy, J. Bellanti, C. E. Buckler, G. Cannellos, P. Carbone, R. M. Chanock, V. DeVita, M. A. Guggenheim, E. Homan, A. Z. Kapikian, R. L. Kirschstein, J. Mills, J. E. Vankirk, and M. Worthington, in: *From Molecules to Man, Perspectives in Virology VII* (M. Pollard, ed.), pp. 198–222, Academic Press, New York (1971).
48. J. J. Nordlund, S. M. Wolff, and H. B. Levy, *Proc. Soc. Exp. Biol. Med.* **133**, 439 (1970).
49. H. B. Levy, G. Baer, S. Baron, C. E. Buckler, C. J. Gibbs, M. J. Iadarola, W. T. London, and J. Rice, *J. Infect. Dis.* **132**, 434–439 (1975).
50. G. M. Baer, J. H. Shaddock, S. A. Moore, P. A. Yager, S. Baron, and H. B. Levy, *J. Infect. Dis.* **136**, 58–62 (1977).
51. E. L. Stephen, M. L. Sammons, W. L. Pannier, S. Baron, R. O. Spertul, and H. B. Levy, *J. Infect. Dis.* **136**, 122–126 (1977).
52. H. B. Levy, W. London,. D. A. Fuccillo, S. Baron, and J. Rice, *J. Infect. Dis.* **133**, A256–A259 (1976).
53. R. H. Purcell, W. T. London, V. J. McAliffe, A. E. Palmer, P. M. Kaplan, H. B. Levy, J. L. Gerin, J. Wagner, H. Popper, E. Lvovsky, and D. C. Wong, *Lancet*, October 9, 1976, 757.
54. H. B. Levy and E. Lvovsky, *J. Infect. Dis.* **173**, 78–81 (1978).
55. H. B. Levy, R. L. Riley, E. Lvovsky, and E. E. Stephen, *Infect. Immun.* **39**, 416–421 (1981).
56. G. Felsenfeld and S. Huang, *Biochem. Biophys. Acta* **24**, 234–242.
57. W. Braun and W. Fershein, *Bact. Rev.* **31**, 83–94 (1967).
58. W. Braun and M. Nakano, *Science* **157**, 819–821 (1967).
59. R. Winchurch and W. Braun, *Nature(London)* **223**, 843–844 (1969).
60. J. R. Schmidtke, Doctoral Dissertation, University of Michigan, Ann Arbor, 1969.
61. A. G. Johnson, J. R. Schmidtke, K. Merritt, and I. Gan, in: *Nucleic Acids in Immunology* (O. J. Plescia and W. Braun, eds.), pp. 379–385, Springer-Verlag, New York (1968).
62. E. L. Stephen, D. E. Hilmas, J. A. Marigrafico, and H. B. Levy, *Science* **197**, 1289–1290 (1977).
63. D. L. Harrington, C. L. Crabbs, D. E. Hilmas, J. R. Braun, C. A. Higbee, F. E. Cole, and H. B. Levy, *Infect. Immun.* **24**, 160–166 (1979).
64. D. Harrington, E. Stephen, J. Peters, and H. B. Levy (unpublished observations).
65. R. J. Klein, E. Klein-Buimoviee, U. Moser, R. Mouch, and J. Hilfenhaus, *Arch. Virol.* **68**, 73–80 (1981).
66. H. B. Levy and P. Barker (unpublished observations).
67. M. A. Chirigos, V. Papademetriou, A. Bartocci, E. Read, and H. B. Levy, *Int. J. Immunopharm.* **3**, 329–337 (1981).

68. E. A. Lvovsky, P. H. Levine, A. Bengali, S. S. Leiseca, J. L. Cicmanec, J. E. Robinson, N. Bautro, H. B. Levy, and R. M. Scott, *Int. J. Radiat. Ocol. Biol. Phys.* **8**, 1721–1726 (1982).
69. S. Feldman, W. T. Hughes, R. W. Darlington, and H. K. Kim, *Antimicrob. Agents Chemother.* **8**, 289–294 (1975).
70. M. A. Guggenheim and S. Baron, *J. Infect. Dis.* **136**, 50–58 (1977).
71. A. I. Freeman, S. Bogger-Goren, M. L. Brecher, and J. A. O'Malley, *J. Interferon Res.* **1**, 457–462 (1981).
72. A. I. Freeman, N. Al-Bussam, J. A. O'Malley, L. Stutzman, S. Bjornsson, and W. A. Carter, *J. Med. Virol.* **1**, 79–93 (1977).
73. N. Kemeny, A. Yagoda, Y. Wang, K. Field, H. Wroblesci, and W. Whitmore, *Cancer* **48**, 2154–2157 (1981).
74. M. L. Sammons, E. L. Stephen, H. B. Levy, S. Baron, and D. E. Hilmas, *Antimicrob. Agents Chemother.* **11**, 80–83 (1977).
75. E. L. Stephen, D. E. Hilman, H. B. Levy, and R. O. Spertzel, *J. Infect. Dis.* **139**, 267–273 (1979).
76. A. S. Levine and H. B. Levy, *Cancer Treat. Rep.* **62**, 1907–1913 (1978).
77. A. S. Levine, M. Sivulich, P. H. Wiernik, and H. B. Levy, *Cancer Res.* **39**, 1645–1650 (1979).
78. A. S. Levine, B. Durie, B. Lampkin, B. G. Leventhal, and H. B. Levy, in: *Augmenting Agents in Cancer Therapy* (Evan M. Hersh *et al.*, eds.), Raven Press, New York (1981).
79. S. E. Krown, D. Kerr, W. E. Stewart II, M. S. Pollack, S. Cunningham-Rundles, Y. Hirshaut, C. M. Pinsky, H. B. Levy, and H. F. Oettgen, in: *Augmenting Agents in Cancer Therapy* (Evan M. Hersh *et al.*, eds.), Raven Press, New York (1981).
80. R. Alexanian, J. Gutterman and H. Levy, in: *Clinics in Haematology* (S. E. Salmon, ed.), pp. 211–220, Saunders, Philadelphia (1982).
81. B. G. Leventhal, H. Kashiman, A. S. Levine, and H. B. Levy, *J. Pediat.* **99**, 614–616 (1981).
82. W. K. Engel, R. Cuneo, and H,. B. Levy, *Lancet* **1978**, 503–504.
83. K. J. Champney, D. P. Levine, H. B. Levy, and M. Lerner, *Infect. Immun.* **25**, 831–837 (1979).
84. L. R. Crane, H. B. Levy, and A. M. Lerner, *Antimicrob. Agents Chemother.* **21**, 481–485 (1982).

16

Functionality and Applicability of Synthetic Nucleic Acid Analogues

Kiichi Takemoto

Abstract. Of the chemistry of synthetic nucleic acid analogues, some of the new aspects of their functionality and applicability were reviewed. The specific interaction was studied particularly on poly-L-lysine derivatives having nucleic acid bases in relation to their molecular weight, their conformation in solutions, and other properties. From a photodimerization study on the model compounds containing thymine bases, intramolecular features of the reaction were elucidated. Graft copolymers of the bases on polyethyleneimine were also considered. New results on the functionality of the supported nucleic acid bases as polymeric reagents were given. The preparation of relating cyclic derivatives of pyrimidine bases, as well as their applicability to the field of organic synthesis, were shown. Owing to the specific properties of the polymers, the studies will find a number of applications, including separation techniques and polymeric drug chemistry.

1. Introduction

Nucleic acids are known to play an important role in both the replicative and the transcriptive functions of the genetic codes. DNA, one of the nucleic acids, for example, has a two-stranded structure consisting of two polynucleotide chains twisted about each other in a double helix, which is stabilized by the special base–base interactions of nucleic acids. The interaction in question among such purine and pyrimidine bases is apparently a common subject of substantial interest among chemists in the fields of organic, macromolecular, and biochemical syntheses.

Over the last ten years, the functional monomers and polymers containing heterocyclic moieties have received much attention, and numerous studies have been devoted to the preparation, polymerization, and the properties of the polymers.[1–8] It has been also pointed out that some of the nucleic acid analogues thus prepared should be important for finding applicability, for example, as interferon inducers in the field of polymeric drug chemistry.

Kiichi Takemoto • Faculty of Engineering, Osaka University, Yamadaoka, Suita, Osaka 565, Japan.

In the present review, emphasis is focused particularly on our recent, systematic work concerning the nucleic acid analogues.Investigations by other groups are referenced also, but somewhat briefly, so far as they are closely concerned with the subjects.

2. Synthesis and Properties of Polyamino Acids Containing Nucleic Acid Bases

A series of the nucleic acid analogues have been prepared and their functionalities have been estimated in relation to special base-pairing properties.[3,5] In these past investigations, however, it was difficult to know the conformation of these polymers in solution, because most of them consisted of a vinyl-type backbone and pendant nucleic acid bases. In this respect, a study of nucleic acid analogues having poly-α-amino acid backbone with definite conformations should be of interest. A systematic work has been done in order to get detailed information about the conformational influence of their backbone nature on the complex formation of the nucleic acid analogues.

2.1. Polymer Synthesis

Poly-L-lysine derivatives having pendant nucleic acid bases were chosen and prepared by two different methods: (1) synthesis of the base-substituted L-lysine derivatives followed by their polymerization using the *N*-carboxy-amino acid anhydride (NCA) method; and (2) polymer reaction of carboxy-ethyl derivatives of the base onto poly-L-lysine.[9]

The first preparation route is shown in Scheme 1.[10] In order to incorporate the base exclusively at the ε-position, the α-amino group of L-lysine was blocked to afford (**1**), which was allowed to react with *p*-nitrophenyl esters[11] in DMSO or DMF solution to give (**2**).

Owing to low solubility, the preparation of their NCAs (**3**) by the Fuchs method, which has been used for preparing alanine derivatives containing the bases, was unsuccessful.[12] The NCAs were, therefore, prepared by the Leuchs method using thionyl chloride.[13] The NCAs thus prepared were allowed to polymerize in DMSO solution by using triethylamine as the initiator. The base-substituted L-lysine derivatives, amide, and ester types of L-glutamic acid derivatives were also prepared.[13]

The poly-L-lysine derivatives containing pendant nucleic acid bases can be prepared alternatively by using the polymer modification reaction (Scheme 2).[14] Carboxymethyl derivatives of the bases were grafted onto poly-L-lysine by using the activated-ester method. Poly-L-lysine was allowed to react in this case in the form of the trifluoroacetate.

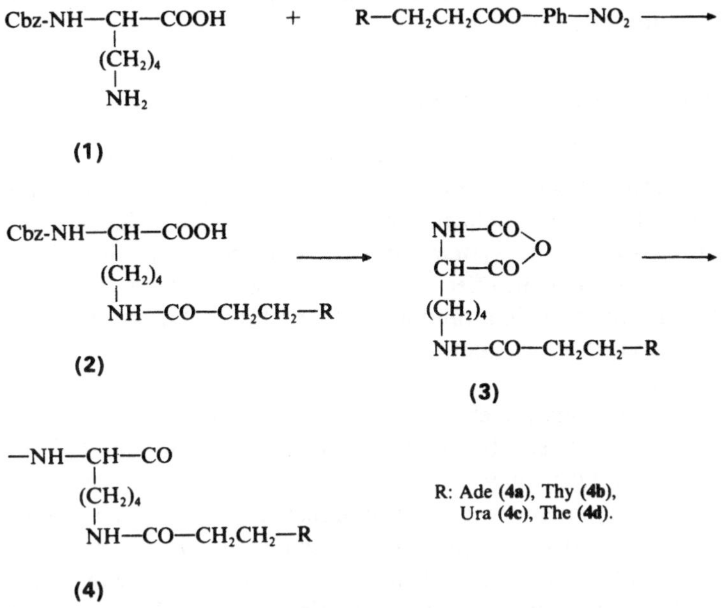

$$\text{Cbz-NH—CH(—(CH_2)_4—NH_2)—COOH} \quad (1) \quad + \quad \text{R—CH_2CH_2COO—Ph—NO_2} \longrightarrow$$

$$\text{Cbz-NH—CH(—(CH_2)_4—NH—CO—CH_2CH_2—R)—COOH} \quad (2) \longrightarrow$$

$$\text{NH—CO—O—CO—CH (cyclic anhydride), CH—(CH_2)_4—NH—CO—CH_2CH_2—R} \quad (3) \longrightarrow$$

$$\text{—NH—CH(—(CH_2)_4—NH—CO—CH_2CH_2—R)—CO} \quad (4)$$

R: Ade (**4a**), Thy (**4b**), Ura (**4c**), The (**4d**).

SCHEME 1

With thymine and uracil derivatives, graft reactions proceed almost quantitatively; in the case of the adenine derivative, its activated ester hardly reacts with poly-L-lysine, and only the copolymer with low adenine content is obtained, probably owing to the instability of the activated ester. Hypochromicity of the copolymer based on nucleic acid bases is negligible for the pyrimidine derivatives and is about 10% for adenine copolymer

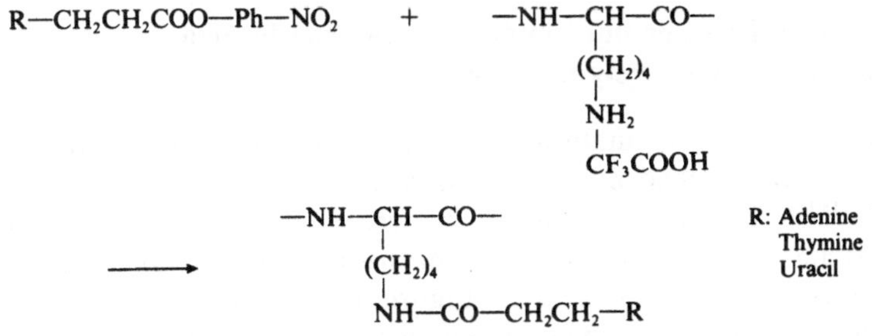

$$\text{R—CH_2CH_2COO—Ph—NO_2} \quad + \quad \text{—NH—CH(—(CH_2)_4—NH_2·CF_3COOH)—CO—}$$

$$\longrightarrow \text{—NH—CH(—(CH_2)_4—NH—CO—CH_2CH_2—R)—CO—}$$

R: Adenine
Thymine
Uracil

SCHEME 2

containing 67% adenine units. This fact suggests the presence of a strong self-association of adenine bases.[15]

2.2. Conformations of the Polymers

Conformations of the polymers were studied by CD (circular dichroism) and optical rotation measurements. Poly-L-lysine is known to be present in disordered, helical, and β-conformation, depending on the temperature, pH of the system, and solvent used. The side chain of the polymer has a significant effect on the backbone conformation. At neutral pH, poly-L-lysine exists in a random coil structure, while at pH above 10 the ε-amino group becomes a neutral form and the polymer undergoes transition to a helical structure.

In order to study the effect of base substituents on the conformation of poly-L-lysine, the CD spectra of the copolymers were measured.[15] From the spectra of poly-L-lysine having adenine units, for example, it was found that the helicity of the polymer increases with increasing pH of the system, due to release of the electrostatic repulsion between positively charged side chains. By adding ethyelene glycol, the helicity tends to increase, which shows that ethylene glycol depresses the electrostatic repulsion between protonated adenine moieties.

Another important factor for the polymer conformation is the solvent effect. The optical rotation measurements and analyses using the Moffitt–Yang equation give the parameter b_0 for the copolymers, which is known to be related to the helix content of poly-α-amino acids. The b_0 value of polycarbobenzoxy-L-lysine is about -550 in DMF or in chloroform solution where the polymer exists in a helical conformation. On the other hand, the value in dichloroacetic acid is nearly zero where the polymer is assumed to be in a random structure. In general, the copolymers having a high content of the bases also tend to exhibit a helical conformation in organic solvents as well as in aqueous solution.

2.3. Polymer–Polymer Interaction between Nucleic Acid Base-Substituted Poly-L-Lysines

The complex formation between the complementary poly-L-lysine derivatives was studied by UV spectroscopy in DMSO and ethyleneglycol mixture.[15,16] The mixing curves between adenine in 67%-containing poly-L-lysine (PLL-A-67) and the one having thymine bases in different contents (PLL-Ts) are shown in Figure 1 as an example. The complex formation between complementary polymers can be clearly observed, and the overall stoichiometry of the complex based on the base units was about 1:2

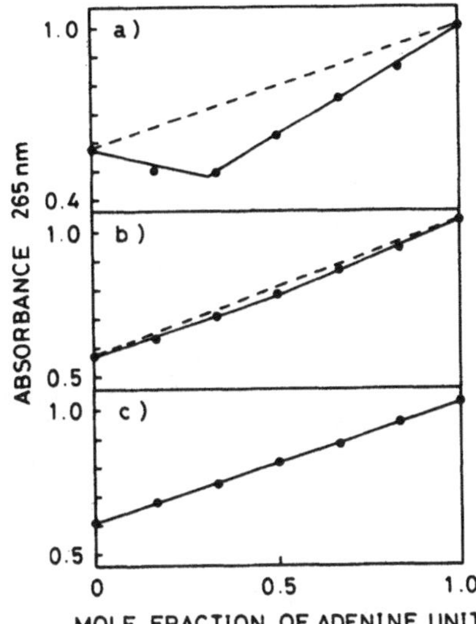

Figure 1. Example of the mixing curves: mixing curves between PLL-A-67 and PLL-Ts in DMSO–ethylene glycol (3/2, v/v). Absorbance in 10-mm cell at 25°C: (a) PLL-T-93; (b) PLL-T-79; and (c) PLL-T-65.[15]

(adenine:thymine). It was also detected that the interaction tends to become weaker with decreasing thymine content of the polymer. It was suggested that a weak interaction between complementary polymers containing a low content of bases may be caused mainly be a low helicity.

The interaction was found to become weaker with raising the temperature from 25 to 90°C, causing dissociation of the polymer complexes. The Moffitt–Yang parameter b_0, however, is constant in the temperature range in question. From these facts, it is concluded that the polymer complexes are formed by specific base pairing between pendant adenine and thymine or uracil units of poly-L-lysine derivatives retaining their helical conformations (Figure 2).[15,16]

Interactions between the low-molecular-weight polymers and between low- and high-molecular-weight polymers were studied, too, but no significant interaction was observed for both systems. The reason may be attributed to the low degree of polymerization and the resulting random coil structures.

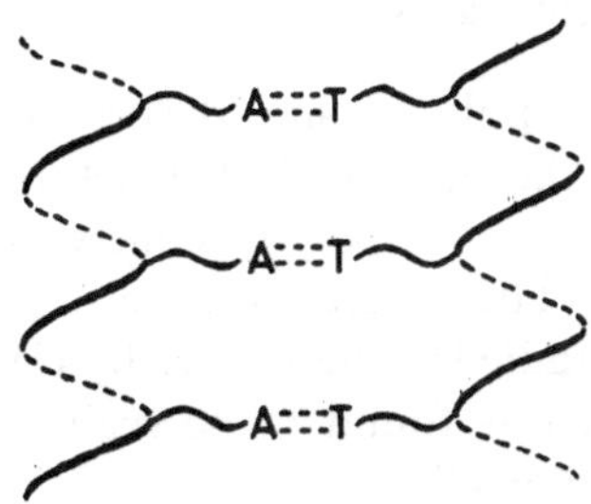

Figure 2. Schematic representation of the interaction between poly-L-lysines having adenine and thymine bases.[15]

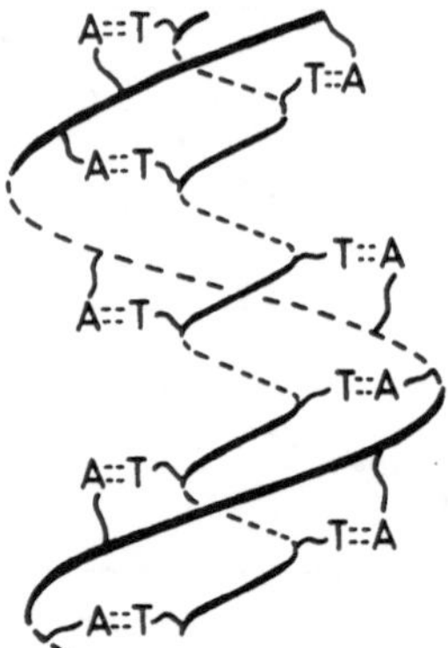

Figure 3. Schematic representation of the interaction between polymethyacryloyloxyethyleadenine and PLL-T-93.[17]

2.4. Polymer–Polymer Interactions between Nucleic Acid Base-Substituted Poly-L-Lysine and Other Synthetic Polymers

In the system of thymine-containing poly-L-lysine and polymethyacryloyloxyethyladenine, the helical conformation of the former polymer was found to be retained even after forming the polymer complex with the latter one. From the 1:1 stoichiometry, the high value of hypochromicity and the helical conformation of the thymine-containing poly-L-lysine, the polymer complex in question was assumed to exhibit a double helical structure as illustrated in Figure 3. In this model, the polymer complex is held together by the specific base pairing between pendant thymine and adenine bases. A similar experiment was also done on the formation of the polymer complex between thymine-containing poly-L-lysine and polymethacryloylaminoethyladenine. It was concluded that both the mutual penetration ability and the compatibility of base–base distances of the polymers are substantial for the formation of the stable polymer complexes by specific base pairing.[17]

2.5. Isopoly-L-Lysine Having Pendant Nucleic Acid Bases

In relation to the case of poly-L-lysine, isopoly-L-lysine derivatives appear to be interesting also, because the latter takes no α-helical structure. The polymer was prepared by the polymer reaction of isopoly-L-lysine with active compounds having nucleic acid bases. The interaction between polymers having complementary bases also was studied in these cases by UV and CD spectroscopies. The specific interaction was found to be realized preferably between the neutral forms of the polymers.[18]

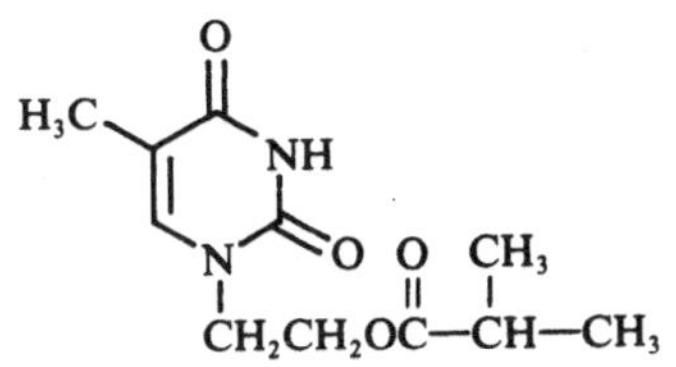

T—M

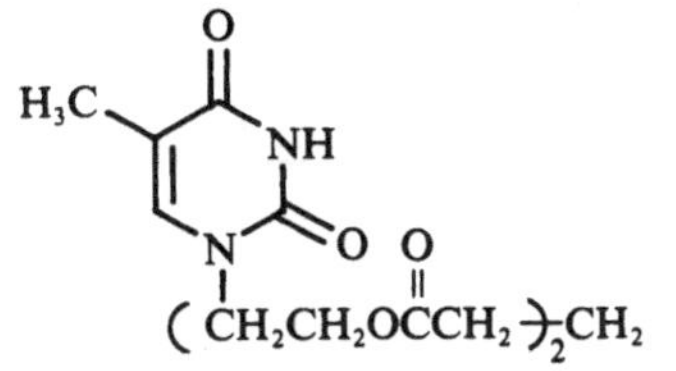

T—T

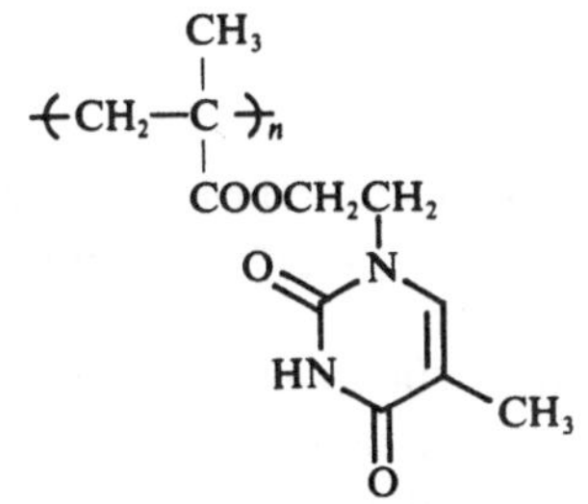

poly(MAOT)

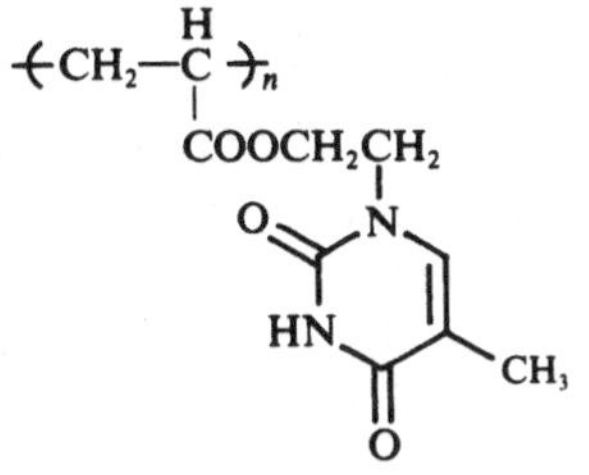

poly(AOT)

Figure 4. Monomer and dimer models and MAOT and AOT polymers.

3. Photochemistry of the Polymers Containing Thymine Bases

In a series of our interaction studies, the photodimerization of thymine bases along the polymer chain was carried out in order to estimate the degree of intramolecular self-association of the bases. The reaction was also studied for monomeric and dimeric model compounds (Figure 4).[19] It is known that cis–syn and trans–syn isomers are formed by photolysis of 1,1′-trimethylene-bis-thymine, TpT, and denaturated DNA. The chemical shift of the 5-methyl proton in our case for the synthetic polymers suggests the formation of syn-type isomers.

3.1. Fundamental Study on Photodimerization

In order to determine how the self-association of thymine bases in thymine derivatives is at the ground state, UV spectra of these compounds were measured.[20] The hypochromicity values for the polymers, poly(MAOT) and poly(AOT), are higher than that for the dimer, T-T, though the difference is small. From the UV spectral data, the self-association was found to increase in the following solvent order: Me_2SO

$$
\begin{array}{ccccccc}
T & \underset{k_1}{\overset{h\nu}{\rightleftharpoons}} & T^{1*} & \xrightarrow{k_{isc}} & T^{3*} & \underset{[T]}{\xrightarrow{k_3}} & \text{Dimer} \\
 & & \downarrow k_2 \ T & & \downarrow k_4 \quad \searrow k_5\ [Q] & & \\
 & & \text{Dimer} & & T \qquad\quad T & &
\end{array}
$$

SCHEME 3

< DMF < Me_2SO–ethylene glycol. The solvent effect on the self-association is closely related to that on the formation of the polymer complexes by specific base pairing.[(21)]

The intrinsic viscosity data indicate that the conformation of the polymer chain is affected by the type of solvent used. The relationship between the self-association of thymine bases and the form of the polymers in the solvents were discussed in detail. Quantum efficiency of the thymine derivatives for the photodimer formation was measured in different solvents.

The photodimerization of thymine derivatives was then studied in the presence of isoprene as the triplet quencher. It was found that the photodimerization of the thymine dimeric model was quenched by isoprene, and the Stern–Volmer plots gave straight lines, while in the case of the polymers no linear relationship was observed. The results appear to indicate that the photodimerization for the former case occurs almost completely from the triplet state, while that for the latter case involves both excited singlet and excited triplet states (Scheme 3).

Photodimerization of acrylic and methacrylic copolymers with different pendant thymine unit content was further studied in DMF solution. The quantum efficiency of thymine base for the photodimerization increased with increasing thymine content in the copolymers. The quenching study revealed that the photodimerization resulted from excited-singlet-state increases with increasing thymine content.[(22)]

3.2. Effect of Complementary Bases

Photodimerization reaction of polyacrylate and polymethacrylate derivatives and the dimer model compound containing thymine bases were also studied in the presence of adenine derivatives in different solvents. The reaction for both polymer and dimer model compounds was found to be quenched by adding adenine derivatives. Base–base interaction in the ground state was studied by UV spectroscopy. The quenching behavior was discussed in terms of the specific interaction beween adenine and thymine bases both in ground and excited states.[23]

The photodimerization reaction of the synthetic polymers containing pendant thymine bases was then studied in polymethyl methacrylate films. The quantum efficiency for the reaction of the thymine bases of polyacrylate and polymethacrylate, fixed in the polymethyl methylacrylate film was measured, and it was found that the value of the film was higher than those measured in solution.[24]

3.3. Photoreaction of the Polymers Having Cyanouracil Units

In relation to the photodimerization behavior of thymine bases, orotic acid has also been known to photodimerize with essentially higher quantum efficiency.[25] We prepared further methacryloyloxy-type polymers having pendant 5- and 6-cyanouracil units, and studied the photoreaction on them. The polymeric 6-cyanouracil derivative (**5a**) was found to show substantially high photoreactivity, while the 5-substituted one (**5b**) had only little reactivity. Further study is now in progress.[26]

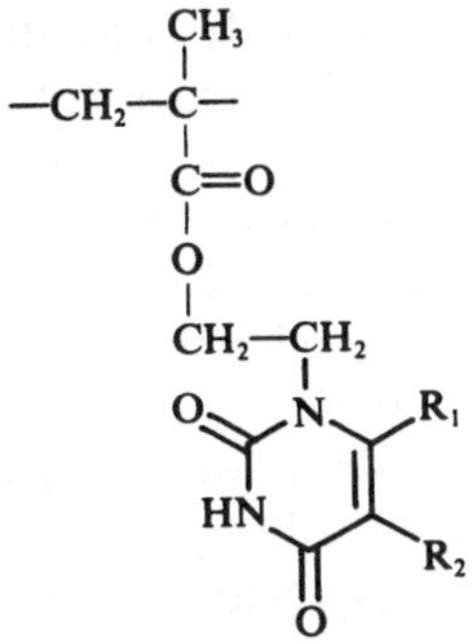

(5)

Polymethacryloyloxy derivates with 5- and 6-cyanouracil; (**5a**) R_1 = CN, R_2 = H; (**5a**) R_1 = H, R_2 = CN.

4. Synthesis and Properties of Polyethyleneimine Derivatives Containing Nucleic Acid Bases

4.1. Polymer Synthesis

In recent years, considerable interest has grown for using natural and synthetic polymers as interferon inducers.[27] It is also known that interferon inducers can prevent or regress certain tumors. In particular, the synethetic, double-stranded complex of poly(I) and poly(C) is effective as an interferon inducer;[28–30] however, it also has a high level of toxicity.[31]

Polyethyleneimine has been shown to be highly effective against ascite tumors.[32] The high toxicity of the polymer is completely eliminated, without significant reduction of its activity, by grafting with monomeric serine and/or histidine residues.[33] Recently, new model polymers with a polyethyleneimine backbone and nucleic acid base derivatives as pending side chains have been prepared[11]; the carboxy derivatives of adenine and thymine were grafted onto linear and branched polyethyleneimine, and the carboxy derivatives reacted with the L-α-amino acid grafted polyethyleneimine to form an amide bond by the *p*-nitrophenyl method.

4.2. Interaction of the Graft Polymers

Interactions of the graft copolymers of adenine and thymine on linear and branched polyethyleneimine with polynucleotides, poly(A) and poly(U), were studied.[34] From the Job plots[35] of the system of polyethyleneimine with adenine (**6**) and that with thymine (**7**), at pH 7.4, for example, formation of the complex was observed, the overall stoichiometry of the latter being 1:1, hypochromicity of the 1:1 complex was about 16%. This value was thought to be the highest among the values reported for the interaction between synthetic polymers with nucleic acid bases. The melting profiles between 25 and 90°C are reversible upon heating and cooling. The results were fully compared with the vinyl-type nucleic acid model polymer.

Subsequently, new model polymers of polynucleotides with linear polyethyleneimine backbones and optically active nucleic acid base derivatives as grafted pendants were prepared.[36] Hypochromic effects were studied systematically on the polymers 2-(thymin-1-yl)propionyl graft polyethyleneimine, and its related monomer and dimer models, and the 2-(adenin-9-yl)propionyl graft, and its monomer and dimer models. This base-stacked conformation in comparison with the corresponding polynucleotides has also been discussed.[37]

CD spectra were measured on a series of polyethyleneimine derivatives having optically active thymine moieties as grafted pendants, and the results were compared with those of the related monomer and dimer model compounds.[38] The CD spectra of the polymer in neutral aqueous solution were different from those of related model compounds, which suggests that the polymer exists in some ordered conformation to allow exciton coupling of π-π^* transition in the base chromophores along the polymer chain. This ordered conformation tends to be random on heating. Similar polyethyleneimine derivatives having pendant optically active thymine moieties, which have a L-proline ring as a spacing group, have also been prepared, and their optical properties have been studied.[39]

Other polymers were prepared by alkylation of polyethyleneimine with chloroethylated adenine and thymine. They interacted with nucleosides to form complementary base pairs. In contrast, the order of their interaction with nucleotides was as follows: purine–purine > purine–pyrimidine > pyrimidine–pyrimidine. The complementary base–base interactions were found not to be appreciably affected by ionic strength.[40]

4.3. Further Studies on Polyethyleneimine Having Thymine Bases

The photodimerization reaction of oligo- and polyethyleneimine derivatives which contain pendant thymine bases in different amounts was also studied in aqueous solution in a wide pH range and in DMF solution.[41] The quantum efficiency was found to increase as thymine units were incorporated in the oligomer and copolymer structure. The result was discussed in terms of the effect of the nearest neighbouring thymine units and singlet-energy migration, particularly in the polymers. A series of oligomer models of polyethyleneimine derivatives having pendant thymine bases was prepared by the reaction of carboxyethyl derivatives of thymine with oligomeric amines using an activated-ester method. The hypochromicity values obtained from UV spectra, and the pKa values obtained from spectrophotomeric titrations were found to depend upon the chain length of the oligomers and the thymine content of the polymers.[42]

Alternatively, oligoethyleneimine derivatives having pendant adenine units were prepared in a similar way as in the case of thymine-containing ones. The hydrochromicity and pKa values were again found to depend linearly on the chain length of the oligomers. The results showed that the intramolecular interaction of adenine bases can be realized less in their protonated forms than in the neutral ones.[43]

A series of polyethyleneimine derivatives and their oligomer models having such pendant bases, but separated by β-alanyl groups as the spacers, were also recently prepared by using an activated-ester method.[44] The result of the spectrophotometric studies was compared with that obtained for the corresponding oligomers without spacers. For the thymine derivatives, the nature of the intramolecular compounds without β-alanyl units, while larger hypochromicities were found for this series. For the adenine derivatives, hypochromicity of the compounds was found to decrease by incorporating the spacer group into the side chains, in contrast to the corresponding thymine systems. It became evident that the intramolecular interaction of adenine bases was due to stacking for the neutral species and to hydrogen bonding and the residual hypochromicity for the protonated ones.

The photodimerization reaction of thymine-base-containing oligomers and polymers of ethyleneimine was further performed by using derivatives in which the thymine bases are separated by β-alanyl units as the spacers.[45] The effects of their molecular weight, aqueous solution pH, as well as the spacing on the photoreaction were measured. Oligo- and polyethyleneimines having α-alanyl group spaced nucleic acid bases were also prepared, and the intramolecular interaction was studied in detail.[46]

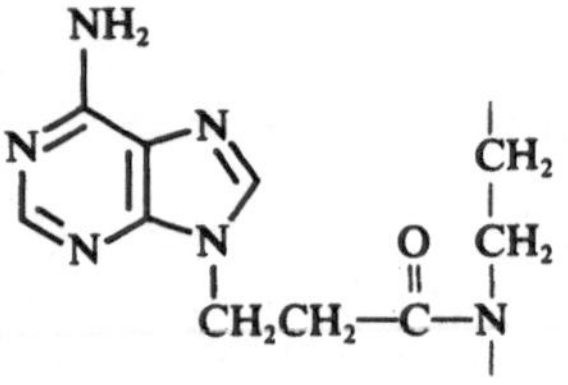

(6) 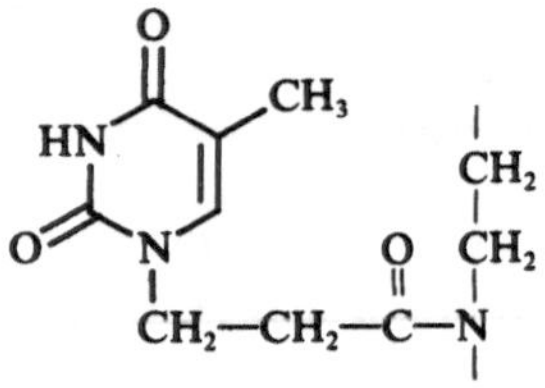 (7)

5. Functionality and Applicability of the Supported Nucleic Acid Bases as Polymeric Reagents

Besides the interest in finding applications for the pendant nucleic acid base synethetic polymers in the medicinal and technochemical fields,[47] one of the promising applications seems to lie in chromatography since the specific interaction realized between nucleic acid bases can be utilized for separating nucleosides, nucleotides, and their purine and pyrimidine bases, in a column packed with a resin containing complementary bases.[48–52] Recently, the resins suitable for chromatography were prepared from the reaction of *p*-chloromethylated polystyrene with the nucleic acid bases in different degrees of substitution:

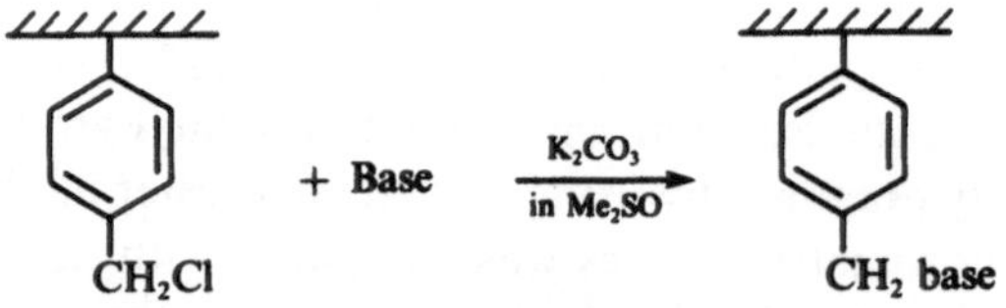

CH_2Cl + Base $\xrightarrow[\text{in } Me_2SO]{K_2CO_3}$ CH_2 base

Base: adenine, uracil

From the measurements of retention times and peak resolution values, it became evident that the specific base–base interaction between the solutes and the resin plays an important role in the separation purpose.[53] Successive work revealed that the silica gel supported bases can separate the complementary nucleic acid bases more efficiently in aqueous solution.[54]

Chloropyrimidine derivatives were found to be useful as the active dehydrating and desulfhydrylating reagents for the preparation of carboiimides, isothiocyanates, esters, lactones, amides or nitriles. In particular, a polystyrene-bound reagent, that is, 4-chloro-2-pyrimidone bound to polystyrene (**8**), was also quite useful due to the advantage of isolating the products readily by using a simple filtration technique[55]:

$$\textbf{(8)} + R\text{—}NH\text{—}\overset{S}{\overset{\|}{C}}\text{—}NH\text{—}R \longrightarrow RN{=}C{=}NR + \textbf{(9)}\ (X = S)$$

$$\textbf{(8)} + R\text{—}NH\text{—}\overset{S}{\overset{\|}{C}}\text{—}S^- \longrightarrow R\text{—}N{=}C{=}S + \textbf{(9)}\ (X = S)$$

$$\textbf{(8)} + C_6H_5COOH + CH_3OH \longrightarrow C_6H_5COOCH_3 + \textbf{(9)}\ (X = O)$$

$$\textbf{(8)} + C_6H_5CH{=}NOH \longrightarrow C_6H_5CN + \textbf{(9)}\ (X = O)$$

$$\textbf{(8)} + C_6H_5NH_2 \longrightarrow C_6H_5CONHC_6H_5 + \textbf{(9)}\ (X = O)$$

The reactivity is related to the structure, and an enhanced ease of lactim–lactam tautomerism in the products is suggested as a driving force of the reactions.

The 2,4-dichloropyrimidine and 6-chloropurine containing polystyrene resins were prepared and their activity was examined.[56] However, their activity as the polymeric reagents was lower than that of the polystyrene-bound 4-chloropyrimidone. The polymeric reagents of oxopyrimidine form can be well regenerated and reused. However, those of thioxypyrimidine, thioxypurine, and oxopurine were found to be regenerated with difficulty.

Synthetic application of 2,4-dichloro-5-nitropyrimidine is also of interest, because of enhanced activity by the inductive effect of the nitro group present. With this reagent, similar reactions were studied.[57]

6. Cyclic Derivatives of Pyrimidine Bases

Cyclonucleosides have been known as valuable intermediates in the synthesis of nucleoside derivatives.[58] One of the typical cyclonucleosides, cyclocytidine, is an intermediate in the synthesis of a carcinostatic nucleoside,[59] 1-β-D-arabinofuranosyl cytosine, and is itself a potent carcinostatic agent.[60] Recently, we found that the cyclic compounds of uracil and thymine (**10**), new bicyclic compounds, can be prepared very easily by the reaction of their *N*-chloroethylated derivatives with 1,8-diaza-bicyclo-(5,4,0)undecene-7 (DBU) even at room temperature; the yield attained over 80% within 5 min.[61]

The cyclic compounds of uracil and thymine are useful as an intermediate for various derivatives of both bases. Thus, the cyclic compounds react with organic and inorganic acids, with ammonia in methanol solution, and with methyl iodide in DMF solution to afford corresponding uracil and thymine derivatives in a high yield. In relation to them, polyacrylic and polymethacrylic derivatives containing different contents of pendant uracil and thymine units were prepared by the polymer reaction of the cyclic compounds with polyacrylic and polymethacrylic acid.[62]

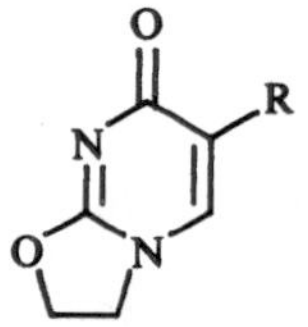

(10)
R = H; uracil
R = CH_3; thymine

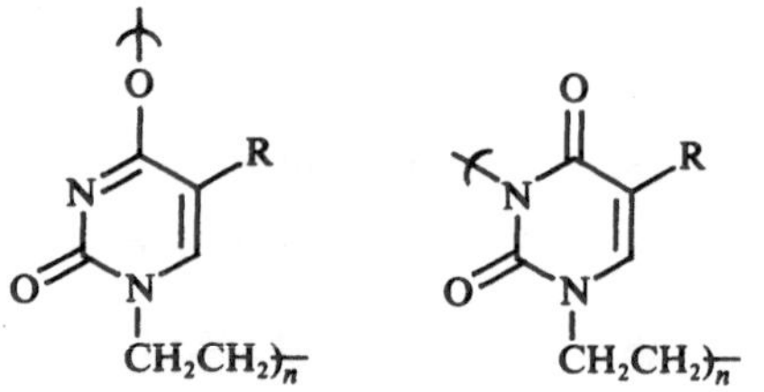

(11)

The cyclic compounds (**10**) could also be polymerized with methyl iodide or methyl *p*-toluenesulfonate as cationic initiators in acetonitrile solution by a ring-opening process. When an excess amount of these initiators was allowed to react with the cyclic compounds, the iodide or the tosylate was obtained. The tosylate was identified by IR and NMR spectra, but a pure compound could not be isolated because of low crystallinity.[63]

The polymerization by a cationic initiator proceeded by ring opening with accompanying isomerization of the pyrimidine ring to afford a polymer in which pyrimidine rings were connected with ethylene between N-1 and N-3 or O-4 in the pyrimidine ring. The structure of these polymers, which contain a uracil or thymine unit in a polymer main chain, was determined by NMR, UV, IR, and mass spectra. The structure was found to be affected by the polymerization temperature.[64]

The ring-opening polymerization of the cyclic compounds was also studied by using anionic initiators, such as sodium methoxide, potassium *tert*-butoxide, and *n*-butyllithium. The polymerization mechanism was discussed in relation to the case of cationic polymerization, and the properties of the polymers obtained were shown.[65]

7. Conclusion

The recent aspects on the chemistry of nucleic acid analogues have been reviewed, focusing on some of the important problems with which we have been concerned. Nucleic acid analogues find application in the biological and pharmaceutical fields, for example, as the polymeric drugs including interferon inducers. A related series by Pitha and co-workers is also noteworthy and has also been reviewed.[7,66,67] The pendant 5-fluorouracil and 6-methylthiopurine polymers of Gebelein and co-workers are also related and have been reviewed.[68,69]

References

1. K. Takemoto, *J. Macromol. Sci., Rev. C5*, 29–102 (1970).
2. M. Imoto and K. Takemoto, *Synthesis 1970*, 173–179.
3. M. Kinoshita, K. Yamauchi, and M. Imoto, *Progr. Polym. Sci. Japan* **7**, 63–106 (1974).
4. K. Takemoto, *J. Polym. Sci., Polym. Symp.* **55**, 105–125 (1976).
5. J. Pitha, *Polymer* **18**, 425–430 (1977).
6. K. Takemoto, in: *Polymeric Drugs* (L. G. Donaruma and O. Vogl, eds.), pp. 103–129, Academic Press, New York (1978).
7. J. Pitha, M. Akashi, and M. Draminski, in: *Biomedical Polymers: Polymeric Materials & Pharmaceuticals for Biomedical Use* (E. P. Goldberg and A. Nakajima, eds.), pp. 271–297, New York, Academic Press (1980).
8. K. Takemoto and Y. Inaki, *Adv. Polym. Sci.* **41**, 1–51 (1981).
9. Y. Inaki, T. Ishiwaka, and K. Takemoto, in: *Modification of Polymers* (C. E. Carraher and M. Tsuda, eds.), pp. 359–370, American Chemical Society Symposium Series No. 121, Washington, D.C. (1980).
10. T. Ishikawa, Y. Inaki, and K. Takemoto, *Polym. Bull.* **1**, 215–220 (1978).
11. C. G. Overberger and Y. Inaki, *J. Polym. Sci., Polym. Chem. Ed.* **17**, 1739–1758 (1979).
12. K. Takemoto, H. Tahara, A. Yamada, Y. Inaki, and N. Ueda, *Makromol. Chem.* **169**, 327–331 (1973).
13. T. Ishikawa, Y. Shigeno, T. Takahara, Y. Inaki, K. Kondo, and K. Takemoto, *Nucl. Acids Res. Suppl.* **5**, 279–282 (1978).
14. T. Ishikawa, Y. Inaki, and K. Takemoto, *Polym. Bull.* **1**, 85–89 (1978).
15. T. Ishikawa, Y. Inaki, and K. Takemoto, *J. Polym. Sci., Polym. Chem. Ed.* **18**, 1847–1856 (1980).
16. T. Ishikawa, Y. Inaki, and K. Takemoto, *J. Polym. Sci., Polym. Chem. Ed.* **18**, 949–958 (1980).
17. Y. Inaki, T. Ishikawa, S. Sugita, and K. Takemoto, *J. Polym. Sci., Polym. Lett. Ed.* **18**, 725–736 (1980).
18. Y. Inaki, Y. Suda, and K. Takemoto, *Polym. J.* **16**, 303–306 (1984).
19. Y. Kita, Y. Inaki, and K. Takemoto, *J. Polym. Sci., Polym. Chem. Ed.* **18**, 427–439 (1980).
20. Y. Kita, T. Uno, Y. Inaki, and K. Takemoto, *J. Polym. Sci., Polym. Chem. Ed.* **19**, 477–485 (1981).
21. M. Akashi, T. Okimoto, Y. Inaki, and K. Takemoto, *J. Polym. Sci., Polym. Chem. Ed.* **17**, 905–916 (1979).

22. Y. Kita, Y. Uno, Y. Inaki, and K. Takemoto, *J. Polym. Sci., Polym. Chem. Ed.* **19**, 1733–1744 (1981).
23. Y. Kita, T. Uno, Y. Inaki, and K. Takemoto, *J. Polym. Sci., Polym. Chem. Ed.* **19**, 3315–3324 (1981).
24. Y. Kita, T. Uno, Y. Inaki, and K. Takemoto, *J. Polym. Sci., Polym. Chem. Ed.*, 2357–2355 (1981).
25. H. E. Johns, *Photochem. Photobiol.* **7**, 633–636 (1968).
26. Y. Inaki, S. Fukunaga, Y. Suda, and K. Takemoto, (to be published).
27. E. DeClercq and T. C. Merigan, in: *Annual Review of Medicine* (A. C. Degraff, ed.), Vol. 21, Palo Alto, California, Annual Reviews, Inc., pp. 17–46 (1970).
28. A. K. Field, A. A. Tytell, and G. P. Lampson, and M. R. Hilleman, *Proc. Natl. Acad. Sci. U.S.A.* **61**, 340–346 (1968).
29. M. M. Nemes, A. A. Tytell, G. P. Lampson, A. K. Field, and M. R. Hilleman, *Proc. Soc. Exp. Biol. Med.* **132**, 776–783 (1969).
30. Y. Inaki and K. Takemoto, *Kaguku (Chemistry)* **31**, 903–914 (1976).
31. M. Absher and W. R. Stinebring, *Nature* **223**, 715–717 (1969).
32. T. D. Perrine, National Institutes of Health, private communication.
33. K. W. Dixon, Ph. D. thesis, The University of Michigan (1974).
34. C. G. Overberger, Y. Inaki, and Y. Nambu, *J. Polym. Sci., Polym. Chem. Ed.* **17**, 1759–1769 (1979).
35. G. J. Thomas and Y. Kyogoku, *J. Am. Chem. Soc.* **89**, 4170–4175 (1967).
36. C. G. Overberger and Y. Morishima, *J. Polym. Sci., Polym., Chem. Ed.* **18**, 1247–1265 (1980).
37. C. G. Overberger and Y. Morishima, *J. Macromol. Sci., Chem.* **A13**, 573–585 (1979).
38. C. G. Overberger and Y. Morishima, *J. Polym. Sci., Polym. Chem. Ed.* **18**, 1267–1277 (1980).
39. C. G. Overberger and Y. Morishima, *J. Polym. Sci., Polym. Chem. Ed.* **18**, 1433–1446 (1980).
40. T. Shimizu, Y. Konishi, and A. Murakami, *Makromol. Chem.* **178**, 2581–2587 (1977).
41. Y. Inaki, Y. Suda, Y. Kita, and K. Takemoto, *J. Polym. Sci., Polym. Chem. Ed.* **19**, 2519–2530 (1981).
42. Y. Inaki, Y. Sakuma, Y. Suda, and K. Takemoto, *J. Polym. Sci., Polym. Chem. Ed.* **20**, 1917–1933 (1982).
43. Y. Sakuma, Y. Inaki, and K. Takemoto, *J. Polym. Sci., Polym. Chem. Ed.* (in press).
44. Y. Sakuma, Y. Inaki, and K. Takemoto, *J. Polym. Sci., Polym. Chem. Ed.* (in press).
45. Y. Suda, Y. Inaki, and K. Takemoto, (to be published).
46. Y. Suda, M. Kono, Y. Inaki, and K. Takemoto, *J. Polym. Sci., Polym. Chem. Ed.* **22**, 2427–2442 (1984).
47. L. N. Blob, V. E. Vengris, P. M. Pitha, and J. Pitha, *J. Med. Chem.* **20**, 356–359 (1977).
48. N. Ueda, K. Nakatani, K. Kondo, K. Takemoto, and M. Imoto, *Makromol. Chem.* **134**, 305–308 (1970).
49. H. Schott and G . Greber, *Makromol. Chem.* **145**, 11–20 (1971).
50. H. Tuppy and E. Küchler, *Biochem. Biophys. Acta* **80**, 669–671 (1964).
51. A. S. Jones, D. G. Parsons, and D. G. Roberts, *Eur. Polym. J.* **3**, 187–198 (1967).
52. Y. Kato, T. Seita, T. Hashimoto, and A. Shimizu, *J. Chromatogr.* **134**, 204–207 (1977).
53. K. Kondo, T. Horiike, and K. Takemoto, *J. Macromol. Sci., Chem. A16*, 793–802 (1981).
54. K. Kondo, K. Hyodo, T. Horiike, and K. Takemoto, *Nucl. Acids Res. Suppl.*, 125–128 (1981).
55. K. Kondo, K. Hyodo, M. Murakami, and K. Takemoto, *Makromol. Chem.* **182**, 3411–3417 (1981).
56. K. Kondo, M. Murakami, and K. Takemoto, *Makromol. Chem.* **184**, 497–503 (1983).

57. K. Kondo, C. Komamura, M. Murakami, and K. Takemoto (to be published).
58. S. S. Cohen, *Progr. Nucl. Acid Res. Mol. Biol.* **5**, 1–88 (1966).
59. W. V. Ruyle and T. Y. Shen, *J. Med. Chem.* **10**, 331–334 (1967).
60. A. Hoshi, F. Kanazawa, K. Kuretani, M. Saneyoshi, and Y. Arai, *Gann* **62**, 145–150 (1971).
61. M. Akashi, H. Futagawa, Y. Inaki, K. Kondo, and K. Takemoto, *Nucl. Acids Res. Suppl.*, 7–10 (1977).
62. Y. Kita, H. Futagawa, Y. Inaki, and K. Takemoto, *Polym. Bull.* **2**, 195–199 (1980).
63. Y. Inaki, H. Futagawa, and K. Takemoto, *Org. Prep. Proc. Int.* **12**, 275–281 (1980).
64. Y. Inaki, H. Futagawa, and K. Takemoto, *J. Polym. Sci., Polym. Chem. Ed.* **18**, 2959–2969 (1980).
65. Y. Inaki, E. Mochizuki, and K. Takemoto (to be published).
66. P. M. Pitha and J. Pitha, *Pharmacol. Ther. A* **2**, 247–260 (1978).
67. J. Pitha, in: *Biomaterials and Organism* (Y. Imanishi, Y. Sakurai, M. Senoo, and K. Takemoto, eds.), pp. 151–162, Kodansha, Tokyo (1981).
68. C. G. Gebelein, R. M. Morgan, R. Glowacky, and W. Baig, in: *Biomedical and Dental Applications of Polymers* (C. G. Gebelein and F. F. Koblitz, eds.), pp. 191–201, Plenum Press, New York (1981).
69. C. G. Gebelein, in: *Biological Activities of Polymers* (C. E. Carraher, Jr. and C. G. Gebelein, eds.), pp. 193–203, American Chemical Society Symposium Series No. 186, Washington, D.C. (1982).

17

Enzyme-Mimetic Polymers

Yukio Imanishi

Abstract. Enzymes are proteins with catalytic groups or proteins which bind or coordinate to small molecules with catalytic action. Enzymes bind substrate prior to reaction to condense and activate the substrate, and provide the bound substrate with suitable reaction field, resulting in an enhanced rate and specificity of reaction. The bound substrate is oriented toward an intramolecular catalytic group for high efficiency and specificity of reaction. Catalytic groups can increase catalytic ability by intramolecular cooperation. These aspects of enzyme catalysis can be reproduced to some extent by synthetic polymers and molecular aggregates. Among investigations on enzyme-mimetic polymers as such, reports published during 1978 and 1982 are reviewed in this article, and the reactivity of enzyme-mimetic polymers is discussed from a viewpoint of intramolecular reaction proceeding along a chain of multifunctional polymer molecules.

1. Introduction

Biological molecules that sustain life activities and control life phenomena are very specific in the recognition of foreign molecules and the acceptance of information, very efficient in the transmittance and conversion of information, and are subjected to environment control. These features of biological molecules are clearly seen in enzymes, and the mechanism of enzyme reaction has been elucidated on a molecular level by molecular biology.

Synthetic polymer materials such as fibers, plastics, rubbers, and films are inevitable to our daily life. The synthesis, structure, property, and function of polymeric compounds have been investigated. Most markedly, functions and properties of polymeric compounds are understandable in relation to the structure, and the method of molecular design which realizes the desired structure or function is also known.

A trend was begun about 25 years ago, in which highly functional polymeric materials could be prepared which contained chemical and biological functionalities. Since the beginning of the 1970s, the molecular design of biomimetic polymers, that is, the development of polymeric

Yukio Imanishi • Department of Polymer Chemistry, Kyoto University, Yoshida Honmachi, Sakyo-ku, Kyoto 606, Japan.

materials having biological activities, has become more important in attempts to overcome social problems, such as the energy crisis and environmental pollution.

Polymer scientists learned about the behavior of biological macromolecules and in particular enzymes, investigated the structure–function relationship of polymeric compounds having enzyme-like activities, and brought about a remarkable progress related to understanding the mechanism of intramolecular reactions of multifunctional polymers. In spite of a great deal of research effort, little success has been achieved with respect to the synthesis and characterization of polymeric materials having enzyme-mimetic abilities.

In the present stage, we should know what has been found by the investigation of enzyme-mimetic polymers, and we must understand what should be done for the future development of the area. The author reviewed early-stage investigations on enzyme-mimetic polymers in a book[(1)] published in 1972, and described the activity of enzyme-mimetic polymers from the viewpoint of intramolecular reactions proceeding along a chain of multifunctional polymer molecules in a review article[(2)] published in 1979. In the present article, subsequent developments in this field will be reviewed. Since the above review article[(2)] was written in 1977, research achievements reported mainly after 1978 will be described in this chapter. Due to a limitation of space, enzyme reactions, catalysts immobilized on the polymer matrix, and micellar catalysts will not be included. Review articles on enzyme-mimetic polymers have recently been written by Westheimer,[(3)] Klotz,[(4)] and Manecke and Storck.[(5)]

2. Polymer Catalysts That Bind the Substrate

In enzyme reactions, like the lock-and-key combination, a substrate (guest) molecule is bound by an enzyme (host) molecule. The driving force for substrate binding is hydrophobic interaction, electrostatic interaction, hydrogen bonding, charge transfer interaction, coordination through metal ion, and so on. In case of a rigid host molecule, the substrate binding is subjected to a stereochemical selection, leading to a substrate selectivity. A bound substrate is sometimes strained and activated for the reaction. In one case host molecules possess only a substrate-binding site, and in other cases they may possess catalytic sites as well. In any case, by the substrate binding, substrate molecules are concentrated in the vicinity of the catalytic site of the host molecule, and a bound substrate is oriented to the catalytic site. Furthermore, desolvation of ions and salt effects occur within the environmental area of the host molecule including hydrophobic or electrostatic atmosphere effects.

2.1. Polymer Catalyst Having a Substrate-Binding and a Catalytic Group

2.1.1. Hydrophobic Binding

As can be easily imagined from the description that an "enzyme is an oil droplet having a hydrophilic cover", substrate binding by enzyme is often carried out by hydrophobic interaction. Based on this idea, an efficient hydrolysis of hydrophobic substrate was attempted using flexible polymers as catalysts; that is, polymers which are equipped with imidazolyl groups and long-chain alkyl or acyl chains. A polymer catalyst based on branched polyiminoethylene (molecular weight 600), in which approximately 10% of the primary amino group was laurylated and approximately 15% imidazolylmethylated, was found by Kiefer *et al.*[6] to be 75 times as effective as a specific enzyme type-II sulfatase (37°C, pH 6.9) in the hydrolysis of 2-hydroxy-5-nitrophenylsulfate (20°C, pH 9.2). However, Kunitake and Sakamoto[7] could not reproduce this striking report in the hydrolysis of the same substrate by laurylimidazolylmethylpolyiminoethylene **[1]** under nearly the same conditions. The reasons of disagreement between the two similar experiments were not reported. In the same paper,[7] it was reported that partially quaternized polyiminoethylene **[2]** hydrolyzed 2,4-dinitrophenylsulfate at 30°C and pH 9.5 according to the Michaelis–Menten kinetics, giving k/K_m = 24.2 liter mol^{-1} s^{-1}. The latter value is larger than any of the second-order rate constants so far reported for the above reaction.

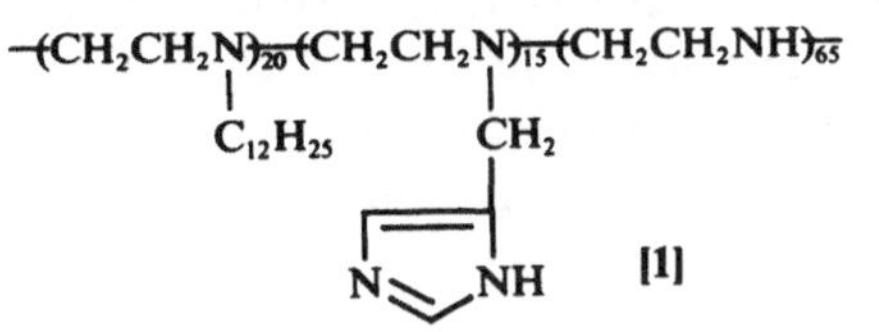

$$-(CH_2CH_2\underset{C_{12}H_{25}}{N})_{20}(CH_2CH_2\underset{CH_2-\text{imidazolyl (N, NH)}}{N})_{15}(CH_2CH_2NH)_{65} \quad [1]$$

$$-(CH_2CH_2\overset{CH_3}{\underset{C_{18}H_{37}}{N^+}})_{16}(CH_2CH_2\overset{CH_3}{N^+})_{84} \quad [2]$$

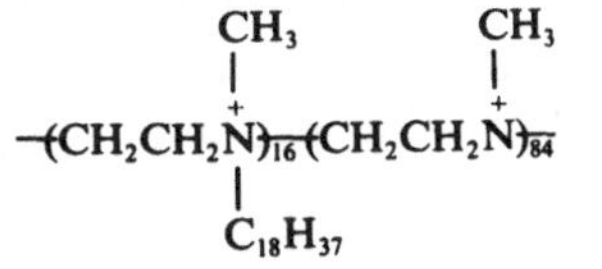

Laurylated polyiminoethylene, to which dialkyl pyridines and 3-[(*N*-methyl-*N*-4-pyridyl)amino]propionic acids are covalently bound, was reported to be highly active as a catalyst for the hydrolysis of nitrophenyl acylates, and the catalysis was found to be nucleophilic.[8] The hydrolysis of alkyl acetate by sulfonated polystyrene or poly(vinyl alcohol) treated with *O*-benzalsulfonic acid was investigated under pressures up to 2000 kg cm^{-2}, and the hydrophobic interaction or hydrogen bonding between the polymer catalyst and the substrate was discussed.[9] Polyiminoethylene, to which lauryl groups and L-histidines are introduced, was an enantiomer-selective catalyst in the hydrolysis of *N*-protected phenylalanine *p*-nitrophenyl ester, in which D-esters were hydrolyzed slightly faster than L-esters.[10] Similarly, copolymers of dodecyl methacrylate and *N*-methacryloyl-L-histidine or

N-methacryloyl-L-histidine methyl ester were an enantiomer-selective catalyst in the hydrolysis of *N*-protected phenylalanine *p*-nitrophenyl ester, in which L-ester was hydrolyzed slightly faster than the D-ester.[11] In these enantiomer-selective catalyses the substrate binding by hydrophobic interaction is essential.

2.1.2. Electrostatic Binding

Acetylcholine esterase is a typical example of an enzyme catalysis in which the substrate binding by electrostatic interaction plays an important role. Polymers having catalytic sites and charged sites have been used as catalysts for the hydrolysis of oppositely charged substrates. Aspirin was hydrolyzed with poly(4-vinylpyridine) lightly cross-linked with divinylbenzene (PVP-Gel), poly(4-vinylpyridine) quaternized by *t*-butyl chloride (*t*-Bu-QPVP, rate of quaternization 90%), and *t*-Bu-QPVP treated with formaldehyde (QPVP-Gel).[12] Apparent first-order rate constants are listed in Table 1. Since the charged polymer catalysts did not accelerate the hydrolysis of aspirin within regions of low pH, the acceleration must have been brought about by electrostatic binding of the anionic substrate by cationically charged polymer catalysts. The particularly effective catalysis by QPVP-Gel may be due to the suitable shape and size of the catalytic site of polymer. This effect is seen from the unusual thermodynamic parameter values obtained.

Copolymers of *p*-vinylthiophenol and acrylic acid were found to be an effective catalyst for the hydrolysis of 3-acetoxy-*N*,*N*,*N*-trimethylanilinium iodide.[13] The reaction proceeded according to the Michaelis–Menten kinetics, but the reaction mechanism was changed by the increase of ionic strength of the medium. Therefore it was shown that a complex of the

Table 1.[a] **Apparent First-Order Rate Constants, *k*, for Aspirin Hydrolysis with Various Catalytic Systems**[b]

	k/d^{-1} at [Aspirin]/mmol liter^{-1}	
Catalytic system	0.3	3.0
Buffer	0.49	0.43
PVP-Gel	—	1.07
t-Bu-QPVP	3.89	2.66
QPVP-Gel	9.76	5.61

[a] Reproduced from Table 1 of Reference 12 by permission of the copyright owner.

[b] [PVP-Gel] = [*t*-Bu-QPVP] = [QPVP-Gel] = 30 mmol liter^{-1}; pH 5.0; temperature 30°C.

negatively charged polymer and the positively charged substrate was subjected to a nucleophilic attack of thiophenolate ion.

2.1.3. Hydrogen Bonding

It has been shown in the cleavage of the β-1,4-glycoside bond of mucopolysaccharide by lysozyme that cooperative hydrogen bondings take place between enzyme and substrate, and that the substrate is bound into a cleft near the catalytic site of enzyme and strained and activated for the reaction. Based on this example, polymer catalysts have been designed that can bind substrates by hydrogen bonding.

Arai and Ogiwara[14–17] and Arai *et al.*[18,19] investigated the hydrolysis of polysaccharides by vinyl copolymers having sulfonic acid groups and alcoholic groups. Hydrolyses by these copolymers are usually faster than those catalyzed by H_2SO_4. Hydrogen bonding of substrate with hydroxy groups of polymer and high concentration of hydrogen ions along the polymer catalyst are important factors in the fast hydrolysis. In the hydrolysis of fibrous cellulose by copolymers of vinyl sulfonic acid and vinyl alcohol, the acceleration by the polymeric acid with respect to H_2SO_4 decreased with increasing density of the fine structure of the substrate.[14] In the hydrolysis of dextrin by copolymer resins of styrene sulfonic acid and vinyl alcohol, the acceleration reached ninefold in comparison with a hydroxy-group-free catalyst.[15] In the hydrolysis of dextrin by block copolymers of styrene sulfonic acid and vinyl alcohol, the reaction proceeded according to the Michaelis–Menten kinetics, provided that the content of vinyl alcohol was high. The ability of substrate binding of the block copolymer was lower than that of the random copolymer, because in the block copolymer there are strong intramolecular interactions between the blocks of vinyl alcohol. However, the catalytic activity of the block copolymer was higher than that of the random copolymer, because in the block copolymer the charge density of vinyl sulfonic acid block was high. In total, the acceleration of hydrolysis was more marked with the block copolymers than with the random copolymers.[16]

In the hydrolysis of amylose by block copolymers of styrene sulfonic acid and vinyl alcohol, substrates bound by hydrogen bonding are reacted more rapidly than unbound substrates. Therefore, the hydrolysis of amylose by the random copolymers takes place more randomly compared with hydrolysis effected by the block copolymers.[17] In the hydrolysis of dextrin by copolymer resins of styrene sulfonic acid and vinyl alcohol, the acceleration was more marked as the content of the vinyl alcohol component increased. Thus, the largest acceleration in comparison with Amberlite 120 B resin was sixfold. Usually in the hydrolysis of saccharides by sulfonic acid resins, polysaccharides are hydrolyzed after being bound by the polymer catalyst, whereas sucrose is not bound by the polymer catalyst and shows only a low degree of acceleration. The difference between polysaccharide and sucrose becomes more marked with increasing content of the vinyl alcohol component in the polymer catalyst.[18] However, the hydrolysis of dextrin by the copolymer resin did not proceed by the Michaelis–Menten kinetics. This is in contrast to the Michaelis–Menten type of hydrolysis observed when soluble copolymers are employed. The inhibition of hydrolysis by the addition of poly(vinyl alcohol) or the copolymer resin of vinyl alcohol and styrene is still indicative of a weak binding of substrate by a poly(vinyl alcohol) sequence in the copolymer resin.[19]

Poly(vinyl alcohol) having a degree of polymerization (DP) of approximately 2000 and containing 2.7 mol % 4(5)-imidazolylmethyl groups hydrolyzed *p*-nitrophenyl acetate according to the Michaelis–Menten kinetics. The Michaelis constant K_m decreased by the addition of alcohol and increased on the addition of boric acid. Therefore, the polymer catalyst binds the substrate by hydrogen bonding.[20] Quinone groups were introduced to porous copolymers of styrene and divinylbenzene, and the product was used as a catalyst for the gas-phase oxidation of 2-propanol by oxygen. The introduction of sulfonic acid groups into the copolymer increased the catalyst activity. The reason for the activity enhancement is that sulfonic acid groups bind the substrate, make the polymer catalyst swell, and inhibit the access of the substrate to quinone groups.[21]

2.1.4. Asymmetric Reactions

The most important feature of enzyme reactions is strict stereospecificity or an enantiomer recognition. So far, the success of the design of enzyme-mimetic polymers has been judged by the rate of acceleration of selected systems in comparison with the catalysis by small molecules. However, stereospecificity is also a critical point of catalysis, and the design of stereospecific polymer catalyst requires more rigorous control of the reaction path. For a preliminary step, asymmetric synthesis and asymmetric selection may be possible with chiral polymer catalyst having substrate-binding groups and catalytic groups. Examples of asymmetric reactions catalyzed by chiral polymer catalysts will be described below.

Yamashita *et al.* investigated the addition of methanol to phenylmethylketene using as the catalyst homopolymers of *N*-benzyl-2-pyrrolidinylmethyl acrylate and its copolymers with methyl acrylate[22] or poly(2-quinuclidinylmethyl acrylate).[23] They observed an asymmetric addition, but did not discuss the mechanism of reaction. Using copolymers of cinchona alkaloid and acrylonitrile, Kobayashi and Iwai[24,25] investigated the addition reactions of methylindanone-2-carboxylate to methylvinylketone, lauryl mercaptan to methyl methacrylate, benzyl mercaptan to β-nitrostyrene, and lauryl mercaptan to β-substituted phenylvinylketone. They observed asymmetric reactions, but the reaction mechanism was not investigated. An asymmetric addition reaction of lauryl mercaptan to isopropenylmethylketone, catalyzed by homopolymers of *d*-bornyl methacrylate and its copolymers with styrene, has been reported.[26]

Yashiro *et al.*[27] investigated the hydrolytic reactions of Z-D- or L-Phe-OPh(NO_2) or Z-D- or L-Val-OPh(NO_2), where Z represents a benzyloxycarbonyl group, employing poly[(*R*)- or (*S*)-2-ethylazirizinoacetic acid] as a catalyst. The (*R*)-catalyst was found to hydrolyze L-substrate preferentially, with the reaction proceeding according to Michaelis–Menten kinetics. Furthermore, the racemic catalyst was found not to discriminate between the D- and the L-substrate.

2.2. Reactions Occurring under Circumstances Regulated by Polymers

2.2.1. Reactions Occurring under the Influence of Nonionic Polymers

Yamazaki *et al.*[28] observed that the Williamson reaction (1) was accelerated by the addition of poly(*N*-vinylpyrrolidone). The reaction was also accelerated by the addition of *N*-methylpyrrolidone, a monomer analogue, but to a much smaller extent than by the polymer.[29] It was shown that the acceleration by poly(*N*-vinylpyrrolidone) is due to the cooperative solvation of sodium ion by pyrrolidone species, leading to an enhanced dissociation of PhONa.

$$PhONa + n\text{-}C_4H_9Br \rightarrow PhO\text{-}(n\text{-}C_4H_9) + NaBr \qquad (1)$$

The addition of polyoxyethylene accelerated the oxidation of *trans*-stilbene and several kinds of nucleophilic substitutions, such as alkylations of potassium acetate and diethyl benzylsodiomalonate, as well as reaction (1).[30] The acceleration is caused by an enhanced dissociation of the ion pair by a cooperative coordination of oxygen atoms in polyoxyethylene with metal cations. The relationship between the reaction rate and the cooperative coordination or the molecular weight of polyoxyethylene was investigated.[31]

Matsumoto and Orihara investigated the additive effect of nonionic polymers on various chemical reactions. The alkaline fading reaction of triphenylmethane dyes was accelerated by the addition of poly(*N*-vinylpyrrolidone).[32] The effect was most marked when the molecular weight of polymer was approximately 2×10^4, and the acceleration was caused by the adsorption and concentration of dyes and alkali to the polymer chain. This reaction was retarded slightly by the addition of poly(vinyl alcohol). Since copolymers of vinyl alcohol and *N*-vinylpyrrolidone also decelerated the reaction, the acceleration by poly(*N*-vinylpyrrolidone) should require rather long sequences of the pyrrolidone residues.[33] The reactions of dinitrobenzene or dinitrofluorobenzene with sodium hydroxide or cyclohexylamine were also accelerated by the addition of poly(*N*-vinylpyrrolidone) or polyoxyethylene, but decelerated by the addition of poly(vinyl alcohol).[34] Alkaline hydrolysis of *p*-nitrophenyl carboxylates was accelerated by the addition of poly(*N*-vinylpyrrolidone) or polyoxyethylene, and decelerated by the addition of poly(vinyl alcohol).[35] Since the acceleration effect becomes increasingly important with increasing chain length of carboxylic acid, the substrate binding by hydrophobic interaction should be important. However, in this case too, the coordination of polymer to metal ions should not be ignored. The same authors[36] investigated the effect of poly(*N*-vinylpyrrolidone) added to various reactions in aqueous solutions and concluded that the acceleration of bimolecular reactions occurs only when both substrate and catalyst are adsorbed by added polymers, and that the acceleration of unimolecular reactions occur only when the adsorbed substrate is activated for the reaction. Furthermore, it was considered that if substrate and catalyst are adsorbed by added polymers and compete for binding sites in a polymer coil, the extent of acceleration is decreased.[37]

It is well known that the reactions of anionic species proceed much more rapidly in organic solvents than in aqueous solutions. The reason for the acceleration is considered to be due to a microenvironmental activation of the reaction site. Decarboxylation of 6-nitrobenzisoxazole-3-carboxylate (2) is a typical example which is strongly accelerated in non-hydrogen-bond-forming solvents. The active site of enzymes is usually situated in a hydrophobic circumstance and scarcely hydrated. Therefore, a major portion of enzymatic activity involving the anionic nucleophilic site might be explained in terms of the desolvation–activation in a hydrophobic circumstance.

$$\text{COO}^- \quad \text{C} \quad \text{N} \quad \text{O} \quad \text{NO}_2 \longrightarrow \left[{}^-\text{O}\cdots\text{C}{=}\text{O} \quad \text{C} \quad \text{N} \quad \text{O} \quad \text{NO}_2 \right] \longrightarrow \text{C}{\equiv}\text{N} \quad \text{NO}_2 \quad \text{O}^- + \text{CO}_2 \qquad (2)$$

Crown ethers strongly bind metal ions and ammonium ions into the ether cavity by the coopertive coordination of oxygen atoms. The binding ability is very great for a certain cation which possesses a size and a shape matching the cavity. Since each crown unit in poly(vinylbenzo-18-crown-6) [3] is specific to sodium ion or potassium ion, the reaction (2) of sodium salt catalyzed by **3** may proceed in a hydrophobic atmosphere in the polymer coil. This situation is caused by the ion-pair formation of the anion with sodium ion captured by the crown ether unit. In fact, the decarboxylation reaction was accelerated by 2×10^3–5×10^3 times according to the charge density of **3**, as shown in Figure 1.[38] The acceleration induced by the ion binding by crown ethers was reported for the decomposition of arenediazonium fluoroborate[39] and the decarboxylation of sodium carboxylates.[40]

$-\!\!\!-\![CHCH_2]_n\!-\!\!\!-$

[3]

As is evident in the above examples, the additive effect of nonionic polymers on chemical reactions can be summarized as follows: a concentration effect through binding of the substrate, and an activation effect on the bound substrate through a microenvironmental effect.

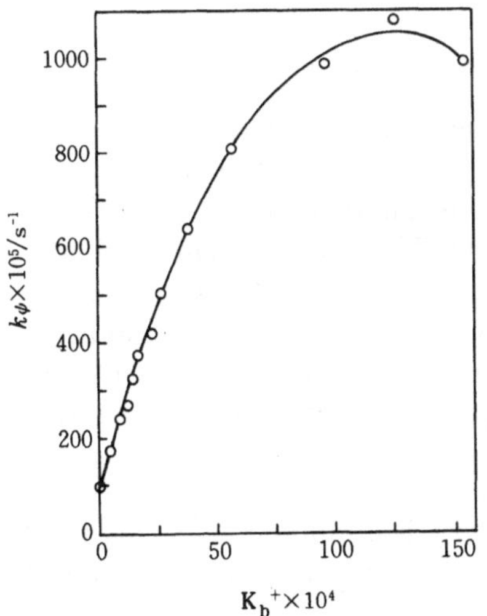

Figure 1. Dependence of the observed rate constant k_ψ of decarboxylation of 6-nitrobenzisoxazole-3-carboxylate in a water/P18C6 mixture as a function of the concentration of K^+ charges bound to P18C6 at 25°C. (Reproduced from Figure 4 of Reference 38 by permission of the copyright owner.)

2.2.2. Reactions in Polysoap

The polymer effect described in the preceding section is similar to the micellar effect. Though all of the micellar effects on chemical reactions cannot be reviewed here, cases involving polymeric detergents will be considered below.

Shinkai *et al.*[41] carried out reaction (2) using poly(4-vinylpyridine) quaternized by octyl(3–15 mol %), dodecyl(22–33 mol %), octadecyl(3–15 mol %), or docosyl(3–10 mol %) groups. The rest of the pyridyl groups were quaternized by ethyl groups. The substrate was taken into the polysoap by hydrophobic interaction and then reacted very rapidly under hydrophobic conditions. Shinkai *et al.* derived "average side-chain length $\bar{L}$," as an index for the contribution of side-chain alkyl groups to the polymer hydrophobicity. According to their parameter, the optimal balance between hydrophobicity and hydrophilicity was obtained when the octyl-quaternized polymer was $\bar{L} = 5.0$.

A complex of bovine serum albumin and tetradecyltrimethylammonium bromide was found to offer an environment for the alkaline hydrolysis of *p*-nitrophenyl hexanoate and was found to strongly accelerate the reaction.[42,43]

A cationic detergent polypeptide, melittin, which is isolated from the venom of honeybee and composed of 26 amino acids, strongly accelerated the hydrolysis of *p*-nitrophenyl laurate[44] The acceleration was attributed to the depression of substrate aggregation and the enhancement of hydroxide ion attack on the dispersed substrate. Interestingly, a complex of melittin with a long-chain carboxylic acid, which is a reaction product, was found to be a more effective catalyst than melittin itself.

2.2.3. Activation of Anionic Nucleophiles in Hydrophobic Environments

In nonenzymatic systems there are several examples showing a wide variation of acid dissociation constants according to the properties of hydrogen bonding. In enzymatic systems, a microenvironmental effect on pK_a of the catalytic site in lysozyme is evident. The side-chain carboxyl group of Glu-35 situates itself in a hydrophobic domain and its pK_a is 6.7. It is undissociated under optimal conditions for the enzyme reaction and acts as a general acid. On the other hand, the side-chain carboxyl group of Asp-52 situates itself in a polar region and its pK_a is 3.8. It is dissociated at the optimal condition for the enzyme reaction and stabilizes a reaction intermediate (carbocation) by the formation of an ion pair. As seen in these examples, an extremely high catalytic activity is attainable by changing the pK_a of ionic catalytic groups through a microenvironmental effect and by desolvation–activation.

It has been shown that cationic micelles greatly increase the reactivity of anionic nucleophiles, such as oximate, hydroxamate, thiolate, and imidazole anion. In this phenomenon, the formation of a nucleophilic ion pair within the hydrophobic environment, a hydrophobic ion pair, was considered to be critical.

Kunitake *et al.*[45] investigated the hydrolysis of *p*-nitrophenyl acetate using as catalyst four kinds of zwitterionic hydroxamates (**4**, **5**) and three kinds of simple hydroxamates as shown below (**6**–**8**). Table 2 presents the

$CH_3(CH_2)_{11}$–Im⁺(R)–CH_2–C(=O)–N(O^-)–CH_2–Ph

[4]

{R = CH_3: C_{12}—MIm$^{\pm}$—HA
{R = H : C_{12}—Im$^{\pm}$—HA

$CH_3(CH_2)_{n-1}$–Py⁺–CH_2–C(=O)–N(O^-)–CH_2–Ph

[5]

n = 8: C_8—Py$^{\pm}$—HA
n = 13: C_{13}—Py$^{\pm}$—HA

$CH_3(CH_2)_{11}$N(O^-)–C(=O)–Ph — C_{12}—BHA **[6]**

$CH_3(CH_2)_{12}$C(=O)–N(O^-)–CH_3 — C_{13}—MHA **[7]**

Ph–CH_2–N(O^-)–C(=O)–Ph — BBHA **[8]**

rate constants of reactions carried out in the presence of an anionic detergent, sodium dodecylsulfate(SDS); a cationic detergent, cetyltrimethylammonium bromide(CTAB); a neutral detergent, polyoxyethylene oleyl ether(POOA);

Table 2.[a] **Second-Order Rate Constants of Acylation in Micellar Systems**[b]

	$k_{a,obs}$ (liter mol^{-1} s^{-1})				
Hydroxamate	Nonmicellar	SDS	CTAB	POOA	DMOG
C_{12}-Im$^+$-HA	12	1.2	1000	270	1200
C_{12}-MIm$^+$-HA	13	0.8	1300	250	1100
C_{13}-Py$^+$-HA	14	1.0	1100	310	1300
C_8-Py$^+$-HA	12	—	320	45	420
C_{12}-BHA	—	0.9	1500	25	75
C_{13}-MHA	—	1.3	1300	32	45
BBHA	12	—	1500	13	23

[a] Reproduced from Table 1 of Reference 45 by permission of the copyright owner.

[b] 30°C; 3 v/v % EtOH–H_2O; μ = 0.01 (KCl); pH 8.90 ± 0.05 (0.01 mol liter^{-1} borate). [PNPA] = 9.46 × 10^{-6} mol liter^{-1}; [catalyst] = (3.05–7.08) × 10^{-5} mol liter^{-1}. Surfactant concentrations are 1 × 10^{-3} mol liter^{-1}, except for [SDS] = 1 × 10^{-2} mol liter^{-1}.

and a zwitterionic detergent, N,N-dimethyl-N-octadecylglycine(DMOG). In the reactions catalyzed by simple hydroxamates, the activation was observed only in cationic CTAB micelles where hydrophobic ion pairs are formed. In the reactions catalyzed by zwitterionic hydroxamates, various degrees of activation were observed in cationic, neutral, and zwitterionic micelles. In all of these micelles an intramolecular hydrophobic ion pair is formed. Micelles contribute to the activation of nucleophiles by generating a hydrophobic environment. The activation by micelles originates from the lowering of the pK_a and the activation of hydroxamate in a hydrophobic environment. The anionic SDS micelle increases the pK_a of a nucleophile, thus leading to a deceleration. A similar phenomenon was observed in the nucleophilic decomposition of dinitrophenyl sulfate by the zwitterionic hydroxamate micelle.[46]

A similar investigation was carried out by Murakami *et al.* Detergents containing a thiol group such as N-hexadecyl-N^{α}-glutaryl-L-cystein amide and N-hexadecanoyl-L-cystein were found to be very efficient catalysts for the hydrolysis of carboxylic acid *p*-nitrophenyl esters, when these detergents formed a comicelle with the cationic detergent CTAB.[47] An efficient catalysis was considered to result from an increased nucleophilicity of the thiol group in the micelle by an electric field effect at the CTAB micelle interface. Furthermore, Murakami *et al.*[48] synthesized other cationic detergents containing a thiol group including N-hexadecyl-N^{α}-(3-trimethylammoniopropionyl)-L-cystein amide bromide and N-dodecyl-N^{α}-(6-trimethylammoniohexanoyl)-L-cystein amide bromide. These detergents, above their critical micelle concentrations, accelerated the hydrolyses of carboxylic acid *p*-nitrophenyl esters. The acceleration of the reaction was considered to result from the lowering of the pK_a of SH groups by the effect of cationic charges in the Stern layer of the micelle and the desolvation of thiolate anions in a hydrophobic environment, the former being more important.

Williams *et al.*[49–51] synthesized microgels of cross-linked poly(methyl methacrylate) by emulsion polymerization, and introduced hydroxamate and carboxylate groups. With increasing degree of ionization of catalytic hydroxamate groups (FB), these functionalized microgels were found to be highly efficient catalysts for the hydrolysis of various carboxylic acid *p*-nitrophenyl esters. For example, for the hydrolysis of *p*-nitrophenyl acetate, methacrylohydroxamic acid gave $k_n = 47$ liter mol^{-1} s^{-1}, which was independent of FB, and a soluble copolymer of methacrylohydroxamic acid and methyl methacrylate gave $k_n = 0.54$–9.03 liter mol^{-1} s^{-1} when FB increased from 0 to 1. On the other hand, the k_n of the microgel catalyst increased from 1.76 to 5×10^4 when FB increased from 0 to 1. The efficient catalysis of the microgel was considered to be due to the dissociation of hydroxamate groups in the anionic microenvironment and the increased nucleophilicity of the hydroxamate anion under a hydrophobic circumstance.

It should be noted here that if (−)-methyl acrylate was copolymerized into the microgel, the resultant microgel was an asymmetric catalyst.[52] In the hydrolysis of Z-Phe-ONp, the (*S*)-isomer was hydrolyzed about twice as fast as the (*R*)-isomer.

2.2.4. Reactions Occurring in the Cavity of Cyclodextrin

It is well known that cyclodextrins include aromatic compounds by hydrophobic interaction. Reactions of substrates included into the cavity of cyclodextrin are variously affected by the host compound, the regiospecificity being most important among the inclusion effects.

The Reimer–Tiemann reaction (3) of *p*-substituted phenols usually yields *o*-formylphenol as a main product. The yield of 2,5-cyclohexadienones is usually less than 10%, and the selectivity is only 20%. The same reaction

$$\text{(X,Y,Z-substituted phenoxide)} \xrightarrow{CHCl_3,\ NaOH} \text{4-}CHCl_2\text{-2,5-cyclohexadienone} + \text{2-CHO phenoxide} \qquad (3)$$

	2,5-cyclohexadienone yield
$Y = CH_3$, $X = Z = H$	28%
$Y = C_6H_5$, $X = Z = H$	12%
$X = Y = Z = CH_3$	15%

under the presence of β-cyclodextrin produces 2,5-cyclohexadienone in an 80% yield, and the selectivity is nearly perfect.[53] Inclusion of dichlorocarbene into the cavity of cyclodextrin and the access of *p*-substituted phenol directing more hydrophobic para position toward the cavity of β-cyclodextrin are the important reasons for the high selectivity. The reaction product is also included in the cavity and protected from side reactions. This is another reason for the high yield. Regiospecificity, also brought about by the inclusion into the cavity of cyclodextrins, was observed in the selective synthesis of 4-hydroxybenzoic acid from phenol and CCl_4[54]; the selective synthesis of 2,4-dihydroxybenzaldehyde from resorcinol and $CHCl_3$[55]; the selective synthesis of 4-hydroxybenzoic acid from phenol and $CHCl_3$[56,57]; the selective one-pot synthesis of 4-hydroxychalcone from phenol, $CHCl_3$, and acetophenone[58]; the selective synthesis of indole-3-aldehyde from indole and $CHCl_3$[59]; and the selective synthesis of 4-hydroxy-3-(2-nitrovinyl)benzoic acid from phenol, $CHCl_3$, and nitromethane and of 3-(2-benzoylvinyl)-4-hydroxybenzoic acid from phenol, $CHCl_3$, and *p*-acetylphenol.[60]

When chlorpromazine was photoirradiated in the presence of

cyclodextrin, the promazine produced by dechlorination was photooxidized to yield promazine sulfoxide.[61] This reaction proceeds in an inclusion state and the product yield depends on the size of cyclodextrin cavities ($\beta > \gamma > \alpha$). The effect of β-cyclodextrin on the photooxidation of the stable carbanion by a singlet oxygen was also reported.[62]

When 2-methylhydronaphthoquinone-1,4 was reacted with allyl, crotyl, methallyl, or prenyl bromide in a dilute alkaline solution containing β-cyclodextrin, vitamin K_1 (or K_2) derivatives were one-pot-synthesized in an excellent yield.[63] Partially charged carbanions are activated in the hydrophobic cavity of cyclodextrin for allylation and the oxidative decomposition by H_2O_2 of the product quinone is retarded by the inclusion. These two factors are responsible for the excellent yield. Additional examples follow. It was shown[64] that the proton migration of *m*- and *p*-nitrophenylazosalicyclic acid is affected by its inclusion into β-cyclodextrin. The selective inhibition by the inclusion into cyclodextrin of the enzymatic hydrolysis of *p*-nitrophenyl-β-D-glucopyranoside was reported.[65] The substituent effect in the alkaline hydrolysis of 7-substituted coumarin changed by the inclusion into cyclodextrins.[66] β-Cyclodextrin acylated with an electrophoric *o*-benzoylbenzoic acid was reported to include benzyl esters of various carboxylic acids and to decompose them by the reduction at the relatively low reduction potentials of the electrophore.[67] Dediazotization of substituted benzene diazonium salts was found to be accelerated by β-cyclodextrin. This reaction proceeded by a radical mechanism irrespective of the nature of the substituent and the atmosphere (N_2 or O_2).[68]

2.2.5. *Polyelectrolyte Catalysis of Ionic Reactions*

Enzymes are usually polyelectrolytes, because some of the component α-amino acid residues possess a dissociative side chain. Therefore, in enzymatic reactions the effect of polyelectrolytes should be involved. Many examples are known of polyelectrolyte effects upon the reaction between charged species.[69] Polyelectrolytes affect not only the substrate distribution through the electrostatic site binding, but also the reaction rate through the electrostatic interaction with the reactants or the activated complex.

Ise and his co-workers attempted to explain the reaction (4) between ionic species A and B, which proceeds through an activated complex X, in terms of the activated complex theory of Brönsted and Bjerrum.

$$A + B \rightleftharpoons X \rightarrow C + D \tag{4}$$

According to this theory, the ratio of reaction rate constants k and

k^* in the presence or absence of polyelectrolyte is given by Eq. (5), where f represents the activity coefficient.

$$k/k^* = f_A\,(f_B/f_X) \tag{5}$$

Manning's theory on polyelectrolytes was applied to various interionic reactions, which had been investigated so far, to evaluate f, and consequently calculated values of k/k^* were shown to agree well with the observed values.[70,71] According to this theory, catalysis of interionic reactions by polyelectrolytes is determined by the stabilization or destabilization of reactants or activated complexes through their interactions with polyelectrolytes. Furthermore, Shikata *et al.*[72] observed that the electron transfer from Fe^{2+} (inner sphere) or V^{2+} (outer space) to Co^{3+} is catalyzed by the addition of polyanion, the acceleration being more marked in the former, and that these reactions were accelerated only slightly by the addition of polycation. These experimental results were explainable qualitatively by the Brönsted–Bjerrum–Manning theory.

Fernandez–Prini[73] explained the polyelectrolyte effect on the bimolecular reaction between equally charged species and the unimolecular reaction by the primary salt effect. Ishikawa[74] explained the acceleration of the reaction between m- and n-valent ions by using the numerical solution of the Poisson–Boltzmann theory.

In addition to this electrostatic factor, polyelectrolytes were found to change f values and therefore affect the reaction rate through the influence on the solvation state of reactants and activted complex by the interaction with solvent. For example, the rate of cyanoethylation reaction (6) was affected by the addition of poly(4-vinylpyridine) quaternized by benzyl chloride.[75] The reaction in aqueous solution was accelerated by the addition of Me_2SO.

$$NH_2CH(R)COO^- + CH_2 = CHCN \rightarrow NCCH_2CH_2NHCH(R)COO^- \tag{6}$$

Acceleration occurs because Me_2SO forms a hydrogen bond with the amino group of phenylalanine to increase its nucleophilicity and stabilizes ammonium groups involved in the transition state by solvation. However, the addition of quaternized poly(4-vinylpyridine) strengthened the interaction between the polymer and Me_2SO to destroy the hydrogen bond between phenylalanine and Me_2SO, leading to a smaller rate of reaction. Therefore k/k^*, which measures the polyion effect, is smaller in mixed solvents than in aqueous solutions at either pH 10.06 or pH 8.48, as shown in Figure 2. In particular, at pH 8.48 the addition of polyion decreased the reaction rate. The alkaline hydrolysis of p-nitrophenyl acetate in 1-hexanol/water mixed solvent was accelerated by the addition of less than a certain amount of quaternized

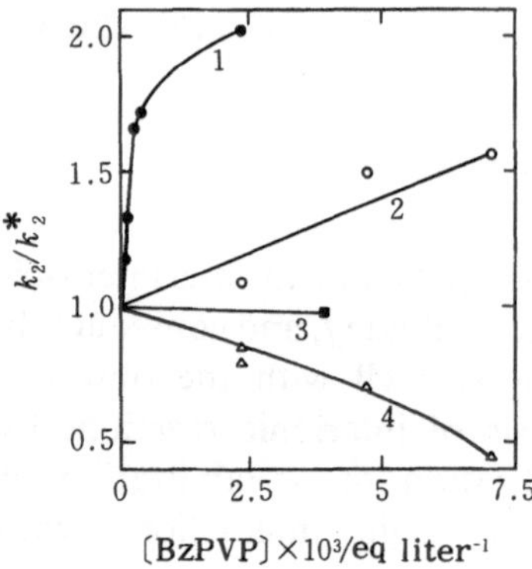

Figure 2. Polyelectrolyte influence on the cyanoethylation of L-phenylalanine at 30°C: [CH_2═CHCN] = 0.2 mol liter^{-1}; [L-phenylalanine] = 1 mmol liter^{-1}. Curve 1 represents pH 10.06 buffer; Curve 2, 50 (v/v)%, pH 10.06 buffer/Me_2SO mixture; Curve 3, pH 8.48 buffer; and Curve 4, 50 (v/v)%, pH 8.48 buffer/Me_2SO mixture. (Reproduced from Figure 1 of Reference 75 by permission of the copyright owner.)

polyiminoethylene.(76) The acceleration is due to the desolvation–activation of hydroxyl ions by the selective hydration of the polyion. The aquation of tris(oxalato)cobalate was greatly accelerated by the addition of Me_2SO or $HCONMe_2$, and the importance of desolvation of H_3O^+ or the activated complex was clearly shown. However, the addition of poly(iminoethylene propionate) made the reaction occur in the vicinity of selectively hydrated polymer chains and decreased the extent of acceleration by the binary solvent.(77)

As described above, polyelectrolytes affect the solvation state of reactants or the activated complex, hence the entropy or enthalpy state, and ultimately the f value in Eq. (5). To investigate the above relationship directly, the determination of activated volume $\Delta V^\ddagger$ by the kinetic investigation of polyelectrolyte-catalyzed reactions under pressure is desired. $\Delta V^\ddagger$ is related to the rate constant k and the pressure p by Eq. (7) and is given by the sum of $\Delta V_1^\ddagger$ and $\Delta V_2^\ddagger$. $\Delta V_1^\ddagger$ represents a contribution from the volume change of reactant accompanied by the change of bonding state in the

$$\Delta V^\ddagger = -RT\partial \ln k/\partial p, \qquad \Delta V^\ddagger = \Delta V_1^\ddagger + \Delta V_2^\ddagger \tag{7}$$

activated complex, and $\Delta V_2^\ddagger$ that accompanied by the change of solvation state of reactants and activated complex. In interionic reactions which are accompanied by the change of valence state in activated complex, $\Delta V_2^\ddagger$ is more important than $\Delta V_1^\ddagger$. Therefore we can consider $\Delta V^\ddagger \fallingdotseq \Delta V_2^\ddagger$.

The aquation reaction (8) of $Co(NH_3)_5Br^{2+}$ by Ag^+ was investigated in the presence of poly(sodium styrene sulfonate) (NaPSS), and the pressure effect on the reaction is shown in Figure 3.(78) Second-order rate constant k_2

$$Co(NH_3)_5Br^{2+} + Ag^+ \rightleftharpoons [Co(NH_3)_5Br^{2+} \ldots Ag^+] \text{ (activated complex X)}$$

$$\xrightarrow{H_2O} Co(NH_3)_5H_2O^{3+} + AgBr \tag{8}$$

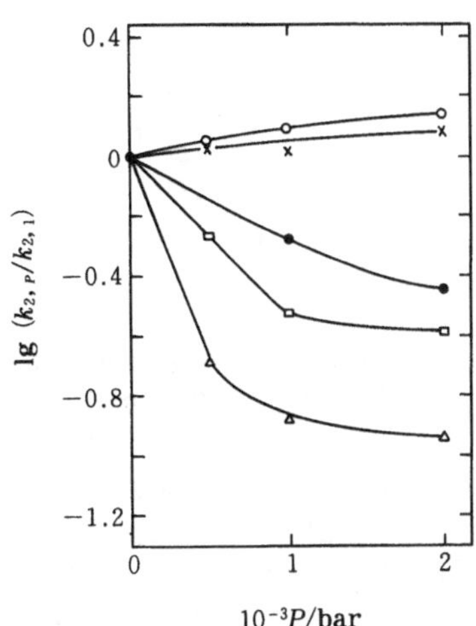

Figure 3. Log ($k_{2,p}/k_{2,1}$) plotted against pressure for the Ag^+-induced aquation of $Co(NH_3)_5Br^{2+}$ in the presence of poly(sodium styrene sulfonate) (NaPSS) and in the absence at 25°C: $[Co(NH_3)_5Br^{2+}]$ = 38 μmol liter^{-1}; $[HClO_4]$ = 117 μmol liter^{-1}; $[AgNO_3]$ = 1 mmol liter^{-1}. (○) represents [NaPSS] = 0 mol liter^{-1} ($k_{2,1} = 5.3 \times 10^{-3}$ liter mol^{-1} s^{-1}); (x), [NaPSS] = 10^{-6} mol liter^{-1} ($k_{2,1} = 9.5 \times 10^{-3}$ liter mol^{-1} s^{-1}); (●), [NaPSS] = 10^{-5} mol liter^{-1} ($k_{2,1} = 3.2 \times 10^{-2}$ liter mol^{-1} s^{-1}); (□), [NaPSS] = 5×10^{-5} mol liter^{-1} ($k_{2,1}$ = 0.48 liter mol^{-1} s^{-1}); and (△), [NaPSS] = 10^{-4} mol liter^{-1} ($k_{2,1}$ = 2.16 liter mol^{-1} s^{-1}). (Reproduced from Figure 5 of Reference 78 by permission of the copyright owner).

in the absence of polyelectrolyte increased with pressure ($k_{2,p} > k_{2,1}$). However, with increasing concentration of added polyelectrolyte, the rate constant decreased with pressure ($k_{2,p} < k_{2,1}$). In other words, $\Delta V^\ddagger$ varies from a negative to positive value by the addition of polyelectrolyte. $\Delta V^\ddagger$ values are shown in Table 3 together with other activation parameters. It should be noted in Table 3 that $\Delta V^\ddagger$, $\Delta S^\ddagger$, and $\Delta H^\ddagger$ all increased with the addition of polyanion resulting in a decrease of $\Delta G^\ddagger$ (the acceleration of reaction). It is most important that the incremental changes of $\Delta S^\ddagger$ and $\Delta V^\ddagger$ brought about by the addition of polyanion are concurrent. The activated complex of reaction (8) is trivalent and more strongly solvated than the starting species. Therefore, the negative value of $\Delta V^\ddagger$ in the absence of polyelectrolyte is expected. The oppositely charged polyelectrolyte weakens the strong

Table 3. Thermodynamic Parameters for the Ag^+-Induced Equation of $Co(NH_3)_5Br^{2+}$ in the Presence of NaPSS at 25°C[a]

[NaPSS] (equiv liter^{-1})	ΔG‡ (kJ mol^{-1})	ΔH‡ (kJ mol^{-1})	ΔS‡ (J K^{-1} mol^{-1})	ΔV‡ (ml mol^{-1})
0	86.2 ± 0.3	32.6 ± 2.0	− 180 ± 8	− 5.3 ± 2
10^{-6}	84.1	51.5	− 109	− 2.3
10^{-5}	81.2	60.7	− 71	16
5×10^{-5}	74.9	81.6	21	30
$(10^{-4})^b$	—	—	—	(76)

[a] Reproduced from Table 2 of Reference 78 by permission of the copyright owner.
[b] Precipitation occurred above this concentration, but disappeared on application of high pressure.

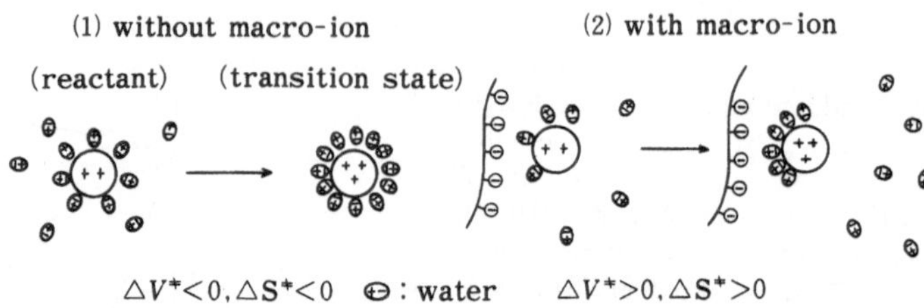

Figure 4. Schematic representation for hydration of ionic solutes and desolvation by macro-ions. (Reproduced from Figure 3 of Reference 78 by permission of the copyright owner).

ion–dipole interaction between the activated complex and water molecules, and liberates many water molecules from the solvation shell to increase $\Delta V^{\ddagger}$. Simultaneously, the randomness of the system increases, leading to the increase of $\Delta S^{\ddagger}$. These situations are illustrated in Figure 4.

The trend in $\Delta V^{\ddagger}$ values, obtained in the experiments under pressure, was very useful in showing that the acceleration of reaction between equally charged ionic species by the addition of oppositely charged polyelectrolyte is caused by not only increased collision frequency due to the concentration of reactants ($\Delta S^{\ddagger}$ term), but also by liberation of water molecules from the activated complex (composite $\Delta S^{\ddagger}$ and $\Delta V^{\ddagger}$ terms). The electrostatic stabilization of activated complex by polyelectrolyte may also contribute to the acceleration of the reaction. Consequently, the effects of polyelectrolyte upon ionic reactions can be summarized as in Figure 5.

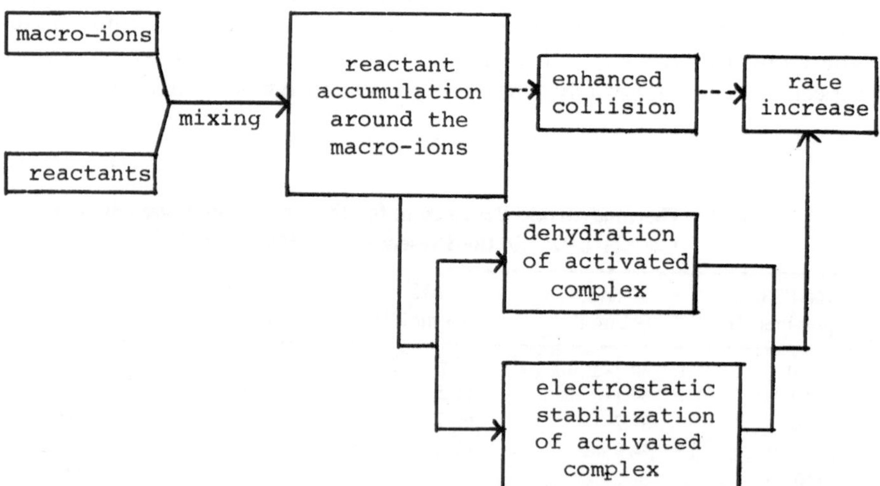

Figure 5. Mechanism of rate increase of ionic reactions caused by macro-ions. (Reproduced from diagram on p. 497 of Reference 78 by permission of the copyright owner.)

The activated complex in the alkaline hydrolyis of $Co(NH_3)_5Br^{2+}$ is monovalent. In this reaction, the charge number and hence the degree of solvation are lower in the activated state than in the initial state. In such a case, either $\Delta S^\ddagger$ or $\Delta V^\ddagger$ is positive, and the addition of polyanion decreased $\Delta S^\ddagger$ and $\Delta V^\ddagger$ and thus the reaction rate.[79]

For reactions between equally charged multivalent ionic species, the reactions between U^{4+} and Tl^{3+} and between $Fe(CN)_6^{4-}$ and $S_2O_8^{2-}$ were investigated. The former reaction was accelerated by poly(sodium ethylene sulfonate) by 10^4 times, and the latter reaction by polycation by 10^5 times.[80] In either case the addition of polyion brought about the increase of $\Delta V^\ddagger$ and $\Delta S^\ddagger$.

Spontaneous aquation of $Co(NH_3)_5Br^{2+}$ proceeds through a dissociative complex, which essentially yields Co^{3+}, whereas that of $Cr(NH_3)_5Br^{2+}$ proceeds through an associative complex, which is not accompanied by the change of charge number. Therefore, the acceleration of the latter reaction by addition of a polyanion was much smaller than that of the former reaction, and the $\Delta V^\ddagger$ of the latter reaction was insensitive to the addition of polyanion.[81]

The polyelectrolyte effect described above derives from a strong interaction between polyions and their counterions, which is closely related with the biphase structure of the polyelectrolyte in solution. Their correlation was briefly reviewed.[82]

2.2.6. *Other Examples of Polyelectrolyte Catalysis*

Other effects related to the addition of polyelectrolytes, besides electrostatic effects, may be considered. The contribution of hydrophobic or charge transfer interactions have been considered as related to the effect of addition of poly(4-vinyl-*N*-alkylpyridinium salt) on the hydrolyses of esters or amides containing the indolyl group[83]; the polyelectrolyte effect on the charge transfer complex between flavin mononucleotides and indole derivatives[84]; effects of polyelectrolytes and their resins on the solvolyses of *t*-BuCl and α-PhEtCl and the deamination of α-$PhEtNH_2$[85]; and the effect of styrene/acrylic acid copolymer latex on the reaction between $Co(NH_3)_5Br^{2+}$ and OH^- and the alkaline fading reaction of crystal violet.[86]

Investigations concerning the cooperative complexation of polyelectrolytes in solution were reported for the fluorescence quenching method,[87] the quenching by low-molecular-weight amphiphilic quenchers of the fluorescence of copolymers containing vinyl naphthalenes and cationic segments,[88] and the acceleration of the excimer formation between bis(α-naphthylmethyl)ammonium chloride and (α-naphthylmethyl)-ammonium chloride by polyelectrolytes.[89]

The oxidation reaction of thiophenols by cobaloximes, which proceeds according to the Michaelis–Menten kinetics, was accelerated through the stabilization of the reaction intermediate by addition of poly(4-vinylpyridine) or 4-vinylpyridine/styrene copolymers.[90] Effects of cationic polyelectrolytes or cationic micelles on the oxidation of $NaHSO_3$ were investigated.[91] Effects of linear aliphatic ionenes having quaternary ammonium groups in the main chain on the alkaline hydrolysis of *p*-nitrophenyl octanoate were reported.[92]

2.3. Stereospecific Micellar Reactions

Investigations employing micellar reactions as models for enzyme reactions are very helpful in elucidating the mechanism of enzyme reactions. However, only a few investigations have been made concerning stereospecific catalysis (which is the most important feature of enzyme reactions). Recently, some examples of highly stereospecific hydrolysis were observed in micelles composed of functionalized detergents.

Moss *et al.*[93] carried out the hydrolysis of Z-L- or D-Ala-L-Pro-ONp in the micelles of $n\text{-}C_{16}H_{33}N^+(Me)_2CH_2CH_2SH \cdot Cl^-$ or $n\text{-}C_{16}H_{33}N^+(Me)_2\text{-}CH_2CH_2NHCOCH(NH_2)CH_2SH \cdot Cl^-$, and obtained $k_{LL}/k_{DL} = 4.33$ and 3.86, respectively. These were the first cases showing stereospecificity values exceeding 3. Moss *et al.*[94] also carried out the hydrolysis of Z-L- or D-Trp-L-Pro-ONp in the micelles of $n\text{-}C_{16}H_{33}N^+(Me)_2CH_2CH_2SH \cdot Cl^-$, and observed k_{LL}/k_{DL} values as high as 5.0. These high stereospecificities were explained in terms of a specific interaction between the substrate and the functional detergent in the micelles. In more detail, the reaction proceeds faster when the methylene chain of the catalytic detergent having a thiol group fits well into a cleft, which in this case is constructed by the L-prolyl residue, the *p*-nitrophenyl group, and the side chain of Ala or Trp residue in the substrate.

A little earlier than this, Ihara[95,96] found that *N*-acylated L-Phe-ONp was hydrolyzed faster than the D-enantiomer by the mixed micelle of *N*-acyl-L-histidine and $Me(CH_2)_{15}N^+Me_3 \cdot Br^-$. The longer the *N*-acyl group of the catalyst, the faster the reaction rate and the higher the enantiomer selectivity, the maximum k_L/k_D being 2.8. It was shown that the enantiomer selectivity was determined by the catalytic acyl transfer to the optically active imidazolyl group.

At nearly the same time, Kuroki and his co-workers were interested in the asymmetric hydrolysis in micelle, and found that Z-D-Phe-ONp was hydrolyzed slightly faster than the L-enantiomer in an asymmetric micelle composed of *N*-dodecyl-*N*-methyl-1-ephedrinium bromide.[97] Later, Ihara's team and Kuroki's team cooperated to investigate the hydrolyses of Z-Phe-ONp and Moc-Phe-ONp in mixed micelles of $Me(CH_2)_{15}N^+Me_3 \cdot Br^-$

Table 4.[a] Apparent Catalytic Rate Constants (k_c) in the Presence of CTAB Micelles[b]

Catalyst	(configuration)	Z-Phe-ONp L	Z-Phe-ONp D	Z-Phe-ONp L/D	Moc-Phe-ONp L	Moc-Phe-ONp D	Moc-Phe-ONp L/D
		k_c (liter mol^{-1} s^{-1})					
Z-His[c]	(L)	111	92.3	1.20	80.7	66.0	1.22
Z-Ala-His	(L–L)	98.9	37.9	2.61	139	32.1	4.33
Z-Val-His	(L–L)	225	56.7	3.97	309	44.7	6.91
Z-Leu-His	(L–L)	473	74.9	6.32	645	52.7	12.2
	(L–D)	78.4	176	2.24[d]	60.9	146	2.40[d]
Z-Phe-His	(L–L)	493	110	4.48	541	74.5	7.26
	(D–L)	202	98.1	2.06	172	74.2	2.32
Z-Try-His	(L–L)	103	67.5	1.53	85.9	47.2	1.82
Z-Lys(Z)-His	(L–L)	84.3	40.2	2.10	78.3	32.1	2.44

[a] Reproduced from Table 1 of Reference 100 by permission of the copyright owner.
[b] At pH 7.30, 0.02 mol liter^{-1} phosphate buffer, and 25°C in the presence of 6.00×10^{-3} mol liter^{-1} CTAB. [Catalyst] = $(0.50\text{–}6.0) \times 10^{-4}$ mol liter^{-1}; [substrate] = 1.0×10^{-5} mol liter^{-1}. The k_c values are calculated by least squares and generally have correlation coefficients > 0.99.

[c] *N*-benzyloxycarbonylhistidine.
Catalyst:

C_6H_5–$CH_2OCONHCH(R)CONHCH(CH_2\text{-imidazolyl})COOH$

Z-Ala-His,	R = CH_3
Z-Val-His,	R = $CH(CH_3)_2$
Z-Leu-His,	R = $CH_2CH(CH_3)_2$
Z-Phe-His,	R = $CH_2C_6H_5$
Z-Try-His,	R = CH_2-(3-indolyl)
Z-Lys-(Z)-His,	R = $(CH_2)_4NHCOOCH_2C_6H_5$

[d] Rate ratios, D/L.
Substrate:

$R^1OCONHCH(R^2)COO$–C_6H_4–NO_2

Z-Phe-ONp,	$R^1 = R^2 = CH_2C_6H_5$
Moc-Phe-ONp,	$R^1 = CH_3$, $R^2 = CH_2C_6H_5$

Surfactant:
$CH_3(CH_2)_{15}N^+(CH_3)_3 \cdot Br^-$ (CTAB)

and dipeptide derivatives containing a histidine (Z-XyZ-His-OH).[98–100] Consequently, as shown in Table 4, in the hydrolysis of Moc-L- or D-Phe-ONp in the mixed micelle of Z-L-Leu-L-His-OH, k_L/k_D was as large as 12.2. Ihara *et al.*[101] compared the hydrolysis of *N*-acyl-L- or D-α-amino acid *p*-nitrophenyl ester by *N*-decanoyl-L- or D-histidine in an asymmetric micelle of *N*-dodecyl-*N*-methyl-1-ephedrinium bromide with that in a simple micelle of $Me(CH_2)_{15}N^+(Me)_3 \cdot Br^-$. The L-substrate was hydrolyzed faster by the L-catalyst, but k_L/k_D was only 1.5–2.7, which was not influenced by the presence of the chiral detergent.

Asymmetric hydrolysis in asymmetric micelle was also investigated at the Kumamoto University. Yamada *et al.*[102] investigated the hydrolysis of ethoxycarbonyl-L- or D-Phe-ONp in a mixed micelle of *N*-lauryl-L- or D-histidine and optically active or achiral detergents, and found $k_{LL}/k_{LD} \doteqdot 1.5$. Furthermore, they investigated the hydrolysis of Z-L- or D-Phe-ONp in the micelle of $Me(CH_2)_{15}N^+(Me)_3 \cdot Br^-$ with a hydroxamic acid derivative of Z-L-α-amino acid or an amino derivative of L-α-amino acid as a catalyst, and found that the maximum k_D/k_L was 2.51 in the former catalysis and 2.65 in the latter catalysis.[103] With the latter catalyst, the asymmetric catalysis disappeared when the distance between the center of chirality and the catalytic site increased. Ohkubo *et al.*[104] found $k_L/k_D = 5.5–5.7$ in the deacylaction of H-$(CH_2)_{n-1}$-CO-L- or D-Phe-ONp ($n = 10–16$) in the mixed micelle of *N*-(*N*-dodecanoyl-L-His)-L-Leu-OH and (*R*)-(+)-*N*-α-MeBzl-*N*,*N*-$(Me)_2C_{18}H_{37}N^+ \cdot Br^-$. This value of k_L/k_D was the highest one at that time so far obtained. Ueoka *et al.*[105] investigated the deacylation reaction of enantiomorphic substrates C_nH_{2n+1}CO-Phe-ONp ($n = 1, 5, 9, 11, 13, 15$) by nucleophiles C_nH_{2n+1}CO-L-His ($n = 1, 7, 11, 15$) in the presence of cationic micelles, such as $Me(CH_2)_{15}N^+(Me)_3 \cdot Br^-$ and $Me(CH_2)_{15}{}^-$-$N^+Bzl(Me)_2 \cdot Br^-$. Deacylation reactions were greatly accelerated by the proximity effect occurring between a nucleophile and a substrate having a suitable acyl chain, through the hydrophobic interaction with the micelle. The rate of acceleration was particularly great for amino acids derivatives, which may have been caused by a promixity effect being strengthened by the hydrophobic interaction between the phenyl group of substrate and the imidazolyl group of nucleophile and the hydrogen bonding between amide groups of substrate and nucleophile. The highest rate of deacylation was observed in the reactions of the substrate having $n = 9$ with the nucleophile having $n = 11$, and the substrate having $n = 11$ with the nucleophile having $n = 15$. The highest value of k_L/k_D was 2.6–2.8. Hydrophobic interaction and hydrogen bonding between a specific pair of substrate and nucleophile should yield a preferential proximity effect, and this might be the case for the present stereoselective and asymmetric selective deacylation reaction. The same group of researchers investigated the stereoselective deacylation reaction of long-chain α-amino acid esters with comicelles of *N*-acyl-L-histidine and cationic detergents, and obtained as the highest asymmetric selectivity the value of 4.4.[106]

In order to investigate the asymmetric selectivity of inverted micelles, Kon-no *et al.*[107] investigated the aminolysis reaction of L- or D-*p*-nitrophenyl-α-methoxyphenyl acetate in the presence of (+)- or (−)-α-phenethylammonium dodecanoate in *n*-hexane. Either when (−)-detergents were in a molecularly dispersed state, or when they formed inverted micelles, the L-ester was hydrolyzed faster by approximately 1.4-fold than the D-ester. Since no asymmetric selectivity was observed when achiral detergents were employed, the asymmetric selectivity was believed to be due to the free energy difference in the formation of enantiomorphic complexes.

2.4. Chemical Reactions in Organized Molecular Assemblies

It was shown in the previous section that a fairly high asymmetric selectivity is observed in micelles and comicelles of chiral substrates and chiral nucleophiles. To those phenomena, explanations were proposed taking into consideration a stereochemical fit between substrate and nucleophile. However, the micelle structure is very versatile, and the details of the reaction mechanism are not clear. On the other hand, monomolecular membranes, bimolecular membranes, and liquid crystals, which are formed by detergents having a specific structure under specific conditions, are molecular assemblies having an organized structure. The change of phase structure by circumstantial conditions can be treated quantitatively. Furthermore, these organized molecular assemblies are similar to biological membranes. For these reasons, the environmental effects on chemical reactions in organized molecular assemblies are investigated.

Stable bimolecular membranes have been obtained only from biological lipids or their derivatives. However, in 1977, Kunitake and Okahata[(108)] found that stable bimolecular membranes are formed from synthetic amphiphilic molecules having two linear alkyl chains with a suitable (C_{10}–C_{20}) length. A review article was published[(109)] on the preparation and structure of synthetic bimolecular membranes and the reaction control using the membrane structure. Kunitake and Sakamoto[(110)] investigated the hydrolyses of *p*-nitrophenyl acetate(PNPA) and palmitate(PNPP) by the chloesteryl ester of imidazolecarboxylic acid **[9]** as a catalyst. In the initial

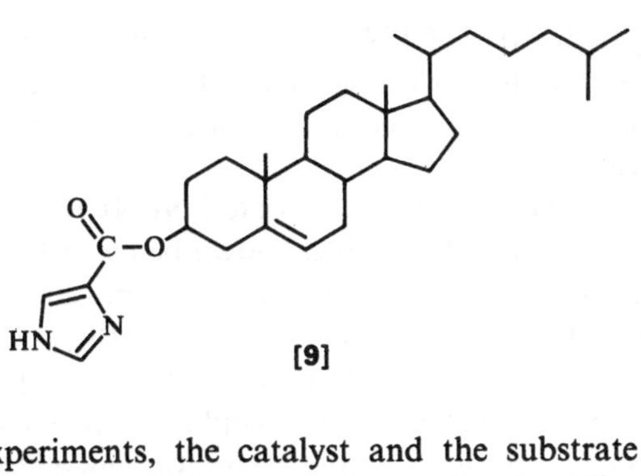

[9]

series, A, of experiments, the catalyst and the substrate were dissolved separately in aqueous solutions containing didodecyldimethylammonium bromide ($2C_{12}N^{+}2C_{1}Br^{-}$), and two stock solutions were added to the buffer solution for initiating the reaction without sonication. In this case, the substrate and the catalyst are bound by different vesicles and intervesicular

Table 5.[a] Effect of Sonication of Mixed Stock Solutions on k_{obsd}[b]

	k_{obsd}/s^{-1}			
	$2C_{12}N^+2C_1Br^-$		CTAB	
	PNPA	PNPP	PNPA	PNPP
Series A[c]	0.53	0.032	0.066	0.12
Series B[d]	0.45	7.8	0.065	0.19

[a] Reproduced from Table 1 of Reference 110 by permission of the copyright owner.
[b] [Substrate] = $(1.2\text{–}1.5) \times 10^{-5}$ mol liter^{-1}; catalyst = $(1.1\text{–}1.2) \times 10^{-4}$ mol liter^{-1}, except for 1.1×10^{-5} mol liter^{-1} in the hydrolysis of PNPP in series B. The observed rate constants were corrected to the values for [catalyst] = 1.0×10^{-4} mol liter^{-1}; $2C_{12}N^+2C_1Br^-$ = 1.0×10^{-3} mol liter^{-1}; CTAB = 2.0×10^{-3} mol liter^{-1}. Conditions: 30°C; pH 9.5 ± 0.1; 0.01 mol liter^{-1} borate buffer; μ = 0.01 (KCl).
[c] Simple mixing of stock solution.
[d] Sonication of mixed stock solutions.

reactions take place. In a second series, B, of experiments, two stock solutions were mixed, sonicated, and added to the buffer solution for initiating the reaction. In this case, the substrate and the catalyst are bound by the same vesicle, and intravesicular reactions take place. The experimental results are shown in Table 5. The K_{obs} for PNPA was independent of sonication, while K_{obs} for PNPP increased by 200-fold by sonication. This finding indicates that the intravesicular reaction proceeds much faster than the intervesicular one. In the latter reaction the intervesicular transfer of catalyst and substrate should be a rate-determining step. The same phenomenon was not observed with CTAB micelle. It was concluded that the difference between intra- and intervesicular reactions is enhanced, when the reactants are firmly held by a vesicle.

When the hydrolysis of PNPP by imidazolecarboxylic amide having long alkyl chains as a catalyst was carried out in a bimolecular membrane of double-chain ammonium-type detergents, the reaction rate was controlled by the phase transition.[(111)]

PNPA and *p*-nitrophenyl nonanoate(PNPN) were hydrolyzed by a series of hydroxamates and imidazoles shown below (**10–15**) as nucleophilic catalysts in the presence of single-, double-, and triple-chain ammonium-type amphiphilic substances.[(112)] As shown in Table 6, the nucleophilic reaction rates were accelerated by some 100 times by the hydrophobic aggregates. Catalytic activities of single-chain nucleophiles C_{12}-BHA and C_{12}-ImAm increased with increasing hydrophobicity of the microenvironment of the ammonium-type aggregate, that is, in the order single < double < triple chain. On the other hand, the chloesteryl ester of imidazolecarboxylic acid was particularly reactive in a bimolecular membrane composed of double-chain detergents. Apparently, a specific binding of cholesteric

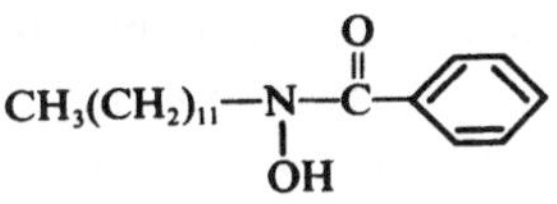

$CH_3(CH_2)_{11}$—N(OH)—C(=O)—C_6H_5

C_{12}—BHA
[10]

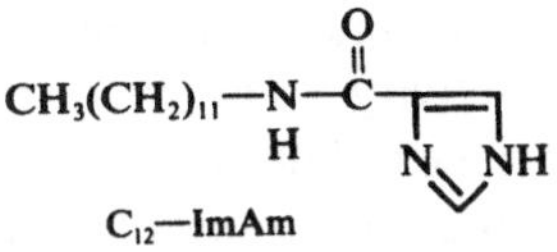

$CH_3(CH_2)_{11}$—NH—C(=O)—Im

C_{12}—ImAm
[11]

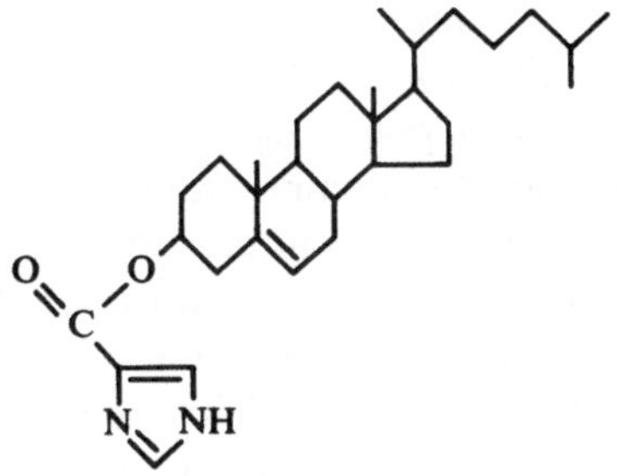

[12] Cholest-Im

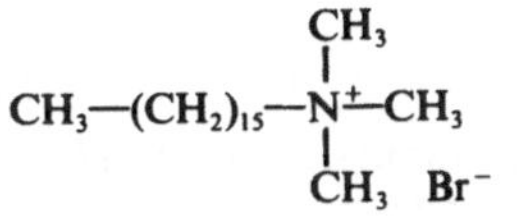

CH_3—$(CH_2)_{15}$—$N^+(CH_3)_3$ Br^-

[13] CTAB
($C_{16}N^+3C_1$)

$[CH_3$—$(CH_2)_{n-1}]_2N^+(CH_3)_2$ Br^-

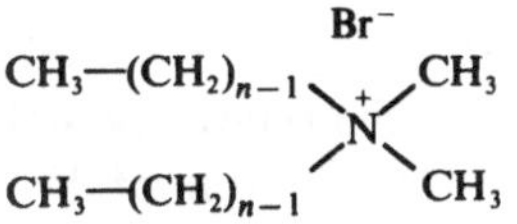

n = 12; $2C_{12}N^+2C_1$ [14]
n = 18; $2C_{18}N^+2C_1$

[15] $[CH_3$—$(CH_2)_7]_3N^+$—CH_3 Cl^-

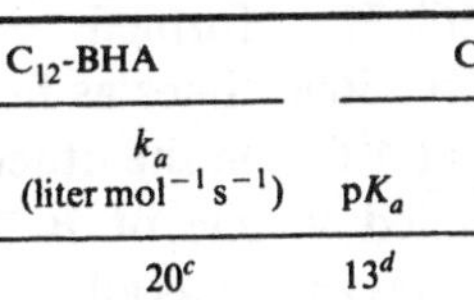

TMAC
($3C_8N^+C_1$)

Table 6.[a] Acidity and Rate Constants for Hydroxamate and Imidazole Nucleophiles[b]

	C_{12}-BHA		C_{12}-ImAm		Cholest-Im	
Ammonium salts (1×10^{-3} mol liter^{-1})	pK_a	k_a (liter mol^{-1} s^{-1})	pK_a	k_a (liter mol^{-1} s^{-1})	pK_a	k_a (liter mol^{-1} s^{-1})
None	9.5[c]	20[c]	13[d]	12[d]	13[d]	12[d]
CTAB ($C_{16}N^+3C_1$)	8.4	2560	10.3	1620	8.9	194
$2C_{12}N^+2C_1$	–	–	9.7	1430	8.7	3980
$2C_{18}N^+2C_1$	–	–	–	–	–	–
TMAC ($3C_8N^+C_1$)[e]	–	–	9.3	5200	–	–

[a] Reproduced from Table 2 of reference 112 by permission of the copyright owner.
[b] 30°C; 3 v/v % EtOH–H_2O; μ = 0.01 (KCl).
[c] *N*-benzylbenzohydroxamic acid was used as the nucleophile. Cited from literature.
[d] *N,N*-dimethyl(4-imidazolecarboxamide) was used as nucleophile. Cited from literature.
[e] From literature.

nucleophile to the bimolecular membrane causes the unusually large amount of catalytic activity. The hydrolysis of *p*-nitrophenyl phosphate by hydroxamate nucleophiles as catalysts was also investigated in the presence of dialkylammonium bimolecular membranes.[113]

Murakami *et al.*[114] investigated the amino transfer reactions between *N*-dodecyl-L-alanine amide and pyridoxal-5′-phosphate in the presence of a single-walled vesicle composed of *N,N*-ditetradecyl-N^{α}-(6-trimethylammoniohexanoyl)-L-histidine amide bromide ($N^+C_5His2C_{14}$). Pyridoxal-5′-phosphate was bound to a polar region of the vesicle by the electrostatic interaction, and *N*-dodecyl-L-alanine amide was bound to a hydrophobic region of the vesicle by the hydrophobic interaction. Despite the polarity of reaction site in the vesicle being similar to that in dioxane/water (7:3) mixed solvent, the reactions in the vesicle were 230 times as fast as those in dioxane/water (7:3) mixed solvent. This remarkable acceleration of the vesicular reaction is due to the hydrophobic reaction site and the suitably polar microenvironment.

Ueoka *et al.*[115] were the first to observe asymmetric reactions in chiral bimolecular membranes. The bimolecular membrane of double-chain-type detergents $(C_mH_{2m+1})_2N^+Me_2 \cdot Br^-$ (m = 12, 14) containing palmitoyl-L-histidine showed a relatively high asymmetric selectivity (k_L/k_D = 3.7–5.6) in the deacylation reaction of H-$(CH_2)_{n-1}$CO-Phe-ONp. The same reaction, catalyzed by comicelles of palmitoyl-L-histidine and $C_{18}N^+Me_3 \cdot Cl^-$, gave a slightly lower asymmetric selectivity (k_L/k_D = 3.5–3.6) than that found for the above micelle. Ohkubo *et al.*[116] investigated on the deacylation of *N*-acylphenylalanine *p*-nitrophenyl ester using bilayer vesicles containing a dipeptide nucleophile, Z-L-Leu-L-His, and double-chain ammonium salt, $(C_{12}H_{25})_2N^+Me_2 \cdot Br^-$ at 10°C, and observed very high asymmetric selectivity (such as k_L/k_D = 80.0) for the *N*-tetradecarylphenylalanine ester.

Murakami *et al.*[117] investigated asymmetric hydrolyses in the presence of micelles on vesicles formed by functional detergents. $Br^- \cdot Me_3N^+(CH_2)_5$CO-His-$NHC_{12}H_{25}$ formed a spheric micelle and hydrolyzed Z-L-Phe-ONp at 20°C 5690 times as fast as alkaline hydrolysis, k_L/k_D being 2.5 at 20°C and 2.7 at 4°C. On the other hand, the hydrolysis of Z-L-Phe-ONp by the single-walled vesicle of $Br^- \cdot Me_3N^+(CH_2)_5$CO-His-$N(C_{12}H_{25})_2$ at 20°C was 22,100 times as fast as alkaline hydrolysis, k_L/k_D being 3.3 at 20°C and 4.4 at 4°C. Furthermore, the acceleration of reaction and the asymmetric selectivity by the single-walled vesicle were controlled by the phase transition of the vesicle.

Water-soluble nitrates of azonaphthol and azophenol dyes **16** and **17**, and chloroacetate **18** form a supersaturated solution at low concentrations such as 10^{-5} mol liter^{-1}, when mixed with the buffer solution. If dust (in air) contaminates the solution, nucleation takes place, resulting in a diluted colloidal dispersion of fine particles. Within the neutral pH region, the homogeneous hydrolysis was very slow. Once colloidal particles are formed, the reaction accelerated by 80–100 times. If nucleation occurs before a considerable extent of hydrolysis takes place in the supersaturated solution, autoacceleration is observed.[118]

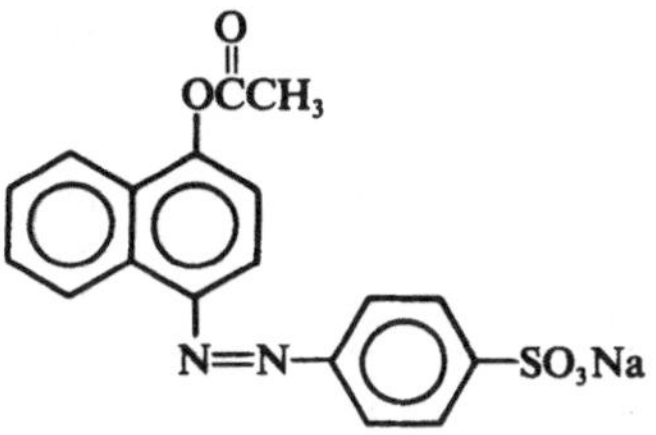

[16]

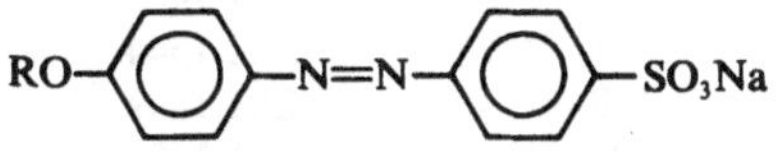

[17] R = $-COCH_3$
[18] R = $-COCH_2Cl$

The displacement reaction (9) of methyl *p*-(dimethylamino) benzenesulfonate(MSE) into a zwitterion goes by the intermolecular methyl transfer. This reaction is controlled by a stacking of MSE, and takes place in a crystalline state but not in a solution. Then, reaction (9) was carried out in metaphase-forming solvents shown in Table 7, and catalysis by liquid crystals was investigated.[119] The reaction temperatures selected were above the

Table 7. Mesomorphic Solvents[a]

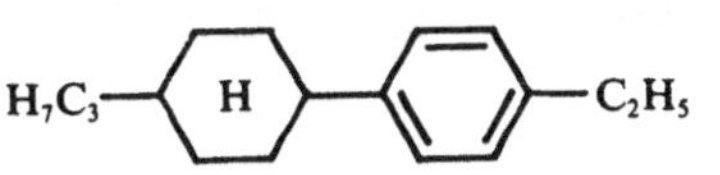 H_7C_3–(H)–(Ph)–C_2H_5; extrapolated clearing point −70°C	S-1484
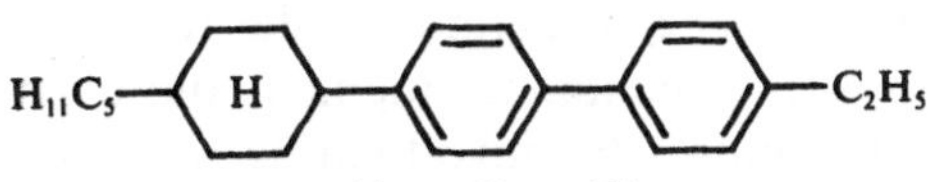 $H_{11}C_5$–(H)–(Ph)–(Ph)–C_2H_5; cry $\overset{34}{\longleftrightarrow}$ s $\overset{146}{\longleftrightarrow}$ n $\overset{164}{\longleftrightarrow}$ i	ZLI-1409
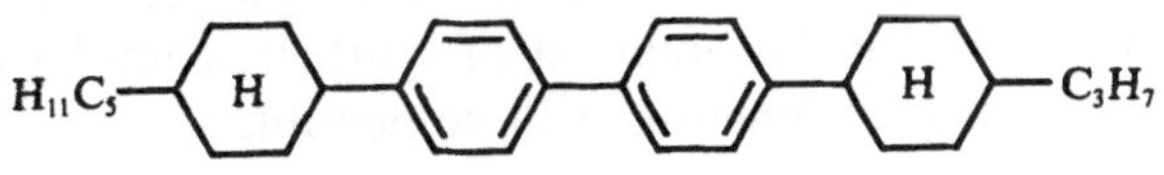 $H_{11}C_5$–(H)–(Ph)–(Ph)–(H)–C_3H_7; cry $\overset{54}{\longleftrightarrow}$ s_1 $\overset{232}{\longleftrightarrow}$ s_2 $\overset{251}{\longleftrightarrow}$ n $\overset{312}{\longleftrightarrow}$ i	ZLI-1544
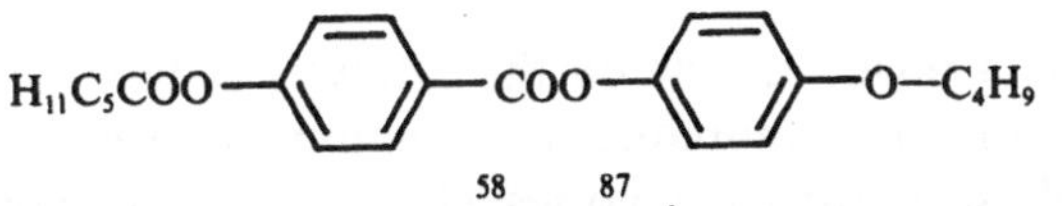 $H_{11}C_5COO$–(Ph)–COO–(Ph)–O–C_4H_9; cry $\overset{58}{\longleftrightarrow}$ n $\overset{87}{\longleftrightarrow}$ i	CBPB

[a] Reproduced from Table 1 of Reference 119 by permission of the copyright owner.

Table 8. Conversion Yield of MSE in Zwitterion[a]

Temperature (°C)	Solvent	Phase	Reaction time, h	Percent product
81	ZLI-1409	s_B	4	7
81	ZLI-1409	s_B	16	50
81	ZLI-1544	s	16	50
100	ZLI-1409	s_B	4	15
100	ZLI-1409	s_B	16	65
100	ZLI-1409	s_B	24	70
100	ZLI-1409/S-1484	s_B-n	24	
100	S-1484	i	24	
81	CBFC	n	24	
50	Ethyl acetate	i	65	

[a] Reproduced from Table II of Reference 119 by permission of the copyright owner.

melting point of MSE (100°C) or below it (81°C). As seen from the experimental results shown in Table 8, the reaction occurs only in smectic liquid crystals. It was proposed that a face-to-face orientation of MSE molecules is possible in smectic liquid crystals, whereas a head-to-tail orientation of MSE molecules occurs in nematic liquid crystals.

$$p\text{-}Me_2NPhSO_3Me \rightarrow p\text{-}Me_3N^+PhSO_3^- \qquad (9)$$

2.5. Polymeric Coenzymes

In complex enzymes, which consist of apoenzyme (protein) and coenzyme, the reaction site is the coenzyme, and a similar reaction proceeds without the protein. Apoenzymes function to augment the efficiency and specificity of the reaction. Therefore, by investigating systems where a coenzyme is combined with a well-designed polymer molecule, the reaction mechanism(s) which characterizes the functions of complex enzymes may be understood, and ultimately the development of synthetic complex enzymes having chemical and biological functions will be possible.

2.5.1. Polymeric Flavins

Bruice[120] wrote a review article on the catalysis of flavin enzymes, in which he classified direct oxidoreductive reactions between flavin and substrate into the following three categories: (1) biological dehydrogenation reaction, (2) electron transfer by the alternation between a reduced state and a radical state; and (3) oxygenation of substrate by triplet oxygen produced by the activation of molecular oxygen. Investigations on polymeric flavins have been carried out mainly on category (1).

Slama *et al.*[121] synthesized flavopapain **19** by alkylation of Cys-25, involving the active site of papain, with a reactive flavin analogue. The same researchers had already shown that the 7-substituted isomer of **19** was an effective catalyst for the oxidation of *N'*-alkyl-1,4-dihydronicotinamide (NADH). However, **19** was found to be more than 10 times as reactive as the 7-substituted isomer, and nearly as reactive as typical native flavin enzymes.

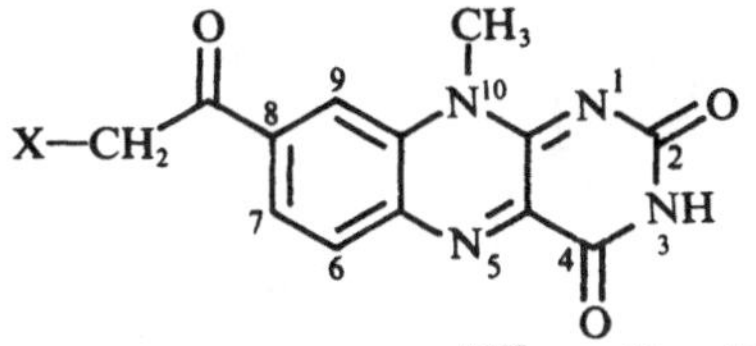

[19] X = Papain-S-

They named this method of producing enzyme-mimetic molecules "chemical mutation." The principle of chemical mutation is to develop synthetic enzymes which possess both the binding specificity of proteins and the characteristic chemical reactivity of a coenzyme. This was done by combining the binding site of the existing protein (papain) with the chemically active coenzyme (flavin). In **19** and the 7-substituted isomer, the carbonyl group at the 8- and 7-positions is important, and it may form a hydrogen bond with the peptide chain of papain to take flavin into a hydrophobic grooving.

Takahashi *et al.*[122] found that the reduced keratin–flavin adenine dinucleotide (FAD) combination converted succinic acid into fumaric acid obeying the Michaelis–Menten kinetics. Kondo *et al.*[123] found that copolymers of 10-(3'-methacryloyloxypropyl)isoalloxazine and styrene accelerated slightly the dehydrogenation of methyl mandelate. They considered that the local increase of the substrate concentration in the neighborhood of isoalloxazine residues, which was caused by the uptake of substrate by the copolymer, should be the origin of the acceleration.

Shinkai *et al.* have done numerous investigations on the reactions catalyzed by flavins embedded in cationic micelles. Shinkai *et al.*[124] carried out the oxidation reaction of thiophenol or 2-mercaptoethanol by 10-ethyl-3-methylisoalloxazine in the presence of a micelle-like polymer. For the latter polymer, poly(2-ethyl-1-vinylimidazole) quaternized by EtBr or lauryl bromide was used. When the micellar polymer having a lauryl content of 28.9 mol % (L-29) or 40.9 mol % (L-41) was present, 10^2–10^5 times the acceleration (as compared with the nonpolymeric system) was observed. On the other hand, the presence of the polymer having a lauryl content of 8.8 mol % (L-9) did not induce the acceleration. The thiolate anion bound into a polymeric domain should be activated by the formation of a hydrophobic ion pair as described in Section 2.2.3.

Shinkai *et al.*[125] synthesized copolymers of styrene derivatives **20**, in which the para positions of the phenyl group are substituted partly by quaternary ammonium groups and partly by a flavin derivative through its

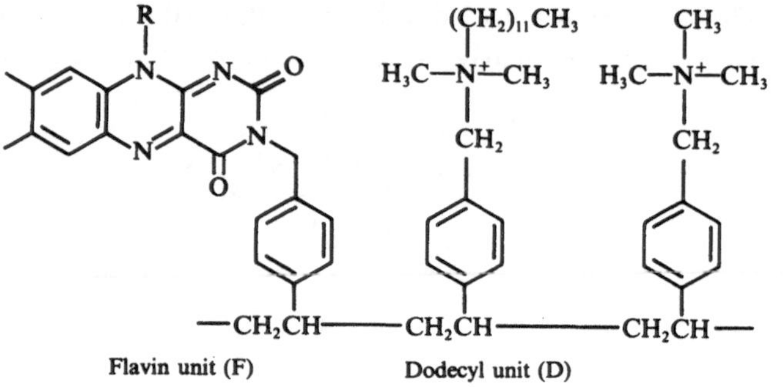

Flavin unit (F) Dodecyl unit (D) [20]

3-position (see the structural formulas **19**). The reduction of NADH by this cationic polymeric flavin was 460–4560 times as fast as the reduction by 3-methyltetra-*O*-acetylriboflavin. The extent of acceleration decreased with increasing the ionic strength. The cationic polymeric flavin did not accelerate the reduction of 1-benzyl-1,4-dihydronicotinamide. These experimental results indicate that the complex formation of the cationic polymeric flavin with NADH by electrostatic interaction is the origin of the large amount of acceleration. Shinkai[126] found that a similar polymeric flavin increased the rate constant for the addition of SO_3^- by 15–37 times and the binding constant by 7–19 times as compared with monomeric flavins. Shinkai[127] observed that a similar polymeric flavin is also a good catalyst for the aerobic oxidation of froin and emphasized the importance of a hydrophobic environment in the oxidation by flavin.

Shinkai *et al.*[128] investigated the reaction of CN^- and aromatic aldehydes or α-keto acids catalyzed by the polymeric flavin **22**. They found that reaction route (10) yields a benzoin condensate predominating by the monomeric flavin, while reaction route (11) yields a benzoic acid predominating by the cationic polymeric flavin. The alteration of the reaction path took place through a rapid oxidation of the intermediate **21**, by the cationic polymeric flavin, which is produced by the rate-determining deprotonation or decarboxylation of the cyanide adduct. This experimental result showed the importance of the microenvironmental effect provided by the polymer on the efficiency of flavin trapping. A similar alteration of the reaction path was also reported with a bis-coenzyme **22**, which contains both flavin and thiazolium ion in one molecule. Shinkai *et al.*[129] found that, in the

$$\text{R—CHO} + \text{CN}^{\ominus} \rightleftharpoons \text{R—CH(O}^{\ominus}\text{)—CN} \rightleftharpoons \underset{[21]}{\text{R—C}^{\ominus}\text{(OH)—CN}} \xrightarrow{\text{RCHO}} \text{R—CO—CH(OH)—R} \xrightarrow{\text{O}_2} \text{R—CO—CO—R} \qquad (10)$$

$$\text{R—CO—CO}_2^{\ominus} + \text{CN}^{\ominus} \rightleftharpoons \text{R—C(OH)(CN)—CO}_2^{\ominus} \xrightarrow{-\text{CO}_2} [21] \xrightarrow{\text{O}_2} \text{R—CO—CN} \xrightarrow{\text{H}_2\text{O}} \text{R—CO}_2^{\ominus} + \text{CN}^{\ominus} \qquad (11)$$

R = 4-chlorophenyl ($-C_6H_4-Cl$), 2,4-dichlorophenyl, C_6H_5

presence of **22** and CTAB, aldehydes were efficiently oxidized into carboxylic acids accompanied with slight acyloin condensates. Therefore, this system can be regarded as a model for pyruvate dehydrogenase.

$C_{12}H_{25}$; $(CH_2)_3$; Br^-

[22]

Shinkai *et al.*[130] found by spectroscopy that quaternized poly(4-vinylpyridine), to which flavin is covalently linked through the 8α-position, possesses a quinone form of flavin. This type of flavin is inactive toward molecular oxygen and has no oxidation ability.

2.5.2. Polymeric Nicotinamides

Shinkai *et al.*[131] synthesized poly[1-(*p*-vinylbenzyl)-3-carbamoyl-pyridinium chloride], **23**. In the CN^- addition to **22**, the second-order rate constant and the equilibium constant were 108 times and 218 times, respectively, as large as those in the CN^- addition to 1-benzyl-3-carbamoyl-pyridiniumchloride. Since the increase of the ionic strength lowered the degree of acceleration by the polymer, **23** seems to produce an ionic environment around the reaction site. Adjacent cationic charges along the

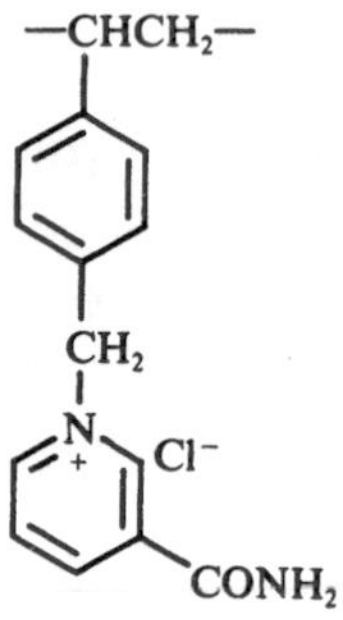

[23]

chain of **23** should destabilize the nicotinamide group and stabilize the CN^- species. This experimental result shows that by the incorporation of cationic species into the cationic polyelectrolyte, its reaction with anionic species is accelerated. This energetically unfavourable incorporation is assisted by the covalent bonding.

Participation of the charge transfer interaction in biological reactions has been indicated, in particular, in reactions involving the NAD^+ coenzyme. Shinkai *et al.*[(132)] showed by spectroscopy that in partially reduced **23** the electron-accepting NAD^+ groups and the electron-donating NADH groups are adjacent in the polymer chain, and that they form charge transfer complexes. Rypáček *et al.*[(133)] synthesized homopolymers of methacrylate carrying a nicotinamide group in the side chain and its copolymers with methacrylate carrying a quaternary ammonium group in the side chain. They showed that the abilities of these homopolymers and copolymers to form charge transfer complexes with iodide were about 10^3 times as great as the corresponding low-molecular-weight model.

It has been shown that, in the asymmetric reduction of active C=C or C=O by using chiral NADH models having polar groups in the chiral 3-carbomoyl moiety, one of the diastereomeric faces of the dihydropyridine group is blocked by the charge transfer interaction with NAD^+. Seki *et al.*[(134)] synthesized a bis-NADH compound which involves L-proline amide units as chiral groups and which is connected by a xylylene unit. This chiral bis-NADH compound underwent the chiral reduction of various C=C and C=O in the presence of $Mg(ClO_4)_2$. In particular, formation in excess of 98.1% of the enantiomer obtained in the reduction of $PhCOCO_2Et$ should be noted. The structure of the C_2-symmetric bis-NADH is proposed to be **24**.

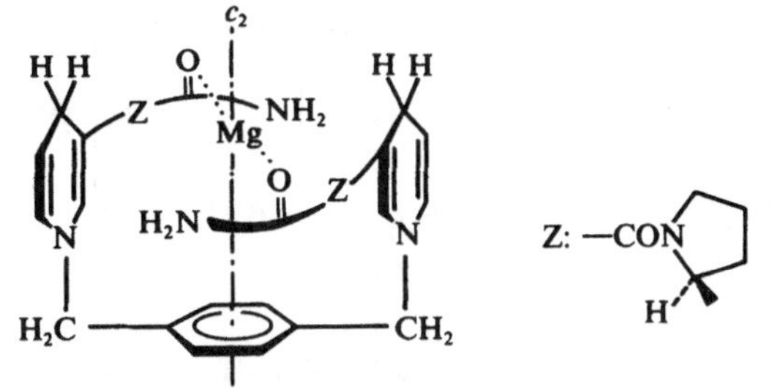

[24]

Shinkai *et al.*[135] synthesized copolymers of 4-vinylpyridine and *N*-(*p*-vinylbenzyl)-1,4-dihydronicotinamide (the monomer of **23** in a reduced state). The reduction of benzil by these copolymers was 4–7 times as fast as that by a monomer analogue, *N*-benzyl-1,4-dihydronicotinamide. Since the reduction catalyzed by the copolymers containing more NADH groups was accelerated by Zn^{2+} or Ni^{2+}, the mechanism of metal-ion assistance was proposed as depicted in **25**.

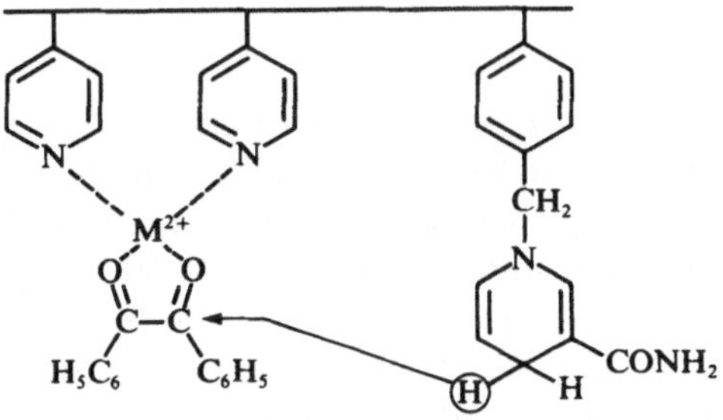

[25]

Murakami *et al.*[136,137] synthesized [20]paracyclophane carrying an NADH group in the benzene ring **26** and that carrying a pyridine-2-carboxylic acid group at C-10 **27**. Compound **26** formed a relatively stable 1:1 complex with Zn^{2+}, in which the NADH group was a ligand for Zn^{2+}. Therefore, the reduction rate of hexachloroacetone by **25** was decreased greatly by the coordination of Zn^{2+}. On the other hand, **26** formed 1:1 and 2:1 complexes with Zn^{2+}. In the 1:1 complex, the NADH group and the pyridyl group coordinate simultaneously to Zn^{2+}. In the 2:1 complex, two molecules of **26** coordinate to Zn^{2+} through their pyridyl groups. The 2:1 complex was seven times as effective as the Zn^{2+}-free catalyst for the reduction. The mechanism of reduction is proposed as depicted in **28**. This catalytic system can be regarded as an NAD^+-dependent alcohol dehydrogenase model.

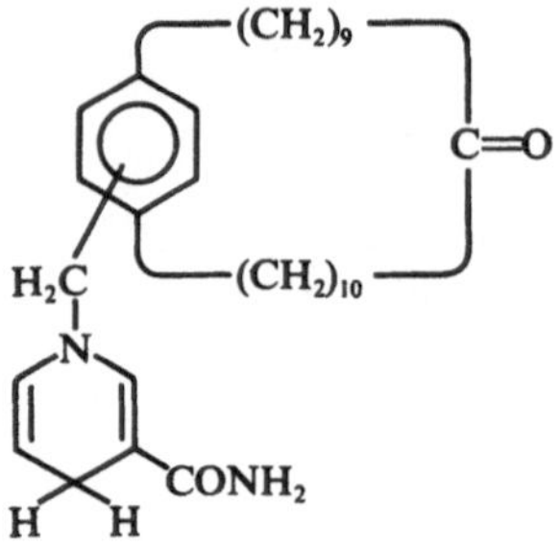

[26]

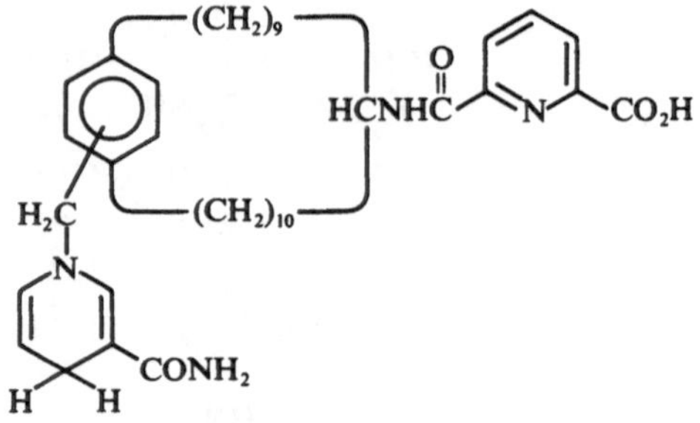

[27]

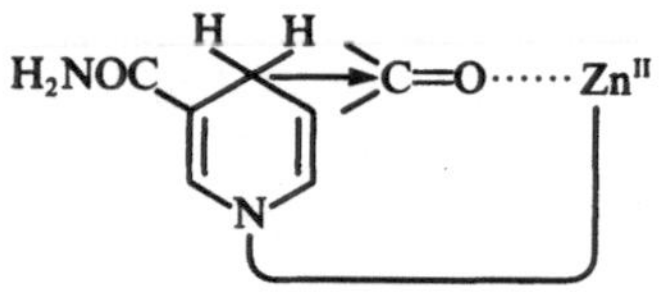

[28]

Kojima *et al.*[138] introduced a nicotinamide group to one of the secondary alcohol groups of β-cyclodextrin. The reduction of ninhydrin by the reduced state of the cyclodextrin derivative proceeded according to the Michaelis–Menten kinetics, and was faster by 40–60 times as compared with that produced by the presence of only a simple NADH. Hirano and Takahashi[139] showed that the reduced keratin and NAD^+ formed a 1:1 complex which oxidized glyceraldehyde after complex formation.

Tsubokawa *et al.*[140] found that insoluble polymers containing a 1-benzyl-1,4-dihydronicotinamide group reduced alloxane and in turn the reduced alloxane reduced insoluble copolymers containing benzoquinone. The oxidation–reduction between two insoluble polymers was attained by alloxane acting as an electron carrier.

2.5.3. *Polymeric Pyridoxal Phosphates*

If one can introduce a chirality to a pyridoxal without damaging the catalytic activity, a novel stereospecific catalyst having an enantiomer selectivity may be developed. Keeping this idea in mind, Kuzuhara *et al.*[141] synthesized (−)-15-formyl-14-hydroxy-2,8-dithia[9](2,5)pyridinophane, **29**. Compound **29** racemized L-glutamic acid approximately 1.3 times as fast as D-glutamic acid. This experimental result reveals that **29** can act as a stereospecific catalyst for racemization.

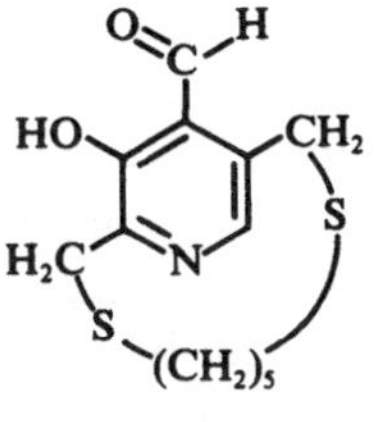

[29]

Breslow *et al.*[142] linked pyridoxamine with a 6-SH derivative of β-cyclodextrin, and investigated the transamination of an α-keto acid with the pyridoxamine-β-cyclodextrin conjugate. There were no differences in the transamination of pyruvic acid, phenylpyruvic acid, and indolepyruvic acid with pyridoxamine. However, aromatic keto acids were transaminated approximately 200 times as fast as pyruvic acid with pyridoxamine-β-cyclodextrin conjugate. In the reaction of an equimolar mixture of indolepyruvic acid and pyruvic acid, tryptophane occupied more than 97% of the initial product. In the reaction of an equimolar mixture of phenylpyruvic acid and pyruvic acid, phenylalanine occupied more than 98% of the initial product. Apparently, the aromatic group of aromatic keto carboxylic acids is bound into the cavity of β-cyclodextrin, and the bound keto acid is transaminated selectivity. Furthermore, some excess of the L-enantiomer was obtained in the product—12% in the case of dinitrophenyltryptophane and 52% in the case of dinitrophenylphenylalanine. This investigation shows the success of artificial transaminase.

Belokon *et al.*[143] synthesized copolymers of N^{α}-5-methacryloylaminosalicylidene-N^{ε}-methacryloyl-(*S*)-lysinate copper(II) and acrylamide or *N*,*N*′-methylenebisacrylamide. By extracting Cu(II) ions from the copolymer, polymer gels containing salicylaldehyde residues and lysyl residues were obtained. The equilibrium constant for the intramolecular aldimine formation was 30 (at pH 7.1)–100 (at pH 9.2) times as large as that for the intermolecular reaction employing low-molecular-weight analogues. The α-NH_2 groups of lysyl residues in the polymer gel form a Schiff base with salicylaldehyde groups in the same gel, and the subsequent reaction of the gel with semicarbazide was 5.3 times as fast as that of a gel from lysyl residues.

Nakano *et al.* investigated the reaction of α-amino acids with pyridoxal phosphate which was bound to quaternized poly(4-vinylpyridine) and activated by metal ions. The decomposition of threonine, yielding acetaldehyde and glycine, by pyridoxal in the presence of Cu^{2+}, Al^{3+}, Fe^{3+}, and VO^{2+} was accelerated with increasing pH.[144,145] This overall acceleration is caused by the acceleration of α-CH abstraction which is due to the increase of OH^- concentration. However, EtBr-quaternized poly(4-vinylpyridine)

(C_2PVP) had little influence on this reaction. The same reaction catalyzed by pyridoxal phosphate was greatly accelerated by addition of partially quaternized poly(4-vinylpyridine). The acceleration should have been induced by the uptake of pyridoxal phosphate by the polycation through the electrostatic interaction, and by the enhanced proton abstraction by the condensed OH^- ions or the general-base catalysis of the pyridyl groups. When the same catalyst was used for the α,β-elimination reaction, similar experimental results were obtained.[(146)] However, in the two kinds of reactions, the details of the effects of added C_2PVP were different. On the other hand, the effect of poly(4-vinylpyridine) partially quaternized by lauryl bromide ($C_{12}PVP$) on the α,β-elimination reaction of tryptophane, yielding indole, by pyridoxal derivative and Cu^{2+} ion was investigated.[(147)] Interestingly, the decomposition of threonine and the α,β-elimination of serine with the pyridoxal phosphate/Cu^{2+} complex were accelerated greatly by C_2PVP, whereas the α,β-elimination of tryptophane was not accelerated. Contrary to this, $C_{12}PVP$ did not influence the former reactions, but greatly accelerated the latter reaction. This difference represents the contribution of hydrophobic interaction involved in the coenzyme–apoenzyme interactions.

2.6. Metalloenzyme Models

The participation of metal ions in enzyme reactions is very important. Metal ions participate in the electron transfer as a catalytic site, activate the catalytic groups, act as a substrate-binding site, or determine the conformation of the apoenzyme. By using metalloenzyme models, the role of apoenzymes in metalloenzymes will be investigated in more detail than by using polymeric coenzymes. Some review articles on metalloenzyme models have been published.[(148,149)]

2.6.1. *Catalase and Peroxidase Models*

The decomposition of H_2O_2 catalyzed by the Fe^{3+} complex of poly(acrylic acid) partially amidated by diethylenetriamine was investigated, and the activation for decomposition by diethylenetriamine units as cofactors was shown.[(150)] Investigations on the decomposition of H_2O_2 were reported for other catalysts, such as the poly(vinyl alcohol)/Cu(II) complex[(151)] and the alginic acid/Cu(II) complex.[(152)]

Barteri *et al.*[(153)] found that *cis*-$[Fe(pmen)(OH)_2]^+$ and *trans*-$[Fe(tetpy)(OH)_2]^+$ supported on poly(L-glutamic acid) were efficient catalysts for the H_2O_2 decomposition at pH 7.8, where pmen and tetpy represent *N,N'*-bis(2-pyridylmethyl)ethylenediamine and 2,2′:6′,2″:6″,2‴-tetrapyridyl, respectively. Barteri *et al.*[(154)] further investigated kinetically the decomposition of H_2O_2 by *trans*-$[Fe(tetpy)(OH)_2]^+$ supported on poly(L-glutamic acid), and discussed the role of the polypeptide matrix for the environmental control of catalysis.

With the above experimental results in mind, and considering the reaction represented by a general equation(12), the stereoselective peroxidase-like activity of the polypeptide-supported catalyst toward a suitable substrate AH_2 was tested.[155,156] Reactions using L(+)-ascorbic acid as the substrate AH_2 and $[Fe(tetpy)(OH)_2]^+$/poly(L-glutamic acid) (FeL) and $[Fe(tetpy)(OH)_2]^+$/poly(D-glutamic acid)(FeD) as the catalysts were investigated. At pH 6.3–7.5, these complexes were found to be an efficient and stereoselective catalyst of reaction (12). It was shown using various complex/polypeptide molar ratios that the reaction occurs through two different mechanisms; one is a catalytic electron-transfer mechanism at the inside of the Fe(III) complex supported on the polypeptide, and the other is a noncatalytic electron-transfer mechanism between the ascorbate anion and H_2O_2. By increasing the molar ratio of the Fe(III) complex against the polypeptide, the α-helix content of the polypeptide (x_α) increased; for example, x_α was 0.70 at the molar ratio 0.20. Under these conditions, the ratio of the catalytic electron-transfer rate constant $k_{FeD}/k_{FeL} = 4.0$. The stereospecificity was found to be entropy control and based on the rigidity of the Fe(III) complex supported on the polypeptide. The environmental control by the chiral Fe(III) complex/polypeptide system was also shown to be operating in the noncatalytic electron-transfer process.

$$AH_2 + H_2O_2 \xrightarrow{\text{Cat.}} A + 2H_2O \tag{12}$$

The x_α of poly(L- or D-glutamic acid) should be increased by lowering pH as well. Barteri *et al.*[155,157] investigated the effects of pH on the reaction of ascorbic acid and H_2O_2 catalyzed by the same Fe(III) complex/polypeptide catalysts. The reaction, which proceeds by a composite catalytic and noncatalytic mechanism, became slower by decreasing pH, but the stereoselectivity increased in both mechanisms. At pH 6.3, k_{FeD}/k_{FeL} was 5.2, and k_D/k_L of the noncatalytic reaction was 1.9. The increase of x_α of the polypeptide support is closely related to increasing stereoselectivity.

2.6.2. Other Metal-Containing Reductooxidase Model

The oxidation of ascorbic acid by molecular oxygen, catalyzed by the multinuclear complex of poly(4-vinylpyridine) partially quaternized by dimethylsulfate and Cu^{2+}, was investigated in detail by Skurlatov *et al.*[158]

Yamamoto and Hayakawa[159] investigated the oxidation of L-ascorbic acid or D-isoascorbic acid by the Cu(II) ion complex with optically active, basic polypeptides, such as poly(L- or D-lysine), poly(L-arginine), and poly(L-histidine). The oxidation reaction by the polypeptide/Cu(II) complexes was slower than that with Cu(II) ions alone. However, the ratio of the reaction rate catalyzed by poly(D-lysine)/Cu(II) and that by poly(L-lysine)/Cu(II) was 1.8–2.8 at pH 5 and 0.7 at pH 10.5, which means that the reactions are stereospecific. The oxidation reactions of homogentisic acid and hydroquinone were investigated using Cu(II), Au(III), and Ag(I) ions supported on some cross-linked chelate resins.[160]

A bicomponent system, consisting of a ferredoxin model compound $[Fe_4S_4(SPh)_4]^{2-}$ and a Mo(IV) complex supported on a partially mercaptomethylated cross-linked polystyrene, was highly catalytically active for the reduction of acetylene by $NaBH_4$. This bicomponent catalyst is useful as a nitrogenase model.[161] Bovine serum albumin coordinated to a Fe_4S_4 cluster showed a hydrogenase-like activity. Its affinity to methyl viologen was similar to hydrogenase, but the hydrogen evolution was only 10^{-5} times as large as that by hydrogenase. Albumin was cleaved by BrCN into fractions and fractionated. The fraction having a molecular weight 4×10^4 coordinated to Fe_4S_4 was more reactive than the fraction having a molecular weight 2×10^4. This experimental finding indicates the importance of the amino acid sequence of apoenzymes involved in the active site.[162]

Using Co^{2+}/poly(styrene sulfonate)/2-(2-pyridylazo)-1-naphthol, the catalytic effect of polyelectrolyte upon the oxidation of the Co(II) chelate by molecular oxygen was investigated. The acceleration of the reaction was explained in terms of the stabilization of the intermediate complex by the coordination to the polyanion.[163] Furthermore, the structural change of the polyanion/Co(II) complex during the process was investigated by the electric dichroism.[164]

2.6.3. Carboxypeptidase Models

Takeishi *et al.* investigated ester hydrolyses by polymer/metal-ion complexes. A poly(acrylic acid)/Ag^+ complex was reactive in the hydrolysis of 2,4-dinitrophenyl vinylacetate, and the activity increased with increasing ratio of Ag^+.[165] In this catalytic system, Ag^+ coordinates to the olefinic bond through charge transfer interactions to concentrate the substrate around the carboxylate groups, which act as intramolecular nucleophiles. However, it was unexpected that an aliphatic carboxylic acid ester, 2,4-dinitrophenyl propanoate, was hydrolyzed rapidly by poly(acrylic acid) in the presence of Ag^+ ions.[166] In this case, the anionic charges of poly(acrylic acid) were neutralized by Ag^+ ions and the hydrophobic interaction between the polymer catalyst and the substrate was enhanced. Another reason for the acceleration is that Ag^+ ions coordinate to the carboxyl oxygen of bound substrates to activate them for the intramolecular nucleophilic attack by the carboxylate groups. When the same substrate was hydrolyzed by acetate ions, such bifunctional catalysis occurs only with difficulty so that addition of Ag^+ ions did not cause the acceleration. The hydrolyses of 2,4-dinitrophenyl isonicotinate and picolinate by partially neutralized poly(methacrylic acid) and poly(acrylic acid) were accelerated by Cu(II), Ni(II), Co(II), and Zn(II) ions.[167] The acceleration by these divalent metal ions was greater than that observed in the hydrolysis by acetate ions. Therefore, the formation of a ternary complex of the polymer, the metal ion, and the substrate was considered. In the ternary complex, the metal ion acts as a template for the nucleophilic attack of the carboxylate anion toward the ligand substrate. The hydrolysis of 2,4-dinitrophenyl acetate by a poly(vinyl alcohol)/Cu(II) complex as a catalyst at pH 5.4–6.3 was accelerated by the addition of KOH.[168] The possibilities of a general-base catalysis or a nucleophilic catalysis by alcoholate groups of the poly(vinyl alcohol) coordinated to Cu(II) ions were experimentally eliminated. It was confirmed that the reaction was catalyzed by OH^- ions, which coordinated to Cu(II) ions bound by poly(vinyl alcohol).

Yamskov *et al.*[169] observed asymmetric selectivity in the alkaline hydrolyses of α-amino acid methyl esters catalyzed by poly(*p*-chlorostyrene) substituted by *S*-(2-aminoethyl)-L-cystein, *S*-methoxycarbonyl-L-cystein, or *N*-methoxycarbonyl-L-valine and coordinated with Cu(II) or Ni(II) ions. The substrate esters are activated for hydrolysis by coordination to metal ions bound by the asymmetric polymers. The same researchers[170] observed the asymmetric hydrolyses of α-amino acid esters catalyzed by polystyrene or polyacrylamide substituted by L-hydroxypyroline and coordinated by Cu(II) or Ni(II) ions.

Lau and Gutsche[171] investigated the aminolysis of acetyl phosphate by triethylenetetramine carrying a nucleophilic group at the *N*-terminal with or without metal ions, and the hydrolysis of acetyl phosphate by oxime triamide with or without metal ions. In the aminolysis reaction, about a threefold increase of the reaction rate was observed, only when Cu(II) ions were present. In the hydrolytic reaction, the coordination of Cu(II) ions caused about a 17-fold increase of the reaction rate, but Zn(II) ions decelerated the reaction by five to six times. The acceleration was considered to be due to the formation of a mixed ligand to which the substrate is bound by forming a four-centered structure. On the other hand, the deceleration was considered to be due to the formation of a mixed ligand to which the substrate is bound by forming a six-centered structure.

Takagishi and Klotz[172] found that the hydrolysis of *p*-nitrophenyl caproate (PNPC) by a modified polyiminoethylene containing imidazolyl groups was greatly accelerated by divalent metal ions. Cu(II) ions brought about the largest acceleration of 20-fold. An explanation for the metal-ion acceleration was that the metal ions coordinated to imidazolyl groups acting as electrophiles to activate the substrate, and that free imidazolyl groups undergo an intramolecular nucleophilic attack toward the activated substrates. Another explanation for the metal-ion acceleration was that the metal ions coordinated to imidazolyl groups assist the attack of OH^- ions from the inner sphere of the metal coordination shell or from the bulk of the solvent.

Ogino *et al.*[(173)] found that the hydrolysis of Z-L-Phe-ONp by *N*-laurylimidazole containing L-2-pyrrolidine methanol in the presence of CTAB micelles was accelerated by 56–401 times by the addition of Zn(II) ions. Furthermore, this catalytic system hydrolyzed the L-substrate more rapidly than the D-substrate. In the absence of Zn(II), $k_L/k_D = 1.14$–2.41, while in the presence of Zn(II), $k_L/k_D = 1.40$–3.92.

Tanihara and Imanishi[(174)] synthesized cyclo(D-Leu-L-Glu-L-His)$_2$ as a hydrolytic enzyme model. This cyclic hexapeptide possesses two D-Leu residues having a hydrophobic side chain, two L-Glu residues having an anionic side chain, and two L-His residues having a nucleophilic side chain. A high catalytic activity was expected for this cyclic hexapeptide by the cooperative interaction among functional side chains. However, the absence of direct interactions between the side chains of L-Glu and L-His residues has been verified. When neutral carboxylic acid esters were hydrolyzed in 20% dioxane/water mixture at 25°C and at pH 7.8, a large catalytic activity was found for long-chain carboxylic acid esters, as shown in Table 9. This cyclic hexapeptide was found to bind the substrate by the hydrophobic interaction and hydrolyze the bounded substrate efficiently. The cyclic hexapeptide was more efficient than the linear tripeptide having the same sequence, Boc-D-Leu-L-Glu-L-His-OMe, where Boc represents a tertiary-butyloxycarbonyl group. Therefore, it was shown that the spatial arrangement of the functional groups, which is maintained by the relatively rigid chain of the

Table 9.[a] **Second-Order Rate Constant k_{cat} (liter mol^{-1} min^{-1}) for the Hydrolysis of $CH_3(CH_2)_{n-2}COONp$**[b]

Catalyst	$n = 2$[c]	$n = 10$[d]	$n = 12$[d]
None ($k_w \times 10^3$, min^{-1})	0.92	0.89	0.24
Imidazole	28	15	1.8
Boc-D-Leu-L-Glu-L-His-OMe	6.2	7.9	9.5
Cyclo(D-Leu-L-Glu-L-His)$_2$	5.8	46	34

[a] Reproduced from Table 1 of Reference 174 by permission of the copyright owner.
[b] In 20% dioxane/H_2O mixture at 25°C and pH 7.8 (KH_2PO_4/NaOH buffer).
[c] In 3% dioxane/H_2O mixture, $[\text{substrate}]_0 = 3.0 \times 10^{-5}$ mol liter^{-1}.
[d] $[\text{Substrate}]_0 = 6.0 \times 10^{-6}$ mol liter^{-1}.

cyclic peptide, should be suitable to promote cooperation between the binding substrate and the nucleophilic catalysis. On the other hand, this cyclic hexapeptide was not very efficient for the hydrolyses of cationically charged substrates in aqueous solutions at 25°C and at pH's of 6.95 and 7.8. Next, *p*-nitrophenyl ester hydrochlorides of leucine and valine, Leu-ONp · HCl and Val-ONp · HCl, respectively, were hydrolyzed by the cyclic hexapeptide. As seen from the experimental results listed in Table 10, the peptide catalysts were not efficient and no asymmetric selection was observed in the hydrolysis carried out in a phosphate-buffered solution at pH 6.95. On the other hand, if the hydrolysis was carried out under the presence of $Cu(ClO_4)_2$ in phosphate-buffered solution at pH 6.04, as seen in Table 10, a 150-fold acceleration (highest) was observed. Furthermore, under the present conditions, the L-esters were hydrolyzed slightly faster than the D-esters. If the hydrolysis was carried out in a citrate-buffered solution at pH 6.01, neither the acceleration nor the asymmetric reaction was observed when $Cu(ClO_4)_2$ was added. In the phosphate-buffered solution, Cu(II) ions coordinate to the side-chain carboxylate groups of the L-Glu residues in the cyclic hexapeptide and trigger the enhanced, asymmetric hydrolysis.

It is advantageous in the use of cyclic peptides as asymmetric catalysts or asymmetric ligands that the activities of cyclic peptides can be investigated in relation to the conformation or the conformational change of the cyclic peptides. By NMR and CD (circular dichroism) spectroscopy cyclo(D-Leu-L-Glu-L-His)$_2$ was shown to take a C_2-symmetric random conformation in aqueous solution.[(175)] When $Cu(ClO_4)_2$ was added to a phosphate-buffered solution of the cyclic hexapeptide at pH 6.95, the conformation of the cyclic

Table 10.[a] Second-Order Rate Constant k_{cat} (liter mol^{-1} s^{-1}) for the Hydrolysis of α-Amino Acid *p*-Nitrophenyl Ester Hydrochloride[b]

	Val-ONp · HCl		Leu-ONp · HCl	
Catalyst	L	D	L	D
None ($k_w \times 10^3$, s^{-1})	3.26	3.31	4.46	4.70
Imidazole	1.2	1.2	2.1	2.0
Boc-D-Leu-L-Glu-L-His-OMe	0.63	0.76	0.93	0.82
Cyclo(D-Leu-L-Glu-L-His)$_2$	0.13	0.26	0.40	0.36
	11.6[c]	9.9[c]	60.6[c]	52.7[c]

[a] Reproduced from Tables III and V of Reference 174 by permission of the copyright owner.
[b] In aqueous solution at 25°C and pH 6.95 (KH_2PO_4/NaOH buffer); [substrate]$_0$ = 6.0×10^{-5} mol liter^{-1}.
[c] In aqueous solution at 25°C and pH 6.04 (KH_2PO_4/NaOH buffer); [substrate]$_0$ = 6.0×10^{-5} mol liter^{-1}; [$Cu(ClO_4)_2$] = 7.8×10^{-4} mol liter^{-1}; [cyclo(D-Leu-L-Glu-L-His)$_2$] = 4.33×10^{-4} mol liter^{-1}.

Figure 6. Corey–Pauling–Koltun molecular model for the cyclo(D-Leu-L-Glu-L-His)$_2$/Cu^{2+} complex: (a) a top view; (b) a side view. [Reproduced from Figure 8 of Reference 175 by permission of the copyright owner (for Figure 6(b))].

hexapeptide changed into a type-II β-turn structure. The molecular model of the cyclo(D-Leu-L-Glu-L-His)$_2$/Cu(II) complex, which is matched best with the experimental findings, was constructed and is shown in Fig. 6. When a Cu(II) ion chelates with two L-Glu carboxylate groups, which situate on one side of the plane of cyclic hexapeptide, a conformational change takes place to yield a pocket on the other side of the plane, which consists of hydrophobic side chains of L-Leu residues and nucleophilic side chains of L-His residues. This pocket is asymmetric and hydrophobic and can be used for substrate binding and intramolecular catalysis.

2.6.4. *Other Metalloenzyme Models*

Wilson and Whitesides[176] considered that the introduction of a catalytically active metal ion to a suitable site in a protein might yield a well-defined steric circumstance around the metal ion. With this idea in mind, they made achiral diphosphine Rh(I) complex supported on various globular proteins. By using these protein-supported Rh(I) complexes, α-acetoamidoacrylic acid was hydrogenated yielding *N*-acetylalanine. When a supported catalyst having an avidin/Rh molar ratio 10 was used, the reaction was fast and the enantiomer excess of the product was more than 40%.

Kmošták and Setínek[177] carried out the gas-phase isomerization of cyclohexene and the dehydration of 1-propanol by sulfonated polystyrene resins, and found that the addition of transition-metal ions yielded some additional effect besides the neutralization effect. The nature of the additional effect could be a participation of the transition-metal ions in the reactions or the influence on the substrate migration inside the resin cross-linked by the metal ions. However, the details are not known.

Watanabe and Imazawa[178] coordinated Co(II) ions to poly(4-vinylpyridine), styrene/4-vinylpyridine copolymer, methyl methacrylate/4-vinylpyridine copolymer, or 4-vinylimidazole/methyl acrylate copolymer. They used these copolymers or a homopolymer as catalysts for the cross-aldol condensation of aldehyde and ketone, and reported the formation of α,β-unsaturated ketones as byproducts. The activity of these catalytic systems was investigated in reference to reactions catalyzed by class-II aldolase.

3. Intramolecular Cooperation between Binding Site and Catalytic Site along a Polymer Chain

One of the reasons for enzymes being 10^3–10^{11} times as efficient as low-molecular-weight catalysts in spite of similar structures of the catalytic sites could be that an extremely efficient intramolecular reaction occurs between a catalytic group and a bound substrate. In other words, a substrate-binding group and a catalytic group in an enzyme are arranged suitably for the intramolecular cooperation. This point has been tacitly taken into account, in Sections 2.4–2.6, in considering the efficiency of enzyme reactions. In the present section, investigations will be reviewed, with the distinct intent to study this intramolecular cooperation.

3.1. Linear Polymer Catalysts

The nucleophilic addition reaction of dodecane thiol to a carbon–carbon double bond of α,β-unsaturated carbonyl compounds, such as isopropenylmethylketone, can be catalyzed by amine bases, as described in Eq. (13):

$$CH_3(CH_2)_{11}SH + CH_2 = CH(CH_3)COCH_3 \xrightarrow{RNH_2} CH_3(CH_2)_{11}SCH_2CH(CH_3)COCH_3 \tag{13}$$

In reaction (13), a thiolate anion, produced by the activation of thiol by amine, undergoes a nucleophilic addition to the double bond, and the produced anion is subsequently protonated. Since either the thiolate anion or the intermediate carbanion forms an ion pair with a protonated amine, either the addition of thiolate anion or the recombination of proton may be stereospecific when an optically active amine is employed as a catalytic base. Reaction (13) is essentially the same as the stereospecific hydration reaction of fumaric acid catalyzed by fumarase yielding (*S*)-malic acid, and the stereospecific addition reaction of ammonia to fumaric acid catalyzed by aspartase yielding L-aspartic acid. To mimic these enzymatic mechanisms, reaction (13) was investigated using as a catalyst the terminal amino group of optically active polypeptides such as poly(L-alanine), poly(γ-benzyl L-glutamate), and poly(β-benzyl L-aspartate), and the product, (1-dodecylthiomethyl)ethylmethylketone, was found to be optically active. Early investigations in this field are reviewed by Inoue.[179]

In order for an asymmetric addition reaction to take place by polypeptide catalysis, substrate-binding groups, as well as the terminal catalytic group, should exist in the same polymer and the binding groups should hold the bound substrate in a specific arrangement. In order for a strict intramolecular cooperation of functional groups to occur, the chain of the catalyst must be rigid to some extent. Polypeptides are more suitable than vinyl polymers in providing suitable rigidity of the chain. The asymmetric addition reactions (13) catalyzed by the terminal amino group of optically active polypeptides have been investigated regarding the reaction mechanism and the influence of the catalyst being a polymer. It was shown that the presence of a small amount of EtOH had a serious influence on the asymmetric addition reactions catalyzed by polyglutamates or poly-aspartates.[180] In the author's opinion, the product of reaction (13) could be racemized in the presence of amines, and alcohols may also affect the racemization process. EtOH also affected the asymmetric addition reaction catalyzed by polyalanine. It was found that the stereospecificity of the reaction was high only when the chiral polyalanine chain was present.[181] This point is very interesting in relation to stereospecific addition reactions catalyzed by enzymes. The enantiomer excess of the product in the polypeptide-catalyzed reaction reached 47%, when a suitable amount of EtOH was added.[182] L-Alanine *N*-propylamide was synthesized as the model for the terminal amino group of poly(L-alanine), and the asymmetric addition reaction and the accompanying isomerization reaction, which are catalyzed by L-alanine *N*-propylamide, were investigated.[183] Amines corresponding to the terminal group of polyalanine or a dipeptide of alanine gave only a low degree of optical yield, but a tripeptide amine and a tetrapeptide amine, which take a β-sheet conformation, gave a much higher optical yield.[184] These experimental results suggest the importance of the highly-ordered structure of enzyme proteins in stereospecific catalysis. Finally, a tertiary amino group was introduced to the *C*-terminal of polyalanine or polyglutamate, and reaction (13) was catalyzed by their presence.[185] The experimental results were opposite to those so far obtained since the low-molecular-weight amines brought about higher stereospecificity than the corresponding polypeptide catalysts. It is possible that a subtle difference between the carboxyl and the amino ends of a polypeptide chain strongly affects the stereospecificity of the reaction.

Juliá *et al.*[186] carried out the asymmetric hydrogenation of several chalcones and various electron-deficient olefins in a triphase system consisting of a catalytic amount of poly[(*S*)-amino acid], water, and toluene, and obtained an optical yield of the product of (highest) 96%. They discussed the effect of molecular structures of the substrate and the catalyst as related to the stereospecificty of the reaction.

Nango *et al.*[187] investigated an asymmetric-selective hydrolysis of α-amino acid active esters catalyzed by polyiminoethylene carrying L-histidine units. In the author's opinion, the asymmetric reaction is not probable with such a flexible chain molecule as polyiminoethylene. Therefore, in the asymmetric selectivity observed under Nango's conditions, polymer association is probably involved.

Hui *et al.*[188] carried out the hydrolysis of *p*-nitrophenyl esters of aliphatic carboxylic acids in a Me_2SO/H_2O (1:1) mixed solvent using amylose having a DP of 350. The amylose, taking an α-helical structure, binds a long-chain ester as cyclodextrins do, and the alcoholic hydroxyl groups catalyze the reaction. In the hydrolysis of *p*-nitrophenyl acetate, amylose had a reactivity nearly equal to that when glucose was present, whereas in the hydrolyses of *p*-nitrophenyl esters of lauric and palmitic acid, the presence of amylose caused a reaction that was 2.8×10^2 times, for the lauric acid, and 5.6×10^3 times, for the palmitic acid, as reactive as when glucose was present.

Guthrie *et al.*(189) synthesized a steroid dimer, α,α′-bis [17β-(4′-imidazolyl)-11-keto-5α-androstan-3β-amino]-*p*-xylene [**30**], and investigated the hydrolyses of 2-nitro-5-carboxyphenyl esters of aromatic carboxylic acids by **30**. In the hydrolysis of 3-(3-phenanthryl)propionic acid ester, **30** was

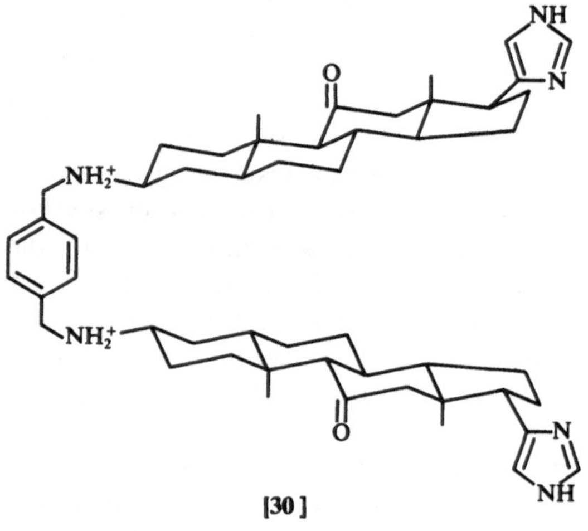

[30]

200 times as reactive as imidazole. For such an efficient catalysis, it is considered that the phenanthrene ring of the substrate is seiged by the two steroidal rings of **30** and the substrate is attacked by the imidazolyl group. Both the catalyst and substrate are restricted from undergoing ready internal rotation, and therefore the geometry of the transition state is depictable. Assuming that the reactions of phenanthryl derivatives and propionic derivatives proceed through a transition state having the same arrangement of the reactive species, the real reaction rate of the former was 3000 times as large as the calculated reaction rate of the latter. The importance of the hydrophobic interaction between the substrate and the catalyst can thus be quantitatively estimated.

3.2. Cyclic Polymer Catalysts

3.2.1. Cyclic Peptides

In Reference 189, the geometry of the transition state for a reaction was assumed and the cooperation between the substrate binding and the intramolecular catalysis was treated. For other linear catalyst molecules treatment of this kind has not been carried out. One of the reasons for the lack of the quantitative or semiquantitative treatment is the undetermined spatial arrangement of functional groups in linear catalyst molecules. Therefore, the use of cyclic peptide as a catalyst seems to overcome the difficulties in the investigation of the intramolecular cooperation of the substrate-binding and catalytic groups.

The author's team investigated the intramolecular cooperation of the binding and catalytic groups in the hydrolyses of aliphatic carboxylic acid esters having different acyl chains using as catalysts cyclic dipeptides consisting of a nucleophilic L-histidine and a hydrophobic D- or L-α-amino acid.[190] An example of the experimental results in the hydrolyses carried out in a 20% dioxane/H_2O mixture at pH 7.8 is shown in Table 11. It should be noted in the table that hydrophobic cyclic dipeptide catalysts cyclo(D-Val-L-His) and cyclo(D-Leu-L-His) are more efficient than imidazole toward hydrolysis of a long-chain carboxylic acid ester, $CH_3(CH_2)_{10}COONp$. Evidently, the importance of the hydrophobic interaction between the catalyst and the substrate is in operation. Another important point is that between the diastereomeric cyclic dipeptides the D–L-type was more efficient than the L–L-type, and the difference became most marked when hydrophobic substrates were involved in the reaction. Evidently, the configuration of the cyclic dipeptide affects the intramolecular nucleophilic attack on the substrate bound by hydrophobic interactions. The conformation of cyclic dipeptides in solution was investigated by the author's team.[191] For example, in cyclo(D-Leu-L-His), two side chains exist on opposite sides with respect to the nearly flat plane of the cyclic dipeptide chain. In cyclo(L-Leu-L-His) the plane of the cyclic dipeptide chain takes a boat-type conformation and two side chains reside on the same side of the plane (see Figure 7). The flagpole–boat conformation seems to be disadvantageous for a reaction requiring intramolecular nucleophilic attack on the bound substrate.

On the other hand, similar hydrophobic, nucleophilic cyclic dipeptides were found to be extremely reluctant in catalyzing the hydrolysis of charged esters of long-chain carboxylic acids.[192] Irrespective of the charge at the terminal of acyl chain or in the alcoholic part, and the nature of the charge being positive or negative, an efficient catalysis effect was not observed. The

Table 11. Second-Order Rate Constants (k_{cat}, liter mol^{-1} min^{-1}) for the Hydrolysis of Aliphatic Carboxylic Acid *p*-Nitrophenyl Esters with Different Acyl Chain Lengths by Hydrophobic Cyclic Dipeptide Catalysts[a]

Catalyst	CH_3COONp	$Ph(CH_2)_2COONp$	$Ph(CH_2)_4COONp$	$CH_3(CH_2)_8COONp$	$CH_3(CH_2)_{10}COONp$
Cyclo(Gly-L-His)	1.5	1.7	3.5	—	1.1
Cyclo(L-Phe-L-His)	1.9	1.9	1.2	0.4	—
Cyclo(D-Phe-L-His)	3.2	3.4	2.5	1.9[b]	0.8[b]
Cyclo(L-Val-L-His)	1.9	1.6	—	1.2[c]	0.11[c]
Cyclo(D-Val-L-His)	3.5	3.7	2.4	3.5[c]	16[c]
Cyclo(L-Leu-L-His)	1.6	1.5	—	1.2[c]	0.15[c]
Cyclo(D-Leu-L-His)	3.2	4.1	2.1	6.3[c]	16[c]
Cyclo(DL-Nle-L-His)	3.7	4.2	3.4	—	—
Imidazole	17.5	14.0	8.9	9.7[c]	4.9[c]

[a] 25°C; pH 7.8; 20% dioxane/water; $[S]_0 = 3.0 \times 10^{-5}$ mol $liter^{-1}$.
[b] $[S]_0 = 6.0 \times 10^{-6}$ mol $liter^{-1}$.
[c] $[S]_0 = 2.0 \times 10^{-6}$ mol $liter^{-1}$.

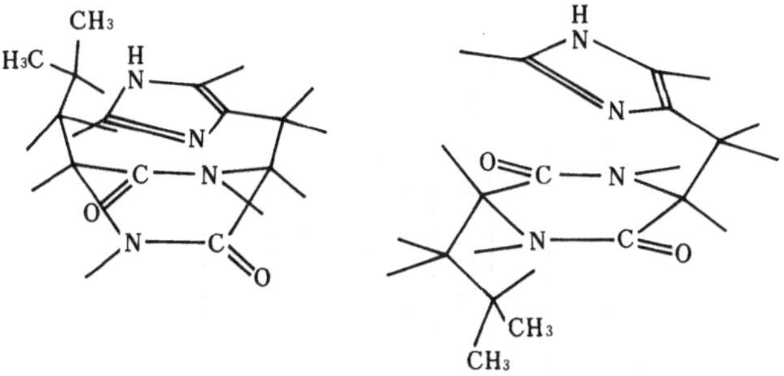

Figure 7. Most probable conformation of cyclic dipeptide in aqueous solution: (a) cyclo(L-Leu-L-His); (b) cyclo(D-Leu-L-His). [Reproduced from Figure 7 of *Biopolymers* **16**, 2217 (1977) by permission of the copyright owner].

hydration of substrate, induced by the charge, may hinder the effective hydrophobic interaction with the catalyst.

The solvolysis of positively charged esters of carboxylic acids was investigated using cyclo(L- or D-Glu-L-His) as a catalyst. This catalyst possesses an anionic L- or D-glutamic acid and a nucleophilic L-histidine.(193) Some of the experimental results are shown in Table 12. Interestingly, the solvolysis of Gly-ONp · HCl was faster in a 42% $(CH_3)_2CHOH/H_2O$ mixed solvent than in a totally aqueous solution, showing the importance of the electrostatic interaction between the substrate and the catalyst. The acceleration of solvolysis induced by a change of solvent composition was observed in the solvolysis of Gly-ONp · HCl, which is subjected to a general-base catalysis, whereas catalysis was not observed in the solvolysis of $Cl^- \cdot H_3N^+(CH_2)_{11}COONp$, when subjected to a nucleophilic catalysis. This observation indicates the importance of the stereochemical fit of the substrate with the catalyst. In a 42% $(CH_3)_2CHOH/H_2O$ mixed solvent, the diastereomeric cyclic dipeptides were almost equally reactive, similar to imidazole, and more reactive than the corresponding linear dipeptides. No experimental findings have been reported for the efficient catalysis by the L–L-type cyclic dipeptides. Therefore, it was indicated that the arrangement of side chains fixed by the cyclic dipeptide chain greatly contributes to the general-base catalysis by the His–imidazolyl group on the substrate bound by the electrostatic interaction.

Although the development of efficient hydrolytic catalysts was accomplished to a certain extent by using hydrophobic or charged cyclic dipeptides, none of them were asymmetric catalysts in regard to the hydrolysis of optically active α-amino acid esters. The asymmetric catalysis

Table 12.[a] **Second-Order Rate Constant (k_{cat}) for the Solvolysis of α- and ω-Amino Acid Active Esters**[b]

	$Cl^-H_3N^+CH_2COONp$		$Cl^-H_3N^+(CH_2)_{11}COONp$	
	k_{cal}/liter mol^{-1} s^{-1}		k_{cal}/liter mol^{-1} min^{-1}	
Catalyst	100% H_2O	42% *i*-PrOH	0.65% MeOH	42% *i*-PrOH; 0.83% MeOH
None (k_w)	1.67×10^{-2} s^{-1}	1.03×10^{-2} s^{-1}	1.44×10^{-3} min^{-1}	0.40×10^{-3} min^{-1}
Imidazole	5.6	3.8	11.7	2.5
Cyclo(L-Glu-L-His)	1.8	3.5	1.6	0.04
Cyclo(D-Glu-L-His)	2.1	3.7	3.6	0.3
Moc-L-Glu-L-His-OMe	1.4	1.9	2.5	0.6
Moc-D-Glu-L-His-OMe	1.8	2.8	1.5	0.3

[a] Reproduced from Table 7 of Reference 193 by permission of the copyright owner.
[b] 25°C; pH 7.8; $[S]_0 = 3.0 \times 10^{-5}$ mol $liter^{-1}$

Table 13.[a] Second-Order Rate Constants (k_{cat}/liter $mol^{-1} s^{-1}$) for the Hydrolysis of L- or D-Val-ONp · HCl and L- or D-Leu-ONp · HCl[b]

Catalyst	L-Val-ONp · HCl	D-Val-ONp · HCl	L-Leu-ONp · HCl	D-Leu-ONp · HCl
None (k_w/s^{-1})	4.79×10^{-3}	5.34×10^{-3}	4.76×10^{-3}	4.35×10^{-3}
Imidazole	1.6 (3.6)	1.7 (3.9)	3.8 (8.6)	3.6 (8.2)
	1.7 (3.9)	1.5 (3.4)	3.5 (8.0)	3.6 (8.2)
Cyclo[L-Glu(L-Leu-OBzl)-L-His]	8.3 (10.2)	7.6 (9.4)	3.8 (4.7)	6.9 (8.5)
	10.3 (12.7)	9.8 (12.1)	2.7 (3.3)	6.1 (7.5)
Cyclo[L-Glu(L-Leu-OH)-L-His]	0	0.6 (0.8)	2.9 (4.0)	4.1 (5.7)
	0	0.7 (1.0)	0.9 (1.3)	3.7 (5.1)
Moc-L-Glu(L-Leu-OH)-L-His-OMe	3.9 (6.5)	4.0 (6.7)	4.3 (7.2)	4.3 (7.2)

[a] Reproduced from Table 5 of Reference 194 by permission of the copyright owner.

[b] In aqueous solution at 25°C and pH 6.95; $[\text{substrate}]_0 = 6.0 \times 10^{-5}$ mol liter^{-1}; $[\text{catalyst}] = (2.56\text{–}3.02) \times 10^{-4}$ mol liter^{-1}; k_{cat} correction for catalytically active imidazolyl functions is given in parentheses.

may take place if another barrier for the asymmetry is introduced to the cyclic dipeptides. Keeping this idea in mind, tripeptide catalysts, cyclo[L-Glu(L-Leu-OBzl)-L-His] and cyclo[L-Glu(L-Leu-OH)-L-His], were synthesized by introducing another α-amino acid to cyclo(L-Glu-L-His).[194] In the hydrolyses of α-amino acid active-ester hydrochlorides with these tripeptide catalysts, the enantiomer recognition was observed as seen in Table 13. In the hydrolyses of Val-ONp · HCl, all catalysts gave the same reaction rate between the D- and L-substrates (within experimental error). However, in the hydrolysis of Leu-ONp · HCl, the D-enantiomer was hydrolyzed about twice as fast as the L-enantiomer with cyclo[L-Glu(L-Leu-OBzl)-L-His] and cyclo[L-Glu(L-Leu-OH)-L-His]. In addition to the asymmetric selectivity, these tripeptide catalysts were more efficient than imidazole. The hydrophobic interaction between the catalyst and the substrate seems to play an important role in either the asymmetric selectivity or the efficiency.

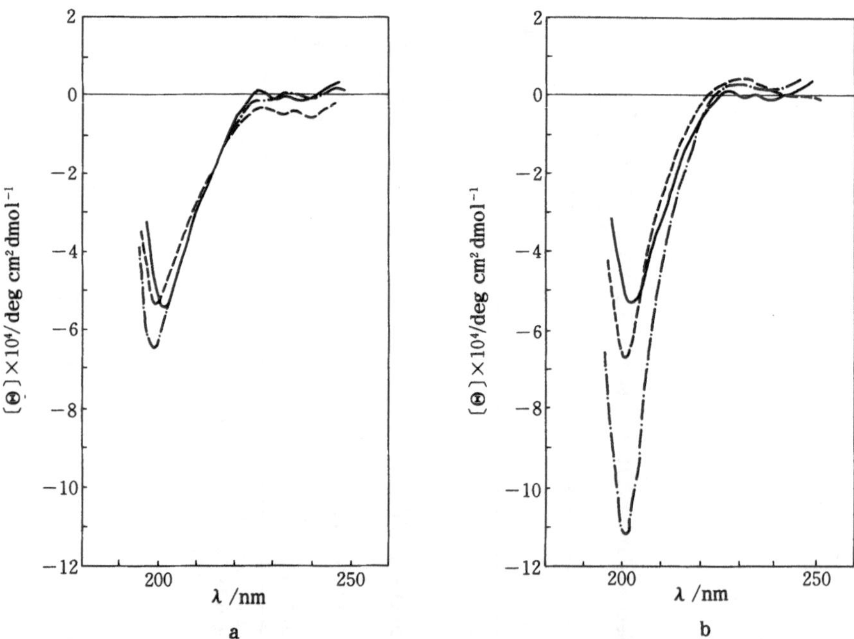

Figure 8. Cd spectra of 1.3×10^{-4} mol liter^{-1} aqueous solution of cyclo[L-Glu(L-Leu-OBzl)-L-His] at pH 6.95 with or without the addition of L- or D-Leu-OEt · HCl. (a) Addition of L-Leu-OEt · HCl; (——), 0, and (– – –) represents [L-Leu-OEt · HCl] = 1.70×10^{-4} mol liter^{-1} and (– · –) represents 9.62×10^{-4} mol liter^{-1}. (b) Addition of D-Leu-OEt · HCl; (——), 0, and (– – –) represents [D-Leu-OEt · HCl] = 1.70×10^{-4} mol liter^{-1} and (– · –) represented 9.88×10^{-4} mol liter^{-1}. (Reproduced from Figure 6 of Reference 195 by permission of the copyright owner].

The solution conformation of these tripeptides was investigated by spectroscopy, and discussed in relation to the asymmetric catalysis.[195] Most importantly, the conformation of the tripeptides in solution was changed drastically by the addition of D-Leu-OEt. The change observed in the CD spectrum of cyclo[L-Glu(L-Leu-OBzl)-L-His] is shown in Figure 8. It is believed that the steric barrier composed of the chiral chain of the cyclic dipeptide and the chiral side chains, as well as the induced fit occurring on binding the substrate, caused the asymmetric hydrolysis.

Inoue *et al.* investigated the asymmetric cyanohydrin synthesis by the addition of HCN to a carbonyl group catalyzed by polypeptide catalysts (see the Section 3.1.). Recently, cyclic dipeptides containing a histidine were used as a catalyst.[196,197] In the synthesis of mandelonitrile from benzaldehyde catalyzed by cyclo(L-His-L-Phe), an enantiomer excess as high as 90% was reported. For the transition state of this reaction, a scheme as depicted in Figure 9 was proposed. We should be able to evaluate the probability of the proposed transition state by spectroscopy, but as yet no experimental work has been done.

3.2.2. Cyclodextrin Catalysis

Cyclodextrins bind substrate selectively into the cavity and one of the hydroxyl groups existing on the rim of the cylindrical wall of the cyclodextrin attacks the bound substrate. Therefore, cyclodextrins are regarded as enzyme models, in which the binding ability and the catalytic function cooperate intramolecularly, and they have been extensively investigated.

The rate of acceleration caused by the inclusion catalysis by cyclodextrins was not very remarkable, but Czarniecki and Breslow[198] considered that greater acceleration could be obtained if an optimum substrate was found. They considered that it is necessary for the acyl part of an active ester to be firmly bound in the cavity, and they chose *p*-nitrophenyl-3-*trans*-ferrocenylpropenoate **31** as a substrate that fits this restriction. β-Cyclodextrin accelerated the reaction of **31** in a 60%

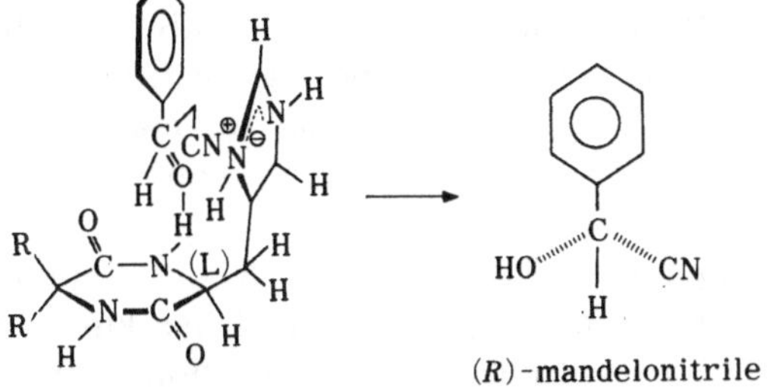

(R)-mandelonitrile

Figure 9. Transition-state model for the stereospecific addition reaction of HCN to benzaldehyde catalyzed by cyclic dipeptide containing L-histidine leading to formation of (*R*)-mandelonitrile. (Reproduced from scheme on p. 585 of Reference 197 by permission of the copyright owner).

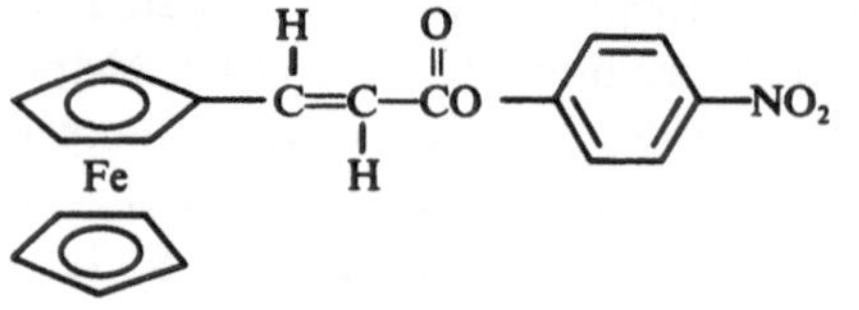

[31]

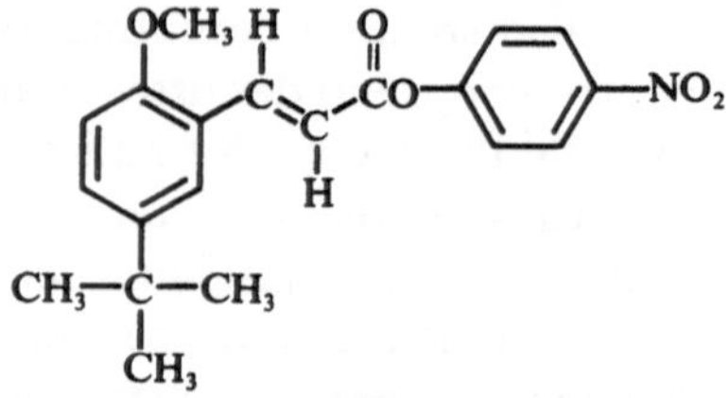

[32]

Me_2SO/H_2O mixture in the phosphate-buffered solution by 51,000 times. The very great acceleration on **31** is based on the fact that the bound **31** can be transformed into a tetrahedral intermediate without changing the state of binding in the cavity. Use of **32** gave a 630-fold acceleration under the same conditions. Compound **32** was included by β-cyclodextrin to the same extent as **31**, but could not go into the tetrahedral intermediate without changing the binding state. In order to eliminate the slight residual freedom of internal rotation in the transition state of **31**, Trainor and Breslow[199] modified **31** into two fused-ring versions **33**. There are two enantiomorphic isomers of **33**. By

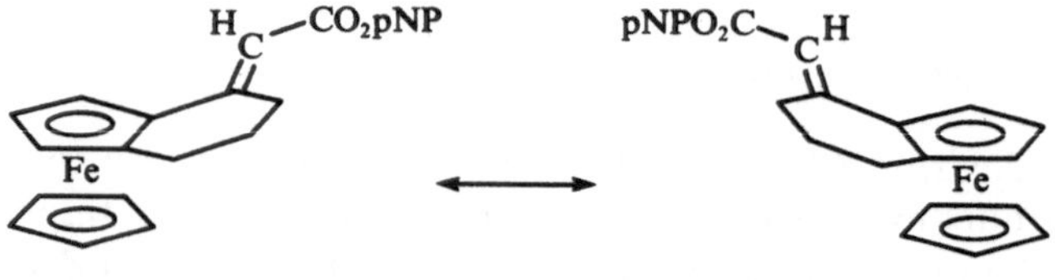

[33]

freezing out the internal rotation, the more reactive isomer reacted 10 times as fast as **31**. The reactivity difference between the two isomers was 20-fold.

In general, the catalysis by cyclodextrin has been considered to be nucleophilic. However, Komiyama and Inoue[200,201] found that the hydrolysis of 2,2,2-trifluoroethyl-4-nitrobenzoate at pH 9.6 proceeded by the mixed general-base (major) and nucleophilic (minor) catalysis by α-cyclodextrin. Komiyama and Hirai[202] investigated the hydrolyses of seven kinds of alkyl benzoates by α-cyclodextrin, and showed the applicability of general-base catalysis to a wide range of alkyl esters.

Komiyama and Bender investigated the cyclodextrin catalysis of the decomposition of thiocarboxylic acid *S*-esters[203] and the hydrolysis of *p*-nitrophenyl 1-adamantylacetic acid,[204] and pointed out the correlation between the hydrolysis of phenyl esters and the time-averaged conformation of the inclusion complex.[205]

Daffe and Fastrez[206] investigated the formation of *N*-acyl-α-amino acids in the hydrolyses catalyzed by α- or β-cyclodextrin of five kinds of 5(4H) oxazolones which have substituents at 2- and 4-positions having different hydrophobicities and bulkiness. It was found that the substituent at the 2-position plays an important role in the stereoselectivity of the reaction.

Motozato *et al.*[207] investigated kinetically the decarboxylation reaction of trichloroacetic acid in the presence of β-cyclodextrin.

Taniguchi *et al.*[208] investigated the pressure effect up to 2 kbar on the hydrolyses of *p*-nitrophenyl acetate and 2-naphthyl acetate by α-, β-, or γ-cyclodextrin at pH 8.3. From the volume change occurring in the formation of the inclusion complex, it was suggested that the inclusion takes place by hydrophobic interaction and the conformation change of the cyclodextrin occurs during the inclusion process. From the experimental values of activated volume, it was shown that both tight and loose inclusion complexes are formed according to various combinations of cyclodextrin and substrate. The tight complex seemed to be more strongly solvated. Makimoto *et al.*[209] investigated the pressure effect on the hydrolyses of three kinds of *m*- and *p*-substituted phenyl acetates by α-cyclodextrin and interrelated the structure of the substrate with the activated volume and the volume change during the binding equilibrium.

It has been alleged that C-3 hydroxyl groups of cyclodextrins are catalytically inactive, because the catalytic activities of cyclodextrins are lost by the methylation of C-2 and C-6 hydroxyl groups. However, Bergeron and Burton[210] carried out static and dynamic NMR investigations on sodium 4-nitrophenolate or sodiuim 2,6-dimethyl-4-nitrophenolate included by dodecakis-2,6-*O*-methylcyclohexaamylose, and verified that the C-3 hydroxyl groups, which are intrinsically catalytically active, become inactive because of the inaccessibility of the bound substrate to the C-3 hydroxyl groups as a result of the methylation of C-2 and C-6 hydroxyl groups.

Breslow and Rideout[211] and Sternbach and Rossana[212] reported that the intramolecular Diels–Alder reaction of molecules, which have both a conjugated diene group and a dienophilic group in the same molecule, was accelerated by cyclodextrins when the two reacting groups take a configuration suitable for the reaction.

From the investigation on the hydrolysis of 3-nitrophenyl acetate by β-cyclodextrin in the presence of a series of substituted imidazoles, it was found that the substituted imidazoles act as general-base catalysts to assist the attack of hydroxyl groups on the bound substrate.[213]

3.2.3. *Functionalized Paracyclophanes*

In higher paracyclophanes the methylene chain and the phenyl groups move freely at room temperature, and the face orientation of phenyl groups is attained statistically. At low temperatures the molecular motions are frozen out so that phenyl groups tend to assume the lateral orientation. In general, higher paracyclophanes include hydrophobic substrates and show a substrate specificity, which is different from that of cyclodextrins. Therefore, functionalized paracyclophanes, which are paracyclophanes having catalytic groups, are inclusion catalysts which are similar to, but different from, cyclodextrins.

Murakami *et al.*[(214)] demonstrated that 10-amino-[20]paracyclophane binds *p*-nitrophenyl carboxylates by a hydrophobic interaction and deacylates them efficiently. The reaction proceeds either by the nucleophilic catalysis of the amino group or by the general-acid catalysis of the ammonium group. Murakami *et al.*[(215)] deacylated various *p*-nitrophenyl carboxylates with synthetic *N*-(4-ethylimidazolyl)-10(11)-oxo[20]paracyclophane-22-carboxamide. This functionalized paracyclophane binds and deacylates hydrophobic esters in solution and micellar states. The binding ability and the deacylation rate are particularly great in the micellar state. The relative acceleration, k_{rel}, of 12 kinds of *p*-nitrophenyl carboxylates by paracyclophane catalyst in solution was shown by Hansch[(216)] to be related to the hydrophobic parameter of the acyl chain, π, by Eq. (14). The numbers in parentheses represented the 95% confidence limit, the number of data points

$$\log k_{rel} = 0.45\ (\pm\ 0.09)\pi - 0.53(\pm\ 0.39) \tag{14}$$

was 11, the correlation factor was 0.968, and the standard deviation was 0.260.

Murakami *et al.*[(217)] synthesized water-soluble azaparacyclophanes having quaternary ammonium groups in the cyclic skeleton and long alkyl branches. One of the above-described azaparacyclophanes, *N*,*N*′,*N*″,*N*‴-tetrakis-(μ-carboxydecyl)-2,11,20,29-tetraaza[3.3.3.3]paracyclophane-3,10,21,28-tetraone, has an octopus-like structure, in which four long alkyl chains and macrocyclic skeleton cooperatively form a strongly hydrophobic field and bind a substrate having a neutral or cationic charge and an appropriate molecular size. By this hydrophobic binding, the hydrolysis of *p*-nitrophenyl 3,5-dimethylcyclohexylacetate proceeded efficiently. Murakami *et al.*[(218)] investigated the hydrophobic binding of substrates and the deacylation reaction on a similar water-soluble azaparacyclophane, *N*,*N*′,(*N*″)-bis-{10-[dimethyl(imidazolylmethyl)ammonio]decyl}-*N*″(*N*′)*N*‴-bis-{10-[dimethyl(hydrogen)ammonio]decyl}-3,10,21,28-tetraoxo-2,11,20,29-tetraza[3.3.3.3]paracyclophane tetrachloride.

Tabushi *et al.*[(219)] hydrolyzed *p*-nitrophenyl, α-naphthyl, and β-naphthyl chloroacetates using a synthetic water-soluble heterocyclophane, *N*,*N*,*N*′,-*N*′,*N*″,*N*″,*N*‴,*N*‴-octamethyl-2,11,20,29-tetraaza[3.3.3.3]paracyclophane-tetraammonium tetrafluoroborate as a catalyst. This heterocyclophane catalyst was more efficient than CTAB micelle, cyclodextrins, and the open-chain analogue, and also substrate selective. From mechanistic investigations, these features were found to originate from a specific substrate binding in which the ammonium ion–oxy anion interaction in the transition state to form a tetrahedral intermediate depends strongly on the structure of substrate.

3.2.4. Functionalized Crown Ethers

It is well known that crown ethers bind metal ions or ammonium ions selectivity according to the extent of the cooperative ion–dipole interactions and the stereochemical fit. It is also well known that the enantiomer-selective inclusion of α-amino acid ester hydrochlorides by crown ether derivatives having chiral barriers takes place and that a stability difference between the enantiomorphic complexes exists. Lehn and Sirlin[220] found that 18-crown-6 having four L-cysteinyl residues **34** binds dipeptide ester hydrochlorides with a high chiral recognition and undergoes an asymmetric thiolysis by the intramolecular nucleophilic catalysis by the thiol groups. For example, Gly-L-Phe-ONp · HBr was hydrolyzed faster than Gly-D-Phe-ONp · HBr by 50-fold in a CH_2Cl_2(97.9%)/MeOH(2%)/H_2O(0.1%) mixture at pH 7.0, and by 90-fold in a CH_2Cl_2 (95 %)/EtOH(5 %) mixture at pH 7.0.

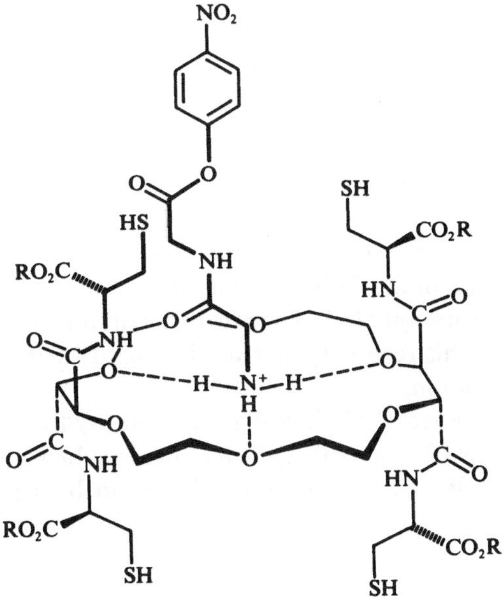

NO_2 HS SH RO_2C NH CO_2R HN O C NH O O C O O H N+ H O H O O C O NH O O C HN O RO_2C CO_2R SH SH

[34]

Chao *et al.*[221] synthesized 18-crown-6 derivatives having (*S*)- or (*R*)-dinaphthyl groups as chiral barriers and introduced a CH_2SH group to the naphthyl groups. By this procedure, enzyme models, with which asymmetric binding sites and nucleophilic catalytic sites are equipped, were designed. Using these enzyme models, the asymmetric thiolysis of many α-amino acid *p*-nitrophenyl ester hydrobromides was investigated. A preliminary report of this investigation was published in 1976 and it was described in detail in another review.[2]

3.3. Intramolecular Catalysis

Orientation of two functional groups is very important in the intramolecular catalysis on the bound substrate. A catalytic group and another group corresponding to substrate or reaction intermediate were covalently linked to one molecule, and the relationship between their orientation and the reactivity was investigated.

Utaka *et al.*[222] synthesized **36** and **37** as models for acetyl α-chymotrypsin and compared their reactivities with the open-chain analogue **35** for the hydrolysis in water at 50°C. The rate constant for the intramolecular general-base-catalysis by the imidazolyl group of **36** was,

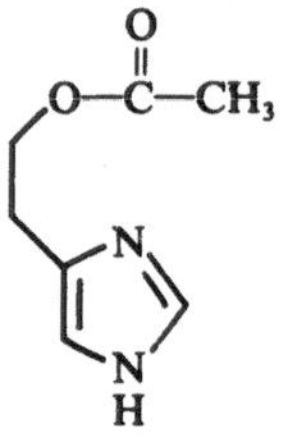

[35] 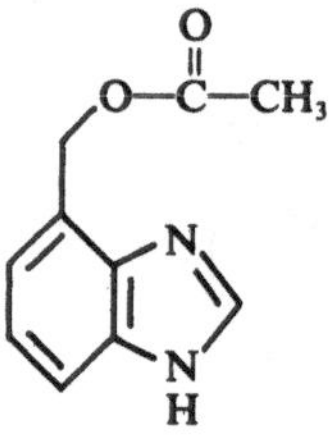 [36] 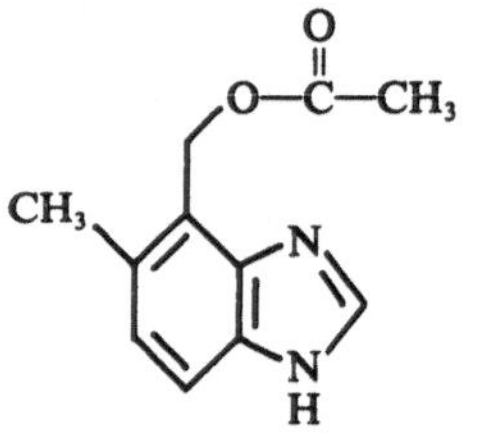 [37]

correcting for the basicity difference and the polar substitution effect, 4.1 times as large as that of **35**. This reactivity increase comes from the freezing out of the internal rotation of **35**. The rate constant for **37** is 1.4 times as large as that for **36**, and this reactivity increase is caused by the restriction of the second internal rotation.

Morris and Page[223,224] investigated the hydrolysis catalyzed by hydroxide ions of *N*-(2-aminoethyl)-6-*endo*-hydroxybicyclo[2.2.1]heptane-2-*endo*-carboxamide **38**. This reaction was found to be a composite type of

HO $CONHCH_2CH_2NH_2$

[38]

reaction where the terminal amino group of **38** is protonated and nonprotonated, and proceeded about 10^9 times as fast as that of a substrate without the hydroxyl group and the ammonium group. This hydrolytic reaction proceeded through a tricyclic lactone intermediate, to which an intramolecular nucleophilic catalysis by hydroxyl group (10^6–10^7-fold acceleration) and an intramolecular general-acid catalysis by ammonium group (150-fold acceleration) contribute. Implication of the present reaction mechanism to that of amide hydrolysis by α-chymotrypsin is discussed.

Roberts and Kanamori[225] found that the hydrolysis of *endo*-5-[4′(5′)-imidazolyl]bicyclo[2.2.1]hepto-*endo*-2-yl *trans*-cinnamate in 42 mol % aqueous dioxane solution was accelerated by 2500 times with the addition of 0.5 mol liter^{-1} sodium benzoate. This acceleration was believed to be due to a carboxylate-assisted process which has been proposed as being responsible for the activities of serine proteinases. In greater detail, the benzoate anion, as if it were Asp-carboxylate of α-chymotrypsin, increases the nucleophilicity of the His-imidazolyl group by hydrogen bonding prior to formation of the transition state, and it stabilizes the imidazolium ion, which is formed in the transition state, by the formation of stronger hydrogen bonding. This reaction mechanism is slightly different from the so-called charge-relay system.

4. Multifunctional and Multiple Polymer Catalysts

It is important when considering the efficiency and the specificity of enzyme reactions, to remember that enzymes are not a single catalyst but a multifunctional or multiple catalyst. In enzyme reactions, more than two catalytic groups can simultaneously or succeedingly attack different sites of a substrate and a multifunctional catalysis occurs. Enzymes are then multifunctional polymer catalysts in which more than two catalytic groups interact intramolecularly with each other and strengthen the activity. One can see a typical example of this in α-chymotrypsin. Points involved in the intramolecular cooperation of more than two catalytic groups along a polymer chain are essentially the same as the points involved in the intramolecular cooperation of a binding group and a catalytic group, which was described in Section 3. Research dealing with multifunctional and multiple polymer catalysts will be described below.

4.1. Bifunctional Catalyses

Kunitake *et al.*[226] synthesized salicylohydroxamic acid **39**, *N*-methyl-salicylohydroxamic acid **40**, and *N*-methylphthalohydroxamic acid **41**, and

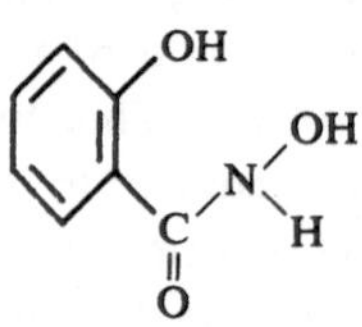

[39] 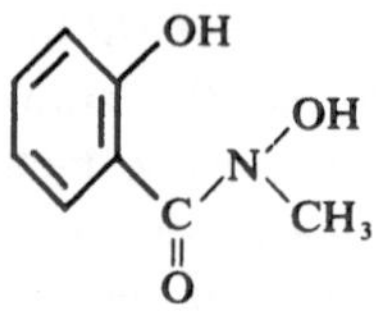 [40] 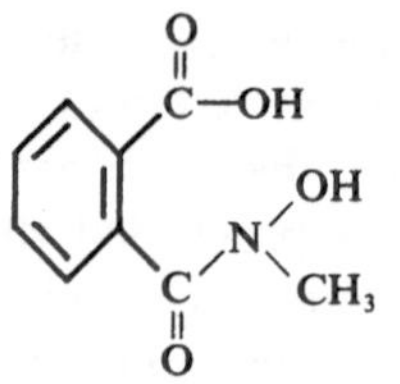 [41]

investigated the acylation reaction of these hydroxamic acids by *p*-nitrophenyl acetate in aqueous ethanol. Both the acidity and nucleophilicity of the hydroxamic acid groups were affected by the nature of orthosubstituent. Specifically, the nondissociated hydroxamic acid group of the **41** monoanion showed a nucleophilic reactivity. Judging from the solvent isotope effect, in which $k_H/k_D = 2.5$, the above situation was caused by the general-base assist by the adjacent carboxylate group, and the acyl-transfer mechanism, as described in reaction (15), can be considered. Hydroxamate anions were nucleophiles in the cases of **39** and **40**.

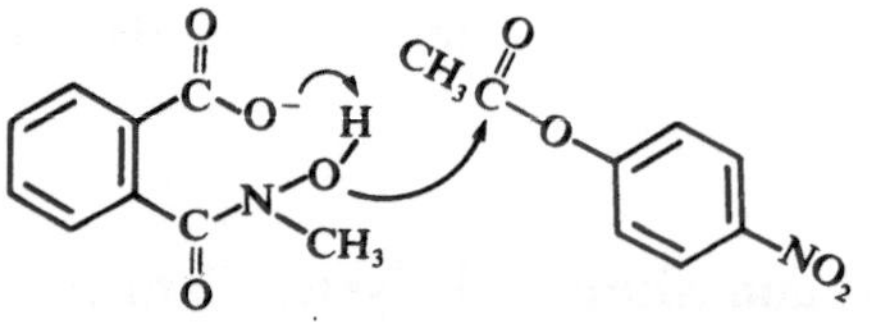

(15)

Hine[227] has investigated the use of bifunctional catalysis for the α-hydrogen exchange reaction of aldehydes or ketones which is involved in decarboxylation of β-keto acids or aldol condensations. Further, Hine and Zeigler[228] investigated the dedeuterization reaction of 3-pentanone-2,2,4,4-d_4 catalyzed by diamines which carry a primary and a tertiary amine in one molecule. The reaction mechanism is represented by Eq. (16). It is particu-

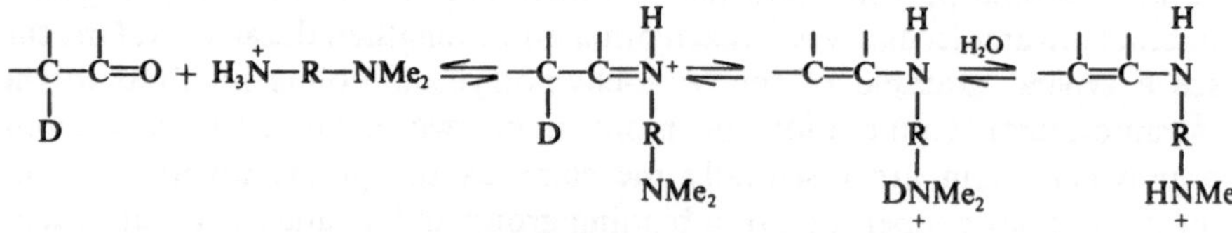

$$-\underset{\mathrm{D}}{\overset{|}{\mathrm{C}}}-\overset{|}{\mathrm{C}}=\mathrm{O} + \mathrm{H_3\overset{+}{N}-R-NMe_2} \rightleftharpoons -\underset{\mathrm{D}}{\overset{|}{\mathrm{C}}}-\overset{|}{\mathrm{C}}=\overset{\mathrm{H}}{\underset{\underset{\mathrm{NMe_2}}{\mathrm{R}}}{\mathrm{N^+}}} \rightleftharpoons -\overset{|}{\mathrm{C}}=\overset{|}{\mathrm{C}}-\overset{\mathrm{H}}{\underset{\underset{\mathrm{D\overset{}{N}Me_2^+}}{\mathrm{R}}}{\mathrm{N}}} \overset{\mathrm{H_2O}}{\rightleftharpoons} -\overset{|}{\mathrm{C}}=\overset{|}{\mathrm{C}}-\overset{\mathrm{H}}{\underset{\underset{\mathrm{H\overset{}{N}Me_2^+}}{\mathrm{R}}}{\mathrm{N}}} \qquad (16)$$

larly interesting that the monoprotonated form of the chiral catalyst **42** is a very effective bifunctional catalyst, and further, it acts as a stereoselective catalyst. Of the two C-2 deuteriums the pro-*S*-deuterium was abstracted 70 times as fast as the pro-*R* deuterium. Therefore, the partially dedeuterized ketone containing more than 80% of the starting dideuterio species was optically active.

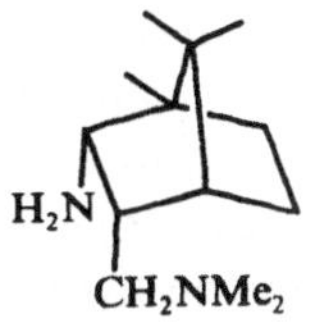

H_2N CH_2NMe_2

[42]

Tachibana[229] investigated the dehydrobromination reaction of 7-bromochloesterol. The 5,7-diene and the 4,6-diene were the main products in the reactions catalyzed by 2- and 3-(hydroxyalkyl)pyridine, respectively. The isomer ratio in the diene product depended on the structure of hydroxyalkylpyridine. The reaction mechanisms are explained in Figure 10.

Kunitake and Sakamoto[230] investigated the hydrolysis of *p*-nitrophenyl acetate using as a catalyst the polysoap-bound bifunctional catalyst, *N*-lauryl-4-imidazolecarbohydroxamic acid **43**. The polysoaps employed were polyvinylimidazole quaternized by dodecyl and ethyl groups and

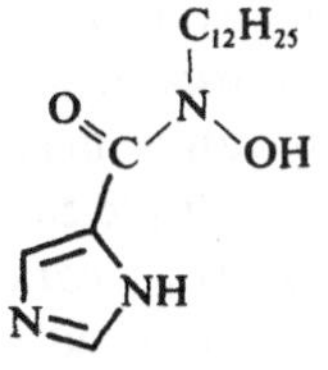

$C_{12}H_{25}$ O C N OH NH N

[43]

RO H N C OH_2 Br H 5,7-diene

RO H N CH_2 O H Br 4,6-diene

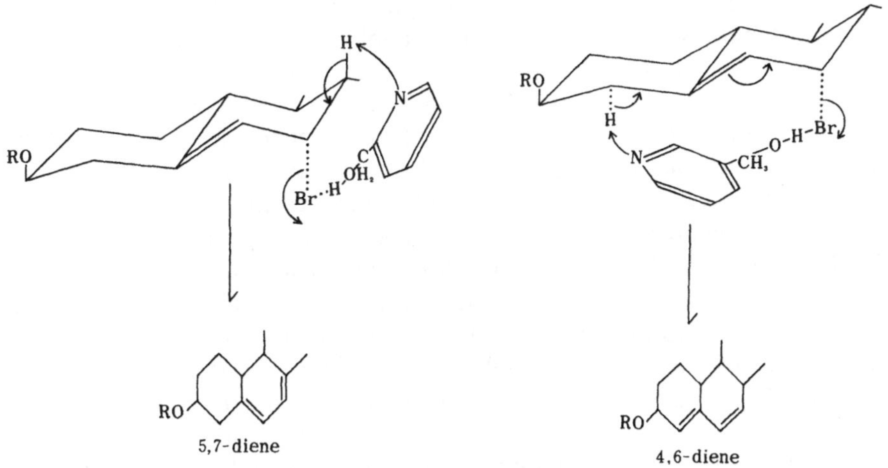

Figure 10. Bifunctional catalysis by 2- or 3-hydroxymethylpyridine for the stereospecific dehydrobromination of esters of 3-β-hydroxy-7-bromo-Δ^5-steroids. (Reproduced from scheme on p. 3085 of Reference 229 by permission of the copyright owner.)

polyiminoethylene partly quaternized by methyl and higher alkyl (C_8, C_{12}, C_{18}) groups. In the reaction the acyl transfer occurs at first from the substrate to the dissociated hydroxamate group of **43**, and subsequently the decomposition of the intermediate by the adjacent imidazolyl group takes place. Since the reaction takes place in the hydrophobic domain of polysoap, the acceleration by hydrophobic environment, as described in Section 2.2.3, was observed. However, the acceleration was greater by cationic CTAB micelle than by the polysoaps. The efficiency of bifunctional catalysts was about 1000 times as large as imidazole.

4.2. Polymeric Multiple Catalysts

When Overberger *et al.* hydrolyzed carboxylic acid active esters using as catalysts poly[4(5)-vinylimidazole] and poly[5(6)-vinylbenzimidazole], a very high catalytic activity was observed in the pH region above 7. This phenomenon was explained in terms of increased catalytic activity due to the intramolecular interaction of neutral or anionic imidazolyl groups, that is, multiple catalysis.[(2)] Since then, the catalysis by copolymers of vinylimidazole and other functional vinyl compounds has been extensively investigated, and the investigations on polymeric multiple catalysis have become very popular. However, recently, in the hydrolysis catalyzed by polymeric imidazoles, Overberger's results were not reproduced or the explanations proposed by Overberger were severely criticized.[(231–234)]

Okamoto[(231)] synthesized poly(N^{α}-methacryloyl-L-histidine and its low-molecular-weight model compound, N^{α}-pivaloyl-L-histidine, and investigated the solvolyses in aqueous ethanol of active esters of α-amino acid and acetic acid using the above imidazole derivative as a catalyst. In the buffer solution at pH's 7.20, 8.25, and 9.31, the polymer catalyst was less effective than the low-molecular-weight catalyst. Furthermore, in none of the solvolytic reactions, was asymmetric selectivity observed.

Mirejovsky[(232)] measured the solvolytic reaction of *p*-nitrophenyl caproate in the presence of polyiminoethylene carrying various amounts of imidazole groups. With increasing ratio of imidazole groups in the polymer catalyst, the second-order rate constant for the reaction decreased and the numbers of imidazolyl groups involved in one catalytic site increased. Such inefficiency of the polymer catalyst was ascribed to the association of imidazolyl groups in the hydrophobic domain of the polymer. Only imidazolyl groups exposed to the surface of polymer coil were available for the catalytic site, and the polar domain near the imidazolyl groups increased the efficiency of hydrolysis. Therefore, the multiple catalysis of imidazolyl groups occurring in the polymer coil was concluded to be improbable.

Lege and Deyrup[233] hydrolyzed *p*-nitrophenyl esters of acetic and caproic acid using partially dodecylated polyiminoethylene (PEI-D-NH_2). The reaction was catalyzed by the nucleophilic primary amino groups in the side chains, and the second-order rate constant k was dependent on pH. The pK_a value of PEI-D-$NH_2 \cdot HCl$ was also dependent on pH. Therefore, the Brønsted relation held between log k and pK_a. The pH dependence in the hydrolysis of *p*-nitrophenyl acetate may be explained by the electrostatic effect upon the nucleophilicity of amines. The same authors[234] obtained the same conclusion regarding the catalysis by the polymer catalysts PEI-D-Im, PEI-D-Pyr, and PEI-D-APyr, which represent imidazole, pyridine, and 2-aminopyridine linked to partially dodecylated polyiminoethylene, respectively. They analyzed the hydrolyses of active esters of carboxylic acids by polymeric bases, which have been reported in literature, and found that the same explanation is applicable. Therefore, they concluded that in these catalyses the cooperation of catalytic groups need not be taken into account.

Nishi and Nakajima[235] hydrolyzed *p*-nitrophenyl acetate by Boc-Asp-βAla-Gly-Ser-βAla-Gly-His-βAla-Gly-OEt, Boc-Asp-βAla-Gly-Ser-βAla-Gly-His-βAla-Gly-OH, cyclo(Asp-βAla-Gly-Ser-βAla-Gly-His-βAla-Gly), and poly(Asp-βAla-Gly-Ser-βAla-Gly-His-βAla-Gly). In any of the peptide catalysts, the His residue constitutes the catalytic site. The detailed mechanism of the catalysis will be hopefully reported soon.

Iwabuchi *et al.*[236] synthesized a homopolymer of 3,4-dihydroxybenzyl methacrylate **44** and its copolymers. The titration curves for these polymers were higher by 20–30% than those for the corresponding low-molecular-weight compounds. The difference could be explained by the interaction of oxidized polymers and electrode.

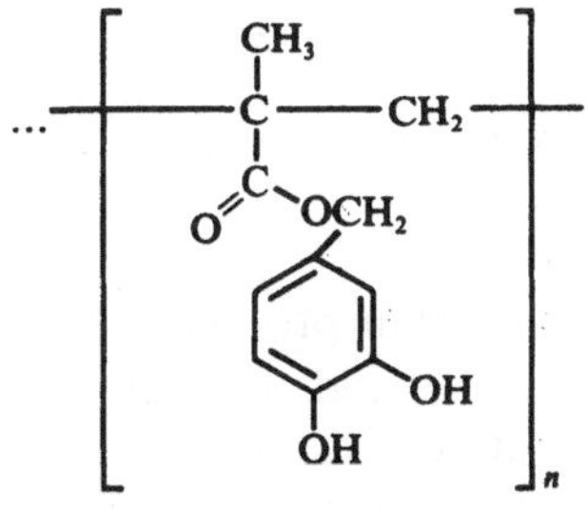

[44]

4.3. Cyclic Multiple Catalysts

Since it became possible to introduce one or two substituents into the hydroxyl groups of cyclodextrins regioselectively, many functionalized cyclodextrins have been synthesized. The functionalized cyclodextrins may be called "synthetic enzymes", in which the substrate-selective binding ability of cyclodextrins and the catalytic actions of newly introduced functional groups are combined.

A cationically charged cyclodextrin mono-(6-trimethylammonio-6-deoxy)-β-cyclodextrin hydrogencarbonate was synthesized. The hydrolyses of anionically charged esters of *o*-, *m*-, and *p*-acetoxybenzoic acid by this functionalized cyclodextrin were accompanied by electrostatic interactions as well as hydrophobic interactions, and differed greatly from reactions catalyzed by β-cyclodextrin.[237]

Eiki and Tagaki[238] found that β-cyclodextrin phosphate was an excellent catalyst for the oxidation by iodine of benzylmethylphosphate in aqueous solution.

Breslow *et al.*[239] synthesized β-cyclodextrin bisimidazole **45** and decomposed a cyclic phosphate of 4-*t*-butylcatechol with **45**. They observed a bell-shaped pH–rate profile and pointed out the cooperative catalysis by a basic imidazolyl group and an acidic imidazolyl group. In this regard, **45** mimics ribonuclease. Furthermore, the reaction catalyzed by **45** was very selective. Compound **45** cleaved only the P-O(1) bond, giving 4-*t*-butyl-

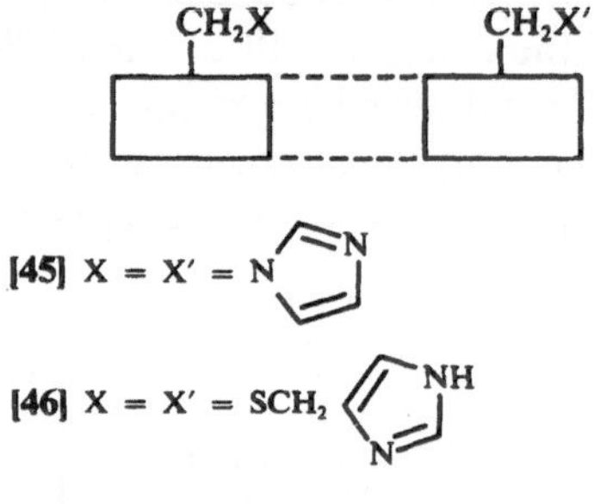

CH_2X CH_2X'

[**45**] X = X' = N(imidazolyl)

[**46**] X = X' = SCH_2(imidazolyl, NH)

catechol-2-phosphate as the main product. Therefore, the reaction scheme shown in Figure 11 was considered. Breslow *et al.*[240] synthesized another cyclodextrin bisimidazole **46** and decomposed the same substrate with **46**. Nearly the same kinetic behaviours were observed as in the case catalyzed by **45**. The rate of acceleration was less marked in the case of flexible **46**. By **46**

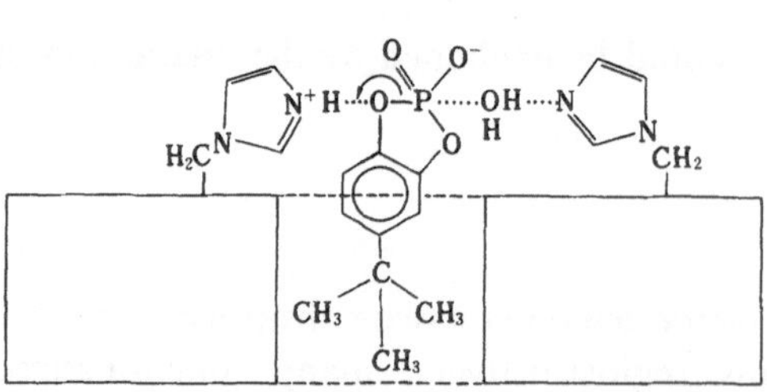

Figure 11. The specific cleavage of cyclic phosphate of 4-*tert*-butylcatechol bound in the cavity of catalyst **45** in the half-protonated state. (Reproduced from Figure 2 of Reference 240 by permission of the copyright owner).

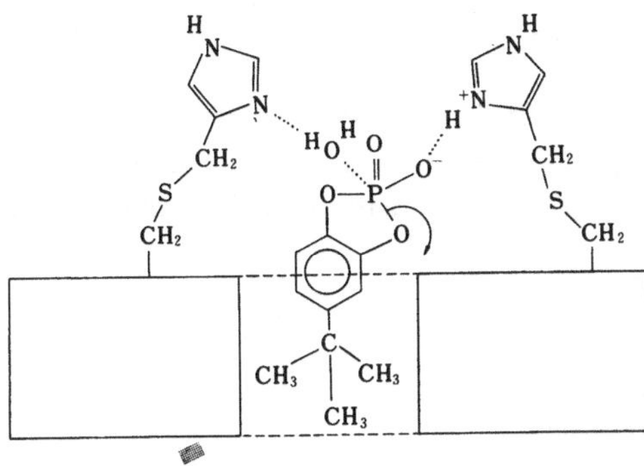

Figure 12. The specific cleavage of cyclic phosphate of 4-*tert*-butylcatechol bound in the cavity of catalyst **46** in the half-protonated state. (Reproduced from Figure 3 of Reference 240 by permission of the copyright owner.)

the P-O(2) bond was cleaved selectively. The reaction mechanism could be explained as depicted in Figure 12.

As was described in Section 3.2.2. with regard to References 198 and 199, the rate of acceleration in the hydrolysis of active esters by β-cyclodextrin has been found to reach as high as 10^6-fold by freezing out the rotational freedom of the substrate. Breslow *et al.*[241] supplied β-cyclodextrin with an intrusive floor by the *N*-methyl- or *N*-ethylamidation of the 6-hydroxyl groups, and found an enormous acceleration of 10^6–10^7-fold, which exceeds the enzymatic acceleration, in the reaction of substrates having hindered internal rotation catalyzed by the functionalized β-cyclodextrin.

Carbonic anhydrase possesses a zinc ion in the active site, which is sieged by three imidazolyl groups. Carbon dioxide is bound to the active site near the zinc ion by hydrophobic interaction. Water coordinates with the zinc ion to be activated for the attack on CO_2. In this process, the general-base catalysis by the imidazolyl group is involved. In order to obtain a model for carbonic anhydrase from β-cyclodextrin, Tabushi *et al.*[242] synthesized **47** and **48**. After a zinc ion was coordinated with each cyclodextrin derivative, the hydration reaction of CO_2 was performed. The reaction rate constants (liter

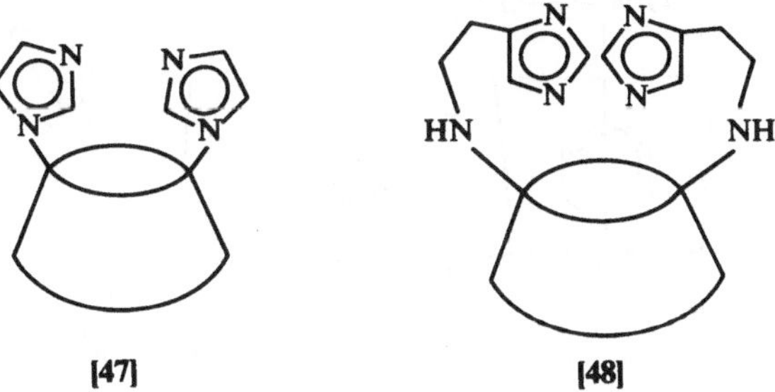

$mol^{-1} s^{-1}$) of the complexes were 16.2 for **47**–Zn^{2+} and 166 for **48**–Zn^{2+}. These values are far smaller than those reported for the enzyme catalysis, but much larger than those reported for reactions catalyzed by the corresponding

bases with the zinc ion. This catalytic system is very useful for designing synthetic enzymes.

Fujita *et al.*[243] *S*-methylated one of 6-hydroxyl groups of β-cyclodextrin and synthesized a capped cyclodextrin, **49**, and investigated the hydrolyses of substituted phenyl acetates. By *S*-methylation, neither the

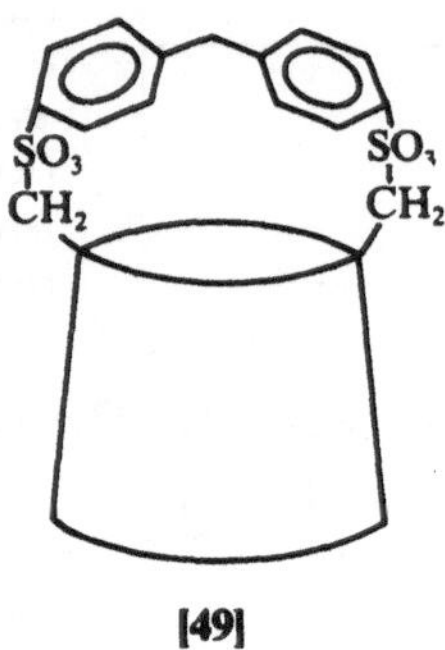

[49]

dissociation constant of the complex, K_d, nor the rate constant for the conversion of the complex, k_c, was affected. On the other hand, by capping, K_d decreased by 300-fold at the maximum and k_c decreased by 10-fold at the maximum. In particular, K_d for the complex of para-substituted esters decreased so drastically that the usual meta-selectivity of β-cyclodextrin changed into para-selectivity.

Ueno *et al.*[244] synthesized a β-cyclodextrin capped with a *cis* or *trans* azobenzene group. The hydrolysis of *p*-nitrophenyl acetate by the azobenezene-capped cyclodextrin was accelerated by irradiation of 320–390 nm light. The kinetic parameters of the reaction are shown in Table 14. The acceleration is caused by a strong substrate binding into the deep cavity with a *cis*-azobenzene bridge **51** after the photoirradiation. The design

hν / Heat

trans-**[50]**
"shallow cavity"

cis-**[51]**
"deep cavity"

Table 14. Kinetic Parameters[a]

Catalyst	$10^3k_2/s^{-1}$	10^2K_m	$10^2(k_2/K_m)$
β-Cyclodextrin[b]	1.09	0.73	15
Trans-(**50**)	1.62	2.34	6.9
Cis(**51**)	0.70	0.19	37

[a] Reproduced from Table 2 of Reference 244 by permission of the copyright owner.
[b] Reported values k_2, 1.15×10^{-3} s^{-1}; K_m, 0.83×10^{-2}; k_2/K_m, 14×10^{-2} [A. Harada, M. Furue, and S. Nozakura, *Macromolecules* **9**, 705 (1976)].

of an enzyme-mimetic catalyst having a photosensitive group would be a promising step in the development of synthetic enzymes.

For cyclic multiple catalysts other than functionalized cyclodextrins, a Cu^{2+} complex of 12,12′-bis-{*N*-[4(5)-imidazolylethyl]amide}-5-oxo-[10,10]-paracyclophane, **52**, which was synthesized by Murakami *et al.*[(245)] will be introduced. The Cu^{2+}–**52** (1:1) complex was an efficient catalyst for the

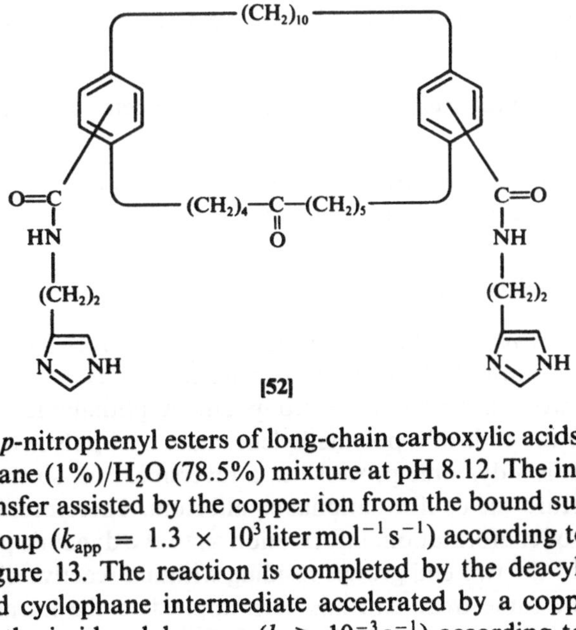

$(CH_2)_{10}$ … $O{=}C$, HN, $(CH_2)_2$, N NH … $(CH_2)_4{-}C(=O){-}(CH_2)_5$ … $C{=}O$, NH, $(CH_2)_2$, N NH

[52]

hydrolysis of *p*-nitrophenyl esters of long-chain carboxylic acids in an EtOH (20.5%)/dioxane (1%)/H_2O (78.5%) mixture at pH 8.12. The initial reaction is an acyl transfer assisted by the copper ion from the bound substrate to an imidazolyl group ($k_{app} = 1.3 \times 10^3$ liter mol^{-1} s^{-1}) according to the scheme shown in Figure 13. The reaction is completed by the deacylation of the monoacylated cyclophane intermediate accelerated by a copper ion coordinated with the imidazolyl group ($k \geqslant 10^{-3}$ s^{-1}) according to the scheme shown in Figure 14.

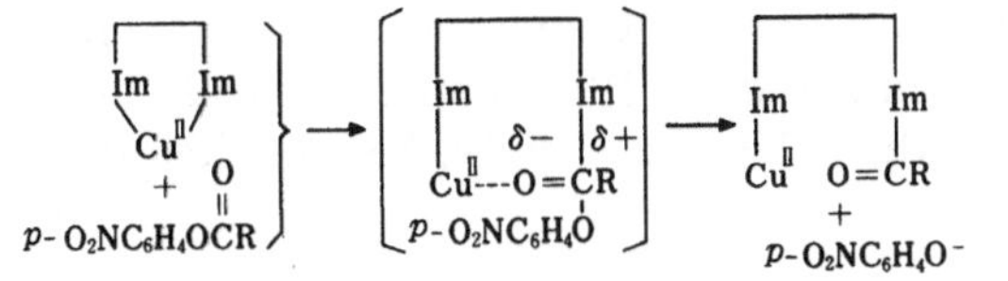

Im Im Cu^{II} + O=CR–O–$C_6H_4NO_2$-p (written: p-$O_2NC_6H_4OCR$) → [Im Im, δ−, δ+, Cu^{II}---O=CR, p-$O_2NC_6H_4O$] → Im Im, Cu^{II} O=CR + p-$O_2NC_6H_4O^-$

Figure 13. Mechanism of copper-ion-assisted acylation of catalyst **52** by carboxylic acid *p*-nitrophenyl ester. (Reproduced from Scheme 2 of Reference 245 by permission of the copyright owner).

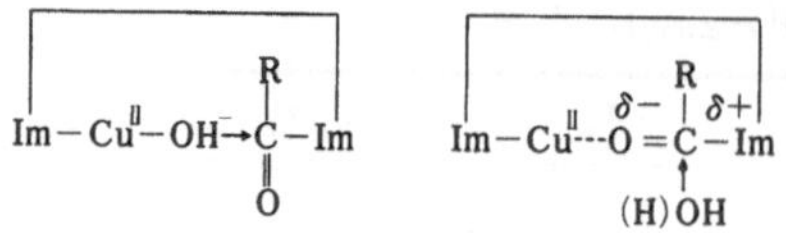

Im−Cu^{II}−OH^{-}→C(R)(=O)−Im   Im−Cu^{II}---O=C(R)(←(H)OH)−Im ($\delta-$, $\delta+$)

Figure 14. Mechanism of copper-ion-assisted deacylation of acylated **52**. [Reproduced from (4) and (5) of Reference 245 by permission of the copyright owner].

5. Intrachain Reactions Proceeding on a Polymer Chain

As is clear from the investigations described in Sections 3 and 4, the participation of the nature of the polymer in the mechanism of enzyme reactions can be treated quantitatively by investigating effects of the nature of the polymer chain on the intramolecular reactions or interactions of a pair of functional groups at both ends of a polymer chain. Furthermore, using these reactions as probes, the chain dynamics of polymers can be studied, which are otherwise very difficult to investigate. Here, relevant literatures will only be listed without describing the details of the investigations.

5.1. Intrachain Reactions among Functional Groups Distributed along a Polymer Chain

On this subject References 246–248 have been published.

5.2. Intrachain Reactions of Pairs of Functional Groups Attached to the Chain Ends–Statistical Treatments

5.2.1. Hydrocarbon Chains

Sisido's work has been published in References 249 and 250. The intrachain reaction dealt with in Reference 250 is an electron-transfer reaction between a phthalimide group and a phthalimide anion radical connected to the ends of a polymethylene chain or a polyoxyethylene chain. This reaction is a dynamic process but a reversible one, so that the reaction does not perturb the conformational equilibrium. Therefore, by using static calculations of conformation, information as to the dynamic process, that is, the lifetime of the two end groups residing within a reactive distance, can be obtained. The same treatment has been applied to the intrachain electron transfer between terminal naphthyl groups.[(251)]

Winnik's work has been published in References 252–255. A review article on the present topic has been written by Winnik.[(256)]

Illuminati's work has been published in References 257–259. A review article on the present topic has been written by Illuminati and Mandolini.[(260)] The work of Nishijima and others has been published in References 261–264.

5.2.2. Polypeptide Chains

Sisido's work has been published in References 265–269. In particular, Reference 266 concerns a photocontrol of polypeptide cyclization and is very interesting. Luisi's work has been published in References 270 and 271. Guillard and Englert[(272)] have also published on this topic.

5.2.3. Polyester Chains

Suter and Mutter[(273)] made a calculation of macrocyclization equilibria for poly(ethylene terephthalate).

5.3. Intrachain Reactions of Pairs of Functional Groups Attached to the Chain Ends—Dynamic Treatments

5.3.1. Hydrocarbon Chains

Winnik's work has been published in References 274–277. A few years ago Shimada and Szwarc investigated the electron-transfer reaction between naphthyl groups or phthalimide groups attached to the ends of a polymethylene chain, and announced that information was obtained on the dynamics of the end-to-end cyclization of a hydrocarbon chain. Winnik[(276)] criticized this announcement suggesting that Shimada and Szwarc measured only a static quantity in their experiment because the intramolecular electron exchange is a conformation-controlled process. However, in the author's opinion, Winnik seems to ignore the difference between reversible and irreversible processes. In an irreversible intramolecular diffusion-controlled reaction, the process for distant terminal groups to approach each other by diffusion is the major contribution. On the other hand, in a reversible intramolecular rate-determining reaction, the collision frequency between relatively close end groups is estimated. Under such conditions the conformational equilibrium is maintained, so that a static treatment can afford information concerning the dynamic quantity.

Nishijima's work has been published in References, 278–280. Other work has been published in References 281–284.

5.3.2. Polypeptide Chains

Sisido *et al.*[(285)] investigated the intrachain energy transfer between terminal groups of a polysarcosine chain. Cerf[(286)] investigated the Brownian motion of terminal groups of an oligopeptide chain. Ibemesi *et al.*[(287)] investigated the intrachain excimer formation along a polyamide chain containing naphthyl groups in the chain.

5.3.3. Polyether Chains

Todesco *et al.*[288] investigated the intrachain excimer formation in bis-(α-naphthylmethyl)ether.

Symbols

- APyr — 2-Aminopyridine
- Boc — Tertiary-butyloxycarbonyl group
- *t*-Bu-QPVP — Poly(-4-vinylpyridine) quaternized with *t*-butyl chloride
- C_2PVP — EtBr-quaternized poly(4-vinylpyridine)
- CTAB — Cetyltrimethylammonium bromide
- DMOG — *N*,*N*-dimethyl-*N*-octadecylglycine
- f — Activity coefficient
- FAD — Flavin adenine dicnucleotide
- FeD — Fe(tetpy)$(OH)_2^+$/poly(D-glutamic acid)
- FeL — Fe(tetpy)$(OH)_2^+$/poly(L-glutamic acid)
- $\Delta G^‡$ — Activated free energy
- $\Delta H^‡$ — Activated enthalpy of reaction
- Im — Imidazole
- ImAm — Imidazolecarboxylic amide
- k — Rate constant
- k/k^* — Ratio of rate constants in the presence or absence of additive
- k_L/k_D — Enantiomer selectivity; asymmetric selectivity
- MSF — Methyl *p*-(dimethylamino)benzene sulfonate
- NADH — *N*′-alkyl-1,4-dihydronicotinamide
- P — Pressure
- PEI — Polyethyleneimine (Polyiminoethylene)
- PEI-D-NH_2 — Partially dodecylated PEI
- pmen — *N*,*N*′-bis(2-pyridylmethyl)ethylene-diamine
- PNPC — *p*-Nitrophenyl caproate
- PNPA — *p*-Nitrophenyl acetate
- PNPP — *p*-Nitrophenyl palmitate
- POOA — Polyoxyethylene oleyl ether
- PVP — Poly(4-vinylpyridine cross-linked)
- Pyr — Pyridine
- $\Delta S^‡$ — Activated entropy
- SDS — Sodium dodecylsulfate
- tetpy — Tetrapyridyl isomers
- $\Delta V^‡$ — Activated volume
- Z — Benzyloxycarbonyl group

References

1. Y. Imanishi, in: Zyūgō Hannōron (T. Saegusa *et al.*, ed.), Vol. 11, pp. 57–257, Kagaku Dojin, Kyoto (1972).
2. Y. Imanishi, *J. Polym. Sci., Macromol. Rev.* **14**, 1 (1979).
3. F. H. Westheimer, *Chemtech 1980*, 748.
4. I. M. Klotz, *Adv. Chem. Phys.* **39**, 109 (1978).
5. G. Manecke and W. Storck, *Angew. Chem.* **90**, 691 (1978).
6. H. C. Kiefer, W. I. Congdon, I. C. Scarpa, and I. M. Klotz, *Proc. Natl. Acad. Sci. U.S.A.* **69**, 2155 (1972).
7. T. Kunitake and T. Sakamoto, *Bull. Chem. Soc. Jpn.* **52**, 2402 (1979).
8. M. A. Hierl, E. P. Gamson, and I. M. Klotz, *J. Am. Chem. Soc.* **101**, 6020 (1979).
9. Y. Taniguchi, N. Sugiyama, and K. Suzuki, *J. Phys. Chem.* **82**, 1231 (1978).
10. M. Nango, H. Kozuka, Y. Kimura, N. Kuroki, Y. Ihara, and I. M. Klotz, *J. Polym. Sci., Polym. Lett. Ed.* **18**, 647 (1980).
11. I. Cho and J.-S. Shin, *Makromol. Chem.* **183**, 2041 (1982).
12. Y. Osada and T. Chiba, *Makromol. Chem.* **180**, 1617 (1979).
13. S. Hayama, M. Takeishi, K. Takahashi, and S. Ninno, *Makromol. Chem.* **181**, 1889 (1980).
14. K. Arai and Y. Ogiwara, *Makromol. Chem.* **177**, 367 (1976).
15. K. Arai and Y. Ogiwara, *Polym. J.* **10**, 99 (1978).
16. K. Arai and Y. Ogiwara, *J. Polym. Sci., Polym. Chem. Ed.* **17**, 4041 (1979).
17. K. Arai and Y. Ogiwara, *J. Polym. Sci., Polym. Chem. Ed.* **18**, 1643 (1980).
18. K. Arai, Y. Ogiwara, and C. Kuwabara, *J. Appl. Polym. Sci.* **25**, 2935 (1980).
19. K. Arai, Y. Ogiwara, and C. Kuwabara, *J. Polym. Sci., Polym. Chem. Ed.* **19**, 1885 (1981).
20. M. Kida and H. Nakano, *Makromol. Chem.* **179**, 2355 (1978).
21. Z. Prokop and K. Setínek, *Collect. Czech. Chem. Commun.* **43**, 1990 (1978).
22. T. Yamashita, H. Yasueda, N. Nakatani, and N. Nakamura, *Bull. Chem. Soc. Jpn.* **51**, 1183 (1978).
23. T. Yamashita, H. Yasueda, and N. Nakamura, *Bull. Chem. Soc. Jpn.* **52**, 2165 (1979).
24. N. Kobayashi and K. Iwai, *J. Am. Chem. Soc.* **100**, 7071 (1978).
25. N. Kobayashi and K. Iwai, *Polym. J.* **13**, 263 (1981).
26. J. H. Liu, K. Kondo, and K. Takemoto, *Angew. Makromol. Chem.* **104**, 209 (1982).
27. N. Yashiro, K. Asakawa, and K. Tsuboyama, *Kobunshi Ronbunshu* **39**, 549 (1982).
28. N. Yamazaki, A. Hirao, and S. Nakahama, *Polym. J.* **7**, 402 (1975).
29. N. Yamazaki, A. Hirao, and S. Nakahama, *J. Polym. Sci., Polym. Chem. Ed.* **14**, 1229 (1976).
30. A. Hirao, S. Nakahama, M. Takahashi, and N. Yamazaki, *Makromol. Chem.* **179**, 915 (1978).
31. A. Hirao, S. Nakahama, M. Takahashi, and N. Yamazaki, *Makromol. Chem.* **179**, 1735 (1978).
32. M. Matsumoto and K. Orihara, *Kobunshi Ronbunshu* **38**, 457 (1981).
33. K. Orihara, *Makromol. Chem., Rapid Commun.* **3**, 7 (1982).
34. K. Orihara, S. Higuchi, and M. Matsumoto, *Kobunshi Ronbunshu* **38**, 769 (1981).
35. K. Orihara and M. Matsumoto, *Kobunshi Ronbunshu* **38**, 775 (1981).
36. K. Orihara, T. Onodera, and M. Matsumoto, *Kobunshi Ronbunshu* **39**, 141 (1982).
37. K. Orihara, T. Onodera, and M. Matsumoto, *Kobunshi Ronbunshu* **39**, 373 (1982).
38. J. Smid, S. C. Shah, A. J. Varma, and L. Wong, *J. Polym. Sci., Polym. Symp.* **64**, 267 (1978).
39. G. D. Hartman and S. E. Biffar, *J. Org. Chem.* **42**, 1468 (1979).
40. D. H. Hunter, M. Hamity, V. Patel, and R. A. Reery, *Can. J. Chem.* **56**, 104 (1978).

41. S. Shinkai, S. Hirakawa, M. Shinomura, and T. Kunitake, *J. Org. Chem.* **46**, 868 (1981).
42. J. Koga, K.-M. Chen, M. Nawata, Y. Yamazaki, and N. Kuroki, *Chem. Lett. 1982*, 663.
43. J. Koga, K.-M. Chen, Y.-J. Lim, and N. Kuroki, *Kobunshi Ronbunshu* **39**, 675 (1982).
44. B. K. McQuoid, A. R. Goldhammer, and E. H. Cordes, *J. Org. Chem.* **41**, 2020 (1976).
45. T. Kunitake, Y. Okahata, S. Tanamachi, and R. Ando, *Bull. Chem. Soc. Jpn.* **52**, 1967 (1979).
46. T. Kunitake and T. Sakamoto, *Bull. Chem. Soc. Jpn.* **52**, 2624 (1979).
47. Y. Murakami, A. Nakano, and K. Matsumoto, *Bull. Chem. Soc. Jpn.* **52**, 2996 (1979).
48. Y. Murakami, A. Nakano, K. Matsumoto, and K. Iwamoto, *Bull. Chem. Soc. Jpn.* **52**, 3573 (1979).
49. R. H. Weatherhead, K. A. Stacey, and A. Williams, *J. Chem. Soc., Chem. Commun. 1979*, 598.
50. K. A. Stacey, R. H. Weatherhead, and A. Williams, *Makromol. Chem.* **181**, 2517 (1980).
51. R. H. Weatherhead, K. A. Stacey, and A. Williams, *Makromol. Chem.* **181**, 2529 (1980).
52. M. G. Harun and A. Williams, *Polymer* **22**, 946 (1981).
53. M. Komiyama and H. Hirai, *Makromol. Chem., Rapid Commun.* **2**, 177 (1981).
54. M. Komiyama and H. Hirai, *Makromol. Chem., Rapid Commun.* **2**, 611 (1981).
55. M. Komiyama and H. Hirai, *Makromol. Chem., Rapid Commun.* **2**, 707 (1981).
56. M. Komiyama and H. Hirai, *Makromol. Chem., Rapid Commun.* **2**, 715 (1981).
57. M. Komiyama and H. Hirai, *Bull. Chem. Soc. Jpn.* **54**, 2053 (1981).
58. M. Komiyama and H. Hirai, *Makromol. Chem., Rapid Commun.* **2**, 733 (1981).
59. M. Komiyama and H. Hirai, *Makromol. Chem., Rapid Commun.* **2**, 757 (1981).
60. M. Komiyama and H. Hirai, *Makromol. Chem., Rapid Commun.* **2**, 759 (1981).
61. K. Uekama, T. Irie, and F. Hirayama, *Chem. Lett. 1978*, 1109.
62. K. Yamada, K. Shigehiro, T. Tomizawa, and H. Iida, *Bull. Chem. Soc. Jpn.* **51**, 3302 (1978).
63. I. Tabushi, K. Yamamura, K. Fujita, and H. Kawakubo, *J. Am. Chem. Soc.* **101**, 1019 (1979).
64. N. Yoshida and M. Fujimoto, *Chem. Lett. 1980*, 231.
65. M. Kodaka, *Kobunshi Ronbunshu* **38**, 447 (1981).
66. K. Uekama, C.-L. Lin, F. Hirayama, M. Odagiri, A. Takadate, and S. Goya, *Chem. Lett. 1981*, 563.
67. C. Z. Smith and J. H. P. Utley, *J. Chem. Soc., Chem. Commun. 1981*, 792.
68. K. Fukunishi, H. Kazumura, H. Yamanaka, M. Nomura, and S. Kojo, *J. Chem. Soc., Chem. Commun. 1982*, 799.
69. N. Ise, *J. Polym. Sci., Polym. Symp.* **62**, 205 (1978).
70. N. Ise and T. Okubo, *Macromolecules* **11**, 439 (1978).
71. K. Mita, T. Okubo, and N. Ise, *J. Chem. Soc., Faraday Trans. 1* **72**, 1033 (1976).
72. M. Shikata, S. Kim, K. Mita, N. Ise, and, in part, S. Kunugi, *Proc. R. Soc. London, Ser. A* **351**, 233 (1976).
73. R. Fernandez–Prini, *J. Chem. Soc., Faraday Trans. 1* **74**, 2460 (1978).
74. M. Ishikawa, *J. Phys. Chem.* **83**, 1576 (1979).
75. K. Yamashita, H. Kitano, and N. Ise, *Macromolecules* **12**, 241 (1979).
76. T. Ishiwatari, T. Okubo, and N. Ise, *Macromolecules* **13**, 53 (1980).
77. M. Sugimura, T. Okubo, and N. Ise, *Macromolecules* **14**, 124 (1980).
78. N. Ise, T. Maruno, and T. Okubo, *Proc. R. Soc. London, Ser. A* **370**, 485 (1980).
79. T. Okubo, T. Maruno, and N. Ise, *Proc. R. Soc. London, Ser. A* **370**, 501 (1980).
80. A. Enokida, T. Okubo, and N. Ise, *Macromolecules* **13**, 49 (1980).
81. N. Ise, T. Okubo, and Y. Yamamura, *J. Phys. Chem.* **86**, 1694 (1982).
82. N. Ise, *Makromol. Chem., Suppl.* **5**, 107 (1981).
83. T. Ishiwatari, T. Okubo, and N. Ise, *J. Polym. Sci., Polym. Chem. Ed.* **18**, 1807 (1980).
84. T. Ishiwatari, T. Okubo, and N. Ise, *J. Polym. Sci., Polym. Chem. Ed.* **18**, 1815 (1980).

85. Y. Nishiyama, I. Moroe, H. Kitano, T. Okubo, and N. Ise, *J. Polym. Sci., Polym. Chem. Ed.* **18**, 3289 (1980).
86. T. Ishiwatari, T. Maruno, M. Okubo, T. Okubo, and N. Ise, *J. Phys. Chem.* **85**, 47 (1981).
87. N. R. Povlova, Yu. E. Kirsh, and V. A. Kabanov, *Vysokomol. Soed.* **21**, 2062 (1979).
88. Y. Morishima, T. Tanaka, Y. Itoh, and S. Nozakura, *Polym. J.* **14**, 861 (1982).
89. N. J. Turro, T. Okubo, C.-J. Chung, J. Emert, and R. Catena, *J. Am. Chem. Soc.* **104**, 4799 (1982).
90. H. Nishikawa, M. Kasai, E. Terada, and E. Tsuchida, *Bull. Chem. Soc. Jpn.* **50**, 3419 (1977).
91. O.-K. Kim, *J. Polym. Sci., Polym. Chem. Ed.* **19**, 287 (1981).
92. V. E. R. Nieves, E. J. Ribaldo, R. Baroud, and F. H. Quina, *J. Polym. Sci., Polym. Lett. Ed.* **20**, 433 (1982).
93. R. A. Moss, Y.-S. Lee, and T. J. Lukas, *J. Am. Chem. Soc.* **101**, 2499 (1979).
94. R. A. Moss, Y.-S. Lee, and K. W. Alwis, *J. Am. Chem. Soc.* **102**, 6646 (1980).
95. Y. Ihara, *J. Chem. Soc., Chem. Commun. 1978*, 984.
96. Y. Ihara, *J. Chem. Soc., Perkin Trans 2, 1980*, 1483.
97. J. Koga, M. Shoshi, and N. Kuroki, *Nihon Kagaku Kaishi 1978*, 1179.
98. Y. Ihara, M. Nango, and N. Kuroki, *J. Org. Chem.* **45**, 5009 (1980).
99. Y. Ihara, R. Hosako, M. Mango, and N. Kuroki, *J. Chem. Soc., Chem. Commun. 1981*, 393.
100. Y. Ihara, N. Kunikiyo, T. Kunimasa, M. Nango, and N. Kuroki, *Chem. Lett. 1981*, 667.
101. Y. Ihara and R. Hosako, *Bull. Chem. Soc. Jpn.* **55**, 1979 (1982).
102. K. Yamada, H. Shosenji, and H. Ihara, *Chem. Lett. 1979*, 491.
103. H. Ihara, S. Ono, H. Shosenji, and K. Yamada, *J. Org. Chem.* **45**, 1623 (1980).
104. K. Ohkubo, K. Sugahara, K. Yoshinaga, and R. Ueoka, *J. Chem. Soc., Chem. Commun. 1980*, 637.
105. R. Ueoka, T. Terao, and K. Ohkubo, *Nihon Kagaku Kaishi 1980*, 462.
106. K. Ohkubo, K. Sugahara, H. Ohta, K. Tokuda, and R. Ueoka, *Bull. Chem. Soc. Jpn.* **54**, 576 (1981).
107. K. Kon-no, M. Tosaka, and A. Kitahara, *J. Colloid Interface Sci.* **79**, 581 (1981).
108. T. Kunitake and K. Okahata, *J. Am. Chem. Soc.* **99**, 3860 (1977).
109. T. Kunitake, *J. Makromol. Sci.-Chem.* **A13**, 587 (1979).
110. T. Kunitake and T. Sakamoto, *J. Am. Chem. Soc.* **100**, 4615 (1978).
111. T. Kunitake and T. Sakamoto, *Chem. Lett. 1979*, 1059.
112. Y. Okahata, R. Ando, and T. Kunitake, *Bull. Chem. Soc. Jpn.* **52**, 3647 (1979).
113. Y. Okahata, H. Ihara, and T. Kunitake, *Bull. Chem. Soc. Jpn.* **54**, 2072 (1981).
114. Y. Murakami, A. Nakano, and K. Akiyoshi, *Bull. Chem. Soc. Jpn.* **55**, 3004 (1982).
115. R. Ueoka, Y. Matsumoto, and Y. Ninomiya, Y. Nakagawa, K. Inoue, and K. Ohkubo, *Chem. Lett. 1981*, 785.
116. K. Ohkubo, N. Matsumoto, and H. Ohta, *J. Chem. Soc. Chem. Commun. 1982*, 738.
117. Y. Murakami, A. Nakano, A. Yoshimatsu, and K. Fukuya, *J. Am. Chem. Soc.* **103**, 728 (1981).
118. R. L. Reeves, S. A. Harkaway, and R. T. Klingbiel, *J. Am. Chem. Soc.* **100**, 3879 (1978).
119. B. Samori and L. Fiocco, *J. Am. Chem. Soc.* **104**, 2634 (1982).
120. T. C. Bruice, *Acc. Chem. Res.* **13**, 256 (1980).
121. J. T. Slama, S. R. Oruganti, and E. T. Kaiser, *J. Am. Chem. Soc.* **103**, 6211 (1981).
122. F. Takahashi, T. Ogasa, M. Hirano, and S. Suzuki, *Bull. Chem. Soc. Jpn.* **49**, 1130 (1976).
123. K. Kondo, K. Takahashi, and K. Takemoto, *J. Polym. Sci., Polym. Lett. Ed.* **15**, 77 (1977).
124. S. Shinkai, R. Ando, and T. Kunitake, *J. Chem. Soc., Perkin Trans. 2, 1978*, 1271.
125. S. Shinkai, S. Yamada, and T. Kunitake, *Macromolecules* **11**, 65 (1978).
126. S. Shinkai, *Makromol. Chem.* **179**, 2637 (1978).

127. S. Shinkai, *J. Polym. Sci., Polym. Chem. Ed.* **17**, 3905 (1979).
128. S. Shinkai, Y. Kusano, and O. Manabe, *Makromol. Chem.* **181**, 1791 (1980).
129. S. Shinkai, T. Yamashita, and O. Manabe, *Chem. Lett. 1981*, 961.
130. S. Shinkai, Y. Kusano, and O. Manabe, *J. Chem. Soc., Perkin Trans. 1, 1980*, 1622.
131. S. Shinkai, K. Tamaki, and T. Kunitake, *J. Polym. Sci., Polym. Lett. Ed.* **14**, 1 (1976).
132. S. Shinkai, K. Tamaki, and T. Kunitake, *Biopolymers* **15**, 1001 (1976).
133. F. Rypáček, M. J. Beneš, J. Drobník, and B. Sedláček, *Collect. Czech. Chem. Commun.* **42**, 648 (1977).
134. M. Seki, N. Baba, J. Oda, and Y. Inoue, *J. Am. Chem. Soc.* **103**, 4613 (1981).
135. S. Shinkai, K. Tanake, T. Ide, and O. Manabe, *J. Polym. Sci., Polym. Chem. Ed.* **17**, 2585 (1979).
136. Y. Murakami, Y. Aoyama, and J. Kikuchi, *J. Chem. Soc., Perkin Trans. 1, 1981*, 2809.
137. Y. Murakami, Y. Aoyama, and J. Kikuchi, *Bull. Chem. Soc. Jpn.* **55**, 2898 (1982).
138. M. Kojima, F. Toda, and K. Hattori, *J. Chem. Soc., Perkin Trans 1, 1981*, 1647.
139. M. Hirano and F. Takahashi, *Bull. Chem. Soc. Jpn.* **50**, 1125 (1977).
140. N. Tsubokawa, T. Endo, and M. Okawara, *J. Polym. Sci., Polym. Chem. Ed.* **20**, 2205 (1982).
141. H. Kuzuhara, M. Iwata, and S. Emoto, *J. Am. Chem. Soc.* **99**, 4173 (1977).
142. R. Breslow, M. Hammond, and M. Lauer, *J. Am. Chem. Soc.* **102**, 421 (1980).
143. Y. N. Belokon, V. I. Tararov, T. F. Savelėva, and V. M. Belikov, *Makromol. Chem.* **181**, 2183 (1980).
144. H. Nakano, T. Goto, and Y. Yamamoto, *J. Polym. Sci., Polym. Lett. Ed.* **18**, 381 (1980).
145. H. Nakano, R. Yamane, O. Sangen, and Y. Yamamoto, *J. Polym. Sci., Polym. Chem. Ed.* **20**, 2335 (1982).
146. H. Nakano, M. Nishioka, O. Sangen, and Y. Yamamoto, *J. Polym. Sci., Polym. Chem. Ed.* **19**, 2919 (1981).
147. H. Nakano, T. Yagi, O. Sangen, and Y. Yamamoto, *J. Polym. Sci., Polym. Lett. Ed.* **20**, 23 (1982).
148. C. Carlini and G. Sbrana, *J. Macromol. Sci.-Chem.* **A16**, 323 (1981).
149. M. Kaneko and E. Tsuchida, *J. Polym. Sci., Macromol. Rev.* **16**, 397 (1981).
150. V. S. Pshezhetskyĭ, S. G. Ikryannikov, T. A. Kuznetsova, and V. A. Kavanov, *J. Polym. Sci., Polym. Chem. Ed.* **14**, 2595 (1976).
151. N. Hojo, H. Shirai, Y. Chujo, and S. Hayashi, *J. Polym. Sci., Polym. Chem. Ed.* **16**, 447 (1978).
152. T. Uragami, N. Niwa, and M. Sugihara, *Kobunshi Ronbunshu* **39**, 669 (1982).
153. M. Barteri, M. Farinella, and B. Pispisa, *Biopolymers* **16**, 2569 (1977).
154. M. Barteri, M. Farinella, B. Pispisa, and L. Splendorini, *J. Chem. Soc., Faraday Trans. 1* **74**, 288 (1978).
155. M. Barteri, B. Pispisa, and M. V. Primiceri, *Biopolymers* **18**, 3115 (1979).
156. M. Barteri and B. Pispisa, *Biopolymers* **21**, 1093 (1982).
157. M. Barteri and B. Pispisa, *Makromol. Chem., Rapid Commun.* **3**, 715 (1982).
158. Yu. I. Skurlatov, V. Ya. Kovner, S. O. Travin, Yu. E. Kirsh, A. P. Purmal, and V. A. Kabanov, *Eur. Polym. J.* **15**, 811 (1979).
159. H. Yamamoto and T. Hayakawa, *Int. J. Biol. Macromol.* **4**, 116 (1982).
160. H. Egawa, T. Nonaka, and N. Kozakura, *Bull. Chem. Soc. Jpn.* **55**, 3536 (1982).
161. N. Oguni, S. Shimazu, Y. Iwamoto, and A. Nakamura, *Polym. J.* **13**, 845 (1981).
162. I. Okura, S. Nakamura, and M. Kobayashi, *Bull. Chem. Soc. Jpn.* **54**, 3794 (1981).
163. A. Yamagishi, *J. Phys. Chem.* **86**, 229 (1982).
164. A. Yamagishi, *J. Phys. Chem.* **86**, 233 (1982).
165. M. Takeishi, S. Niino, and S. Hayama, *Makromol. Chem.* **177**, 1225 (1976).
166. M. Takeishi, T. Watanabe, S. Fujii, S. Niino, and S. Hayama, *Makromol. Chem.* **180**, 1099 (1979).

167. M. Takeishi, T. Watanabe, S. Niino, and S. Hayama, *J. Polym. Sci., Polym. Chem. Ed.* **18**, 3081 (1980).
168. M. Takeishi, M. Ueda, and S. Hayama, *Makromol. Chem., Rapid Commun.* **2**, 167 (1981).
169. I. A. Yamskov, B. B. Berezin, L. A. Belchich, and V. A. Davankov, *Makromol. Chem.* **180**, 799 (1979).
170. I. A. Yamskov, B. B. Berezin, L. A. Belchich, and V. A. Davankov, *Eur. Polym. J.* **15**, 1067 (1978).
171. H.-P. Lau and C. D. Gutsche, *J. Am. Chem. Soc.* **100**, 1857 (1978).
172. T. Takagishi and I. M. Klotz, *Biopolymers* **18**, 2497 (1979).
173. K. Ogino, I. Tomita, K. Machiya, and W. Tagaki, *Chem. Lett. 1982*, 1875.
174. M. Tanihara and Y. Imanishi, *Polym. J.* **7**, 499 (1983).
175. M. Tanihara and Y. Imanishi, *Polym. J.* **7**, 509 (1983).
176. M. E. Wilson and G. M. Whitesides, *J. Am. Chem. Soc.* **100**, 306 (1978).
177. S. Kmošťák and K. Setínek, *Collect. Czech. Chem. Commun.* **46**, 2354 (1981).
178. K. Watanabe and A. Imazawa, *Bull. Chem. Soc. Jpn.* **55**, 3208 (1982).
179. S. Inoue, *Adv. Polym. Sci.* **21**, 77 (1976).
180. H. Fukushima and S. Inoue, *Makromol. Chem.* **177**, 2617 (1976).
181. K. Ueyanagi and S. Inoue, *Makromol. Chem.* **177**, 2807 (1976).
182. K. Ueyanagi and S. Inoue, *Makromol. Chem.* **178**, 235 (1977).
183. K. Ueyanagi and S. Inoue, *Makromol. Chem.* **178**, 375 (1977).
184. K. Ueyanagi and S. Inoue, *Makromol. Chem.* **179**, 887 (1978).
185. S. Inoue and Y. Kawano, *Makromol. Chem.* **180**, 1405 (1979).
186. S. Juliá, J. Guixer, J. Masana, J. Rocas, S. Colonna, R. Annuziata, and H. Molinari, *J. Chem. Soc., Perkin Trans. 1, 1982*, 1317.
187. M. Nango, H. Kozuka, Y. Kimura, N. Kuroki, Y. Ihara, and I. M. Klotz, *J. Polym. Sci., Polym. Lett. Ed.* **18**, 647 (1980).
188. Y. Hui, S. Wang, and X. Jiang, *J. Am. Chem. Soc.* **104**, 347 (1982).
189. J. P. Guthrie, P. A. Cullimore, R. S. McDonald, and S. O'Leary, *Can. J. Chem.* **60**, 747 (1982).
190. Y. Masuda, M. Tanihara, Y. Imanishi, and T. Higashimura, *Bull. Chem. Soc. Jpn.* **58**, 497 (1985).
191. M. Tanihara, T. Hiza, Y. Imanishi, and T. Higashimura, *Bull. Chem. Soc. Jpn.* **56**, 1155 (1983).
192. K. Kawaguchi, M. Tanihara, and Y. Imanishi, *Polym. J.* **15**, 97 (1983).
193. M. Tanihara and Y. Imanishi, *Int. J. Biol. Macromol.* **4** 193 (1982).
194. M. Tanihara, Y. Kikuchi, and Y. Imanishi, *Int. J. Biol. Macromol.* **4**, 297 (1982).
195. Y. Kikuchi, M. Tanihara, and Y. Imanishi, *Int. J. Biol. Macromol.* **4**, 305 (1982).
196. J. Oku and S. Inoue, *J. Chem. Soc., Chem. Commun. 1981*, 229.
197. J. Oku, N. Ito, and S. Inoue, *Makromol. Chem.* **183**, 579 (1982).
198. M. F. Czarniecki and R. Breslow, *J. Am. Chem. Soc.* **100**, 7771 (1978).
199. G. L. Trainor and R. Breslow, *J. Am. Chem. Soc.* **103**, 154 (1981).
200. M. Komiyama and S. Inoue, *Chem. Lett. 1979*, 1101.
201. M. Komiyama and S. Inoue, *Bull. Chem. Soc. Jpn.* **53**, 3334 (1980).
202. M. Komiyama and H. Hirai, *Chem. Lett. 1980*, 1251.
203. M. Komiyama and M. L. Bender, *Bull. Chem. Soc. Jpn.* **53** 1073 (1980).
204. M. Komiyama and S. Inoue, *Bull. Chem. Soc. Jpn.* **53** 3266 (1980).
205. M. Komiyama and H. Hirai, *Chem. Lett. 1980*, 1471.
206. V. Daffe and J. Fastrez, *J. Am. Chem. Soc.* **102**, 3601 (1980).
207. Y. Motozato, Y. Furuya, T. Matsumoto, and T. Nishihara, *Bull. Chem. Soc. Jpn.* **53**, 2578 (1980).
208. Y. Taniguchi, S. Makimoto, an . K. Suzuki, *J. Phys. Chem.* **85**, 3469 (1981).
209. S. Makimoto, K. Suzuki, and Y. Taniguchi, *J. Phys. Chem.* **86**, 4544 (1982).

210. R. J. Bergeron and P. S. Burton, *J. Am. Chem. Soc.* **104**, 3664 (1982).
211. R. Breslow and D. C. Rideout, *J. Am. Chem. Soc.* **102**, 7816 (1980).
212. D. D. Sternbach and D. M. Rossana, *J. Am. Chem. Soc.* **104**, 5853 (1982).
213. M. Akiyama, T. Ohmachi, and M. Ihjima, *J. Chem. Soc., Perkin Trans. 2, 1982*, 1511.
214. Y. Murakami, J. Sunamoto, H. Kondo, and H. Okamoto, *Bull. Chem. Soc. Jpn.* **50**, 2420 (1977).
215. Y. Murakami, Y. Aoyama, M. Kida, and A. Nakano, *Bull. Chem. Soc. Jpn.* **50**, 3365 (1977).
216. C. Hansch, *J. Org. Chem.* **43**, 4889 (1978).
217. Y. Murakami, A. Nakano, R. Miyata, and Y. Matsuda, *J. Chem. Soc., Perkin Trans. 1 1979*, 1669.
218. Y. Murakami, A. Nakano, K. Akiyoshi, and K. Fukuya, *J. Chem., Soc., Perkin Trans. 1 1981*, 2800.
219. I. Tabushi, Y. Kimura, and K. Yamamura, *J. Am. Chem. Soc.* **103**, 6486 (1981).
220. J.-M. Lehn and C. Sirlin, *J. Chem. Soc., Chem. Commun. 1978*, 949.
221. Y. Chao, G. R. Weisman, G. D. Y. Sogah, and D. J. Cram, *J. Am. Chem. Soc.* **101**, 4948 (1979).
222. M. Utaka, M. Takatsu, and A. Takeda, *Bull. Chem. Soc. Jpn.* **50**, 3276 (1977).
223. J. J. Morris and M. I. Page, *J. Chem. Soc., Chem. Commun. 1978*, 591.
224. J. J. Morris and M. I. Page, *J. Chem. Soc., Perkin Trans. 2 1980*, 1131.
225. J. D. Roberts and K. Kanamori, *Proc. Natl. Acad. Sci. U.S.A.* **77**, 3095 (1980).
226. T. Kunitake, Y. Okahata, R. Ando, and S. Hirotsu, *Bull. Chem. Soc. Jpn.* **49**, 2547 (1976).
227. J. Hine, *Acc. Chem. Res.* **11**, 1 (1978).
228. J. Hine and J. P. Zeigler, *J. Am. Chem. Soc.* **102**, 7524 (1980).
229. Y. Tachibana, *Bull. Chem. Soc. Jpn.* **51**, 3085 (1978).
230. T. Kunitake and T. Sakamoto, *Polym. J.* **11**, 871 (1979).
231. Y. Okamoto, *Nihon Kagaku Kaishi 1978*, 870.
232. D. Mirejovsky, *J. Org. Chem.* **44**, 4881 (1979).
233. C. S. Lege and J. A. Deyrup, *Macromolecules* **14**, 1629 (1981).
234. C. S. Lege and J. A. Deyrup, *Macromolecules* **14**, 1634 (1981).
235. N. Nishi and B. Nakajima, *Int. J. Biol. Macromol.* **4**, 281 (1982).
236. S. Iwabuchi, T. Nakahira, Y. Ikebukuro, T. Takada, and K. Kojima, *Makromol. Chem.* **182**, 2715 (1981).
237. Y. Matsui and A. Okimoto, *Bull. Chem. Soc. Jpn.* **51**, 3030 (1978).
238. T. Eiki and W. Tagaki, *Chem. Lett. 1980*, 1063.
239. R. Breslow, J. B. Doherty, G. Gillot, and C. Lipsey, *J. Am. Chem. Soc.* **100**, 3227 (1978).
240. R. Breslow, P. Bovy, and C. L. Hersh, *J. Am. Chem. Soc.* **102**, 2115 (1980).
241. R. Breslow, M. F. Czarniecki, J. Emert, and H. Hamaguchi, *J. Am. Chem. Soc.* **102**, 762 (1980).
242. I. Tabushi, Y. Kuroda, and A. Mochizuki, *J. Am. Chem. Soc.* **102**, 1152 (1980).
243. K. Fujita, A. Shinoda, and T. Imoto, *J. Am. Chem. Soc.* **102**, 1161 (1980).
244. A. Ueno, K. Takahashi, and T. Osa, *J. Chem. Soc., Chem. Commun. 1981*, 94.
245. Y. Murakami, Y. Aoyama, and M. Kida, *J. Chem. Soc., Perkin Trans. 2 1980*, 1665.
246. N. Platé and O. V. Noah, *Adv. Polym. Sci.* **31**, 133 (1979).
247. H. Miyakawa and N. Saito, *Polym. J.* **10**, 27 (1978).
248. A. D. Litmanovich, *Eur. Polym. J.* **16**, 269 (1980).
249. M. Sisido and K. Shimada, *Macromolecules* **12**, 790 (1979).
250. M. Sisido and K. Shimada, *Macromolecules* **12**, 792 (1979).
251. M. Sisido, *J. Am. Chem. Soc.* **99**, 7785 (1977).
252. D. S. Saunders and M. A. Winnik, *Macromolecules* **11**, 18 (1978).
253. D. S. Saunders and M. A. Winnik, *Macromolecules* **11**, 25 (1978).
254. U. Maharaj, M. A. Winnik, B. Dors, and H. J. Schäfer, *Macromolecules* **12**, 905 (1979).

255. A. Mar, S. Fraser, and M. A. Winnik, *J. Am. Chem. Soc.* **103**, 4941 (1981).
256. M. A. Winnik, *Chem. Rev.* **81**, 491 (1981).
257. C. Galli, G. Giovannelli, G. Illuminati, and L. Mandolini, *J. Org. Chem.* **44**, 1258 (1979).
258. C. Galli, G. Illuminati, and L. Mandolini, *J. Org. Chem.* **45**, 311 (1980).
259. G. Illuminati, L. Mandolini, and B. Masci, *J. Am. Chem. Soc.* **103**, 4142 (1981).
260. G. Illuminati and L. Mandolini, *Acc. Chem. Res.* **14**, 95 (1981).
261. Y. Hatano, M. Yamamoto, and Y. Nishijima, *J. Phys. Chem.* **82**, 367 (1978).
262. T. Kanaya, Y. Hatano, M. Yamamoto, and Y. Nishijima, *Bull. Chem. Soc. Jpn.* **52**, 2079 (1979).
263. M. Furue and S. Nozakura, *Bull. Chem. Soc. Jpn.* **55**, 513 (1982).
264. R. Bird, G. Griffiths, G. F. Griffiths, and C. J. M. Stirling, *J. Chem. Soc., Perkin Trans. 2 1982*, 579.
265. M. Sisido, Y. Kanazawa, and Y. Imanishi, *Biopolymers* **20**, 653 (1981).
266. M. Sisido and Y. Imanishi, *Biopolymers* **20**, 665 (1981).
267. M. Sisido, T. Shimizu, Y. Imanishi, and T. Higashimura, *Biopolymers* **19**, 701 (1980).
268. M. Sisido, H. Ito, and Y. Imanishi, *Biopolymers* **21**, 1597 (1982).
269. M. Sisido and Y. Imanishi, *Biopolymers* **21**, 1613 (1982).
270. P. Skrabel, V. Rizzo, A. Baici, F. Bangerter, and P. L. Luisi, *Biopolymers* **18**, 995 (1979).
271. A. Baici, V. Rizzo, P. Skrabal, and P. L. Luisi, *J. Am. Chem. Soc.* **101**, 5170 (1979).
272. R. Guillard and A. Englert, *Biopolymers* **15**, 1301 (1976).
273. U. W. Suter and M. Mutter, *Makromol. Chem.* **180**, 1761 (1979).
274. M. A. Winnik, T. Redpath, and D. H. Richards, *Macromolecules* **13**, 328 (1980).
275. A. E. C. Redpath and M. A. Winnik, *J. Am. Chem. Soc.* **102**, 6869 (1980).
276. M. A. Winnik, *J. Am. Chem. Soc.* **103**, 708 (1981).
277. A. E. C. Redpath and M. A. Winnik, *J. Am. Chem. Soc.* **104**, 5604 (1982).
278. S. Ito, M. Yamamoto, and Y. Nishijima, *Bull. Chem. Soc. Jpn.* **54**, 35 (1981).
279. S. Ito, M. Yamamoto, and Y. Nishijima, *Bull. Chem. Soc. Jpn.* **55**, 363 (1982).
280. T. Kanaya, K. Goshiki, M. Yamamoto, and Y. Nishijima, *J. Am. Chem. Soc.* **104**, 3580 (1982).
281. C. Cuniberti and A. Perico, *Eur. Polym. J.* **16**, 887 (1980).
282. N.-C. C. Yang, S. B. Neoh, T. Naito, L.-K. Ng, D. A. Chernoff, and D. B. McDonald, *J. Am. Chem. Soc.* **102**, 2806 (1980).
283. M. Van der Auweraer, A. Gilbert, and F. C. De Schryver, *J. Phys. Chem.* **85**, 3198 (1981).
284. M. Migita, T. Okada, N. Mataga, Y. Sakata, S. Misumi, N. Nakashima, and K. Yoshihara, *Bull. Chem. Soc. Jpn.* **54**, 3304 (1981).
285. M. Sisido, Y. Imanishi, and T. Higashimura, *Macromolecules* **12**, 975 (1979).
286. R. Cerf, *Biopolymers* **18**, 731 (1979).
287. J. A. Ibemesi, J. B. Kinsinger, and M. A. El-Bayoumi, *J. Polym. Sci., Polym. Chem. Ed.* **18**, 879 (1980).
288. R. Todesco, J. Gelan, H. Martens, J. Put, and F. C. De Schryver, *J. Am. Chem. Soc.* **103**, 7304 (1981).

18

Bioactive Carboxylic Acid Polyanions

Raphael M. Ottenbrite

Abstract. Synthetic polymers with carboxylic acid functionality have been found to elicit a broad range of biological activity. Among these activities are inhibitory effects against tumors, viruses, bacteria, and fungi. The distribution of these polymers, by ^{14}C studies, indicate that they are oriented to the reticuloendothelial system. The mode of antitumor activation is related to their ability to activate a population of macrophage cells. The unique aspect of these polycarboxylic acid polymer-activated macrophages is that they become cytotoxic or cytostatic for tumor cells, such as Lewis lung and Ehrlich ascites, while they have no apparent effect on normal cell populations. Cell cycle analysis shows that polycarboxylic acid-activated macrophages shift the cellular population from the DNA synthesis and division phases to the resting phase. Clinical studies of these types of polymers have only been reported for the phase I program.

1. Introduction

Natural polyanions, such as heparin, heparinoids, and other polysaccharide-type macromolecules, produce a variety of biological responses in the host. In the past 20 years, several synthetic polyanions have been investigated for a comparison of their physiological properties. One of the most widely studied is the carboxylic acid containing polymers.(1) Some of the biological responses elicited by these synthetic polymers include interferon induction, antiviral activity, tumor cell cytotoxicity, and anticlotting effects.

These carboxylic acid polymers seem to behave similarly to certain proteins, glycoproteins, and polynucleotides which modulate a variety of biological responses related to host defense mechanisms. They have been reported to induce interferon production,(2) to modify reticuloendothelial function,(3) and to invoke immunoadjuvant,(4) antiviral,(5) and antineoplastic activity.(6) Although most of the activity observed is prophylactic, the prolonged protective properties exhibited by these materials makes them

Raphael M. Ottenbrite • Department of Chemistry and the Massey Cancer Center, Virginia Commonwealth University, Richmond, VA 23284.

potential clinical candidates. Presently, the fundamental role of these polyanions in controlling host resistance to a variety of diseases is being investigated. A major interest lies in modes of effecting macrophage activation and tumor cell changes effected by "activated" macrophages.

2. Types of Carboxylic Acid Polymers Evaluated for Biological Activity

2.1. Homopolymers

These polymers were the first to be evaluated. The most extensively studied homopolymer for biological drug application is poly(acrylic acid). It was found to give 100% protection against semliki forest virus[7] and a 55% reduction in pox count for vaccina virus infections.[8] Although poly(acrylic acid) is a poor interferon inducer,[9] it is capable of effecting a 25% increase in survival time against lethal doses of Mengo virus,[10] and a 32% increase in survival time for animals subjected to sarcoma 180.[11]

Poly(methacrylic acid) is less effective than poly(acrylic acid) as an interferon inducer and against viral infection. However, when administered 24 hours before infection, poly(methacylic acid) does exhibit significant inhibition of Sindbis and vesicular viruses.[12] Poly(ethacrylic acid), recently prepared by Tirrell,[13] is much more effective against Lewis lung carcinoma than any other homopolymer tested.[14] This polymer elicits macrophages with higher 5′ nuclotidase levels than the macrophage stimulant thioglycollate.

The homopolymer of maleic anhydride was found to be effective against Friend leukemia.[15] It induced phagocytic activity but did not depress the reticuloendothelial response experienced with other synthetic polyanions. The most significant effect is that supernantant fluids of the peritoneal exudate cells, from mice pretreated with poly(maleic acid), transferred protection against Friend leukemia.[16]

Although poly(itaconic acid) has been used industrially, it has not been extensively evaluated biologically. An increase of 45% has been reported in the life span of mice inflicted with ascites sarcoma 180.[11]

2.2. Copolymers of Maleic Anhydride

Maleic anhydride is a unique comonomer since it undergoes copolymerization to form a 1 : 1 alternating copolymer sequence with most alkenes and dienes.[17] This behavior has been attributed to the ability of this agent to act as an electron acceptor to form π-complexes with other π-systems.

One of the most widely studied synthetic material for biological activity has been the copolymer prepared from divinyl ether and maleic anhydride. During the course of polymerization, two types of ether rings are formed along the polymer backbone; a six-membered (pyran) and a five-membered (furan) ring. Originally, it was thought that only six-membered rings were formed and the polymer was called pyran. In the literature, it is also referred to by the acronym DIVEMA (divinyl ether–maleic anhydride) and MVE (maleic anhydride–vinyl ether). Pyran was first synthesized by Butler in 1960,[18] but submitted for NIH screening about the same time by both Butler[18] and by Brewlow.[19] Initial studies showed that pyran had interesting activity and was subsequently designed as NSC 46015 by the National Cancer Institute.

In fact, this polymer exhibited a broad spectrum of biological activities which stimulated investigations in many areas. Pyran has the following characteristics: (1) it is a mild inducer of interferon;[20–23] (2) it is active against a number of viruses,[20–28] some of these are foot-and-mouth diseases, encephalomyocarditis, vesicular stomatitis, polyoma, Rauscher leukemia, Friend leukemia, Moloney sarcoma, and Mengo;[3] it is an antifungal agent[29] and an antibacterial agent;[29–31] and (4) it stimulates immune responses.[29–42] Pyran has also been shown to be an effective anticoagulant as well as an agent for removing plutonium from the liver.[43] It is currently being evaluated as

Table 1. Effect of Maleic Anhydride Copolymers Against Ascites Tumor Cells[a]

	1000–10,000 MW	10,000–30,000 MW
Polymer	Percent control[b]	Percent control
Saline	100	100
Pyran	30	49
MA-co-acrylic Acid	58	118
MA-co-styrene	7	45
MA-co-methracyclic acid	60	93
MA-co-allyl phenol	16	86
MA-co-allyl succinic anhydride	42	87
MA-co-1,3-dioxepin	30	94
MA-co-isobutenyl succinic anhydride	61	80

[a] 10^6 Ehrlich ascites cell were inoculated i.p.

[b] Five days after inoculation, animals were sacrificed and total peritoneal exudate cells were counted with a hemocytometer.

Table 2. Effect of 2,3-Dicarboxynorborn-5-ene Copolymers against Ehrlich Ascites Tumor

Copolymer	1000–10,000 MW Percent control	10,000–30,000 MW Percent control
Control	100	100
Pyran	30	49
Maleic anhydride	7	108
Acylic acid	22	106
Vinyl acetate	34	41
Vinyl alcohol	32	56

an immunoadjuvant in the treatment of L1210.[25,43] Pyran is effective as a tumor growth inhibitor for Lewis lung and Erhlich ascites carcinoma.[45]

Since pyran appears to have considerable variability in structure, we have synthesized and evaluated a number of other maleic anhydride copolymers as tumor inhibitors. The results of these evaluations against Ehrlich ascities are present in Tables 1 and 2. These results indicate that although all the carboxylic acid polymers tested were active, those with a large hydrophobic group were as good or better than pyran copolymer. Several other maleic anhydride copolymers, such as vinyl acetate, cyclohexyl-1,3-dioxepin, ethylene, and allyl urea, have been evaluated against Lewis lung carcinoma and are reported in Table 3. Our present synthesis and evaluation program should provide data on 20 to 30 additional copolymers of maleic anhydride.

Table 3. Lewis Lung Tumor Cell Culture Cytotoxicity by Carboxylic-Acid-Polymer-Activated Macrophage

Macrophage activation	Percent cytotoxicity: Morphological assay	Percent cytotoxicity: Tritium release
Normal	15	10
Thioglycolate	2	—
C. parvum	85	18
Pyran	80	15
Poly(maleic anhydride-co-styrene)	2	5
Poly(maleic anhydride-co-ethylene)	18	9
Poly(maleic anhydride-co-allyl urea)	8	11
Poly(maleic anhydride-co-4-methyl-2-pentenone)	70	30
Poly(maleic anhydride-co-cyclohexyl-1,3-dioxepin)	80	28
Poly(itaconic acid-co-styrene)	85	21
Poly(ethacrylic acid)	40	14

2.3. Other Carboxylic Acid Copolymers

Generally copolymerization of monomers not involving maleic anhydride produces copolymers in which the monomers add randomly or in block sequence. Hodnett *et al.*[11] prepared five copolymers of acrylic acid and isobutyl vinyl ether which produced good results against sarcoma 180 ascities tumor in mice. These researchers also prepared and evaluated the copolymer of acrylic acid and β-(*N*,*N*-dimethylamino)ethylmethacrylate which was relatively nontoxic (LD_{50} = 600 mg/kg) and gave a 45% increase in survival time against sarcoma. Poly(acrylic acid-co-isooctyl vinyl ether) was very toxic and less effective against this tumor.

We have prepared low-molecular-weight poly(itaconic-co-styrene) which appears to have an alternating structure. This material activates peritoneal macropohages to be cytotoxic to Lewis lung carcinoma similar to pyran and *C. parvum* bacterium.[45] We have also prepared a number of copolymers of 2,3-dicarboxylnorborn-5-ene which were evaluated against Erhlich ascites tumors. These polymers were found to be as effective as pyran and the maleic anhydride copolymers (Table 2).

2.4. Carboxylic Acid–Half-Amide and Imide Polymers

Data indicated that tumor inhibition may be a function of the distribution and density of the carboxylic acids on the polymer chain. The first study by Regelson[46] involved poly(acrylic acid), poly(methacrylic acid), and poly(acrylamide) with sarcoma 180. The carboxylic acids were most effective giving 60–70% inhibition to tumor growth, whereas poly(acrylamide) was ineffective.

It was then decided to study the combination of amide and carboxylic acid groups[46] by postreaction modification of preformed copolymers. This study involved the treatment of poly(ethylene-alt-maleic anhydride) and other anhydride polymers with ammonia. The half-amide–carboxylic acid and the half-amide–carboxylate ammonium salt were tested for sarcoma 180 inhibition and were found to be similar in activity to the original dicarboxylic acid polymer. The corresponding diimide similar to poly(acrylamide) was significantly less effective. Conversely, poly(propylene-alt-maleic anhydride) and poly(isobutylene-alt-maleic anhydride) showed very little activity against sarcoma 180, while their half-amide–half-acids exhibit 65 and 75% inhibition, respectively.

In a study of low molecular weight poly(ethylene-alt-maleic acid) (EMA) and the corresponding half-amide–half-acid (EMAA) the latter material was much more effective. For example, for tumors, such as Krebs, EMA = 34% and EMAA = 55% inhibition; for L1210, EMA = 12.5%

Table 4. Antitumor Activity of Some Carboxyl-Containing Polymers and Copolymers[1]

NSC Number	Polymer or Copolymer[a]	Tumor[a]	T/C (%)
D59196	DVE-MA	LE	122
D59199	Poly(1,4-pentadiene-co-MA)	SA	49
D59200	(2-Vinyloxy-2-ethoxyl-benzaldehyde-co-MA)	CA	55
84645	Poly(DVSO$_2$-methacrylic acid)	LE	101
86469	1,4-Pentadiene-co-MA $[\eta] = 0.18$	LE	95
84650	1,4-Pentadiene-co-MA $[\eta] = 0.26$	LL	43
99425	1,4-Pentadiene-MA-BrCCl$_3$ (telomer)	LE	146
99426	1,4-Pentadiene-MA-BrCCl$_3$ (telomer)	LE	97
99427	DCE-MA-BrCCl$_3$ (telomer)	LE	107
104304	4-Vinylcyclohexane-co-MA	WM	92
119165	Furan-co-itaconic anhydride	LE	97
119166	Furan-co-MA	LE	105
119167	Furan-co-MA-(half-amide)	LE	105
119168	Furan-co-itaconic anhydride (half-amide)	LE	137
133788	DVE-co-citraconic anhydride	LE	103
133789	β-Chloroethylvinylether-co-citraconic anhydride	LE	117
133790	β-Chloroethylvinylether-co-MA	LE	119
133791	2,5-Dihydrofuran-co-MA	LE	110
133792	2-Methyl-4,5-dihydrofuran-co-MA	LE	107
148129	Isoprene-co-MA	PS	118
148130	2-Methylenenorbornene-co-MA	LE	93
148131	2-Ethylidenenorbornene-co-MA	LE	98
148132	2-Vinylnorbornene-co-MA	LE	106
148133	Butadiene-co-MA	LR	109

[a] DVE = divinyl ether; MA = maleic anhydride; LL = Lewis lung carcinoma; LE = L1210 lymphoid leukemia; CA = adenocarcinoma 755; SA = sarcoma 180; WM = Walker carcinosarcoma 256; and PS = P388 lymphocytic leukemia.

and EMAA = 65% inhibition; and for carcinoma 755, EMA was stimulatory but EMAA = 58% inhibition was observed.[46] The effects of other half-amide–half-acid polymers as antitumor agents are listed in Table 4.

Recently, Fields *et al.* have prepared a low molecular weight ethylene–maleic anhydride copolymer derivatized to contain both half-amide half-carboxylate salt and imide functions.[47,48] The synthesis was carried out by preparing a low-molecular-weight alternating copolymer of ethylene and maleic anhydride. The anhydride groups of the copolymer were converted to half-amide, half-ammonium carboxylate salt functions with a liquid ammonia–acetone mixture. This ammoniated copolymer was then converted to the partial imide by heating a xylene slurry under reflux. The product, coded "NED 137," was recovered by filtration and vacuum drying and was shown to have 14–25 wt % of the succinimide rings. The NED 137 copolymer was evaluated for biological activity against several transplantable tumors. It was found to have relatively low acute toxicity in mice and rats with an LD_{50} approximately 2500 mg/kg body weight. It inhibited Lewis lung carcinoma and several other murine solid tumors.

NED 137 has been potent as a tumor inhibitor and in preventing metastases of a methylcholanthrene-induced carcinoma of the bladder (FBCa) in F344 rats.[(49)] This tumor is known to metastasize to the lung within one week of tumor implantation. Animals treated with NED 137 at 30 mg/kg showed prolonged survival as compared to control animals. All the treated animals were found to be free of pulmonary metastases when autopsied, while all the control animals had extensive pulmonary metastases.

The effect of NED 137 as an adjuvant to surgical excision of the tumor was also examined. The treated animals were observed for tumor recurrence and survival times after excision *vs.* untreated control animals. Tumor recurrence was 100% in the control animals with subsequent death by 35 days after surgical excision. Autopsy of the control animals indicated massive lung metastases. The rats treated with NED 137, however, showed no local

Table 5. Antitumor Activity of Some Poly(Divinylether-co-*N*-Substituted Maleimides)[(1)]

NSC Number	Monomer or Comonomers[a]	Tumor[a]	*T/C* (%)
77033	DVE-*N*-phenylmaleimide	SA	71
84651	DVE-maleimide	LE	88
85652	DVE-*N*-methylmaleimide	SA	114
84653	DVE-*N*-ethylmaleimide	SA	113
84654	DVE-*N*-*n*-propylmaleimide	SA	106
85655	DVE-*N*-i-propylmaleimide	LE	100
85656	DVE-*N*-*n*-butylmaleimide	SA	112
84657	DVE-*N*-i-butylmaleimide	SA	82
84658	DVE-*N*-benzylmaleimide	LE	105
84659	DVE-*N*-*β*-naphthylmaleimide	SA	109
148134	DVE-*p*-carboxyphenylmaleimide	PS	110
217996	DVE-*N*-morpholinomethylmaleimide (10 : 1)	LE	100
266062	DVE-*N*-maleimide of *d,l*-alanine	PS	117
266063	DVE-maleimide of *d,l*-phenylalanine	PS	101
266064	DVE-maleimide of *d,l*-methionine	PS	112
266066	DVE-maleimide of *d,l*-leucine	PS	125
266067	DVE-maleimide of glycine	PS	109
266226	DVE-maleimide of 1-phenylalanine	PS	105

[a] DVE = divinyl ether; LE = L1210 lymphoid leukemia; SA = sarcoma 180; and PS = P388 lymphocytic leukemia.

recurrences of tumor after tumor excision. Autopsy after 60 days indicated that NED 137-treated animals were free of pulmonary metastases.[39] Indefinite survival in these animals could be obtained with repeated administration of the drug. It was noted that the experimental animals showed no acute or chronic toxicity with NED 137.

Subsequently, using the bladder carcinoma FBCa, the antitumor activity of NED 137 was compared with a series of known immunoadjuvants. These were bacilli Calumette-Guerin, *Coreynebacterium parvum*, and pyran.[40] These adjuvants were administered after excision of tumor implants. A single intraperitoneal injection of NED 137 at 30 mg/kg body weight prolonged survival beyond 60 days with no evidence of recurrent or metastatic disease, whereas with the other adjuvants, the animals survived an average of 30–40 days with 100% local recurrence and a 60–90% incidence of pulmonary metastases. The effects of several other polymers containing the maleimide structure as antitumor agents are listed in Table 5.

3. Effect of Molecular Weight and Structure of Polycarboxylic Acid Polymers on Biological Activity

It has been clearly demonstrated by a number of investigations that the molecular weights (MW) of synthetic polycarboxylic acid polymers has significant effects on the therapeutic index of these materials. For example, Mück *et al.*[12] found that poly(acrylic acid) below 5000 MW was ineffective against viruses such as encephalomyocarditis and Colombia SK. Optimum activity was found between 6000 and 15,000 MW with 25,000 being too toxic for the dose levels administered.

Pyran fractions prepared with narrow polydispersity and MW ranging from 2500 to 32,000 were prepared and evaluated by Breslow.[19] Molecular weights up to 15,000 stimulated the reticuloendothelial system, whereas the higher MW fractions caused a suppression. Greater liver damage, inhibition of drug metabolism, and sensitization to endotoxin were also reported to occur as the molecular weight of pyran increased. These findings were confirmed and elaborated on by Ottenbrite *et al.*[50] A sample of pyran XA124-177 was separated into 1000–10,000 and 10,000–30,000 MW fractions. The parent materials caused hepatosplenomegly, inhibition of microsomal enzymes, and sensitization to endotoxin, whereas the lower molecular weight fractions greatly diminished all of these toxic effects. Furthermore, the pyran fractions exhibited similar activity against Lewis lung carcinoma. The antiviral activity against encephalomyocarditis was decreased to one-third for the high MW fractions and no activity was observed for the low MW fraction.

Table 6. Effect of Molecular Weight and Counterions of Pyran on Acute Toxicity and Sensitivity to Bacterial Endotoxin[42]

	LD_{50}		Endotoxin LD_{50}	
Polymer MW	Na^+ salt	Ca^{2+} salt	Na^+ salt	Ca^{2+} salt
12,500	112	–	24.0	–
15,500	98	190	7.0	10
21,300	94	–	0.8	–
52,600	86	170	0.5	0.5

Recently, Munson *et al.*[42] found that the calcium salt of pyran was less toxic than the sodium salt. Both salts showed marked differences in acute toxicity, but insignificant differences were observed for sensitization to bacterial endotoxin (Table 6). Concurrently, we have investigated the effect of molecular weight on some other polycarboxylic acid polymers.[51] Several molecular-weight fractions of pyran, poly(acrylic acid-co-maleic acid) (PAAMA), poly(maleic acid) (PMA), and poly(acrylic acid-co-3,6-endoxo-1,2,3,6-tetrahydrophthalic acid) (PAATHP) were evaluated for activity with Lewis lung tumor and encephalomyocarditis virus (Table 7). Although three of the polymers significantly reduced the growth of the primary tumor, these results did not produce a coincident increase in life span (1LS). For example, PMA exhibited a consistent reduction of tumor size of > 70%, but only an 1LS of 15% compared to pyran with 80% reduction of tumor size and 40% 1LS compared to untreated animals. The antiviral effects were found to be very susceptible to molecular weight. In all cases, the 1000–10,000 MW fraction was totally ineffective. The activity increased with molecular weight with maximum effectiveness occurring with polymers have > 50,000 MW.

Although high molecular weight polycarboxylic acid polymers induced many undesirable toxicologic difficulties, many of these problems are moderated in the lower molecular weight fractions (Tables 6 and 7). Further evidence that molecular weight plays a significant role in the biological activity of polycarboxylic acid copolymers is also illustrated in Tables 1–3.

Recently, we have evaluated a series of polycarboxylic acid polymers that differed in charge density, molecular weight, lipophilicity, and chain rigidity for their ability to induce activation of macrophages. To determine the level of macrophage activation induced by these polymers, we evaluated peritoneal exudate cells that were elicited by intraperitoneal administration of polymers for their cytotoxicity to Lewis lung tumor cells. In addition, an ectoenzyme analysis on the lysates of these cell populations was determined.[55]

The ability of polymer-activated macrophages to be cytotoxic to Lewis

Table 7. Effect of Molecular Weight on Biological Activity

Polymer size	Polymer			
	Pyran[a] (%)	PAAMA[b] (%)	PMA[c] (%)	PAATHP[d] (%)
		Inhibition of tumor size		
Whole polymer	78	74	75	30
(increased life span)	(40)	(33)	(15)	(< 10)
1,000– 10,000	89	—	74	22
10,000– 30,000	84	80	72	—
30,000– 50,000	74	—	76	—
50,000–100,000	70	78	74	30
		Antiviral encephalomyocarditis protection		
Whole polymer	89	90	30	25
1,000– 10,000	0	0	0	0
10,000– 30,000	30	29	0	—
30,000– 50,000	58	54	< 10	—
50,000–100,000	86	80	26	38
		Acute toxicity (LD_{50})		
Whole polymer	74	110	120	150
1,000– 10,000	120	—	160	> 200
10,000– 30,000	115	> 200	150	—
30,000– 50,000	95	—	140	—
50,000–100,000	84	180	135	180
		Endotoxin sensitization		
Whole polymer	0.12	1.0	15	3.0
1,000– 10,000	15	—	> 20	> 10
10,000– 30,000	15	—	> 20	> 10
30,000– 50,000	—	> 20	> 20	> 3
50,000–100,000	—	> 20	> 20	> 3

[a] Poly(divinyl ether-co-maleic anhydride).
[b] Poly(acrylic acid co-maleic anhydride).
[c] Poly(maleic anhydride).
[d] Poly(acrylic acid-co-3,6-endoxo-1,2,3,6-tetrahydrophthalic anhydride).

lung cells in culture were determined by ^3H release and morphological assays. A good correlation was obtained for both assays with poly(itaconic acid-co-styrene) (IAS), poly(cyclohexyl-1,3-dioxepin-co-maleic acid) (CDA-MA), and poly(4-methyl-2-pentenone-co-maleic acid) (MP-MA) which demonstrated the greatest induction of tumoricidal macrophages.[(55)]

Survival studies of mice that were subcutaneously implanted with Lewis lung cells indicate that CDA-MA and MP-MA are the most effective antitumor agents (Table 8). CDA-MA enhanced survival at all doses and MP-MA was most effective at 100 mg/kg. Only two out of five mice receiving IAS survived three days after administration due to acute toxicity; however, those two survived more than 90 days. LS was totally ineffective.

Table 8. Enhanced Survival of Mice Inflicted with Lewis Lung Carcinoma and Treated with Polycarboxylic Acid Polymers

	Dosage (mg/kg body weight)		
Agent[a]	25	50	100
CDA-MA	130[b]	160[b]	173[b]
MP-MA	95	115	158[b]
IAS	130[b]	120[b]	80[b]
LS-MA	102	107	98
Pyran	—	142[b]	—
C. parvum[c]	130[b]	—	—

[a] CDA-MA: poly(cyclohexyl-1,3-dioxepin-alt-maleic anhydride); MP-MA: poly(4-methyl-2-propenone-alt-maleic anhydride); IAS: poly(itaconic acid-alt-styrene); LS-MA: poly(styrene-alt-maleic anhydride).

[b] Indicates groups where mice survived more than 90 days.

4. Macrophage Activation by Polycarboxylic Acid Polymers

Macrophage cells are generally considered to be involved only with the phagocytosis and the degradation of antigenic materials. In recent years, however, the role of macrophages in the immune system has been undergoing considerable reevaluation. Consequently, macrophages have been found to have a function in the generation of humoral response[52] and antibody production.[53]

Phenomenally, under suitable conditions, macrophages can be transformed into an activated state that is characterized by cell enlargement with undulating membranes (see Figure 1). These "activated macrophages" are both qualitatively and quantitatively different from normal macrophages. Several biologic reticuloendothelial stimulants such as bacilla Calumette-Guerin, *C. parvum*, and *Toxoplasma gondii* are known to enhance macrophage function and to induce resistance to tumor growth.[54]

Synthetic polycarboxylic acids have also been found to be potent activators of macrophages as well. Wc have recently reported that several polycarboxylic acid polymers are capable of activating macrophages to tumoricidal function.[55] A unique finding among all the polycarboxylic acid polymer-activated macrophage populations is that they are cytotoxic or cytostatic to tumor cells, such as Ehrlich ascites and Lewis lung, while they have no apparent effect on normal cells.[56,57] Presently, the basis for this discrimination is unknown; it is assumed, however, that the activated macrophage can recognize some feature in a tumor cell that is not found in a normal cell. Furthermore, it is not clear whether cytotoxicity and cytostasis involve the same macrophage function.

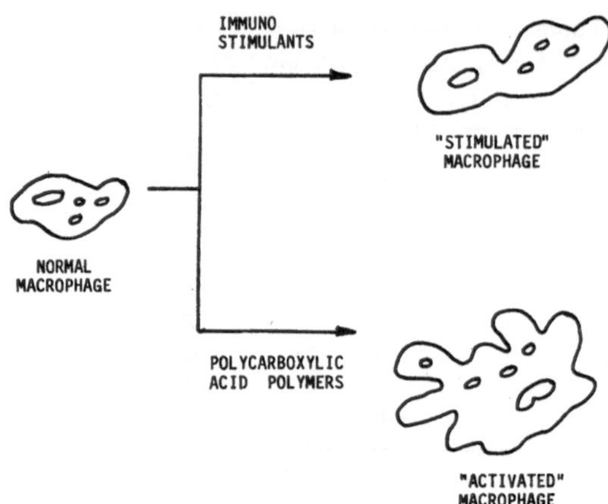

Figure 1. Comparison of "Normal", "Stimulated", and "Activated" Macrophages.

Although the mechanism of macrophage activation or the tumoricidal activity of macrophage is not understood, it has been observed that Lewis lung cells cultured with activated macrophage exhibit reduced cell division.[58] It was found that tumor cells contained 50% less DNA per cell. This depletion of DNA eventually led to cell death since the tumor cells no longer had the appropriate amount of DNA to prepare the cell for division. Cell cycle analysis showed that the *in vitro* Lewis lung cells' distribution was 15% in the resting phase (G_1), 52% in the DNA synthesis phase (S), and 33% in the cell division phase (G_2M).[59] After culturing with polycarboxyl acid "activated" macrophage, the tumor cell distribution changed to 71% in the resting phase (G_1), to 22% in the cell division stage (G_2M), and in the DNA synthesis stage to only 4%. The shift of S to the G_1 phase was not seen with tumor cells cultured with normal macrophage or normal cells incubated with activated macrophages.

In vivo studies of Ehrlich ascites tumor cells before and after administration of pyran copolymer also showed a significant decrease in mitotic activity and an increase in cells in the resting phase. It appears from these results that activated macrophages may induce tumor cells with lower DNA content which eventuates their destruction.

5. Distribution of Polyanions in the Host

The mode of administration of an agent can have a significant effect on its activity based on the barriers and carriers that are encountered. For example, polycarboxylic acid polymers are totally ineffective if given orally, but several interesting biological events are triggered when administered intravenously or intraperitoneally. Studies have been carried out to evaluate the organ distribution of ^{14}C and ^{3}H labelled poly(acrylic acid) administered intravenously[60] as well as C^{14} labelled pyran administered intravenously[61] and intraperetoneally.[62]

Mück *et al.*[60] polymerized isopropyl acrylate-2,3-^{14}C and isopropyl acrylate-2,3-^{3}H using anionic initiators. These isotactic polymers were fractionated and hydrolyzed to poly(acrylic acid). These radiolabeled polymers were injected intravenously into mice in 40 and 100 mg/kg doses. About two-thirds of the radioactive material was excreted within the first two days and 10% still remained in the body after nine weeks. After 4 h, very little was found in the blood (0.2 μg/g). The main distribution was throughout the kidneys, liver, spleen, and bones. Although the peak concentrations in these organs occurred on day 14, approximately 90% of the maximum radioactivity in these areas was observed within 4 h after injection. Very little was found in the muscles or the brain. The only organ that showed any difference in the concentration of radioactive material with molecular weight was the spleen where a direct relationship was observed.

^{14}C-pyran was administered intravenously (i.v.) by Munson *et al.* (unpublished results)[61] and intraperitoneally (i.p.) by Papamatheakis *et al.*[62] The bulk of the ^{14}C was excreted within 24 h and less than 0.2% of the residual activity remained in the blood. The rest of the ^{14}C material was distributed throughout the reticuloendothelial system with most of the radioactivity residing in the liver and the spleen. Peak levels in the liver were found five days after i.v. injection and two days after i.p. injection. The peak levels of ^{14}C-pyran in the spleen occurred on day 1 and day 5 for i.p. and i.v., respectively. Similar to poly(acrylic acid), the spleen levels were less than those found in the liver.

No detectable ^{14}C material was observed at any time in the brain after i.p. administration, whereas after i.v. administration small amounts (0.2%) were detected. The concentration in the lungs 24 h after i.p. injection was small compared to the liver and spleen, while i.v. injection resulted in 4% of the residual ^{14}C activity in the lung which decreased to 1% on day 7 and continued to decline. The high activity in the lung after i.v. injection is probably due to the lung capillary beds being encountered soon after administration.

The kinetics of the distribution of radiolabeled polycarboxylic acid polymers by i.v. and i.p. injection indicate a difference in the time for organs to achieve peak levels. These patterns, however, demonstrate that polycarboxylic acid polymers have a reticuloendothelial distribution and concentrate primarily in the liver and the spleen.

6. Cellular Uptake of Polycarboxylic Acid Polymers

From the distribution data, it has been rationalized that most of the polycarboxylic acid polymers accumulate in the organs associated with the reticuloendothelial system. The central question is how these synthetic macromolecules active the macrophages. The present information indicates that they are taken into the macrophage cells where the process of activation takes place.

Generally, the pinocytosis of a substance can occur in two ways. Absorptive pinocytosis is a low-energy process whereby the material passes through the cell membrane and can be determined by uptake of ^{198}Au by the macrophage. Fluid-phase pinocytosis is a high-energy mechanism whereby a volume of fluid near the surface of the cell is engulfed.

Very few studies have been concerned with whether polycarboxylic acid polymers were actually internalized by macrophages with the exception of those done by Pratten *et al.*[(63–67)] They observed that ^{14}C-pyran and ^{125}I-labeled pyran were rapidly accumulated *in vitro* by rat peritoneal macrophages. The uptake of pyran was 100-fold more rapid than ^{125}I-poly(vinyl pyrrolidine). Uptake was inhibited at 4°C as well as in the presence of 2,4-dinitrophenol. Since the rate of uptake was consistent with an adsorptive pinocytosis mechanism, it was concluded that macrophage activation is due to internalization of the polymer rather than binding to the cell surface.

The rate and mechanism of pinocytosis is related to the nature and physical characteristics of the substance being pinocytized. For example, enhanced positive charge and hydrophilicity increase pinocytic uptake,[(63,66)] as well as the inclusion of substrate into a liposome. A study is currently underway to evaluate all three of these characteristics for a number of specifically designed polycarboxylic acid polymers.[(68)]

7. Clinical Effects of Polycarboxylic Acid Polymers

Drug–water solubility is essential for systemic administration, since an injection into the bloodstream of undissolved particles can cause "colloidoclasmic shock" leading to major clinical toxicity. Most polyanions, at a physiologic pH, are water soluble and may be distributed in a living host by means of blood circulation, cellular transport by mobile phagocytic cells, and by absorption on cell surfaces.

Pyran copolymer was intravenously administered to advanced cancer patients who were no longer responding to other treatment modalities. Survival was estimated after one month of treatment of patients who were taken off other chemotherapy or radiation for at least two weeks previously.

The polymer was given to 62 patients. The limiting toxicity dose was governed by thrombocytopenia and fever which was experienced by half of the patients. Clinical studies were ceased due to these and other side effects until the recent development of less toxic pyran. The new pyran drug has a narrower polydispersity and lower molecular weight. Furthermore, it is being administered as a more compatible calcium salt interperitoneally rather than intervenously. Two phase I studies on this agent have been completed and phase II evaluations are in progress.

The other polycarboxylate that is being studied clinically is NED 137, a half-acid–half-amide polymer with imide groups. It is prepared from low molecular weight (1000–2000 MW) poly(maleic anhydride-alt-ethylene). The clinical study consisted of 212 patients with a variety of tumor types, tumor burdens, and prior therapeutic treatments. The efficacy of NED appears to depend on the tumor type and extent of the disease. It seems that patients with minimal disease residue were more responsive to NED; however, the lack of appropriate control groups does not allow for any meaningful conclusions. Similarly, longer life spans were reported for patients with colorectal and pancreatic tumors than with SFU-BCG regimens; however, a more definitive and valid study must be conducted before meaningful conclusions can be drawn.

8. Summary

It has been well documented that polycarboxylic acid polymers have significant biological activity. Their most interesting property at the present time is their ability to act as immunostimulants and immunoadjuvants. These properties make them potential drugs for many diseases and biological disorders. The fact that they can elicit activated macrophages to be cytotoxic to tumor cells and not normal cells is especially important. Present research is being carried out to determine the mechanism(s) of activation by these synthetic materials.

Acknowledgments

The author wishes to thank the NIH for Grant AI-15612 for partial support of this work, as well as the Massey Cancer Center. I also wish to thank A. Kaplan, K. Kuus, A. Munson, and W. Regelson for their help.

References

1. L. G. Donaruma, R. M. Ottenbrite, and O. Vogl, *Anionic Polymer Drugs*, Wiley, New York (1980).

2. A. M. Kaplan, R. M. Ottenbrite, W. Regelson, R. Carchman, P. Morahan, and A. Munson in *Handbook of Cancer and Immunology*, Vol. 5, H. Walters (ed.), Garland Publishing Co., New York (1978), p. 135.
3. A. E. Munson, W. Regelson, W. Lawrence, and W. R. Wooles, *J. Reticuloendothel. Soc.* **7**, 285–375 (1970).
4. L. G. Baird and A. M. Kaplan, *Cell Immunol.* **20**, 167–176 (1975).
5. P. S. Morahan, W. Regelson, and A. E. Munson, *Antimicro. Agents Chemother. 1972*, 16–22.
6. P. S. Morahan and A. M. Kaplan, *Int. J. Cancer 1976* 82; S. J. Mohr, M. A. Chirigos, F. S. Fuhrman, and J. W. Pryor, *Cancer Res. 1975*, 2750–3654.
7. A. Billiau, J. Desmyter, and P. DeSomer, *J. Virol.* **5**, 321 (1970).
8. P. DeSomer, E. Declerca, A. Billiau, E. Schonnean, and M. Claesn, *J. Virol.* **2**, 878 (1968).
9. T. C. Merigan and M. S. Finkelstein, *Virology* **35**, 363 (1968).
10. A. Billian, J. Muyembe, and P. DeSomer, *Nature* **232**, 183 (1971).
11. E. M. Hodnett, J. Amirmoazzami, and J. Tien Hai Tai, *J. Med. Chem.* **21**, 652 (1978).
12. K. F. Mück, H. Rolly, and K. Burg, *Makromol. Chem.* **178**, 2773 (1977).
13. D. A. Tirrell, private communication.
14. R. M. Ottenbrite, K. Kuus,. and A. Kaplan, *Polym. Prepr., Am. Chem. Soc., Div. Polym. Chem.* **24**, 25 (1983).
15. R. M. Ottenbrite, E. Goodell, and A. E. Munson, *Polymer* **18**, 461 (1977).
16. A. E. Munson, J. M. Yeager, S. E. Loveless, and R. M. Ottenbrite, *J. Reticuloendothel. Soc.* **18**, 406 (1975).
17. B. M. Culbertson and B. C. Trivedi, *Maleic Anhydride*, Plenum Press, New York (1982).
18. G. B. Butler, *J. Polym. Sci.* **48**, 279 (1960).
19. D. S. Breslow, *Pure Appl. Chem.* **46**, 103 (1976).
20. T. C. Merigan, *Nature* **214**, 416 (1967).
21. T. C. Merigan, *New Engl. J. Med. 1967* , 277–1283.
22. T. C. Merigan, Ciba Foundation Symposium on Interferon, G. W. Wolstenholme and M. O'Connor, eds., J. A. Churchill Ltd., London, England (1967), p. 50.
23. T. C. Merigan and W. Regelson, *N. Engl. J. Med.*, 277, 1283 (1967).
24. E. DeClereq and T. C. Merigan, *Arch. Intera. Med. 1970*, 126–194.
25. S. J. Mohr, M. A. Chirigos, F. S. Fuhrman, and J. W. Pryor, *Cancer Res.* **35**, 3750 (1975).
26. P. S. Morahan and A. M. Kaplan, *Int. J. Cancer* **17**, 82 (1976).
27. W. Regelson, *Adv. Exp. Med. Biol.* **1**, 315 (1967).
28. T. C. Merigan and M. S. Finkelstein, *Virology* **35**, 363 (1968).
29. E. DeClereq and T. C. Merigan, *J. Gen. Virol.* **5**, 359 (1969).
30. W. Regelson, A. Munson, and W. Wooles, *International Symposium on Standards of Interferon and Interferon Inducers*, London (1969) and *Symposium Series Immunobiological Standards*, Vol. 14, pp. 227–236, Karger, Basel, New York (1970).
31. F. F. Pindak, *Infect. Immunol.* **1**, 217 (1970).
32. D. J. Givon, J. P. Schmidt, R. J. Ball, and F. F. Pindak, *Antimicrob. Agents Chemother.* **1**, 80 (1972).
33. J. Y. Richmond, *Infect. Immunol.* **3**, 249 (1971) and *Arch. Ges. Vi. Rusfersch.* **36**, 232 (1972).
34. G. B. Schuller, P. S. Morahan, and M. J. Snodgrass, Tenth National Meeting of the Reticulo Society, Abstract 28 (1973).
35. C. H. Campbell and J. Y. Richmond, *Infect. Immunol.* **7**, 199 (1973).
36. W. Regelson and A. E. Munson, *Ann. N.Y. Acad. Sci. 1970*, 173–831.
37. Y. Shamash and B. Alexander, *Biochim. Biophys. Acta.* **1**, 449 (1969).
38. M. A. Kapusta and J. Mendelson, *Arthritis Rheum.* **12**, 463 (1969).
39. D. W. Baxter, M. W. Rosenthal, and A. Lindenbaum, Abstracts of the 21st Annual Meeting of the Radiation Research Society, St. Louis, Missouri, (April 29, 1973);

N. E. Egan, G. S. Kalesperus, E. S. Moretti, and J. J. Russel, Annual Report, Division of Biological Medical Research, Argonne National Laboratory (1972), pp. 121–125.

40. W. Regelson, Biologically active water-soluble polymers, in: *Polymer Science and Technology* (N. M. Bikales, ed.), Vol. 2, pp. 161–177, Plenum Press, New York (1973).
41. T. J. Leavitt, T. C. Merigan, and J. M. Freeman, *Am. J. Dis. Child.* **121**, 43 (1971).
42. A. E. Munson, K. L. White, and P. Klykken, *Cancer Res.* **16**, 329 (1981).
43. J. S. Mohr and M. A. Chirigos, *Progress Cancer Research Therapy*, Raven Press, New York (1977), p. 421.
44. R. M. Ottenbrite, in: *Biological Acitivities of Polymers* (C. E. Carraher and G. G. Gebelein, eds.), American Chemical Society Symposium Series 186, Washington, D.C. (1982).
45. R. M. Ottenbrite, K. Kuus, and A. M. Kaplan, Symposium of Polymers in Medicine, Porto Cervo, Sardinia (1982).
46. W. Regelson, S. Kuhar, M. Tumis, J. Fields, J. Johnson, and E. Gluesenkamp, *Nature* **186**, 778 (1960).
47. J. E. Fields, S. S. Asculai, and J. H. Johnson, Monsanto Company, St. Louis, Missouri, U.S. Patent 4,255,537 (March 10, 1981).
48. R. E. Falk, K. Makowka, N. A. Nossal, J. A. Falk, J. E. Fields, and S. S. Asculae, *Brit. J. Surg.* **66**, 861–863 (1979).
49. R. E. Falk, L. Makowka, N. A. Nossal, L. E. Rotstein, and J. A. Falk, *Surgery* **88**, 126 (1980).
50. R. M. Ottenbrite, E. Goodell, and A. Munson, *Polymer* **18**, 461 (1977).
51. R. M. Ottenbrite, The antitumor and antiviral effects of polycarboxylic acid polymers, in: *Biological Activities of Polymers* (C. E. Carraher, Jr. and C. G. Gebelein, eds.), p. 205, American Chemical Society Symposium Series 186, Washington, D.C. (1982).
52. E. R. Unanue, *Adv. Immunol.* **15**, 95 (1972).
53. I. N. Abdour and M. Richter, *Adv. Immunol.* **12**, 202 (1970).
54. R. M. Schultz, J. D. Papamatheakis, and M. A. Chirigos, *Cell. Immunol.* **29**, 403 (1977).
55. K. Kuus, R. M. Ottenbrite, and A. Kaplan, *Fed. Proc.* **42**, 3364 (1982).
56. A. M. Kaplan, P. S. Morahan, and W. Regelson, *JNCI* **54**, 989 (1975).
57. R. P. Harml and B. Zbar, *JNCI* **54**, 989 (1975).
58. M. S. Melter, R. W. Tucker, and A. C. Breuer, *Cell. Immunol.* **17**, 30 (1975).
59. A. M. Kaplan, K. M. Connolly, and W. Regelson, in: *The Host Invader Interplay* (A. van Vendborshe, ed.), p. 479, Elsevier, Netherlands (1980).
60. K. F. Mück, O. Christ, and H. M. Kellner, *Macro. Chem.* **178**, 2785 (1977).
61. A. Munson (unpublished data).
62. J. D. Papamatheakis, R. M. Schultz, M. A. Chrigos, and J. G. Massicot, *Cancer Treat. Rept.* **62**, 1845 (1978).
63. M. K. Pratten, R. Duncan, H. C. Cable, R. Schhe, H. Ringsdorft, and J. B. Lloyd, *Chem. Biol. Inter.* **35**, 319 (1981).
64. M. K. Pratten and J. B. Lloyd, *Biochem. J.* **1980**, 567 (1979).
65. T. Koolstra, M. K. Prattan, and J. B. Lloyd, *Bioscience Repts.* **1**, 587 (1981).
66. M. K. Pratten, P. C. Millard, and J. B. Lloyd, *Mioscience Repts.* **1**, 125 (●●●●).
67. R. M. Ottenbrite and J. Sunamoto (unpublished results).
68. R. M. Ottenbrite, K. Kuus, and A. L. Kaplan, IUPAC, Macromolecular Symposium Proceedings, p. 366 (1982).

19

Polymeric Antitumor Agents on a Molecular and Cellular Level

Klaus Dorn, Gerhard Hoerpel, and Helmut Ringsdorf

1. Introduction

During the last 80 years polymer science developed from H. Staudinger's controversial hypothesis of the existence of macromolecules to highly sophisticated applications such as polymers in medicine. Within the field of pharmacologically active polymers, the area of polymeric antitumor agents is a new and attractive field, where polymer scientists try to combine the advantages of macromolecular systems with the requirements of tumor therapy. That is to say, that polymer chemists have to enter neighboring fields, for example, the rapidly developing area of the life sciences. This interdisciplinary thinking may lead to new concepts, which could serve as guidelines in the thicket of facts arising from the innumerable attempts to contribute to the solution of the cancer problem. The present survey does not review the whole field of polymeric antitumor agents, but summarizes facts and speculations fitting into the hypothesis of antitumor agents on a molecular and cellular level.

1.1. Definition: Molecular Level—Cellular Level

Within the meaning of this definition most of the present antitumor agents may be regarded as drugs on a molecular level. A certain concentration of molecules is required inside the cell to damage for example, the DNA, or inhibit metabolic pathways and thus provide cytotoxicity. The nature of action leading finally to cell death has a chemical or biochemical character in its first step.[1] For example, cyclophosphamide[2] (CP) (3) is known to cross-link DNA irreversibly (alkylating agent). Methotrexate[3] (MTX) (9), an antimetabolite, inhibits the enzyme dihydrofolate reductase. Subsequently, in both cases cells are unable to survive due to damages on a

Klaus Dorn, Gerhard Hoerpel, and Helmut Ringsdorf • Johannes Gutenberg-Universität, Fachbereich Chemie, Institut für Organische Chemie, Joh.-Joachim-Becher-Weg 18-22, D-6500 Mainz, West Germany.

molecular level. Both CP and MTX are clinically established antitumor agents against a wide spectrum of human cancers.

In nature, however, one may observe another strategy to combat antigenic particles, that is, transformed cells. The immune system recognizes the antigen. Subsequently, antibodies or, alternatively, killer cells destroy the tumor cell membrane. Membrane damage, that is, the destruction of the compartment character of the cell, is a physical phenomenon at least in its initial steps. The internal medium is no longer separated from the cell's environment, thus leading to cell death. Systems which destroy a tumor cell as a whole in this manner, may be called antitumor agents acting on a cellular level.

1.2. Tumor Cells and Tumor Therapy

The target of any antitumor agent is tumor cells. It must be considered, however, that there is no such thing as "the tumor cell." More than 100 human tumor species are known, all significantly different from each other (cf., e.g., References 4 and 5). In addition, each tumor species consists of several cell subpopulations, which differ, for example, in cell surface structure, that is, antigenicity.[6]

Tumor cells are derived from normal cells. Although there are a number of reported hypotheses concerning carcinogenesis (cf., e.g., References 7 and 8), we still lack information about the molecular and cellular mechanisms involved during the transformation of a normal cell to a tumor cell. The consequences of cell transformation, however, are known in more detail.[9,10] The biochemical mutations of DNA structure induce errors in DNA replication and subsequently alterations in cell metabolism, cell structure, and the behavior of cells in tissues.[11] Table 1 summarizes some differences between normal and neoplastic cells.

It is unquestionable that transformed cells are insufficiently recognized by the immune system. This may be due to the insufficient antigenicity of cancer cells (i.e., the differences between normal and transformed cells are not severe enough) or an immune defect of the body. With regard to both points, experiments are reported to overcome the inadequate immune response by, for instance, polymer-mediated stimulation of the macrophages (see Section 2.3.4). The question how to mimic the immune response may be rather speculative, but nevertheless leads to interesting results (see Section 3). In general, tumor therapy encloses radiation treatments, surgery, and chemotherapy. Whereas radiation therapy and surgery have optimal prognoses when the tumor is localized, chemotherapy treatments are administered more frequently when the tumor is spread over the whole body (fast, proliferating, and metastasizing tumors).[11] Ideally, chemotherapy

Table 1. Characteristics of Tumor Cells as Compared with Normal Cells

Properties	Ref.
CELLULAR METABOLISM AND INTRACELLULAR ENVIRONMENT	
1. Increased content of methylated nucleosides	[14]
2. Different enzyme patterns	[15,16]
– Increased activity of DNA- synthesizing and decreased activity of DNA-catabolizing enzymes	[15,16]
– Lack of asparagine-synthetase (several types of leukemia)	
– Increased activity of proteolytic lysosomal enzymes	[18]
– Strongly decreased activty of Mn^{2+}-superoxide dismutase	[19,20]
3. Higher need for exogenous Zn^{2+}	[21]
4. Lower Ca^{2+} and higher K^+-concentrations	[22]
5. Lower pH of cytoplasm (after injection of glucose)	[23,24]
MEMBRANE STRUCTURE AND PROPERTIES	[25–30]
6. Higher rate of endocytosis	[31,32]
7. Altered phospholipid contents	[28]
8. Different glycoproteins (lectin receptors) of the membrane	[33–36]
9. Tumor associated surface antigens	[37,30]
10. Higher amounts of contractile proteins (actin) altering the shape of cancer cells	[38]
CELL GROWTH AND BEHAVIOR OF CELLS IN TISSUES	
11. Different cell cycle	[39]
12. Impaired cell proliferation	[40]
13. Lower cohesion in tissues	[41]
14. Lower contact inhibition	[41]
15. Metastasizing activity	[42,43]
16. 'Invasiveness' = active penetration into other tissues	[41]
17. Increased locomotion ability	[38]

should not damage the tumor cells, but retransforms them to normal cells.[12] Although this aim seems to be rather speculative, there is one interesting experiment reported. Zimmermann and Sutter[13] found that addition of a new neuronal hormone, the nerve growth factor (βNGF), to malignant neurones results in a retransformation of the neoplastic cells to normal cells monitored by changing shape and functions of the cells.

In reality, chemotherapeutic agents are designed to damage tumor cells irreversibly. The relatively moderate selectivity of most of the established anticancer drugs is based on the high mitotic activity of the rapidly proliferating tumor cells (see 11 in Table 1), compared to normal cells. Unfortunately, some cell types like liver or lymph system, which are also dividing rapidly, are sensitive to anticancer agents in a similar manner leading to severe and known side effects. Taking advantage of the differences between neoplastic and normal cells to direct anticancer agents selectively to the tumor cells, seems to be a promising approach. The asparaginase treatment of certain leukemias,[11] requiring an extracellular asparagine supply, has already been used successfully. Improvements coming from the higher endocytotic rate of certain cancer cells or the expression of tumor-associated antigens will be discussed later (Section 2.4.1). Yet one must consider that although the differences known so far may be used in clinical tumor therapy, they provide only a narrow basis for selective cancer chemotherapy.

Many approaches have been made to increase the canceroselectivity (and thus reduce toxicity) of established compounds rather than synthesizing or isolating new drugs. As early as the beginning of the century, Paul Ehrlich[44] proposed targeting to affected tissues by means of appropriate receptors. A better understanding about these receptors comes from investigations on the characteristics of cell morphology and cell function, as well as the mechanisms of cell–cell recognition in particular.[45] In order to direct the anticancer agents to the desired cell receptors, scientists placed their hopes in the use of carriers, both low molecular weight and polymeric, which should increase the selectivity of drugs up to the ultimate canceroselectivity.

1.3. The Carrier Concept

In order to direct chemotherapeutic drugs to their desired place of action, attempts have been made to combine the drug with tumor-cell-specific transport moieties such as hormones.[46,47] Thus the qualities of two compounds (tumor cell specificity and cytotoxicity) should be combined.

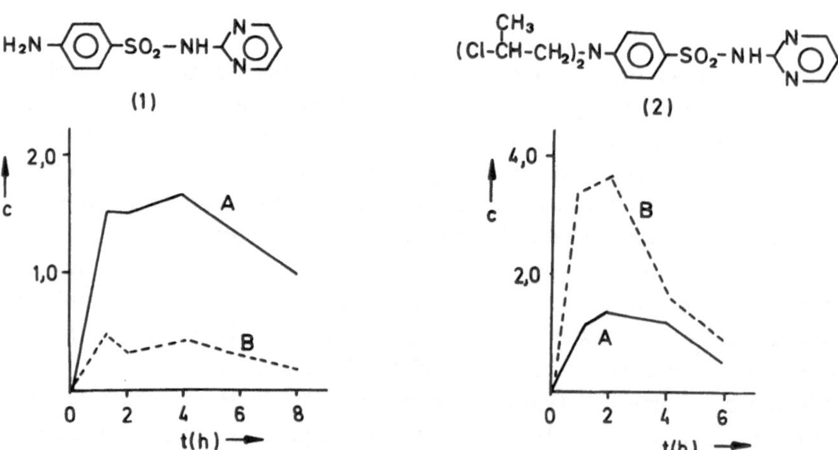

Figure 1. Different cell uptake of sulfadiazines (1) and (2) by liver and tumor cells (Walker tumor in rats). *c*: μg of compound per mg tissue; *t*: time after intraperitonal injection (i.p.); A: tumor cells; B: liver cells.[50]

Often, the combination of two characteristic properties met with substantial difficulties, that is, did not lead to the desired combination of advantages.[48,49] Frequently, the structural alteration of each of the two compounds leads to a complete loss in activity. In one classical experiment, Conners *et al.*[50] investigated the influence of substituents on the tumor-cell-specific uptake of sulfadiazine. In Figure 1 the uptake of sulfadiazine (1) by Walker tumor tissue and liver tissue compared to the modified sulfadiazine (2) is shown. The modified sulfadiazine (2), mainly taken up by liver cells, failed to show the cancerospecificity of free sulfadiazine. Conclusively, the dramatic structural change of the low-molecular-weight carrier by attaching it to the alkylating lost system, results in a complete loss of its targeting properties.

An approach to overcome this problem may be the use of high-molecular-weight drug carriers. In these systems, the ratio carrier/drug can be varied to a large extent so that the carrier still retains its properties. Unlike low-molecular-weight carriers, such as sulfadiazine or hormones, the homing property is only one aspect with respect to macromolecular carriers. Other features are solubilizing and detoxifying properties, or depot properties (changes in body distribution). Three different types of high-molecular-weight drug carriers are discussed at present: cells,[51] liposomes,[52] and polymers [53-59] (see Section 2).

In the following we would like to focus our attention on one of these carriers, which has advanced to the point where first successes can be seen, namely, polymers.

2. Antitumor Agents on a Molecular Level

2.1. Model for a Polymeric Drug Carrier

Figure 2 illustrates the general model for a suitable polymeric drug carrier, where different structural units are linked to the polymeric bone.[60] The model has the following properties:

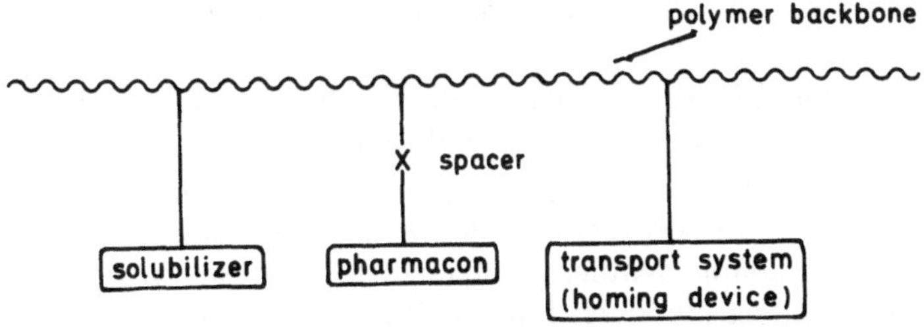

Figure 2. Model of pharmacologically active polymers.[69]

1. The solubilizing unit, sometimes the backbone itself, enables intravenous (i.v.) application of the whole system. Moreover, parameters, such as hydrophilicity, hydrophobicity, charge, and density, influence pharmacokinetics and pharmacodynamics of the drug.[61]
2. The backbone can be varied in biostability or molecular weight, in copolymer composition, and in the ability to provide certain domains (e.g., hydrophobic bags). Specific molecular weight ranges are essential for membrane permeation (retention properties of the blood–brain barrier,[63,64] and stimulation of endocytotic uptake).[65]
3. The drug may be linked to the carrier via a bioerodible (hydrolyzable) or biodegradable bond (spacer length and stability). The pharmacon, which is often derived from one of the established antitumor agents, is fixed either as an active metabolite or as a masked species (transport form).
4. Finally, the "homing device" acts as a vector to direct the whole transport system to the desired target tissue. The realization of the homing device appears to be the most speculative and questionable approach of the polymeric carrier concept.

The great variety of possible structural manipulations on the polymeric carrier, as well as the special separation of the different components, admits positive alterations in pharmacological properties, for example, body distribution, transmembrane penetration, biotransformation, and also pharmacodynamics. The main potential advantages of a polymeric carrier may be summarized as:

- Decreased toxity of the drug (e.g., by masking the cytotoxic species.[66,67]
- Sustained intracellular release of the active species (see Section 2.3.2).
- Increased selectivity by affinity chemotherapy (see Section 2.4).

In contrast to the advantages of polymer-specific properties mentioned so far, there are important disadvantages, objections, and problems[68–70]:

- Biocompatibility of water-soluble polymers has so far only been tested in a few isolated cases.[70] We still lack broader experiences with hemolytic activity, pyrogenicity osmotic properties, and interaction with plasma constituents under *in vivo* conditions.
- Nondegradable synthetic polymers can be used in order to obtain depot effects.[71,72] A long-lasting systematic storage of high-molecular-weight compounds can, however, not be accepted from

pharmacological and toxicological viewpoints. Therefore the development of biodegradable polymers is becoming more important. Oligomers or polymers of low molecular weight, which can be excreted, may also be used.[70]

- There have been few systematic investigations of the body distribution of polymers and its change with time.[61]
- Little attention has been given to immunological reactions against pharmacologically active polymers.[73]
- Synthetic polymers investigated so far are not resorbed by the gastrointestinal tract[74] – a fact which decisively limits the choice of appropriate drug formulations.

Nothing could be said in favor of polymeric pharmaca if they were just "me-too-drugs"[75] – drugs which can do a job that low-molecular-weight pharmaca can do just as well. Therefore, it is reasonable to concentrate efforts on the development of pharmacologically active polymers in such areas where low-molecular-weight drugs have failed or have only given insufficient therapeutic results. One such area is the chemotherapy of cancer. There is no doubt that up until now, the use of polymeric antitumor agents has only been moderately successful. However, investigations carried out so far give clear-cut evidence that these drugs open up new possibilities which could not be realized by the use of low-molecular-weight drugs.

2.2. How Do Macromolecules Enter Cells?

Much attention has been given to the process of cell uptake of polymeric drugs acting on a molecular level (i.e., References 65 and 76–78). But before a polymeric drug conjugate arrives at the cell wall, it has to overcome several different barriers following i.v. injections[11] (e.g., reticuloendothelial system, cell wall, and cell membrane). Restrictions to the size and structure of the polymeric system, with regard to the above-mentioned barriers, have been discussed elsewhere [70,79] and will not be dealt with in this survey. In a first approximation, the cell membrane can be considered as the most important barrier.

2.2.1. Polymer–Membrane Interaction

There is accumulating evidence that much of the behavior of a living cell, including cell–cell recognition as well as adhesion functions and cell uptake, is mediated by the cell surface.[80,81] As a result, attention is focused on the processes involved during the interaction of natural and synthetic macromolecules with cell surfaces and subsequent cell uptake.[82] The primary process of adsorption at the cell membrane may be discriminated into a specific and an unspecific one. The former includes the receptor-mediated recognition process; the latter is just an adsorption via electrostatical or hydrophobic interaction with the lipid moiety of the bilayer.

A great understanding of the so-called unspecific adsorption process can be reached by investigating interactions of synthetic and natural macromolecules[83–85] with model membranes such as monolayers and liposomes, as well as biomembranes such as erythrocytes. Tirrell studied systematically the adsorption of synthetic polymers like DIVEMA (divinyl ether–maleic anhydride copolymer) with the membrane systems mentioned above.[86,87] He observed that the adsorption of the anionic DIVEMA at the polar membrane surface is promoted by the presence of Ca^{2+} ions, a bivalent cation. On the other hand, DIVEMA is known to display an antitumor activity via macrophage stimulation (see Section 2.3.4). This stimulation mechanism is induced by a Ca^{2+} flux. whereas polyanions like DIVEMA act as mediators.[88] Some authors have speculated that DIVEMA may function as a macromolecular ionophore.[89] From Tirrell's experiments, however, there was evidence that the DIVEMA–Ca^{2+} complex is not inserted into the membrane, thus not being able to act as ionophore. It is definitely rather speculative to predict biological actions from pure adsorption measurements with model membranes; but it may be regarded as a stimulation of the adsorption mechanism at macrophages, thus giving insight into some molecular processes.

The nature of interaction may not only be electrostatic, as it was claimed for highly charged polymers like DIVEMA and the cationic poly-L-lysine (PLL),[90] but also hydrophobic interaction forces may play an important role. These were studied using incorporation experiments of membrane proteins,[28,91] such as the insertion of the hydrophilic–hydrophobic glycophorin into synthetic model membranes.[83] For this incorporation or fusion mechanism, it is important that the polymer (synthetic or natural) forms amphiphilic domains, at least in its tertiary structure (see Section 2.3.5).

A more specific process of recognition between a polymer and a cell membrane is the receptor-mediated type of adsorption. It has long been known that the cell surface exposes a great variety of cell-specific receptors (e.g., glycoproteins) which regulate not only the cell–cell recognition, but also the specific adsorption of macromolecules.[92] Subsequently, a certain molecule to be transported is released (transport proteins) or a particular function is initiated (e.g., transmembrane signaling induced by hormones).[93] How one can take advantage of the different cell surfaces with respect to tumor affinity chemotherapy will be discussed later (see Sections 2.4).

An instructive model demonstrating the receptor-mediated adsorption

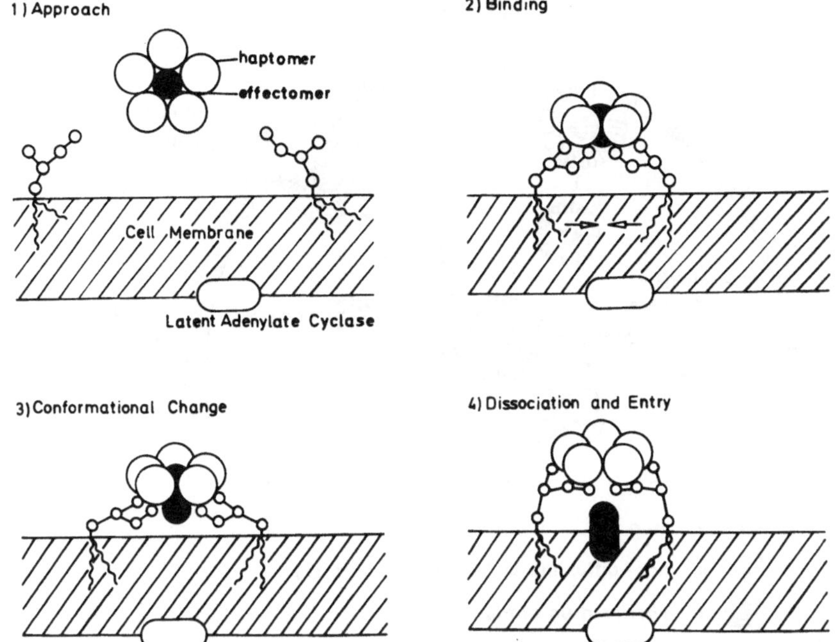

Figure 3. Hypothetical mechanism for the binding of cholera toxin to the cell surface as a model for receptor-mediated cell uptake.[94]

at the cell membrane with subsequent function is the interaction of cholera toxin with G_{M1} ganglioside (Figure 3).[94] Only one molecule of the toxin is required to kill the cell. The toxin is known to consist of two parts.[95] Part A, the so-called effectomer, is responsible for the toxic action. Part B is a certain lectin able to recognize carbohydrate structures on the cell surface (haptomer). The processes involved can be separated into several phases:

1. Recognition of the cell surface (Part B).
2. Binding to the receptor (Part B).
3. Transmission of the effectomer through the membrane (Part A and B?).
4. Activation of internal metabolic pathways (Part A).

If part A is separated from part B they do not produce any toxic effects.

The adsorption step is often followed by an internalization or cell uptake of the whole particle via endocytosis.

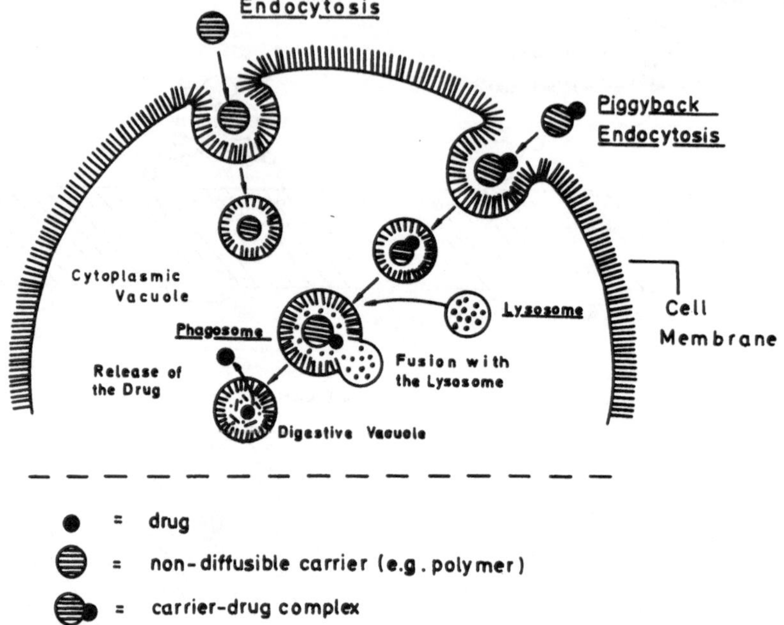

Figure 4. Schematic representation of cell uptake by endocytosis and piggyback endocytosis.[104]

2.2.2. *Cell Uptake of Polymers Via Endocytosis*

While low-molecular-weight compounds can cross the membrane barrier differently (diffusion, active transport), high-molecular-weight systems are able to enter cells by endocytosis only,[76] as shown in Figure 4.

Endocytosis[96] regulates the quantal cellular uptake of exogenous molecules from the cell's environment via plasma-membrane derived vesicles and vacuoles. The idea of polymer-mediated uptake of membrane-impermeable compounds [cf. piggyback endocytosis (de Duve)] is one important postulation in the model of polymeric antitumor agents on a molecular level, since the drug should act after internalization via damage of essential intracellular functions, which is the meaning of our definition.

Endocytosis has been divided into two categories: phagocytosis or "eating" and pinocytosis or "drinking."[96] Phagocytosis occurs by close contact of a segment of a plasma membrane to large particles (i.e., those visible by light microscopy) to the surface of the particle, including the surrounding fluid. Pinocytosis describes the uptake of smaller particles dissolved in the surrounding fluid (proteins, lipoproteins, hormones, antibodies, toxins, and low-molecular-weight solutes). Endocytosis is a temperature-dependent process requiring "cellular energy" like ATP. The rate at which a macromolecule is captured via endocytosis depends on both the morphology of the endocytotic vacuole (surface properties) and the nature of the material to be internalized. With respect to pinocytosis, which can be considered to represent the main route for a macromolecule to be transported into a cell, one may distinguish among three mechanisms dependent on rate and amount of uptake,[97] described in the following as a series of decreasing endocytosis rates:

1. *Stimulated Pinocytosis*: Often accompanied by a receptor recognition mechanism, thus achieving substrate selectivity in pinocytosis.
2. *Adsorptive Pinocytosis*: Known from highly charged polymers binding to the cell surface unspecifically, as mentioned above.
3. *Nonadsorptive Pinocytosis*: Exclusively dependent on the endocytotic activity of a cell, that is, the "drinking rate" of the cell and by which neither substrate selectivity nor incorporation of high amounts are achieved.

Lloyd and co-workers[98,99] investigated the pinocytotic uptake of potential drug carriers and solubilizers, such as charged polymers (vinylamine and DIVEMA) compared to a neutral plasma expander polyvinylpyrrolidone (PVP), by the yolk sac system and macrophages. Whereas the PVP was captured entirely by nonadsorptive pinocytosis, the uptake of the cationic vinylamine copolymer as well as the anionic DIVEMA was more efficient, probably because they adsorb at the cell surface. DIVEMA failed to stimulate pinocytosis,[100] contradictory to expectations from some earlier data.[101,102]

A stimulated uptake mechanism is reported to occur, for example, with CON A,[103] a plant lectin able to bind specifically to oligosaccharides of the cell surface. CON A is a tetravalent lectin, which induces receptor capping or so-called "coated pits" facilitating subsequent engulfment of the membrane. However, after succinylation of CON A, leading to a dimer and thus being unable to form coated pits, the molecule is entirely captured by nonadsorptive pinocytosis. These examples should demonstrate that by choosing a certain polymer as a drug carrier, one may influence the mechanism and thus the rate and amount of endocytosis uptake.

The special case of polymer-mediated drug uptake may be called piggyback endocytosis.[104] Figure 4 illustrates the model where a pharmacon with insufficient penetration into cells is attached to a macromolecule known to be efficiently internalized. Assuming that some tumor cells indeed have a higher endocytotic activity (see Table 1), a first step in drug selectivity could be achieved by that mechanism.

The fate of a pinocytotic vesicle, that is, in particular the process by which the drug ultimately enters the cytoplasm after leaving the vesicle, is a subject of controversy.[105] One process was reported by de Duve and Trouet (see Figure 4): the pinocytotic vesicle fuses with a lysosomic vesicle to build the digestive vacuole. Substrates for the lysosomic enzymes are biodegradable units (peptides, esters, amides, etc.). Polymeric drug carriers taken up via piggyback endocytosis and releasing the drug only after degradation of the spacer or the carrier may be called lysosomotropic carriers (de Duve *et al.*[106]).

2.3. Polymeric Carriers

There are two main approaches to designing polymeric antitumor agents:

1. Utilization of polymers showing anticancer properties themselves.[107–110]
2. Fixation of anticancer drugs to polymeric carriers, which may either be biologically inert[109] (like plasma expanders) or which belong to the former group. In the following we would like to focus attention on some selected examples and also problems,[111] to illustrate the different aspects of polymer fixation.

2.3.1. Solubilization and Detoxification of Drugs via Polymer Fixation

The lipophilic character and the severe toxicity of a great deal of the antitumor agents mentioned already in Section 1.2 are the main dose-limiting properties, in particular with the i.v. administration route. In the following example, the fixation of cyclophosphamide [Cytoxan®, 2-bis(2-chlorethyl)-amino-1,3,2-oxazaphosphorinane-2-oxide] (3) to polymeric carriers is described, illustrating the circumventions finally leading to the less toxic, water-soluble polymeric antitumor agent.

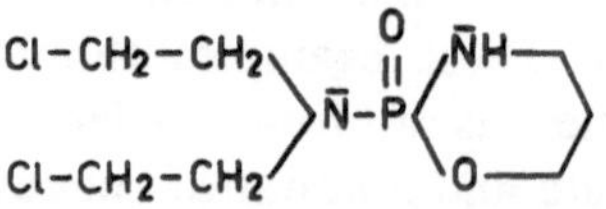

$Cl-CH_2-CH_2$ / $Cl-CH_2-CH_2$ $\rangle\bar{N}-P(=O)$ (NH, O ring) (3)

Cyclophosphamide (CP) (3)[2] may be regarded as the inactive transport form of the ultimate alkylation species *N,N*-bis(2-chlorethyl) phosphoric acid diamide (4). According to the model of drug carriers discussed above, CP was covalently linked to a water-soluble polymethacrylate system (5a,b).[112]

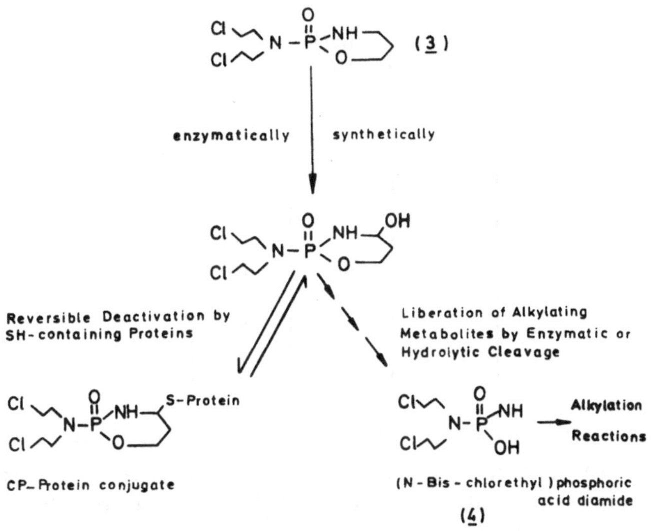

Figure 5. Metabolic pathway of cyclophosphamide (CP).[113,115]

However, the polymeric antitumor agent thus obtained produced only weak cytostatic effects in tumor-bearing animals. A better understanding of the polymeric drug and its failure to produce antitumor activity comes from investigations concerning the metabolic pathway of CP (Figure 5).[113–116]

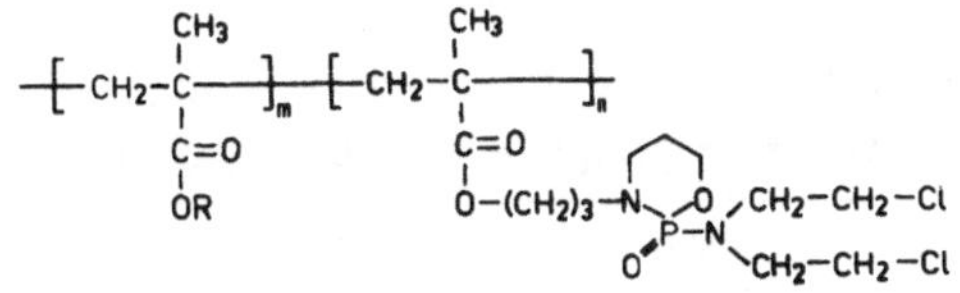

(5a) $R = -(CH_2)_2-SO-CH_3$ (5b) $R = -(CH_2)_2-\overset{\oplus}{N}(CH_3)_3\ Cl^{\ominus}$

CP (3) itself is inactive *in vitro* and must be hydroxylated *in vivo* by liver enzymes in order to liberate the ultimate alkylating species (Friedman acid) (4). The CP-containing polymer, however, was inactive. Even if, for example, CP is released from the polymer within the tumor cell, it would not damage the DNA, for according to its metabolic pathway, it has to interact with the liver enzymes first. The problem was that the polymer (5) indeed releases the prodrug (CP), but this prodrug cannot liberate the alkylating metabolite at the site of action.

The approach, synthesizing prodrugs which are more labile, leads to the fixation of sulfidoderivatives of CP[117–119] (6a–f), which are hydrolyzed very

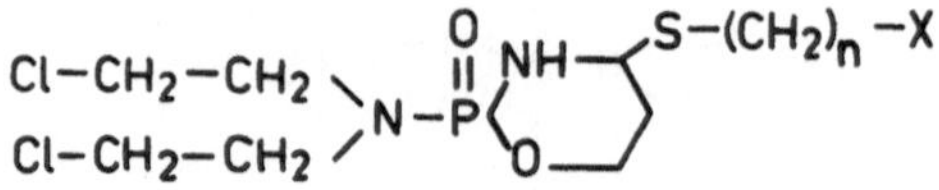

$$\mathrm{(Cl{-}CH_2{-}CH_2)_2N{-}P(=O)}\text{-ring(NH, O)}{-}\mathrm{S{-}(CH_2)_n{-}X} \quad (\underline{6})$$

6	n	X
a	2	$-CH_2OH$
b	5	
c	10	
d	2	$-COOH$
e	5	
f	10	

rapidly due to the nature of both spacer and polymeric backbone.[120] The conjugation of lipophilic CP-sulfido derivative (6e) to a water-soluble dextran carrier, a nontoxic, biodegradable polysaccharide, leads to a completely water-soluble system (7). The conjugate results in increasing

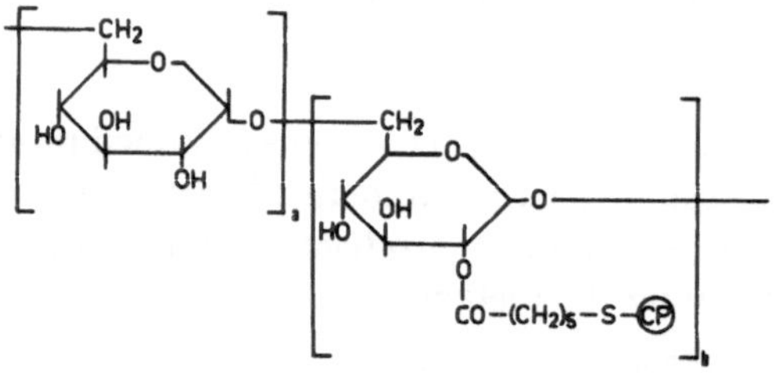

(7) [dextran unit]$_a$–[dextran unit with O–CO–$(CH_2)_5$–S–CP]$_b$

Table 2. *In Vivo* and *in Vitro* Data[a] of Some Polymeric Cyclophosphamide Conjugates

Derivative	Cytotoxicity EC_{50}[b] (μg/ml)	Acute/toxicity DL_{50}[c] (mg/ml)	Antitumor activity DC_{50}[d] (mg/ml)	Therapeutic index(TI)[e]
CP—S—(CH_2)—COOH(6e)	13	96	3.0	32
Dextran—CP (7)	4.2	141	1.4	100
Dextran	—	Not toxic	No effect	—
PEO—CP[f]	25	83	No effect	—
PEO	—	Not toxic	No effect	—
4—HO—CP	2	100	1.0	

[a] The pharmacological investigations were performed by N. Brock and J. Pohl, ASTA-Werke Bielefeld, West Germany.
[b] Tumor cells, YA-ASTA, EC_{50}: Dose producing cytotoxicity at 50% of the species.
[c] Animals, rats, DL_{50}: Lethal dose producing acute toxicity on 50% of the animals.
[d] Tumor, L5222, DC_{50}: Curative dose producing a curative effect on 50% of the animals.
[e] $TI = DL_{50}/DC_{50}$.
[f] Conjugate of polyethyleneoxide and (6e).

antitumor activity,[121] demonstrated in the lower curative dose (DC_{50} value) for CP equivalents in Table 2. Simultaneously, the lethal doses(LD_{50} values) are increased, showing a lower toxicity compared to the corresponding low-molecular-weight drug. This experiment illustrated the solubilizing and detoxifying capacity of the dextran polymer. Other conjugates have been synthesized[119b] with sometimes impressive, yet somewhat disappointing results. From this experiment uncharged as well as zwitterionic polymers appear to be suitable for drug carriers rather than positively or negatively charged macromolecules.

The problem of polymeric transport forms, that is, masking potency of synthetic, water-soluble polymers, was investigated in more detail by Schnee and Ringsdorf using hapten–polymer conjugates.[120]

In designing polymeric transport forms of low-molecular-weight agents, one has to consider not only that the active metabolite is liberated from the backbone, but also at which rate this process occurs.

2.3.2. Sustained Release by Polymer Fixation

The utilization of polymeric drugs was originally aimed to achieve depot effects, that is, the controlled slow release of a pharmacon into the body[71,72,122–125] thus obtaining a prolonged action within the therapeutic range. Although sustained-release preparations are mainly based on insoluble polymer systems (microspheres, microcapsules, polymer gels, implants), one may improve the therapeutic efficiency of drugs by utilizing soluble carriers providing a prolonged release of the active agent. In addition, the prolonged release seems to affect the part of polymer properties which presently can be manipulated best. This is possibly due to the great variety of hydrolyzable or biodegradable spacers, the topic of the following paragraph. Molz and Ringsdorf investigated daunomycin–polymethacrylate conjugates[126] (8a–c) linked via hydrolyzable hydrazon spacers. They

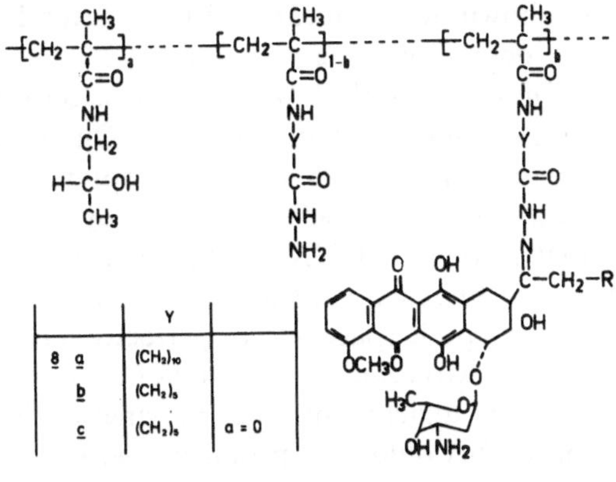

	Y	
8 a	$(CH_2)_{10}$	
b	$(CH_2)_5$	
c	$(CH_2)_5$	a = 0

Table 3. Effect[a] of Polymeric Daunomycin (DAU) Conjugates *in Vivo*[b] Compared to Free Drug

Compounds		Equivalent dose DNR (mg/kg)	Weight difference[c] (g)	Increase in life span (ILS)[d] (%)
Copolymer	*8a*	31.0	−0.1	43
		15.5	−0.5	37
Copolymer	*8b*	31.0	−0.5	56
		15.5	−0.1	53
Homopolymer	*8c*	31.0	0.9	51
		15.5	1.6	51
Daunomycin	*8*	4[e]	−2.0	17
		2	2.1	19

[a] The pharmacological experiments were performed by Rhone-Poulenc S.A., France.
[b] Drug treatment i.p. on day 0; tumor inoculation: 10^6 P388 cells on day 1 into $B_6D_2F_1$ mice (20 g).
[c] Weight difference between day 9 and day 0.
[d] Increase in life span: ILS = (T − C)/C × 100. T = life span of treated animals and C = life span of untreated animals.
[e] Maximal dose of daunomycin.

showed that the increase in life span following polymer conjugate application was superior to that of free daunomycin (DNR), both administered one day before tumor inoculation (Table 3). In this example the polymer acts as a drug depot, gradually releasing free DNR. A more sophisticated mechanism of sustained release was realized by Trouet and his co-workers, namely, the lysosomotropic carrier.

2.3.3. The Lysosomotropic Drug Carrier

The concept of lysosomotropic carriers with respect to cancer chemotherapy was shown to be meaningful by several authors.[127–130,76] According to its definition (see Section 2.2.2), Trouet designed polymer-linked antitumor agents[131] with drugs selectively cleavable by intracellular enzymes, but undegradable in serum. One instructive example were DNR–albumin complexes with different peptide spacers. Only the spacers with three and four amino acids are cleaved by lysosomal enzymes *in vitro*. Correspondingly, polymer–DNR conjugates containing such spacers produced an increase in life span superior to free DNR. The authors also proved the tri- and tetrapeptide spacers to be perfectly stable in serum. Thus, side rections arising from extracellular liberation of the free drug are avoided, resulting in a reduced toxicity of the polymeric drug preparation. In addition, this approach would be a promising possibility to increase the selectivity of antitumor agents if not only lysosomal-enzyme-specific spacers, but also "tumor-specific spacers," could be designed.

Shen and Ryser[77] reported *in vitro* experiments of methotrexate (MTX) linked to a biodegradable polyamino acid (poly-L-lysine, PLL) (9).

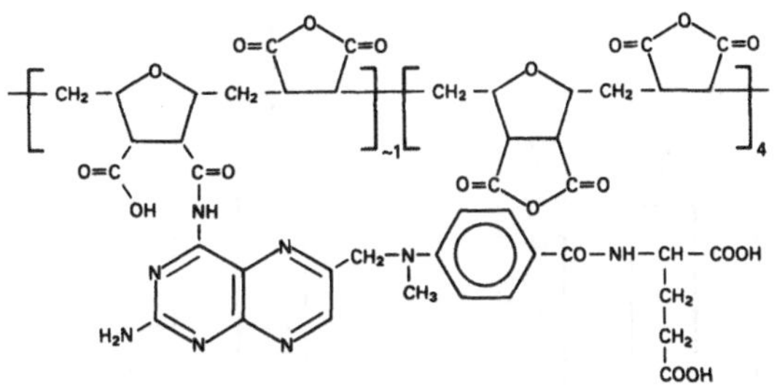

DIVEMA-MTX, NSC NO. 282447
MTX Content: 25.2%
$\bar{M}_n$ 24,000

(9)

The PLL–MTX conjugate was also proved to be cleaved by lysosomal enzymes. In addition, the PLL enabled the uptake of polymer-bound MTX into Chinese hamster ovarian cancer cells, resistant to free MTX. These two examples suggest that the controlled cleavability of the polymer–pharmacon bond via lysosomal enzymes plays a decisive role in the use of more cancer-specific polymeric antitumor agents.

After this discussion of controlled cleavable active metabolites of anticancer drugs from carriers, we would now like to concentrate on the bioactivity of the polymers themselves and in particular the immunogenicity.

2.3.4. *Immunoadjuvant Properties of Polymers*

The observation that highly charged polymers, both polycations[132,110] and polyanions,[54] can activate macrophages increased the hope that polymers might be used to stimulate immune response against tumor cells. Unlike bacteria or viruses, cancer cells are generally not antigenic. On the other hand, most of the antitumor agents are immunosuppressive. The attempts at linking those drugs to immunostimulating carriers were guided by the following conditions: that the carrier would overcome the immunosuppressive properties of the low-molecular-weight drug and, in addition, that the carrier would activate the macrophages. It may be worthwhile to note that, according to our definition, polymer-mediated immune activation may be regarded as a link between actions on the molecular and cellular level.

While the immunoadjuvant properties in general, for example, of DIVEMA, were frequently reported (see Reference 133), there is one experiment of a DIVEMA–MTX conjugate (9) where the activation of macrophages was studied.[134–137] The results indicate that the immunopotentiating effect of

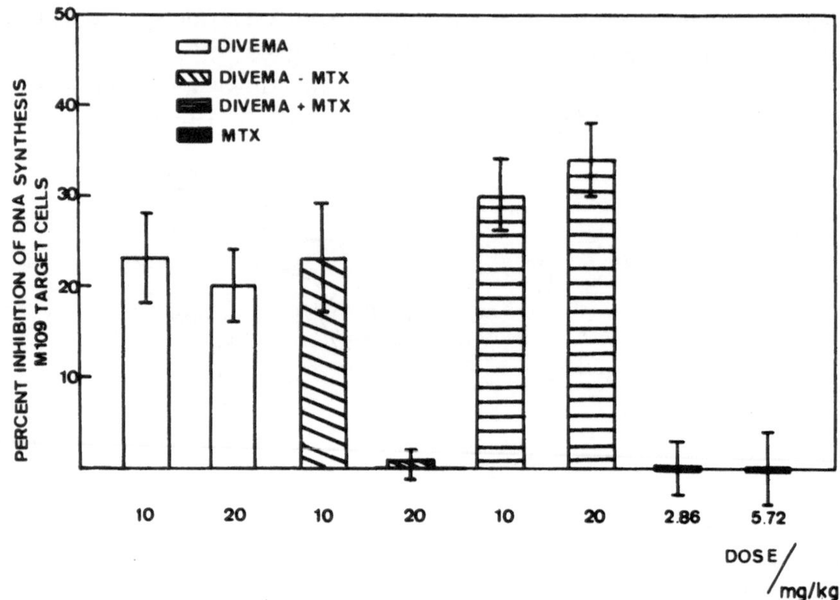

Figure 6. Effect of DIVEMA and DIVEMA–MTX (conjugate of DIVEMA and methotrexate) on the cytostatic activity of BALB/c peritoneal macrophages. Drugs at the indicated doses were given i.p. on day 0. Macrophages were harvested six days later and tested for their ability to inhibit M109 cell DNA synthesis.[135]

DIVEMA can be maintained in the DIVEMA–MTX complex. In this experiment, the cytostatic effects of mouse peritoneal macrophages harvested six days after drug treatment were compared (Figure 6). While the MTX did not activate the macrophages, almost all polymer preparations were efficient stimulators measured by inhibition of DNA synthesis in M109 target cells. The inhibitory effect of high DIVEMA–MTX-dose-treated macrophages was insufficient. This was probably due to the immunosuppressive properties of the MTX exceeding, at high doses of the conjugate, the carrier quality.

Unfortunately, these findings cannot be generalized, that is, transferred to other DIVEMA–drug conjugates.[138] Thus, corresponding CP-containing DIVEMA conjugates produced completely unexpected high toxicities,[120] which is unquestionably due to a polymer effect, but unfortunately in the wrong direction.

Apart from immunogenic properties of drug carriers in nature, the transport of, for example, hydrophobic molecules, is mediated by amphiphilic transport proteins and in particular by micelle-forming lipoprotein complexes. In order to achieve more biogenic drug carriers, the question arises as to how to mimic these transport proteins.

2.3.5. *Amphiphilic, Micelle-Forming Polymeric Drug Carriers*

The general scheme of polymeric drug carriers, described in Section 2.1, may act as an instructive model in order to imagine the structural composition of units, different in properties and desired functions. According to the amino acid sequence in proteins, a series of different units along a polymer chain may be called the primary structure. Thus the model represents the primary structure of polymeric drugs. However, little attention has been given to the conformation or tertiary structure of synthetic polymers, since most of them occur in the random coil conformation.

Klesse and Ringsdorf created an amphiphilic polymer[139](10), which arranges in "polymer micelles" spontaneously in aqueous solution, as illustrated schematically in Figure 7. Polymer micelles may be regarded as compartments with hydrophobic cores and hydrophilic shells.

As an example, polyethyleneimine (PEI), a hydrophilic water-soluble

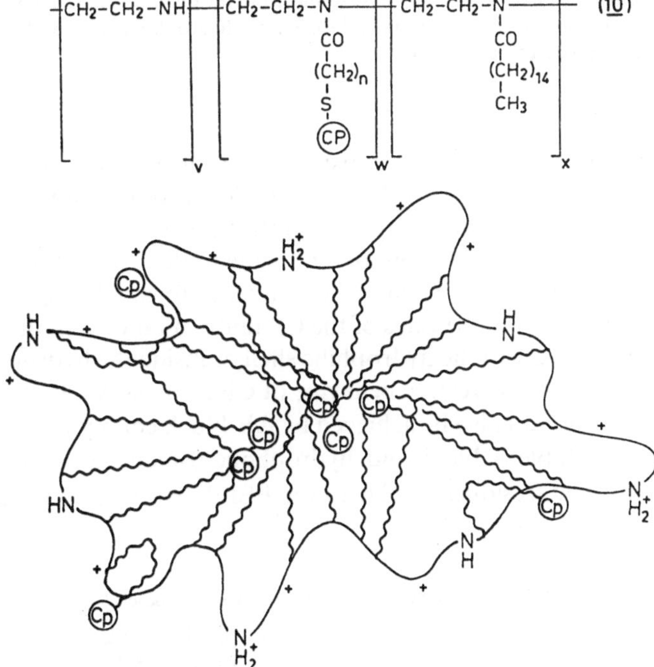

Figure 7. The micelle model—structure and conformation of amphiphilic micelle-forming ethyleneimene polymers.

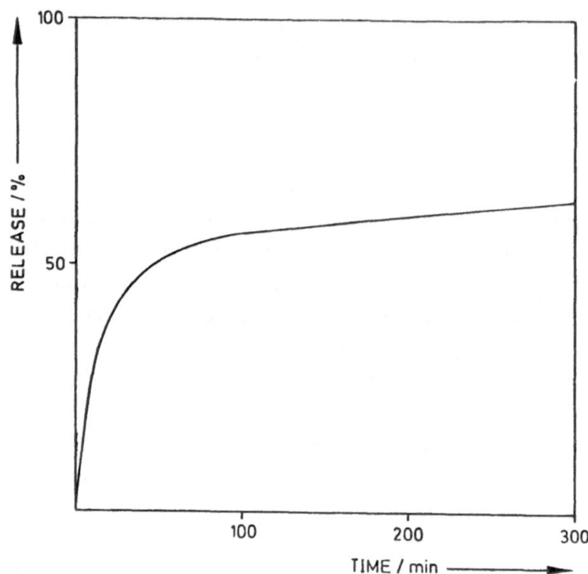

Figure 8. Biphasic release kinetics of drug from a cyclophosphamide-containing micelle-forming polyethyleneimine. The rapid phase results from the liberation of that CP, placed on the hydrophilic shell of the micelle. In contrast, the CP in the hydrophobic core of the micelle results in a slow-release phase (cf. Figure 7).

polymer, forms micelles upon statistical fixation of hydrophobic palmitic acid units to the backbone. Additional reaction with CP-sulfido derivatives (6) with different hydrophobic spacer lengths leads to a more or less effective insertion of CP into the hydrophobic core, due to the hydrophobicity (spacer length) of CP. The theoretical model of a micelle-forming drug carrier could be easily checked by release rates of the CP units. As shown in the model, CP may be bound closely to the hydrophilic shell or inside the hydrophobic core. This leads to a biphasic release behavior of CP as demonstrated in Figure 8. Micelle formation can also be achieved by AB-block copolymers[140] (11a,b) with a water-soluble A-block and hydropholic B-block containing both CP and palmitic acid residues similar to the PEI derivatives. A variation from

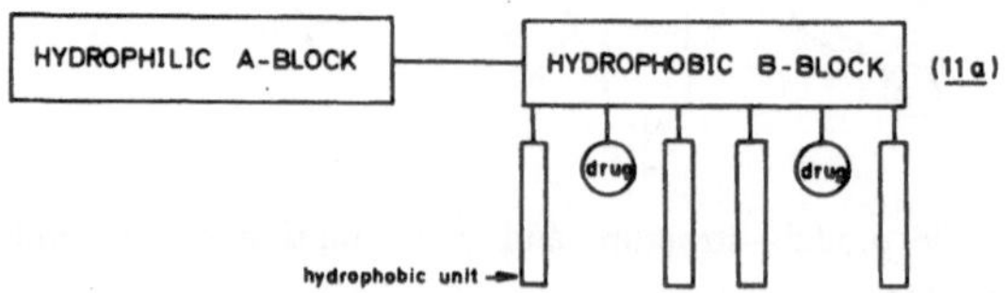

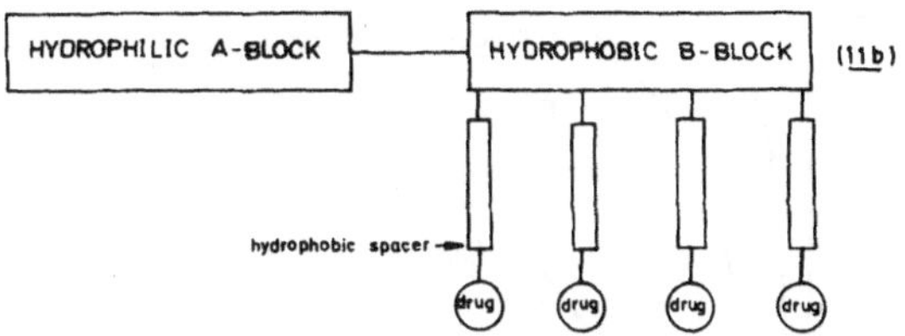

an amphiphilic, statistical copolymer described above to an amphiphilic block copolymer can certainly avoid the biphasic release behavior. In the following example, the A-block is a nontoxic polyethylene oxide (PEO) excretable from the body within several days. The B-block consists of a biodegradable poly-L-amino acid (12a,b). Thus, by polymer fixation the

$$CH_3\left[O-CH_2-CH_2\right]_a NH\left[CO-\underset{\underset{NH_2}{|}}{\underset{(CH_2)_4}{\underset{|}{CH}}}-NH\right]_b\left[CO-\underset{(CH_2)_4-NH-CO-(CH_2)_{10}-S-\text{CP}}{\underset{|}{CH}}-NH\right]_c \qquad (12a)$$

$$CH_3\left[O-CH_2-CH_2\right]_a NH\left[CO-\underset{(CH_2)_4-NH_2}{\underset{|}{CH}}-NH\right]_b\left[CO-\underset{(CH_2)_4-NH-CO-(CH_2)_{10}-S-\text{CP}}{\underset{|}{CH}}-NH\right]_c\left[CO-\underset{(CH_2)_4-NH-CO-(CH_2)_{14}-CH_3}{\underset{|}{CH}}-NH\right]_d \qquad (12b)$$

relatively labile CP-sulfido derivatives can be stabilized due to their insertion into the hydrophobic cores of the polymeric micelles. The liberation rate of the active metabolite 4-hydroxy-cyclophosphamide can be varied within a broad time spectrum from minutes to several hours with respect to the half-time ($T_{1/2}$) (see Figure 9).

In vitro studies[141] gave evidence that the CP-containing block copolymer acts as an intracellular depot for the active metabolite of CP. Figure 10 demonstrates the time course of DNA interstrand cross-linking in L1210 cells treated either with the low-molecular-weight drug or with the block copolymer. The analysis of DNA interstrand cross-links by DNA alkaline elution assay[142] turned out to be an appropriate method by which to measure the action of alkylating agents quantitatively. The cross-linking frequency *vs.* time shows a maximum for the low-molecular-weight drug at 4–5 h postincubation time, whereas the peak maximum correlated to the

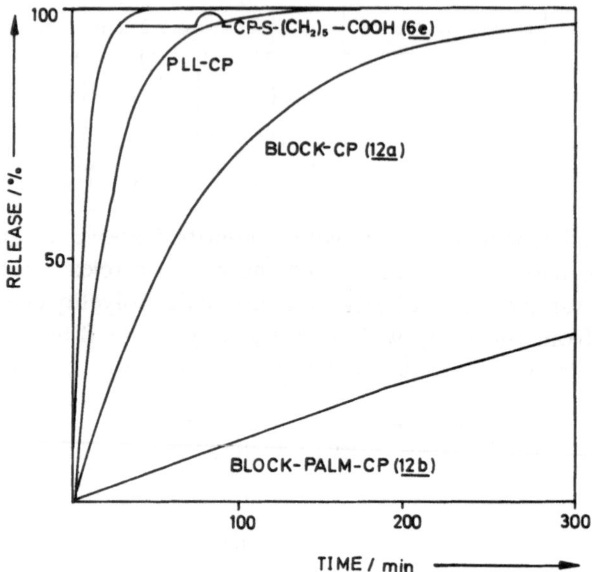

Figure 9. Variation of the release kinetics of cyclophosphamide due to the hydrophobic nature of the amphiphilic carrier compared to a low-molecular-weight CP-sulfido derivative. (6e): low-molecular-weight drug, water soluble; (PLL-CP): CP conjugate with poly-L-lysine, water soluble; (12a): CP-containing block copolymer (block-CP), micelle forming; (12b): CP-containing block copolymer with additional palmitic acid residues (block-Palm-CP), micelle forming.

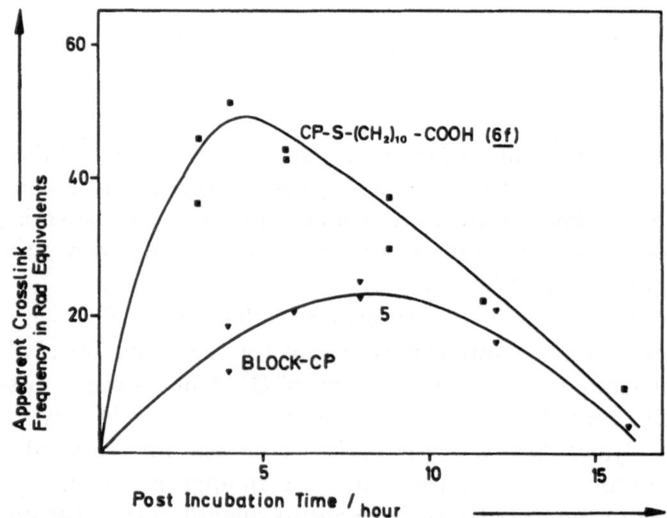

Figure 10. Time course of DNA cross-linking induced by the block-copolymer-cyclophosphamide conjugate (block-CP) (12a) compared to the corresponding low-molecular-weight drug (6f).

Table 4. *In Vivo* Antitumor Activity of Micelle-Forming Cyclophosphamide-block-copolymer Conjugates[a]

Drug	CP-equivalent dose (mmol/kg)	ILS (%)	Cures
Cyclophosphamide (3)[b]	0.95	160	6/10
CP—S—$(CH_2)_{10}$—COOH (6f)	1.26	14	0/6
Block—CP (12a)	0.16	43	1/6
	0.25	86	1/6

[a] These studies were performed in collaboration with L. C. Erickson, K. W. Kohn, and D. S. Zaharko, National Cancer Institute, NIH, Bethesda, Maryland. 10^5 L1210 cells were inoculated 24 h prior to drug treatment into BDF_1 mice.

[b] These data obtained in earlier studies[(119b)] are for comparative purposes.

polymer. Both are shifted to longer postincubation times (8 h) and are widened, indicating a cell uptake prior to sustained release of the active metabolite 4HO-CP. Preliminary *in vivo* studies pointed out the usefulness of micelle-forming, amphiphilic polymers not only in stabilizing the active metabolite, but also in increasing the life span of L1210 tumor-bearing mice. Even at reduced CP-equivalent doses the polymer drug conjugate showed a higher (five times) activity compared to the corresponding low-molecular-weight drug (see Table 4).

Amphiphilic transport systems with hydrophobic cores and hydrophilic shells are well known from nature.[(143)] For example, lipoproteins transport hydrophobic cholesterol esters in the hydrophilic blood serum. So far, the natural lipoproteins and their synthetic analogues are comparable to the point that a defined domain structure (hydrophobic domains in aqueous environment) leads to a certain function. But in addition to the hydrophobic solubilizing ability, the lipoproteins are able to recognize specific receptors on the cell surface and subsequently undergo cell uptake. The simulation of this process would therefore be a further step to realizing higher drug selectivity. In general, the question of selectivity is not a black or white problem. This means that the benefit of chemotherapeutic agents is not based on an exclusive canceroselective agent. All examples discussed in the Section 2.3 exhibit more or less canceroselectivity, not the ultimate tumorspecificity.

In the last decade, however, an accumulating number of experiments have been reported, dealing with more specific homing groups, such as antibodies, recently stimulated by the advantages of the cloning techniques. According to Wilchek's suggestion,[(114)] the use of homing devices in chemotherapy may be called affinity chemotherapy.

2.4. Affinity Chemotherapy

The growing interest in affinity chemotherapy is documented in numerous papers as well as reviews (see References 143–146) concerning this topic. Besides the application of the various receptor-active components, much attention is focused on the usefulness of both polyclonal and monoclonal antibodies as homing devices in cancer chemotherapy.

2.4.1. Monoclonal Antibodies—The Ultimate Homing Device?

As discussed in Section 1.2, tumor cells exhibit tumor-specific properties, in particular, exposing tumor-associated antigens on their cell surface. Antibodies, which are directed towards those antigens, are able to recognize the corresponding tumor cells. Those antibodies, generally raised in rabbits, goats, or mice and subsequently purified, are still polyclonal, that is, they bind to several cell populations. The specific recognition of only one population is realized by monoclonal antibodies, which are available by cloning techniques introduced during the last two years.[147]

Besides their potential application in tumor diagnosis, antibodies may be useful in anticancer chemotherapy.[148] Antibodies, for instance, were linked to toxins (see Section 2.2) achieving impressive therapeutic results.[145–151] In general, toxins are more effective than conventional drugs; that is, often one molecule of toxin is sufficient to kill a cell. Therefore, a conjugate of toxin effectomer (A chain) and an antibody, called immunotoxin, appears to be the ideal drug carrier. However, there are two main disadvantages in the use of toxins or their effectomers. First, the haptomer (B chain) provides a point of interaction with the cell and thus a binding possibility in addition to the one to be conferred by the antibody. Second, if the selectivity of the antibody is not exclusive for the desired cell type (cross reactions are also observed with monoclonal antibodies), the side effects would be much more severe than with conventional anticancer drugs.

In the following, the combination of conventional drugs with antibodies recognizing tumor cells will be described. Apart from problems with the preparation and isolation of pure tumor-specific antibodies, there are two possible routes to load an antibody with drugs. The direct fixation of antitumor agents to the antibody is limited by two factors:

1. The amount of functional groups (often lysine residues) along the polypeptide (= antibody) chains.
2. Increasing fixation degree leads to destruction of the recognition determinant.

These difficulties are avoided by linking drugs to polymer (B), which

Table 5. ***In Vivo***[a] **Antitumor Effects of Dextran–Daunomycin Conjugates to Mouse Monoclonal Antibodies (Dau-dex-KH$_{3\text{-}4,10}$) Compared to Polyclonal Purified Antibodies (Dau-dex-antiYAC).**[152]

Treatment[b]	Equivalent dose Dau (mg/kg)	Content of antibody (mg/kg)	Survival (%)
Phosphate-buffered Saline (PBS)	–	–	0
Daunomycin (8)	12	–	60
	17	–	60
Dau-dex-KH$_{3\text{-}4,10}$[c]	12.5	100	20
	20	200	83
Dau-KH$_{3\text{-}4,10}$	12.5	250	0
Dau-dex-anti-YAC	12.5	50	100
Dau anti-YAC	12.5	100	60
Anti-YAC	–	100	20

[a] 10^5 YAC cells in 0.5 ml PBS were injected subcutaneously into A/J mice (10 weeks old).
[b] Treatment was given i.v. on day 3.
[c] KH$_{3\text{-}4,10}$: anti-YAC IgM obtained from a cloned mouse–mouse hybridoma.

itself is attached to the antibody (A) achieving thus an A–B block copolymer. Ideally, there is only one linkage between the drug-loaded polymer simulating a toxin effectomer, and the antibody, thus avoiding chemical modification at the recognition center. Moreover, in a block copolymer there is a chance of immiscibility of the two blocks when they consist of unflexible, unpenetrable chains and therefore behave like two boats drifting behind each other in water.

Ruth Arnon[152] described a series of detailed experiments using such a polymer-antibody complex. The system described consists of the drug daunomycin attached to the dextran carrier and an antibody. The tumor system is YAC lymphoma induced by the Moloney virus in A/J mice.[153] Two kinds of antibodies were tested: purified ones of Ig fraction of goat antisera (anti-YAC), as well as monoclonal anti-YAC antibodies (KH$_{3\text{-}4}$). The latter showed high specificity *in vitro*. It was disappointing, yet interesting, to recognize (see Table 5) that the conventional goat anti-YAC antibodies were much more effective than the corresponding conjugates with the monoclonal ones. The experiments may suggest that the monoclonal antibodies are not as useful as expected for, in spite of their high specificity, they are only able to recognize a few binding sites on the cell surface. In contrast to this, the polyclonal anti-YACs, while less specific, react with a larger number of cell-surface receptors, thus being more potent in binding higher amounts of drug to the cell. Moreover, another reason might be that there is indeed a highly specific monoclonal antibody, but the tumor is polyclonal with respect to the cell subpopulations (see Section 1.2). Possibly in the future a "cocktail" of monoclonal antibodies will prove to be the optimal reagent.

A further disadvantage in using monoclonal antibodies is the difficulty in their availability. Although the techniques in cloning cells are pursued routinely, much effort is necessary to select the hybridoma-producing antibodies with the ultimate specificity and also to produce them in suitable amounts.

2.4.2. Other Routes to Realize Selective Drug Action

In the following, compounds will be discussed which exhibit tumor selectivity yet often show no cytotoxicity. A comprehensive summary of "tumor localizers" is given by Conners.[154] In particular, hormones were discussed as directing anticancer agents to tumor tissues. The usefulness of steroid hormones attached to polymers as homing devices was recently investigated.[155] The conjugate was proved to accumulate in steroid receptor-containing tumor cells.

Lectins and Glycoproteins comprise a promising class of compounds with relatively high tumor specificity (for comprehensive reviews, see References 156 and 157). Lectins are proteins or glycoproteins found throughout nature, for example, in plants, invertebrates, and even in vertebrates,[158] but nevertheless exhibit specific affinity to sugar residues. Lectin–sugar interaction plays an important role in cell–cell recognition.[80] Moreover, recent investigations on the function of endogenic lectins, with respect to tumor cells, resulted in a greater understanding of tumor immunology, that is, the problems associated with the organotropic metastasis of certain tumors.[158] On the other hand, glycoproteins were used to mark liver-associated lectins.[159] Both routes, the recognition of tumor-associated sugar moieties by lectins and the direction of glycoproteins to tumor-membrane-bound lectins, may be practiced in tumor diagnosis[160,161] and also in affinity cancer chemotherapy.

Attempts of coupling anticancer agents like mitomycin, adriamycin,[162] and methotrexate[163] to Con A, a lectin from jack beans, have already been published. Yet one must note that the direct coupling of drugs to lectins may be accompanied with a loss in specificity, similar to the antibodies mentioned above. According to the model of pharmacologically active macromolecules, it appears worthwhile to bind the drug to the biopolymer via a synthetic polymer bridge.

Lectins, glycoproteins, and antibodies are most likely to yield useful therapeutic results when used as homing devices in the development of new anticancer agents.[164] The first successes in experimental cancer chemotherapy justify further attempts within the concept of affinity chemotherapy.

2.5. From Research Lab to Clinic?

The concept of using polymeric rather than low-molecular-weight antitumor drugs proved to be useful in more systematical drug design. There is accumulating evidence that polymers do decrease the toxicity of antitumor drugs, while maintaining or even increasing the anticancer potency by the various polymer-specific properties mentioned above. In particular, in the area of affinity drugs some encouraging approaches have been described. Despite impressive therapeutical results in experimental cancers and recently even in the treatment of clinical tumors, it has to be emphasized that polymeric antitumor agents are far from being applicable in a broad spectrum of human tumors. The relevent problems and questions have already been discussed. The answers may come from a better understanding in the interdisciplinary field of both the tumor cell and the structure-bioactivity relationship of the drug carrier, especially the molecular mechanisms, involved in the process of drug action on a molecular level. The immunological aspects, in particular, the activation of the body's immune response, plays an important role in the molecular process. However, it has already been pointed out that one might think of other approaches to overcome the usually insufficient immune reaction, such as simulating the body's immune response. How this process may be regarded as a drug action on a cellular level will be discussed in Section 3.

3. Polymeric Antitumor Agents on a Cellular Level

3.1. Death of a Tumor Cell—How To Duplicate this Event

Of all parts of a cell which change during the transition from a normal to a malignant cell (carcinogenesis), the biomembrane is affected to a great extent (see Table 1). Many signals controlling cell growth, mitosis, and cell uptake originate in the membrane. It is therefore not surprising that minor changes in the cell surface have major consequences for cell behavior. In this context it is very impressive to see what happens if tumor cells fail to circumvent the immune defense of the body.[165]

The tumor-associated antigens are recognized by an attacking lymphocyte. This corresponds to the principle of affinity chemotherapy described in Section 2.4, but in this case on a cellular level. The subsequent destruction of the tumor cell is not a molecular process (as defined in Section 1), but is the consequence of the membrane destruction after contact of the T-lymphocyte with a tumor cell and thus an action on a cellular level. This mechanism of destruction by a membrane destabilization is not understood in all details; but several possibilities are discussed.[166–168]

The basis of all hypotheses, however, is destruction on a purely physical level. Bearing this in mind one wonders whether antitumor agents on a cellular level can be conceived and whether they would be able to mimic such a process successfully.[169]

A possible starting point for solving this problem could be the interaction of a tumor cell with a synthetic vesicle, which would have to fulfill the following criteria:

1. Possibility of cell-specific recognition.
2. Ability to destabilize the tumor cell membrane.
3. Similar or higher stability than the tumor cell membrane.

All three points are largely unsolved problems in the investigation of the use of liposomes in medicine described in the literature. Attempts to mimic the above-described interaction of a lymphocyte with liposomes from natural or synthetic phospholipids will lead to the destruction of the liposome rather than of the malignant cell. The vesicle undergoes endocytosis, or its membrane is destabilized.[170] This is of course not what one expects from an antitumor agent on a cellular level. Even less is known about the problems of cell-specific recognition (see Section 3.3.1c) and of cell membrane opening (see Section 3.3.1b). Membrane-destroying compounds would also destroy the liposomes. It is essential therefore to increase the stability of synthetic liposomes. In the following we will describe current considerations on, and experiments designed for, the creation of models of biomembranes and cells with high and variable stability.

3.2. Stable Synthetic Bilayer Membranes via Polymerization

The well-known membrane model of Singer and Nicholson (Figure 11) pictures a double layer formed by a lipid matrix and proteins, for examples, enzymes as informational and functional units "floating" in this lipid matrix.[171] However, this representation is oversimplified: Vesicles from natural phospholipids have the same lipid matrix, but are rather unstable; on the other hand some biomembranes contain only 25% lipids.[172] Obviously, nature finds additional means for creating membranes of high stability: Besides the "integral" membrane proteins (which are embedded in the hydrophobic part of the membrane), there are peripheral proteins, only found in the hydrophilic part of the membrane. Some of these peripheral proteins act as a support, because they are associated with several integral proteins. A well-known example is spectrin situated at the inside of the erythrocyte membrane.[173]

How can a stabilization of biomembranes be achieved synthetically?

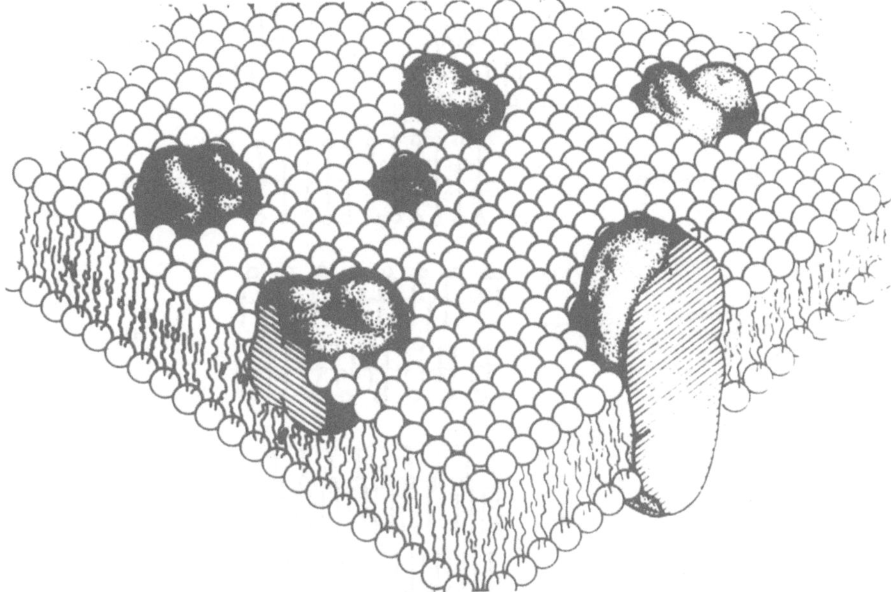

Figure 11. Fluid mosaic model of biological membranes after Singer and Nicholson.[171]

The attempt to mimic a support similar to the spectrins seems unlikely, for very little is known, as yet, about the interactions between the peripheral and integral proteins.

The most convenient and realistic attempt seems to be a method which uses membrane lipids for stabilization. One experiment of this type (though having a different purpose) has been described by Khorana *et al.*[174] These authors incorporated lipids carrying photoreactive groups into a membrane and could (by irradiation) covalently fix proteins to these lipids. Another method with much greater potential uses polyreactions of lipids carrying polymerizable groups.

3.2.1. Polymerizable Lipids

If one intends to synthesize polymerizable lipids to build up membranes of high stability, the polymerizable group can be introduced into different parts of the lipid molecule, that is, into the hydrophilic head group or into the alkyl chain (Figure 12).[175] All four principles in Figure 12 result in systems which have properties differing from those of cell membranes.

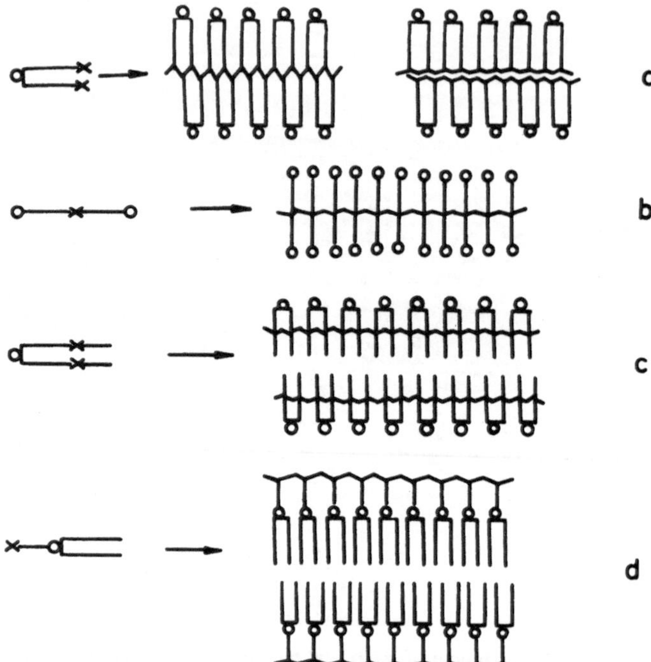

Figure 12. Possible ways to synthesize polymeric model membranes (x = polymerizable group): (a)–(c) polymerization with preservation of head-group properties and (d) polymerization with preservation of chain mobility. (For examples of appropriate monomers, see Table 6.)

Methods (a)–(c) have no influence on the head groups and hence preserve physical properties (such as charge, charge density, etc.), but change the fluidity of the hydrophobic chains. In case (d) the fluidity is not affected, but there is no free choice of head groups. All polymeric lipid systems will show an increase in viscosity and a decrease in the lateral mobility of the molecules.

All four possibilities shown in Figure 12 for the formation of polymerizable lipids have been realized synthetically.[175] Kunitake[176a] was able to show that simple molecules like dialkylphosphates or dialkyldimethylammonium salt can form liposomes. Acrylic and methacrylic groups [type (a) and (d) in Table 6], as well as dicetylene and diene groups [type (b) and (c)], have been used as polymerizable groups. Some of the synthesized polymerizable lipids are listed in Table 6.

Besides polymerization, another type of polyreaction, the polycondensation, can be applied to stabilize membrane systems. Recently, Fukuda *et al.*[177] described polyamide formation in monolayers at the gas–water

Table 6. Several Examples of Polymerizable, Liposome-Forming Lipid Analogues

Type Compound		References
a)		
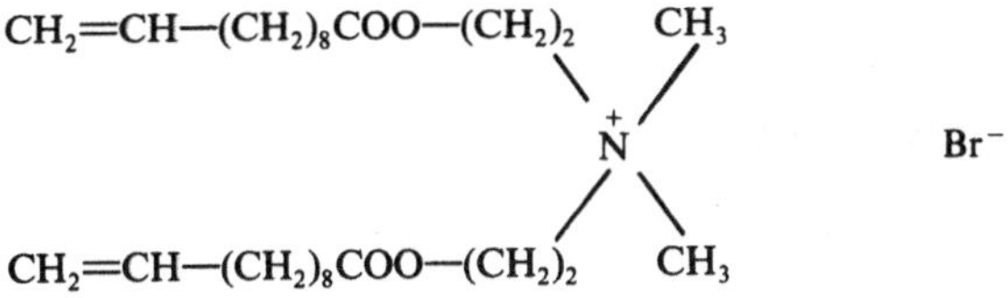 $CH_2{=}CH{-}(CH_2)_8COO{-}(CH_2)_2 \quad CH_3$ $\overset{+}{N}$ $\quad Br^-$ $CH_2{=}CH{-}(CH_2)_8COO{-}(CH_2)_2 \quad CH_3$	(13)	188
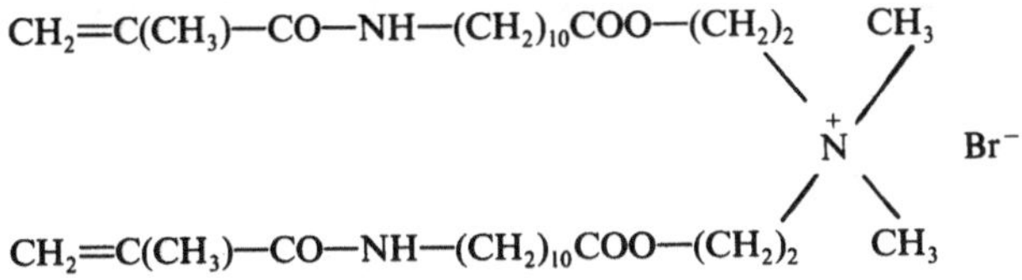 $CH_2{=}C(CH_3){-}CO{-}NH{-}(CH_2)_{10}COO{-}(CH_2)_2 \quad CH_3$ $\overset{+}{N}$ $\quad Br^-$ $CH_2{=}C(CH_3){-}CO{-}NH{-}(CH_2)_{10}COO{-}(CH_2)_2 \quad CH_3$	(14)	175
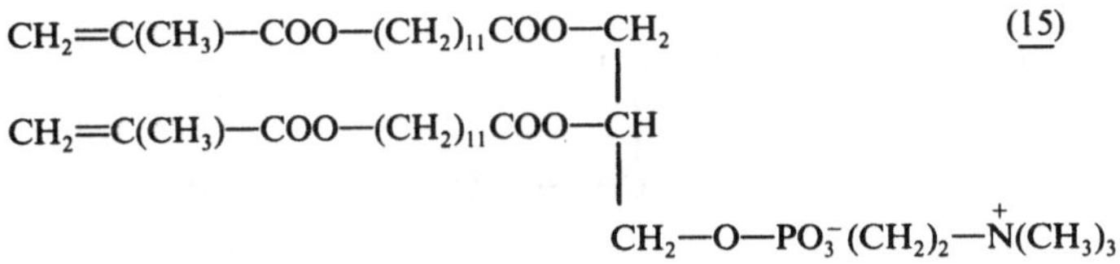 $CH_2{=}C(CH_3){-}COO{-}(CH_2)_{11}COO{-}CH_2$ $CH_2{=}C(CH_3){-}COO{-}(CH_2)_{11}COO{-}CH$ $CH_2{-}O{-}PO_3^-(CH_2)_2{-}\overset{+}{N}(CH_3)_3$	(15)	189
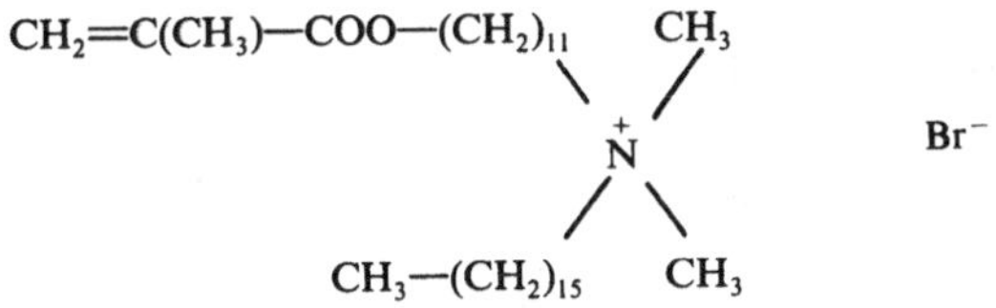 $CH_2{=}C(CH_3){-}COO{-}(CH_2)_{11} \quad CH_3$ $\overset{+}{N}$ $\quad Br^-$ $CH_3{-}(CH_2)_{15} \quad CH_3$	(16)	183
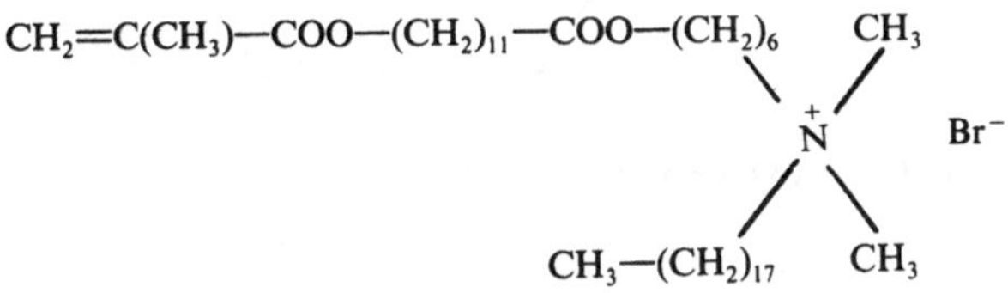 $CH_2{=}C(CH_3){-}COO{-}(CH_2)_{11}{-}COO{-}(CH_2)_6 \quad CH_3$ $\overset{+}{N}$ $\quad Br^-$ $CH_3{-}(CH_2)_{17} \quad CH_3$	(17)	187
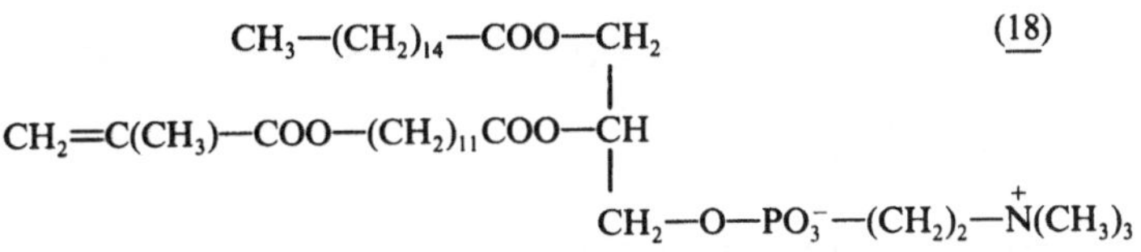 $CH_3{-}(CH_2)_{14}{-}COO{-}CH_2$ $CH_2{=}C(CH_3){-}COO{-}(CH_2)_{11}COO{-}CH$ $CH_2{-}O{-}PO_3^-{-}(CH_2)_2{-}\overset{+}{N}(CH_3)_3$	(18)	189

(Continued)

Table 6. (*Continued*)

Type Compound		References
b) 		
$HOOC{-}(CH_2)_8{-}C{\equiv}C{-}C{\equiv}C{-}(CH_2)_8{-}COOH$	(19)	175
$HO{-}(CH_2)_9{-}C{\equiv}C{-}C{\equiv}C{-}(CH_2)_9{-}OH$	(20)	175
$H_2O_3P{-}O{-}(CH_2)_9{-}C{\equiv}C{-}C{\equiv}C{-}(CH_2)_9{-}O{-}PO_3H_2$	(21)	185
c)		
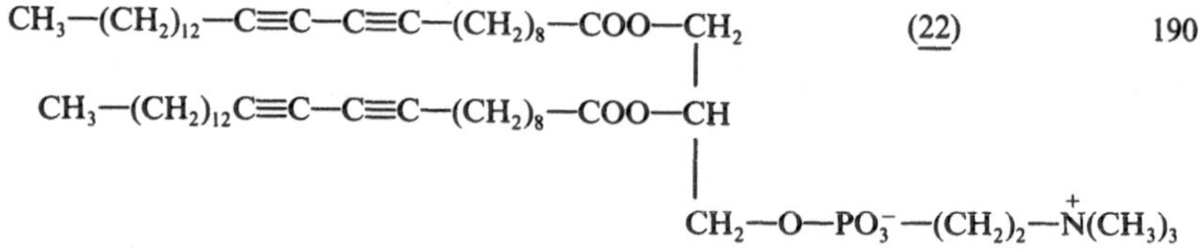 $CH_3{-}(CH_2)_{12}{-}C{\equiv}C{-}C{\equiv}C{-}(CH_2)_8{-}COO{-}CH_2$ $CH_3{-}(CH_2)_{12}C{\equiv}C{-}C{\equiv}C{-}(CH_2)_8{-}COO{-}CH$ $CH_2{-}O{-}PO_3^-{-}(CH_2)_2{-}\overset{+}{N}(CH_3)_3$	(22)	190
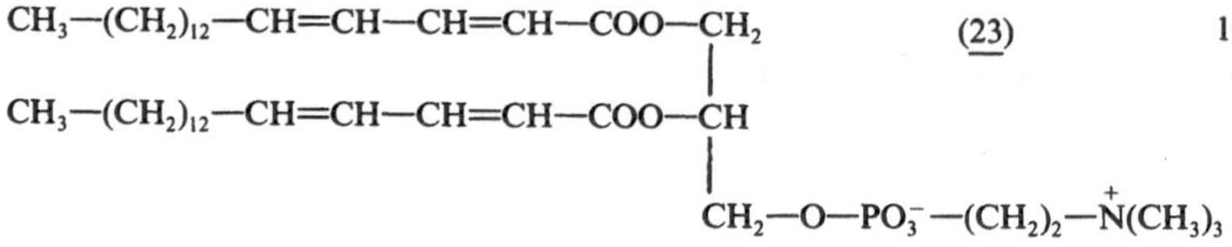 $CH_3{-}(CH_2)_{12}{-}CH{=}CH{-}CH{=}CH{-}COO{-}CH_2$ $CH_3{-}(CH_2)_{12}{-}CH{=}CH{-}CH{=}CH{-}COO{-}CH$ $CH_2{-}O{-}PO_3^-{-}(CH_2)_2{-}\overset{+}{N}(CH_3)_3$	(23)	190
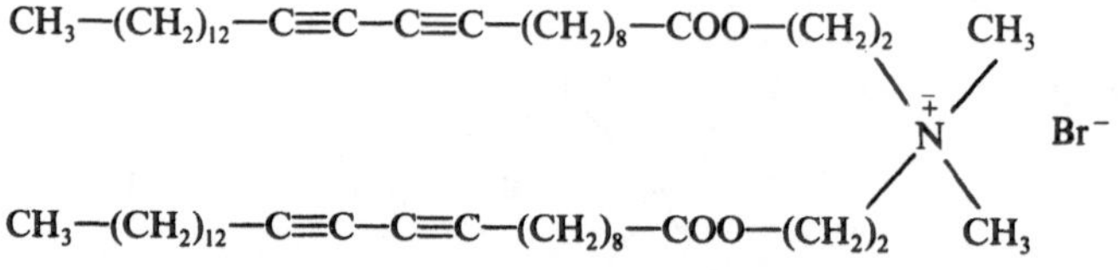 $CH_3{-}(CH_2)_{12}{-}C{\equiv}C{-}C{\equiv}C{-}(CH_2)_8{-}COO{-}(CH_2)_2$ CH_3 $\overset{+}{N}$ Br^- $CH_3{-}(CH_2)_{12}{-}C{\equiv}C{-}C{\equiv}C{-}(CH_2)_8{-}COO{-}(CH_2)_2$ CH_3	(24)	191
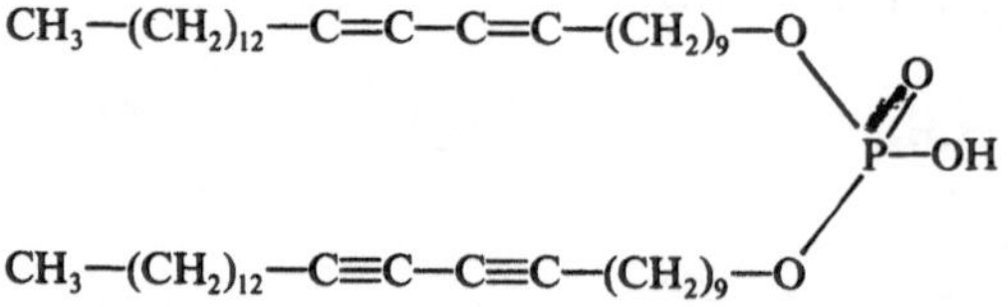 $CH_3{-}(CH_2)_{12}{-}C{=}C{-}C{=}C{-}(CH_2)_9{-}O$ $P(=O){-}OH$ $CH_3{-}(CH_2)_{12}{-}C{\equiv}C{-}C{\equiv}C{-}(CH_2)_9{-}O$	(25)	192
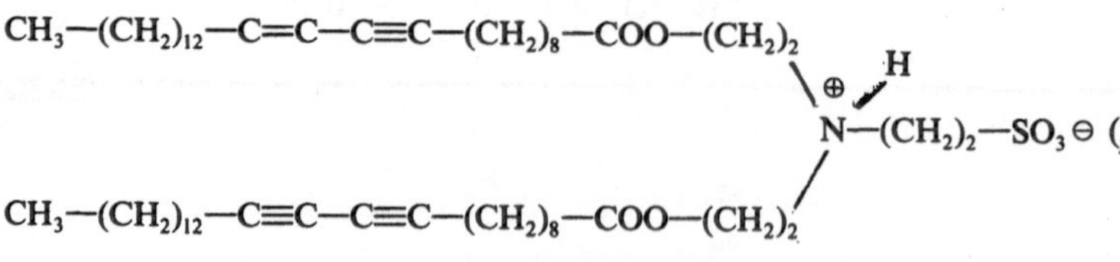 $CH_3{-}(CH_2)_{12}{-}C{=}C{-}C{\equiv}C{-}(CH_2)_8{-}COO{-}(CH_2)_2$ $\overset{\oplus}{N}(H){-}(CH_2)_2{-}SO_3^{\ominus}$ $CH_3{-}(CH_2)_{12}{-}C{\equiv}C{-}C{\equiv}C{-}(CH_2)_8{-}COO{-}(CH_2)_2$	(26)	193

Table 6. (*Continued*)

Type Compound		References
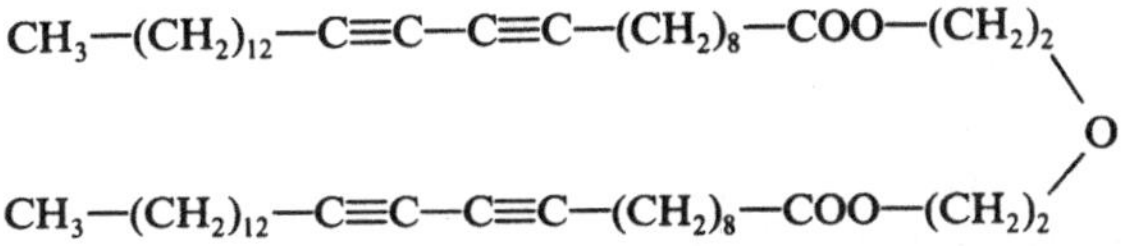 $CH_3-(CH_2)_{12}-C\equiv C-C\equiv C-(CH_2)_8-COO-(CH_2)_2-O-(CH_2)_2-OOC-(CH_2)_8-C\equiv C-C\equiv C-(CH_2)_{12}-CH_3$	(27)	191
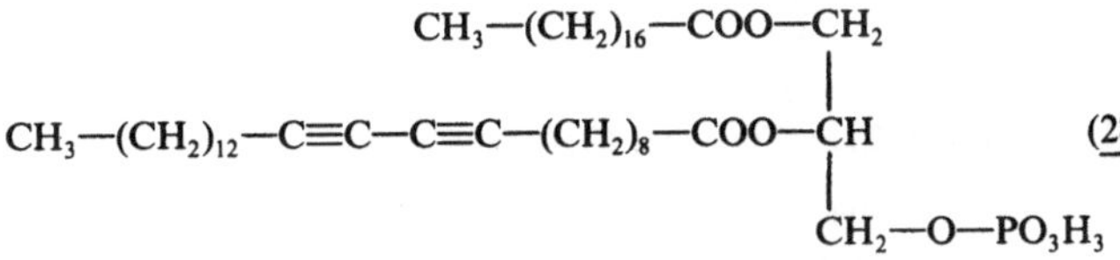 $CH_3-(CH_2)_{16}-COO-CH_2$; $CH_3-(CH_2)_{12}-C\equiv C-C\equiv C-(CH_2)_8-COO-CH$; $CH_2-O-PO_3H_3$	(28)	194
d)  ═══O—X		
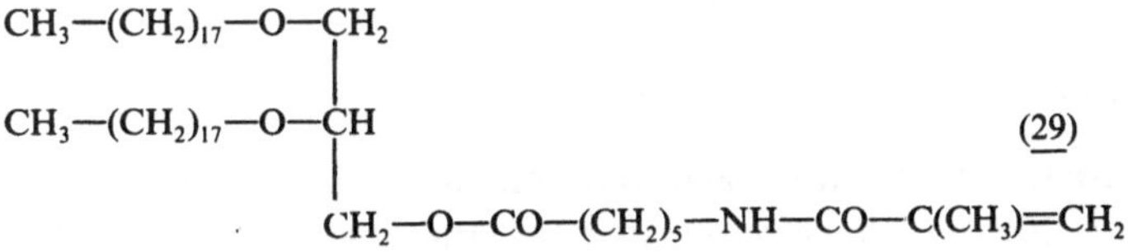 $CH_3-(CH_2)_{17}-O-CH_2$; $CH_3-(CH_2)_{17}-O-CH$; $CH_2-O-CO-(CH_2)_5-NH-CO-C(CH_3)=CH_2$	(29)	175
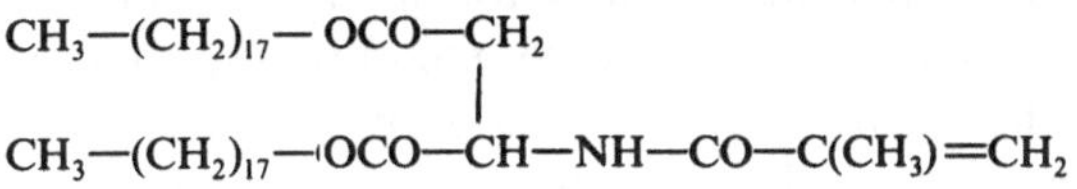 $CH_3-(CH_2)_{17}-OCO-CH_2$; $CH_3-(CH_2)_{17}-OCO-CH-NH-CO-C(CH_3)=CH_2$	(30)	175
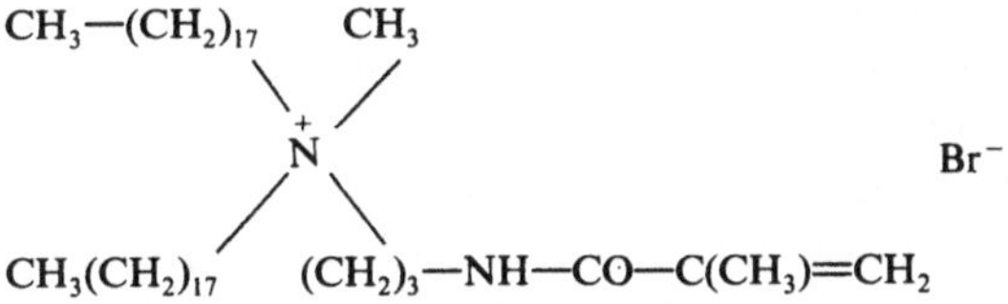 $CH_3-(CH_2)_{17}$, CH_3, $CH_3(CH_2)_{17}$, $(CH_2)_3-NH-CO-C(CH_3)=CH_2$ on $\overset{+}{N}$ Br^-	(31)	187
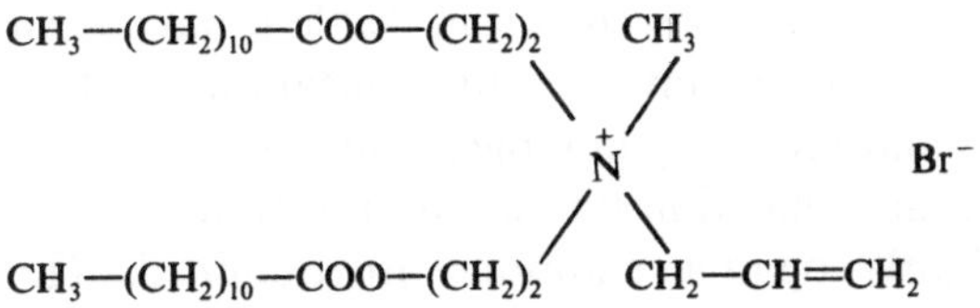 $CH_3-(CH_2)_{10}-COO-(CH_2)_2$, CH_3, $CH_3-(CH_2)_{10}-COO-(CH_2)_2$, $CH_2-CH=CH_2$ on $\overset{+}{N}$ Br^-	(32)	188

interface. Long-chain esters of glycine and alanine were polycondensed to yield *nonoriented* polyamide films of polyglycine and polyalanine.

The self-condensation of esters of long-chain amino acids (33), (34). (35), (36), however, can be used to prepare *oriented* polypeptide films and vesicles in order to stabilize model membranes via polycondensation.

$$CH_3{-}(CH_2)_{15}{-}\underset{\displaystyle NH_2}{\underset{|}{CH}}{-}CO{-}O{-}CH_3$$

(33)

$$CH_3{-}(CH_2)_{15}{-}\underset{\displaystyle NH_2}{\underset{|}{CH}}{-}CO{-}O{-}(CH_2)_{21}{-}CH_3$$

(34)

$$CH_3{-}(CH_2)_{23}{-}\underset{\displaystyle NH_2}{\underset{|}{CH}}{-}CO{-}O{-}CH_3$$

(35)

$$CH_3{-}(CH_2)_{23}{-}\underset{\displaystyle NH_2}{\underset{|}{CH}}{-}CO{-}O{-}(CH_2)_{21}{-}CH_3$$

(36)

Compared to the polymerizable lipids based on acrylates, butadienes, and diacetylenes mentioned above, these polypeptide-forming monomers provide the additional advantage of being biodegradable.

3.2.2. *Polymeric Liposomes*

Liposomes are the closest approach to biomembranes. They are closed, spherical structures having an aqueous interior and one or several lipid double layers[178] (see Figure 13). Such vesicles can be formed from synthetic amphiphiles and membrane extracts. Reconstituted membranes, that is, liposomes of cell membrane constituents, contain nearly all components of cell membranes (lipids, proteins, glycolipids, etc.). Vesicles of synthetic lipids or lipid analogues have a much simpler composition and, depending on the method of formation, are double- or multilayered with different sizes.[179]

The most common methods of preparing liposomes[179] are by the ultrasonication of lipid suspensions in water, by the injection of methanolic or etheral solutions of lipid into water, by the dialysis of surfactant–lipid mixtures, or by the handshaking of lipid films on glass surfaces in water. The polymerizable lipids and lipid analogues shown in Table 6 have been transformed into liposomal solutions mainly by ultrasonication of their crystal suspensions.[175,180] Small and relatively homogeneous vesicles with a single bilayer are formed after long sonication times. On filtration through a millipore filter, clear or slightly opaque solutions are obtained. The monomeric liposomes are relatively unstable – like vesicles from normal lipids, their solutions turn turbid after a few days.

The clear monomeric solutions of methacrylic, diene, and diacetylene lipids and surfactants can be polymerized by UV irradiation. In the case of the diacetylene lipids, the transitions from monomeric to polymeric liposomes can be observed visually and spectroscopically by the color change which results from formation of a conjugated polymer backbone.[180] Electron

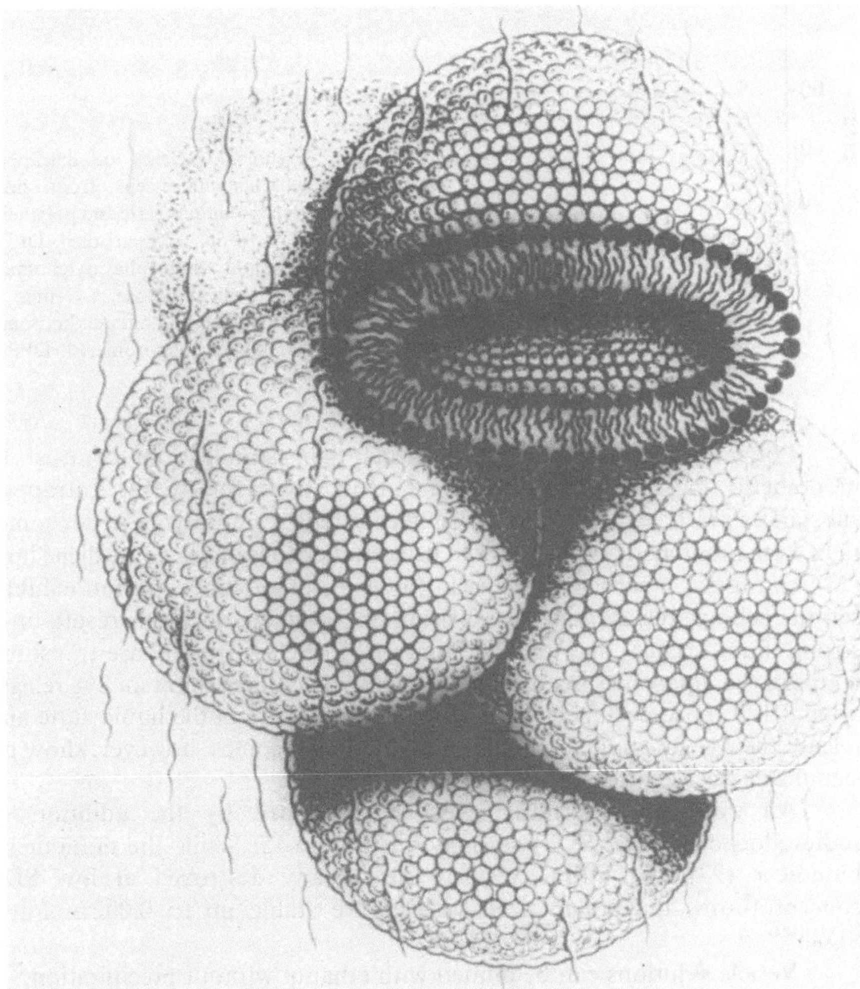

Figure 13. Schematic representation of unilamellar liposomes (i.e., consisting of one bilayer).

microscopy can provide direct evidence that the polymers still possess a liposomal structure.[180] The structure remains unchanged during polymerization, as can be seen by a comparison with the corresponding monomer solutions.[176,181–183] The size distribution of monomeric and polymeric vesicles is essentially the same as that shown for liposomes of the diene lipid (23) by GPC measurements.[184]

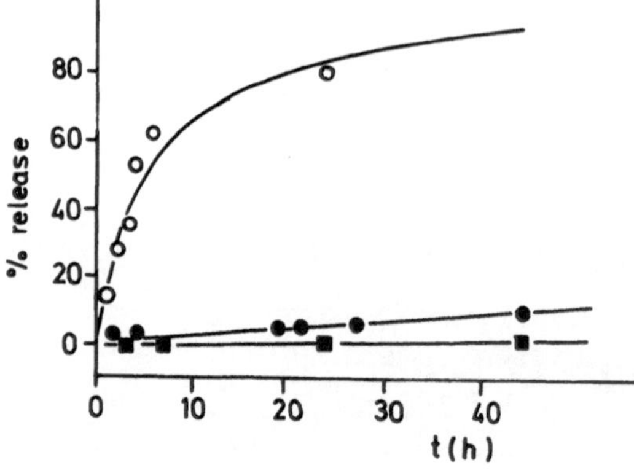

Figure 14. Release of entrapped 6-carboxyfluorescein from liposomes of monomeric and polymeric (23). For comparison: DPPC (dipalmitoylphosphatidylcholine). R = percent release; t = time in hours; (○) monomeric liposome; (■) polymeric liposome; (●) DPPC.

Polymerization of liposomes effects their stability. In contrast to monomeric liposomes, the polymers remain stable for weeks. Entrapped substances are released to a much smaller extent from polymeric liposomes than from monomeric ones. This had been studied in the case of the diene lipid (23): entrapped 6-carboxyfluorescein (6-CF) in high concentration exhibits self-quenching; release into the surrounding aqueous medium results in a strong fluorescence due to the dilution.[59] Below the phase-transition temperature, liposomes made from dipalmitoyllecithin exhibit an 8% release after 40 h. Liposomes made from monomeric (23) are in the liquid state and release the dye much more rapidly. Polymeric liposomes, however, show no significant release after 40 h (Figure 14).[184]

An even more striking effect is produced by the addition of sodiumdodecylsulfate (SDS) as shown in Figure 15. While the monomeric butadiene (23) and DPPC vesicles are already destroyed at low SDS concentrations, the polymerized vesicles are stable up to 0.002 mol/liter SDS.[184]

Vesicle solutions can be diluted with ethanol without precipitation.[184] The solution of polymeric liposomes of (24) are stable in 80% ethanol for weeks. This could also be shown for the case of the polymerized methacrylic lipid (16).[185]

The increased membrane stability described above is combined with the presence of a polymer chain (in the membrane itself or on its surface) introducing increased viscosity and thus reduced flexibility into the membrane system. How does this reduced membrane mobility affect one of the most vital biomembrane properties, that is, the phase transition, which is necessary to provide biocompatibility?

Whether polymerized artificial biomembrane systems are too rigid to show a phase transition strongly depends on the type of polymerizable lipid

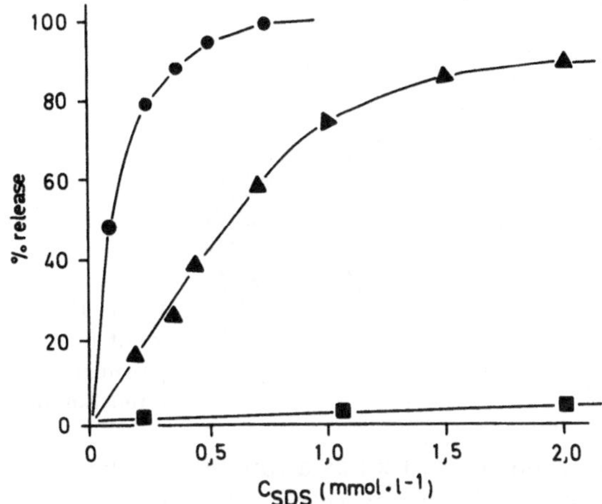

Figure 15. SDS induces release of 6-carboxyfluorescein from liposomes of monomeric and polymeric (23) compared to DPPC: (●) monomeric liposome; (▲) DPPC; (■) polymeric liposomes.

involved in membrane polymerization. Especially in the case of the diacetylene lipids a loss of phase transition can be expected due to the formation of the rigid conjugated polymeric backbone. This was demonstrated by DSC (differential scanning calorimetry) measurements with the diacetylene-containing taurin derivative (26). Figure 16 illustrates the

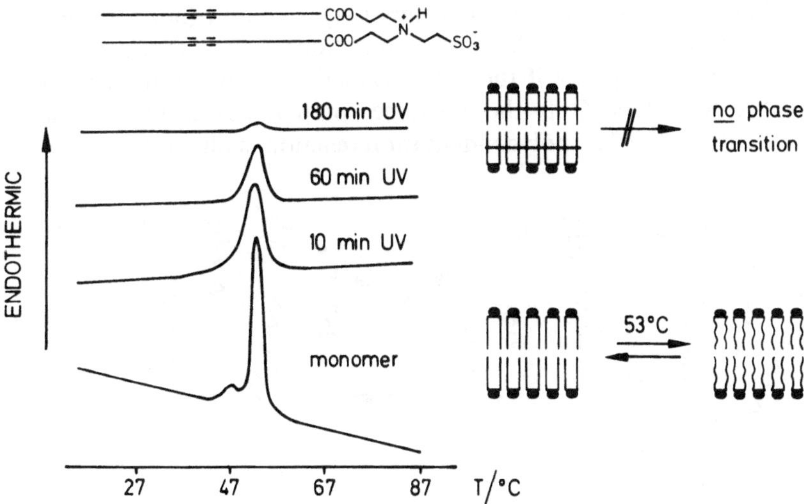

Figure 16. Phase-transition behavior of (26) as a function of polymerization time.

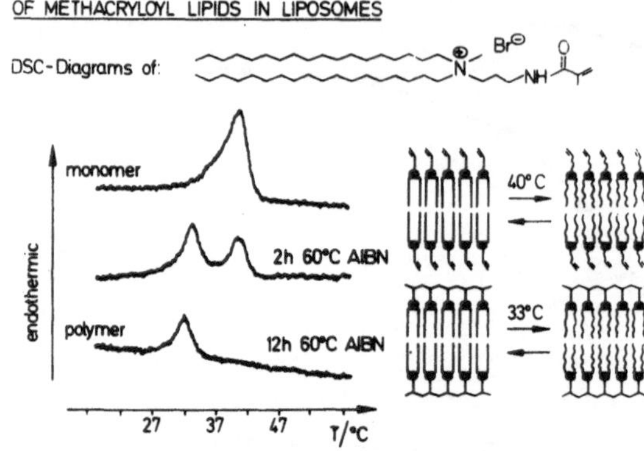

Figure 17. Phase-transition behavior of an amphiphilic molecule (31) polymerizable in the head group as a function of polymerization time.

phase-transition behavior of (26) as a function of the polymerization time. The pure monomeric liposomes show a transition temperature of 53°C. During the polymerization a decrease in phase-transition enthalpy indicates a restricted mobility of the polymerized hydrocarbon core. Moreover, the phase transition finally disappears after complete polymerization of the monomer.[186]

In contrast to this the phase transitions of polymeric liposomes are retained if the polymer chain is more flexible or located at the surface of the vesicles instead of in the hydrophobic core. The polymerized vesicles of the methacrylamide (31), for instance, show a phase-transition temperature which is even slightly lower than the one of the corresponding monomer vesicles,[187] as shown in Figure 17. This could be explained by a disordering influence of the polymer chain on the head-group packing.

The long-chain amino acid esters (34), (35) and (36) form liposomes on sonication in water under acidic conditions. Liposomes prepared from (34) and (36) precipitate if the aqueous medium is neutralized by titration with 0.02 M NaOH. Only the liposomes prepared from (35) are stable in basic solutions.[195] The polycondensation reaction is illustrated in Figure 18.

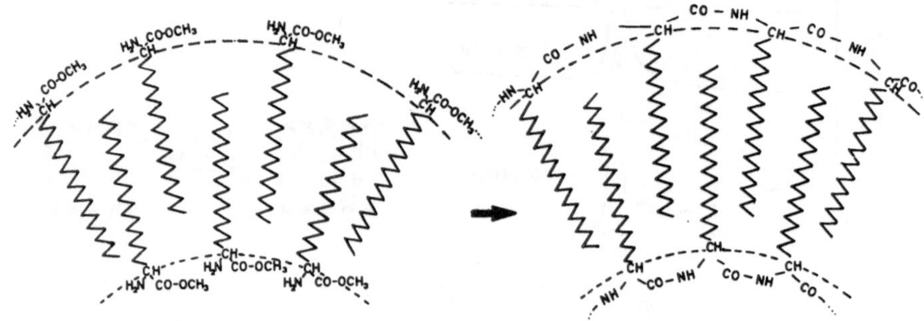

Figure 18. Polycondensation reaction in liposomal bilayers.

Polycondensation reactions in liposomal bilayers are not catalyzed at all and simply occur due to the high packing density of the reactive groups and their orientation in the layers. This orientation requirement can be demonstrated by the fact that all investigated amino acid esters show no measurable polycondensation in solution. Polycondensation in liposomes leading to oriented polyamides represents a new route for stabilizing model membranes under mild conditions. In addition it can be expected that the polypeptide vesicles can be enzymatically cleaved, thus representing a group of stable, but biodegradable, polymeric liposomes.

3.3. How To Improve Biological Functionality

So far it can be shown that stable liposomes can be prepared by polymerization of lipids. These vesicle systems, however, are still far removed from a real biocompatible carrier model, not yet showing any typical biological membrane properties, such as selective opening enzymatic activity, surface recognition, and the ability to undergo fusion.

There are two principally different ways to approach this problem of polymeric biomembrane models. One can begin with a completely synthetic system and introduce natural lipids and membrane proteins to increase biological similarity. On the other hand, one can take a completely natural system, which could be a living cell itself, and introduce synthetic components into the cell membrane, for example, polymerizable lipids. These two aspects will be discussed in more detail.

3.3.1. Synthetic Route to Biofunctionality

In this section emphasis will be placed on the preparation of polymeric biological membrane models using purely synthetic components or membrane constitutents which are isolated in pure form from natural systems, that is, one completely avoids the application of any kind of natural membrane segments prearranged by a living system.

Systems within this category are, for instance, mixed membranes of polymerizable and natural lipids or isolated membrane proteins reincorporated into polymerizable lipids. Both the hydrolysis of the natural lipid component by enzymes added to the vesicle preparation or the remaining enzymatic activity of the reincorporated membrane protein in its new environment can be used to test the biological behavior of these functionalized membrane systems. Further biological criteria can be introduced into polymerizable membranes using lipids with recognizable hydrophilic head groups providing surface recognition by proteins.

3.3.1a. Incorporation of Membrane Proteins into Polymerizable Membranes. The high-performance transport specificity of biomembranes solely depends on the presence of membrane proteins embedded in the lipid matrix. On the other hand, however, most membrane proteins cease to function in the absence of lipids. So in order to introduce biological transport abilities into artificial membrane systems, protein–lipid interactions are of vital interest; that is, how is the activity of membrane proteins affected if they are placed into a polymeric environment?

As an example of an asymmetric membrane-integrated protein, the synthetase complex (ATPase) (*Rhodospirillum rubrum*) was incorporated into liposomes of the diacetylene taurine (26).[186] The protein consists of a hydrophobic membrane-integrated part (F_0) and a water-soluble moiety (F_1) carrying the catalytic site of the enzyme. The purified ATPase synthetase complex is almost completely inactive. Activity is substantially increased in the presence of a variety of amphiphiles.

For soybean lecithin and the taurin diacetylene (26), the maximum enzyme activity is obtained at 500 lipid molecules per enzyme molecule. Using soybean lecithin, the ATPase activity is increased eightfold compared to a fivefold increase in the presence of the taurine derivative.

The incorporation of the ATPase into monomeric and polymeric diacetylene liposomes is achieved by simple incubation. There is approximately a twofold activity increase in the polymeric liposome

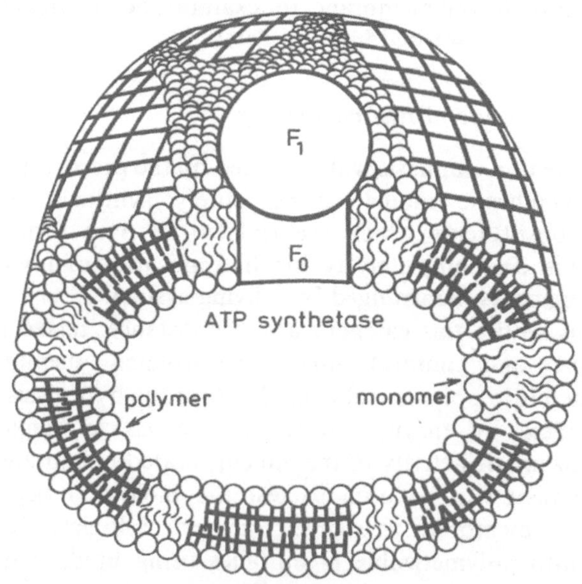

Figure 19. Incorporation of F_0F_1 ATPase in polymeric liposomes.

compared to the monomer which can be ascribed to a structural change in the bilayer organization during polymerization. DSC data indicate residual "monomeric domains" in the polymerized liposomes, so that the ATPase is most probably embedded in these domains, stabilized by the polymer matrix shown schematically in Figure 19.

These polymeric proteoliposomes have a considerably higher long-term stability and activity than, for instance, ATPase-containing soybean lecithin liposomes. Their lifetime is practically only determined by the bacterial growth rate in the particular sample.

3.3.1b. Selective Openings of Mixed Polymeric Liposomes. Several methods are possible to open up closed polymeric membrane compartments selectively to release incorporated substances, as illustrated in Figure 20. In order to carry out these releasing procedures, the polymeric vesicle has to contain destabilizable areas comparable to cork stoppers on bottles. These sensitive areas could eventually be opened by variation of the pH,[198] or enzymatic processes.

Such an enzymatic process, for instance, is the hydrolysis of a natural phospholipid membrane component by phospholipase A_2, schematically

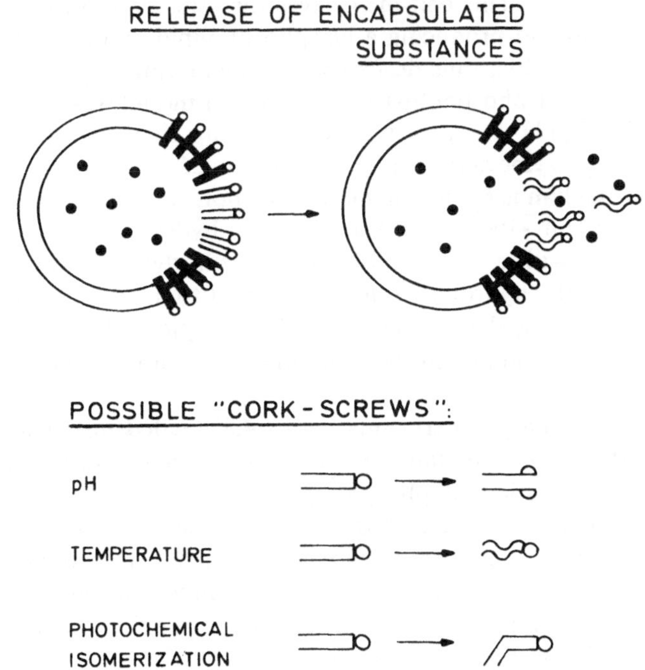

Figure 20. "Corked" liposomes—possibilities to open selectively polymeric membrane compartments.

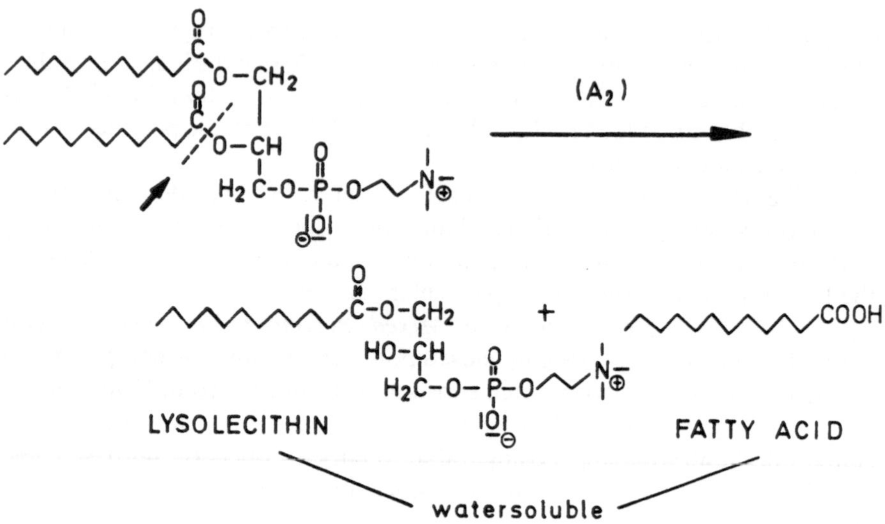

(A$_2$)

LYSOLECITHIN

FATTY ACID

watersoluble

Figure 21. Action of phospholipase A_2 on L-α-lecithins.
60

shown in Figure 21. This enzyme cleaves the ester bond in position 2 of a natural phospholipid producing a lysophospholipid and a fatty acid which are both water soluble, thus destroying the membrane.

The action of phospholipase A_2 on mixed monolayers of natural and polymerizable lipids was investigated, revealing that the main parameter influencing the enzyme activity is the miscibility of the lipid components and not whether the film is being polymerized or not. In both cases—miscible or immiscible system—the enzyme is able to hydrolyze the natural lipid component, but with a considerable difference in the hydrolyzing rate. In a miscible system the hydrolysis rate is decreased to a very great extent. This could be explained by the fact that each natural lipid molecule is surrounded by nonhydrolyzable lipid and therefore the action of phospholipase A_2 may be hindered.

First preliminary experiments with 6-CF-loaded mixed liposomes of DPPC and the diacetylene taurine (14)[(199)] show only a very slow release of 6-CF after the addition of phospholipase A_2. Reasons for this might be head-group interactions decreasing the enzyme activity or the multi-lamellarity of the vesicle preparation used. Further investigations will be carried out with "giant" liposomes (visible under the light microscope) which, besides being uni- or oligolamellar, provide higher enclosure percentage and low surface curvature.

3.3.1c. Surface Recognition of Polymeric Liposomes. If polymeric vesicle systems are to be used as selective carriers for membrane-destroying

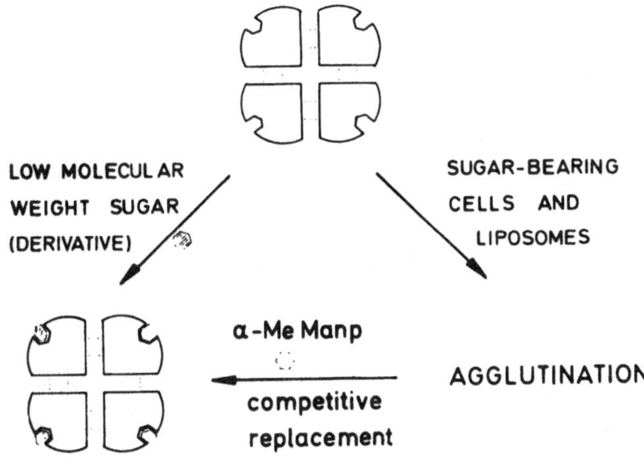

Figure 22. Lectin–sugar interactions. Lectins—proteins with sugar-binding sites.

agents, they have to be able to recognize their desired target. Biomembranes are capable of recognizing target proteins, such as antibodies, resulting, for instance, in heavy blood-cell agglutination. These recognizable abilities are in general performed by hydrophilic head groups of special lipid components in the membrane.

In order to imitate biological surface recognition with polymeric membrane systems recognizable natural lipids can either be incorporated into polymerizable membranes or polymerizable lipids have to be synthesized carrying recognizable head groups. For first simple recognition experiments, sugars were chosen as recognizable head groups. Sugars can be recognized by lectins (phytohaemagglutinins), which are proteins with specific binding sites for sugars and are mainly found in plants. Concanavalin A (Con A) for instance (see Figure 22) is able to bind α-mannopyranoside and α-glucopyranoside. It is a tetramer with subunits, each having one binding site. If a low-molecular-weight sugar is added to an aqueous lectin solution, the binding sites of the protein are saturated without observing a turbidity increase. If the sugar moiety, however, is part of an aggregate, such as a liposome or a cell, these aggregates are agglutinated and precipitation occurs.[200,201] Adding a low-molecular-weight sugar to these agglutinated aggregates will result in a competitive replacement of the aggregates by the low-molecular-weight sugar. Thus, the precipitate will be redissolved leaving the original liposomes or cell solution next to the saturated lectin.

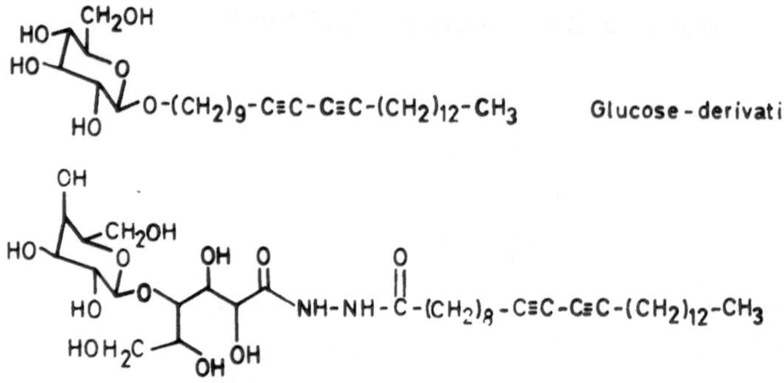

$O-(CH_2)_9-C{\equiv}C-C{\equiv}C-(CH_2)_{12}-CH_3$ Glucose-derivative

$NH-NH-C(=O)-(CH_2)_8-C{\equiv}C-C{\equiv}C-(CH_2)_{12}-CH_3$ **Lactose**-derivative

Figure 23. Polymerizable glycolipids.

With these interactions in mind, glycolipids with different sugar head groups were synthesized carrying polymerizable diacetylene functionalities in the hydrophobic chain (Figure 23).[185]

When an aqueous suspension of the glucose derivative is sonicated for 15 min at 50°C, a clear, colorless solution is obtained, which can be polymerized at 0°C by UV irradiation. During the polyreaction the usual color change via blue to red takes place as already observed for different liposome-forming diacetylenes.

On addition of the lectin Con A, within a few seconds agglutination occurs and a red precipitate next to a colorless supernatant is formed as schematically shown in Figure 24. This agglutination is reversible. On addition of α-methyl mannopyranoside, the red precipitate is redispersed giving a clear red solution of the original polymeric liposomes. This agglutination does not occur with vesicles not containing sugar moieties at their surface. These experiments show that polymeric vesicular systems can be recognized by target proteins, indicating that these functionalized compartments could be potentially used as a homing device for membrane hydrolyzing agents.

3.3.1d. Biofunctionality via Fusion Processes. In contrast to the "synthetic" route to polymeric carrier models, which involves combining all kinds of natural and synthetic membrane-forming substances, the method discussed in the following will use intact natural cell membranes and modify them by incorporating polymerizable "accessories." The basic procedure applied for this incorporation process is the fusion of two membrane systems, that is, living cell and polymerizable vesicle. Fusion methods, however, have been applied successfully to cell–cell fusion. So, in order to fuse a polymerizable vesicle with a natural cell membrane, investigations are necessary to determine whether particular fusion methods can be used on giant liposome–liposome fusion itself before proceeding to cell–liposome fusion.

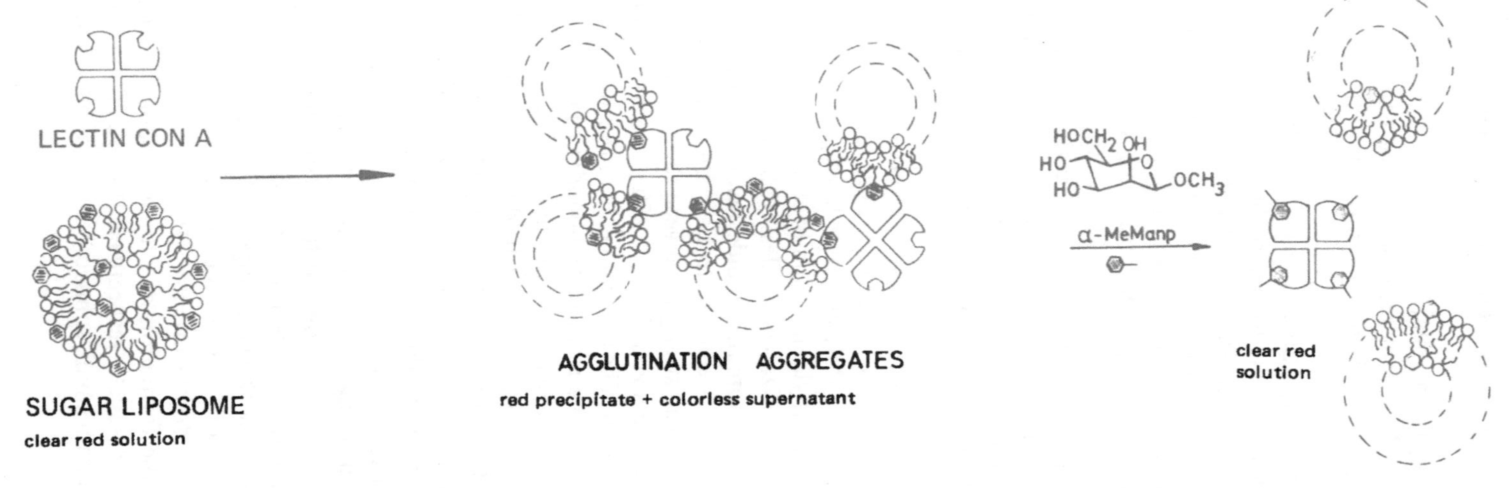

Figure 24. Interactions between Con A and sugar-bearing liposomes (reversible aggregation) and competitive replacement of sugar-bearing liposomes by α-methylmannopyranoside.

Among the many fusion methods known,[202–204] a very selective method recently described by Zimmermann *et al.* proved to be applicable for these purposes. Essentially this method is based on subjecting cell membranes – in which high-intensity electrical fields occur naturally – to a short external electric field pulse of comparable intensity.[205] Under these conditions the membrane breaks down locally (dielectric breakdown), becomes permeable, and can be fused with an adjuvant membrane. Contact of membrane compartments to be fused can be achieved by dielectrophoresis.[206]

The preparation of "giant" liposomes[207] made it possible to apply this electrical-field-induced fusion to the fusion of vesicles from natural and polymerizable lipids. Since in dielectrophoresis the net force pulling vesicles together is proportional to the volume of the vesicles, only large vesicles can be fused. Using these large vesicles, the fusion process can be followed in a phase-contrast light microscope. This is in contrast to several other fusion techniques,[208–210] which use small, submicroscopic vesicles (< 100 nm) proving fusion only indirectly.

Natural lipids used for fusion experiments were mainly phospholipids with different chain lengths – saturated and unsaturated – and mixtures thereof with cholesterol.[211] Polymerizable lipids, butadiene derivatives with a lecithin (23), and a dimethylammoniumbromide head group were also used.

In Figure 25 polymerizable vesicles made from a butadiene lipid with the cationic dimethylammoniumbromide head group and cholesterol (1 : 1 mixture) are oriented between electrodes. By establishing an optimum membrane contact, the field strength between the electrodes causes flattening of the vesicles in the contact zone [Figure 25(a)]. Fusion is initiated by a field pulse of 20–50 μs duration. The intermingling of the membrane occurs within a fraction of a second. The membrane boundary between the two vesicles disappears, forming an oval as shown in Figure 25(b). A completely spherical liposome is formed after turning off the field [Figure 25(c)]. If the vesicles are unilamellar the whole fusion process is completed within 1 s. In the case of multilamellar vesicles, the fusion time is extended to about 3–6 s. Fusion times for cells are usually in the range of minutes[204] to hours.

During the fusion process the relative surface area decreases with increasing volume indicating a loss in membrane material (about 20% in the demonstrated example). In analogy to the fusion process of protoplasts, it can be assumed that the excess lipid is removed in the form of small, submicroscopic vesicles as shown in Figure 26. The electrical breakdown in the membrane contact zone leads to the formation of several pores, in which lipid molecules are randomly oriented [Figure 26(b)]. The molecules now reorient leading to the formation of submicroscopic vesicles and the new membrane of the fused vesicle, respectively [Figure 26(c)]. Thus, fused giant liposomes should contain small, submicroscopic vesicles. The validity of this

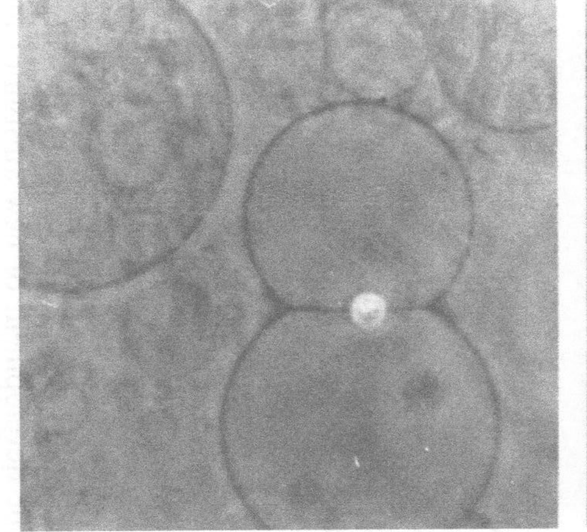

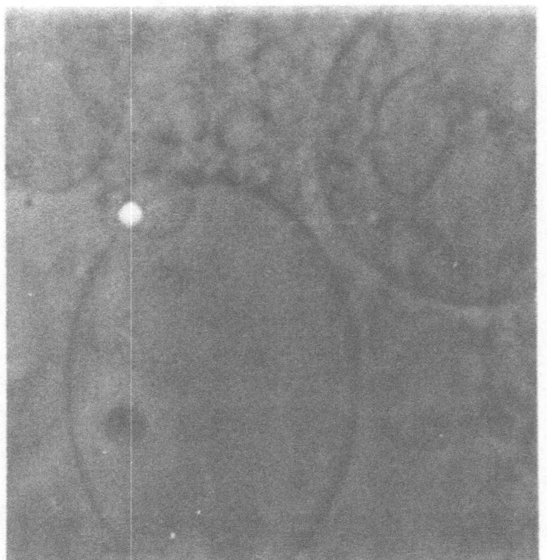

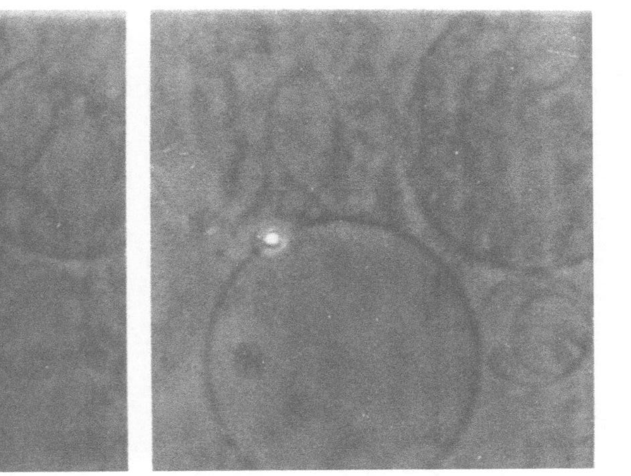

Figure 25. Phase-contrast photographs of the fusion process of lipid vesicles, prepared from a 1 : 1 mixture of lipid A/cholesterol; diameter of the vesicles, 37 and 45 μm, respectively. (a) Pearl-chain formation in an alternating electric field of 200 V/cm; (b) elongated fused liposome, 1 s after application of a pulse of 3 kV/cm strength and 30-μs duration; and (c) spherical new vesicle after turn-off of the alternating electric field new diameter, 51 μm.

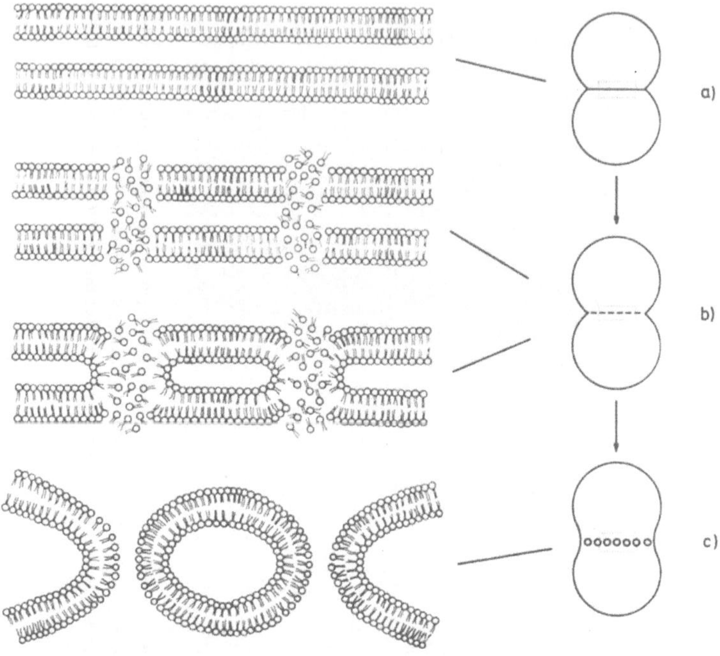

Figure 26. Proposed mechanism of the fusion process of liposomes resulting in the formation of small unilamellar vesicles at the interface: (a) lipid bilayers in contact; (b) pores generated by electric breakdown and lipid reorientation; and (c) formation of membrane bridges and, in turn, of submicroscopic vesicles.

could possibly be proved by using fluorescence-labeled lipids for liposome fusion.

Preliminary experiments to fuse the described giant liposomes with Friend cells show problems in the simultaneous orientation of vesicles and cells in the 0.3 *M* mannit solution, necessary as an isotonic medium for the cells. The induced dipole of cells (necessary to provide dielectrophoresis) is much larger than that of the vesicles, leading primarily to cell–cell contacts and thus to cell–cell fusion. Using vesicles filled with electrolyte or applying multilamellar vesicles in cell–liposome fusion may overcome this problem.

4. Conclusion

In this paper we have attempted to report on the possible contributions of polymer science to the development of new antitumor agents. We have proceeded from the viewpoint of a polymer chemist—which is necessarily a limited one.

Our objective was to point out some solved and unsolved problems and to speculate on possible further solutions. We are aware of the fact that the joint treatment of two entirely different fields (namely, the development of polymeric antitumor agents on a molecular level and first attempts to construct a stable cell model), may seem rather arbitary to the reader. Indeed, development of the former has advanced to a point where first successes are seen—together with the corresponding difficulties. One can surely say that this field is a promising one, and needs further intensive research. In the case of cell models and hypothetical antitumor agents on a cellular level, only preliminary steps have been undertaken and it remains to be seen whether this concept is a realistic one. Both approaches have one crucial point in common: targeting to tumor cells in the human body still remains an unsolved problem.

The cancer problem, however, is such an urgent one that we cannot give up the empirical or rational search for antitumor agents. If this search is performed in cooperation with chemists, pharmacologists, immunologists, and molecular biologists, it should help in gaining more detailed information about tumor cells. Thus, development of antitumor agents can help to characterize and treat tumors. The reader may consider that a great deal of the concepts presented here is speculative—he or she is correct. We do feel, however, that this particular area of research merits stimulation and further collaborative investigation.

References

1. E. S. Greenwald, *Cancer Chemotherapy*, 2nd Edition, Medical Examination Publishing Company, Inc., Flushing, N.Y. (1973).
2. D. L. Hill, *A Review of Cyclophosphamide*, Thomas Springfield, Ill. (1975).
3. R. B. Livingston and S. K. Carter, Methotrexate, in: *Single Agents in Cancer Chemotherapy*, pp. 131–172, Plenum Press, London (1970).
4. K. W. Brunner and G. A. Nagel, *Internistische Krebstherapie*, 2nd Edition, Springer, Berlin/Heidelberg/New York (1979).
5. R. E. La Fond, *Cancer, The Outlaw Cell*, American Chemical Society, Washington D.C. (1978).
6. Baldwin, Tumor associated antigens, *Z. Krebsforsch.* **89**, 1 (1977).
7. R. Holliday, A new theory of carcinogenesis, *Brit. J. Cancer* **40**, 513 (1979).
8. M. S. Melzer, *Eur. J. Cancer* **16**(1), 15 (1980).
9. D. Schmähl, *Maligne Tumoren*, 3rd Edition, Editio Cantor, Aulendorf (1981).
10. W. W. Franke, in: *Neoplastic Transformation: Mechanisms and Consequences* (H. Koprowsky, ed.), Dahlem Workshop Reports, Verlag Chemi, Weinheim (1977).
11. A. S. Sartorelli and D. G. Johns, *Handbook of Experimental Pharmacology*, Vol. 38, Parts 1 and 2, Springer, Berlin (1974/1975).

12. See Reference 39, p. 36.
13. A. Zimmermann and A. Sutter, personal communication.
14. C. Cheng, *J. Pharm. Sci.* **61**, 645 (1972).
15. G. Weber, *N. Engl. J. Med.* **296**, 486 and 541 (1977).
16. G. Weber, *Adv. Enzyme Regul.* **11**, 79 (1973).
17. C. Tan, *Hosp. Prac.* **7**, 99 (1972).
18. A. R. Poole, *Lysosomes in Biology and Pathology*, Vol. 3, Elsevier, North-Holland, Amsterdam (1973), p. 303.
19. L. W. Oberley and G. R. Buettner, *Cancer Res.* **39**, 1141 (1979).
20. G. M. Bartoli, S. Bartoli, T. Galeotti, and E. Bertoli, *Biochim. Biophys. Acta* **620**, 205 (1980).
21. C. P. Li, *Anti cancer Agents Recently Developed in the Peoples' Republic of China*, DHEW Publication No. NIH 74, Washington, D.C. (1974).
22. A. K. Brewer, *Am. Lab.* **5**, 12 (1973).
23. C. H. Lo, V. J. Cristofalo, H. P. Morris, and S. Wienhouse, *Cancer Res.* **28**, 1 (1968).
24. Z. P. Papanastassiou, R. J. Bruni, E. White, and P. L. Levins, *J. Med. Chem.* **9**, 725 (1966).
25. J. C. Robbins and G. L. Nicolson, in: *Biology of Tumors: Surfaces, Immunology and Comparative Pathology* (F. F. Becker, ed.), Vol. 4, p. 3, Plenum Press, New York (1975).
26. G. L. Nicolson and G. Poste, *N. Engl. J. Med.* **295**, 197 (1976).
27. G. L. Nicolson, *Biochim. Biophys. Acta* **457**, 57 (1976).
28. D. F. H. Wallach, *Membrane Molecular Biology of Neoplastic Cells*, Elsevier, Amsterdam (1975).
29. G. Poste, *Biochem. Soc. Trans.* **8**, 695 (1980).
30. G. L. Nicolson, *Biochim. Biophys. Acta* **458**, 1 (1976).
31. H. Busch, E. Fujiwara, and D. C. Firszt, *Cancer Res.* **21**, 371 (1961).
32. C. H. Sutton and N. H. Becker, *Ann. N.Y. Acad. Sci.* **159**, 497 (1969).
33. L. A. Smets, *Biochim. Biophys. Acta* **605**, 93 (1980).
34. G. L. Nicolson, *Int. Rev. Cytol.* **39**, 89 (1974).
35. M. M. Burger, *Proc. Fed. Am. Soc. Exp. Biol.* **32**, 91 (1973).
36. M. F. Bramwell, *Biochem. Soc. Trans.* **8**, 697 (1980).
37. D. A. Weiss, *Current Topics Microbiol. Immunol.* **89**, 1 (1980).
38. M. Bessis, G. A. Dunn, H. Felix, G. Haemmerli, G. Isenberg, E. F. Lüscher, M. Mareel, R. Rungger-Brändle, U. Gröschel-Stewar, P. Sträuli, and K. E. Wolfarth-Bottermann, *Eur. J. Cancer* **16**, 1 (1980).
39. W. E. G. Müller, *Chemotherapie von Tumoren*, Verlag Chemie, Weinheim (1975).
40. See References 4, p. 69.
41. R. Süss, V. Kinzel, and J. D. Scribner, *Krebs-Experimente und Denkmodelle*, Springer, Berlin (1970), p. 85.
42. G. Poste and J. J. Fidler, *Nature* **283**, 139 (1980).
43. G. L. Nicolson, *Spect. Wissenschaft* **1979**, 44.
44. P. Ehrlich, in: *Readings in Pharmacology* (L. Shuster, ed.), p. 231, Little, Brown, Boston (1962).
45. V. T. de Vita and H. Busch, *Methods in Cancer Research*, Vol. 16, Part A, Academic Press, New York (1979).
46. C. E. Tananis, M. G. McLoughlin, and P. C. Walsh, *Cancer Treat. Rep.* **61**, 265 (1977).
47. P. G. Riches, E. D. Gilby, S. M. Sellwood, R. Wilkinson, and I. Konyves, *Eur. J. Cancer* **13**, 873 (1977).
48. W. C. J. Ross in: *Handbook of Experimental Pharmacology* (A. C. Sartorelli and D. G. Johns, eds.) pp. 33–51, Springer, Berlin/Heidelberg/New York (1974).
49. S. Neidle, *Nature* **268**, 196 (1977).

50. C. Abel, T. A. Conners, W. C. J. Ross, N. H. Nam, H. Hoellinger, and I. Pichat, *Eur. J. Cancer* **9**, 49 (1973).
51. (a) U. Zimmermann and G. Pilwat, *Z. Naturforsch.* **C 31**,732 (1976). (b) U. Zimmermann, F. Riemen, and G. Pilwat, *Biochem. Biophys. Acta* **436**, 475 (1976). (c) M. Furusawa, M. Yamaizumi, T. Nishimura, T. Ushida, and Y. Okada, *Methods Cell. Biol.* **14**, 73 (1976). (d) S. J. Updike, R. T. Wakamiya, and E. N. Lightfood, Jr.,*Science* **193**, 681 (1976). (e) D. A. Tirell and B. E. Ryman, *Biochem. Soc. Trans.* **4**, 677 (1976).
52 (a) H. K. Kimelberg and E. J. Mayhew, *CRC Crit. Rev. Toxicol.* **6**, 25 (1978). (b) G. Gregoriadis and A. C. Allison. *Liposomes in Biological Systems*, Wiley, New York (1980). (c) D. Papahadjopoulos, *Ann. N.Y. Acad. Sci.* **308**(1978). (d) R. E. Pagano and J. N. Weinstein, *Ann. Rev. Biophys. Bioeng.* **7**, 435 (1978) (gives a critical review of possible liposome–cell interactions, including experimental techniques).
53. L. G. Donarauma and O. Vogl, *Polymeric Drugs*, Academic Press, New York/San Francisco/London (1978).
54. L. G. Donaruma, R. M. Ottenbrite, and O. Vogl, *Anionic Polymeric Drugs*, Wiley, New York (1980).
55. L. G. Donaruma, *Progr. Polym. Sci.* **4**, 1 (1974).
56. O. Vogl and D. Tirrell, *J. Macromol. Sci. Chem.* **13**, 415 (1979).
57. D. S. Zaharko, M. Przybylski, and V. T. Olivera, in: *Methods in Cancer Research* , Vol. 16, pp. 347–380, Academic Press, New York (1979).
58. M. Szekerke and J. S. Driscoll, *Eur. J. Cancer* **13**, 529 (1977).
59. L. Gros, H. Ringsdorf, and H. Schupp, *Angew. Chem. Int. Ed. Engl.* **20**, 305 (1981).
60. H. Ringsdorf, *J. Polym. Sci., Polym. Symp.* **51**, 135 (1975).
61. (a) K. F. Mück, o. Christ, and H. M. Kellner, *Makromel. Chem.* **178**, 2785 (1977). (b) V. Hofmann, H. Ringsdorf, and E. Schaumlöffel, *Makromol. Chem.* **181**, 351 (1980). (c) L. Sprincl, J. Exner, V. Sterba, and J. Kopecek, *J. Biomed. Mater, Res.* **10**, 953 (1976). (d) G. Lindblad and J. Falk, *Infusionstherapie* **3**, 301 (1976).
62. S. I. Rapoport, *Blood–Brain Barrier in Physiology and Medicine*, Raven Press, New York (1976).
63. G. Hecht, *Ther. Ber.* **10**, 264 (1956).
64. E. Gundmann, *Fortschr. Staublungenforsch.* **2**, 223 (1967).
65. H. J. P. Ryser, *Science* **390** (1968).
66. J. Kalal, J. Drobnik, J. Kopecek, and J. Exner, *Brit. Polym. J.* **10**, 111 (1978).
67. E. Hurwitz, R. Maron, A. Bernstein, M. Wilchek, M. Sela, and R. Arnon, *Int. J. Cancer* **21**, 747 (1978).
68. J. Kalal, J. Drobnik, J. Kopecek, and J. Exner, in: *Polymeric Drugs* (L. G. Donaruma and O. Vogl, eds.), p. 131, Academic Press, New York (1978).
69. J. Kopecek, *Makroimol. Main Lect. Int. Symp.* **27**, 305 (1981).
70. J. Kalal, *Makromol. Chem. Suppl.* **2**, 215 (1979(.
71. A. C. Tanguary and R. E. Lacey, *Controlled Release of Biologically Active Agents*, Plenum Press, New York and London (1974).
72. R. J. Kostelnik, *Polymeric Delivery Systems*, Gordon & Breach, New York/London/Paris (1978).
73. R. Schnee, Immun- und Endocytoseverhalten Synthetischer Polymerer, Dissertation, Mainz, West Germany (1979).
74. P. Ferruti, M. C. Tanzi, and F. Vaccaroni, *Makromol. Chem.* **180**, 375 (1979).
75. E. J. Artiens, *Drug Design*, Vol. 1, Academic Press, New York (1971).
76. E. D. Korn, *MTP Int. Rev. Sci. Biochem. Ser. I* **2**, 1 (1975).
77. W. C. Shen and H. J. P. Ryser, *Biochem. Biophys. Res. Commun.* **102**, 1048 (1981) and further references cited therein.
78. R. Duncan, J. B. Lloyd, and J. Kopecek, *Biochem. Biochem. Biophys. Res. Commun.* **94**(1), 284 (1980).

79. (a) Z. A. Cohn and E . Parks, *J. Exp. Med.* **125**, 213 (1967). (b) H,. J. P. Ryser, *Nature* **215**, 934 (1967). (c) H. J. P. Ryser, M. P. Gabathuler, and A. B. Roberts, *Biomembranes* **2**, 197 (1971). (d) C. J. van Oss, C. F. Gillman, and A. W. Neumann, *Phagocytotic Engulfment and Cell Adhesiveness,* Marcel Dekker, New York (1975). (e) T. Kooistra, A. M. Duursma, J. M. W. Bouma, and M. Gruber, *Biochim. Biophys. Acta* **587**, 282 (1980). (f) J. Quinkert, M. A. Leroy-Houyet, A. Trouet, and P. Baudhuin, *J. Cell Biol.* **82**, 644 (1979).
80. J. L. Denburg, *Adv. Comp. Physiol. Biochem.* **7**, 106 (1978).
81. J. Becker *et al., Cell Surface Carbohydrate Chemistry,* R. E. Harmon (Ed), Academic Press, San Francisco (1978).
82. S. C. Silverstein, *Transport of Macromolecules in Cellular Systems*, report of the Dahlem Workship, April 24–28, 1978, Abakon Verlagsgesellschaft, Berlin (1978).
83. F. Macritchie, *Adv. Protein Chem.* **32**, 283 (1978).
84. J. D. Morrisett, R. L. Jackson, and A. M. Golto, *Biochim. Biophys. Acta* **272**, 93 (1977).
85. D. J. Rejngnoud, S. Lind-Katz, H. Hauser, and M. C. Phillips, *Biochemistry* **21**, 2977 (1982).
86. D. A. Tirrell, *Polym. Prepr., Am. Chem. Soc., Div. Polym. Chem.* **22**, 393 (1981).
87. L. K. Mawaha and D. A. Tirrell, in: *Biological Activities of Polymers* (C. E. Carraher, Jr. and C. G. Gebelein, eds.), pp. 163–175, American Chemical Society Symposium Series No. 186, Washington, D.C. (1982).
88. W. Regelson, *J. Polym. Sci., Polym. Symp.* **66**, 539 (1979).
89. R. J. Fiel, E. Mark, and H. I. Levine, *Amionic Polymeric Drugs*, Wiley, New York (1980).
90. H. J. P. Ryser, *Nature* **215**, 934 (1967).
91. H. Utsumi, B. D. Tunggal, and W. Stoffel, *Biochemistry* **10**, 2385 (1980).
92. I. G. Goldstein, *Adv. Carbohydr. Chem. Biochem.* **35**, 128 (1978).
93. M. Achtman and L. Jarett, Group Report, *Adv. Comp. Physiol. Biochem.* **7**, 134 (1978).
94. R. O. Brady and P. H. Fishman, *Adv. Comp. Physiol. Biochem.* **7**, 69 (1978).
95. S. Olsnes, *Adv. Comp. Physiol. Biochem.* **7**, 103 (1978).
96. S. C. Silverstein, R. M. Steinman, and Z. A. Cohn, *Ann. Rev. Biochem.* **46**, 669 (1977).
97. W. S. Sly, *Adv. Comp. Physiol. Biochem.* **7**, 266 (1978).
98. R. Duncan, M. K. Pratten, H. C. Cable, H. Ringsdorf, and J. B. Lloyd, *Biochem. J.* **194** (1980).
99. S. J. Freeman and J. B. Lloyd, *Biochem. Soc. Trans.* **8**, 434 (1980).
100. M. K. Pratten, R. Duncan, H. C. Cable, R. Schnee, H. Ringsdorf, and J. B. Lloyd, *Chem. Biol. Interactions* **35**, 319 (1981).
101. P. S. Morahan and A. M. Kaplan, *Int. J. Cancer* **17**, 82 (1976).
102. R. M. Schultz, J. D. Papametheakis, J. Luetzeler, and M. A. Chirigos, *Cancer Res.* **37**, 3338 (1977).
103. N. Sharon and H. Lis, *Science* **177** (4053); 949 (1972).
104. A. J. Sbarra, W. Shirly, and W. A. Bardawil, *Nature* **194**, 255 (1962).
105. R. M. Steinman, J. M. Silver, and Z. A. Cohn, *Adv. Comp. Physiol, Biochem.* **7**, 167 (1978).
106. C. de Duve, T. de Varsy, B. Poole, A. Trouet, P. Tulkens, and F. van Hoof, *Biochem. Pharmacol.* **23**, 2495 (1974).
107. R. E. Falk, L. Makowka, L. E. Rotstein, J. A. Falk, N. Nossal, and U. Ambus, in *Progress Cancer Research Therapy* (E. Mittersch, M. A. Chirigos, M. J. Mastrangelo, eds.), Vol. 16, p. 295 (1981).
108. R. E. Falk, E. Makowka, N. Nossal, J. A. Falk, J. E. Fields, and S. S. Asculai, *Brit. J. Surg.* **66**, 861 (1979) and British Patent No. GB 2040951 A.
109. R. Bierling, G. D. Wolf, and B. Böhmer, *Naturwissenschaften* **67**, 366 (1980) and DOS 3026447, Vol. 12, p. 3026447, (July 1980/February 4, 1982).

110. L. J. Arnold, A. Dagan, J. Gutheil, and N. O. Kaplan, *Proc. Natl. Acad. Sci. U.S.A.* **76**, 3246 (1979).
111. J. Kopecek, *Polym. Med.* **7**(3), 191 (1977).
112. H. G. Batz, H. Ringsdorf, and H. Ritter, *Makromol. Chem.* **175**, 2229 (1974).
113. N. Brock and H. J. Hohorst, *Naturwissenschaften* **49**, 610 (1962).
114. N. Brock and H. J. Hohorst, *Z. Krebsforsch.* **88**, 185 (1977).
115. T. A. Connors, P. J. Cox, P. B. Farmer, and M. Jarmann, *Biochem. Pharmacol.* **23**, 115 (1974).
116. P. J. Cox, B. J. Philips, and P. Thomas, *Cancer Res.* **35**, 3755 (1975).
117. G. Peter, T. Wagner, and H. J. Hohorst, *Cancer Treat. Rep.* **60**, 429 (1976).
118. T. Hirano, W. Klesse, and H. Ringsdorf, *Makromol. Chem.* **180**, 1125 (1979).
119. (a) T. Hirano, W. Klesse, J. H. Heusinger, G. Lambert, and H. Ringsdorf, *Tetrahedron Lett.* **1979**, 883. (b) T. Hirano, H. Ringsdorf, and D. S. Zaharko, *Cancer Res.* **40**, 2263 (1980).
120. L. Gros, H. Ringsdorf, R. Schnee, J. B. Lloyd, E. Rüde, H. U. Schorlemmer, and H. Stötter, American Chemical Society Symposium Series No. 195, pp. 7, 84, Washington, D.C. (1982).
121. G. Hörpel and H. Ringsdorf (to be published).
122. C. G. Overberger, H. Ringsdorf, and B. Avchen, *J. Med. Chem.* **8**, 862 (1962).
123. H. Ringsdorf, A. G. Heisler, F. H. Müller, E. H. Graul, and W. Rüther, in: *Biological Aspects of Radiation Protection* (T. Sugahara and O. Hug, eds.), Igaku Shoin Ltd., Tokyo (1971).
124. D. S. Breslow, *Pure Appl. Chem.* **46**, 102 (1976).
125. R. Langer, D. S. T. Hirsch, and L. Brown, American Chemical Society Symposium Series No. 186, Washington, D.C. (1982), p. 95.
126. P. Molz, Dissertation, Mainz, West Germany (1982).
127. A. Trouet, *Bull. Mem. Acad. R. Med. Belg.* **135**, 261 (1980).
128. A. Trouet, P. Pirson, R. Steiger, M. Masquelier, R. Baurain, and J. Gillet, *WHO Bull.* **59**, 449 (1981).
129. W. C. Shen and H. J. P. Ryser, *Biochem. Biophys. Res. Commun.* **102**, 1048 (1981).
130. A. Bernstein, E. Hurwitz, R. Maron, R. Arnon, M. Sela, and M. Wilchek, *J. Natl. Cancer Inst.* **60**, 379 (1978).
131. A. Trouet, M. Masquelier, R. Baurain, and D. Deprez-De Campeneere, *Proc. Natl. Acad. Sci. U.S.A.* **79**, 626 (1982).
132. M. Sela and E. Katechalski, *Adv. Protein Chem.* **14**, 391 (1959).
133. L. G. Baird and A. M. Kaplan, *Anionic Polymeric Drugs*, Wiley, New York (1980).
134. M. Przybylski, E. Fell, H. Ringsdorf, and D. Zaharko, *Makromol. Chem.* **179**, 1719 (1978).
135. M. Przybylski, D. S. Zakarko, M. A. Chirigos, R. H. Adamson, R. M. Schultz, and H. Ringsdorf, *Cancer Treat. Rep.* **62**, 1837 (1978).
136. M. Przybylski, W. P. Fung, H. Ringsdorf, R. H. Adamson, and D. Zaharko, 69th Annual Meeting, American Association of Cancer Research (1978), Proceedings, Vol. 19, p. 2.
137. W. P. Fung, M. Przybylski, H. Ringsdorf, and D. Zaharko, *J. Natl. Cancer Inst.* **62**, 1261 (1979).
138. G. B. Butler, *J. Macromol. Sci. Chem.* **A13**(3), 351 (1979).
139. W. Klesse, Synthese und Untersuchung von Konjugaten des 4-Hydroxy-Cyclophosphamid mit Mizellaren und nicht Mizellaren Polymeren Trägern, Dissertation, Mainz, West Germany (1981).
140. G. Hörpel and H. Ringsdorf (to be published).
141. L. C. Erickson, G. Hörpel, K. W. Kohn, H. Ringsdorf, and D. S. Zaharko (in press).

142. K. W. Kohn, R. A. G. Ewig, L. C. Erickson, and L. A. Zwelling, in: *DNA-Repair* (E. C. Friedberg and P. C. Hanawalt, eds.), Vol. I, Part B, Chapter 29, p. 379, Marcel Dekker, New York/Basel (1981).
143. J. C. Osborne and H. B. Brewer, *Adv. Protein Chem.* **31**, 253 (1977).
144. M. Wilchek, *Makromol. Chem. Suppl.* **2**, 207 (1979).
145. E. Goldberg, L. G. Donaruma, and O. Vogl, *Targeted Drugs*, Wiley, New York (1983).
146. G. Gregoriadis, J. Senior, and A. Trouet, *Targeting of Drugs*, Nato Advanced Study Institutes Series, Series A: Life Sciences, Vol. 47, Plenum Press, New York and London (1982).
147. C. Milstein, Monoclonal antibodies, *Sci. Am.* **10** (1980).
148. E. Haber and R. M. Krause, *Antibodies in Human Diagnosis and Therapy*, Raven Press, New York (1977).
149. S. Olsnes, *Nature* **290**, 84 (1981).
150. E. Vitetta, K. A. Krolick, M. Miyama-Inaba, W. Cushley, and J. W. Uhr, *Science* **219**, 644 (1982).
151. D. C. Edwards, P. E. Thorpe, and A. J. S. Davies, *Targeted Drugs*, Wiley, New York (1983), p. 83.
152. R. Arnon, *Targeted Drugs*, Wiley, New York (1983), p. 31.
153. E. Klein and G. Klein, *J. Natl. Cancer Inst.* **32**, 547 (1964).
154. T. A. Conners, *Targeted Drugs*, Wiley, New York (1983), p. 97.
155. H. Ritter, Synthetische Polymere als Potentielle Antitumor-Chemotherapeutika, Dissertation, Mainz, West Germany (1974).
156. G. L. Nicolson, *Biochim. Biophys. Acta* **458**, 1 (1976).
157. M. Burger, *Adv. Cancer Res.* **20**, 1 (1974).
158. G. Uhlenbruck, Lektine, toxine immunotoxine, *Naturwissenschaften* **68**, 606 (1981).
159. L. Fiume, C. Busi, A. Mattioli, P. G. Balboni, G. Barbanti-Brodano, and Th. Wieland, *Targeted Drugs*, Wiley, New York (1983), p. 1.
160. C. M. Sturgeon, *Trends Biolog. Sci.* **4**(6), 121 (1979).
161. R. A. Neumann, *J. Natl. Cancer Inst.* **63**, 1339 (1979).
162. E. P. Goldberg, *Polym. Prepr., Am. Chem. Soc., Div. Polym. Chem.* **20**(1) 362 (1979).
163. P. Tsuruo, T. Yamori, S. Tsukagoshi, and Y. Sakurai, *Int. J. Cancer* **26**, 655 (1980).
164. P. Franchimont and P. F. Zangerle, *Eur. J. Cancer* **13**, 637 (1977).
165. L. J. Old, *Sci. Am.* **236**(5), 62 (1977).
166. (a) G. Poste and A. C. Allison, *Biochim, Biophys. Acta* **300**, 421 (1973). (b) J. A. Lucy, *J. Reprod. Textiles* **44**, 193 (1975).
167. S. Sell, *Immunologie, Immunpathologie und Immunität*, Verlag Chemie, Weinheim (1977), p. 199.
168. (a) D. Papahadjopoulos, W. J. Vail, K. Jacobson, and G. Poste, *Biochim. Biophys. Acta* **394**, 483 (1975). (b) D. Papahadjopoulos, S. Hui, W. J. Vail, and G. Poste, *Biochem. Biophys. Acta* **448**, 245 (1976).
169. (a) G. Gregoriadis and G. Poste, in: *Liposomes on Biological Systems* (G. Gregoriadis and A. C. Allison, eds.), pp. 36 and 107, Wiley, New York (1980). (b) F. Szoka, K. Jacobson, and D. Papahadjopoulos, *Biochim. Biophys. Acta* **551**, 295 (1979).
170. G. Gregoriadis and A. C. Allison in: *Liposomes on Biological Systems* (G. Gredoriadis and A. C. Allison, eds.), pp. 34, 105, and 340, Wiley, New York (1980).
171. S. J. Singer and G. L. Nicolson, *Science* **175**, 720 (1972).
172. F. S. Sjoestrand, in: *Chemotherapy of Cancer Dissemination and Metastasis* (S. Carattini and C. Franchi, eds.), p. 4, New York (1973).
173. S. J. Singer, in: *Cell Membranes—Biochemistry, Cell Biology and Pathology* (G. Weissman and R. Clairborne, eds.), p. 44, New York (1975).
174. C. M. Gupta, C. E. Costello, and H. G. Khorana, *Proc. Natl. Acad. Sci. U.S.A.* **76**, 3139 (1979).

175. A. Akimoto, K. Dorn, L. Gros, H. Ringsdorf, and H. Schupp, *Angew. Chem.* **93**, 108 (1981); *Angew. Chem. Int. Ed. Engl.***20**, 90 (1981).
176. (a) T. Kunitake and Y. Okahata, *Chem. Lett.* **1977**, 1337. (b) T. Kunitake, *Macromol. Sci. Chem.* **13**, 587 (1979).
177. K. Fukuda, Y. Shibasaki, and H. Nakahara, *J. Macromol. Sci., Chem.* **15**, 999 (1981).
178. A. D. Banghham, M. M. Standish, and J. C. Watkins, *J. Mol. Biol.* **13**, 238 (1965).
179. H. C. Huang, *Biochemistry* **8**, 344 (1969).
180. H.-H. Hub, B. Hupfer, H. Koch, and H. Ringsdorf, *Angew. Chem.* **92**, 962 (1980).
181. (a) B. Pearce, *Trends Biochem. Sci. Pers. Ed.* **1980**, 131. (b) C. D. Ockleford and A. Whyte, *Coated Vesicles* Cambridge Univ. Press, London (1980).
182. Y. Okahata and T. Kunitake, *J. Am. Chem. Soc.* **101**, 5231 (1979).
183. S. L. Regen, B. Czech, and A. Singh, *J. Am. Chem. Soc.* **102**, 6638 (1980).
184. H. Schupp, Ph.D. Thesis, University of Mainz, West Germany (1981).
185. H. Bader, H. Ringsdorf, and J. Skura, *Angew, Chem. Int. Ed. Engl.* **20**, 91 (1981).
186. N. Wagner, K. Dose, H. Koch, and H. Ringsdorf, *FEBS Lett.* **132**, 313 (1983).
187. K. Dorn, Ph.D. Thesis, University of Mainz, West Germany (1983).
188. P. Tundo, D. J. Kippenberger, P. L. Klahn, N. E. Prieto, T. C. Jao, and J. H. Fendler, *J. Am. Chem. Soc.* **104**, 456 (1982).
189. S. L. Regen, A. Singh, G. Oehme, and M. Singh, *Biochem. Biophys. Res. Commun.* **101**, 131 (1981).
190. B. Hupfer, H. Ringsdorf, and H. Schupp, *Makromol. Chem.* **182**, 247 (1981).
191. D. Day, H.-H. Hub, and H. Ringsdorf, *Isr. J. Chem.* **18**, 325 (1979).
192. K. Aliev (unpublished results).
193. H. Koch and H. Ringsdorf, *Makromol. Chem.* **182**, 255 (1981).
194. H.-H. Hub, B. Hupfer, H. Koch, and H. Ringsdorf, *Angew, Chem. Int. Ed. Engl.* **19**, 938 (1980).
195. T. Folda, L. Gros, and H. Ringsdorf, *Makromol. Chem. Rapid Commun.* **3**, 167 (1982).
196. M. B. Yatvin, W . Kreutz, B. A. Horwitz, and M. Shinitzky, *Science* **210**, 1253 (1980).
197. J. N. Weinstein, R. L. Magin, M. B. Yatvin, and D. S. Zaharko, *Science* **204**, 188 (1979).
198. K. Kano, Y. Tanaka, T. Ogawa, M. Shimomura, J. Okahata, and T. Kunitake, *Chem. Lett.* **1980**, 421.
199. R. Büschl, personal communication.
200. I. J. Goldstein, ed., *Carbohydrate Protein Interaction*, American Chemical Society Symposium Series No. 88, Washington D.C. (1979).
201. J. H. Fendler, *Acc. Chem. Res.* **13**, 7 (1980).
202. G. Poste and G. L. Nicholson, *Cell Surface Reviews*, Vol. 5, North-Holland, Amsterdam (1979).
203. N. R. Ringertz and R. E. Savage, *Cell Hybrids*, Academic Press, New York (1976).
204. U. Zimmermann, P. Scheurich, G. Pilwat, and R. Benz, *Angew. Chem. Int. Ed. Engl.* **20**, 325 (1981).
205. U. Zimmermann, J. Schultz, and G. Pilwat, *Biophys. J.* **13**, 1005 (1973).
206. P. Scheurich, U. Zimmermann, and H. Schnabl, *Plant Physiol.* **62,** 849-853 (1981).
207. H.-H. Hub, U. Zimmermann, and H. Ringsdorf, *FEBS Lett.* **140**, 254 (1982).
208. D. Papahadjopoulos, S. Hui, W . J. Vail, and G. Poste, *Biochim. Biophys. Acta* **448**, 245 (1976).
209. G. Poste and J. L. Nicholson, eds, *Membrane Fusion*, Elsevier/North-Holland, Amsterdam (1978).
210. J. Wilschut, N. Duezguenes, and D. Papahadjopoulos, *Biochemistry* **20**, 3126 (1981).
211. R. Büschl, H. Ringsdorf, and U. Zimmermann, *FEBS Lett.* **150**, 38 (1982).

20

Biological Activities of *cis*-Dichlorodiamineplatinum II and Its Derivatives

Charles E. Carraher, Jr., William J. Scott, and David J. Giron

Abstract. Malignant neoplasms are the second leading cause of death. In 1964 Rosenberg and co-workers discovered that *cis*-dichlorodiamineplatinum II prevented select bacteria from dividing. This led to the licensing, in 1980, of Platinol and brought about the synthesis and characterization of numerous derivatives of *cis*-dichlorodiamineplatinum II. The structural requirements, mode of activity, animal and human toxicity, and antineoplastic effects of such derivatives are reviewed. Due to the rapidity with which antineoplastic platinum compounds are filtered through the kidney, a high level of renal, and related, damage occurs. One method of lowering this toxicity, which may also permit prolonged controlled release of the active platinum-containing compounds, is the placement of the platinum into polymers.The rational synthesis and biological and physical characterization of these polymeric platinum-containing polymers is discussed.

1. Introduction

A number of basic chemical investigations have not been done related to the synthesis and chemistry of *cis*-dichlorodiamineplatinum II, *cis*-DDP, derivatives, because most investigations did not involve chemists and because the major impetus for studying *cis*-DDP derivatives involves biological, not physical or chemical, characterization of these compounds. Conversely, more is known about the biological properties, including mode of transport and sites of concentration, than for any other organometal-containing group of compounds. Platinum-containing compounds exhibit typical types of toxicities including the broad classification known as heavy-metal toxicity. Understanding the state of affairs regarding derivatives of *cis*-DDP may assist us in understanding the role, mechanism, and pathways of other

Charles E. Carraher, Jr., William J. Scott, and David J. Giron • Departments of Chemistry and Microbiology and Immunology, Wright State University, Dayton, OH 45435.

metal-containing drugs including metal-containing polymers which are active through a controlled-release mechanism. This chapter reviews findings related to *cis*-DDP and its derivatives and ends with results derived from polymers containing *cis*-DDP derivatives. (The *cis*-DDP is also known as Platinol and *cis*-platin in the literature.)

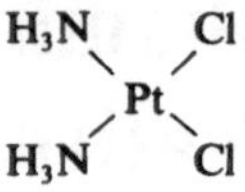

$$\begin{array}{ccc} H_3N & & Cl \\ & \diagdown \; \diagup & \\ & Pt & \\ & \diagup \; \diagdown & \\ H_3N & & Cl \end{array}$$

Cis-DDP

(1)

2. General Discussion

Malignant neoplasms are the second leading cause of death in the United States. Until the late 1960s little effort was made towards the synthesis and characterization of antineoplastic agents, except for totally organic compounds. In 1964 Barnett Rosenberg and co-workers discovered by chance that bacteria failed to divide, but continued to grow giving filamentous cells in the presence of platinum electrodes.[1] The cause of this inhibition to cell division was eventually traced to small concentrations (about 10 ppm) of *cis*-dichlorodiamineplatinum II (*cis*-DDP) and *cis*-tetrachlorodiamine platinum IV (*cis*-TCP). This ability to inhibit bacterial cell division prompted their trial as antitumor agents. Both compounds were found to be active against sarcoma 180 and leukemia L1210 test tumors in mice with *cis*-DDP having greater activity than *cis*-TCP.[2,3] In fact, *cis*-DDP is active against a wide range of tumors in animals[4] and man,[5] including several which are particularly resistant to treatment by other techniques, such as ovarian and testicular carcinomas.

The success of platinum compounds has catalyzed the synthesis and characterization of a number of organometallic compounds, focusing on derivatives of *cis*-DDP, for antitumor activity.

Continued study of the effects of platinum compounds, both on animals and on man, has indicated that their positive attributes are coupled with a high level of negative side effects which may be quite serious or even lethal (e.g., see References 5–9). Major complications include gastrointestinal, hematopoietic, immunosuppressive, auditory, and renal dysfunction, with the latter two being by far the most serious. One group has also shown an allergic reaction in a man after a number of doses of *cis*-DDP.[10]

The gastrointestinal effects are displayed as acute nausea and vomiting, normally lasting four to six hours. Because the anorexia associated with this may be cumulative, a subject's appetite may decrease even as the subject's tumor shrinks.[5] Thus, it has been suggested by some researchers that psychopharmacological drugs with cyproheptadine or Δ^9-tetrahydrocanabinol be employed to diminish these effects and to stimulate an appetite.[11,12] The hematopoietic effects are normally seen as leukopenia occasionally complicated with mild thrombocytopenia. Because they are not well understood, the immunosuppressive effects are currently being studied.

During early treatment with *cis*-DDP, cumulative and irreversible hearing losses begin to occur in the high range (4000–8000 Hz). With continued treatment, losses in the spoken range (1000–4000 Hz) can develop, leading eventually to total deafness.[5,13] Deafness has been found to indicate that death is imminent.

Renal damage, also cumulative and eventually irreversible, is the most important side effect.[5,9,14] It limits both the size of a dose and the length of treatment. This problem occurs because soon after application, a large percentage of the *cis*-DDP is filtered from the blood and excreted through the urine. Many studies have shown that a large proportion of the platinum which does remain in the body is found in the kidneys.[15] Other problems are continuing to be discovered and dealt with (e.g., see References 15 and 16).

The potency of such activity may be measured in a number of ways. The three most prevalent are the therapeutic index (TI), the percentage increased life span (% ILS), and the treated-to-control tumor ratio (TC). In the first case, the ratio of the LD_{50} (minimum dose required to kill 50% of the test animals) to the ID_{90} (minimal dose required to cause a 90% tumor regression) is calculated. A high TI implies an effective drug. The % ILS simply relates the relative lengthening of the life span of test animals under treatment with respect to control animals. Finally, the TC is expressed as the ratio of the tumor weights of treated animals to the tumor weights of control animals. In this case, a small value implies an effective drug. Because the data on the relative antineoplastic activity of platinum compounds may employ any of these indices, they will all appear in this chapter.

3. Structural Requirements

Most of the current efforts related to the synthesis of platinum compounds for antitumor activity are aimed at new compounds exhibiting antitumor activity comparable to *cis*-DDP, but with decreased toxicity or greater therapeutic properties allowing decreased dosage level. Such efforts have produced a wide variety of compounds, a number of which show antitumor activity.[16–18] From studying the structural features of such compounds, several features have emerged. Active compounds are (1) neutral, (2) contain two inert and two labile ligands, and (3) must have the ligands cis to each other.

One of the first points studied was the geometric arrangement of

Table 1. Antitumor Activity of Some Platinum Complexes

Platinum complex	Therapeutic index	Tumor	Tumor/Control weight ratio	Tumor	References
cis-$(NH_3)_2PtCl_2$	8.1	ADJ/PC6A mice	1	Sarcoma 180 mice	5, 6
trans-$(NH_3)_2PtCl_2$	1.0	ADJ/PC6A mice	85	Sarcoma 180 mice	5, 6
cis-$(NH_3)_2PtBr_2$			30	Sarcoma 180 mice	6
trans-$(NH_3)_2PtBr_2$			110	Sarcoma 180 mice	6
cis-$(H_3CNH_2)_2PtCl_2$			25	Sarcoma 180 mice	6
trans-$(H_3CNH_2)_2PtCl_2$			100	Sarcoma 180 mice	6
cis-(NH)$_2PtCl_2$	21.7	ADJ/PC6A mice			5
trans-(NH)$_2PtCl_2$	1.0	ADJ/PC6A mice			5

Table 2. Toxicity and Efficacy of Platinum Complexes

Platinum complex	LD_{50}	ID_{90}	Tumor	References
cis-$(NH_3)_2PtCl_2$	13.0	1.6	Sarcoma 180 mice	5, 6
trans-$(NH_3)_2PtCl_2$	27.0	27	Sarcoma 180 mice	5, 6
cis-(⌐NH)$_2$PtCl$_2$	56.0	2.6	Sarcoma 180 mice	5, 6
trans-(⌐NH)$_2$PtCl$_2$	18.0	18	Sarcoma 180 mice	5, 6

ligands. Where trans compounds exist, they seem to be less active in comparison to the cis isomer (e.g., see Table 1). Early antineoplastic platinum complexes were characterized as having a pair of inert amino ligands and a pair of labile chloride ligands, which were believed to behave as leaving groups. As suggested by the well-known trans effect, a chloride ligand is more reactive in the trans isomer, with *trans*-DDP hydrating about four times faster[(19)] and undergoing amination 10 times faster[(20)] than *cis*-DDP. This greater reactivity might imply a lowered reaction specificity for the trans isomer. Thus, even though the two isomers are of approximately equal toxicity, as shown in Table 2, their therapeutic levels differ vastly.[(20–22)] This difference might be due to the reaction of the trans isomer with various constituents of the body prior to reaching the tumor site. Distribution and excretion studies showed *cis*-DDP to be excreted much faster initially.[(23)] However, within five days the levels of platinum derived from the cis and trans isomers were comparable with about 20% retention. Even at this point the distribution of the two compounds differed radically—platinum levels from *trans*-DDP remaining high in plasma at all times, while levels from *cis*-DDP fall off markedly.[(15,23)] This might be explained by suggesting that *trans*-DDP reacts with some constituents in the blood, remaining there for some time, while *cis*-DDP reacts somewhat later, thus being readily filtered from the blood by the kidneys.

It has also been postulated that the difference in the TI of the two isomers may be due to the mode of action of the drug. Since cis compounds are more likely to form chelates, the activity of *cis*-DDP may be due to its chelation with biological material, such as ribonucleic acid (RNA), deoxyribonucleic acid (DNA), or proteins, at the tumor site.

Activity of the platinum compound is highly dependent on the leaving group. Generally, as shown in Table 3, a change from the chloride acts to raise the TC of the compound.[(22,23)] If the leaving groups are too reactive, such as with the diaquo complex, the drug will not chelate with the proper cellular material to show antineoplastic activity.

More recently, antitumor activity has been observed with compounds containing bidentate leaving groups, such as malonate, sulfate, or nitrate

Table 3. Efficacy of *cis*-$Pt(NH_3)_2X_2$

X	X_2	TC	Tumor	References
Cl^-		1	Sarcoma 180 mice	6, 7
NO_3^-		54	Sarcoma 180 mice	6, 7
NO_2^-		99	Sarcoma 180 mice	6, 7
Br^-		30	Sarcoma 180 mice	6, 7
I^-		110	Sarcoma 180 mice	6, 7
	$^-O_2CCO_2^-$ (oxalate)	9	Sarcoma 180 mice	6
	$^-O_2CCH_3CO_2^-$ (malonate)	7	Sarcoma 180 mice	6
	Cl^-, Br^-	11	Sarcoma 180 mice	7
	Cl^-, I^-	14	Sarcoma 180 mice	7

(e.g., see References 24–27). Because these ligands should be released more slowly than chloride, due to their bidentate nature, a different mechanism of replacement must be sought to remain consistent with the requirement of ligand replacement being at about the rate of chloride ligands. The activity of these compounds is probably due to enzymatic cleavage of Pt–O bonds or to cleavage or oxidation of other portions of the ligand.[28] Though either process would lead to rapid hydrolysis, cleavage of the Pt–O bond seems most likely in view of the fact that the toxicity of diamineoxalato platinum II is closely related to that of oxalate alone.[22]

The amine ligands have few basic requirements. First, the amine ligand should be inert towards replacement in the body, as is almost always the case. Second, because they do not exert a trans effect on the leaving groups, it has been found that aliphatic amines work well.[22,23] Third, the aliphatic substitution has a large effect on the water solubility of the drug.[29] As the amine chain becomes longer, the resulting compound becomes less soluble.[23,29] Because these drugs are generally delivered *via* intravenous injection, aqueous solubility is important for transportation of the drug to the tumor site,[30] while insoluble compounds tend to remain at the injection site.[23,24] The aliphatic chain may be modified to make the compound more water soluble. Thus, *cis*-dichlorobis(2-aminoethanol)platinum II (DEIP) **(2)** is very soluble in water.[23] This leads, however, to very high dosage requirements, presumably due to rapid excretion of the drug. In fact, DEIP is not effective at 15 times the level used for *cis*-DDP.[23]

```
HO—CH2—CH2—NH2     Cl
               \   /
                Pt
               /   \
HO—CH2—CH2—NH2     Cl
```

(2)

The final structural requirement is that these compounds must be neutral. This is a common requirement for drugs which exert their effects within the cell. Because of the nonpolar nature of the cell membrane, compounds must be neutral to pass through it. One exception to the requirement of neutrality for platinum drugs is a class of antineoplastic platinum complexes called "platinum blues" (e.g., see References 31 and 32). However, electron microscopy studies show that the platinum blues do not tend to pass through the membrane, and may act by a different mechanism.[33] Though the "platinum blues" appear to be the exception, such studies should serve as a warning against describing structural requirements that are too limited regarding compounds to be synthesized for antitumor applications.

Relatively small changes in structure appear to lead to large changes in activity.[18] The absence of the use of donor atoms other than nitrogen may be due to a lack of synthesis and testing of such compounds rather than such compounds actually being inactive. Thus other less similar compounds should be synthesized and tested to resolve this question.

4. Mode of Activity

There are a number of theories proposed for the activity of *cis*-DDP and related compounds. Interaction between platinum-containing compounds and biological molecules has been known for some time and studied by a number of investigators.

Most antineoplastic agents act on the cell to block nucleic acid or protein biosynthesis as shown in Figure 1 (see e.g., References 34 and 35). These drugs are essentially antigrowth drugs. Because of the lack of biological differences between tumor cells and normal cells, those which multiply at the greatest rate will normally realize the greatest kill rate from the drug. Unfortunately, tumor cells do not necessarily multiply at the greatest rate. Accordingly, these drugs are highly toxic to organs characterized by rapid cell division, such as bone marrow and the gastrointestinal tract.

As mentioned earlier, active complexes have a pair of leaving groups in the cis position. It is believed that due to the relatively high chloride concentration in the blood serum, the leaving groups remain in position and the molecule remains neutral.[36] In this relatively lipid soluble form, the molecule may then enter the cell, which contains about 25 times less chloride.[37] At this point the dichloro complex may readily become hydrated forming the diaquo complex.

When *cis*-DDP is dissolved in water, hydrolysis occurs with the chloride atoms being substituted by water or atoms derived from water. The half-life for the formation of the diaquo complex at 37° C is 1.7 h with an activation enthalpy of about 20 kcal/mol.[38]

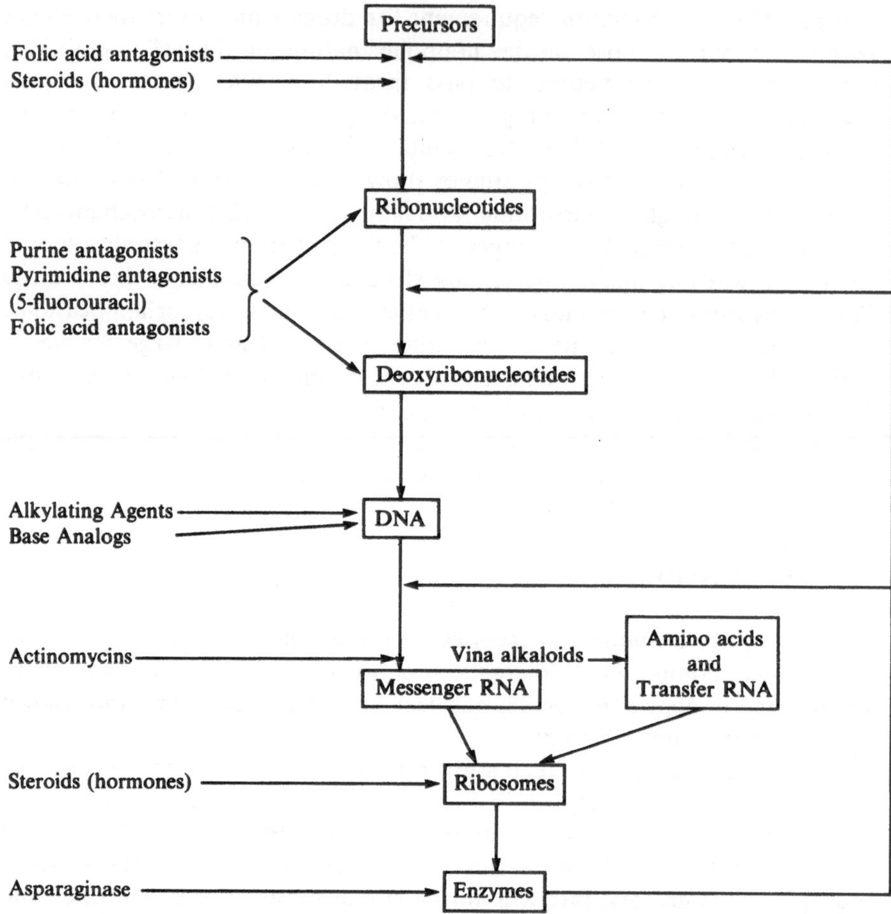

Figure 1. Possible modes of action of antitumor drugs on cell biosynthesis. The general cell biosynthetic pathway is shown in boxes linked by arrows. The portion of the pathway disrupted by the antitumor drugs is shown by arrows leading into the pathway.

$$\underset{\textbf{(3)}}{\begin{matrix} H_3N & & NH_3 \\ & Pt & \\ Cl & & Cl \end{matrix}} \xrightarrow{H_2O} \underset{\textbf{(4)}}{\begin{matrix} H_3N & & NH_3 \\ & Pt & \\ H_2O & & Cl \end{matrix}} \xrightarrow{H_2O} \underset{\textbf{(5)}}{\begin{matrix} H_3N & & NH_3 \\ & Pt & \\ H_2O & & OH_2 \end{matrix}}$$

The exact form of the aquated species is dependent on the pH of the surrounding liquid as depicted below. Further, these aquated species readily oligomerize predominantly existing as the monomer, a hydrogen-bridged centrosymmetric dimer and a hydroxo-bridged trimer with the proportion pH dependent.[38] At low pH's the major monomeric form is the diaquo form and at high pH's, the dihydroxy form. The formation of the oligomers occurs most readily around a pH of 7.

$$\underset{\textbf{(6)}}{[Pt(NH_3)_2(H_2O)_2]} \underset{}{\overset{pK_a = 5.5}{\rightleftharpoons}} \underset{\textbf{(7)}}{[Pt(NH_3)_2(H_2O)(OH)]} \overset{pK_a = 7.4}{\rightleftharpoons} \underset{\textbf{(8)}}{[Pt(NH_3)_2(OH)_2]}$$

The activity of the monomeric, dimeric, and trimeric species was studied by injecting each oligomer into animals. While the exact forms may change in the body fluids, it can give a clue as to their relative activities. All three are more toxic (lower LD_{50}'s) than *cis*-DDP.[39–41] The dimer and trimer did not display anticancer activity at any dose level, while the monomer is active at dose levels below the toxic level.

These results may be relevant to the cytotoxicity of normal cells containing large amounts of *cis*-DDP, such as kidney cells where the oligomerization reaction may be rapid. This may also provide a rationale for the lower toxicity (but maintained efficacies) of *cis*-DDP given as a divided dose over 24 h or as a slow infusion over 24 h.[38] Thus injection of *cis*-DDP is followed by rapid excretion (half-life in man of 1.5 h), where 95% remains in the original *cis*-DDP form since uptake may be limited on a time basis with the remaining (less than 5%) platinum-containing forms being of the more toxic dimer and trimer form which are also washed through the body. Other factors also suggested include that the *cis*-DDP-derived moieties act both as characteristic *cis*-DDP and as "platinum blues" depending on the site of activity, and so on. Thus the situation becomes more complicated as more information is gathered, both with regard to mechanism(s) of action and even the active species.

The fact that cis compounds are active led early workers to postulate that the platinum drugs act through chelation to some cellular material. It was soon suggested that platinum complexes act in a manner similar to bifunctional alkylating agents, such as nitrogen mustards.[3] This hypothesis was based on the similarity of the results of treatment of cells with the two drugs. Both are capable of stopping cell division and can form giant cells of mammalian tissue and filamentous bacteria cells.[6]

Bifunctional alkylating agents act as relatively nonselective electrophiles, reacting with various cellular material. The cellular component most sensitive to attack is DNA, with substitution occurring at N-7 of guanine, N-3 of adenine, and N-1 of adenine and cytosine in decreasing order of importance.[35] The alkylating agents can act to form interstrand cross-links, inhibiting DNA synthesis.[42,43]

Such lesions have been demonstrated on the use of *cis*-DDP by Roberts and Pascoe[44,45] by labeling separate DNA strands. HeLa cells were grown in the presence and absence of 5-bromo-2′-deoxyuridine and radioactive thymidine and then treated with mustard gas, *cis*-DDP or a blank. The blank showed a heavy fraction, containing the brominated moieties, and a light fraction. Cells treated with either alkylating agent, however, also showed a third fraction of intermediate density. This was believed to be due to the cross-linking of a light strand to a heavy strand.

In order to cross-link between guanine bases at the N-7 position, as is believed to commonly occur for alkylating agents, such as the mustards, the leaving groups must be at least 8 Å apart. Because the chloride ligands are only 3.3 Å apart, the same cross-link cannot occur. Many alternate reaction sites have been suggested, such as the amino nitrogen of adenine.[46,47] However, this, along with the fact that platinum complexes have been shown to inhibit a singly stranded bacteriophage, as well as the doubly stranded, also led workers to look for a new lesion site.

It has been suggested that the most important bonding to DNA may be the formation of intrastrand lesions, either by chelation to one base or by interbase bonding.[3] Thomson and Mansy studied the reaction of *cis*- and *trans*-DDP with the dinucleotide adenosine-3′p5′-adenosine (ApA) using circular dichroism.[47] At room temperature ApA has a stacked conformation, which is destabilized upon warming. Reaction with *cis*-DDP caused the dinucleotide to retain the stacked conformation over a wider temperature range. In contrast, reaction with *trans*-DDP led to complete unstacking of the dimer. The added temperature stability was due to the linking of the adenosine groups into the stacked position by *cis*-DDP. Two dinucleotides were thought to have reacted with *trans*-DDP, thereby forcing the unstacked structure. Comparative UV spectra of reacted and unreacted nucleosides indicated that the cis compound could chelate to adenosine or cytidine, as well as monofunctionally bind to guanosine.[48]

An alternate site proposed by a number of researchers is the formation by the *cis*-DDP of a chelate on the N7-06 sites of guanine.[45–51] Such a complex fits the suggested geometry and involves the 06 guanine site, suggested to be a likely site for alkylation leading to carcinogenesis. Thus, *cis*-DDP-complexed guanine could be present unrepaired in a cell deficient in the necessary repair enzyme until DNA replication has occurred with a consequent mispairing of thymidine and guanine.[52] A second replication will give about one of every four such sites with a T-A pairing where previously a G-C pair existed. This latter suggestion has been confirmed using the Ames tester strains[53] and mammalian cells,[54] though other hypotheses may also lead to the same prediction. Further, the chemical evidence for the existence of this complex is not strong.[55] All nucleophilic sites of guanine are accessible to attack by the platinum-containing moiety and all such reaction products will probably be found.

Other evidence suggests that binding to adjacent guanine bases in the same DNA strand is important (e.g., see Reference 56). Following initial binding to guanine bases, DNA is locally denatured, exposing additional sites for chelation and subsequent cross-linking. Evidence for local denaturation, unwinding, and a shortening of the DNA is found in electron micrographs of platinated DNA double helixes and other physical characterization techniques (e.g., see Reference 57).

More recently several base sequences have been identified as being important sites of binding (e.g., see References 58 and 59). Thus, investigation of the nonsense mutants from *E. coli* revealed that 70% of these mutations were derived from GC-TA or GC-AT substitutions at sites where the guanine is part of a GAG or GCG sequence.[59] Such studies are also relating binding sites with DNA locations where the incidence of spontaneous mutagenesis is high. Such sites are called "hot spots."

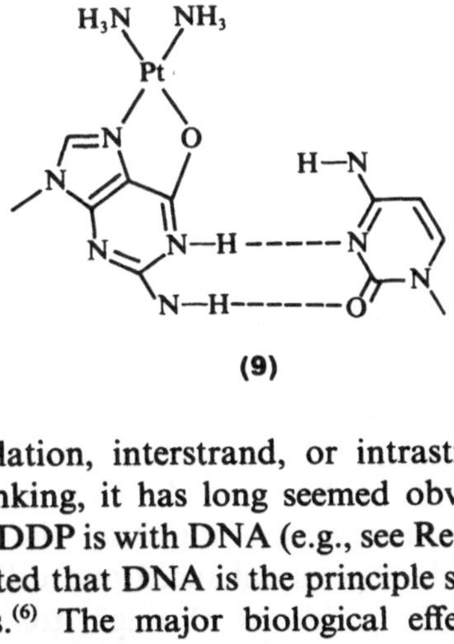

H_3N NH_3 Pt N O H—N N N N—H-----N N N—H------O N

(9)

Whether by chelation, interstrand, or intrastrand cross-linking, or DNA–protein cross-linking, it has long seemed obvious that the primary cellular reaction of *cis*-DDP is with DNA (e.g., see Reference 60). Howle and co-workers demonstrated that DNA is the principle site of action at normal dosage concentrations.[6] The major biological effect of *cis*-DDP is the inhibition of DNA synthesis and the major reaction product is a *cis*-DDP–DNA complex. Inhibition occurs at levels below that necessary for either RNA or protein inhibition.

As further proof of the importance of the DNA lesion, a number of workers studied the effects of the platinum interaction on intercalation agents. Intercalation agents are normally planar aromatic molecules which stack between nucleic acid moieties. For example, 9-aminoacridine increases the base separation from 3.4 to 6.7 Å.[61] Reaction with *cis*-DDP or dichloro(ethylenediamine) platinum II (DENP) led to a decrease in the amount of intercalation which could occur.[62,63] In fact, it has been suggested that there is a one-to-one relationship between this effect and the number of molecules of platinum incorporated into the DNA chain at low platinum levels.[63] This decrease is theoretically caused by Pt–DNA bonding, most likely cross-linking.[62,63]

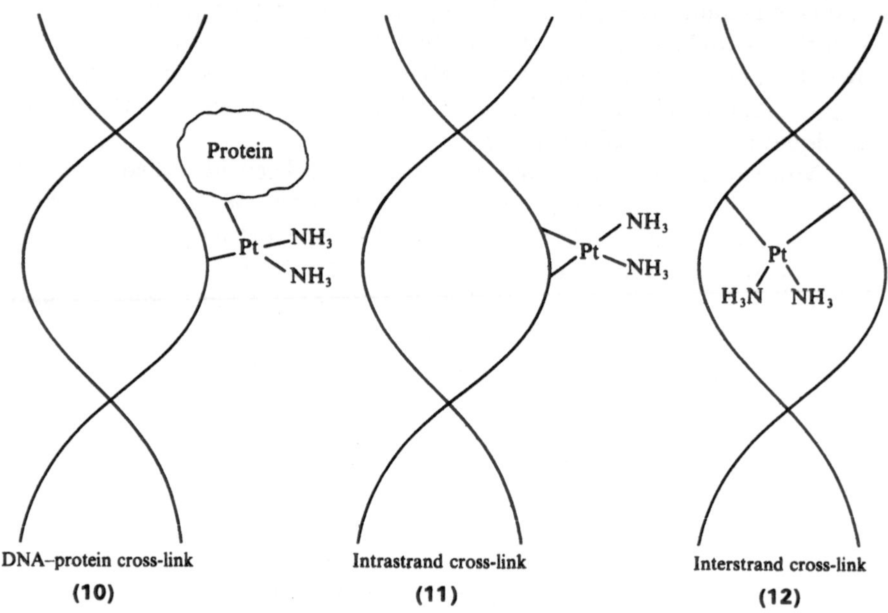

DNA–protein cross-link (10) Intrastrand cross-link (11) Interstrand cross-link (12)

Roberts has studied the repair of DNA reacted with *cis*-DDP.[64] The envisioned repair process consists of degradation of a section of the DNA strand, resynthesis of that section, and finally, resealing of that section into the DNA strand. During replication, a gap is thought to be left opposite the damaged portion of the DNA strand. This gap is filled through a postreplication repair process. By inhibiting the postreplication repair process, chromosome damage is enhanced and cellular survival lessened. Such inhibition may be achieved through treatment with nonlethal concentrations of trimethylxanthine or caffeine. Indeed, treating cells with *cis*-DDP and caffeine greatly increased the lethal effects of the *cis*-DDP and the number of chromosomal aberrations. Studies showed that the molecular weight of daughter DNA was reduced proportionately with the increasing platinum concentration at constant caffeine levels.[64]

Thus, a massive amount of information has been accumulated to support the view that interaction with DNA represents the mode of action of platinum complexes. The platinum acts to modify the DNA and correspondingly inhibit its use as a template for synthesis by DNA polymerase. How the DNA is modified, whether mono- or bifunctionally, whether chelated, inter-, or intrastrand cross-linked, or any combination of these effects remains undefined.

Even the theory of DNA modification does not explain all of the observed facts. Hoeschele and Van Camp found that mice with tumors excrete *cis*-platinum compounds more slowly than do nontumored mice.[65] Also, there seems to be a tumor-specific toxicity, as shown by the difference between the toxicity to the body and the therapeutic level or toxicity to the tumor. This, in spite of the fact that platinum complexes are not selectively absorbed into the tumor,[65] led some researchers to suggest a totally different mode of action.

Rosenberg has hypothesized that the platinum complex somehow enhances the antigenicity toward tumor cells. The tumor cells are then attacked by the body as an immunologic response.[66,67] It was shown that mice cured of sarcoma 180 tumors with *cis*-DDP rejected all further attempts to transplant another tumor into their bodies. Additionally, Reslova transplanted sarcoma 180 tumor tissue pretreated with platinum complexes into mice and found that few grew to sizable tumors (13%), while control tumors grew readily (90%).[67,68] A second, nontreated tumor, was transplanted into those animals given the pretreated tumor and again only a small number grew (38%) compared to control animals (86%). These results indicated that the animals had developed immunity after exposure to pretreated tumor tissue.

Also, Conran and Rosenberg showed that *cis*-DDP, which produces almost 100% cures of sarcoma 180 in Swiss mice, caused only 40% cures when administered with hydrocortisone, an immunodepressant.[69] In a second series of experiments, they showed that *cis*-DDP, which does not produce a cure of sarcoma 180 in BALB/c mice, caused 50%–100% cures when administered with zymosan, an immunostimulant. While the meaning of the therapy with zymosan is speculative, the hydrocortisone results seem to show again the importance of the host defenses in the antitumor action of *cis*-DDP.

Khan and Hill[71] demonstrated that this effect is not simply the immune system working in conjunction with *cis*-DDP. Among other results, they found that *cis*-DDP inhibited the antibody formation of sheep red blood cells in mice and inhibited the graft-vs.-host reaction in mice.[70–72] These results indicate immunosuppression. They also found that malonato(1,2-diaminocyclohexane)platinum II (MCCP) (**13**) acts as an immunosuppressive agent at high doses. However, at low doses this complex acts as an immunostimulant. These results led Rosenberg to hypothesize that platinum drugs increase the antigenicity of a tumor cell sufficiently so even a suppressed immune system can destroy the cell.

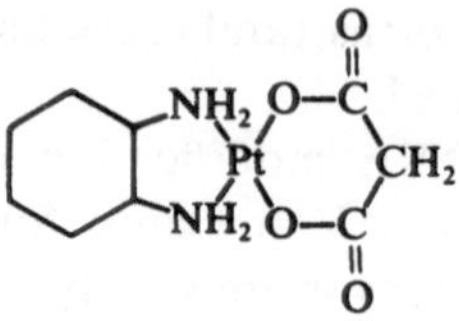

(13)

5. Animal and Human Toxicity

Before a new drug may be made available for public use, it must pass through the four developmental stages described in Table 4 (e.g., see Reference 73). After assessing the evidence of the activity of *cis*-DDP (**1, 2**) the National Cancer Institute (NCI) sponsored preclinical testing of the drug. In these studies the formation, administration mode, and toxicity of the platinum complex was determined using animals. Following this step, the first phase of clinical trials, the establishment of human parameters, took place. This phase of study is limited to patients with advanced, solid tumors, which are resistant to all currently available therapy. Because of the goals of phase I clinical testing and the advanced state of the patients' disease, a favorable response during treatment, while gratifying and possibly important, is not required for advancement to phase II testing. This step is taken after evaluating the toxicity data generated from phase I and animal tests.

Many studies were conducted to determine the distribution of platinum drugs in test animals.[74–76] Usually the drug was followed by atomic absorption analysis of platinum,[15,77] or by detection of radiolabeled platinum.[65,78,79] However, Taylor and co-workers followed the metabolism of DENP using carbon-14.[75] Their assumption that the amine ligand remains with the platinum was later justified when they radiolabeled both platinum and carbon.

Table 4. Development Stages of a Drug

I.	Synthesis and antitumor characterization discovery of the compound and its antitumor activity
II.	Preclinical testing
	Determination of the toxicity, administration mode, and dose level using animals
III.	Clinical testing
	Phase I: Determination of toxicity at given dose levels
	Phase II: Antitumor activity screening
	Phase III: Screening for general use
IV.	Introduction into medical practice

Table 5. Distribution of *cis*-DDP after 24 Hours[a]

	Nontumored rat[b]	Nontumored beagle[c]	Nontumored shark[d]	Nontumored mouse[e]	Tumored mouse[e]
Blood		0.27[f]			
Liver	2.89	1.17	0.21	7.10	7.9
Spleen		0.43		1.10	0.08
Pancreas		0.66			
Testes	0.32				
Kidneys	11.61	2.15	1.97	2.10	3.8
Adrenals		1.08			
Heart		0.26			
Lungs		0.80		0.06	0.56
Brain		0.06		0.04	0.06
Muscle		0.25			
Fat		0.04			
Tumor					2.1
GI tract				15.0	11.0
Uterus		1.53	1.24		

[a] All values in micrograms per gram wet weight.
[b] Reference 67.
[c] Reference 8.
[d] Reference 67.
[e] Values are high value of a range given; Reference 44.
[f] Plasma.

Table 6. Distribution of *cis*-DDP as a Percentage of Dose[a]

	Nontumored rat	Tumored rat
Blood	2.43	2.45
Skin	10.56	10.57
Thyroid	0.01	0.01
Liver	3.52	3.52
Spleen	0.27	0.17
Stomach	0.19	0.21
Testes	0.10	0.09
Kidneys	2.92	2.80
Adrenals	0.02	0.01
Heart	0.07	0.06
Lungs	0.29	0.29
Brain	0.04	0.05
Muscle	5.37	4.92
Bone	5.37	4.58
Marrow	0.01	0.02
Tumor	—	0.46

[a]Taken from Reference 68.

Results of the distribution studies summarized in Table 5 show that the drug concentrates in similar organs for various animals. The liver and kidney continually show the highest levels of platinum. Similarly, the highest percentage of administered drug remains in these organs, as well as in the blood, skin, and muscle (Table 6). The latter three have high percentages of administered platinum due to their high body mass and not due to their high concentration.

Platinum concentrations in tumor tissue are no higher than in the liver or kidneys.[(78)] Additionally, presence of a tumor does not seem to cause any additional retention of the drug (Table 6), contrary to the early findings of Hoeschele and Van Camp.[(65)] Because these drugs seem to have some specificity for killing tumor cells (the TI is greater than 1) not related to the concentration, Rosenberg has used these findings, as noted earlier, to support the hypothesis that platinum drugs act to stimulate the immune system of the body.[(4)]

Time-dependent concentration studies have shown 50–60% of the platinum to be excreted through the urine within four hours of administration.[(15)] Initial plasma levels fell quite rapidly with excretion, but within 4 h the platinum loss slowed considerably. Litterst *et al.* suggested that there was a fast and a slow phase to clearance of the drug.[(15)] The latter phase is so prolonged that platinum remains in the blood 12 days later. Probably, the fast phase represents clearance of the unbound drug, whereas the slow phase represents release of the drug, which was bound at some time to biological material, such as albumin.

Nephrotoxicity was the dose-limiting effect as expected due to the rapid excretion of *cis*-DDP through the urine and the high concentration of platinum found in the kidneys.[14] The damage was cumulative and eventually irreversible.[5] Renal damage was characterized by tubular necrosis and eventually led to death.[14] Limits on the dose size and the length of treatment due to nephrotoxicity seriously limit the usefulness of *cis*-DDP as a drug.[80]

Though the liver contained high concentrations of platinum, and it is known that the liver diverts a small amount of the platinum into the bile, no damage to the liver was seen.[28] This is possibly due to the entrance of the drug after the drug is already bound to some biological material.

Gastrointestinal effects are displayed as acute nausea and vomiting, normally lasting four to six hours.[5] This is a common side reaction of antineoplastic agents and probably results from destruction of intestinal flora. This has been reported in all cases of treatment with platinum compounds. Because anorexia associated with this effect may be cumulative, a patient's appetite may decrease even as the patient's tumor shrinks.[71] Thus, it has been suggested by some researchers that psychopharmacological drugs like cyproheptadine or Δ^9-tetrahydrocannabinol be employed to diminish these effects and stimulate appetite.[11,12,81]

Renal and gastrointestinal toxicity had been predicted prior to clinical studies, however, with the onset of phase I testing another major problem was discovered. In a small number of patients, treatment with *cis*-DDP caused cumulative and irreversible hearing losses in the high range (4000–8000 Hz). With continued treatment, losses in the spoken range (1000–4000 Hz) developed, leading eventually to total deafness.[14,81] Studies later showed that this side effect also occurred in rhesus monkeys. It appeared up to three months after intramuscular injection of *cis*-DPP and was accompanied by hair losses in the inner ear.[13]

Late in the phase I studies, Khan *et al.* reported an allergic reaction to *cis*-DDP after a number of doses of the drug was given to a patient.[82] This group later showed that allergic reactions could be induced in rats.[72] Like most antitumor drugs, *cis*-DDP is an immunosuppressive agent.[70] However, this suppression seemed to be short-lived. Thus, the drug persisted beyond the suppression and caused hypersensitivity. This side reaction supported the theory that the drug somehow stimulates the immune system of the body.

Finally, *cis*-DDP exhibited hematopoietic toxicity characterized by leukopenia (destruction of white blood cells) and thrombocytopenia (destruction of platelets).[81] Though cumulative, these effects are mild and seem to be reversible. They probably result from reaction of the drug with the bone marrow.

Table 7. Antitumor Activity of *cis*-DDP in Phase I Studies[a]

Carcinoma	Number of patients	Less than 50% regression (%)	Greater than 50% regression (%)
Endometrial carcinoma	2	0	100
Testicular carcinoma	27	44	56
Hepatoma	2	50	50
Head and neck	11	55	45
Squamous cell carcinoma	12	58	42
Bladder and urethra	19	63	37
sarcomas	12	67	33
Kidneys	3	67	33
Ovary	58	71	29
Non-Hodgkin's lymphoma	16	75	25
Lung	4	75	25
Leukemia	43	91	9
Stomach and colon	10	100	0
Breast	7	100	0
Pancreas	2	100	0
Prostate	1	100	0

[a] Data taken from References 79–81 and 83–88; data includes a phase II study (Reference 85).

6. Antineoplastic Effects

Along with the toxicity data from phase I studies came a certain amount of information on the antitumor activity of *cis*-DDP. As seen in Table 7, partial or complete responses (greater than 50% tumor regression) are found for a number of tumor types.[83–92] This table, however, must be interpreted with caution. A number of important factors are not shown, such as the mode of *cis*-DDP administration, criteria of patient eligibility, patient history (earlier chemotherapy and advancement of the disease), and definition of tumor remission. Certain numbers of Table 7 may also be inflated due to the authors publishing the same data in more than one article.

Higby and co-workers found *cis*-DDP extremely potent in the chemotherapy of testicular tumors, as evidenced by a 30–50% complete remission rate, lasting a median of three months.[83] Most surprisingly, only about 15% of the patients treated showed no response.

Studies by Wiltshaw on the treatment of ovarian carcinoma, while not impressive numerically, also showed a significant response to tumors which are in an advanced state.[84] Because it has been found that once the ovarian carcinoma develops resistance to one alkylating drug, improvement due to

Table 8. Recent Treatment of Human Cancers Employing *cis*-DDP

Cancer type and/or location	References[a]
Choriocarcinoma	93
Squamous	94–97
Interstitial	98
Marine bladder	99
Bladder	100
Myelogenous leukemia	101
Ovarian	102–107
Nonseminomatous testicular	108
Testicular	109–111
Endodermal sinus	112
Urothelial	113

[a] While there are numerous references, those listed are typically reviews or contain review sections.

other drugs is rare, Wiltshaw's work suggested that *cis*-DDP might be a valuable addition to the regimen of drugs used in the treatment of this disease.

Again, it should be noted that phase I studies were not meant to develop data on the usefulness of the drug, but on toxicity. Accordingly, such favorable responses were quite encouraging. However, despite the apparent success of the drug, by 1973 few researchers were working with *cis*-DDP due to its prohibitive kidney toxicity.[80,85,86] In most cases, the therapeutic dose was simply too close to the toxic dose to allow use.

Recent efforts have concentrated on difficult cancers and cancers in difficult-to-treat areas. Table 8 cites recent uses of *cis*-DDP in the treatment of human cancers.

7. Toxicity Minimization

The hydration technique of Cvitkovic *et al.*[114] represented the first effort to overcome the toxicity of *cis*-DDP. In this method, the patient was given 1 to 2 liters of fluid overnight, then *cis*-DDP was administered along with mannitol, a diuretic. By flushing the drug along with large amounts of water past the kidneys, renal toxic effects were substantially lowered. This method allowed the administration of 3–4 mg of *cis*-DDP per kilogram body weight, about 10 times the normal dosage, while lowering damage to the kidneys. Though the antitumor effects of *cis*-DDP at this high dosage seemed to be improved, nausea and auditory losses were still observed and were probably enhanced.[115–117]

A common method of minimizing the toxicity of antineoplastic agents is to administer them along with a number of other antineoplastic agents at reduced dose levels. Thus, *cis*-DDP has been administered in combination with adriamycin (ADM), cyclophosphamide (CTX), bleomycin (BLM), vinblastine (VLB), and actinomycin D (ACD), among other agents.

Merrin reported a 50% complete remission rate of advanced testicular tumors when *cis*-DDP was administered in a three-part program along with BLM, VLB, ACD, and vincristine.[117] Einhorn and Furnas reported 100% partial or complete remission of testicular tumors in 39 patients with a combination of *cis*-DDP, VLB, and BLM.[118] Bruckner *et al.*,[119] as well as Kwong and Kennedy,[120] have shown less spectacular, though quite good, response rates on genitourinary tumors using *cis*-DDP in combination with ADM. Other recent efforts are given in References 121–123.

A final method of lowering the toxicity of chemotherapy using platinum is the synthesis of new compounds showing equal or enhanced activity and toxicity in mice, as alluded to during the discussion of structural requirements.[22–24,29,124,125] This effort has continued for some time and some derivatives of *cis*-DDP, such as MCCP (**4**), are now beginning to enter phase I studies.[126–128] The use of polymeric derivatives of *cis*-DDP is described in the following sections.

8. Rationale

The preference of including platinum-containing moieties within polymers compared with delivery *via* smaller molecules is clearly a debatable point, but there is sufficient evidence and potential to justify at least preliminary studies of such situations. The use of biological and synthetic macromolecules for delivery of tumor-suppressant drugs is well known and has in some cases proved to be advantageous (e.g., see References 129–131 and other references in this book). Specifically, advantages for employing polymeric derivatives of *cis*-DDP may include: (1) restricted biological movement, (2) controlled release, and (3) increased probability of critical attachment.

Heavy metal toxicity related to the presence of large quantities of *cis*-DDP derivatives in the circulatory system (such as renal failure) is well established. Chain lengths of about 100 and greater are typically prevented from easy movement through biological membranes. Thus the location of the platinum drug can be somewhat restricted, for instance, from the kidney, decreasing damage to the kidney-associated organs, when the drug is contained in a polymer.

Studies by the authors have established that most metal-containing polymers undergo hydrolysis when wet. A Me_2SO solution of polymeric derivatives of *cis*-DDP can act as a controlled-release agent, releasing therapeutic quantities of the active drug.

Finally, if attachment—interstrand, intrastrand, or otherwise—to DNA is essential for antitumor activity and multiple attachments are required or advantageous (i.e., more than one attached platinum compound per DNA), then the fact that the platinum is itself an integral portion of a polymer is advantageous since the probability that additional attachments will be made on the same strand and adjacent strands is high after the first attachment.

9. Synthesis

Fortunately, nature has supplied a ready route to the synthesis of both cis and trans derivatives of the plus two oxidation state of platinum, palladium, and nickel. The following equations describe the synthesis of both cis and trans derivatives of *cis*-DDP. The trans effect is dominant for both platinum and palladium so that the derived cis or trans product is synthesized to almost (generally less than 0.1%) the total exclusion of the other isomer. Inclusion of the platinum-containing moiety in a polymer matrix can occur through a polymerization reaction or through reaction with a preformed polymer.

$$[PtCl_4]^{2-} \; (\mathbf{14}) + RNH_2 \longrightarrow cis\text{-}[Pt(NH_2R)_2Cl_2] \; (\mathbf{15})$$

(14) **(15)**

cis-Product

$$[Pt(NH_2R)_4]^{+2} \; (\mathbf{16}) + Cl^- \longrightarrow trans\text{-}[Pt(NH_2R)_2Cl_2] \; (\mathbf{17})$$

(16) **(17)**

trans Product

Allcock and co-workers[132–136] typically synthesized phosphazene derivatives of *cis*-DDP from reacting K_2PtCl_4 (hereafter called the tetrachloroplatinate) or $PtCl_2$ with the appropriate phosphazene, in organic media or with K_2PtCl_4 in aqueous media yielding two types of products—neutral, square planar platinum complexes (as **18**) or tetrachloroplatinate salts as $[H_2N_4P_4(NHCH_3)_8]^{+2}[PtCl_4]^{-2}$. Phase-transfer agents were employed for some syntheses. The platinum-containing products can be made water soluble and typically were designed to degrade hydrolytically releasing the platinum-containing moiety.

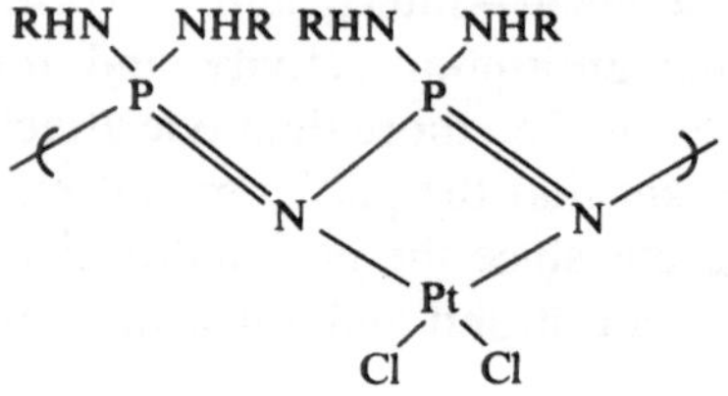

(18)

We are currently studying the synthesis of *cis*-DDP derivatives from the reaction of water-soluble polymers such as polyvinyl amine and polyethyleneimine. A major portion of the resulting polymer product is soluble in Me_2SO consistent with preferential internal chelating (as **19**) with only a limited amount of cross-linking.[137] The reaction system consists of bringing together two aqueous solutions, one containing the tetrachloroplatinate and the other the nitrogen-containing polymer. Future work includes reaction of tetrachloroplatinate and other appropriate platinum-containing reactants with basic nucleic acids and proteins, liposomes, and other nitrogen-containing natural and synthetic polymers.

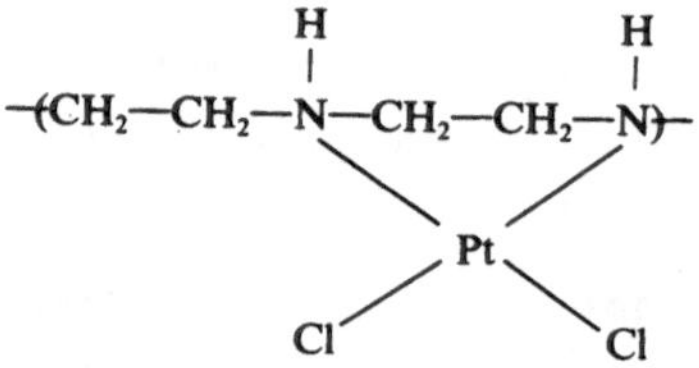

(19)

Inclusion of platinum through polymer formation reactions is analogous to the general reaction scheme employed in the synthesis of monomeric derivatives of *cis*-DDP.[138–144] Aqueous solutions containing equal molar amounts of the diamine (generally 1 mmol in 50 ml of water) and tetrachloroplatinate (also 1 mmol in 50 ml of water) are mixed together and the resultant solution mixed employing a stir plate. The polymer precipitates from the reaction mixture and is collected by suction filtration or centrifugation.

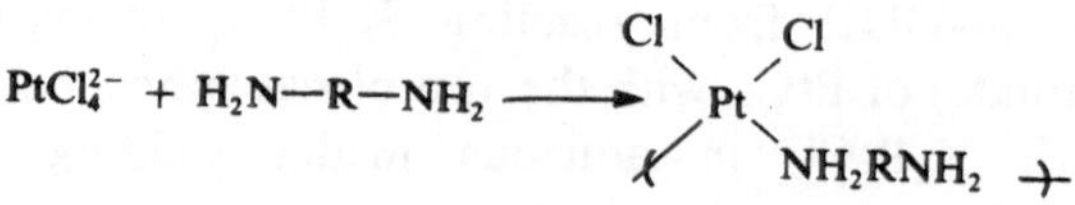

$$PtCl_4^{2-} + H_2N{-}R{-}NH_2 \longrightarrow \left[\mathrm{Pt}(Cl)_2{-}NH_2RNH_2\right]$$

(20)

Other sequences have been developed allowing the ready synthesis of a number of polymeric derivatives of *cis*-DDP. Some of these routes are described in the following equations:

$$PtCl_4^{2-} \longrightarrow PtI_4^{2-} \xrightarrow{H_2NRNH_2} \left[Pt(I)_2-NH_2-R-NH_2 \right]$$

(21)

$$\textbf{(21)} \xrightarrow[\text{2) NaCl}]{\text{1) AgNO}_3} \left[Pt(Cl)_2-NH_2-R-NH_2 \right]$$

(22)

The reaction of the tetrahaloplatinate with a typical diamine such as 1,6-diaminohexane follows the order I > Br > Cl (Table 9). Thus, timewise it is advantageous to employ the synthetic procedure described directly above.

The reaction has been expanded to include tetrachloropallidate and such dinitrogen-containing reactants as urea, thiourea, hydrazines, hydrazides, purines, and pyrimidines. A distinct reaction rate trend exists such that a decrease in the electron density on the nitrogen results in a decreased reaction rate as described in Table 10. Further, the reaction rates of the tetrachloroplatinate with substituted derivatives of phenylhydrazine follows the rate trend predicted by Hammett sigma values.[44]

Molecular weight is directly related to the molar ratio of reactant groups. Thus the reaction between the tetraiodioplatinate and 1,6-diaminohexane for a molar ratio of 0.9985:1.000 led to a weight-average number of repeating units of 2600, whereas a ratio of 0.794:1.000 lead to a weight-average number of repeating units of 6.3; both values are near that which would be calculated from a simple statistics model.[139]

Generally, the platinum polymers are more soluble than the analogous palladium products presumably due to the greater size and polarizability of the platinum. The platinum polymers are typically soluble in dipolar aprotic

Table 9. Approximate Observed PtX_4^{-2} Reaction Times with HDA

PTX_4^{-2}	Approximate reaction time
Chloride	48 h
Bromide	6 h
Iodide	5 min

Table 10. Approximate (to 50% yields) Reaction Rates with Tetrachloroplatinate

Nitrogen-containing species	Reaction time
Hydrazines	1 h
Diamines	2 days
Dithioureas	3 days
Diureas	> 5 days

solvents such as DMSO, HMPA, and TEP and some even soluble in simple organic solvents such as chloroform. After solution in DMSO, the products are relatively stable for several weeks. They remain in solution on addition of water, but do undergo hydrolysis releasing the platinum-containing moiety. Thus such polymers may act like "platinum blues" rather than intact derivatives of *cis*-DDP in animal studies.

$$PtCl_4^{2-} + H_2N{-}\overset{\overset{\large X}{\|}}{C}{-}NH_2 \longrightarrow \left(\!\!-\,Pt(Cl)_2{-}NH_2{-}\underset{\underset{\large X}{\|}}{C}{-}NH_2\,\!\!-\right)$$

(23)

$$PtCl_4^{2-} + H_2N{-}\overset{\overset{\large X}{\|}}{C}{-}R{-}\overset{\overset{\large X}{\|}}{C}{-}NH_2 \longrightarrow \left(\!\!-\,Pt(Cl)_2{-}NH_2{-}\underset{\underset{\large X}{\|}}{C}{-}R{-}\underset{\underset{\large X}{\|}}{C}{-}NH_2\,\!\!-\right)$$

(24)

$$PtCl_4^{2-} + H_2H{-}NHR \longrightarrow \left(\!\!-\,Pt(Cl)_2{-}NH_2{-}\underset{\underset{\large R}{|}}{N}H\,\!\!-\right)$$

(25)

$$PtCl_4^{2-} + \text{uracil} \longrightarrow \left(\!\!-\,Pt(Cl)_2{-}\text{uracil}\,\!\!-\right)$$

(26)

10. Physical and Structural Characterization

The condensation products of forms **19–26** have been well characterized utilizing light-scattering photometry, IR, pyroprobe–MS, TG–MS, NMR (some), TG, DSC, UV–VIS–near-IR, and complete elemental analyses. Following is a description of items that may be useful.

The infrared spectra of the polymeric products typically are similar to the spectra of the nitrogen-containing reactant itself. The major difference is the appearance of a medium intensity band between 300–330 cm^{-1} attributed to the Pt–Cl stretch and two bands in the 450–550 cm^{-1} region attributed to the presence of the Pt–N bond. Some authors have noted that the cis and trans isomers can be differentiated through the appearance of one (trans) or two (cis) bands in this region. Other authors note that the bands may be weak and that the absence of two detectable bands in this region means little. This latter argument is supported by crystal studies.

A second ploy typically employed to differentiate between cis and trans isomers is ultraviolet-visible spectroscopy. The tetrachloroplatinate itself exhibits four bands in the 150–400 nm region (177, 210, 255, and 302). Trans products generally exhibit only three bands within this region, while cis products exhibit four. The *cis*-DDP exhibits bands at 240, 273, 331, 372, whereas *trans*-DDP exhibits bands at 268, 317, and 367 (all values in nanometers). The bulk of the polymers synthesized by the authors also show four bands within this region. Even so, such information should be used only as collaborative evidence and not employed as conclusive evidence of the cis or trans geometry because infrared spectroscopy typically does not yield three or four well-defined bands with at least one of the bands being weak. The major structural consideration regarding cis or trans geometry is the trans effect which often appears to persevere in spite of geometrical constraints.

The TG for the products typically show at least two kinetically dependent stability plateaus, the first occurring after major loss of the nitrogen-containing moiety (about 175–260°C), while the second corresponds to the loss of the halide ligands. After about 600–800°C, only elemental platinum remains as a residue, making TG an ideal technique for platinum determination. The platinum appears to act as a combustion catalyst resulting in a "clean burn."

While valuable data have been obtained employing NMR, the general lack of solubilities in the 5–10% range preclude accurate NMR data collection.

The use of coupled pyroprobe and TG–MS permits the ready identification of both impurities and variations in products synthesized at different times. Such combinations are important in biological testing, particularly if the response is particularly dependent on the exact structure of the product and presence of impurities.[145]

11. Biological Characterization

Several phosphazene carrier polymers have undergone stage I (Table 4) testing. These polymers showed tumor inhibitory activity in initial anticancer screening tests against mouse P 388 lymphocytic leukemia and in the Ehrlich ascites tumor regression test.[135]

Products derived from the condensation of tetrahaloplatinates with nitrogen-containing reactants have undergone both stage I and II testing (forms **19–26** emphasizing products of form **20**). Summaries are as follows:

1. The so-called "windows of activity" established for the monomeric derivatives of *cis*-DDP appear to hold for the polymeric derivatives of *cis*-DDP. Activity is typically aliphatic > aromatic and Cl > I against a wide variety of tumorous cell lines.
2. The polymers show distinct crossover concentration levels with regard to inactivity to activity. For aliphatic diamine-chloro derivatives this crossover level is about 30 μg/ml, whereas in hydrazine derivatives this crossover is increased to about 60 μg/ml. Almost without exception, the *cis*-DDP derivatives show good inhibition to all tested tumors – L929, HeLa, WISH, Detroit, and so on.
3. The polymeric derivatives are, compared to *cis*-DDP, less toxic to mice. Thus mice are able to tolerate repeated (once a day for 30 days) near-lethal (if in the form of *cis*-DDP, itself) injections of the polymeric products. Live-animal injections are done utilizing the polymer dissolved in Me_2SO which are then diluted (ninefold or greater) by addition of Hank's solution or distilled water.
4. In tests related to extending the lifetimes of mice injected with lethal doses of a cancer, the majority of the products extended life with one increasing the life span over 150%.
5. There does not exist a relationship between activity toward bacteria and cancer cell lines. Generally, the polymeric derivatives are mild, at best, antibacterial agents.
6. The majority of polymeric derivatives of *cis*-DDP exhibit good cell differentiation favoring inhibition of transformed cells over the same normal cell line.
7. The majority of the polymers show good antiviral activity at concentration levels (as 10 μg/ml) well below the point where

antitumoral inhibition is found. Polio-I-type virus is inhibited in the presence of HeLa cells at a concentration of 10 μg/ml, whereas antitumoral activity begins at 30 μg/ml for these cells. It is believed that certain cancers are virally related. Thus the use of the polymeric derivatives of *cis*-DDP may prove effective against these cancers without harm to the healthy or cancerous cells.

Present work is now divided between use of the polymers as antitumoral agents and as antiviral agents. Preliminary studies are positive in regard to using these polymers against hard to control viruses.

8. The most active polymers studied to date are ones where the nitrogen-containing moiety is also an antitumor agent. There appears to be a synergistic effect in employing such moieties together.
9. Preliminary live-animal tests are consistent with at least some of the activity being of a controlled-release variety.

In summary, the polymeric platinum-containing derivatives of *cis*-DDP typically exhibit wide antitumoral activity; are relatively nontoxic; show good cell differentiation; and are both antitumoral and antiviral agents.

Putrescine spermidine and spermine are naturally occurring polymers which contain large numbers of amine-functional groups. These polymers are often referred to as natural polyamines[146] and can be associated with tumor growth.[147,148] Raised levels of such polyamines are associated with cell growth, such as embryogenesis, hepatic regeneration, and location, as well as solid and haematological tumors.[149] Their function as growth promoters in neoplastic processes has been described.[150] The findings that polyamine levels are raised in the urine and sera of cancer patients have catalyzed work towards using the concentration levels of these polyamines as biochemical markers for cancer detection.[151–153] The variation in concentration levels of these polyamines has been used to determine a patient's response to radiation and chemotherapy.[154,155]

The presence of these natural polyamines and their regulating enzymes during the cancer process appear to be closely related. This is supported by the work of O'Brien and Diamond who demonstrated that increased levels of the polyamines were intrinsic to tumor promotion *in vitro*, suggesting that normal and transformed cells differ in their control of polyamine biosynthesis.[156]

The biological synthesis of such polyamines related to cancer increases during the growth of cancer cells and typically precedes increases in RNA, DNA, and protein synthesis.[146] Putrescine, in elevated amounts, appears to be related to the growth fraction of a tumor and elevated levels of spermidine appear related to the cell-loss factor.[146] The fact that polyamines can influence cancer therapy has been demonstrated with hyperthermia, where their presence increases heat-mediated cell death.[157,158] Recently, Roizin-Towle showed, using V79 hamster cells *in vitro*, that the presence of putrescine and spermine reduced the cytotoxicity of *cis*-DDP.[146]

The interaction between *cis*-DDP and the natural polyamines may be analogous to that found for polymers synthesized by Carraher and co-workers where the cytotoxicity of the synthetic platinum-containing polyamines is reduced in comparison to *cis*-DDP itself. Further, it may establish a rationale for some of the antitumor activity of the platinum polyamines. While the use of natural polyamines reduces the toxicity of *cis*-DDP, the synthetic platinum polyamines exhibit good antitumoral activity along with reduced toxicity toward the test animal.

Several additional findings are possibly related. First, methotrexate, cytosine arabinoside and 5-azacytidine (all antitumor agents) were shown to reduce polyamine levels in the spleens of leukemic mice suggesting a relationship between their antitumoral activity and polyamine concentration.[159] (In general, though, most chemotherapeutic drugs appear not to affect polyamine levels, possibly due to a lack of testing related to polyamine levels.) Second, Rupniak and co-workers showed that transformed cells lacked polyamine growth regulatory mechanisms and may not be subject to the normal restraints of polyamine biosynthesis.[160] Thus traditional *in vitro* studies involving transformed cells may not give a complete picture of drug action since the natural polyamines are absent.

A major effort is now beginning concerning use of a nature-like polymer modified through reaction with *cis*-DDP or the tetrachloroplatinate (e.g., see References 161 and 162). For example Strommen and Peticolas synthesized products derived from the interaction of the ribonucleotide polycytosine (poly C) with *cis*- and *trans*-diamminedichloroplatinum II.[161] Both platinum compounds chelate the N-1 position of the cytosine residue, while the trans form also appears to react with the phosphate moiety. Studies involving poly(guanine–cytosine) (poly G–C) and coupled strands involving poly G–C are also popular (e.g., see Reference 162). While the majority of such studies are aimed at determining the sites of platinum attachment, several involved biological testing with the results that such platinum-containing polymers exhibited decent antitumoral activity.

The analogous products involving natural polymers, as DNA, RNA, and proteins, have been studied for some time and again the products themselves can exhibit antitumoral activity (e.g., see References 163–169).

There are numerous studies underway involving both synthetic and natural polymers containing derivatives of *cis*-DDP, each showing promise in describing the actions of *cis*-DDP in combating cancer, with many exhibiting anticancer activity meriting further study.

Abbreviations Used in Chapter

Å	Angstrom (10^{-8} cm)
ACD	Actinomycin D
BLM	Bleomycin
c-DDP	*cis*-Dichlorodiamineplatinum II
c-TCP	*cis*-Tetrachlorodiamineplatinum IV
CTX	Cyclophosphamide
DEIP	*cis*-Dichlorobis(2-aminoethanol)platinum II
DENP	*cis*-Dichloro(ethylenediamine)platinum II
Hz	Hertz, frequency
% ILS	Percentage increased life span
IR	Infrared spectroscopy
MCCP	Malonato(1,2-diaminocyclohexane)platinum II
MS	Mass spectroscopy
NMR	Nuclear magnetic resonance spectroscopy
Poly C	Polycytosine
Poly G–C	Coupled strands of polyguanine and polycytosine
TC	Treated-to-control tumor ratio
TG	Thermogravimetric analysis
TI	Therapeutic index
UV	Ultraviolet spectroscopy
VLB	Vinblastine

References

1. B. Rosenberg, L. Van Camp, and T. Krigas, *Nature* (*London*) **205**, 698 (1965).
2. B. Rosenberg, L. Van Camp, J. Trosko, and V. Mansour, *Nature* (*London*) **222**, 385 (1969).
3. B. Rosenberg, *Platinum Metals Rev.* **15**, 42 (1971).
4. B. Rosenberg, *Cancer Chemother. Rep., Part 1* **59**, 589 (1975).
5. J. Gottlieb and B. Drewinko, *Cancer Chemother. Rep., Part 1* **59**, 621 (1975).
6. J. Howle, G. Gale, and A. B. Smith, *Biochem. Pharm.* **21**, 1465 (1972).
7. J. Hill, E. Loeb, A. MacLellan, N. Hill, A. Khan, and J. King, *Cancer Chemother. Rep., Part 1* **59**, 647 (1975).
8. P. Kamalakar, A. Freeman, D. Higby, H. Wallace, and L. Sinks, *Cancer Treat. Rep.* **61**, 835 (1977).
9. J. Ward, D. Young, K. Fauvie, M. Wolpert, R. Davis, and A. Guarino, *Cancer Treat. Rep.* **60**, 1675 (1976).
10. A. Khan, J. Hill, W. Grater, E. Loeb, A. MacLellan, and N. Hill, *Cancer Res.* **35**, 2766 (1975).
11. J. Holland, J. Rowland, and M. Plumb, *Cancer Res.* **37**, 2425 (1977).
12. S. Sallan, N. Zinberg, and E. Frei, *New Engl. J. Med.* **293**, 795 (1975).
13. S. Stadnicki, R. Fleischman, U. Schaeppi, and P. Merriman, *Cancer Chemother. Rep. Part 1* **59**, 467 (1975).

14. J. Ward and K. Fauvie, *Toxical Appl. Pharm.* **38**, 535 (1976).
15. C. Litterst, T. Gram, R. Dedrick, A. Leroy, and A. Guarino, *Cancer Res.* **36**, 2340 (1966).
16. K. P. Lewis and W. D. Medina, *Cancer Treat. Rep.* **64**(10–11), 1162 (1980).
17. M. S. Aapro and D. S. Alberts, *Cancer Chemother. Pharmacol.* **7**(1), 11 (1981).
18. R. Speer, H. Ridgway, L. Hall, D. Stewart, K. Howe, D. Lieberman, D. A. Newman, and J. Hill, *Cancer Chemother. Rep. Part 1* **59**, 629 (1975).
19. M. Tucker, C. Colvin, and D. Martin, *Inorg. Chem.* **3**, 1373 (1964).
20. C. Colvin, R. Gunther, L. Hunter, J. McLean, M. Tucker, and D. Martin, *Inorg. Chim. Acta* **3**, 487 (1968).
21. T. Conners, M. Jones, W. Ross, P. Braddock, A. Khokharard, and M. Tobe, *Chem. Biol. Interact.* **5**, 415 (1972).
22. M. Cleare and J. Hoeschele, *Platinum Metals Rev.* **17**, 2 (1973).
23. M. Cleare and J. Hoeschele, *Bioinorg. Chem.* **2**, 187 (1973).
24. P. Schwartz, S. Meischen, G. Gale, L. Atkins, A. Smith, and E. Walker, *Cancer Treat. Rep.* **61**, 1519 (1977).
25. K. Inagaki, Y. Kidani, K. Suzuki, and T. Tashiro, *Chem. Pharm. Bull.* (*Tokyo*) **28**(8), 2286 (1980).
26. K. Okamoto, M. Noji, T. Tashiro, and Y. Kidani, *Chem. Pharm. Bull.* (*Tokyo*) **29**(4), 929 (1981).
27. P. Ribaud, D. P. Kelsen, N. Alcock, G. E. Garcia, P. Dubouch, C. C. Young, F. Muggin, J. Burchenal, and G. Mathe, *Recent Results Cancer Res.* **74**, 156 (1980).
28. B. Rosenberg, *Inorganic and Nutritional Aspects of Cancer* (G. Schrauzer, ed.), Chapter 10, pp. 129–150, Plenum Press, New York (1971).
29. M. Tobe and A. Khokhar, *J. Clin. Hematol. Oncol.* **7**, 114 (1977).
30. R. Dedrick, D. Zonharko, R. Bender, W. Bleyer, and R. Lutz, *Cancer Chemother. Rep., Part 1* **59**, 795 (1975).
31. J. Davidson, P. Faber, R. Fischer, S. Mansy, J. Persie, B. Rosenberg, and L. Van Camp, *Cancer Chemother. Rep., Part 1* **59**, 287 (1975).
32. B. Rosenberg, *Naturwissenschaften* **60**, 399 (1973).
33. M. Clear, *J. Clin. Hematol. Oncol.* **7**, 1 (1977).
34. T. Connors and W. Ross, *Advances in Antimicrobial and Antineoplastic Chemotherapy*, Volume 3, University Park Press, Baltimore (1972), p. 771.
35. A. Korolkovas and J. Burckhalter, *Essentials of Medicinal Chemistry*, Wiley, New York (1976), p. 564.
36. M. C. Lim and R. B. Martin, *J. Inorg. Nucl. Chem.* **38**, 1911 (1976).
37. T. Ruch and H. Patton, eds., *Physiology and Biophysics*, 12th Edition, Saunders, Philadelphia (1974), Vol. 2, pp. 466–467.
38. B. Rosenberg, *Cancer Treat. Rep.* **63**(9–10), 1433 (1979).
39. B. Lippert, C. Lock, and B. Rosenberg, *Inorg. Chem.* **16**, 1525 (1977).
40. R. Faggiani, B. Lippert, and C. Lock, *J. Am. Chem. Soc.* **99**, 777 (1977).
41. R. Faggiani, B. Lippert, and C. Lock, *Inorg. Chem.* **16**, 1192 (1977); **17**, 1941 (1978).
42. R. Goldacre, A. Loveless, and W. Ross, *Nature* **163**, 667 (1949).
43. P. Brookes and F. Lawley, *Biochem. J.* **80**, 496 (1961).
44. J. Roberts and J. Pascoe, *Advances in Antimicrobial and Antineoplastic Chemotherapy*, Volume 2, p. 249, University Park Press, Baltimore (1972).
45. J. Roberts and J. Pascoe, *Nature* (*London*) **235**, 282 (1972).
46. J. Drobnik and P. Horacek, *Chem. Biol. Interact.* **7**, 223 (1973).
47. A. Thomson and S. Mansy, *Advances in Antimicrobial and Antineoplastic Chemotherapy*, Vol. 2, p. 199, University Park Press, Baltimore (1972).
48. S. Mansy, B. Rosenberg, and A. Thomson, *J. Am. Chem. soc.* **95**, 1633 (1973).
49. J. P. Macquet and T. Theophanides, *Inorg. Chim. Acta* **18**, 189 (1976).

50. D. Goodgame, I. Jeeves, and F. Phillips, *Biochim. Biophys. Acta* **378**, 153 (1975).
51. J. Dehand and J. Jordanov, *J. Chem. Soc., Chem. Commun. 1976*, 598.
52. A. E. Pegg, *Nature* **274**, 182 (1978).
53. D. J. Beck and J. E. Fisch, *Mut. Res.* **77**(1), 45 (1980).
54. L. A. Zwelling, K. Kohn, and T. Anderson, *Proc. Am. Assoc. Res.* ASCO, **19**, 233 (1978).
55. B. Rosenberg, *Cancer Treat. Rep.* **63**(9–10), 1433 (1979).
56. A. D. Kelman and H. J. Peresie, *Cancer Treat. Rep.* **63**(9–10), 1445 (1979).
57. G. L. Cohen, W. R. Bauer, J. K. Barton, and S. L. Lippard, *Science* **203**, 1014 (1979).
58. G. L. Cohen, J. A. Ledner, W. R. Bauer, H. M. Ushay, C. Caravana, and S. J. Lippard, *J. Am. Chem. Soc.* **102**, 2487 (1980).
59. J. Brouwer, P. van de Putte, A. J. Fichtinger-Schepman, and J. Reedijk, *Proc. Natl. Acad. Sci. U.S.A.* **78** (II), 7010 (1981).
60. J. Roberts, *Recent Results Cancer Res.* **48**, 79 (1974).
61. A. Blake and A. Peacocke, *Biopolymers* **6**, 1225 (1968).
62. I. Roos and M. Arnold, *J. Clin. Hematol. Oncol.* **7**, 374 (1977).
63. J. Macquet and J. Butour, *J. Clin. Hematol. Oncol.* **7**, 469 (1977).
64. H. Van den Berg, H. Fravol, and J. Roberts, *J. Clin. Hematol. Oncol.* **7**, 349 (1977).
65. J. Hoeschele and L. Van Camp, *Advances in Antimicrobial and Antineoplastic Chemotherapy*, Vol. 2, p. 241, University Park Press, Baltimore (1972).
66. B. Rosenberg, *J. Clin. Hematol. Oncol.* **7**, 231 (1977).
67. B. Rosenberg, *Advances in Antimicrobial and Antineoplastic Chemotherapy*, Vol. 2, p. 101, University Park Press, Baltimore (1972).
68. P. Conran, *Recent Results Cancer Res.* **48**, 134 (1974).
69. P. Conran and B. Rosenberg, *Advances in Antimicrobial and Antineoplastic Chemotherapy*, p. 235, Vol. 2, University Park Press, Baltimore (1972).
70. A. Khan, *Recent Results Cancer Res.* **48**, 131 (1974).
71. A. Khan and J. Hill, *Advances in Antimicrobial and Antineoplastic Chemotherapy*, Vol. 2, p. 259, University Park Press, Baltimore (1972).
72. A. Khan, K. Wakasugi, B. Hill, D. Richardson, J. Disahato, and J. Hill, *J. Clin. Hematol. Oncol.* **7**, 787 (1977).
73. S. Carter and M. Goldsmith, *Recent Results Cancer Res.* **48**, 137 (1974).
74. J. Toth-Allen, Ph.D. Thesis (1970) as abstracted in *Diss. Abst., Inter. Ed. B* **31**, 6445 (1971).
75. D. Jones, A. Robins, and D. Taylor, *Advances in Antimicrobial and Antineoplastic Chemotherapy*, Vol. 2, p. 229, University Park Press, Baltimore (1972).
76. D. Taylor, J. Jones, and A. Robins, *Recent Results Cancer Res.* **48**, 125 (1974).
77. C. Litterst, C. Torres, and A. Guarino, *J. Clin. Hematol. Oncol.* **7**, 169 (1977).
78. W. Wolf and R. Manaka, *J. Clin. Hematol. Oncol.* **7**, 79 (1977).
79. W. Wolf and R. Manaka, *Recent Results Cancer Res.* **48**, 128 (1974).
80. I. Krakoff, *J. Clin. Hematol. Oncol.* **7**, 604 (1977).
81. I. Krakoff and A. Lippman, *Recent Results Cancer Res.* **48**, 183 (1974).
82. A. Khan, J. Hill, W. Grater, E. Loeb, A. MacLellan, and N. Hill, *Cancer Res.* **35**, 2766 (1975).
83. D. Higby, H. Wallace, D. Albert, and J. Holland, *J. Urology* **112**, 100 (1974).
84. E. Wiltshan and B. Carr, *Recent Results Cancer Res.* **48**, 178 (1974).
85. J. Hill, E. Loeb, A. MacLellan, N. Hill, A. Khan, and J. Koglerg, *Recent Results Cancer Res.* **48**, 145 (1974).
86. Z. Dienstbier, O. Andrysek, and J. Zamecnik, *Recent Results Cancer Res.* **48**, 191 (1974).
87. A. Lippman, C. Helson, L. Helson, and I. Krakoff, *Cancer Chemother. Rep., Part 1* **57**, 191 (1973).
88. A. Russof, R. Slayton, and C. Perila, *Cancer* **30**, 145 (1972).
89. E. Wiltshan and T. Kroner, *Cancer Treat. Rep.* **60**, 55 (1976).

90. D. Higby, H. Wallace, and J. Holland, *Cancer Chemother. Rep., Part 1* **57**, 459 (1973).
91. J. Hill, E. Loeb, A. Pardue, A. Khan, N. Hill, J. King, and R. Hill, *J. Clin. Hematol. Oncol.* **7**, 681 (1977).
92. A. Yagoda, R. Watson, H. Grabstald and W. Whitmore, *Proc. Am. Assoc. Cancer Res.* **17**, 296 (1976).
93. J. B. Schlaer, C. P. Morrow, and A. D. Depetrillo, *Am. J. Obstet. Gynecol.* **136**(8), 983 (1980).
94. J. H. Glick, L. M. Zehngebot, and S. G. Taylor, *Am. J. Otolaryngol.* **1**(4), 306 (1980).
95. Z. Steiger, R. Franklin, R. F. Wilson, L. Leichman, I. Asfaw, G. Baishanpayan, J. C. Rosenberg, J. J. Loh, A. Dindogru, H. Seydel, J. Hoschner, P. Miller, T. Knechtges, and V. Vaitkevicius, *Am. Surg.* **47**(3), 95 (1981).
96. R. Arriagada, T. LeChevalier, and B. I. Sillet, *Bull. Cancer (Paris)* **68**(2), 163 (1981).
97. M. Schroder and H. W. Von Heyden, *H* **29**(7), 225 (1981).
98. S. Davis, N. A. DiMartino, and G. Schneider, *Cancer* **47**(2), 425 (1981).
99. M. S. Soloway and C. E. Cox, *Trans. Am. Assoc. Genitourin. Surg.* **71**, 8 (1979).
100. K. R. Kedin, C. Gibbons, and L. Persky, *J. Urol.* **125**(5), 655 (1981).
101. K. A. Foon, F. Naiem, C. Yale, and R. P. Gale, *Leuk. Res.* **3**(63), 171 (1979).
102. R. C. Wallach, C. Cohen, H. Bruckner, B. Kabakow, G. Deppe, and L. Ratner, *Obstet. Gynecol.* **55**(3), 371 (1980).
103. R. C. Young, D. D. von Hoff, P. Gormley, R. Makuch, J. Cassidy, D. Howser, and J. M. Bull, *Cancer Treat. Rep.* **63**(9–10), 1539 (1979).
104. E. Wiltshaw, S. Subramarian, C. Alexopoulos, and G. H. Barker, *Cancer Treat. Rep.* **63**(9–10), 1545 (1979).
105. S. B. Lele, M. S. Pives, and J. J. Barlow, *Gynecol. Oncol.* **8**(1), 74 (1979).
106. D. V. Razis, E. Poulakou, A. Petounis, G. Papadimitriou, P. Kosmides, and G. Delides, *Oncology* **39**(4), 205 (1982).
107. C. J. Williams, B. Mead, A. Arnold, J. Green, R. Buckanan, and M. Whitehouse, *Cancer* **49**(9), 1778 (1982).
108. H. G. Taylor, W. A. Brown, W. M. Butler, M. D. Weltz, J. L. Berenberg, D. G. McCleod, J. E. Fowler, R. E. Stutzman, and J. Blom, *Cancer* **48**(5), 1110 (1981).
109. S. D. Williams and L. H. Einhorn, *Adv. Intern. Med.* **27**, 531 (1982).
110. L. H. Einhorn, *Cancer Res.* **41**(9), 3275 (1981).
111. T. F. Reynolds, D. Vugrin, E. Cvitkovic, E. Cheng, D. W. Braun, M. A. Ohehir, M. E. Dukeman, W. F. Whitmore, and R. B. Golby, *Cancer* **48**(4), 888 (1981).
112. C. G. Julian, J. M. Barrett, R. L. Richardson, and F. A. Greco, *Obstet. Gynecol.* **56**(3), 396 (1980).
113. A. Yagoda, *Cancer Treat. Rep.* **63**(9–10), 1565 (1979).
114. D. Hayes, E. Cvitkovic, R. Golby, E. Scheiner, and I. Krakoff, *Proc. Am. Assoc. Cancer Res.* **17**, 169 (1976).
115. K. Chary, D. Higby, E. Henderson, and K. Swingerton, *Cancer Treat. Rep.* **61**, 367 (1977).
116. K. Chary, D. Higby, E. Henderson, and K. Swingerton, *J. Clin. Hematol. Oncol.* **7**, 633 (1977).
117. C. Merrin, *Proc. Am. Assoc. Cancer Res.* **17**, 243 (1976); **18**, 298 (1977).
118. L. Einhorn and B. Furnas, *J. Clin. Hematol. Oncol.* **7**, 662 (1976).
119. H. Bruckner, C. Cohen, G. Deppe, B. Kabakow, R. Wallach, E. Greenspal, S. Gusberg, and J. Holland, *J. Clin. Hematol. Oncol.* **7**, 619 (1977).
120. R. Kwong and B. Kennedy, *Proc. Am. Assoc. Cancer Res.* **18**, 317 (1977).
121. J. A. Mabel, P. C. Merker, M. L. Sturgeon, I. Wodinsky, and K. I. Geran, *Cancer* **42**(4), 1711 (1978).
122. M. H. Amer, R. M. Izbick, V. K. Vaitkerleins , and M. Al Sarraf, *Cancer* **45**(2), 217 (1980).

123. H. Takita, F. Edgerton, P. Marabella, D. Conway, and S. Harguindey, *Cancer* **48**(7), 1528 (1981).
124. M. Cleare, *Coord. Chem. Rev.* **12**, 349 (1974).
125. R. Speer, H. Ridgway, L. Hall, D. Stewart, K. Howe, D. Lieberman, and J. Hill, *Wadley Med. Bull.* **5**, 19 (1975).
126. R. Speer, H. Ridgway, L. Hall, A. Newman, K. Howe, D. Stewart, G. Edwards, and J. Hill, *Wadley Med. Bull.* **5**, 335 (1975).
127. Y. Kidani, K. Indgaki, R. Saito, and S. Tsukagoshi, *J. Clin. Hematol. Oncol.* **7**, 197 (1977).
128. E. Loeb, J. Hill, A. Pardue, N. Hill, A. Khan, and J. King, *J. Clin. Hematol. Oncol.* **7**, 701 (1977).
129. G. Rowland, G. O'Neil, and D. Davis, *Nature* **255**, 487 (1975).
130. H. Ringsdorf, *Midland Macromolecules Meeting* (H. Elias, ed.), Marcel Dekker, New York (1978).
131. H. J. Ryser, *Nature* **215**, 934 (1967).
132. H. Allcock, R. Allen, and J. O'Brien, *Chem. Commun. 1976*, 717.
133. H. Allcock, *Science* **193**, 1214 (1976).
134. H. Allcock, *Polym. Prepr., Am. Chem. Soc., Div. Polym. Chem.* **18**, 857 (1977).
135. H. Allcock, *Organometallic Polymers* (C. Carraher, J. Sheets, and C. Pittman, eds.), pp. 283–288, Academic Press, New York (1978).
136. H. Allcock, R. Allen, and J. O'Brien, *J. Am. Chem. Soc.* **99**, 3984 (1977).
137. C. Carraher, C. Admu-John, and J. Fortman (unpublished results).
138. C. Carraher, D. J. Giron, I. Lopez, D. R. Cerutis, and W. J. Scott, *Org. Coatings Plast. Chem.* **44**, 120 (1981).
139. C. Carraher, W. J. Scott, J. A. Schroeder, and D. J. Giron, *J. Macromol. Sci., Chem. A15*(4), 625 (1981).
140. C. Carraher, *Org. Coatings Plast. Chem.* **42**, 428 (1980).
141. C. Carraher, T. Manek, D. Giron, D. R. Cerutis, and M. Trombley, *Polym. Prepr.* **23**(2), 77 (1982).
142. C. Carraher and A. Gasper, *Polym. Prepr., Am. Chem. Soc., Div. Polym. Chem.* **23**(2), 75 (1982).
143. C. Carraher, *Biomedical and Dental Applications of Polymers* (i.e., C. G. Gebelein and F. Koblitz, eds.), Chapter 16, Plenum Press, New York (1981).
144. C. Carraher, D. J. Giron, T. Manek, and D. Blair (unpublished results).
145. C. Carraher, H. M. Molloy, M. L. Taylor, T. O. Tiernan, and W. J. Scott (unpublished results).
146. L. Roizin-Towle, *Brit. J. Cancer* **43**, 378 (1981).
147. R. G. Ham, *Biochem. Biophys. Res. Commun.* **14**, 34 (1964).
148. P. E. Duffy, R. Defendini, and L. T. Kremzner, *J. Neuropath, Exp. Neurol.* **30**, 678 (1971).
149. A. V. Hospattankar, S. H. Advani, N. R. Vaidya, S. E. Electricwalla, and B. M. Braganca, *Int. J. Cancer* **25**, 463 (1980).
150. D. H. Russell, ed., *Polyamines in Normal and Neoplastic Growth*, Raven Press, New York (1973).
151. D. H. Russell, *Nature* **233**, 144 (1971).
152. S. I. Harik and C. H. Sutton, *Cancer Res.* **19**, 5010 (1979).
153. L. J. Marton, J. G. Vaughn, I. A. Hawk, C. C. Levy, and D. H. Russell, *Polyamines in Normal and Neoplastic Growth* (D. H. Russell, ed.), p. 367, Raven Press, New York (1973).
154. G. M. Durie, S. E. Salmon, and D. H. Russell, *Cancer Res.* **37**, 214 (1977).
155. D. H. Russell, B. G. H. Durie, and S. E. Salmon, *Lancet* **II**, 797 (1975).
156. T. G. O'Brien and L. Diamond, *Cancer Res.* **37**, 3895 (1977).
157. E. Ben-Hur and E. Riklis, *Cancer Biochem. Biophys.* **4**, 25 (1979).

158. E. Gerner and D. H. Russell, *Cancer Res.* **37**, 482 (1977).
159. O. Heby and D. H. Russell, *Polyamines in Normal and Neoplastics Growth* (D. H. Russell, ed.), Raven Press, New York; *Cancer Res.* **40**, 293 (1980).
160. H. T. Rupniak and P. Dieter, *Cancer Res.* **40**, 293 (1980).
161. D. P. Strommen and W. L. Peticolas, *Biopolymers* **21**, 969 (1982).
162. B. Malfoy, B. Hartmann, and M. Leng, *Nucl. Acids. Res.* **9**, 5659 (1981).
163. J. P. Alix-Alain, L. Bernard, M. Manfait, P. K. Ganguliad, and T. Theophanides, *Inorg. Chim. Acta* **55**, 147 (1981).
164. G. L. Cohen, J. Ledner, W. R. Baner, H. M. Usay, C. Caravana, and S. J. Lippardy, *J. Am. Chem. Soc.* **102**, 2487 (1980).
165. V. Kleinwaechter, *Stud. Biophys.* **81**, 85 (1980).
166. B. Malfoy, B. Hartmann, J. P. Macquet, and M. Leng, *Cancer Res.* **41**, 4127 (1981).
167. A. J. Fornace and D. S. Seres, *Mut. Res.* **94**, 277 (1982).
168. A. Eastman, *Biochem. Biophys. Res. Commun.* **105**, 869 (1982).
169. N. P. Johnson, *Biochem. Biophys. Res. Commun.* **104**, 1394 (1982).

21

Iron-Complexing Bioactive Polymers

Anthony Winston

Abstract. Polymers bearing functional groups possessing a special ability to bind iron are under consideration as potential drugs for treating iron overload caused by iron poisoning and by diseases such as Cooley's anemia. The functional groups of greatest current interest include hydroxamic acids, catechols, and phenols. Model compounds include naturally occurring desferrioxamine-B, a *tris*-hydroxamic acid, and enterobactin, a *tris*-catecholate. Attachment of the functional groups to polymers enhances the iron chelating ability by the "chelate effect" caused by the polymers holding the groups in close proximity to each other. The determination of the stability constants of a series of hydroxamic acid polymers has confirmed the existence of the chelate effects in such cases. The *in vivo* activities of the polymers were determined by means of a mouse screen designed to measure the ability of the drug to remove iron from iron-overloaded mice. Bioassays of only a few polymers of the hydroxamic acid type have been reported. The polymers removed iron from iron-overloaded mice to an extent approaching that of the standard desferrioxamine-B. Additional polymers of the hydroxamic acid type, as well as catechol and phenolic types, are under consideration or are being tested. These results should be available soon.

1. Introduction

This chapter will deal with the development of polymeric materials either designed to bind iron(III) selectively and specifically, or that bear functional groups that would be expected to possess an especially high affinity for iron(III). Since we are concerned primarily with iron(III), the functional groups included are especially prone to favor this particular oxidation state of iron. Although iron-binding polymers have been developed for both medical and nonmedical applications, medical uses will be emphasized. However, polymers for nonmedical purposes are included since the structure and synthetic methods are similar and there is the potential that these polymers might be adapted to medical purposes when iron chelation is important.

Anthony Winston • Department of Chemistry, West Virginia University, Morgantown, WV 26506.

2. Medical Problems of Iron Overload

Medical interest in the development of iron chelating drugs is due almost entirely to their potential use in removing iron from patients experiencing iron overload. Following the discovery of certain highly active naturally occurring iron chelators in the 1950s, iron chelation therapy has become an accepted means for such treatment.

There are two major causes of iron overload – iron poisoning and a disease called β-thalassemia, or Cooley's anemia. Iron poisoning is a problem most often arising in small children through the inadvertent ingestion of iron preparations. Before the introduction of iron chelation therapy in the early 1960s, such conditions were often fatal. Now, through the use of a powerful iron chelator, iron poisonings are treated with an extraordinary degree of success. The clinical manifestations of iron poisoning and the results of various treatment protocols using iron chelation therapy have been reviewed.[1] β-Thalassemia (Cooley's anemia) is a genetic disorder, rare in the United States, but widely distributed throughout the Mediterranean area, the Middle East, India, and Southeast Asia.[2,3] The disease, appearing largely in persons with Greek, Italian, or Oriental decent, is characterized by an inability to synthesize adequate amounts of the β-chain of hemoglobin. Since excess α-chains cannot form soluble tetramers, precipitation occurs in the red cell percursors leading to their death and to the condition of anemia.[4]

Two mRNA mutations have been identified as causing β-thalassemia, a form of thalassemia in which the β-globin chains are not formed. In a Chinese patient an AAG to UAG mutation converts the β^{17Lys} codon to an amber termination codon.[5] Similarly, in a Sardinian patient, a CAG to UAG mutation converts β^{39Gln} to a terminator.[6] Both of these disorders would cause premature termination of partially formed β-globin.

Because of the inability to synthesize the β-chain of hemoglobin, the only effective treatment of β-thalassemia is to administer blood transfusions throughout life. Such continual transfusions introduce large quantities of iron, which, if not removed, accumulate and form deposits in the liver, spleen, heart, and other vital organs. Death is usually by cardiac failure. To remove iron and to prevent its accumulation, a naturally occurring iron chelator, desferrioxamine-B (DFO)(1), is administered. DFO reduces iron levels by forming a stable soluble iron complex, which is eliminated in the urine and stool. Because DFO clears rapidly from the plasma, the most effective route of administration is by frequent slow subcutaneous infusion. Because of these various problems associated with the use of DFO, there is still a need for new iron chelators that can be administered easily and that will remove iron rapidly and efficiently.[4] The vast amount of literature concerning β-thalassemia and its treatment has been adequately cited in a number of reviews.[1–4,7–12]

3. Naturally Occurring Iron Chelators

The history of naturally occurring iron chelators began in 1952 when Neilands[13] isolated a brown iron-containing material from *Ustilago sphaerogena*, which he named ferrichrome. The structure of ferrichrome was later found to be the *tris*-hydroxamic acid iron complex (**2**).[14,15] In subsequent years a variety of other hydroxamic acids and their iron complexes were isolated and identified including the medically important desferrioxamine-B, (**1**),[16–18] also a *tris*-hydroxamic acid. A variety of ferrichromes and ferrioxamines have been found in nature, each differing from one another by slight modifications in structure. Another class of iron chelators in which catechol groups comprise the iron-complexing ligand is represented by enterobactin (**3**), a *tris*-catecholate. Several other members of this series, bearing only one or two catechols, have also been identified. The chemistry of siderophores, a generic name used for all these iron chelators, has been the subject of a number of excellent reviews.[19–25]

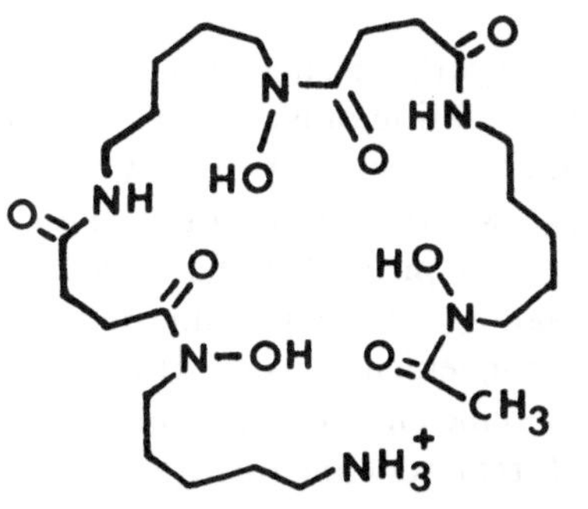

1 Desferrioxamine B

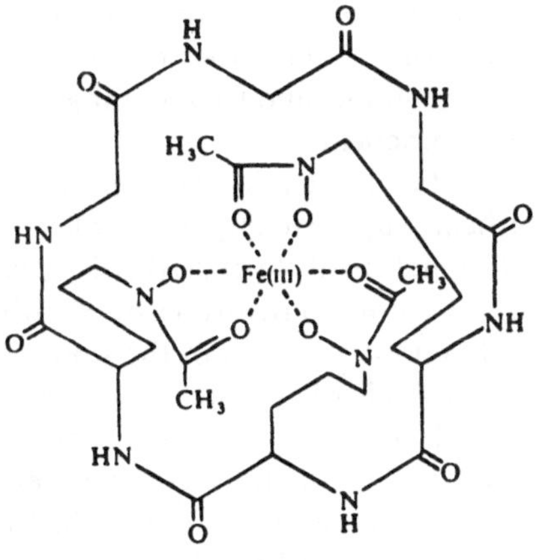

2 Ferrichrome

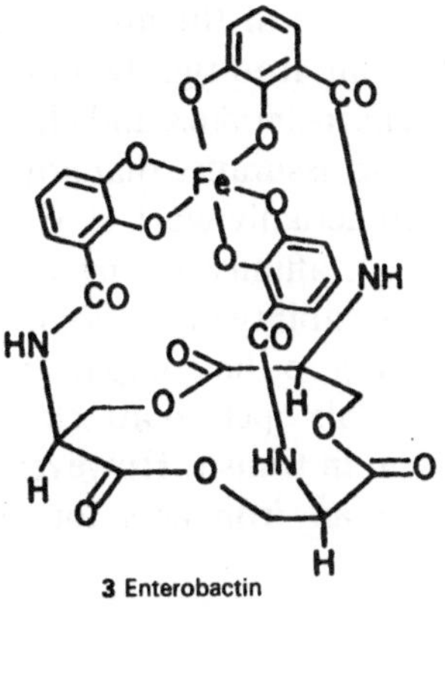

3 Enterobactin

In 1963 the newly discovered compound, desferrioxamine-B (DFO), was found to be a potent agent for eliminating iron in cases of hemochromatosis and iron poisoning.[26] Soon thereafter the drug was found safe for use in humans and clinical investigation began. Because of its high activity and low toxicity in treating iron overload and iron poisonings, DFO was selected by Ciba Pharmaceutical Company for commercial production. The material, produced by fermentation, is available as the methane sulfonate salt under the name Desferal.®

4. Programs for the Development of New Iron Chelators

Although DFO therapy is effective in removing large quantities of iron rapidly, there are some drawbacks in its use. One of these is the short plasma residence time, about 30 min, which causes a significant reduction in the efficiency of DFO to remove iron. To counteract this rapid plasma clearance, the chelator is often administered by means of a portable pump that is worn by the patient and which continuously administers a controlled amount of the drug.[27] On the other hand, the pump is often psychologically unacceptable, and frequent injections throughout the day are painful and not very practical.

For these reasons a program designed to synthesize new iron chelators is being supported by the Cooley's Anemia Foundation of New York and the National Institute of Arthritis, Diabetes, and Digestive and Kidney Diseases of the National Institutes of Health. The object is to devise structures that not only remove iron from the body, but also possess properties that provide a better mode of administration. A definite advantage would be achieved if the new drug could be given orally, the most convenient route. Also, if the plasma survival time could be extended, the efficiency of drug utilization would be increased, and thus the frequency of treatment and quantity of drug required would be reduced.

Advances have been made in the area of designing and synthesizing chemical structures for binding iron tightly and specifically, and possessing the appropriate solubility characteristics and chemical stability for drug use. On the other hand, there is no assurance that any such compound, no matter how carefully designed, will actually work *in vivo*. Although the compound must, of course, have a high affinity for iron, there is no simple relation between this property and the ability of the compound to remove excess iron from living systems.[28] Adding to the complexity of the problem is the idea that there are several iron pools open to attack by an iron chelator, such as transferrin, ferritin, and iron in transit. However, knowledge as to which of these pools is available to an iron chelator such as DFO is not well understood. Also, there is evidence that the rat and mouse screens, used to provide a preliminary evaluation of the drug, do not always provide a measure of the behaviour of the drug in humans. For example, rhodotorulic acid, which appeared promising on the basis of animal tests, nevertheless produced painful reactions when used in humans.[29] Clearly, there is much to be learned about iron transport, the manner in which iron chelators really work, and the effect of structure on adverse physiological reaction.

The design of new iron chelators has been directed largely toward mimicking the naturally occurring siderophores, such as desferrioxamine and enterobactin, by inclusion of hydroxamic acids, catechols, and phenols into a variety of structures in order to produce compounds having exceptionally high stability constants for iron, close to or higher than that of desferrioxamine. Since, a high iron-binding constant does not ensure high activity *in vivo*,[28] many compounds will have to be prepared and screened biologically in order to find a satisfactory balance of chemical and physiological properties.

5. Development of Polymeric Iron Chelators for Iron Chelation Therapy

5.1. Hydroxamic Acid Type

5.1.1. Early History—Nonmedical Applications

Hydroxamic acids, the functional group responsible for the iron-binding activity of DFO, have been known since 1869, when Lossen[30] reported the reaction between hydroxylamine and diethyloxylate. This general reaction of an organic ester with hydroxylamine is the basis of the synthesis of most hydroxamic acids. Reaction of hydroxylamine with acid chlorides or anhydrides may not be entirely satisfactory due to the possibility of further reaction of the hydroxamic acid to give the *O*-acyl derivative.

The first hydroxamic acid polymer appears to have been made by Coffman[31] who in 1946 reported the reaction of hydroxylamine with a maleic anhydride copolymer. Some years later a similar study by Cocea *et al.*,[32] confirmed these findings and reported that hydroxamic acid polymers formed complexes with Fe^{3+}, Cu^{2+}, Cd^{2+}, and UO_2^{2+}.

The reaction of polyacryloyl chloride with hydroxylamine was reported by Vranken and Smets[33] to give a hydroxamic acid polymer. However, this polymer was not isolated, but was converted immediately to the isocyanate polymer by means of the Lossen rearrangement.

Ion exchange resins based on hydroxamic acid groups have been prepared from Amberlite IRC-50 (Rohm and Haas) by conversion of the carbonyl groups to the acid chloride,[34,35] or to the ester,[36] followed by treatment with hydroxylamine. Used in a column, the modified resin showed a significant increase in the retardation of various metal ions. Recently, a hydroxamic acid ion exchange resin, prepared from Chelex 100 (Bio-Rad) was shown to be capable of reducing the iron level of water supplies to less than 1 ppm.[37]

Kern and Schulz[38] prepared hydroxamic acid polymers by treating poly(methyl methacrylate) with hydroxylamine and made the interesting observation that these polymers were generally insoluble, except under conditions of high dilution or in the presence of a large excess of iron(III). The effect of excess iron on solubility will be discussed later in this chapter. Another report by polymeric hydroxamic acids is concerned with copper chelation instead of iron.[39]

Vinyl monomers bearing hydroxamic acids have been reported to polymerize under a variety of conditions. Acrylo-, methacrylo-, crotono-, and cinnamohydroxamic acids, prepared by reactions of the corresponding esters with hydroxylamine, polymerized on initiation with peroxides, azo compounds, and UV light.[40] Acrylohydroxamic acid was produced in only 8% yield from the acid chloride and hydroxylamine.[41] An earlier report of the synthesis of this monomer from ethyl acrylate and hydroxylamine was incorrect, the only product being the Michael addition adduct of hydroxylamine to the vinyl unsaturation.[42] Performed under proper conditions, the Michael addition can be suppressed and the acrylohydroxamic acid isolated.[43]

An unusual method for preparing hydroxamic acid polymers was used by Gasparini[44] who condensed hydroxamic acid derivatives of phenol and catechol with formaldehyde. The resulting resins were said to be useful as ion exchange resins.

Polyhydroxamic acids can also be prepared from polyacrylonitrile by either of two methods—reaction with hydroxylamine to give the amidoxime followed by hydrolysis, or by hydrolysis of the nitrile group to the amide followed by treatment with hydroxylamine. Early work in this area was reported by Schouteden[45,46,48] and Schouteden and Herbots.[47] More recent reports by Vernon and Eccles[49,50] show the usefulness of cross-linked polymers of this type as ion exchange resins for separating and isolating several metal ions. Vernon[49] has reviewed the procedures for converting polyacrylonitrile to polyhydroxamic acids. Patents have also been issued on this procedure.[51,52]

5.1.2. Graft Copolymers of Desferrioxamine-B

The first concerted effort to synthesize iron-binding polymers

Table 1. Polymer–DFO Grafts[a]

Grafts	Percent DFO grafted
Polyacrolein	57
Poly(acrylyl chloride-co-styrene)	30
Poly(methacrylyl chloride-co-styrene)	53
Poly(methacrylyl chloride-co-styrene–co-vinyl pyrrolidone)	60
Chloromethylated polystyrene	68
Chloromethylated Amberlite XAD-2	28
Chloromethylated Amberlite XAD-4	33

[a] Reference 55.

specifically for medical use was made by Ramirez and Andrade[53] in a project designed to develop a means for treating acute iron poisoning in cases where excessively high doses of DFO would be undesirable. The idea was to graft DFO onto a polymer framework and to then use the graft copolymer to remove iron from the blood by extracorporeal ion exchange. Three grafting reactions were used, all involving the free amino group of DFO: reaction of the amine with the aldehyde or acetal groups of polyacrolein to give the Shiff base, reaction with acid chloride copolymers to give amides, and reaction with chloromethylated styrenes. Grafting efficiencies ranged from 30 to 60%.[54,55] The base polymers used and the results of these studies are shown in Table 1.

Although the grafting reactions were successful, there have been no further developments in the synthesis or medical applications of DFO grafted polymers since the results of Table 1 were reported in 1976.

5.1.3. Synthetic Polymers with Controlled Spacing of Hydroxamic Acid Groups

A substantial body of evidence shows that a strong chelate effect is responsible for the high stability of the iron(III) complexes of ferrioxamine-B, ferrichrome, enterobactin, and other siderophores. In Table 2 the iron(III) stability constants of several hydroxamic-acid-type siderophores are compared with that of acetohydroxamic acid. In the case of ferrioxamine, the natural high affinity of the hydroxamic acid for iron(III) is augmented by the chelate effect arising as a result of the 9-atom spacing between hydroxamic acids in the DFO molecule. Molecular models show that this spacing is about the right length to permit the three hydroxamic acids to fit the octahedral requirements of the iron(III) without strain. The chelate effect is even more pronounced in desferrioxamine-E, a cyclic analogue of desferrioxamine-B in which the hydroxamic acid groups are in an even better position to complex iron(III). The stability constant in this case is over 10,000 times that of acetohydroxamic acid. For desferrichrome,

Table 2. Formation Constants of Iron Chelators[a]

Chelator	Log *K*
Acetohydroxamic acid	28.2
Desferrioxamine-B	30.6
Desferrioxamine-E	32.5
Desferrichrome	29.1

[a] References 56–58.

although the spacing distance between hydroxamic acids is 10 atoms, the stereochemistry is apparently not as favorable for forming the iron(III) complex, hence the stability constant is somewhat lower.

Using the naturally occurring trihydroxamic acids as models, Winston *et al.*[59,60] reasoned that polymers bearing hydroxamic acid groups on side chains would have a strong chelate effect if the length of the side chains were sufficient to allow three neighboring hydroxamic acids to fit the octahedral coordination sphere of the iron(III) without strain. Such polymers should have an enhanced iron-binding capability over other polyhydroxamic acids that did not have this designed arrangement. Although a variety of hydroxamic acid polymers have been synthesized in the past, none had been constructed with the idea of providing a spacing distance favorable for intramolecular complex formation.

In order to see the effect of hydroxamic acid group spacing on the stability constant of the iron(III) complex, Winston and Kirchner[61] prepared a series of polymers in which the hydroxamic acid groups were spaced by 9, 11, and 13 atoms (designated as P-9, P-11, and P-13). Later, a P-15 analogue was prepared.[62] The polymers were synthesized by the procedure shown in Figure 1.

The first step was to introduce the spacer amino acids through the Schotten–Bauman reaction with methacryloyl chloride. The terminal carboxyl group was then activated by conversion to the *N*-hydroxysuccinimide (NHS) ester through condensation using dicyclohexylcarbodiimide[14] (DCC). Polymerization was initiated by free-radical means to give insoluble polymers. Treatment of the polymers with methylhydroxylamine gave the water-soluble hydroxamic acid polymers. The polymers were purified by dialysis (8000 cutoff Spectrapor membrane tubing) and isolated by freeze drying.

In order to confirm that the hydroxamic acid (HA) groups of the polymers would form the normal 3 : 1 complex with iron, the compositions of the complexes were analyzed by the mole ratio method using the

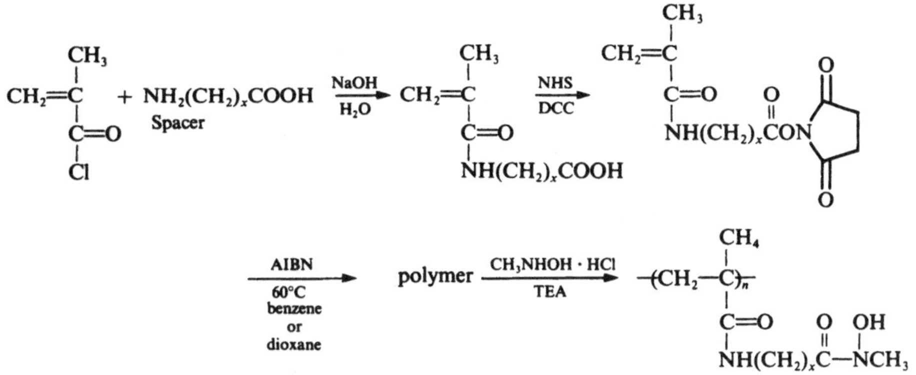

$$CH_2{=}C(CH_3){-}C({=}O){-}Cl + NH_2(CH_2)_xCOOH \text{ (Spacer)} \xrightarrow[H_2O]{NaOH} CH_2{=}C(CH_3){-}C({=}O){-}NH(CH_2)_xCOOH \xrightarrow[DCC]{NHS} CH_2{=}C(CH_3){-}C({=}O){-}NH(CH_2)_xCON(\text{succinimide})$$

$$\xrightarrow[\text{60°C benzene or dioxane}]{AIBN} \text{polymer} \xrightarrow[TEA]{CH_3NHOH \cdot HCl} {-}(CH_2{-}C(CH_4))_n{-} \text{ with side chain } C({=}O){-}NH(CH_2)_xC({=}O){-}N(OH)CH_3$$

P-15 spacer = $NH_2CH_2CONHCH_2COOH$

NHS = *N*-hydroxysuccinimide
DCC = dicyclohexylcarbodiimide
TEA = triethylamine
AIBN = azobis (isobutyronitrile)

P-9 $x = 1$
P-11 $x = 2$
P-13 $x = 3$

Figure 1. Reactions for the synthesis of hydroxamic acid polymers with controlled spacing.

characteristic absorption band at 430–450 nm. The results are shown in Figure 2. The sharp intersections in the region of 0.3 Fe/HA indicate stable complexes at the 3 : 1 ratio. Molecular models show that the spacings of the HA units are sufficiently large to permit complete utilization of hydroxamic acids in forming intramolecular iron(III) complexes. Although the sharp intersection of Figure 2 indicates that the complexes are highly stable, the mole ratio method is not sensitive to small changes in stability constants.

In order to reveal the effect of hydroxamic acid spacing on the magnitude of the chelate effect, the stability constants for the iron(III) complexes were determined. The method involved the measurement of the competition between the polymer and EDTA or DTPA for iron as described by Anderegg *et al.*[(58)] and by Winston and Kirchner.[(61)] The results are shown in Table 3.

The stability constants of the iron–polymer complexes are in the same range as those of the naturally occurring siderophores ferrioxamine and ferrichrome, and indicate that the hydroxamic acid units are utilized completely in forming the 3 : 1 HA–Fe complexes. The side chains are of sufficient length to give the chelate effect, but differences occur because of different steric arrangements. The fact that the stability constant changes significantly with changing side-chain length indicates that the 3 : 1 iron complex is formed through three neighboring hydroxamic acid units. If this were not the case, we would not expect much influence from side-chain length

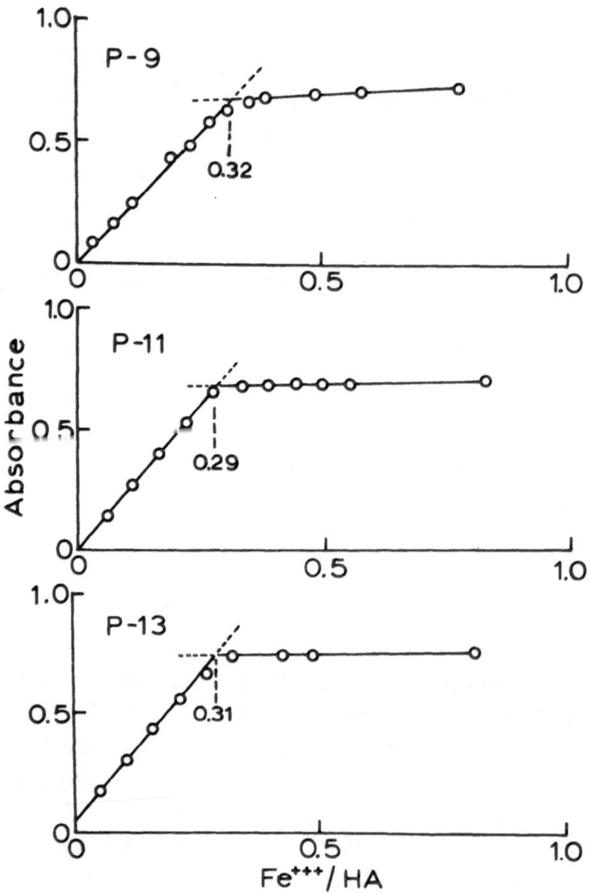

Figure 2. Mole ratio plots for iron(III) complexation of hydroxamic acid polymers.

Table 3. Stability Constants for Fe–HA Complexes

Polymer	$K \times 10^{-29}$	Log K
P-9[a]	0.42	28.6
P-11[a]	4.86	29.7
P-13[a]	2.24	29.4
P-15[b]	1.0	29.0
DFO[a]	27.9	30.4
DFO[c]	31.6	30.5
FCHO[c,d]	1.12	29.0

[a] Reference 61.
[b] Reference 62.
[c] Reference 58.
[d] FCHO = Ferrichrome.

and we would also expect the complex to cross-link through the ferric ion and become insoluble.

The stability constants clearly show that the order of stability is Fe(P-11) > Fe(P-13) > Fe(P-15) > Fe(P-9). Molecular models show that in order for the 3 : 1 iron complex to form using neighboring hydroxamic acid units, the side chains must be long enough to wrap around the iron in the required octahedral arrangement. In the model of P-11, the side chains can do this without strain, but for P-9 the fit is not nearly as good and considerable strain is encountered, especially for the attachment of the third side chain onto the iron. In the case of P-13, which has ample side-chain length to form the complex, the lower stability is attributed to the greater entropy due to the longer side arms. Thus, the optimum side-chain length is the 11-atom spacing of P-11. In comparing the polymer with the natural siderophores, the stability constant of the P-11 complex is about one-fifth that of the ferrioxamine complex, but about four times greater than that for the ferrichrome complex.

Another question is the selectivity of the polymers for iron over other metals. A solution of Fe(P-11) was prepared and treated with Hg^{2+}, Pb^{2+}, Al^{3+}, Ca^{2+}, and Mg^{2+}. Of these, only aluminum complexes sufficient amounts of polymer, displacing iron, to have any effect on the intensity of the absorbance of the iron complex. From a measure of the competition between iron and aluminum for the polymer, the stability constants for the aluminum complexes were determined, Table 4.

Although all three polymers are selective for iron, polymer P-11 is more selective by a factor of 10 than either P-9 or P-13. Thus, the side-chain length and the hydroxamic-acid-group spacing are important factors not only in complex stability but also in metal selectivity. It is also interesting that in the case of aluminum, the complex with P-13 is now the more stable of the three.

Although in dilute solution the red-brown iron–polymer complexes are crystal clear, in concentrated solution addition of iron causes an immediate precipitation of a red-brown complex. The interesting point is that when additional iron is added to this mixture, the precipitate dissolves over a period of a few minutes to form a clear, red-brown solution. This

Table 4. Stability Constants for Al–Hydroxamic Acid Polymer Complexes[a]

Polymer	$K_{Al\text{-}P} \times 10^{-29}$	$K_{Fe\text{-}P}/K_{Al\text{-}P}$
P-9	0.008	54.5
P-11	0.00695	699
P-13	0.0296	75.7

[a] Reference 61.

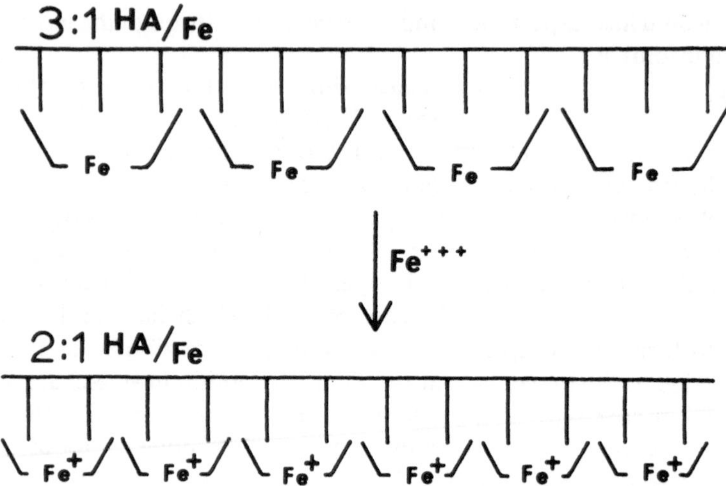

Figure 3. Conversion of an insoluble 3 : 1 to a soluble 2 : 1 iron(III)–hydroxamic acid (HA) complex in the presence of excess iron.

phenomenon, originally observed by Kern and Schultz,[38] was mentioned earlier in this review. A semiquantitative measurement indicates that the precipitate forms at a 3 : 1 HA–Fe ratio where the clear solution is reached at a 2 : 1 ratio. The λ_{max} changes during the process from 430 to 460–470 nm. This effect is not observed on addition of other metal ions or of acids. This phenomenon can be interpreted on the basis of a conversion of the insoluble 3 : 1 complex to a soluble 2 : 1 complex as shown schematically in Figure 3.

The 2 : 1 HA–Fe complex, being a polyelectrolyte, would be expected to be soluble. The λ_{max} of 460–470 nm is intermediate between 430 nm, characteristic of the 3 : 1 complex, and 520 nm, characteristic of the purple 1 : 1 complex[22] of iron with acetohydroxamic acid. Even in the presence of a large excess of iron, λ_{max} for the polymer–iron system never approached the 520 nm of the 1 : 1 complex.

These results indicate that two hydroxamic acids on neighboring side chains form a tight complex with iron, whereas the third hydroxamic acid is less firmly attached and is capable of displacement to form the 2 : 1 species. Molecular models reveal that in order for the third neighboring hydroxamic acid to become attached to the iron in the octahedral arrangement, the third side chain must reach around to the opposite side of the iron to form the final coordinate linkage. This is clearly a less accessible position than the other five, which are more easily reached from the polymer backbone. Hence a lower stability for this third linkage is reasonable.

Hydroxamic acid polymers synthesized without spacer units (the P-3 series **4** and **5**) form iron complexes that are much less soluble than those in which the hydroxamic acids are separated by spacer units (P-9 to P-15).

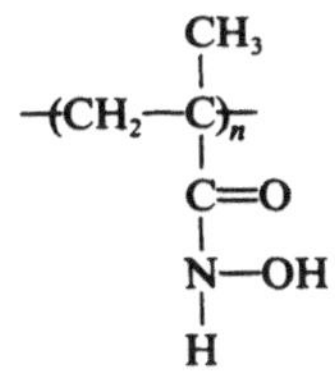

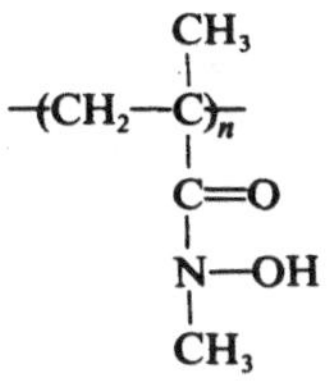

$$\text{-(}CH_2\text{-}C(CH_3)(C(=O)N(OH)H)\text{)}_n\text{-} \qquad \text{-(}CH_2\text{-}C(CH_3)(C(=O)N(OH)CH_3)\text{)}_n\text{-}$$

P-3 series

(4) **(5)**

If the hydroxamic acids are situated close together as in the P-3 series, the spacing distance is insufficient to permit intramolecular complexation to occur between three neighboring hydroxamic acid groups. Instead, the iron forms complexes with hydroxamic acids from different chains to produce highly cross-linked insoluble complexes.

At the other end of the scale, hydroxamic acid polymers have been prepared in which the hydroxamic acid groups have been spaced far apart on the polymer chain.[88] These polymers were prepared by copolymerization of acrylamide with the monomer derived from β-alanine (Figure 1, $n = 2$) to give copolymers with from 0.2 to 2 mol % hydroxamic acid units. Addition of iron to aqueous solutions of these copolymers causes either intermolecular or intramolecular cross-linking depending on the concentration of the copolymer. In concentrated solution, addition of iron causes an increase in viscosity, clear evidence of intermolecular cross-linking. In dilute solution iron causes a decrease in viscosity, a result of intramolecular cross-linking, where the most probable reaction is for the iron to complex with hydroxamic acid groups from the same polymer molecule, causing it to assume a more compact shape. Figure 4 illustrates these effects. In either case, addition of excess iron converts the 3:1 or 2:1 complexes to the 1:1 Fe–HA complex, breaking up the cross-links and restoring the viscosity to its former range. Since the viscosity of the intermolecular cross-linked complex decreases with time, the thermodynamics favors the soluble intramolecular cross-linked form.

5.2. Phenol Type

Recognizing the excellent chelating ability of ethylene-1,2-bis(2-hydroxyphenyl)glycine (EHPG), **6**, for iron(III), Dawson *et al.* investigated the construction of a polymer bearing 2-hydroxyphenylglycine units for use as a potential chelator for treating iron overload.[63] The synthetic method was to alkylate the amine groups of a random 60:40 mol % vinyl

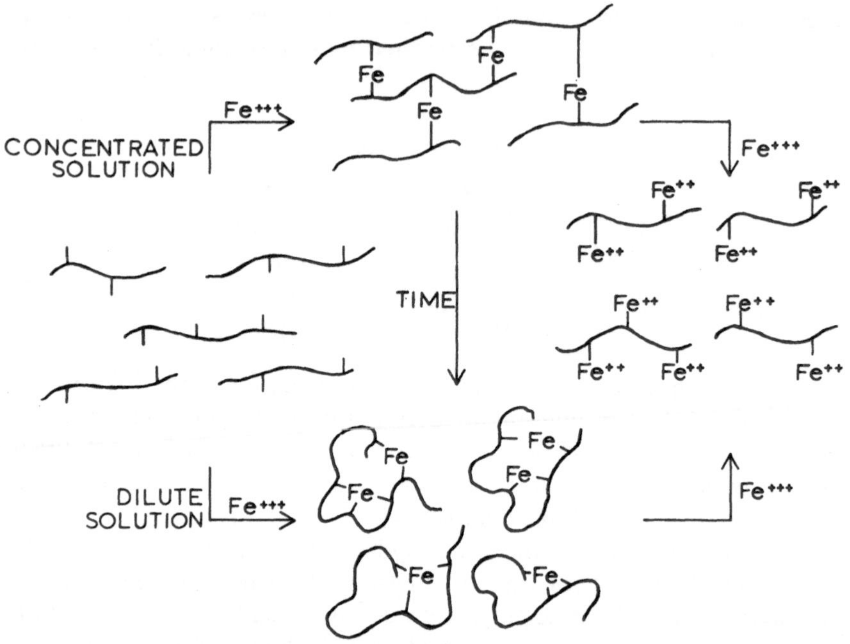

Figure 4. Schematic representation of the effect of iron on hydroxamic acid–acrylamide copolymers (Reference 88).

amine–vinyl sulfonate copolymer[64] with arylbromoacetate followed by hydrolysis of the ester and the groups protecting the phenolic hydroxyls (Figure 5). The UV spectrum of the polymer and the visible spectrum of the polymer–iron complex indicated that the polymer possessed some of the structural features of the model EHPG, **6**. On the other hand, the rate of iron

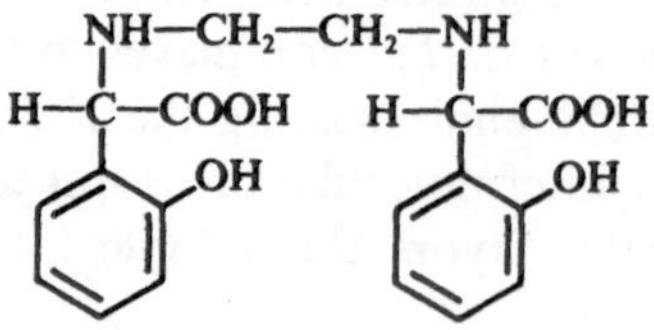

6

complexation of the polymer was considerably slower than that of the EHPG model. Since 43% of the amines of the 60 mol % vinyl amine copolymer were derivatized, only about one-quarter of the total monomer residues will bear 2-hydroxylphenylglycine groups. Thus, it would seem unlikely that the derivatized copolymer would possess a sufficient number of 2-hydroxyphenylglycine units on adjacent positions to provide a useful number of

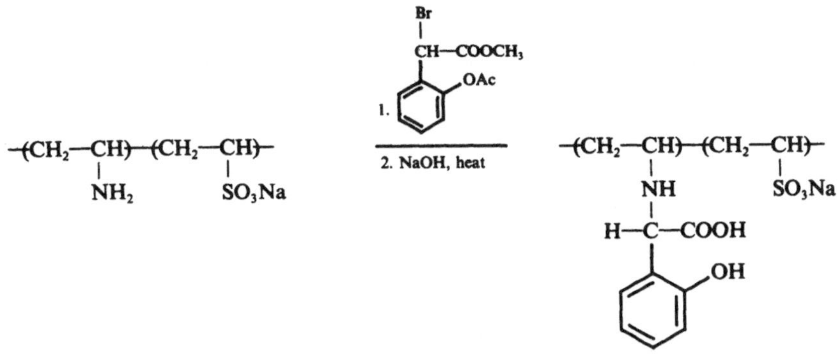

$-(CH_2-CH)-(CH_2-CH)-$ with NH_2 and SO_3Na side groups

1. $BrCH(COOCH_3)C_6H_4OAc$
2. NaOH, heat

$-(CH_2-CH)-(CH_2-CH)-$ with $NH-CH(COOH)-C_6H_4OH$ and SO_3Na side groups

Figure 5. Method for preparing polymers bearing 2-hydroxyphenylglycine groups (Reference 63).

bis-structural units similar to the EHPG model. Without such neighborliness, most of the 2-hydroxyphenylglycine units will be acting independently. This, together with the low mobility introduced by the polymer chain, would cause a low rate of iron chelation.

5.3. Catechol Type

The naturally occurring enterobactin **3** serves as the model for *tris*-catechol iron-binding structures. Enterobactin, first isolated by Neilands from *Salmonella typhimurium*,[65] promotes iron uptake by the organism in much the same way as desferrioxamine or desferrichrome.[20–22,24,25] An iron(III) binding constant of enterobactin of $\log K = 52$[66] reveals an unusually stable iron complex.

A number of *tris*-catecholes have been prepared[67–69] in an effort to mimic the binding ability of enterobactin and to produce compounds having potential use as bioactive iron chelators.

In an approach directed toward polymeric iron chelators, Dawson *et al.* constructed polymers with side chains bearing catechol groups.[63,70] Starting with a 60 : 40 vinyl amine–vinyl sulfonate copolymer,[64] the free amine groups of the copolymer were acylated using the acid chlorides of the catechols. Two types of polymers were prepared, one in which the side chains consisted of single catechols and the other in which the side chains were sequences of three catechol units with spacing distances designed to provide strong chelate effects to maximize iron binding (Figure 6). The proportion of available amino groups functionalized in these reactions ranged from 9 to 25%. The UV spectra of the copolymers were consistent with the proposed structures. The copolymers bound iron readily as evidenced by the broad absorption band centered around 500 nm for the iron complex.

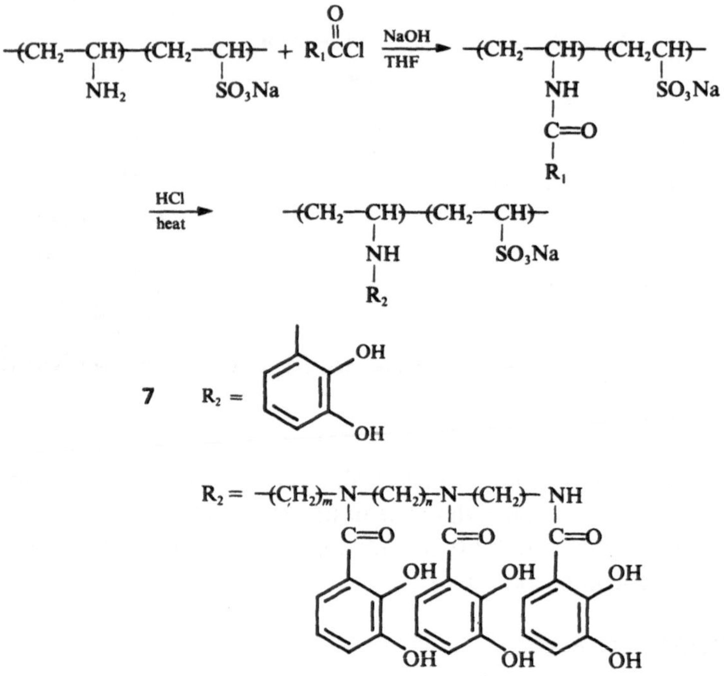

$$-(CH_2-\underset{NH_2}{CH})-(CH_2-\underset{SO_3Na}{CH})- + R_1\overset{O}{\overset{\|}{C}}Cl \xrightarrow[THF]{NaOH} -(CH_2-\underset{\underset{\underset{R_1}{C=O}}{NH}}{CH})-(CH_2\underset{SO_3Na}{CH})-$$

$$\xrightarrow[heat]{HCl} -(CH_2-\underset{\underset{R_2}{NH}}{CH})-(CH_2-\underset{SO_3Na}{CH})-$$

7 $R_2 =$ 2,3-dihydroxyphenyl (OH, OH)

$R_2 = -(CH_2)_m-\underset{C=O}{N}-(CH_2)_n-\underset{C=O}{N}-(CH_2)_o-\underset{C=O}{NH}$ (each C=O bearing a 2,3-dihydroxyphenyl group: OH, OH)

R_1 = Same as R_2 except ortho hydroxyls protected as the acetone ketal

	m	*n*	*o*
8	3	3	3
9	5	3	3
10	5	4	3
11	5	4	4

Figure 6. Catechol polymers (References 63 and 70).

Another iron-binding catechol-type polymer was prepared by Fair.[71] In contrast to the previously described cases where only a fraction of the available sites were functionalized, this synthesis was designed to ensure that all active sites on the polymer would be occupied by catechol units. The synthesis (see Figure 7) involved the polymerization of a vinyl monomer bearing the dimethyl ether of 2,3-dihydroxybenzohydrazide followed by demethylation. Elemental analysis and a mole ratio study of the iron complex of the catechol polymer confirmed complete demethylation. By similar techniques the 3,4-dihydroxybenzohydrazide polymer was also prepared.

Both the 2,3 and 3,4 polymers readily form complexes with iron. The properties of the two iron complexes are different in that the 2,3-polymer

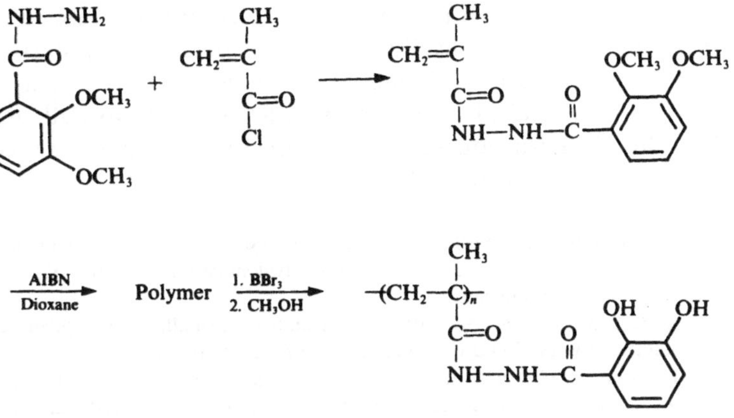

Figure 7. Synthesis of Poly(N'-methacryloyl-2,3-dihydroxylbenzohydrazide) (Reference 71).

complex is soluble over a wide pH range, whereas the 3,4-polymer complex is not. The difference in solubility of the two complexes can be accounted for on the basis of a higher probability of the iron to form a cross-linked structure in the case of the 3,4 polymer as contrasted with the 2,3 polymer which appeared to prefer an intramolecular complex involving only catechols from the same chain. This difference in the manner of complexation can be attributed to the difference in the stereochemistry of the two polymers. In the 2,3 polymer the hydroxyl groups are directed toward the side of the phenyl ring and are in a good position to complex an iron that is already complexed with a neighboring catechol (intramolecular complexation). On the other hand, the two hydroxyls of the 3,4 polymer are directed away from the polymer chain and are not favorably disposed to form intramolecular complexes. Hence, the complex forms between catechol units of different chains, causing cross-linking and resulting in insolubilization. A similar situation exists in the case of hydroxamic acid polymers, the iron complexes of which are insoluble when the side-chain length becomes too short to permit intramolecular complex formation. The stability constant of the iron(III) complex was found to be $\log K = 40$, a value in reasonable agreement with other *tris*-catechols.

6. Selection of Iron Chelating Polymers for Potential Use in Iron Chelation Therapy

A wide variety of iron chelating compounds have become available over the last few years as a result of the synthetic effort being applied to the

Cooley's anemia program. It is hoped that one or more of these compounds will eventually emerge as a leading candidate for continued development, with the ultimate objective of finding, or at least approaching, what is considered to be the ideal drug for this purpose. The overall properties desired for this ideal drug are well stated in the following passage taken from an article by Elmer B. Brown.[72]

1. It must be efficient and selective for chelating Fe^{3+}. This implies high affinity for iron while ignoring various competing divalent cations (Ca, Mn, Cu, Mg, etc.) of biologic importance.
2. The drug must have accessibility to find iron from a small, slowly replenished, chelatable iron pool and not interfere with heme synthesis.
3. After iron is chelated, it should be promptly excreted in a nontoxic form—preferably in the urine to facilitate measurement. Enhanced biliary iron excretion via the feces is not precluded and may even be desirable.
4. Iron removal should be from multiple tissues at risk of toxic damage, not just from the relatively inert storage sites. This requirement stresses the need for iron removal from the heart, liver parenchyma, and endocrine organs that show the greatest effects of iron toxicity in man.
5. During the process of iron excretion, there must be no redistribution of iron from nontoxic storage depots to more vulnerable tissues such as the heart.
6. The ideal chelating drug should be inexpensive since many patients with iron storage disorders, such as those with thalassemia, require lifelong treatment.
7. The drug should be easy to take—preferably by mouth—to reduce the problems of compliance.
8. The ideal drug should have no immediate or long-term toxicity.
9. The ideal drug should permit a net removal of iron from patients with pre-existent iron overload and prevent iron buildup and related tissue damage in patients at risk of iron overload.

Another set of criteria for the ideal chelating drug was expressed in biochemical terms by Jacobs[73]:

1. Removes iron from transferrin
2. Enters cells freely
3. Chelates intracellular iron
4. Inhibits iron uptake by ferritin
5. Does not inhibit iron uptake by heme
6. Does not inhibit DNA synthesis
7. Chelate leaves cells freely
8. Chelate not utilized for other metabolic needs
9. Chelate not utilized by microorganisms
10. Chelate rapidly excreted

The design and synthesis of efficient and selective chelating agents for iron can be rationally approached by the chemist using well-established principles of theoretical and synthetic organic chemistry.[74,75] On the other hand, the biochemical factors are extremely complex and not well understood, and for this reason are much more difficult to incorporate into the structure of the iron chelator. Thus, as with all drugs, a bioassay is essential to see just how the drug will behave *in vivo*.

In order to see whether or not a particular compound is a reasonable candidate drug, some knowledge of iron-binding capability is important. The formation constant (stability constant) of the complex is a measure of this capability and is usually expressed in the form of $\log K$. In order to be considered as a potential drug for iron chelation therapy, the compound must have a reasonably high $\log K$.

Although values of $\log K$ are useful for comparing ligands having similar functionalities, they are less useful in comparing different types of ligands. In order to compare compounds of different ligand types, pM values are used. The pM is defined as $-\log[Fe^{3+}]$ in a solution at the physiological pH of 7.4 and made up to contain 1×10^{-6} M in iron(III) and 1×10^{-5} M in ligand.[66,76] As an example, in the case of a *tris*-hydroxamic acid, which loses three protons on complexation, the equation for the stability constant, Eq. (2), is combined with Eq. (4), which defines the proton dissociation constant of the ligand. The resulting equation, Eq. (5), is then solved for $[Fe^{3+}]$ and converted to the pM value, Eqs. (6) and (7).

$$Fe^{3+} + L^{3-} \rightleftarrows FeL \tag{1}$$

$$K_{FeL} = \frac{[FeL]}{[Fe^{3+}][L^{3-}]} \tag{2}$$

$$H_3L \rightleftarrows 3H^+ + L^{3-} \tag{3}$$

$$K_{H_3L} = \frac{[H^+][L^{3-}]}{[H_3L]} \tag{4}$$

$$K_{FeL} = \frac{[FeL][H^+]^3}{[Fe^{3+}][H_3L][K_{H_3L}]} \tag{5}$$

$$[Fe^{3+}] = \frac{[FeL][H^+]^3}{K_{FeL}[H_3L][K_{H_3L}]} \tag{6}$$

$$pM = -\log[Fe^{3+}] \tag{7}$$

In the case of other types of iron–ligand complexes, such as the catechols, the general procedures are similar, but the particular calculations and equations will differ, depending on the stoichiometry of the complex and the number of ionizable protons of the ligand. Another way of expressing the ability of a ligand to sequester iron is the "effective" stability constant ($\log K_{eff}$) which permits a comparison of complex stabilities over a wide pH range and in the presence of interfering ions. The calculation of $\log K_{eff}$ has been reviewed by Pitt and Martell[(74)] and Martell.[(75)]

7. Iron Chelating Ability of Polymeric Iron Chelators

Over the past 10 years, 300 to 400 compounds have been examined as potential candidates for iron chelating drugs. The iron-binding capabilities in terms of $\log K$ and pM of many of these compounds are known and have been the subject of extensive reviews.[(67,74–76)] Of these compounds, only a few are polymeric in nature. In Table 5 the values for $\log K$ and pM of the polymeric iron chelators are compared with each other and with standards representative of each ligand type.

All of the polymeric iron chelators possess high values of $\log K$ indicating very stable iron chelates. These high stabilities arise in part from the powerful chelate effect caused by the polymer chain holding the ligands in close proximity to each other and by this means increasing the iron stability constant. Thus, there is no doubt that polymeric iron chelators can be designed to possess chelate effects approaching those of model compounds, and that these polymers are viable candidates for use as iron chelating drugs.

Transferrin, a buffer protein for iron,[(73,78)] absorbs excess iron in times of plenty and releases iron during periods of deficiency. Any compound with a pM value greater than that of transferrin (Table 5) is capable of removing iron from transferrin, and an ability to remove iron from transferrin is generally considered to be an essential feature of any iron chelator that is to work *in vivo*.

The ability of iron chelators to remove iron from transferrin can be measured spectrophotometrically. The results of such measurements show that in spite of the high iron-binding constant of many chelators, the rate of transfer of iron to the chelator is often slow.[(67,79)] However, the addition of low-molecular-weight chelators, such as citrate, nitrilotriacetate, and ascorbic acid, can promote a rapid transfer of iron from transferrin to the chelator. The supposition is that the secondary chelator forms a ternary complex with transferrin and ferric iron and aids in the removal of the iron from transferrin.[(78)]

Table 5. Comparison of the Iron Chelating Ability of Polymeric Iron Chelators

Compound	Log *K*	p*M*	References
Hydroxamic acid type			
Desferrioxamine-B	30.6	26.6	56–58
Desferrioxamine-E	32.5	27.7	58
Desferrichrome	29.1	25.2	58
P-9	28.6	25.7	61
P-11	29.7	26.6	61
P-13	29.4	26.4	61
P-15	29.0	26.1	62
Catechol type			
Enterobactin	52	35.6	66,76
Catechol	43.7	15	77
3,4-Dihydroxyphenylacetic acid	43.9	15	77
2,3-Dihydroxy-*N*,*N*′-Dimethylbenzamide	39.8	15	77
7		25.4	70
8		24.5	70
9		24.7	70
10		25.1	70
11		25.1	70
12	40	26.9	71
Protein type			
Transferrin		23.6	76

Since the p*M* values are greater than that of transferrin, the polymers should be able to remove iron from transferrin. Kinetic problems might arise as indicated above, but this effect can probably be countered by the addition of an appropriate secondary chelator.

8. Bioassays of Iron Chelating Drugs

Two bioassays have been developed to test the ability of iron chelating drugs to capture iron and to remove it from living systems. One involves the use of Chang cells, and the other employs either rats or mice that have been brought to a state of iron overload through hypertransfusion with red blood cells.

In the Chang-cell model, developed largely by Jacobs *et al.*,[73,80–82] the Chang cells are incubated in a medium containing iron transferrin, whereupon the iron becomes incorporated into the ferritin within the cell. However, when a powerful iron chelator is present, the iron is captured by the chelator and is not incorporated into ferritin. This ability of a chelator to capture iron in transit between transferrin and ferritin is considered to be a sensitive measure of intercellular iron chelation.

The mouse screen was brought to its present state largely by Dr. Harris Rosenkrantz, Mason Research Laboratory, Worcester, Massachusetts and by Dr. Robert W. Grady, Cornell University Medical Center and Rockefeller University, New York. The goal of the mouse screen is to measure directly the ability of an iron chelating drug to remove iron *in vivo*.[28,82,83] At the same time, any adverse effects of the drug on the mouse can be noted. The essential features of the screen are as follows:[28]

1. Hypertransfusion of mice with canine red blood cells.
2. Daily injections (i.p.) of the iron chelator.
3. Daily collection of urinary and fecel pools.
4. Sacrifice of mice and dissection of liver and spleens.
5. Reduction of livers, spleens, urine, and fecel matter to a form suitable for iron analyses.
6. Measurement of iron content by atomic absorption.
7. Statistical analysis of data and comparison with controls.

Controls include nontransfused nontreated mice, transfused vehicle-treated mice, and transfused mice treated with desferrioxamine-B, the standard against which all new drugs are compared.

The results are compared in tables which show changes in the iron contents of the spleen, liver, feces, and urine. To be effective the iron chelator should cause spleen and liver iron to decrease and fecel and urine iron to increase. Of these, the urine iron appears to be the most significant indication of activity. The reports also include any observed toxic response that could render the drug unsuitable for use in humans. The results of the screening of several hundred iron chelators have appeared in a number of pages and reviews.[28,29,74,82,84]

9. Results of Bioassays of Polymeric Iron Chelators

The results of bioassays of several polymeric iron chelators have appeared in the literature, but these include only the hydroxamic acid type. New polymers are being synthesized and reports on these structures, together with bioassay data, are expected in the near future. The new systems will include hydroxamic acid, catechol, and possibly phenol types of polymers. Preliminary results of the mouse screen of several hydroxamic acid polymers are given in Table 6.

The most significant feature of Table 6 is the urine iron level, which for

Table 6. Bioassay Results of Polymeric Hydroxamic Acid Iron Chelators

Drug	Dose (mg/kg)	Toxicity	Iron content (percent change from controls)[a]				References
			Spleen	Liver	Feces	Urine	
DFO	250	None	+ 2	− 24	+ 22	+ 226	[a]
P-11,[b] $[\eta]^c = 0.92$	100	None	− 17	− 24	+ 44	+ 138	85
Same	300	None	− 20	− 52	−	+ 380	28
Same	500	None	− 41	− 49	− 3	+ 331	85
P-11, $[\eta]^c = 2.18$	500	None	− 35	− 12	− 12	+ 631	85
P-3 (**5**)	375	Toxic[d]	− 46	− 6	− 80	+ 73	85

[a] Mouse screen, Mason Research Institute, Worcester, Mass.
[b] See Figure 1 for structure.
[c] Intrinsic viscosity, DMF–H_2O.
[d] Very toxic, 5/10 survivors.

P-11 is comparable to that of the standard DFO and much greater than that of many simple and complex synthetic and natural hydroxamic acids.[28] Although spleen, liver, and fecal iron can be somewhat erratic, the decrease in spleen and liver iron for P-11 appears to be significant. The higher-molecular-weight P-11 ($[\eta] = 2.18$) appears to be a somewhat better iron chelator, but this result must be confirmed by further tests. P-3 is clearly toxic, in marked contrast to P-11 which shows no toxic signs.

Although the synthetic approach to bioactive hydroxamic acid polymers was to achieve a strong chelate effect by appropriate spacing of hydroxamic acids, there is yet another means by which a polymeric iron chelator might have some advantage over smaller molecules. This idea is concerned with the residence time of the chelator in the plasma. It has been well established that desferrioxamine-B therapy is severely hampered by the rapid depletion of DFO in the plasma, about 30 min for almost complete clearance.[86] Also the efficiency of DFO in removing iron is considerably better when the drug is administered by a slow infusion rather than by periodic injections.[87] This improved efficiency of slow infusion is consistent with the idea that the pool of chelatable iron is established only slowly.

If the loss of DFO from plasma is a result of simple diffusion processes, the polymer itself might provide a distinct advantage merely because its high molecular weight and larger size would retard diffusion through membranes. If the loss of DFO were due to metabolic degradation, it is possible that the polymeric chelator might not serve as a suitable substrate for the enzyme and thus be more resistant to degradation. In either event, the lifetime in the plasma might be extended sufficiently to permit the capture of more iron from the slowly forming chelatable iron pools.

A dynamic factor involving changes in chain conformation on complexation might also be involved. The uncomplexed polymer is highly solvated and exists in an expanded state as a random coil. On complexation with iron, the polymer contracts to occupy a smaller volume. Evidence for this is that when a dilute aqueous solution of hydroxamic acid acrylamide copolymer (containing 1–2 mol % hydroxamic acid units) is treated with iron, the viscosity of the solution decreases significantly, indicating a contraction through the formation of intramolecular cross-links involving iron–hydroxamic acid complexes (Figure 4).[88] When we consider that passage through the membrane depends not only on the molecular weight, but also on the molecular volume, it is not unreasonable to suppose that a polymer of some select molecular weight might be blocked by the membrane when in the form of a large uncomplexed random coil, but easily pass through the membrane while in the contracted complexed state. If such a process could work *in vivo*, the polymer would be of a great advantage. The uncomplexed drug would remain in the plasma because of its expanded size and would have sufficient opportunity to capture iron from the slowly forming iron pools. After complexation with iron, the polymer would occupy a considerably smaller volume and could then diffuse out rapidly.

The high toxicity of the P-3 polymer is in marked contrast to the P-11 series, where little or no toxicity is evident. Since the two polymers are functionally similar, the difference must be due to the difference in molecular architecture. With the 11-atom spacing between hydroxamic acid groups, P-11 can form an intramolecular complex with iron and thus remain water soluble. On the other hand, the 3-atom spacing of the hydroxamic acids of P-3 does not allow three neighboring groups to position themselves about a single iron atom to form the complex. Thus, the iron complex forms intermolecularly to produce a highly cross-linked polymer which is completely insoluble in water. This rapid cross-linking and precipitation of P-3 in the presence of iron would have obvious serious physiological consequences, and it is not surprising that a high order of toxicity is observed in this case. This rather interesting behavior of P-3 would indicate that iron chelators, whether or not they are polymeric, should at least be designed in ways that would not lead to intermolecular iron chelation and possible precipitation. Otherwise, a toxic reaction might merely represent an adverse effect from the cross-linking and not have anything to do with any inherent toxicity of the structure itself.

Polymer P-11 has also been subjected to the Chang-cell screen. The results indicate that P-11 "inhibits iron incorporation into ferritin with either a reduction of ferritin synthesis or cellular iron content."[(73)]

10. Organizations Supporting the Development of New Iron Chelators

Because of the great influence on the rapid development of the science and application of iron chelators to biology and medicine it is appropriate to recognize the following organizations: the National Institute of Arthritis, Diabetes, and Digestive and Kidney Diseases of the National Institutes of Health for organization of the Cooley's Anemia Program under the National Cooley's Anemia Control Act of 1972; the National Heart, Lung, and Blood Institute for their interest throughout the program and for their sponsorship of certain phases of the program; and the Cooley's Anemia Foundation of New York for their interest and sponsorship of certain phases of the program.

References

1. H. S. Waxman and E. B. Brown, Clinical usefulness of iron chelating agents, in: *Progress in Hematology*(E. B. Brown and C. V. Moore, eds.), Vol. VI, pp. 338–373, Grune & Stratton (1969).
2. A. I. Chernoff, Blood, *J. Hematol.* **14**, 899 (1959).
3. T. F. Necheles, D. M. Allen, and H. E. Finkel, *Clinical Disorders of Hemoglobin Structure and Synthesis*, Meredith Corporation, New York (1969).
4. D. J. Weatherall, The iron loading anemias, in: *Development of Iron Chelators for Clinical Use* (A. E. Martell, W. F. Anderson, and D. G. Badman, eds.), pp. 3–12, Elsevier/North Holland, New York (1981).
5. J. C. Chang and Y. W. Kan, *Proc. Natl. Acad. Sci. U.S.A.* **76**, 2886 (1979).
6. R. F. Trecartin, S. A. Leibhaber, J. C. Chang, K. Y. Lee, Y. W. Kan, M. Furbetta, A. Anguis, and A. Cao, *J. Clin. Invest.* **68**, 1012–17 (1981).
7. B. Modell, Advances in the use of iron chelating agents for the treatment of iron overload, in: *Progress in Hematology* (E. B. Brown, ed.), Vol. XI, pp. 267–312, Grune & Stratton, New York (1979).
8. W. F. Anderson, Iron chelation in the treatment of Cooley's anemia, in: *Inorganic Chemistry in Biology and Medicine* (A. E. Martell, ed.), pp. 251–261, American Chemical Society Symposium Series 140, Washington, D.C. (1980).
9. E. C. Zaino and R. H. Roberts, eds., *Chelation Therapy in Chronic Iron Overload*, Symposia Specialists Inc., Miami, Fla. (1977).
10. Assessment of Cooley's Anemia Reseach and Treatment, DHEW Publication No. (NIH) 79-1653 (March 1979).
11. W. F. Anderson and M. C. Hiller, eds., Development of Iron Chelators for Clinical Use, U.S. Department of Health, Education and Welfare, DHEW Publication No. (NIH) 77-994, Bethesda, Md. (1975).
12. D. J. Weatherall and J. B. Clegg, *The Thalassaemia Syndromes*, 3rd Edition, Blackwell, Oxford (1981).
13. J. B. Neilands, *J. Am. Chem. Soc.* **74**, 4846 (1952).
14. S. J. Rogers, R. A. Warren, and J. B. Neilands, *Nature* **200**, 167 (1963).
15. W. Keller-Schierlein and B. Maurer, *Helv. Chim. Acta* **52**, 603 (1969).
16. H. Bickel, G. E. Hall, W. Keller-Schierlein, V. Prelog, and E. Vischer, *Helv. Chim. Acta* **43**, 2129 (1960).
17. H. Bickel, H. Keberle, and E. Vischer, *Helv. Chim. Acta* **46**, 1385 (1963).
18. V. Prelog and A. Walser, *Helv. Chim. Acta* **45**, 631 (1962).
19. W. Keller-Schierlein, V. Prelog, and Z. Zahner, Siderochrome (Naturliche Eisen(III)-trihydroxamat-Komplexe), in: *Fortschritte der Chemie Organischer Naturstoffe* (L. Zechmeister, ed.), pp. 279–322, Springer-Verlag, Wien (1964).
20. J. B. Neilands, Microbial iron transport compounds (siderochromes), in: *Inorganic Biochemistry* (G. Eichhorn, ed.), pp. 167–202, Elsevier, Amsterdam (1973).
21. J. B. Neilands, *Struct. Bond.* **11**, 145 (1972).
22. J. B. Neilands, *Struct. Bond.* **1**, 59 (1966).
23. H. Maehr, *Pure Appl. Chem.* **28**, 603 (1971).
24. J. B. Neilands, Microbial iron transport compounds (siderophores), in: *Development of Iron Chelators for Clinical Use* (W. F. Anderson and M. C. Hiller, eds.), pp. 5–44, U.S. Department of Health, Education and Welfare, DHEW Publication No. (NIH) 77-994, Bethesda, Md. (1975).
25. J. B. Neilands, Microbial iron transport compounds (siderophores) as chelating agents, in: *Development of Iron chelators for Clinical Use* (A. E. Martell, W. F. Anderson, and D. G. Badman, eds.), pp. 13–31, Elsevier/North-Holland, Amsterdam (1981).

26. S. Moeschlin and U. Schnider, *New Engl. J. Med.* **269**, 57 (1963).
27. R. D. Propper and D. G. Nathan, The use of desferrioxamine and "the pump", in: *Chelation Therapy in Chronic Iron Overload* (E. C. Zaino and R. H. Roberts, eds.), pp. 17–35, Symposia Specialists, Miami, Fla. (1977).
28. C. G. Pitt, G. Gupta, W. E. Estes, H. Rosenkrantz, J. J. Metterville, A. L. Crumbliss, R. A. Palmer, K. W. Nordquest, K. A Sprinkle Hardy, D. R. Whitcomb, B. R. Byers, J. E. L. Arceneaux, C. G. Gaines, and C. V. Sciortino, *J. Pharm. Exp. Therap.* **208**, 12 (1979).
29. R. W. Grady, C. M. Peterson, R. L. Jones, J. H. Graziano, K. K. Bharguva, V. A. Berdoukas, G. Kokkini, D. Loukopoulous, and A. Cerami, *J. Pharm. Exp. Therap.* **209**, 342 (1979).
30. H. Lossen, *Justus Liebigs Ann. Chem.* **150**, 314 (1869).
31. D. D. Coffman, U.S. Patent 2,402,604 (1946): *Chem. Abstr.* **40**, 52942^9 (1946).
32. E. Cocea, M. Grigoras, and M. Tutoveanu, *Bull. Inst. Politeh. Iasi* **11**, 159 (1965); *Chem. Abstr.* **64**, 19,800a (1966).
33. M. Vrañcken and G. Smets, *J. Polym. Sci.* **14**, 521 (1954).
34. J. P. Cornaz and H. Deuel, *Experientia* **10**, 137 (1954).
35. J. P. Cornaz, K. Hutschneker, and H. Deuel, *Helv. Chim. Acta* **40**, 2015 (1957).
36. G. Petrie, D. Locke, and C. E. Meloan, *Anal. Chem.* **37**, 919 (1965).
37. A. Winston, C. R. Jenkins, W. Lerdthusnee, and S. J. Masten, Development and Evaluation of Ion-Exchange Resins for Removal of Specific Metals in Water Treatment Information Report 17, Water Research Institute, West Virginia University, Morgantown, West Va. (1981).
38. W. Kern and R. C. Schulz, *Angew. Chem.* **69**, 153 (1957).
39. M. Hatano, Y. Nose, T. Nozawa, and S. Kambara, *Kogyo Kagaku Zasshi* **69**, 571 (1966); *Chem. Abstr.* **65**, 15532g (1966).
40. A. Peden, H. Smith, and E. J. Vickers, Brit. Patent 887,175 (1962); *Chem. Abstr.* **57**, 3630a (1962).
41. F. Becke and C. Mutz, *Chem. Ber.* **98**, 1322 (1965).
42. L. W. Jones and L. Neuffer, *J. Am. Chem. Soc.* **39**, 659 (1917).
43. H. Smith, British Patent 852,176 (1960); *Chem. Abstr.* **55**, 9284i (1961).
44. G. M. Gasparini and S. Vomero, *Com. Naz. Energy Nucl., RT/CHI*, **70**, 25 (1970); *Chem. Abstr.* **74**, 112752f (1971).
45. F. Schouteden, *Makromol. Chem.* **27**, 246 (1958).
46. F. Schouteden, *Chim. Ind. (Paris)* **79**, 749 (1958); *Chem. Abstr.* **52**, 21116e (1958).
47. F. Schouteden and J. A. Herbots, Belgian Patent 560,782 (1958); *Chem. Abstr.* **53**, 5739f (1959).
48. F. Schouteden, *J. Soc. Dyers Colour.* **75**, 309 (1959).
49. F. Vernon and H. Eccles, *Anal. Chim. Acta* **82**, 369 (1976).
50. F. Vernon and H. Eccles, *Anal. Chim. Acta* **83**, 187 (1976).
51. C. A. Fetscher, U.S. Patent 3,154,499 (1964); *Chem. Abstr.* **62**, 4882d (1965).
52. C. A. Fetscher and S. A. Lipowski, U.S. Patent 3,345,344 (1967); *Chem. Abstr.* **67**, 109273a (1967).
53. R. S. Ramirez and J. D. Andrade, *J. Macromol. Sci., Chem.* **A7**, 1035 (1973).
54. R. S. Ramirez and J. D. Andrade, *Polym. Prepr., Am. Chem. Soc., Div. Polym. Chem.* **15**, 391 (1974).
55. R. S. Ramirez and J. D. Andrade, *J. Macromol. Sci., Chem.* **A10**, 309 (1976).
56. G. Schwarzenbach and K. Schwarzenbach, *Helv. Chim. Acta* **46**, 1390 (1963).
57. G. Anderegg, F. L'Eplattenier, and G. Schwarzenbach, *Helv. Chim. Acta* **46**, 1400 (1963).
58. G. Anderegg, F. L'Eplattenier, and G. Schwarzenbach, *Helv. Chim. Acta* **46**, 1409 (1963).
59. A. Winston and E. T. Mazza, *J. Polym. Sci., Polym. Chem. Ed.* **13**, 2019 (1975).

60. A. Winston and G. R. McLaughlin, *J. Polym. Sci., Polym. Chem. Ed.* **14**, 2155 (1976).
61. A. Winston and D. Kirchner, *Macromolecules* **11**, 597 (1978).
62. D. V. P. R. Vasaprasad, J. Rosthauser, and A. Winston, *J. Polym. Sci., Polym. Chem. Ed.* **22**, 2131 (1984).
63. M. I. Dawson, I. S. Cloudsdale, C. S. Tyson, S. Le Valley, and W. R. Harris, Progress toward the synthesis of polymerically bound chelating agents for iron(III) and the development of a new assay method for determining iron chelator effectiveness, in: *Development of Iron Chelators for Clinical Use* (A. E. Martell, W. F. Anderson, and D. G. Badman, eds.), pp. 201–209, Elsevier/North-Holland, Amsterdam (1981).
64. D. J. Dawson, D. O. Otteson, P. C. Wang, and R. E. Wingard, Jr., *Macromolecules* **11**, 320 (1978).
65. J. R. Pollack and J. B. Neilands, *Biochem. Biophys. Res. Commun.* **38**, 989 (1970).
66. W. R. Harris, C. J. Carrano, S. R. Cooper, S. R. Sofen, A. E. Avdeef, J. V. McArdle, and K. E. Raymond, *J. Am. Chem. Soc.* **101**, 6097 (1979).
67. K. N. Raymond, V. L. Pecoraro, and F. L. Weitl, Design of new chelating agents, in: *Development of Iron Chelators for Clinical Use* (A. E. Martell, W. F. Anderson, and D. G. Badman, eds.), pp. 165–187, Elsevier/North-Holland, Amsterdam (1981).
68. M. C. Venuti, W. H. Rastetter, and J. B. Neilands, *J. Med. Chem.* **22**, 123 (1979).
69. W. H. Rastetter, T. J. Erickson, and M. C. Venuti, *J. Org. Chem.* **46**, 3579 (1981).
70. M. I. Dawson, R. L. -S. Chan, I. S. Cloudsdale, and W. R. Harris, *Tetrahedron Lett.* **22**, 2739 (1981).
71. D. L. Fair, Iron Binding Catechol Polymers, Dissertation submitted in partial fulfillment of Ph.D. degree, West Virginia University, Morgantown, W. Va. (1981).
72. E. B. Brown, Candidate chelating drugs: Where do we stand?, in: *Development of Iron Chelators for Clinical Use* (A. E. Martell, W. F. Anderson, and D. G. Badman, eds.), pp. 47–59, Elsevier/North-Holland, Amsterdam (1981).
73. A. Jacobs, Screening for iron chelating drugs, in: *Development of Iron Chelators for Clinical Use* (A. E. Martell, W. F. Anderson, and D. G. Badman, eds.), pp. 39–46, Elsevier/North-Holland, Amsterdam (1981).
74. C. G. Pitt and A. E. Martell, The design of chelating agents for the treatment of iron overload, in: *Inorganic Chemistry in Biology and Medicine* (A. E. Martell, ed.), pp. 279–312, American Chemical Society Symposium Series 140, Washington, D.C. (1980).
75. A. E. Martell, The design and synthesis of chelating agents, in: *Development of Iron Chelators for Clinical Use* (A. E. Martell, W. F. Anderson, and D. G. Badman, eds.), pp. 67–104, Elsevier/North-Holland, Amsterdam (1981).
76. K. N. Raymond, W. R. Harris, C. J. Carrano and F. L. Weitl, The synthesis, thermodynamic behavior and biological properties of metal-iron-specific sequestering agents for iron, in: *Inorganic Chemistry in Biology and Medicine* (A. E. Martell, ed.). pp. 313–332, American Chemical Society Symposium Series 140, Washington, D.C. (1980).
77. W. R. Harris and K. N. Raymond, *J. Am. Chem. Soc.* **101**, 6534 (1979).
78. T. Emery, *Amer. Scient.* **70**, 626, 1982.
79. S. Pollack, P. Aisen, F. D. Lasky, and G. Vanderhoff, *Brit. J. Haematol.* **34**, 231 (1976).
80. G. P. White, A. Jacobs, R. W. Grady, and A. Cerami, *Brit. J. Haematol.* **33**, 486 (1976).
81. G. P. White, A. Jacobs, R. W. Grady, and A. Cerami, *Blood* **48**, 923 (1976).
82. R. W. Grady and A. Jacobs, The screening of potential iron chelating drugs, in: *Development of Iron Chelators for Clinical Use* (A. E. Martell, W. F. Anderson, and D. G. Badman, eds.), pp. 133–164, Elsevier/North-Holland, Amsterdam (1981).
83. A. Cerami, R. W. Grady, and C. M. Peterson, New iron chelators, in: *Chelation Therapy in Chronic Iron Overload* (E. C. Zaino and R. H. Roberts, eds.), pp. 37–52, Symposia Specialists Inc., Miami, Fla. (1981).

84. C. G. Pitt, Structure and activity relationships of iron chelating drugs, in: *Development of Iron Chelators for Clinical Use* (A. E. Martell, W. F. Anderson, and D. G. Badman, eds.), pp. 105–131, Elsevier/North-Holland, Amsterdam (1981).
85. A. Winston, J. Rosthauser, D. Fair, J. Bapasola, and W. Lerdthusnee, Design of polymeric iron chelators for treating iron overload in Cooley's anemia, in: *Biological Activities of Polymers* (C. E. Carraher, Jr. and C. G. Gebelein, eds.), pp. 107–117, American Chemical Society Symposium Series 186, Washington, D.C. (1982).
86. M. R. Summers, A. Jacobs, D. Tudway, P. Perera, and C. Ricketts, *J. Haematol.* **42**, 547 (1979).
87. R. D. Propper and D. G. Nathan, Use of desferrioxamine and "the pump", in: *Chelation Therapy in Chronic Iron Overload* (E. C. Zaino and R. H. Roberts, eds.), pp. 17–35, Symposia Specialists Inc., Miami, Fla. (1981).
88. J. W. Rosthauser and A. Winston, *Macromolecules* **14**, 538 (1981).

Note added in press: Since the completion of the writing of this review, the following papers have been prepared by the author and have appeared or will appear shortly in the journals cited.

A. Winston, Bioactive Hydroxamic Acid Polymers for Iron Chelation, *Polymer News* **10**, 6 (1984). A review.

P. Desaraju and A. Winston, Synthesis of a Polyhydroxamic Acid by Interfacial Polymerization, *J. Polym. Sci., Polymer Letters Ed.* (in press).

A. Winston, D. V. P. R. Varaprasad, J. J. Metterville, and H. Rosenkrantz, Evaluation of Polymeric Hydroxamic Acid Iron Chelators for Treatment of Iron Overload, *J. Pharmacol. Exp. Ther.* (in press). A report of a series of hydroxamic acid polymers prepared by the author and associates and tested for bioactivity in removing iron from iron overloaded mice.

22

Biological Activities and Medical Applications of Metal-Containing Macromolecules

Charles E. Carraher, Jr.

Abstract. The potential and actual applicability of metal-containing polymers as drugs in medical and nonmedical applications is described. Synthetic variations are presented focusing on the ability to generate a wide variety of biologically active polymers emphasizing a coupling of desired biological activity and polymer structure. A number of modes of activity are presented and discussed in view of published data.

1. Introduction

The use of metal- and organometallic-containing polymers in medical applications is widespread focusing on siloxane polymers and to a lesser degree on polyphosphazenes. This work deals with the use of these polymers as medical materials in applications such as biomedical implants as catheters, blood pumps, and breasts. Here we will concentrate on a new, emerging area – the use of organometallic polymers as drug delivery (controlled-release or direct) agents. A number of topics will be covered including philosophy, targeting, synthesis, and modes of action.

2. Philosophy

The toxicity as well as therapeutic value of metals is well known. The interaction of metal ions with biological macromolecules such as proteins and nucleic acids is a continuing area of research. The appearance of metal-containing macromolecules in the human body is extensive, including the metals of iron (transferrin), molybdenum (xanthine oxidase), vanadium

Charles E. Carraher, Jr. • Department of Chemistry, Wright State University, Dayton, OH 45435.

Table 1. General Biological Uses for Metal-Containing Drugs

Metal	Medical usage
Au	Arthritis
Ag	Antiseptic agent, prophylactate
As, Sb	Bactericides
Bi	Skin injuries, alimentary diseases, diarrhea
Co	Vitamin B_{12}
Cu	Fungicide, insecticide, algicide
Ga	Antitumor agent
Hg	Antiseptic
Li	Manias
Mn	Parkinsonism, fungicide
Pt	Antitumor agents
Rb	Substitute for K in muscular dystrophy; protective agent against adverse effects of heart drugs
Ru, Rh, Pd, Os	Antitumor agents (experimental)
Sn	Fungicide, bactericide
Ta, Si	Inert medical applications as gauzes, implants
Tl	Poison
Zn	Fungicide

(hemovanadin), zinc (carbonic anhydrase), and copper (hepatocuprein). The use of organometallic medicinals is widespread and includes merbromine (mercurochrome; mercury), meralein (mercury; antiseptic), silver sulfadizine (prophylactic treatment for severe burns), arsphenamine (antimalarial; arsenic), 4-ureidophenylarsonic acid [therapy of ameblasis, tryparsamide (Gambian sleeping sickness; arsenic], and antimony dimercaptosuccinate (schistosome). Table 1 contains additional uses.

Tables 1–3 contain a summary of useful general information regarding the biological activities of metals, metal oxides, metal salts, and, where available, organometallic compounds. Table 2 contains a general listing of toxicity on a 1 (nontoxic) to 5 (quite toxic) scale and carcinogenicity also on a 1 to 5 scale. Regarding the carcinogenicity scale, a value of 1 indicates that the compounds are not cancer causing; 2 indicates there might be some potential through animal testing that a very few select compounds can cause an increase in tumor activity; and 3 and above indicate that there is good evidence that at least select metal-containing compounds can cause cancer in humans. A value of 3 is given for uranium since there is adequate evidence that long-term exposure will induce cancer after a prolonged incubation period (as long as five to six decades); however, there is little evidence that short-term exposures will induce cancer). Where possible, biological half-lives are given in Table 2. Real values are most dependent on the location of the material. For example, mercury vapor was inhaled and whole-body counter readings gave average half-times (or lives) for mercury clearance of 1.7 days for the lung, 21 days for the head, 43 days for the chest, and 64 days for the kidney.[55] Thus a general average of one month for the mercury half-life is reported in Table 2. Table 2 also contains state-of-the-art analysis techniques for each metal emphasizing techniques employed in the analysis of biological samples.

Toxicity is dependent on the mode of administration and on the form of the material. Thus zirconium compounds exhibit an LD_{50} of several grams per kilogram of body weight when administered orally, with the toxicity increasing to 20-fold when the same compounds are administered intraperitoneally. Toxicity is almost always greater when a compound is accepted into the body as a dust or mist and is greater when administered in a solution and/or in a soluble form.

While the values listed in Table 2 and properties given in Table 3 are general, varying with the compound, it is of interest to note the coincidence of a property as the particular metal is varied from being present in an inorganic salt to being contained within an organometallic compound. Thus, while given biological responses vary in degree, as the nature of the metal-containing compound changes, the kind of response is often similar, emphasizing the importance of the metal. This is only a generalization and each compound must be specifically investigated.

In general organometallic compounds are more toxic than the metal salts which are generally more toxic than the metal oxide. Organometallics have an increased tendency to locate in fatty tissues and organs with toxicity decreasing as the size of the alkyl chain increases, that is, the toxicity is $CH_3 > C_2H_5 > C_3H_7 > C_4H_9 > C_5H_{11}$, and so on. Also of interest is the potential inclusion of trace amounts of metals, as the presence of iron, silicon, and copper as trace impurities in even highly purified aluminum compounds.

Information given in Table 3 may act to signal both sites and mode of activities which may be utilized in the design of a specific drug. For example, chromium tends to seek the spleen and exhibits a high affinity for complexing pyrophosphates. Thus a chromium-containing compound might be considered to correct an excess of pyrophosphates in the spleen. The information given in Table 3 is to be utilized as only a "key" to a generation of creative possibilities and the reader is encouraged to seek out full reviews of these metals, including recent NIOSH reports.

3. Synthesis

Objectives in the design of chemotherapy drugs include that the drugs

Table 2. Toxicities: Analytical Techniques and Related Values for Selected Metals

Metal	Carcinogenicity	Toxicity	Biological half-life	Average exposure limits (TLV values) (mg/m^3)	Preferred analysis technique(s)	Lower limit for detection (g/ml)	References for determination
Al	1	2		10^0	Many	10^0	1
Sb	1	2		10^1	Spectrochemical	10^{-1}	2
As	2	2		10^{-2}	Chemical, AA[a]		3
Ba	1	5			AA, spectrographic	10^{-1}	4, 5
Be	4	4		10^{-1}	AA	10^{-2}	6
Bi	1	1		10^1	AA	10^0	7
Cd	2	2		10^{-2}	AA	10^{-1}	8, 9
Cs	1	1	40 days		AA	10^{-2}	10
Cr	2	2	60 days	10^{-4}	Chemical	10^0	11, 12
Co	1	2		10^1	Ion exchange	10^{-3}	13
Cu	1	1 (excluding mists, dusts)		10^0	AA	10^{-1}	14
Ga	1	1	2 weeks		AA	10^0	10
Ge	1	1		10^{-1}	Chemical	10^0	15
Au	1	1	1 month		AA	10^{-1}	16
In	1	5	4 weeks	10^{-1}	Spectrographic	10^0	17
Fe	1	—		10^1	Spectrophotometric	10^0	18, 19
Pb	2	3		10^{-1}	AA	10^0	1, 20
Li	1	3		10^{-5}	AA	10^{-1}	14
Mg	1	1		10^1	AA	10^{-2}	14
Mn	1	3	2 months	10^0	AA	10^2	14
Hg	1	5 (organometallic)	1 month	10^{-2}	Many	10^1	1
Mo	1	2	2 weeks	10^1	Chemical	10^{-1}	21
Ni	4	3		10^{-1}	AA	10^{-1}	22
Nb	1	2	3 months		Chemical	10^{-2}	23
Pt	1	2	1 week	10^{-3}	AA	10^0	14, 25, 26

Table 2. (*Continued*)

Ru	1	2	1 month	10^{-3}	Spectrographic	10^{0}	24
Rh	1	2		10^{-3}	AA	10^{0}	14
Os	1	2			Spectrophotometric	10^{0}	24, 27
Pd	1	2			AA	10^{0}	25, 26
Rb	1	1			AA	10^{-3}	28, 29
Ag	1	3 (salts)		10^{-1}	AA	10^{-2}	30
Sr	1	2		10^{0}	AA	10^{0}	31
Ta	1	1	4 years (solid) 3 months (soluble)	10^{0}	MS	10^{0}	32
Tl	1	4		10^{-1}	AA, polarographic	10^{-1}	33, 34
Th	5	2			Chemical	10^{0}	35
Sn	1	3		10^{-1}	AA, optical	10^{-2}	36
Ti	2	2			AA, H_2O_2	10^{0}	37
W	1	2	2 months	10^{0}	AA	10^{2}	38
U	3	3		10^{-1}	Neutron act.	10^{2}	1, 39, 40
V	1	2	2 days	10^{0}	AA	10^{-3}	41
Zn	1	2		10^{0}	AA	10^{-2}	42
Zr	1	1		10^{0}	AA, emission spec.	10^{-2}	14, 43, 44

[a] AA = AAS = atomic absorption spectroscopy.

Table 3. General Biological Activities of Metal and Metal-Containing Compounds

Metal	Results of acute toxicities	Results of chronic toxicities	Metabolism and mode of action
Al	Increase in blood glucose, decrease in liver glycogen-soluble forms; insoluble—little or none; alkyls—very reactive with tissue.		Reacts with phosphates and is excreted.[1]
Sb	Gastrointestinal disturbances; eosinophilia, myocardial failure; nausea, vomiting, diarrhea.	Pneumonitis, fatty degeneration of liver, decreased white blood cells.	Concentrates in liver, red blood cells, thyroid, parathyroid; interference with cellular metabolism through combination with sulfhydryl groups in respiratory enzymes.
As	Violent gastroenteritis, vomiting, headache, myocardial failure, anuria, coma, edema, anemia.	Skin, mucous membranes, gastrointestinal and nervous system damage.	Widely distributed in tissues; highest in liver, kidney; combines with SH-containing materials such as cysteine and glutathione; moderated by presence of Se and S.
Ba	Affects central nervous system; hemorrhaging in stomach, intestines, kidneys.		Soluble – permeates gastrointestinal tract into the bloodstream; deposited in muscles, lungs; none in brain, heart, hair; binds with proteins.
Be	Pneumonitis, chest pain, granulomata, pulmonary tumors, bone sarcoma, rickets.	Pneumonitis, granulomata.	Poorly absorbed in gastrointestinal tract; attracted to skeleton; transported to all tissues; storage long term; affects lysosomes, destroys cells.
Bi	Small granulomatous lesions.	Gingivitis, ulcerative stomatitis.	Concentrates in all organs, kidney affinity.
Cd	Powerful emetic, nausea, vomiting, diarrhea, headache.	Pulmonary edema dyspnea, proteinuria, loss of sense of smell, cough, anemia; damage to liver and kidneys.	Long lived, concentrates in liver and kidney; increase in prorainent connective-tissue fiber bundles; thickening of subpleural elastic layer; deposition of fibrous and proteoglycan elements in

			interstitial space; binds sulfate and carbonyl of proteoglycan, amino and amide groups of fibrillar unit, occupies site of other metals normally held by proteoglycan fibrillar; interferes with formation of intramolecular bonds through interaction with Cu in lysyloxidase.
Cs	Seldom, if ever.	Seldom, if ever.	Similar to other Group IA, widely distributed; most tenacious retention in muscle; low concentration in blood.
Cr	Dermatitis, nasal, lung, and larynx inflammation; toxicity $Cr^{+6} > Cr^{+3}$ and Cr^{+2} forms; anuria.	Kidney, pulmonary damage; hyperemia, enlarged liver.	Concentrates in spleen, bone, kidney, liver; reacts with enzymes, great affinity for pyrophosphate; interacts with nucleic acids.[15]
Co	Diarrhea, paralysis, lowers blood pressure.	Pulmonary disease.	Essential trace element; exact variety of physiologic activity varying with Co concentration; involved in regulation of sulfhydryl concentration.
Cu	Suppression of urine, jaundice, hypertension.	Injury to liver, kidney, spleen.	Essential trace metal; involved with oxidative enzymes and other enzymes.[45–47]
Ga	Anorexia, nausea, vomiting.	Decrease in lymphopenia, skin rash.	Mostly absorbed by skeleton or rapidly excreted in urine.
Ge including hydrides	Low except as gases, then no affect to blood or nervous system; respiratory depression.	Low except as gases.	Affect on water balance leading to fall in blood pressure, dehydration, hemoconcentration and hypothermia.
Au	Low.	Papular eruption, erythema nodosum, allergic contact, allergic contact purpura pityriasis rosea.	Absorbed into blood, then transported by plasma, taken to all tissues, mainly kidneys, liver, skin, hair, nails.

(continued)

Table 3. (*Continued*)

In	Convulsive motions, nosebleed, increased reflexes.	Weight loss, pulmonary edema, necrotizing pneumonia, blood damage, liver and kidney damage.	Damage to most organs, tissue distribution uniform; varies with form and administration route; direct calcifier.
Fe	Orally—not toxic; introduced directly into bloodstream— instantaneously toxic, causing respiratory failure, anorexia, oligodipsia, oliguria, alkalosic.	Benign pneumoconiosis.	Essential metal; uptake by the exogenous of excess electrons donated by ferric reductase in the mitochondrial membrane.[48]
Lanthanides (general)	Labored and depressed respiration, increased prothrombin.[49,50]	Some liver damage.[49]	Protein precipitation and complexing.
Pb	Fatigue, loss of sleep; constipation, colic, anemia, neuritis; metallic taste; diarrhea.	Emulative poison—anemia, central nervous system disorders, renal disorders, reproductive disorders.	Mitotic abnormalities in bone-marrow cells, chromosomal aberrations, affinity for nucleic acids, effect on immune mechanism, inhibits red-cell pigment heme.
Li	Toxicity directly related to amount of Na inverse relationship; nausea, vomiting, abdominal pain.	Azotemia, nonreversible toxicity, gastroenteritis, reduced body temperature, atoxin, giddiness, muscular weakness.	Not absorbed through human skin; soluble salts absorbed from gastrointestinal tract; evenly distributed in organs; interacts with neurohormones.
Mg	Low; leukocytosis.	Low.	Essential metal; involved in neuromuscular conduction of skeletal and cardiac muscle, activates many enzymes; catalyzes reactions with ATP; essential for integrity of cell mitochondria.[51]
Mn	Pulmonary edema; toxicity Mn^{+2} > Mn^{+3} in general.	Psychotic diseases, brain disorders, Mn disease, pulmonary disease, pleuritis, pneumonia.	Essential metal, enzyme action, intestinal uptake from lumenal transfer to mucosal surfaces.[1,52]
Hg	Gastrointestinal erosion, elevated temperature, shallow respiration, general malaise, chest pain, pharyngitis, nausea.	Psychic and emotional disturbance, tremors.	Retained in kidney, liver; organics—brain, bound to erythrocytes; inorganic—plasma; affinity for sulfhydryl—cont. enzymes; general binding to enzymes.

Mo	Low (+4), higher (+6); loss of appetite, weight; diarrhea, muscular coordination.	Loss of appetite, listlessness, diarrhea, reduced growth rate.	Seeks all tissues, bone; bound to red blood cells, plasma proteins; antagonism with Cu; aid in retention of F in bone and soft tissue of old rats; small amounts increase antibody formation.
Ni	Mild, nonspecific.	Contact and atopic dermatitis; allergic sensitization; cancers of the lung and nasal sinuses.	Essential; widespread – liver, larynx, kidney, heart, lung, skin, intestine; metabolism regulated by strict homeostatic mechanisms; can cross human placenta; binds to RNA, may depolymerize RNA, bound to phosphate and bases; binds to proteins changing configuration and activity; inhibits ATPase and RNA polymerase.
Nb	Urination and defecation decreased.	Glycosuria, lower urinary protein.	Interfering with activating, respiration metal enzyme cofactors such as ATP.
Pd	Tonic and clonic convulsions.	Proteinuria.	Induced hemolysic and albuminuria; concentrates in kidneys, lungs, bone marrow, spleen, muscle, with damage to kidneys, liver and bone marrow.
Pt	Vomiting, diarrhea, bloody stools, epileptiform convulsions, coma, heart action delayed.	Loss of hearing, renal damage, asthma.	Rapid clearance from gastrointestinal tract, very limited passage through blood–brain barrier; binds to cell walls, nucleic acids, and proteins; concentrates highest in kidney, liver, ovary, and uterus; immunosuppressive.
Os	Injurious to eyes.	Irritation of mucous membranes, semicomatose condition, pulmonary embarrassment.	Binds to DNA and RNA.
Ru	Injurious to eyes and lungs.	Injuries to eyes and lungs.	Antitumoral ability involves interference with mitochondrial transport of Ca. Also found for Pt; immunosuppressive.

(continued)

Table 3. (*Continued*)

Rb	Mostly low; hemorrhages, adhesions; congested, cyanotic lungs.	Mostly low; greater excitability.	Alters heart-muscle contractions; often replaces K; rapidly absorbed from intestine and distributed chiefly to muscle.[1]
Ag	Low; only airborne potential problem for Ag itself, salts; are often highly toxic to most microorganisms. Irritating, kills skin, mucous membranes, eyes.	Argyria, bronchitis.	Silver, itself—none; salts—absorbed from respiratory and gastrointestinal tracts; widely distributed throughout body.[53]
Sr	Respiratory failure.	Histological changes in lungs, liver, kidneys, spleen; hyperemin, hemorrhage.	Omnipresent in human tissues; causes disorder in mineral metabolism; stored in bone; association with and often substitution of Sr for Ca; preferential excretion of Sr over Ca by kidney; crosses placental barrier; interacts with phosphate, Mg, and K; can activate and enhance transmitter release.[1]
Ta	Largely nontoxic; pneumoconiosis, dystrophic changes in lungs and parenchymatous organs.	Largely nontoxic; dust-hypertrophic focal emphysema; bronchia epithelial hyperplasia.	Concentrates in kidney, bone; soluble—excreted rapidly; tracheobronchial passage initially followed by a prolonged alveolar clearance phase.
Tl	Hemorrhages, headache, abdominal pain; polyneuritis, epilation, gastrointestinal symptoms; encephalopathy and retrobulbar neuritis; delirium, hallucinations—all appear to be delayed responses.	Alopecin, loss of hair, epilation, degeneration of nerve cells, axons, and myelin sheaths.	Concentrates in kidney, spleen, lung, and brain; absorbed through digestive tract and skin, then distributed throughout body; no inhibition of nonspecific enzyme system; exciter of preganglionic parasympathetic nervous system; diminished actions of adrenaline and acetylcholine; typical heavy metal poisoning.[1]
Th	Generally low; abnormal leukocytes.[54]	Tumors—squamous and transitional cell types most usual; long latent period	Hydrolyzed to form chain; binds to bovine cortical bone glucoprotein,

			cont. species; binds chondroitin sulfate protein, possible interaction of Th with Zn.
Sn (including organics)	Diarrhea, paralysis, degeneration in epithelium of the renal proximal tubules, headache, muscular weakness.	Benign pneumoconiosis.	Concentrates in liver, lung, kidneys; complexes protein; ultimately excreted in urine; depressed hemoglobin and serum Fe; inducer of hemeoxygenase in kidney.(1)
Ti	Lowering of blood pressure; respiratory distress; generally nontoxic.	Increase in leukocyte count; possible tumors by Cp_2TiCl_2.(1)	Inhibits serum alkaline phosphatase; complexes with tyrosinase; capable of substituting from V, Fe, Co, Ni, Zn.
W	Quite different = soluble > insoluble forms; respiratory problems, blood-stained oral and nasal discharges, necrosis of skin and mucosa.	Hyperplastic lymph nodes, thickening of alveolar walls, perivascular infiltration by lymphocytes.	Concentrates in lungs, kidneys; potential inhibitor of Mo; may replace Mo; acts at enzyme sites often at SH groups.
U	Severe kidney degeneration.	Cancer after many years.	Concentrates in liver, kidney; cell membrane; UO_2^{2+} complexes with most ionic Lewis bases, including those in proteins and nucleic acids (phosphate); inhibits mucosal transfer of glucose and galactose.(1,39,40)
V	Pulmonary edema, irritants to mucous membranes, coughing.	Fatty degeneration of liver, pneumonia, dilated alveoli, tracheitis.	Increases oxidation of fatty acids of phospholipids; limited lower plasma cholesterol; reacts with SH cont. proteins.(1)
Zn	Damage to buccal and gastroenteric mucous membranes; fever, nausea, vomiting, cramps, diarrhea.	Damages pancreas; anemia.	Present as a metalloenzyme, helps control CO_2 exchange; metal–protein complexes—as RNA polymerase, superoxide dismutase, carboxypeptidase, and isocitric dehydrogenase.
Zr	Low; depression, decreased activity.	Low.	Interacts to charge concentration of essential metals.

have (1) good specificity (differentiation), (2) good activity, (3) a long duration of activity, and (4) a wide concentration range of biological activity. Thus variability and specificity are key objectives. Metal-containing polymers, because of the variety of metal size, chemical environment, and electronic structure, offer a wide variability of biological activities and the opportunity for good specificity.

Metals can be incorporated into polymers through numerous routes the major ones including addition, coordination, and condensation reactions. The general area of organometallic polymers has been recently reviewed.[56] Here the discussion will focus on appropriate biomedical aspects.

Most metals exist in aqueous soluble cationic forms which are readily complexed by a host of Lewis Acids, including monomeric difunctional species (such as bis-1,2-amino acids, bis-diamines, and bis-1,2-dioximes; a listing of such general classes is contained in Table 3 in Reference 56) and polymers (such as polyethylene amine, polyvinyl alcohol, polyacrylic acid, dextran, cellulose, proteins, nucleic acids, and polyphosphates). Much is known about the structure and stability of such complexes through analytical attempts to determine these metals employing wet chemical, chromatographic, and other techniques. Metal coordination polymers can also be synthesized through use of preformed metal complexes polymerized through functional groups where the actual polymer-forming step can be a condensation or addition reaction.

Possibly the least effective route with regard to product variability is the addition route. Basically the metal-containing moiety is contained in either or both the X and Y. When X = Y, the polymer is called a homopolymer; when X ≠ Y, the polymer is called a copolymer.

$$\underset{\mathbf{1}}{\mathrm{H(X)C{=}CH_2}} + \underset{\mathbf{2}}{\mathrm{H(Y)C{=}CH_2}} \longrightarrow \underset{\mathbf{3}}{\mathrm{\text{-}(CH_2\text{-}CHX)_n\text{-}(CH_2\text{-}CHY)_m\text{-}}}$$

A wide number of varying structures can be synthesized including blocks, blends, alternating copolymers, tripolymers, and composites. For all of these, the variability as to the nature of X and Y has been severely limited. Thus a number of ferrocene, manganese carbonyl, and organostannane esters have been employed, but each presents the chemist with its own polymerization problems. Thus far only a limited number of metal-containing homopolymers have been synthesized due to the electronic nature of the metal-containing moiety. Further, the bulk of the presently employed metal-containing units effectively shield the metal from ready biological access with the metal "covered" by biologically neutral, inert organic groupings. Even so, success has already been achieved in select areas. The use of organostannane esters contained in addition polymers is described elsewhere in this chapter.

Synthesis of most of the metal-containing condensation polymers can be considered in Lewis acid–base terms as described below, where A has been R_3Sb, R_3As, R_3Bi, R_2Pb, R_2Sn, R_2Ge, R_2Si, R_2Ti, R_2Zr, R_2Hf, R_2Mn, and MoO_2 and Y—B—Y has been a hydrazine, hydrazide, urea, amine, oxime, amidoxime, diol, dithio, salt of a diacid, sugar, cellulose or if contained on a polymer the Lewis base has contained, an alcohol, amine, sulfate, sulfonate, salt of an acid, and amidoxime.

$$\underset{\mathbf{4}}{X - A - X} + \underset{\mathbf{5}}{Y - B - Y} \rightarrow \underset{\mathbf{6}}{-(A - B)-}$$

For reaction to occur, the Lewis bases must be in what can be referred to as an "active form". There should be a match between the number of "active" Lewis-base sites such that the final structure of the product will be linear or cross-linked, depending on the particular intent. For instance, xanthene dyes typically exhibit three Lewis bases (two phenols and one carboxylic acid), but through a wide pH range (often from about 2–13) only two of these Lewis bases are active-forming "polydyes" of form **7** from condensation with Group IVA and IVB organodihalides giving a linear product. A cross-linked product would result from the use of a Lewis acid or base containing an "active" functionality greater than 2.

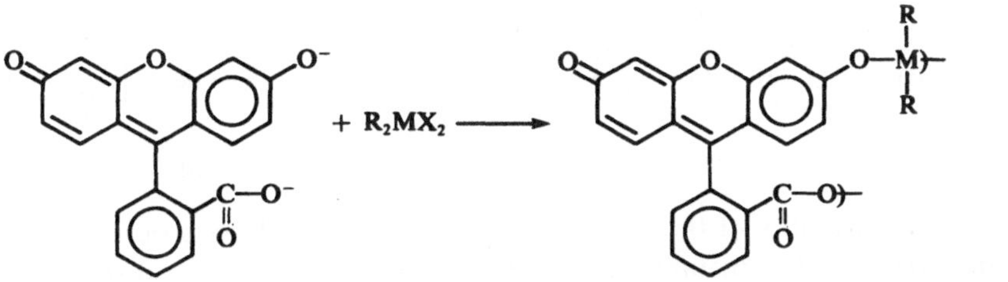

+ R_2MX_2 ⟶

7

The Lewis base can contain like functions as in stilbestrol (*trans*-4,4′-dihydroxystilbene; **8**)

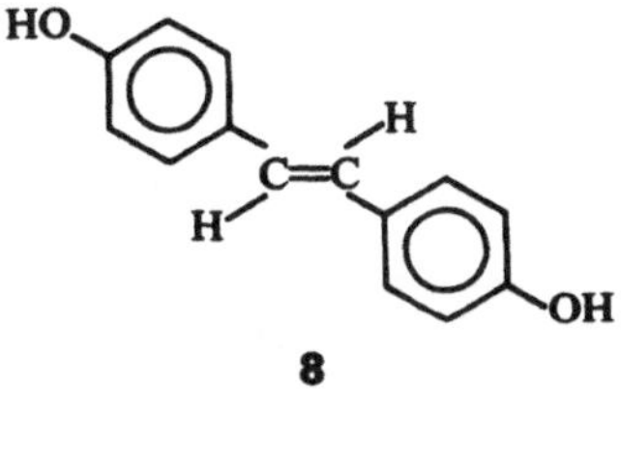

8

or dissimilar functions as in the case of xanthene dyes, *p*-aminobenzoic acid (vitamin $\mathbf{B}_x$; antrickettsial; **9**) and salicylic acid (analgesic and antirheumatic; **10**)

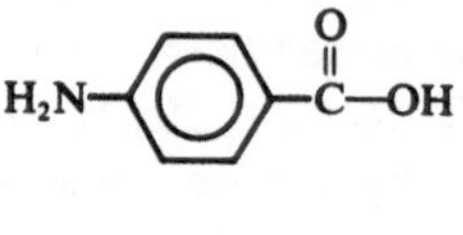

9

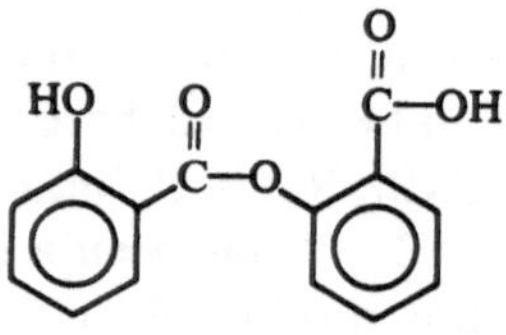

10

or chemically converted from one active form to another as in the case of progesterone (important progestin; **11**).

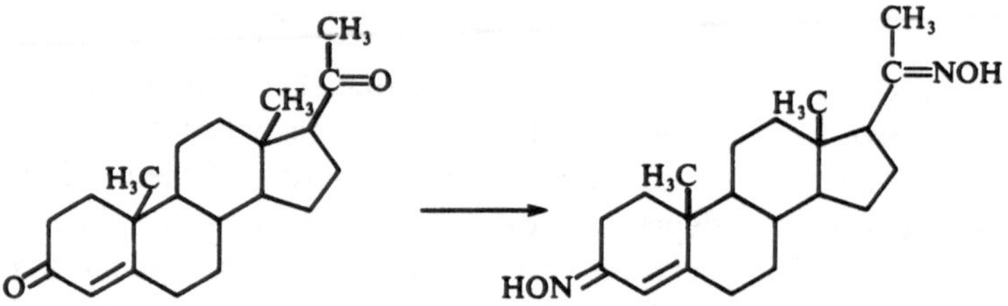

11 **12**

The Lewis base has also been utilized as a metal carrier with polymers containing the metals Co, Fe, Ru, and Rh (**13**, **14**).

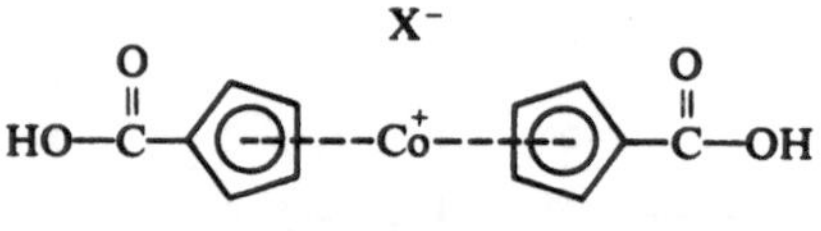

13 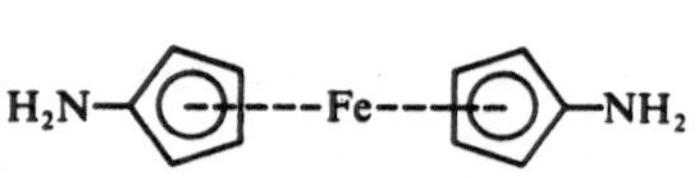 **14**

Desirability of the location of the drug also will vary depending on its intended use. For instance, the drug may be an integral part of the polymeric backbone as in **17–23** or within a side chain near the backbone.[16]

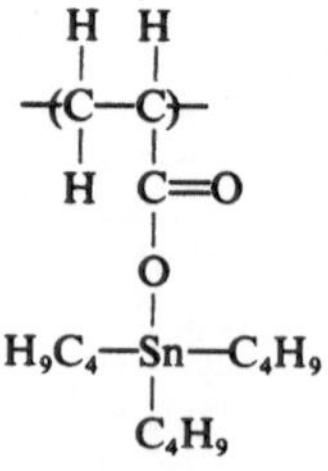

15 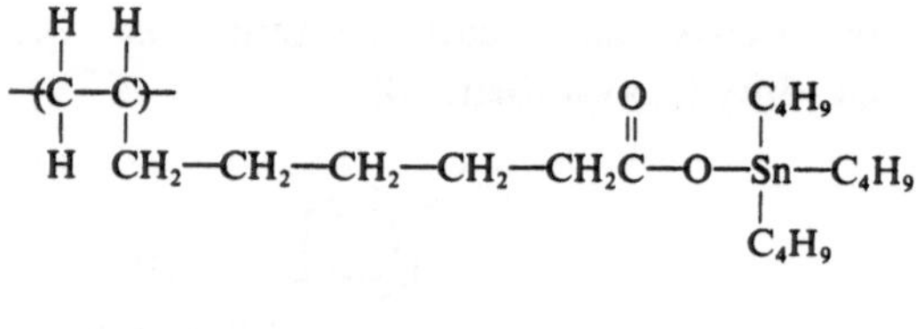 **16**

The desired portion to be delivered (i.e., the drug) may be either or both the metal-containing moiety or the organic comonomer. For instance, our purpose in the synthesis of a number of Group IVB polymers containing steroids (**17**) and derivatives of vitamin K (**18**), is to use such polymers as delivery agents for the nonmetal comonomer portion utilizing the relatively biologically inactive Group IVB Cp_2M moiety.[56]

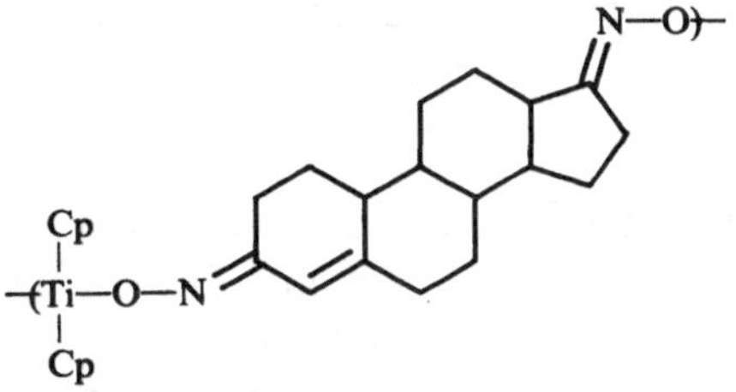

17 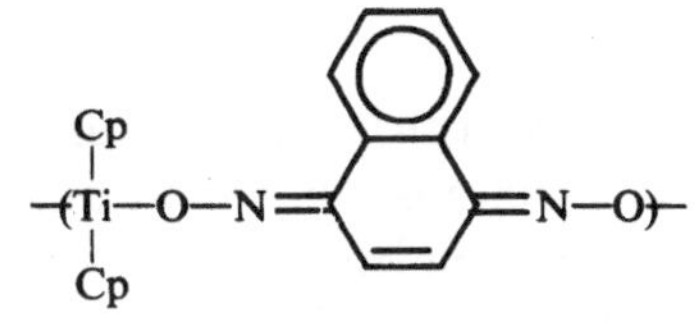 **18**

Polymer **19** was synthesized to deliver both the manganese moiety, which is an essential metal, and the pyrimethamine portion (**19**), which is an antimalarial and antimicrobial agent.[57] Here the delivery of both portions may be advantageous not necessarily through "toxic" routes but through "assisting" routes. Contrary to this would be the synthesis of **20** where the kynurenic acid portion is added to assist in overcoming a vitamin B deficiency with the antimony moiety added as a toxin or modifier of enzymes containing thiol groups.

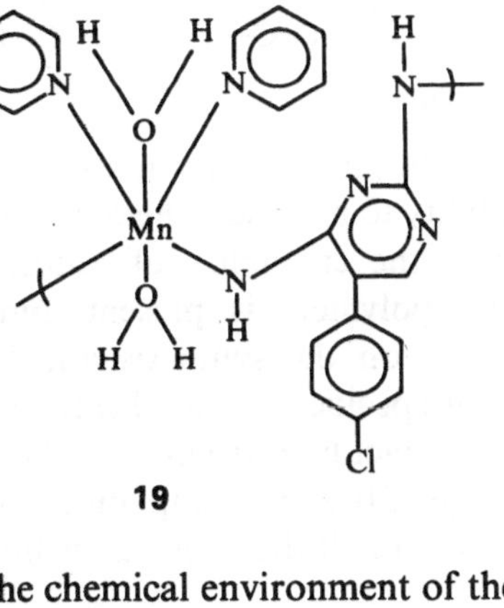

19 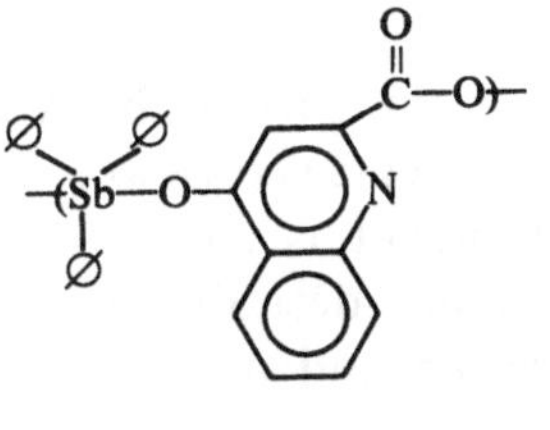 **20**

The chemical environment of the drug can be varied as to its intended use. The chemical (for hydrolysis, enzyme, and the like, reaction) and physical (size, hydrolysis, solubility) environment about the polar linkage can be unobstructed and the hydrophobic character minimal or through use of hydrophobic blocking groups the polar linkage can be sterically hindered and highly hydrophobic, and so on.[58,59] Thus both intended preferential biological location of the drug (blood, lipid, and the like, compatability) and rate of drug release (from fast to the polymers themselves acting as the delivered agent) can, in theory, be built into a polymer.

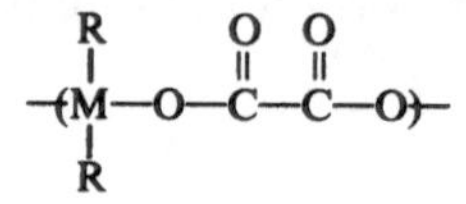

$$-\!\!(\underset{R}{\overset{R}{M}}-O-\overset{O}{\overset{\|}{C}}-\overset{O}{\overset{\|}{C}}-O)\!\!- \qquad -\!\!(\underset{R}{\overset{R}{M}}-O-\overset{O}{\overset{\|}{C}}-C_6(CH_3)_4-\overset{O}{\overset{\|}{C}}-O)\!\!-$$

21 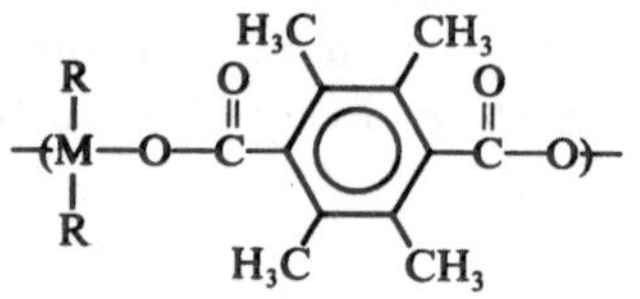 **22**

Many of the employed metal-containing Lewis-acid reactants undergo certain reactions for which analogous products are not formed through attempted condensation with acid chlorides utilizing mild reaction conditions. Thus Cp_2TiCl_2 reacts with salts of dicarborylic acid giving titanium-containing polyesters, whereas the reaction with organic diacid chlorides does not yield the analogues condensation product under similar reaction conditions.[58,59] This is an important consideration if a drug such as **9**, **10**, and **13**, which contains carboxyl groups, is to be delivered utilizing a condensation monomer or polymer.

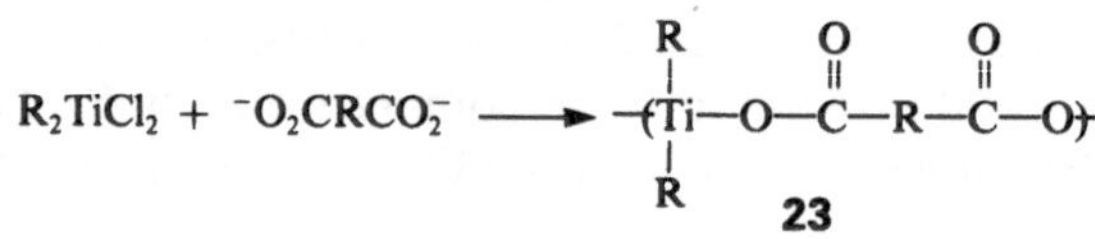

$$R_2TiCl_2 + {}^{-}O_2CRCO_2^{-} \longrightarrow -\!\!(\underset{R}{\overset{R}{Ti}}-O-\overset{O}{\overset{\|}{C}}-R-\overset{O}{\overset{\|}{C}}-O)\!\!-$$

23

$$ClOC-R'-COCl + {}^{-}O_2CRCO_2^{-} \not\longrightarrow -\!\!(\overset{O}{\overset{\|}{C}}-R'-\overset{O}{\overset{\|}{C}}-O-\overset{O}{\overset{\|}{C}}-R-\overset{O}{\overset{\|}{C}}-O)\!\!-$$

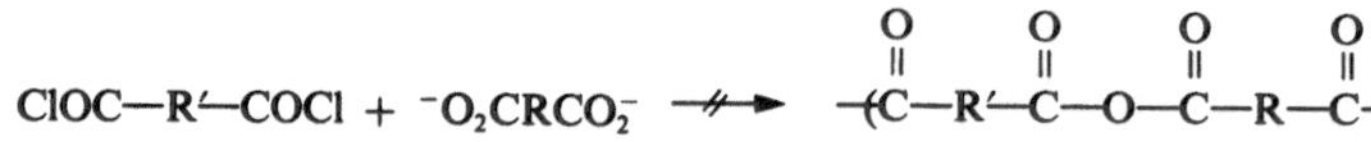

24

In this section the synthetic ability to place metals in known chemical environments is reviewed. While the ability to place any metal in any desired environment is still not accomplished, the capability of tailoring many potentially desirable metal-containing polymers is present and rapidly increasing and it is time to concentrate on the synthesis and biological characterization of specific polymers for specific drug applications.

Biological activity is dependent on exact fine structure of the polymer, molecular weight and distribution; nature of the end group; mode and site of delivery; and so on—a Pandora's box where all that remains is hope and a lot of perspiration.

While little is known concerning the particular biological activities of metal-containing polymers much more is known about the biological activities of the metals themselves, oxides, salts, and certain organometallic compounds. For a first approximation it may be useful to consider the gross biological activity of such metal-containing compounds, allowing extrapolation for potential activity when the metal is present in a polymer as cited in the previous section. Thus, almost all tin-containing monomeric and polymeric substances inhibit some bacteria and it is reasonable to assume that a new tin-containing product should also exhibit some antibacterial properties.

4. Targeting

The area of selective targeting of drugs is currently the weakest link in the overall drug delivery system. Work is involved in the use of coupled drugs (insulin tends to make cell walls more permeable, thus it is being investigated in the delivery of anticancer drugs), embedding in synthetic and natural "homing devices" (liposome-embedded drugs), and in the design of drugs with both the targeting and drug activity built in. While some search for the "magic bullet" which will send a drug directly to a specific site has been done, this search has only started to deal with liposomes, hemaproteins, and so on. Of more immediate use is targeting by size. Containment within the general circulatory system occurs for many polymers with chain lengths of about 10^2.

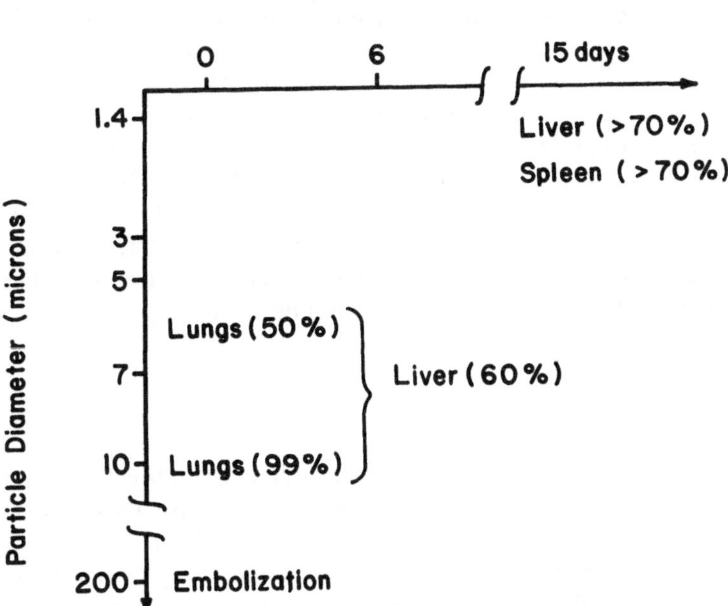

Figure 1. Preferential targeting by particle size.

Figure 1 contains a general scheme relating containment by select organs as a function of size and time regarding drug administration through intravenous injection. An estimation of the chain length needed to achieve such targeting is roughly calculated utilizing the relationship that the root-mean-square end-to-end distance is equal to the square root of the number of units times the root mean square of the unit's length. For nylon-66, a chain length of 10^5 corresponding to a molecular weight of about 2×10^7 is needed. While such polymers can be synthesized, it is easier to synthesize microcapsules of the desired size which contain the desired drug.

Several comments are appropriate regarding recent microencapsulation activity. First, regarding liposome, and other related systems, while early work was quite heartening, later studies have shown that for many drugs much of the drug is not "trapped" by the liposome and the resulting liposome complex often lacks desired stability. Thus, much recent work has involved placing polymeric shields about the liposome complexes—that is, microencapsulation.

Second, while there is much emphasis on the use of synthetic polymers as coating agents, more emphasis should be made utilizing materials derived from the natural sector [as dextran, xanthon, poly(beta-hydroxybutyrate)]. Encapsulation can be accomplished with synthetic materials which are only metabolized with difficulty and which have long lifetimes (excess of five years), yet many requirements call for more prompt delivery. Capsules can be derived from natural sources which can deliver, on call, its "fill" within several minutes to months without leaving an unmetabolized residue.

As previously noted, metals distribute themselves as oxides and salts within the body in known locations through (primarily) known routes. Information regarding metal-containing monomeric compounds is rapidly becoming available through many sources, including the environmental, hygiene, medical, industrial, and nutritional sectors. Judicious choices of metals, polymer size, and degradative modes and rates may allow the use of metal-containing polymers as targeting agents themselves for delivery of either or both the metal or associated organic moiety.

5. Modes of Bioactivity

A bioactive polymer could interact directly or indirectly with another molecule or an active site to elicit a biological response. (An indirect mode of action would involve a chemical and/or physical interaction with some other chemical species which would then make the polymer biologically active.) The bioactive moiety could be (1) the entire polymer chain, (2) segments within the polymer chain, (3) oligomeric degradation products of the polymer chain, or (4) a low-molecular-weight (monomeric) fragment derived from the polymer. Modes (3) and (4) would be termed as activity via controlled release of the bioactive agent or species, while modes (1) and (2) would be a more direct biological activity of the polymer itself. All these modes (1–4) could occur via direct or indirect interactions.

An additional mode utilizes the polymer as a matrix material for the embedding of the biologically active agent. These systems are called embedded or encapsulated systems. In these systems the biologically active agent is embedded within the polymer matrix and the agent is slowly released by a diffusion process or through the decomposition of the polymer within or about the target species. The polymer-embedded agent can be either a reservoir system, in which the agent is coated or surrounded by the polymer, or it can be a monolithic system where the drug is distributed fairly uniformly throughout the polymer matrix.

Insertion of metal-containing moieties can bring about a variety of biological responses – enhanced, decreased, different, or longer activities. The condensation of triphenylarsenic dichloride with diamines results in high polymers which exhibit their greatest activity during the active growth of bacteria, whereas the triphenylarsenic dichloride itself shows similar inhibition under both maintenance and growth cycles.[60] (The primary activity of the arsenic dichloride may be due to its "acid chloride" nature and not the presence of the arsenic – thus this may not be a fair comparison.) Further the toxicity of the polymer is highly specific and where it occurs it shows in excess of a twofold enhanced activity (on a per arsenic atom basis) over the monomeric arsenic dichloride itself.

The complexation of the uranyl ion by polyacrylic acid, polysodium acrylate, and diacid salts greatly reduces the toxicity of the uranyl moiety towards bacteria.[61] Tin-containing derivatives of polyvinylacetate exhibit remarkably long inhibition to mildew and rot-causing agents and against degradation of wood situated in contact with the ocean for periods in excess of five years, long after the corresponding monomeric tin esters and tin oxides cease their biological inhibitory nature.[62]

Tin-containing compounds, almost without exception, are inhibitory towards select bacteria and fungi. The extent, longevity, and particular bioorganisms inhibited are dependent on the nature of the compound and accessibility of the microorganism towards the tin-containing product. For polymers this includes the mode of delivery (solubilized, semisolubilized, or solid), type of connective bond (hydrophopic, hydrophilic, etc.), chemical environment presented by the surrounding moieties, concentration of tin-containing moiety, tin derivatives, and accessibility (cross-link density, etc.).

Subramanian and co-workers have achieved remarkable degradative resistence from the *in situ* polymerization of tin-containing esters in wooden poles and planks.[63] Such impregnated tin products have shown total stability when exposed to seaside environments and to the ocean itself for times in excess of five years. By comparison, untreated wood is degraded before one year, whereas other wood treated with a variety of treatments including the commonly accepted treatments were largely degradated prior to a three years' exposure period. Similar results have been found for the use of cross-linked tin-containing coatings for ship hulls.[63,64]

Each of these applications are superior to nonpolymeric applications in terms of longevity, amount required, and environmental impact. Considering the latter, barnacle repulsion is achieved at surface tin concentrations below ppb, well within possible acceptable environmental limits. Plants and sea creatures survive with no negative signs within centimeters of test objects (panels, posts, hulls).

For such applications the tin-containing moieties can be considered as residing within matrices which allow some moisture to enter, causing limited hydrolyzation followed by controlled exit of a tin-containing, inhibitory moiety. In nonpolymeric applications, much of the bioactive material is rapidly washed away creating a biologically unsatisfactory high concentration of bioactive material, which is subsequently washed away, leaving behind a largely unprotected material. At least two mechanisms for tin release have been suggested.

First, hydrolysis followed by release of the bioactive tin-containing moiety with subsequent bioactivity is the active mechanism for a majority of situations. A second mechanism involves the active participation of the bioorganism. The organism can release an enzyme, or a similar substance, which inadvertently (or otherwise) releases the tin from its site, unintentionally committing suicide. An alternative is for the organism itself to become involved in the release of the tin-containing moiety. While such bioorganism involvement may occur in select cases, it has yet to be conclusively proven. A simple test might indicate the role such a pathway plays. In situations where the active tin-containing moiety is known, inhibition of both the polymeric material and active material toward a number of bioorganisms can be studied. If activity concerned only the physical release of the active tin-containing moiety, then inhibition of specific species should be similar for the two systems. Contrary to this, if species involvement is necessary for at least some situations, then the two systems should show divergency with regard to species inhibited.

Studies related to the size of the bioactive species will be complicated for live-animal studies since: (1) the drugs will experience a wide variety of chemical and biological environments making it difficult to follow real chain-length activities without actual isolation and characterization of the various chain units (actual location and isolation of metal-containing drugs should be easier than for nonmetal-containing drugs because of the presence of the metal); and (2) the actual activity and size of the active chain will probably be dependent on the mode and site of drug delivery.

For cell, bacterial, fungi, and topical studies activity can be directly studied as a function of chain size and nature of the active moiety through synthesis of the appropriate products followed by biological testing of each product. While the procedure appears straightforward, the accumulation of unambiguous facts will be time-consuming and difficult. Thus, even for simpler systems, it appears that detailed, unambiguous studies will be few and will concentrate only on those diseases and drugs where spectacular progress is being made due to the complexity of biological systems and the wide variety of potential biological species and modes of activity. The rationale behind the synthesis of many of the potential drugs by us is to supply a host with a metabolizable portion within the polymer, which when metabolized, will release a moiety toxic to the host.

Recently, we reported the synthesis of titanium-containing polyether esters derived from the condensation of Cp_2TiCl_2 with a wide variety of xanthene dyes.[65] More recently we effected the synthesis of analogous tin-containing xanthene dyes through condensation of organostannane dihalides with xanthane salts.[66] These products will be called tin polydyes. The xanthene dyes utilized to form the polydyes are typically utilized as cell-coloring agents by microbiologists, biochemists, and so forth, and as such, polymers containing these dyes may be readily accepted by various organisms, effectively delivering the potentially toxic stannane. Further, mercurochrome, a xanthene dye, unlike the other utilized dyes, is toxic to many organisms because of the presence of mercury.

A number of tin polydyes derived from xanthene dyes were tested against the bacterial species *Escherichia coli* (C600) and *Pseudomonas aeruginosa* (7430). As expected, the greatest toxicity comes from the polydye derived from mercurochrome, probably due to at least the toxicity of the dye portion itself. Most of the compounds exhibited some inhibitory nature consistent with the idea that toxic moieties can, in select situations, be "activated" through coupling with moieties already known to be "biologically acceptable."

This theme of biological acceptability is emphasized in the synthesis of a number of arsenic V polyamines including a number of pyrimidines. The compound derived from condensation of triphenylarsenic dichloride with 4,6-diamino-2-mercaptopyrimidine was evaluated.[60] It was chosen for study because of the presence of the thiol group, which is both a potential modifying moiety towards the toxicity of arsenic and a typically desirable metabolite for a number of biological organisms—thus encouraging both selected toxicity and metabolism.

Inhibition tests in nutrient broth and saline solution were undertaken.

Inhibition was greatest for the nutrient broth mixtures being consistent with active metabolism being involved in inhibition of the *Ps. fluorescens* (concentration range 27.5–1.71 μg of polymer studied). The inhibition of triphenylarsenic dichloride is 50% or less than that of the polymer and is similar in both nutrient broth and saline solutions. Thus combining the arsenic with the thiopyrimidine as a polymer appears to be advantageous in inhibiting the growth of at least certain strains of *Ps. fluorescens.*

Similar studies were carried out employing polysaccharides containing organostannane moieties. The polysaccharides were utilized as both a source of inexpensive feedstock and as a biologically known, acceptable class of compounds. The tin-containing polysaccharides were synthesized through condensation of mono- and dihaloorganostannanes with the polysaccharide hydroxyls through formation of tin polyether linkages.[61,67–70] The tin-containing polysaccharides show good inhibition towards a wide range of bacteria. Inhibition is generally of polysaccharide source, but is dependent on the nature of the tin derivatives in the expected manner, that is, $CH_3 > C_2H_5 > C_3H_7 > C_4H_9 > C_8H_{17}$.

To assess whether inhibition is due to an inability of the organisms to metabolize the modified polysaccharide or due to the actual toxicity of the products themselves, studies relating the growth in a liquid medium were undertaken. *Trichoderma reesei* and *Chaetominum globosum* readily metabolize polysaccharides, including cellulose derived from cotton. Compared to growth of the fungi in solutions not containing tin-modified cellulose, all the compounds inhibit the two fungi. Growth inhibition in the dextrose-containing media indicates that inhibition is the result of the toxicity of the modified-cellulose compounds of the fungi rather than merely an inability of the fungi to degrade the test compounds.

The results are indicative of the applicability of such tin-containing polysaccharides being employed for the retardation of fungi-related rot and mildew. Questions such as duration and mechanism of fungal inhibition by these materials have yet to be answered, but the ability of certain test organisms to metabolize select tin-containing polysaccharides is shown.

References

1. H. E. Stokinger, *Patty's Industrial Hygiene and Toxicology,* 3rd Edition (G. Clayton and F. Clayton, eds.), Vol. 2A, Chapter 29, New York (1981).
2. R. E. Kinser, R. G. Keenan, and K. E. Kupel, *Am. Ind. Hyg. Assoc. J.* **26**, 249 (1965).
3. *NIOSH Criteria Document for Occupational Exposure to Inorganic Arsenic—New Criteria—1975,* NIOSH, Washington D.C., pp. 100–110.
4. *NIOSH Manual of Analytic Methods–General Procedure for Metals,* No. 173, NIOSH, Washington, D.C.

5. R. J. Grabowski and R. C. Unice, *Anal. Chem.* **30**, 1374 (1958).
6. M. L. Taylor, *Anal. Lett.* **1**, 735 (1968).
7. R. E. Kinser, *Am. Ind. Hyg. Assoc. J.* **27**, 260 (1966).
8. G. Lehnert, *Brit. J. Ind. Med.* **26**, 156 (1969).
9. L. Friberg, *Cadmium in the Environment*, 2nd edition, CRC Press, Cleveland, Ohio (1974).
10. Perkin-Elmer, *Anal. Memos* (September 1976).
11. A. M. Baetjer, C. Damron, and V. Budacz, *Arch. Ind. Health* **20**, 136 (1959).
12. H. J. Cahnmann and K. Bisch, *Anal. Chem.* **24**, 1341 (1952).
13. A. M. Kabiel, *Appl. Spectrosc.* **22**, 183 (1968).
14. *NIOSH Manual of Analytic Methods*, 2nd Edition, NIOSH, Washington, D.C. (1977).
15. E. R. Shaw and J. F. Corwin, *Anal. Chem.* **30**, 134 (1958).
16. T. Groenewald, *Anal. Chem.* **40**, 863 (1968).
17. R. E. Kinser, R. G. Keenan, and R. E. Kupel, *Am. Ind. Hyg. Assoc. J.* **26**, 249 (1965).
18. P. Collins and H. Diehl, *Anal. Chem.* **31**, 1692 (1959).
19. R. G. Reynolds and J. L. Monkman, *Am. Ind. Hyg. Assoc. J.* **23**, 415 (1962).
20. H. E. Stokinger, *J. Occup. Med.* **17**, 108 (1975).
21. B. F. Quin and R. R. Brooks, *Anal. Chim. Acta* **74**, 75 (1975).
22. W. F. Sunderman, Chairman and Chief contributor, 'Nickel,' Rep. Comm. Med. Biol. Effects Environ. Pollutants, National Academy of Sciences, Washington, D.C. (1975).
23. H. A. Schroeder and J. J. Balassa, *J. Chron. Dis.* **18**, 229 (1965).
24. G. H. Ayres and H. J. Belknap, *Anal. Chem.* **29**, 1536 (1937).
25. M. F. Pera and H. C. Harder, *Clin. Chem.* **23**, 1245 (1977).
26. A. H. Jones, *Anal. Chem.* **48**, 1472 (1976).
27. E. L. Steele and J. H. Yoe, *Anal. Chem.* **29**, 1622 (1957).
28. M. Punta, *Detection and Determination of Trace Elements,* Ann Arbor Science Pub., Ann Arbor, Mich. (1971).
29. O. L. Wood, *Biochem. Med.* **3**, 458 (1970).
30. B. L. Carson and I. C. Smith, *Silver, an Appraisal of Environmental Exposure*, Midwest Research Institute Report No. 3 (July 16, 1975).
31. C. A. Helsby, *Anal. Chim. Acta* **69**, 259 (1974).
32. M. L. Jacobs, *Evolution of Spark Source Mass Spectrometry in the Analysis of Biologic Samples*, NIOSH Research Report, HEW Publications No. (NIOSH) 75-186 (May 1975).
33. F. Amore, *Anal. Chem.* **46**, 1597 (1974).
34. A. Fitzek, *Acta Pharmacol. Toxicol.* **36**, 187 (1975).
35. H. G. Petrow and C. D. Strehlow, *Anal. Chem.* **39**, 265 (1967).
36. *NIOSH Criteria for a Recommended Standard for Occupational Exposure to Organotin Compounds*, NIOSH, Washington, D.C. (November 1976).
37. J. L. Leone, *J. Assoc. Off. Anal. Chem.* **56**, 535 (1973).
38. *NIOSH Criteria for a Recommended Standard for Occupational Exposure to Tungsten and Cemented Carbide*, NIOSH Publication No. 77-127 (September 1977).
39. D. A. Becker and P. D. LaFleur, *Anal. Chem.* **44**, 1508 (1972).
40. E. Cordfunke, *The Chemistry of Uranium*, Elsevier, New York (1969).
41. Criteria for Recommended Standard for Occupational Exposure to Vanadium, NIOSH Publication No. 77 (August 1977).
42. R. E. Allan, *Am. Ind. Hyg. Assoc. J.* **29**, 469 (1968).
43. E. E. Campbell, *Am. Ind. Hyg. Assoc. J.* **20**, 281 (1969).
44. V. A. Fassel and R. N. Kniseley, *Anal. Chem.* **46**, 1110A (1974).
45. W. Mertz, *Physiol. Rev.* **49**, 163 (1969).
46. J. Peisach, ed., *The Biochemistry of Copper*, Academic Press, New York (1966).
47. I. H. Scheinberg and I. Sternlieb, *Pharm. Rev.* **12**, 355 (1960).

48. C. J. Gubler, *Science* **123**, 87 (1956).
49. J. G. Graca, *Arch. Environ. Health* **5**, 437 (1962).
50. D. W. Bruce, *Toxicol. Appl. Pharmacol.* **5**, 750 (1963).
51. W. E. C. Wacker and B. L. Vallee, *N. Engl. J. Med.* **259**, 431 and 475 (1958).
52. G. C. Cotzias, *Adv. Neurol.* **2**, 265 (1973).
53. W. R. Buckley, C. F. Oster, and D. W. Fassett, *Arch. Dermatol.* **92**, 697 (A65).
54. E. D. Hutchinson, U.S. Atomic Energy Commission Research Division Report No. UR-561 (January 1960).
55. J. S. Hursh and T. W. Clarkson, *Arch. Environ. Health* **21**, 302 (1976).
56. C. Carraher, *J. Chem. Ed.* **58** (11), 91 (1981).
57. C. Carraher and L. P. Torre, *Macromolecular Solutions* (R. B. Seymour and G. A. Stahl, eds.), Chapter 6. Pergamon Press, Elmsford, New York (1982).
58. C. Carraher, V. Foster, H. M. Molloy, and J. Schroeder, *Org. Coat. Plast. Chem.* **41**, 203 (1979) and unpublished results.
59. C. Carraher and J. L. Lee, *J. Macromol. Sci., Chem.* **A9** (2), 191 (1975).
60. C. Carraher, *J. Polym. Sci., A-1* **9**, 3661 (1971).
61. C. Carraher, W. Moon and T. Langworthy, Polymer Preprints, **17**, 1 (1976).
62. C. Carraher, D. J. Giron, D. R. Cerutis, W. Bert, R. S. Venkatachalam, T. Gehrke, S. Tsuji, and H. S. Blaxall, *Biological Activities of Polymers* (C. Carraher and C. G. Gebelein, eds.), Chapter 2, American Chemical Society, Washington, D.C. (1982).
63. D. M. Anderson, J. A. Mendoza, B. K. Garg, and R. V. Subramanian, *Biological Activities of Polymers* (C. Carraher and C. G. Gebelein, eds.), Chapter 3, American Chemical Society. Washington, D.C. (1982).
64. R. V. Subramanian and K. N. Somasekhavan, *Advances in Organometallic and Inorganic Polymer Science* (C. Carraher, J. Sheets, and C. Pittman, eds.), Chapter 2, Dekker, New York (1982).
65. R. V. Subramanian, B. K. Garg, and J. Corredor, *Organometallic Polymers* (C. Carraher, J. Sheets, and A. C. Pittman, eds.), Chapter 19, Academic Press, New York (1978).
66. C. Carraher, R. Schwarz, J. Schroeder, and M. Schwarz, *Interfacial Synthesis, Vol. III, Recent Advances* (C. Carraher and J. Preston, eds.), Chapter 6, Dekker, New York (1982).
67. C. Carraher, R. S. Venkatachalam, T. O. Tiernan, and M. L. Taylor, *Org. Coat. Appl. Polym. Sci. Proc.* **47**, 119 (1982).
68. C. Carraher, J. Schroeder, C. McNeely, D. Giron, and J. Workman, *Org. Coat. Plast. Chem.* **40**, 560 (1979).
69. C. Carraher, D. Giron, W. Woelk, J. Schroeder, and M. Feddersen, *J. Appl. Polym. Sci.* **23**, 1501 (1979).
70. C. Carraher, J. Schroeder, W. Venable, C. McNeely, D. Giron, W. Woelk, and M. Feddersen, *Additive for Plastics* (R. Seymour, ed.), Vol. 2, p. 81, Academic Press, New York (1978).

Index